Lecture Notes in Networks and Systems 1966

The series "Lecture Notes in Networks and Systems" publishes the latest developments in Networks and Systems—quickly, informally and with high quality. Original research reported in proceedings and post-proceedings represents the core of LNNS.

Volumes published in LNNS embrace all aspects and subfields of, as well as new challenges in, Networks and Systems.

The series contains proceedings and edited volumes in systems and networks, spanning the areas of Cyber-Physical Systems, Autonomous Systems, Sensor Networks, Control Systems, Energy Systems, Automotive Systems, Biological Systems, Vehicular Networking and Connected Vehicles, Aerospace Systems, Automation, Manufacturing, Smart Grids, Nonlinear Systems, Power Systems, Robotics, Social Systems, Economic Systems and other. Of particular value to both the contributors and the readership are the short publication timeframe and the world-wide distribution and exposure which enable both a wide and rapid dissemination of research output.

The series covers the theory, applications, and perspectives on the state of the art and future developments relevant to systems and networks, decision making, control, complex processes and related areas, as embedded in the fields of interdisciplinary and applied sciences, engineering, computer science, physics, economics, social, and life sciences, as well as the paradigms and methodologies behind them.

Indexed by SCOPUS, EI Compendex, INSPEC, WTI Frankfurt eG, zbMATH, SCImago.

All books published in the series are submitted for consideration in Web of Science.

For proposals from Asia please contact Aninda Bose (aninda.bose@springer.com).

Alvaro Rocha · Carlos Ferrás · Luis Loo
Editors

Information Technology & Systems

ICITS 2026, Volume 2

Springer

Editors
Alvaro Rocha
ISEG
University of Lisbon
Lisbon, Portugal

Carlos Ferrás
Facultade de Geografía e Historia
University of Santiago de Compostela
Santiago de Compostela, Spain

Luis Loo
Universidad Tecnológica de Honduras
San Pedro Sula, Honduras

ISSN 2367-3370 ISSN 2367-3389 (electronic)
Lecture Notes in Networks and Systems
ISBN 978-3-032-27537-0 ISBN 978-3-032-27538-7 (eBook)
https://doi.org/10.1007/978-3-032-27538-7

This Springer imprint is published by the registered company Springer Nature Switzerland AG
The registered company address is: Gewerbestrasse 11, 6330 Cham, Switzerland

Introduction

This volume comprises papers written in English and accepted for presentation and discussion at the 2026 International Conference on Information Technology and Systems (ICITS 2026). The conference was jointly organized by the Universidad Tecnológica de Honduras (UTH) and the Information and Technology Management Association (ITMA). ICITS 2026 took place on Roatán Island, Honduras, from February 16 to 18, 2026.

The 2026 International Conference on Information Technology and Systems (ICITS 2026) serves as an international forum for researchers, academics, and practitioners to present and discuss recent innovations, emerging trends, research findings, practical experiences, and contemporary challenges in the broad field of Information Technology and Systems. The conference fosters interdisciplinary dialogue and promotes collaboration among scholars and professionals from diverse geographical and scientific backgrounds.

The ICITS 2026 Program Committee was composed of a multidisciplinary group of 197 experts in Information Systems and Technologies. All submitted papers underwent a rigorous double-blind peer-review process, ensuring the scientific quality, originality, and relevance of the accepted contributions. The conference covered the following main thematic areas:

A) Information and Knowledge Management.
B) Organizational Models and Information Systems.
C) Software and Systems Modeling.
D) Software Systems, Architectures, Applications, and Tools.
E) Multimedia Systems and Applications.
F) Computer Networks, Mobility, and Pervasive Systems.
G) Intelligent and Decision Support Systems.
H) Big Data Analytics and Applications.
I) Human–Computer Interaction.
J) Ethics, Computers, and Security.
K) Health Informatics.
L) Information Technologies in Education.
M) Media, Applied Technology, and Communication.

ICITS 2026 received 298 submissions from 36 countries, reflecting the truly international character of the conference. The accepted papers presented at the conference are published in this volume by Springer and in RISTI. The proceedings will be submitted for indexing in major international databases, including Web of Science (WoS), Scopus, EI-Compendex, and Google Scholar, among others.

We express our sincere appreciation to all those who contributed to the success of ICITS 2026, including authors, reviewers, Program Committee members, workshop organizers, session chairs, sponsors, and the local organizing team. Their commitment, professionalism, and support were fundamental to maintaining the high scientific standards and international recognition of the conference.

Organization

Honorary Chairs

Álvaro Rocha	ISEG, University of Lisbon, Portugal
Hojjat Adeli	The Ohio State University, USA
Javier Enrique Mejía	Universidad Tecnológica de Honduras, Honduras
Rajeev Kanth	University of Turku, Finland

Scientific Committee Chair

Carlos Ferrás Sexto	University of Santiago de Compostela, Spain

Local Organizing Committee

Arely Ponce	Universidad Tecnológica de Honduras, Honduras
Kelly Salgado	Universidad Tecnológica de Honduras, Honduras
Doris Vanessa Puentes	Universidad Tecnológica de Honduras, Honduras
Diego Chacón	Universidad Tecnológica de Honduras, Honduras
Luís Loo	Universidad Tecnológica de Honduras, Honduras

Scientific Committee

Carlos Ferrás Sexto	University of Santiago de Compostela, Spain
Abel Suing	Universidad Técnica Particular de Loja, Ecuador
Alan Ramírez Noriega	Universidad Autónoma de Sinaloa, Mexico
Alcides Alvear	Sistema Universitario Ana G. Méndez - UAGM, Puerto Rico
Alexandra González-Eras	Universidad Técnica Particular de Loja, Ecuador
Álvaro López-Martín	University of Malaga, Spain
Ana Dinis	Polytechnic Institute of Cávado and Ave, Portugal
Ana Paula Silva	Polytechnic University of Castelo Branco, Portugal
Ana Rita Calvão	Universidade de Aveiro, Portugal
Ana Sofia Duque	Polytechnic Institute of Viseu, Portugal
Anabela Moreira Bernardino	Polytechnic of Leiria, Portugal

Andre Kawamoto	Universidade Tecnológica Federal do Paraná, Brazil
Angel Torres Toukoumidis	Universidad Politécnica Salesiana, Ecuador
Angélica Victoria Guillén Pinargote	Univerisdad Técnica de Manabí, Ecuador
Ania Cravero	Universidad de La Frontera, Chile
Ankit R. Patel	University of Minho, Portugal
António Brandão	Metro do Porto, SA, Portugal
António José Gonçalves Fernandes	Instituto Politécnico de Bragança, Portugal
Aracelly Fernanda Nuñez Naranjo	Universidad Indoamérica, Ecuador
Arlindo Silva	Polytechnic University of Castelo Branco, Portugal
Arnulfo Alanis	Tecnologico Nacional de México, Mexico
Áurea Sousa	University of Azores, Portugal
Bertil P. Marques	GILT/ISEP/IPP, Portugal
Bhavya Paliwal	Amazon Web Services, USA
Bruno F Gonçalves	Instituto Politécnico de Bragança, Portugal
Carlos Grilo	Polytechnic Institute of Leiria, Portugal
Carlos Vaz de Carvalho	Instituto Politécnico do Porto, Portugal
Celia Rafael	Politécnico de Leiria, Portugal
Célio Marques	Polytechnic University of Tomar, Portugal
Cesar Collazos	Universidad del Cauca, Colombia
Christian Trefftz	USA
Claudia Cappelli	UERJ - Universidade do Estado do Rio de Janeiro, Brazil
Claudio Cilli	"Sapienza" University of Rome, Italy
Dacyr Gatto	UNINOVE, Brazil
Daniela Borissova	Bulgarian Academy of Sciences, Bulgaria
Denis Cereno Moreno	Universidad Tecnologica de Panama, Panama
Dulce Mourato	ISTEC - Higher Advanced Technologies Institute, Portugal
Eddie Galarza	Universidad de las Fuerzas Armadas - ESPE, Ecuador
Eduardo Marques	Universidade da Madeira, Portugal
Elisabete Paulo Morais	Instituto Politécnico de Bragança, Portugal
Elisabete Vieira	University of Aveiro, Portugal
Eugénia Bernardino	Polytechnic of Leiria, Portugal
Felix Freitag	Universidad Politécnica de Catalunya, Spain
Fernando Bobillo	University of Zaragoza, Spain
Fernando Moreira	Universidade Portucalense, Portugal
Fernando Ribeiro	Polytechnic Institute of Castelo Branco, Portugal
Filipa Ferraz	University of Minho, Portugal

Filipe Caldeira	Polytechnic Institute of Viseu, Portugal
Filipe Cardoso	Instituto Politécnico de Santarém, Portugal
Filipe Fidalgo	Instituto Politécnico de Castelo Branco, Portugal
Francisco Antunes	University of Beira Interior and INESC Coimbra, Portugal
Francisco José García Tartera	UCM, Spain
Gabriel Alberto García-Mireles	University of Sonora, Mexico
Gabriel Guerrero Contreras	University of Cádiz, Spain
Gerardo Arceo Moheno	Universidad Juarez Autonoma de Tabasco, Mexico
Gerardo Matturro	Universidad ORT Uruguay, Uruguay
Giancarlo Sanchez Atuncar	Cesar Vallejo University, Peru
Hugo Sanjurjo-González	University of Deusto, Spain
Inês Messias	Instituto Politécnico de Santarém, Portugal
Iván Armuelles Voinov	Universidad de Panamá, Panama
João Tomaz Simoes	Polytechnic University of Tomar, Portugal
Javier Medina	Universidad Distrital Francisco José de Caldas, Colombia
Joan-Francesc Fondevila-Gascón	Blanquerna-URL, EUM-UdG, Euncet-UPC, EAE Business School, UPF, CECABLE, Spain
João Gondim	University of Brasilia, Brazil
Jorge Barreiros	Polytechnic of Coimbra - ISEC, Portugal
Jorge Buele	Universidad Indoamérica, Ecuador
José Antonio Ogosi Auqui	Universidad César Vallejo, Peru
José Braga de Vasconcelos	Universidade Lusófona, Portugal
Jose Carlos Metrôlho	Polytechnic Institute of Castelo Branco, Portugal
Jose Luis Vázquez Noguera	Universidad Nacional de Asunción, Paraguay
José Marcelo Juca Aulestia	Universidad Técnica Particular de Loja, Ecuador
Jose Ramon Lopez	Universidad Autonoma de Sinaloa, Mexico
José Ribeiro	Polytechnic of Leiria, Portugal
José Rosado	ISEC-IPC, Portugal
José Salgado	Instituto Politecnico Guarda, Portugal
Juan Antonio Meza Fregoso	Universidad Autónoma de Baja California, Mexico
Juan D'Amato	UNCPBA, Argentina
Juan Jesus Ojeda-Castelo	University of Almeria, Spain
Juan M. Santos-Gago	University of Vigo, Spain
Juan Pacheco Torres	Universidad César Vallejo, Peru
Julio Cesar Mello-Román	Universidad Nacional de Asunción, Paraguay
Julio Ernesto Quispe_Rojas	Universidad Nacional Pedro Ruiz Gallo, Peru
Karim Guevara Puente de la Vega	Universidad Nacional de San Agustin de Arequipa, Peru

Khutso Lebea	University of Johannesburg, South Africa
Laura Guerra	Pontificia Universidad Catolica del Ecuador, Ecuador
Laura Magali Chamba Rueda	Universidad Técnica Particular de Loja, Ecuador
Ledyis Lisbeth Jiménez González	Universidad Bolivariana del Ecuador, Ecuador
Leila Weitzel	Fluminense Federal University, Brazil
Leonilde Reis	Instituto Politécnico de Setúbal (ESCE/IPS), Portugal, Portugal
Lilia Carpio-Jiménez	Universidad Técnica Particular de Loja, Ecuador
Lizbeth Suarez	Universidad Bolivariana del Ecuador, Ecuador
Lorena Siguenza	Universidad de Cuenca, Ecuador
Lornel Rivas	Universidad Andina del Cusco, Peru
Luis Álvarez Sabucedo	University of Vigo, Spain
Luis Marques	ISEC, Polytechnic University of Coimbra, Portugal
Luis Silva Rodrigues	ISCAP, Polytechnic University of Porto, Portugal
Luz Angela Rocha Salamanca	Universidad Distrital Francisco José Caldas, Colombia
Luz Maria Hernández Cruz	Universidad Autónoma de Campeche, Mexico
Luz Sussy Bayona Ore	Universidad Nacional Mayor de San Marcos, Peru
Mahdieh Zakizadeh	TUWien, Austria
Manuel Caeiro Rodríguez	Universidade de Vigo, Spain
Manuel Sousa Pereira	Instituto Politécnico de Viana do Castelo, Portugal
Manuel Tupia	Pontificia Universidad Catolica del Peru, Peru
Marco Quintana	UIDE, Ecuador
Marcos Vinícius de Souza Toledo	Instituto Federal de Minas Gerais, Brazil
María Chávez Chuquimango	Universidad Peruana de Ciencias Aplicadas, Peru
Maria Cristina Marcelino Bento	UNIFATEA, Brazil
Maria do Rosário Bernardo	Universidade Aberta e INESCTEC, Portugal
Maria Elisabete Neves	Politécnico de Coimbra- CBS|ISCAC, Portugal
Maria Isabel Ribeiro	Instituto Politécnico de Bragança, Portugal
Maria João Ferreira	Universidade Portucalense, Portugal
Maria José Sousa	ISCTE - Instituto Universitário de LIsboa, Portugal
María Teresa García-Álvarez	University of A Coruna, Spain
Mário Cruz	Polytechnic of Porto, Portugal
Mário Negas	Universidade Aberta, Portugal
Maritza Elisabet Aguirre Munizaga	Universidad Agraria del Ecuador, Ecuador
Marla TGeresinha Barbosa Geller	CEULS/ULBRA, Brazil
Michal Kvet	Žilinská univerzita v Žiline, Slovakia
Miguel Arcos-Argudo	Universidad Politécnica Salesiana, Ecuador

Mikel Ugando Peñate	Pontificia Universidad Católica del Ecuador, Ecuador
Mónica Aresta	University of Aveiro, Portugal
Mónica Costa	Polytechnic Institute of Castelo Branco, Portugal
Narasimha Rao Vajjhala	University of New York Tirana, Albania
Naveen Kumar	Chandigarh University, India
Nelson Rocha	Universidade de Aveiro, Portugal
Nilthon Dennis Paredes Quezada	Gerencia Regional de Educación La Libertad, Peru
Noel García-Díaz	Tecnológico Nacional de México, Mexico
Nuno Costa	IPLeiria, Portugal
Nuno Melão	Instituto Politécnico de Viseu, Portugal
Olger Gutierrez Aguilar	Universidad Nacional de San Agustín, Peru
Orlando Belo	University of Minho, Portugal
Oscar Carlos Medina	Universidad Tecnológica Nacional – Facultad Regional Córdoba, Argentina
Osvaldo Dias Silva	Universidade dos Açores, Portugal
Osvaldo Santos	Instituto Politécnico de Castelo Branco, Portugal
Pablo Payró Campos	Universidad Juárez Autónoma de Tabasco, Mexico
Patrícia Leite	Instituto Politécnico Cavado e Ave, Portugal
Patricia Quiroz-Palma	Uleam, Ecuador
Paulo Batista	CIDEHUS, Portugal
Paulo Rurato	Universidade Fernando Pessoa, Portugal
Pedro Isaias	Universidade Aberta, Portugal
Pedro Martins	Polytechnic University of Viseu, Portugal
Pedro Sousa	University of Minho, Portugal
Peixoto Hugo	University of Minho, Portugal
Petr Doucek	Prague University of Economics and Business, Czechia
Rene Garzozi-Pincay	Universidad Estatal Península de Santa Elena, Ecuador
Rocío González-Sánchez	Rey Juan Carlos University, Spain
Rodolfo Bojorque	Universidad Politécnica Salesiana, Ecuador
Rosa Maria Reis	Instituto Superior de Engenharia do Porto, Portugal
Rui Cruz	INESC-ID/IST-UL, Portugal
Rui Humberto Pereira	ISCAP/CEOS.PP, Portugal
Rui Quaresma	Universidade de Évora, Portugal
Sam Michael Espinoza Vidaurre	UPT, Peru
Samuel Silva	University of Aveiro, Portugal
Sara Balderas-Díaz	Universidad de Cádiz, Spain

Sara Dias-Trindade	Faculty of Arts of the University of Porto, Portugal
Sarat Piridi	Hanwha Qcells, India
Sergio Araya Guzmán	Universidad del Bío-Bío, Chile
Shahzad Ashraf	Gachon University, Pakistan
Sri Tejaswi Gattupalli	Cisco, USA
Tagleorge Silveira	ISTEC Porto, Portugal
Tomasz Kisielewicz	Warsaw University of Technology, Poland
Veronica Martinez Lazcano	Universidad Autónoma del Estado de Hidalgo, Mexico
Veronica Segarra Faggioni	UTPL, Ecuador
Verónica Vasconcelos	ISEC - IPC, Portugal
Víctor Manuel Cornejo Aparicio	Universidad Nacional de San Agustín de Arequipa, Peru
Vitor Gonçalves	Instituto Politécnico de Bragança, Portugal
Waqas Haider Bangyal	Kohsar University Murree, Pakistan
Wagner Tanaka Botelho	Federal University of ABC, Brazil
Zenaide Silva	Universidade Federal do Sul e Sudeste do Pará, Brazil

Contents

Information and Knowledge Management

Socio-technical Modeling of Software Development Team Productivity: A Systematic Literature Review of System Dynamics Applications 3
Luz Marcela Restrepo-Tamayo, Gloria Piedad Gasca-Hurtado, and Natalia Gómez-Velásquez

Optimizing the Balance Between Knowledge Acquisition and Cognitive Load in Virtual Reality-Based Education for Children with Special Needs 16
Stefan Stoev, Katia Rasheva-Yordanova, Daniela Borissova, and Anna Lekova

Bibliometric Overview of Early Intervention in the Educational Context 27
Verónica Castro, Lennis Pérez, Ángel Cornejo, Maribel Puchaicela, and Jessica Cahuate

The Discourse on Political Polarization in the Peruvian Context Through Lima's Digital Media 44
Hugo Gonzalez Aguilar

The Level of English Language Learning According to the Common European Framework of Reference for Languages in Students of Health-Related Programs at a Private University 52
Pilar Valeria Zapana-Aguilar, Kevin Mario Laura-De La Cruz, Domingo Nicolás Pérez-Yufra, Miliam Quispe-Vargas, Gabriela Manzur-Vera, and Cynthia Karín Rodriguez-Palumbo

Cybersecurity Audit-Based Vulnerability Assessment and Risk Mitigation for the Edutainment Platform 62
Omar G. Bravo-Quezada, Pedro J. Orellana-Jaramillo, Jonathan M. Morán-Velasco, and Ledys L. Jiménez-González

Accounting Education and Entrepreneurship: Keys to Transforming Education in the 21st Century 74
Liliana Marisol Jumbo Mena, Margoth Iriarte-Solano, Marcelo Juca-Aulestia, and Edwin José Cabrera-Paucar

Accurate and Rapid Time-Synchronization Using Single Camera Image 85
Masanori Sugimoto

Key Competitive Advantage Factors in the Commercialization of Non-timber Forest Products (NTFPs) 97
Roosa Balk and Jyri Vilko

Designing a Smart Corporate Governance Prototype for Wellness Tourism Ecosystems: An ICT-Driven Approach 108
Lady Fernández-Mora, Efraín Ortiz-Pabón, Daniela Flórez-Villamil, Sara Juliana Peña-Gómez, Juan David Hernández-Rodríguez, María Andrea Méndez-Beltrán, and Nestor Armando Nova-Arévalo

Enhancing Sales Forecast Accuracy for New Fashion Items: Leveraging Online Trend Signals and Historical Data 120
Vera Miguéis and Nelson Alves

Use of Artificial Intelligence in Clinical Image Analysis to Improve Patient Diagnosis and Treatment 131
Renato M. Toasa, Karen Stephany Cordóva-Vera, Aracelly Núñez-Naranjo, and Victor Cusco-V

Use of ICT in Education: Descriptive Analysis Based on Secondary Data from Ecuador 141
María José Carrera Yánez

Patterns of Technology Utilization and Academic Dishonesty in Secondary Education 156
Juan Carlos Torres-Diaz, Pablo Vicente Torres-Carrión, Mónica Hinojosa Becerra, and Diana Rivera-Rogel

Applying a Quantitative Method to Analyze the Cybersecurity Risk a Portuguese Hospital 169
Ana Luísa Teixeira, Ana Patrícia Borges, Dário Carreira, Victor Costa, João Pavão, and Nelson Pacheco Rocha

Artificial Intelligence in Teacher Training: Technological, Ethical, and Pedagogical Challenges for Critical Education 181
Elza Mesquita, Bruno F. Gonçalves, and Maria Raquel Patrício

Word Embedding to Analyze Innovation Projects in Higher Education Institutions 192
Carlos Tandazo-Malla and Ruth Reátegui

Artificial Intelligence as Digital Innovation for Social Responsibility: A Bibliometric Exploration with a Focus on SMEs 208
Lizbeth Suarez-Morales, Cristhian Solorzano, and Roxana Chiquito

Artificial Intelligence and the Human Dimension in Predicting Payment Character in Savings and Credit Cooperatives 220
Lizbeth Suarez-Morales, Cristhian Solorzano, and Roxana Chiquito

Reading Assistant for Visually Impaired People Implemented in Raspberry Pi with 3D Casing Made from Recycled Materials 232
Elizabeth Salazar-Jácome, Rolando Salazar-Paredes, Gustavo Guaigua-Albarracín, and Javier De la Torre-Guzmán

Exploring Digital Metaliteracy Through a Bibliometric Lens: Intergenerational Perspectives in Education 245
Eric Le Fur

Unraveling the Macroeconomic Determinants of Economic Growth 257
Ferdina Paththamesthrige, Gayani Hawpe Liyanage, Uthpala Punchihewa, Danish Khokhar, Daniyal Jawad, and Jyri Vilko

Digital Literacy through Robotics: A Fablab-Based Pathway to Strengthen Programming Skills and Academic Achievement in Secondary Education 269
Estefany Nayeli Quispe Tica, Bryan Antony Quinta Ccosi, and Luis Daniel Haro Laque

Innovation in Neuromuscular Electrostimulation: An Integrative Review and Case Study 278
Manuel Au-Yong-Oliveira, Suzete Carvalho, and Ana Palma-Moreira

Artificial Intelligence in Management: Transforming Decision-Making, Operations, and Organizational Efficiency 290
Ricardo Marcão, Vasco Santos, Filipa Martinho, Rodrigo Nascimento, and Bablu Kumar Dhar

The Impact of Transformational Leadership on Organizational Performance 297
Tiago Costa, Ricardo Marcão, Vasco Santos, Filipa Martinho, and Isabel Reis

Consumer Preferences on Natural Food Products: Perspectives on NTFP Demand 306
Vihtori Hänninen and Jyri Vilko

Flipped Classroom and Gamification for Training Engineers 317
Aracelly Núñez-Naranjo, María-Stefanie Vasquez-Peñafiel, Mauricio Carvache-Franco, Orly Carvache-Franco, Aleksandra Minchenkova, and Wilmer Carvache-Franco

Usability Evaluation of Cross-Platform Mobile Development Frameworks 331
Jairo Pozo, William Montalvo, and Rodrigo Tufiño

Smart Irrigation in Miniature Roses: A Case Study with IoT and Agriculture 4.0 341
Blanca Topon-Visarrea, Adriana Guevara-Arteaga, and Miryam Johana Salcedo-Quishpe

Personalization of Educational Assessments: An Artificial Intelligence Perspective 350
Alexander Puentes-Castro and Diana Janeth Lancheros-Cuesta

Breaking Barriers: Technology as a Tool for the Educational Inclusion of Students with Visual Impairments 360
Víctor Manuel Zamudio García and Glendamira Serrano Franco

Semantic Resizing of Neutral Emotions in Multi-label Classification Systems with SBERT and Cascade Models 369
Christian Salamea-Palacios, Juan Vasquez-Avila, Javier Viñanzaca-Figueroa, and Jairo Sacoto-Cabrera

Agentic AI for Homicide Prediction: A Modular Framework for Geospatial Risk Analysis 381
Luis Loo, Kenny Joseph Cooper, José García, Selvin Maldonado, Aldenis E. Miranda, and José Irias

Audiovisual Narratives and the Audience: The Case of the Ministry of Time ... 392
Cristián Londoño-Proaño

Enhancing Legal Transparency: Interpretable Legal Document Classification with Hybrid BERT-RF Models 403
Paulo Henrique B. Rodrigues, Daniel Alves Silva, Fabio Lucio L. de Mendonca, Edna Dias Canedo, Georges Daniel Amvame Nze, and Geraldo Pereira Rocha Filho

Bibliometric Review on the Contributions of Artificial Intelligence to Energy and Climate Sustainability: A Thematic Study of Scientific Literature 414
Mikel Ugando-Peñate, José Luis Espín-Lopez, Mishell Alejandra García-Acosta, Reinaldo Armas-Herrera, Angel Alexander Higuerey-Gómez, and Ángel Ramón Sabando-García

Information Technologies in Education

Development of Technical and Transversal Skills Through Educational Robotics Using Construction Models and the Lego Serious Play (LSP) Methodology in University Students 429
Fabricio Marcillo, Willian Ocampo, Franklin Carrasco, Rodolfo Córdova, Aide Zumba, and Fausto Orozco

University Students' Perceptions of the Use of Virtual Reality to Improve Verbal Reasoning: A Case Study with Delightex 436
Mariela Tapia-Leon, María Merchán Gavilanez, Gerardo Benavides-Montenegro, and Sergio Luján-Mora

Basketball 3-Point Shot Detection Using GCN-Pose Estimated Keypoints for the Downstream AQA Task 446
Bupe Chindongo and Tevin Moodley

Performance Evaluation of Predictive Methods for Electric Vehicle Investment in Ecuador 458
Renato Arcos, Renato M. Toasa, and Mario Pérez

Contribution of the PLS Methodology in the Field of Information and Knowledge Management 473
Mikel Ugando-Peñate, José Luis Espín-Lopez, Mishell Alejandra García-Acosta, Reinaldo Armas-Herrera, Angel Alexander Higuerey-Gómez, and Ángel Ramón Sabando-García

Media Representations in Emergency Contexts: A Bibliometric Analysis of Humanitarian Crises 484
Angel Torres-Toukoumidis, Moisés Pallo Chiguano, Santiago Castro, and Julio Merchán Romero

Integration of Blockchain and Artificial Intelligence in Accounting Audit: Transparency, Challenges, and Opportunities in Ecuador 495
Carolina Paola Jaramillo Bayas, William Ortega Naranjo, and Lorena Elizabeth Muquis Tituaña

STEM Studies, SDGs and the Role of Women in Ecuadorian Universities 506
Samanta Zabala-Cáceres, Elizabeth Salazar-Jácome, Karen Merizalde-Torres, and Doris Calle-Landázuri

From Attitude to Intention: Modeling Gen Z Mobile Banking Adoption with an Extended UTAUT2 518
Marco Antonio Nolasco-Mamani, Norma C. Velásquez Rodríguez, Renza L. Gambetta Quelopana, Ana Noelia Martinez Valdivia, Ivette Eneida Atencio-Iturry, and Sam M. Espinoza Vidaurre

Trust-Aware Adaptive Framework for Malware Classification with Feedback Patching 540
Iliyan Barzev and Daniela Borissova

Disinformation in Peruvian Political Discourse in Digital Media 552
Hugo Gonzalez Aguilar

Sonification of Centrality Metrics in Layered Graphs 559
José Luis Gómez Ramos, Arturo Corona Ferreira, Carlos Arturo Custodio Izquierdo, Carlos González Zacarías, María Alejandrina Almeida Aguilar, and Eduardo Cruces Gutiérrez

Big Data Analytics and Unsupervised Learning for Digital Payments Adoption in Ecuador. 569
Miguel Angel Peñarreta, Andres Ulloa, and Reinaldo Herrera

Implementation of AI Tool Accessibility in Nursing Procedures 582
Toapanta Vizuete Kevin Joel, Vallejo Burbano Alejandra Jadira, and Nasimba Yautibug Lizbeth Alexandra

Taxonomy of Hierarchical Federated Learning for Intrusion Detection Systems in IoT 589
Jonathan Navarro-Alfaro and Adrian Lara

Integrating Traditional and Emerging Technologies for the Professional Learning of Prospective Mathematics Teachers 602
Jhony Alexander Villa-Ochoa

Interactive NFC System: Multisensory Platform for Early Learning Based on VEX 123 614
Luis Ulcuango, Rubén Jaramillo, Darío Guamán, and Mario Guamán

Mobile Banking as a Tool for Financial Inclusion in Adults in the City of Loja, Year 2025 626
Gloria del Carmen Palacio Valdivieso and Ivanova Quituizaca Songor

Prototype of Smart Wristbands with QR Code for Pre-hospital Care of People with Intellectual Disabilities ... 637
Angela Gabriela Valencia Berrezueta, Ana Isabel Uyaguari Morocho, Sandra El Khori El Khouri, and Carlos Francisco Encalada Ochoa

Use of Social Media in Education: Perceptions of Students and Teachers in Their Integration ... 648
Roger Daniel Ponce Rodríguez, Ana C. Romero, and María Eugenia Sandoval Sanabria

High-Fidelity Simulation for Learning to Manage Tachyarrhythmias in Medical Emergency Students ... 657
Ledys L. Jiménez-González, Wellington Gerardo Abril Valarezo, María Elena Villafuerte Pucha, César Ariel Quesada Gutiérrez, and Carlos Francisco Encalada Ochoa

Technological Solution for the Prevention of Operational Risks in Entities of the Financial Sector ... 669
Marcos Ronceros, Christopher Perez, Juan Mansilla-Lopez, and Ricardo Loza

Key Variables for Teacher Performance and the Development of Autonomous Learning Management in Adolescents ... 679
S. Hinojosa Nuñez, S. Hinojoza Talavera, E. Gonzales-Inca, K. Ríos-Vera, and O. Gutierrez-Aguilar

Corporate Social Responsibility and Organizational Sustainability: An ESG-and-AI Framework for Better Performance ... 694
Carmen Machado, Ricardo Marcão, Vasco Santos, Filipa Martinho, and António Dionísio

Using Portable SDR Systems to Teach Spectrum Monitoring and RF Reconnaissance Concepts ... 705
Bryan Chanataxi-Nacimba, Fernando Caicedo-Altamirano, Janneth Moreta-Changoluisa, Diego Alomoto-Robles, and David Rivas-Lalaleo

Instructional Designer Competencies: A Systematic Literature Review ... 718
Margarida Sampaio and Guilhermina Lobato Miranda

The Use of Photography, Through the Smartphone, for the Development of Creative Competence in the Education ... 731
Silvia Cáceres-Bazalar and María Chávez-Chuquimango

Photographic Language and Peruvian Gastronomy in the Digital Era 742
Génesis Quispe-Coronel and María Chávez-Chuquimango

Integrating Artificial Intelligence and Self-reports for Emotional Recognition and Management in Primary and Secondary Education 753
Eduardo Ñaupa Yapura, Antony Elio Ayansi Huisa, Anthony Dennis Oyola Saloma, Daniel Hachircana Méndez, Yadira Melina Zapata Ttito, and Diana Salas Sanchez

Author Index 763

Information and Knowledge Management

Socio-technical Modeling of Software Development Team Productivity: A Systematic Literature Review of System Dynamics Applications

Luz Marcela Restrepo-Tamayo(✉), Gloria Piedad Gasca-Hurtado, and Natalia Gómez-Velásquez

Universidad de Medellín, Cra. 87 #30-65, Medellín, Antioquia, Colombia
{lmrestrepo,gpgasca,nagomez}@udemedellin.edu.co

Abstract. System Dynamics (SD), as a simulation paradigm, provides a suitable analytical framework for modeling productivity in software development teams, enabling the representation of the interrelationships between technical, human, and organizational factors that impact performance. However, limited access to simulation models presents an obstacle, as they rely on specialized, licensed software that restricts their widespread application in the software development industry. Given this problem, the objective of this study is to examine the current state of research on SD applications for analyzing software development team productivity. Thus, a Systematic Literature Review was conducted using Scopus and Web of Science from 2014 to 2024. The analysis of 20 studies reveals an evolution towards models that integrate both technical and non-technical factors, with a predominance of industrial applications. However, significant limitations are noted, including a lack of reproducibility and inadequate documentation of data management capabilities. Thus, more accessible SD tools that integrate technical and social dimensions, provide validated models with full documentation, and offer robust data management capabilities are required to democratize these analytical methodologies in the software development industry.

Keywords: System Dynamics · Productivity · Simulation · Software Development · Systematic Literature Review

1 Introduction

System Dynamics (SD) is a simulation paradigm suitable for representing complex phenomena characterized by nonlinear relationships, feedback, and time delays [1]. In software development, this approach provides an analytical framework for modeling team productivity, enabling the integration and analysis of the interrelationships between technical and non-technical factors that influence team performance [2].

Despite SD's potential to capture the systemic nature of software development and optimize human resources for timely delivery [3], a gap persists between theory and practice due to limited accessibility from tool costs and complexity, poor reproducibility from

A. Rocha et al. (Eds.): ICITS 2026, LNNS 1966, pp. 3–15, 2027.
https://doi.org/10.1007/978-3-032-27538-7_1

incomplete documentation, and insufficient integration of technical and non-technical factors [4].

SD-based simulation models rely on specialized software such as Vensim, Stella, or Powersim [5], which entail licensing costs, steep learning curves, and limited interoperability [6]. These technical and economic barriers have restricted their use mainly to academia and well-resourced companies, excluding small and medium enterprises that could benefit from simulation-driven decision-making [4].

Although SD has been applied to various aspects of software development, no comprehensive analysis examines its current state, modeled factors, or the accessibility and reproducibility of approaches—limiting practitioners' understanding of available solutions, relevant factors, and required tool characteristics. Given this scenario, it is necessary to review the available scientific literature through a Systematic Literature Review (SLR) [7] to identify the characteristics of current SD applications in software engineering and determine the attributes an innovative tool should incorporate to overcome the identified limitations.

To address these gaps, this study makes three key contributions: (1) provides a classification of technical and non-technical factors considered in current SD models for software development, (2) evaluates the accessibility, reproducibility, and practical applicability of existing SD approaches, and (3) identifies the essential characteristics that future SD tools should incorporate to democratize their use in organizational contexts.

This SLR aims to help bridge the gap between SD's theoretical potential and its practical use in software development. The review analyzed peer-reviewed literature indexed in Scopus and Web of Science (WoS) [7]. The findings show a shift toward socio-technical SD models, predominantly in industrial settings, but constrained by limitations in reproducibility and documentation.

This paper is structured as follows: Sect. 2 presents the research method; Sect. 3 presents the results; Sect. 4 discusses the findings; and finally, Sect. 5 concludes. This study fills a critical gap in the literature by providing the first comprehensive analysis of SD applications in software development productivity, establishing a foundation for developing more accessible, practical analytical tools for the software industry.

2 Methodology

This SLR follows the guidelines established by Kitchenham and Charters [7], widely used in software engineering. The methodology comprises three phases: (1) planning the review, (2) conducting the review, and (3) reporting the review.

2.1 Planning the Review

Following Kitchenham's guidelines [7], the planning phase began with establishing the need for this SLR and defining the research protocol.

- Research question: Based on the study's objective, the following research question was formulated: *How do existing SD applications in software development address the factors that impact team productivity, and what features do these applications exhibit regarding accessibility and reproducibility?*

- Data source: Scopus and WoS were selected because they are referential databases that aggregate literature from multiple publishers, ensuring comprehensive and unbiased coverage across diverse publication venues [8].
- Search string: The same search string was used in both databases to ensure consistent retrieval. To broaden the capture of studies, the exact search string applied was: *"system dynamics" AND ("software development" OR "software engineering") AND (productivity OR performance OR effectiveness).* We selected broader domain terms ("software development" and "software engineering") rather than more specific terms ("software teams" or "development teams") to ensure comprehensive capture of relevant literature.
- Temporal scope: The searches covered 2014–2024 to capture recent developments in software engineering while allowing enough time to observe trends in SD tool evolution and adoption.
- The inclusion criteria are studies published between 2014 and 2024 (CI1), studies related to software engineering (CI2), and studies related to SD (CI3).
- Exclusion criteria include documents not available (CE1), articles not written in English (CE2), articles from another field of knowledge (CE3), duplicated studies (CE4), literature reviews (CE5), and studies without a simulation model (CE6).
- The quality criteria are quantity and quality of bibliographic references (CC1), methodological clarity (CC2), and use of quantitative data (CC3). Each criterion is rated 1 if it is met and 0 if it is not. Two members of the research team evaluate each selected study, identifying whether it meets each quality criterion. Those studies that meet at least two of the requirements are retained. If the two reviewers disagree on whether a study should be retained, the third team member takes the final decision.
- Data extraction strategy: Articles that meet the quality criteria will be read in their entirety. From the reading, general data (title, authors, year of publication, keywords, DOI, and journal) and data related to the research question (technical factors, non-technical factors, technological tool used, tool characteristics, application context, limitations) are extracted.
- Data synthesis: A qualitative summary [8] will be made, presenting frequencies of findings related to the research question.

2.2 Conducting the Review

The review execution comprised study identification, selection, quality assessment, and data extraction.

- Study identification: The search was conducted during the first week of March 2025 using the predefined search string in the selected databases.
- Study selection: The study selection followed a PRISMA-structured process, moving from initial identification to screening, eligibility, and final inclusion. The flow of records through each stage is summarized in Table 1.

- Quality assessment: Table 2 summarizes the consolidated results and confirms that all studies met the minimum inclusion threshold.

Table 1. The selection process for primary studies according to PRISMA.

Phase	Identification	Screening	Eligibility	Included
Records	139 total: Scopus 107; WoS 32	73 after filters	20 full-text assessed	20 studies included
Exclusions	–	53 total: CE3 (21), CE4 (11),CE1 (11), CE6 (5), CE5 (5)	0 excluded	–

Table 2. The selection process for primary studies.

Criterion	Description	Evaluator 1	Evaluator 2
CC1 – References	Adequate bibliographic support	20 (100%)	19 (95%)
CC2 – Methodology	Clear methodological description	19 (95%)	20 (100%)
CC3 – Quantitative data	Report of quantitative data supporting results	16 (80%)	17 (85%)
Studies meeting ≥2 criteria	Threshold for inclusion	20 (100%)	20 (100%)

Two independent reviewers conducted the screening and quality assessment. Inter-rater agreement was calculated using Cohen's kappa. The observed agreement was $P_o = 0.97$, while the agreement expected by chance was $P_e = 0.82$, yielding $\kappa = 0.81$, which represents substantial agreement according to the Landis & Koch scale [9]. Discrepancies were discussed and resolved by consensus. Since all studies met at least two of the criteria, they were included in the data extraction process. The selected studies are listed in Table 3.

Table 3. Selected studies.

ID	Title	Main focus	Ref.
S01	Helping teams to help themselves: An industrial case study on interdependencies during sprints.	Simulation of agile teams	[10]
S02	Capability erosion dynamics	Erosion of capacities	[11]
S03	Using system dynamics to teach about dependencies, correlation, and systemic thinking on the software process workflows	COCOMO II Models	[12]

(continued)

Table 3. (*continued*)

ID	Title	Main focus	Ref.
S04	An empirically validated simulation for understanding the relationship between process conformance and technology skills	Training vs. processes	[13]
S05	Effects of Funding Fluctuations on Software Development: A System Dynamics Analysis	Knowledge management	[14]
S06	Modeling the Dynamics of an Agile Scrum Team in the Development of a Single Software Project	Sprinting behavior	[15]
S07	Design and management of software development projects under rework uncertainty: a study using system dynamics	Complexity and rework	[16]
S08	When to automate software testing? Decision support based on system dynamics: An industrial case study	Test automation	[17]
S09	Productivity of Software Development Teams Based on Social and Human Factors: A Dynamic Conceptual Model.	Social and human factors	[18]
S10	Systems dynamics model for decision support in risk assessment in software projects	Project success/failure	[19]
S11	Characterizing relationships for system dynamics models supported by exploratory data analysis: A conceptualizing approach about the meeting diversity in student software projects	Communication and productivity	[20]
S12	A study of software development project cost, schedule, and quality by outsourcing to low-cost destination	Distributed equipment	[21]
S13	A hybrid simulation model of individual and team performance in software project environment	Human behavior in CMMI	[22]
S14	When to automate software testing? A decision-support approach based on process simulation	Productivity in testing	[23]

(*continued*)

Table 3. (*continued*)

ID	Title	Main focus	Ref.
S15	The effect of onsite-offshore work division on project cost, schedule, and quality for re-engineering projects in Indian outsourcing software industry	Offshore equipment	[24]
S16	System dynamics modeling to manage performance based on scope change for software development projects	Schedule pressure	[25]
S17	A Dynamic Approach to Performance for Software Development Projects	Human capital	[26]
S18	System dynamics modelling of the impact of agile practice on the quality of continuous delivery projects	TDD and culture	[27]
S19	Change-Oriented Open-Source Software Process Simulation	Change management	[28]
S20	A Dynamic Model of Individual and Collective Learning Amid Disruption	Organizational learning	[29]

3 Results

Following Kitchenham's recommendations for data synthesis [7, 30], we qualitatively analyzed the extracted data, categorizing technical and non-technical factors, assessing tool characteristics and accessibility, and identifying research patterns and gaps.

3.1 Technical and Non-technical Factors in SD Models

The technical factors were grouped into six categories derived from the content analysis, reflecting key methodological and evaluation aspects of the software development process (Table 4).

Table 4. Technical aspects in selected studies

Category and representative factors	Related studies
Project management and methodologies: Sprint planning, retrospectives, SCRUM, productivity metrics, quality assurance	S01, S04, S06, S07, S08, S10, S13, S16, S17, S18
Tools and technology: JIRA, version control systems, automation tools, and infrastructure	S01, S08, S14

(*continued*)

Table 4. (*continued*)

Category and representative factors	Related studies
Development and architecture: Software design, code review, system architecture, integration	S02, S05, S12, S19
Quality and testing: Automated testing, defect detection, test coverage, rework	S02, S06, S08, S18, S19
Measurement and metrics: COCOMO II, performance metrics, productivity indicators	S03, S11, S14, S15, S20

Technical factors show advanced modeling of software processes. Project management (S01, S06) links planning to team performance; quality/testing (S08, S14) reveals feedback between automation and long-term productivity; and measurement (S03, S20) incorporates systemic and behavioral effects. Most are combined with communication elements, as seen in nine studies.

Complementarily, the non-technical dimension of software development encompasses the human, social, and organizational aspects that influence project success. Table 5 organizes these factors into six categories that capture the complexity of human and organizational interactions in software development.

Table 5. Non-technical aspects in selected studies.

Category and representative factors	Related studies
Communication and collaboration: Effective communication, team coordination, and interdepartmental collaboration	S02, S06, S08, S11, S12, S15, S18, S19
Team management and leadership: Leadership, staff turnover, resource allocation, and team management	S02, S04, S05, S10, S12, S16
Knowledge and competencies: Knowledge transfer, staff experience, and learning processes	S05, S08, S12, S14, S15, S17, S20
Human and psychological aspects: Motivation, stress, mood, and emotional factors	S01, S04, S09, S11, S13, S19
Organizational culture: Corporate culture, resistance to change, and organizational adaptation	S04, S08, S13, S15, S18
Stakeholder management: Customer satisfaction, user involvement, expectation management	S01, S02, S10, S16

Non-technical factors in SD models include communication (S11, S18), linking collaboration to productivity; knowledge management (S05, S17, S20), showing how expertise distribution drives learning; human factors (S09, S13), adding psychological influences; and organizational culture (S04, S18), modeling transformation and resistance. These factors are usually integrated with technical elements. The interdisciplinary nature of software development is reflected in the co-occurrence of multiple factor categories within the same studies, as shown in Table 6.

Table 6. Co-occurrence of factors in selected studies.

Related aspects	Related studies
Project management + Communication	S01, S04, S06, S08, S11, S16, S18, S19
Quality/Testing + Human Aspects	S02, S06, S08, S13, S16, S18
Tools + Competencies	S01, S08, S14, S15, S17
Methodologies + Organizational culture	S04, S13, S15, S18
Measurement + Equipment management	S05, S12, S14, S20

Co-occurrence patterns show a mature view of software development as a socio-technical system: project management and communication links link management effectiveness to information flow. At the same time, quality and human aspects highlight the role of decision-making in testing. These patterns indicate a shift toward integrated systemic models.

3.2 Characteristics and Accessibility of SD Tools and Models

Regarding modeling software, Vensim was the most frequently used tool (n = 8, 40%), followed by Isee Systems products (Stella or iThink, n = 2, 10%), and AnyLogic (n = 1, 5%). Eight of the remaining articles did not specify the commercial software used. Additionally, one study mentioned developing an SD simulator but provided no access information. Regarding the availability of the simulation models reported in the captured studies, only 3 (15%) provided full access to the models, either through explicit equations, download links, or detailed specifications in the text. Most studies (n = 13, 65%) presented partial information, typically limited to stock-and-flow diagrams without corresponding equations. Four articles (20%) did not specify their simulation models. In two studies, although the text mentioned that resources were available online, they were not accessible.

We also evaluated whether studies reported simulated data management capabilities, which support parameter changes, scenario comparison, and result validation. Among the 20 studies, 17 (85%) did not provide this information, 1 (5%) mentioned it without evidence, and 2 (10%) were conceptual models in which it was not applicable.

4 Discussion

The temporal analysis of SD-based simulation model publications shows a 2016 peak followed by a decline, suggesting a shift from fundamental research to technological maturity and commercialization [31]. This transition from academic to industrial implementation [32], driven by established commercial platforms and proprietary developments [33], aligns with the high concentration of industrial applications (60% of reviewed studies), indicating real demand, practical problem-solving, and technical-economic feasibility in productive software contexts.

On the other hand, the presence of studies generally framed in project management suggests that the applications of simulation models are not limited to specific methodologies but rather address fundamental principles and concepts of project management that are applicable regardless of the methodological approach adopted [34]. These results suggest the need to complement this analysis with patent records, software databases, and industrial gray literature to obtain a more complete perspective of the current state and evolution of the field.

Results from studies on factors considered by SD models reveal the socio-technical nature of software development, as evidenced by studies that consider both technical and non-technical factors (Table 5). This trend contrasts with previous approaches that prioritized exclusively technical aspects [35, 36], suggesting an evolution towards more integrated perspectives that consider both technical competencies and organizational and human skills [37].

Including communication, collaboration, and team management in software development analysis underscores the importance of human factors in project success. This observation is evident in studies of agile methodologies, where the co-occurrence of technical processes with social dynamics suggests that the effectiveness of these practices depends on the effective integration between both dimensions [38, 39], reinforcing the need for multidisciplinary skills, educational programs that incorporate organizational psychology and team management, and organizational frameworks that strengthen human capacities and optimize resources.

The construction of an SD model typically includes a causal diagram, a stock-and-flow diagram, the mathematical equations that support the model, and the parameter and input values [1]. However, the limited availability of complete models (15%) represents a barrier to reproducibility and validation of results in the field of SD applied to software development. This lack of methodological transparency hinders the replication of studies and the incremental construction of knowledge, which are fundamental aspects of scientific progress in this discipline.

Complementarily, the widespread absence of information on simulated data management (85%) represents a methodological limitation that compromises the reproducibility and reusability of the proposed models [40]. This lack suggests that researchers either do not consider or do not adequately report the technical capabilities required for iterative validation and sensitivity analysis, essential to the robustness of SD models. The lack of documentation of these functionalities makes it difficult to assess the suitability of the tools for future research and practical applications [41].

4.1 Implications for Practice and Research

Professionally, results support creating open-source SD tools for software teams with validated, fully documented models in public repositories, integrating simulation into project management platforms, investing in technical training, and establishing metrics to assess their impact on productivity and quality. Academically, SD models should include non-technical factors and be validated through longitudinal studies using software quality indicators, with reproducibility as a core scientific criterion.

For the software industry, the findings indicate the need to actively contribute to shared scientific knowledge by publishing case studies with real data, funding university-industry collaborative research focused on accessible tools, and adopting open standards for the exchange of simulation models. Technology companies should lead the development of integrated ecosystems that combine SD tools with existing development platforms, facilitating the mass adoption of these tools to improve software quality on an industrial scale systematically.

4.2 Limitations and Threats to Validity

This SLR has several limitations that should be considered. This review relied on Scopus and WoS due to institutional access restrictions that prevented the use of IEEE Xplore and the ACM Digital Library. Although these two databases index major software engineering venues and provide broad multidisciplinary coverage, together they index the majority of journal and conference publications in System Dynamics and software engineering. Consequently, the risk of missing highly relevant studies is reduced but cannot be fully eliminated. Claims regarding coverage and completeness have been moderated accordingly.

The search was limited to two academic databases, potentially excluding gray literature, patents, and unpublished commercial developments that could provide a more complete perspective of the current state of SD tools in software development. Although the search string was validated through preliminary exploration, it may not have captured studies using alternative terminology.

Selection and quality assessment relied on subjective criteria, potentially biasing study classification despite reviewers' consensus. Data extraction relied on published information, which may omit tool capabilities or implementation details, and factor categorization depended on reviewers' interpretation of authors' descriptions.

Finally, the findings reflect academic research and may not accurately represent the state of industrial practice. The 10-year time frame (2014–2024) may not capture the most recent developments that have not yet been formally published or are undergoing peer review.

5 Conclusions

This SLR, based on publications between 2014 and 2024, demonstrates that the development of accessible SD-based technological tools for productivity analysis of software development teams must integrate both technical and non-technical factors, provide properly documented and validated simulation models, and offer robust capabilities for managing simulated data. Overcoming the identified access and reproducibility limitations is a crucial first step toward democratizing the use of these analytical methodologies and systematically realizing their full potential to enhance productivity and quality in the software development industry.

This study contributes to scientific knowledge by systematizing the current state of the art of SD applications in software development, providing a classification of technical and non-technical factors considered in existing models. The identification of co-occurrence patterns across different categories of factors provides an empirical basis for designing more comprehensive models that capture the inherent complexity of software development. For future research, we identify the need to complement this analysis with patent records, software databases, and industrial gray literature to gain a more complete perspective of SD-based simulation tools.

References

1. Sterman, J.D.: Business Dynamics: Systems Thinking and Modeling for a Complex World. Mc Graw Hill (2000)
2. Restrepo-Tamayo, L.M., Gasca-Hurtado, G.P., Valencia-Calvo, J.: Simulation of software development team productivity incorporating social and human factors: a system dynamics model. IEEE Access. **13**, 24604–24621 (2025). https://doi.org/10.1109/ACCESS.2025.3527330
3. Li, W., Fang, C.C.: Applying a system dynamics approach for decision-making in software testing projects. PLoS One. **20**, e0323765 (2025). https://doi.org/10.1371/JOURNAL.PONE.0323765
4. Ferreira Franco, E., Hirama, K., Carvalho, M.M.: Applying system dynamics approach in software and information system projects: a mapping study. Inf. Softw. Technol. **93**, 58–73 (2018). https://doi.org/10.1016/j.infsof.2017.08.013
5. Bala, B., Mohamed, F., Mohd, K.: System dynamics. In: Modelling and Simulation. Springer Nature (2017). https://doi.org/10.4324/9780203112694-14
6. Bureš, V.: Comparative analysis of system dynamics software packages. Int. Rev. Modell. Simul (IREMOS). **8**, 245–255 (2015). https://doi.org/10.15866/IREMOS.V8I2.5401
7. Kitchenham, B., Charters, S.: Guidelines for performing Systematic Literature reviews in Software Engineering. Keele, Staffs, UK (2007). https://doi.org/10.1145/1134285.1134500
8. Noblejas, C.J., Rodríguez, A.P.: Information retrieval and visualization in web of science and Scopus: a practical approach. Investigación Bibliotecológica: Archivonomía, Bibliotecología e Información. **28**, 15–31 (2014). https://doi.org/10.1016/S0187-358X(14)70907-4
9. Cruzes, D.S., Dyb, T.: Research synthesis in software engineering: a tertiary study. Inf. Softw. Technol. **53**, 440–455 (2011). https://doi.org/10.1016/j.infsof.2011.01.004
10. Klünder, J., Kortum, F., Ziehm, T., Schneider, K.: Helping teams to help themselves: an industrial case study on interdependencies during sprints. In: Lecture Notes in Computer Science (Including Subseries Lecture Notes in Artificial Intelligence and Lecture Notes in Bioinformatics), pp. 31–50. Springer Verlag (2019). https://doi.org/10.1007/978-3-030-05909-5_3
11. Rahmandad, H., Repenniing, N.: Capability erosion dynamics. Strateg. Manag. J. **37**, 649–672 (2015). https://doi.org/10.1002/smj
12. Dugarte-Peña, G.L., Sánchez-Segura, M.I., de Amescua, A., Medina-Domínguez, F., Armenia, S.: Using system dynamics to teach about dependencies, correlation and systemic thinking

on the software process workflows. IET Softw. **15**, 351–364 (2021). https://doi.org/10.1049/sfw2.12031
13. Matalonga, S., Solari, M., Feliu, T.S.: An empirically validated simulation for understanding the relationship between process conformance and technology skills. Softw. Qual. J. **22**, 593–609 (2014). https://doi.org/10.1007/s11219-013-9214-2
14. Trammell, M.T.I., Moulton, A., Madnick, S.E.: Effects of funding fluctuations on software development: a system dynamics analysis. EMJ – Eng. Manag. J. **28**, 71–85 (2016). https://doi.org/10.1080/10429247.2016.1155390
15. Ching, P.M., Mutuc, J.E.: Modeling the dynamics of an agile scrum team in the development of a single software project. In: IEEE International Conference on Industrial Engineering and Engineering Management. 2019, pp. 386–390 (2018). https://doi.org/10.1109/IEEM.2018.8607430
16. Khatun, M.T., Hiekata, K., Takahashi, Y., Okada, I.: Design and management of software development projects under rework uncertainty: a study using system dynamics. J. Decis. Syst. **32**, 265–288 (2023). https://doi.org/10.1080/12460125.2021.2023257
17. Sahaf, Z., Garousi, V., Pfahl, D., Irving, R., Amannejad, Y.: When to automate software testing? Decision support based on system dynamics: an industrial case study. In: ACM International Conference Proceeding Series, pp. 149–158. Association for Computing Machinery (2014). https://doi.org/10.1145/2600821.2600832
18. Restrepo-Tamayo, L.M., Gasca-Hurtado, G.P., Machuca-Villegas, L.E., Morillo-Puente, S.: Productivity of software development teams based on social and human factors: a dynamic conceptual model. In: Lecture Notes in Networks and Systems, pp. 13–23. Springer Science and Business Media Deutschland GmbH (2022). https://doi.org/10.1007/978-3-031-04829-6_2
19. Lopes, D.S., Braga, J.L., Resende Filho, M.A.: Systems dynamics model for decision support in risk assessment in software projects. J. Softw. Evol. Process. **27**, 976–989 (2015). https://doi.org/10.1002/smr.1754
20. Kortum, F., Klünder, J., Schneider, K.: Characterizing relationships for system dynamics models supported by exploratory data analysis : a conceptualizing approach about the meeting diversity in student software projects. In: Proceedings of the International Conference on Software Engineering and Knowledge Engineering, SEKE, pp. 608–614. Knowledge Systems Institute Graduate School (2017). https://doi.org/10.18293/SEKE2017-143
21. Mishra, D., Mahanty, B.: A study of software development project cost, schedule and quality by outsourcing to low cost destination. J. Enterp. Inf. Manag. **29**, 454–478 (2016). https://doi.org/10.1108/JEIM-08-2014-0080
22. Alshammri, M., Qin, S.: A hybrid simulation model of individual and team performance in software project environment. In: 22nd International Congress on Modelling and Simulation, MODSIM 2017 (2017)
23. Garousi, V., Pfahl, D.: When to automate software testing? a decision-support approach based on process simulation. J. Softw. Evol. Process, 272–285. John Wiley and Sons Ltd (2016). https://doi.org/10.1002/smr.1758
24. Mishra, D., Mahanty, B.: The effect of onsite-offshore work division on project cost, schedule, and quality for re-engineering projects in Indian outsourcing software industry. Strategic Outsourcing. **7**, 198–225 (2014). https://doi.org/10.1108/SO-06-2014-0010
25. Hiekata, K., Khatun, M.T., Chavy-Macdonald, M.A.: System dynamics modeling to manage performance based on scope change for software development projects. In: Advances in Transdisciplinary Engineering, pp. 675–684. IOS Press BV (2019). https://doi.org/10.3233/ATDE190177
26. Díaz Vega, M.I., Ospina, M.J.: A dynamic approach to performance for software development projects. J Manag. **39**, 77 (2023). https://doi.org/10.25100/cdea.v39j77.12579

27. Akerele, O.: System dynamics modelling of the impact of agile practice on the quality of continuous delivery projects. Innov. Syst. Softw. Eng. **14**, 183–208 (2018). https://doi.org/10.1007/s11334-017-0296-z
28. Zhang, X., Wang, X., Kang, Y.: Change-oriented open source software process simulation. IEEE Access. **6**, 70145–70163 (2018). https://doi.org/10.1109/ACCESS.2018.2880998
29. Anderson, E.G., Lewis, K.: A dynamic model of individual and collective learning amid disruption. Organ. Sci. **25**, 356–376 (2014). https://doi.org/10.1287/orsc.2013.0854
30. Brereton, P., Kitchenham, B.A., Budgen, D., Turner, M., Khalil, M.: Lessons from applying the systematic literature review process within the software engineering domain. J. Syst. Softw. **80**, 571–583 (2007). https://doi.org/10.1016/j.jss.2006.07.009
31. Park, M., Leahey, E., Funk, R.J.: Papers and patents are becoming less disruptive over time. Nature. **613**, 138–144 (2023). https://doi.org/10.1038/s41586-022-05543-x
32. Dedehayir, O., Steinert, M.: The hype cycle model: a review and future directions. Technol Forecast Soc Change. **108**, 28–41 (2016). https://doi.org/10.1016/J.TECHFORE.2016.04.005
33. Arora, A., Belenzon, S., Patacconi, A.: The decline of science in corporate R&D. Strateg. Manag. J. **39**, 3–32 (2018). https://doi.org/10.1002/SMJ.2693;REQUESTEDJOURNAL:JOURNAL:10970266;WGROUP:STRING:PUBLICATION
34. Kerzner, H.: Project Management: a Systems Approach to Planning, Scheduling, and Controlling. John Wiley & Sons (2017)
35. Sommerville, I.: Ingenieria del Software. Adisson-Wesley (2011)
36. Sadowski, C., Zimmermann, T.: Rethinking Productivity in Software Engineering. Springer, New York (2019)
37. Hidellaarachchi, D., Grundy, J., Hoda, R., Madampe, K.: The effects of human aspects on the requirements engineering process: a systematic literature review. IEEE Trans. Softw. Eng. **48**, 2105–2127 (2022). https://doi.org/10.1109/TSE.2021.3051898
38. Chagas, A., Santos, M., Santana, C., Vasconcelos, A.: The impact of human factors on agile projects. In: Proceedings - 2015 Agile Conference, pp. 87–91. Agile. 2015 (2015). https://doi.org/10.1109/AGILE.2015.11
39. Das, S., Gary, K.: Challenges and success factors in large scale agile transformation—a systematic literature review. In: ITNG 2024: 21st International Conference on Information Technology-New Generations, pp. 405–416 (2024). https://doi.org/10.1007/978-3-031-56599-1_52
40. Barlas, Y.: Model validation in system dynamics. In: Proceedings of the 1994 International System Dynamics Conference, pp. 1–10 (1994)
41. Dooley, J.F.: Software Development, Design and Coding. Apress, Galesburg, Illinois, USA (2017). https://doi.org/10.1007/978-1-4842-3153-1

Optimizing the Balance Between Knowledge Acquisition and Cognitive Load in Virtual Reality-Based Education for Children with Special Needs

Stefan Stoev[1(✉)], Katia Rasheva-Yordanova[1], Daniela Borissova[1,2], and Anna Lekova[3]

[1] University of Library Studies and Information Technologies, Sofia, Bulgaria
{s.stoev,k.rasheva}@unibit.bg, daniela.borissova@iict.bas.bg
[2] Institute of Information Technologies of Bulgarian Academy of Sciences, Sofia, Bulgaria
[3] Institute of Robotics of Bulgarian Academy of Sciences, Sofia, Bulgaria
a.lekova@ir.bas.bg

Abstract. The integration of Virtual Reality (VR) into education has emerged as a transformative approach for creating immersive and engaging learning environments, particularly for children with special educational needs. VR technologies enable the visualization of complex concepts, support individualized learning, and foster social and emotional development. However, prolonged exposure to virtual environments and inappropriate content complexity may lead to cognitive overload, reducing concentration, increasing fatigue, and negatively impacting learning outcomes. Cognitive Load Theory (CLT) provides a theoretical basis for balancing knowledge acquisition with mental effort, yet formal optimization frameworks addressing this balance in VR-based education remain underexplored. The article proposes a bi-objective mathematical model aiming to deter-mine the optimal duration and complexity of VR-based educational sessions for children with learning difficulties. The model defines an objective function that maximizes knowledge acquisition while minimizing cognitive load, supported by analytical expressions linking session time, content complexity, learning out-comes, and learner fatigue. Numerical simulations demonstrate that effective learning outcomes are achieved within session durations of 15–20 min and medium content complexity, beyond which cognitive load disproportionately in-creases. The contribution of the article is related to the field of education by offering a formalized framework for optimizing VR learning experiences in special education. The results highlight the potential of adaptive approaches to enhance learning effectiveness while ensuring cognitive well-being.

Keywords: Virtual Reality · Cognitive Load · Special Educational Needs · Optimization Model · Adaptive Learning

A. Rocha et al. (Eds.): ICITS 2026, LNNS 1966, pp. 16–26, 2027.
https://doi.org/10.1007/978-3-032-27538-7_2

1 Introduction

Virtual Reality (VR) is emerging as one of the most promising technologies for transforming the educational process, particularly for children with special educational needs. It provides immersive, interactive, and personalized learning environments that create new opportunities for visualizing complex concepts, developing practical skills, and designing adaptive scenarios tailored to the individual cognitive and emotional characteristics of learners [1, 2].

Research shows that VR environments can enhance social and emotional skills through simulated situations, stimulate self-regulation, and promote active engagement in the learning process [3, 4]. Particularly important is the ability to personalize content and adjust the workload, ensuring that children with special needs can participate effectively without being overwhelmed [5, 6].

Despite these advantages, VR poses challenges related to cognitive load, the duration of learning sessions, and the complexity of the content [7]. Excessive time in a virtual environment or overly complex material can lead to fatigue, decreased concentration, and reduced learning effectiveness [8]. Cognitive Load Theory provides a valuable framework for managing this balance, emphasizing that effective learning de-pends on the careful regulation of the complexity and amount of information in accordance with the individual characteristics of learners [9–11].

A review of the scientific literature shows that numerous adaptive systems and methods for measuring and regulating cognitive load have been proposed [12]. However, most existing approaches focus primarily on monitoring or adjusting cognitive load, rather than on formally optimizing the balance between knowledge acquisition and cognitive effort. To address this limitation, the present study develops a mathematical model that determines the optimal duration and content complexity of VR learning sessions for children with special educational needs.

Although VR has been widely applied in inclusive education, existing studies tend to focus either on its pedagogical benefits [2, 5] or on the measurement of cognitive load in virtual environments [8, 12]. Recent systematic reviews further highlight that no formal optimization frameworks have been proposed that jointly consider knowledge acquisition and cognitive load, particularly in special education contexts [13, 14]. This clearly demonstrates the need for models that integrate learning outcomes, content complexity, and cognitive constraints in VR-based instruction for children with special educational needs.

In response to this gap, the proposed model aims to maximize knowledge acquisition while simultaneously minimizing cognitive load and the risk of fatigue. The theoretical framework builds on the principles of Cognitive Load Theory and educational psychology, while validation has been carried out through numerical simulations determining optimal training parameters.

The present article is organized into three sections. The first section provides a review of the existing achievements in the studied field. The second section presents the developed mathematical model. The third section describes the conducted experiments and their results. The conclusion outlines the main findings and future research directions.

2 Theoretical Background and Related Work

2.1 Virtual Reality in Education

The use of VR in education is regarded as a significant innovation that creates engaging and immersive learning environments. These environments support active learning and the development of practical skills by providing new opportunities for visualizing com-plex concepts and enhancing learner motivation [15]. Meta-analyses show that VR technologies enhance satisfaction and learning outcomes across various educational contexts [13]. Similar results have also been confirmed by other studies, demonstrating significant improvement in learners' cognitive and social skills [14].

2.2 Applications of VR for Children with Special Educational Needs

VR technology demonstrates particularly strong potential in the education of children with special educational needs [7]. By creating adaptive scenarios, it can address learners' individual cognitive and emotional difficulties [16]. The ability to personalize content and control stimuli makes VR an effective tool for overcoming the limitations of traditional education [5].

Research shows that VR facilitates social interaction, develops emotional regulation, and supports self-directed learning, while simultaneously reducing learning barriers for children with autism, dyslexia, or intellectual disabilities [17, 18].

2.3 Cognitive Load Theory

Cognitive Load Theory (CLT) provides a valuable framework for optimizing the learning process by controlling the amount and complexity of information. According to Sweller, working memory is a limited resource, and its overload leads to reduced learning effectiveness [10]. Additional studies show that cognitive load is influenced by the content, the duration of the learning session, and the individual characteristics of the learner [9]. In the context of VR, this theory is particularly relevant, as the immersive nature of the environment can lead to sensory overload [19]. Dynamic measurement and adaptation of cognitive load in VR show that it is possible to create learning scenarios that maintain an optimal balance between challenge and effective learning [4, 20].

2.4 Optimization Methods in VR-Based Education

Despite the widespread application of VR in education, few studies propose formal optimization models that simultaneously take into account both knowledge acquisition and cognitive load. In this context, Fu et al. [21] propose a multi-objective optimization that minimizes cognitive load while simultaneously maximizing learning outcomes. Additional methods include the implementation of quality functions combined with neural networks to adapt resources according to learners' individual needs.

Existing results indicate that VR can serve as an effective tool for inclusive education, but the optimization of the learning process requires a systematic approach that balances cognitive efficiency and educational outcomes. The present study positions itself in this direction by proposing a formalized mathematical model for determining the optimal duration and complexity of VR learning sessions for children with special needs.

3 Mathematical Model

The goal of the formulated mathematical model is to determine the optimal combination of time spent in the VR learning environment and the complexity of the instructional content in order to achieve maximum knowledge acquisition with minimal cognitive load on learners. Therefore, to define the balance between acquired knowledge and fatigue/load, the following bi-criteria problem is formulated:

$$\max_{T,S} Z(T,S) = \begin{cases} \max(K(T,S) = \alpha.\ln(1+T).\ln(1+S)) \\ \min\left(F(T,S) = b.T^2 + c.S^2\right) \end{cases} \tag{1}$$

Subject to:

$$0 \le T \le T_{max} \tag{2}$$

$$0 \le S \le S_{max} \tag{3}$$

The formula for cognitive load *F*(*T*, *S*) includes two components: the first (bT^2) reflects the increase of fatigue over time, while the second (cS^2) represents the load resulting from the complexity of the content. The parameters *b* and *c* are weighting co-efficients that balance the influence of the two factors on the overall load. Where:

- *T* is the duration of the learning session (in minutes);
- *S* is the complexity of the instructional content (evaluated on *a* scale or through quantitative indicators);
- α is an empirical sensitivity coefficient that scales the acquired knowledge *K*(*T*, *S*) and determines the relative importance of knowledge in relation to cognitive load. Its value is defined through experimental data;
- *Tmax* and *Smax* are the maximum allowable values.

  The values of the parameters were determined after preliminary analysis and sensitivity testing of the model. Parameter α acts as a scaling factor of the knowledge acquisition function, reflecting the relative weight of learning outcomes. Parameter *b* balances the temporal growth of cognitive fatigue, while *c* calibrates the contribution of content complexity to the overall load. Their values are chosen to ensure a realistic representation of the learning process, aligned with both theoretical assumptions of Cognitive Load Theory and empirical observations reported in related literature.

The formulated problem aims to maximize the overall educational effectiveness *Z*(*T*, *S*), defined as a function of the acquired knowledge *K*(*T*, *S*) and the load *F*(*T*, *S*).

The function *K*(*T*, *S*) describes the growth of knowledge with increasing time and complexity, while also accounting for the diminishing effect when these factors reach excessively high values.

The formulation of the objective functions is directly grounded in Cognitive Load Theory and its core principles. The logarithmic growth of K(T,S) reflects the diminishing returns associated with extended exposure or increasing complexity, as described in CLT's notion of intrinsic and extraneous load [10]. Similarly, the quadratic structure of F(T,S) operationalizes the well-established observation that cognitive fatigue grows non-linearly with time-on-task and sensory stimulation [9]. The proposed mathematical formulation encodes the theoretical constraints imposed by the cognitive architecture, ensuring that the model corresponds to empirically validated cognitive mechanisms.

4 Experimental Validation

To validate the formulated bi-criteria problem, numerical testing was conducted based on parameters corresponding to educational sessions with children with special needs. To solve the problem, an ε-constraint method was used, which can be considered as interactive either a priori or a posteriori, depending on the stage at which the decision-maker expresses their preferences [22]. This method was selected as the most appropriate due to its ability to efficiently identify compromise solutions between conflicting objectives, taking into account both the maximization of acquired knowledge and the limitation of cognitive load.

The parameter ranges used in the numerical validation were informed by preliminary pedagogical observations from VR-assisted activities with children aged 5–7 years in preschool settings. These pilot observations were conducted in collaboration with teachers, involved no data collection, and focused on short exploratory VR tasks de-signed to support attention and basic conceptual understanding. All procedures followed institutional ethical guidelines, and no personally identifiable information was recorded. The VR content included simple spatial navigation and object interaction tasks appropriate for preschool cognitive development.

4.1 Input Data for Numerical Testing

The constraints used in solving the formulated problem (1)–(3) are as follows:

- Duration of the learning session (T): 5 to 30 min.
- Complexity of the instructional content (S): evaluated on a scale from 1 to 5.
- $bb = 0.05$ and $cc = 0.5$;
- $\alpha\alpha = 10$.

The obtained max values for the acquired knowledge K(T,S) and min values for cognitive load F(T,S) under different combinations of time T and complexity S are shown in Table 1.

It is noticeable from Table 1 that the acquired knowledge $K(T, S)$ increases with the duration of the learning session and the complexity of the content, but the growth rate slows down at higher values. The graphical representation of knowledge acquisition as a function of time and complexity is shown in Fig. 1. The comparison highlights the role of the coefficient α as a scaling parameter that reflects the relative importance of knowledge.

The cognitive load F(T,S), calculated as a quadratic function, increases significantly faster, particularly during longer sessions or with higher complexity. This relationship is illustrated in Figs. 2 and 3.

Using the ε-constraints method to solve the multi-objective problem (1)–(3) requires successively solving the objective functions given in (1). The ε-constraints technique

Table 1. Input Data for Numerical Testing with Parameters and $\alpha = 10$

Time, (min)	Complexity	max K (T,S)	min $F(T,S) = b.T^2 + c.S^2$
		Acquired knowledge	Cognitive load
10		26.34357	7.0
15		30.46000	13.25000
20	2	33.44750	22.0
25		35.79385	33.25000
30		37.72621	47.0
10		33.24189	9.500000
15		38.43624	15.75000
20	3	42.20604	24.50000
25		45.16681	35.75000
30		47.60517	49.50000
10		38.59264	13.00000
15		44.62309	19.25000
20	4	48.99970	28.00000
25		52.43704	39.25000
30		55.26789	53.00000
10		42.96452	17.50000
15		49.67812	23.75000
20	5	54.55052	32.50000
25		58.37725	43.75000
30		61.52879	57.50000

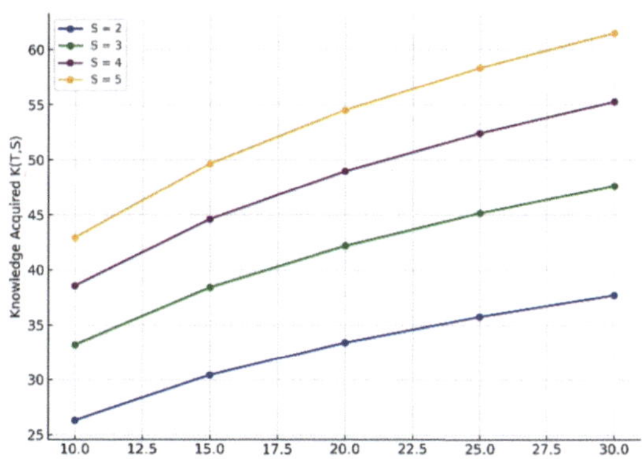

Fig. 1. Knowledge acquisition K(T,S) as a function of session duration and content complexity

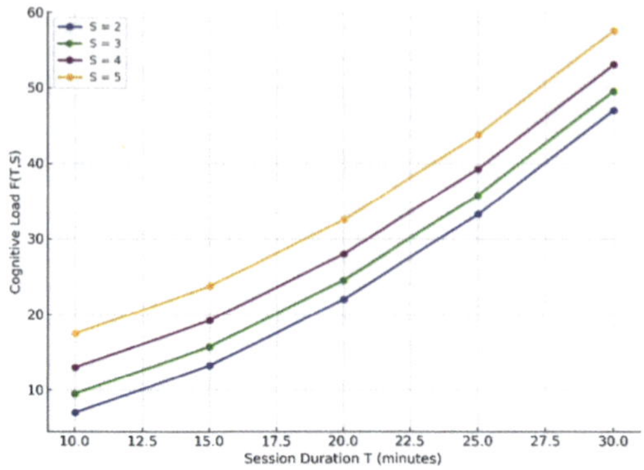

Fig. 2. Cognitive load F(T,S) as a quadratic function of session duration T and content complexity S.

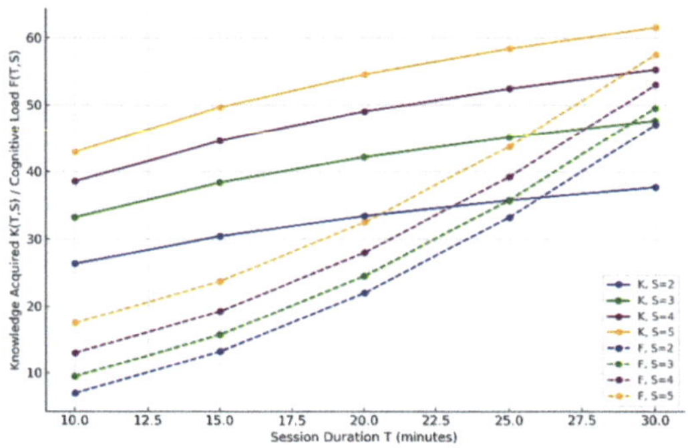

Fig. 3. Combined representation of knowledge acquisition and cognitive load for time duration between 10 and 30 min.

transforms a multi-objective problem into a single-objective one by optimizing one primary objective while constraining the remaining objectives to be less than or equal to a certain value. In the first step, we solve a single-objective problem to maximize $K(T, S)$ as follows:

$$\max \ \mathrm{K(T, S)} = 10. \ln(1 + \mathrm{T}). \ln(1 + \mathrm{S}) \tag{4}$$

Subject to:

$$15 \leq \mathrm{T} \leq 30 \tag{5}$$

$$3 \leq \mathrm{S} \leq 5 \tag{6}$$

The solution determines $K = 61.52879$, $T = 30$ and $S = 5$. On the next step, we determine restriction for K expressed as 20% bellow of maximum or $K > 49.223032$. As the problem is bi-objective, the final Pareto optimal solution is obtained on the second step, solving the single-objective problem to minimize $F(T, S)$ as follows:

$$\min \ \mathrm{F(T, S)} = 0.05.\mathrm{T2} + 0.5.\mathrm{S2} \tag{7}$$

Subject to:

$$15 \leq T \leq 30 \quad (8)$$

$$3 \leq S \leq 5 \quad (9)$$

$$K \geq 49.223032 \quad (10)$$

The solution of this task determines the value for the objective function (7) equal to 23.26637 ($F = 23.26637$), $K = 49.22303$ under $T = 15$ and $S = 4.902321$.

When decreasing the restriction for K by up to 30% of the maximum value, i.e., $K \geq 43.070153$, and replacing it in (10), the final Pareto-optimal solution is achieved at $T = 15$ and $S = 3.727646$, respectively, $K = 43.07015$ and $F = 18.19767$.

If $K \geq 36.917274$ (corresponding to 40% below the maximum), the optimal solution is obtained at T = 15 and S = 3, with $K = 38.43624$ and F = 15.75.

Table 2 summarizes the discrete optimal solutions obtained under different knowledge acquisition constraints, derived from the application of the proposed model.

Table 2. Optimal Discrete Solutions under Different Knowledge Acquisition Constraints.

Parameter Constraints (T, S)	K < 20% Kmax	K < 30% Kmax	K < 40% Kmax
15 < T < 30 3 < S < 5	T = 15 S = 4.902321	T = 15 S = 3.7276	T = 15 S = 3
20 < T < 30 3 < S < 4	T = 20 S = 3.272750	T = 20 S = 3	T = 20 S = 3

The obtained discrete optimal solutions are graphically illustrated in Fig. 4, where their distribution in the T–S and K–F spaces is clearly visualized.

The graphical representation demonstrates that the model yields discrete Pareto points, reflecting the trade-off between acquired knowledge and cognitive load. Under stricter knowledge constraints (K < 20% of Kmax), the optimal solutions correspond to higher complexity and shorter duration, whereas under more moderate conditions (K < 40% of Kmax), the solutions shift towards lower complexity and longer duration. This highlights that the optimal balance depends both on the imposed constraint and on the parameters of the learning session.

Figure 5 presents the Pareto front derived from the ε-constraint method. The curve clearly demonstrates that improvements in knowledge acquisition come at the cost of increased cognitive load, and vice versa. The highlighted points correspond to the three scenarios, showing how the decision-maker can select the most appropriate compromise depending on the educational context and learner's needs.

The comparison across the three scenarios demonstrates how relaxing the knowledge acquisition constraint leads to more balanced solutions. Scenario 1 (80% of maximum

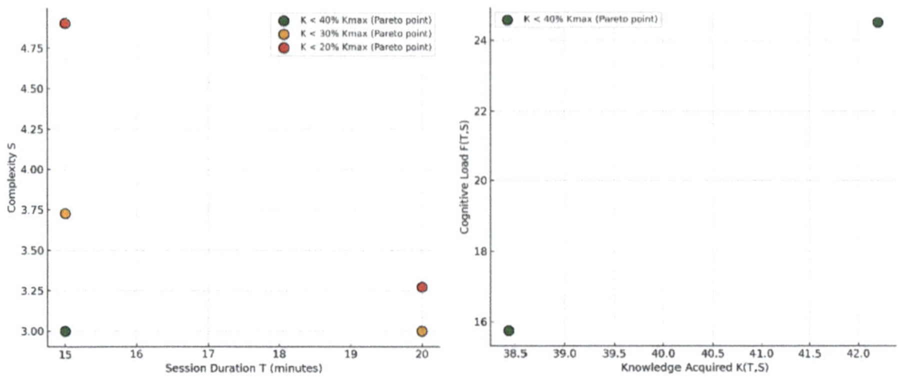

Fig. 4. Discrete Pareto Points in T–S and K–F Spaces

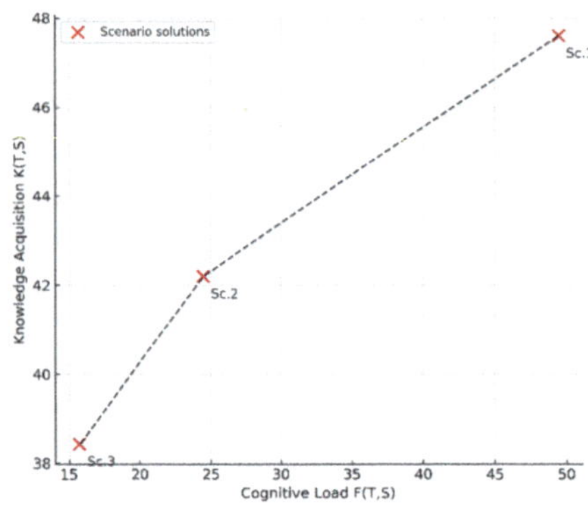

Fig. 5. Pareto-optimal points for knowledge acquisition and cognitive load

knowledge) ensures relatively high learning outcomes but requires higher complexity levels, which in turn increase cognitive load. Scenario 2 (70% of maximum) offers a compromise by maintaining solid knowledge acquisition while reducing the cognitive burden through lower complexity. Scenario 3 (60% of maximum) minimizes cognitive load and maintains learner comfort, though it achieves comparatively lower knowledge levels. Together, these scenarios highlight the flexibility of the optimization model and its ability to support informed decisions depending on whether the priority is to maximize knowledge, minimize cognitive effort, or strike a balance between the two.

5 Conclusion

This article presents a mathematical optimization model aimed at achieving a balance between knowledge acquisition and cognitive load in the education of children with special educational needs in a virtual environment. The proposed model considers two key factors of the learning process—session duration and content complexity—and demonstrates how their optimization leads to improved educational effectiveness.

The numerical modeling results show that excessive increases in VR session duration or content complexity do not yield proportional gains in knowledge acquisition but significantly increase cognitive load. The optimal parameters identified by the model suggest conducting learning sessions lasting between 15 and 20 min with medium content complexity (S $\approx$ 3), which provides the best balance between effectiveness and cognitive comfort.

The study is limited by the simulated nature of the validation and the relatively narrow parameter space derived from pilot observations. The model does not yet incorporate factors such as sensory sensitivity or prior knowledge, which may influence cognitive load dynamics. These limitations highlight the need for broader empirical testing in diverse educational contexts.

Future research will focus on the empirical validation of the model in real educational settings with a broader group of learners. The model will also be extended by incorporating individual learner characteristics such as concentration level, motivation, and emotional state. Another direction for development is the integration of artificial intelligence mechanisms to enable dynamic adaptation of learning sessions in real time according to the learner's current condition.

Acknowledgements. These research findings are supported by the National Scientific Research Fund, Project № КП-06-Н67/1.

References

1. Ajitha, S., Seema, S., Hotagi, H.S.: Guiding special need children using AR/VR TECHNOLOGY. In: 2023 7th International Conference on Computation System and Infor-Mation Technology for Sustainable Solutions (CSITSS), pp. 1–6. IEEE, Bangalore, India (2023). https://doi.org/10.1109/CSITSS60515.2023.10334086
2. Chec a-Domene, L., García-Martínez, I., Gavín-Chocano, Ó., Prieto, M.G.V.: Augmented and virtual reality as a teaching resource to attend to the diversity of students with special educational needs: a systematic review. Eur. J. Spec. Needs Educ. **39**(5), 709–728 (2023). https://doi.org/10.1080/08856257.2023.2282247
3. Poonia, R.C., Saudagar, A.K.J., Prabu, P., Khan, M.B.: Enhancing cognitive accessibility and social interaction in special education through AMLE. J. Stat. Manag. Syst. **27**(5), 1045–1063 (2024). https://doi.org/10.47974/JSMS-1348
4. Stoev, S., Rasheva-Yordanova, K., Bankovska, M., Getova, I.: Analysis of the applicability of VR in the education of children with disabilities. In: 2024 47th MIPRO ICT and Electronics Convention (MIPRO), pp. 364–368. IEEE, Opatija, Croatia (2024). https://doi.org/10.1109/MIPRO60963.2024.10569844
5. Chițu, I.B. et al.: Exploring the opportunity to use virtual reality for the education of children with dis-abilities. Children. **10**, 436 (2023). https://doi.org/10.3390/children10030436
6. Maigeldiyeva, Z., Alexandrova, N., Matagova, K.: Inclusive education in the era of sustainable development. E3S Web of Conferences. **451**, 05015 (2023). https://doi.org/10.1051/e3sconf/202345105015
7. Rasheva-Yordanova, K., et al.: A comprehensive review and analysis of virtual reality scenarios in speech and language therapy. TEM J. **14**(2), 1895–1907 (2025). https://doi.org/10.18421/TEM142-85

8. Albus, P., Seufert, T.: The modality effect reverses in a virtual reality learning environment and influences cognitive load. Instr. Sci. **51**, 545–570 (2023). https://doi.org/10.1007/s11251-022-09611-7
9. Young, J.Q., van Merriënboer, J.J.G., Durning, S., van der Cate, O.: Cognitive load theory: implications for medical education: AMEE guide no. 86. Medical Teacher. **36**(5), 371–384 (2014). https://doi.org/10.3109/0142159x.2014.889290
10. Sweller, J.: Element interactivity and intrinsic, extraneous, and germane cognitive load. Educ. Psychol. Rev. **22**, 123–138 (2010). https://doi.org/10.1007/s10648-010-9128-5
11. Nikolova, T., Staneva, A., Rasheva-Yordanova, K., Toleva-Stoimenova, S., Stanev, D.: Illustration in education with a focus on reducing cognitive strain. In: 2025 MIPRO 48th ICT and Electronics Convention, pp. 753–758. IEEE, Opatija, Croatia (2025). https://doi.org/10.1109/MIPRO65660.2025.11132072
12. Zhang, L. et al.: Cognitive load measurement in a virtual reality-based driving system for autism interven-tion. IEEE Trans. Affect. Comput. **8**(2), 176–189 (2017). https://doi.org/10.1109/TAFFC.2016.2582490
13. Yu, Z., Xu, W.: A meta-analysis and systematic review of the effect of virtual reality technology on users' learning outcomes. Comput. Appl. Eng. Educ. **30**(5), 1470–1484 (2022). https://doi.org/10.1002/cae.22532
14. Huang, W., Roscoe, R.D., Craig, S.D., Johnson-Glenberg, M.C.: Extending the cognitive-affective theory of learning with media in virtual reality learning: a structural equation modeling approach. J. Educ. Comput. Res. **60**(4), 807–842 (2021). https://doi.org/10.1177/07356331211053630
15. Noguchi, Y., Kusakawa, Y., Shimasaki, H., Miwa, C.: Effects of early clinical exposure using virtual reality on learning: a text-mining analysis study. Hong Kong J. Occup. Ther. **38**(1), 24–32 (2025). https://doi.org/10.1177/15691861251327632
16. Ardai, E., Vámos, T., Papp, G., Berencsi, A.: Virtual reality therapy in special needs education. Gyermeknevelés Tudományos Folyóirat. **10**(2–3), 259–271 (2022). https://doi.org/10.31074/gyntf.2022.3.259.271
17. Ullah, F., Johnson, T.A.L.: Customizable virtual reality gamification for enhanced teaching and learning for students with dyslexia. Int. J. Teaching Learn Educ. **3**(5), 45–62 (2024). https://doi.org/10.22161/ijtle.3.5.4
18. Sobota, B., Korecko, S., Mattová, M.: Web-based 3D virtual environments utilization in primary and secondary education of children with multiple impairments. Electronics. **12**(13), 2792 (2023). https://doi.org/10.3390/electronics12132792
19. Thomay, C., et al.: Towards cognitive load-based decision making in VR training. In: Proceedings of the 2023 IEEE 2nd International Conference on Cognitive Aspects of Virtual Reality (CVR), pp. 23–28. IEEE, Veszprém, Hungary (2023). https://doi.org/10.1109/CVR58941.2023.10395602
20. Rodenburg, D., et al.: Dynamically adaptive simulation based on expertise and cognitive load. In: Proceedings of the 2018 IEEE Games, Entertainment, Media Conference (GEM), pp. 1–8. IEEE, Galway, Ireland (2018). https://doi.org/10.1109/GEM.2018.8587618
21. Fu, Q., Lv, J., Zhao, Z., Yue, D.: Research on optimization method of VR task scenario resources driven by user cognitive needs. Information. **11**(2), 64 (2020). https://doi.org/10.3390/info11020064
22. Mavrotas, G.: Effective implementation of the ε-constraint method in multi-objective mathematical programming problems. Appl. Math. Comput. **213**(2), 455–465 (2009). https://doi.org/10.1016/j.amc.2009.03.037

Bibliometric Overview of Early Intervention in the Educational Context

Verónica Castro(✉), Lennis Pérez, Ángel Cornejo, Maribel Puchaicela, and Jessica Cahuate

Instituto Superior Tecnológico Vida Nueva, Quito, Ecuador
veronica.castro@istvidanueva.edu.ec

Abstract. This study gathers information on the relationship between emotional intelligence and early intervention. A qualitative methodology with a documentary approach was used, applying bibliographic and bibliometric techniques. The article poses the following research question: What early intervention strategies are effective for developing emotional intelligence? The relevance of early intervention in education is highlighted, where the identification and treatment of socio-emotional disorders are essential. The results indicate that these interventions must involve the entire educational community to ensure a sustainable impact on student well-being, highlighting the need for specific training for educators in emotional competencies. In addition, innovative pedagogical strategies that promote a safe and collaborative learning environment are presented.

Keywords: Emotional Intelligence · Early Intervention · Comprehensive Development · Emotional Education · Psychological Well-being

1 Theoretical Reference

1.1 Emotional Intelligence

In the field of education, emotional intelligence has become established as a key factor in understanding how individuals interact with themselves and their environment [1]. Their study not only identifies the internal mechanisms that regulate affective responses, but also provides tools for strengthening social and academic competencies [2].

In particular, at an early age, such as between 11 and 12 years old, the development of emotional intelligence is crucial for the integral training of the student, since it provides the basis for the proper management of emotions and the construction of healthy interpersonal relationships [3].

Emotional intelligence is understood as the ability to perceive, value and express emotions accurately; to induce and generate feelings which benefit thought; to understand and know emotions and finally to control and manage emotions for personal development [1].

The development of emotional intelligence in school contributes to the integral training of students by facilitating proper stress management, peaceful resolution of conflicts

A. Rocha et al. (Eds.): ICITS 2026, LNNS 1966, pp. 27–43, 2027.
https://doi.org/10.1007/978-3-032-27538-7_3

and building healthy interpersonal relationships [4]. It also promotes a positive learning environment in which students feel valued and confident to express their ideas and emotions [2].

This set of emotional skills promotes not only individual well-being, but also academic and social success. For example, recent research has shown that greater emotional intelligence in teachers and students is associated with higher intrinsic motivation and better performance in various areas, including academic writing [2, 5].

Teachers also play a crucial role in promoting emotional intelligence, as by strengthening their own emotional skills they are better prepared to guide and support their students [3]. The implementation of teaching strategies that encourage empathy, self-regulation and effective communication creates a favourable school environment for academic and personal development [3].

Early intervention through structured programs has been shown to be effective in fostering these skills, as it allows students to acquire adaptive coping strategies and improve their capacity for emotional self-regulation [4].

Moreover, in populations with specific needs, such as adolescents with neuromotor disabilities, emotional intelligence and coping strategies become even more important to promote their wellbeing and social adaptation [6].

1.2 Emotional Regulation and Empathy

Emotional regulation refers to the ability of individuals to identify, understand and manage their emotions appropriately, adjusting their affective responses according to the demands of context and personal goals [3]. In education, this skill is essential as it enables students to maintain control in stressful situations, manage frustration with academic challenges and establish balanced interpersonal relationships [2].

Empathy is understood as the ability to recognize and understand other people's emotions, thoughts and perspectives, which facilitates the development of positive social bonds and pro-social behaviors [5]. In students aged 11–12, a stage characterized by significant emotional changes, empathy is a protective factor that promotes cooperation, respect for diversity and conflict prevention [4].

Several studies highlight that emotional regulation and empathy are closely related and mutually reinforcing [3, 5]. Intervention programmes aimed at strengthening these skills have shown improvements in the school climate and academic performance [4].

1.3 Socio-emotional Competencies

Socio-emotional competencies comprise a set of skills, attitudes and values that enable students to recognize and manage their emotions, relate positively and make responsible decisions in different contexts [2, 4].

These skills are essential not only for emotional well-being, but also for academic performance and school life. Among the most relevant are self-awareness, which involves identifying and understanding one's own emotions [4], and self-regulation, understood as the ability to control impulses and manage stress effectively [1]. Empathy is also essential, as it facilitates understanding of other people's emotions and fosters healthy and respectful relationships [5]

Social skills such as effective communication, collaboration and the peaceful resolution of conflicts create a harmonious school environment and encourage teamwork [3]. Responsible decision-making, understood as the ability to act ethically and assess consequences, is particularly important in pre-adolescence [6]. The development of these skills strengthens resilience, that is, the ability to adapt positively to adverse situations [2], which is key to meeting educational and personal challenges.

Integrating these skills into the school curriculum promotes inclusive, motivating and more student-involved environments [4]. They also have a preventive effect on problems such as bullying or early school leaving, strengthening self-efficacy and a sense of belonging [5]. Finally, in contexts of functional diversity, these skills are even more relevant, since they promote the inclusion, autonomy and integral well-being of the student [6], making possible a more equitable and quality education.

2 Early Intervention

2.1 Emotional Well-Being

Emotional well-being is a multidimensional construct that comprises psychological balance and the ability to experience positive emotions, as well as effectively managing negative emotions and stress [7]. This concept goes beyond the mere absence of mental disorders, involving aspects such as life satisfaction, self-esteem, and the ability to maintain healthy interpersonal relationships. The relevance of emotional well-being lies in its direct impact on physical health, academic performance and overall quality of life.

Emotional regulation, on the other hand, is the ability to identify, understand and modify one's own emotional responses, favouring a state of balance and well-being. Strategies such as mindfulness (mindfulness), cognitive restructuring and social skills training contribute to improving emotional regulation, reducing the impact of stress and promoting greater emotional stability [8]. Adequate regulation of emotions is especially important in high-pressure contexts, such as academic or work environments.

Another factor that has received increasing attention in recent literature is self-pity, defined as the ability to treat oneself with kindness and understanding at times of failure or suffering [9]. Self-pity has been associated with lower levels of anxiety and depression, and a significant increase in emotional well-being. This construct promotes a healthy relationship with oneself, facilitating the adaptive coping of everyday difficulties.

The recent global health crisis triggered by the COVID-19 pandemic has highlighted the critical importance of emotional well-being in public health [10]. Uncertainty, social isolation and loss of routine have significantly increased levels of stress and emotional distress in the general population, underlining the urgent need to implement interventions that foster coping skills and resilience.

Psychological interventions based on mindfulness, cognitive-behavioral therapy and psycho-educational programs have been shown to be effective in improving emotional well-being in different age groups and contexts [11]. These interventions enhance emotional self-regulation, promote positive cognitive changes and strengthen social networks, facilitating better adaptation to personal and social challenges.

2.2 Cognitive Maturity

The neurobiological development of cognitive maturity is strongly linked to the progressive maturation of the prefrontal cortex, a brain region responsible for executive functions. These include planning, inhibition, selective attention and cognitive flexibility, which are essential elements for informed thinking and behaviour regulation. Neuroscientific research has shown that the prefrontal cortex continues to develop until about 25 years of age, which explains the variations in complex decision-making ability and emotional control during adolescence and early adulthood [12].

Metacognition, defined as the ability to reflect and control one's own cognitive processes, is an essential pillar of cognitive maturity [14]. This aspect includes skills such as strategic planning, monitoring of one's own learning and critical evaluation of results achieved. Metacognition promotes autonomy and allows individuals to adjust their learning strategies to optimize outcomes.

Cognitive maturity is closely related to emotional regulation, as the ability to properly manage emotions directly influences the quality of decision-making and the resolution of social conflicts [13]. Recent studies suggest that the efficient integration of cognitive and emotional processes is essential for ethical and responsible behavior.

The precise measurement of cognitive maturity is carried out through psychometric tools that evaluate dimensions such as executive functions, metacognition and logical reasoning [12]. The differentiation between chronological age and cognitive age makes it possible to identify students who need personalized educational interventions, ensuring that pedagogical strategies respond to their real abilities and not only to their age.

Enhancing cognitive maturity has a direct impact on the integral formation of the individual. In addition to improving academic performance, it strengthens autonomy, adaptability and competence to face contemporary challenges. Developing advanced cognitive skills also contributes to emotional well-being, mental health and building strong personal and social identities.

2.3 Neurobiological Development and Executive Functions

The neurobiological development of cognitive maturity is strongly linked to the progressive maturation of the prefrontal cortex, a brain region responsible for executive functions. These include planning, inhibition, selective attention and cognitive flexibility, which are essential elements for sophisticated thinking and behavioural regulation. Neuroscientific research has shown that the prefrontal cortex continues to develop until about 25 years of age, which explains the variations in complex decision-making ability and emotional control during adolescence and early adulthood [12].

Brain plasticity during these stages allows exposure to enriching educational experiences and stimulating environments to significantly enhance cognitive development. Therefore, the educational and social context becomes a key factor which may accelerate or hinder the acquisition of cognitive maturity.

2.4 Identification of Needs

Identifying needs in the school setting is the first step towards designing effective intervention strategies aimed at the emotional development of students. This process involves a systematic analysis of individual, family, school and community factors that may influence the emotional well-being of the child. Several studies agree that a comprehensive needs assessment can personalize pedagogical and psychosocial strategies, thus facilitating meaningful emotional learning [18].

In the context of basic education, particularly for 11- to 12-year-old s, emotional needs are often related to managing emotions such as anger, frustration and anxiety, as well as the need for social belonging and acceptance. The lack of skills in these aspects can lead to disruptive behaviors, poor school performance and difficulties in school life [22].

La participación del equipo docente, los orientadores y la familia es crucial para lograr una detección eficaz. El enfoque colaborativo permite reconocer no solo los síntomas, sino también las causas contextuales subyacentes, como conflictos familiares, falta de redes de apoyo o eventos estresantes. Además, la identificación temprana permite evitar la cronificación de síntomas y planificar intervenciones preventivas con enfoque inclusivo, equitativo y centrado en el estudiante [19].

La implementación de procesos de observación sistemática dentro del aula se convierte en una herramienta valiosa para identificar patrones emocionales y conductuales relevantes. Docentes capacitados pueden detectar señales sutiles de malestar emocional, como aislamiento, irritabilidad o falta de motivación, lo cual permite una intervención oportuna antes de que los síntomas se intensifiquen [23].

Por otra parte, promover espacios de diálogo y escucha activa con los estudiantes permite que ellos mismos expresen sus emociones, inquietudes y necesidades. Estos espacios fortalecen la relación escuela-estudiante y contribuyen a generar un entorno emocionalmente seguro, en el que los niños se sientan comprendidos y valorados, lo que favorece el desarrollo de su inteligencia emocional [24].

The involvement of the teaching team, counsellors and family is crucial for effective screening. The collaborative approach allows recognizing not only symptoms, but also underlying contextual causes such as family conflicts, lack of support networks or stressful events. In addition, chronic identification can prevent the chronification of symptoms and plan preventive interventions with an inclusive, equitable and student-centred approach [19].

The implementation of systematic observation processes within the classroom becomes a valuable tool for identifying relevant emotional and contextual patterns. Trained teachers can detect subtle signs of emotional distress, such as isolation, irritability or lack of motivation, which allows for timely intervention before symptoms intensify [23].

On the other hand, promoting spaces of dialogue and active listening with students allows them to express their own emotions, concerns and needs. These spaces strengthen the school-student relationship and contribute to generating an emotionally safe environment in which children feel understood and valued, which favors the development of their emotional intelligence [24].

The use of age-appropriate socio-emotional surveys has shown good results in collecting useful information on student well-being. Simple but structured instruments, applied periodically, allow to visualize changes over time, contributing to a more precise and continuous emotional monitoring [25].

Working together with families ensures a more complete view of the child's needs, as many emotional factors originate in the home. Permanent communication between school and family strengthens the support network of the student and allows joint strategies to be designed to accompany him in an integral way [26].

In addition to the above-mentioned strategies, continuing training of teachers on mental health and socio-emotional development issues is essential for improving responsiveness to identified needs [16].

It should be stressed that, according to recent research, the identification of emotional needs must be continuous and not limited to a single point in the school cycle. Tools such as systematic observations, short interviews and emotional self-assessment questionnaires have been shown to be effective in identifying emotional variations over time. In this sense, the use of validated instruments allows not only to detect difficulties, but also to make visible emotional strengths that can be enhanced within the classroom [17].

The implementation of early warning protocols that allow action on specific signals such as drastic changes in behavior, persistent demotivation or frequent conflicts with peers is essential to prevent the aggravation of symptoms [27].

Fostering a school culture where emotional care has the same value as academic achievement should be a priority in institutional policies. Emotional well-being as a transversal axis in pedagogical practices and school management has been highlighted as a determining factor in the integral development of the student [21].

2.5 Emotional and Behavioral Difficulties

Emotional and behavioral difficulties in middle childhood, particularly in students aged 11 to 12, are complex manifestations that reflect alterations in emotional development, affective regulation, and social interaction. These difficulties may be expressed through anxiety, depression, isolation, aggressiveness, impulsiveness or low tolerance for frustration [20].

According to recent studies, the increase in cases of emotional distress in school age is due to multiple factors, including academic stress, bullying, family instability and exposure to violent content on digital media. These conditions directly affect students' learning, school motivation and sense of belonging [21].

Psycho-educational approaches based on emotional intelligence have proved highly effective in addressing these difficulties. Programs such as RULER and PATHS foster skills of self-regulation, empathy, conflict resolution and self-control. Through classroom-based interventions, with active participation of teachers and families, there is a significant reduction in risk behaviors and an increase in social and academic adaptation [15].

It is essential to implement a continuous assessment of the emotional state of trainees through the use of structured observation, interviews, psycho-emotional surveys and regular meetings with guidance teams. These practices contribute to a positive and healthy school culture, allowing for proactive action based on scientific evidence [16].

The 11 to 12 age group represents a critical period in psychosocial development as students begin to build their emotional identity and autonomy. Difficulties not addressed at this stage may intensify in adolescence, leading to more complex and persistent disorders. It is therefore essential that the school act as a protective environment, where healthy emotional expression and respect for diversity of emotions are promoted [20].

In this sense, teachers have a fundamental role as agents of emotional support and communication. Their ability to identify early signs of behavioral or emotional changes is key to timely referral to specialized services or classroom-adapted interventions. Teacher training in emotional intelligence not only improves the school climate, but also strengthens the relationship between students and educators by promoting a more inclusive and empathetic environment [15]. In addition, dysfunctional family dynamics also impact student behavior within the school environment. Lack of affective communication in the home, absence of significant figures or excessive permissiveness can increase a child's emotional vulnerability. These conditions require a joint approach between the school and the family, in order to establish coherent and sustainable strategies that favour the integral development of the child [26].

3 Methodology

The methodology adopted in this research is based on a qualitative, exploratory approach using documentary and bibliometric techniques. This approach allows a more holistic and enriching understanding of the phenomenon under study by integrating qualitative data from the object investigated. Considering the existing problem, a detailed analysis of articles related to emotional intelligence and early intervention was carried out.

Bibliometrix, an open-source tool developed in R that offers comprehensive scientific mapping analysis, with the advantage of being flexible and compatible with other statistical and graphical packages [30], stands out for collecting information. In addition, VOSviewer stands out for its ability to visualize extensive bibliometric maps intuitively and easily interpreted [29]. For this reason, information extracted from the Scopus database was used to process the data (Table 1).

Table 1. Equation for information extraction

Equation	Number of documents found in Scopus
(("emotional intelligence") AND ("intervention strategies"))	264

The research focuses on emotional education focused on developing self-control and self-esteem as essential conditions to prevent negative emotions and cultivate happiness [27].

3.1 Analysis of Cooccurrence

A co-occurrence analysis was generated, the purpose of which is to identify the frequency with which certain keywords appear jointly within the selected set of documents.

This type of analysis is fundamental to the detection of patterns in scientific production, since it makes it possible to observe how terms are grouped according to their inter-connection and relevance within a field of study. In this particular case, the results obtained reflect the thematic structure of the area investigated, providing a clear vision on the most relevant concepts and their relationship with other themes within the corpus analyzed.

The following image shows the results of the analysis, where it is possible to visualize the connections established between the keywords and how they are grouped in different clusters or research nuclei. In this analysis the keywords "emotional intelligence", "child", "cognition", "cross-sectional study" and "emotions" are highlighted.

In addition, the VOSViewer tool shows a heat map showing the development of concepts over the years; as shown in the following image, the most recent concepts worked on are: "mental health", "cross-sectional study", "physicological stress" and "university" (Fig. 1).

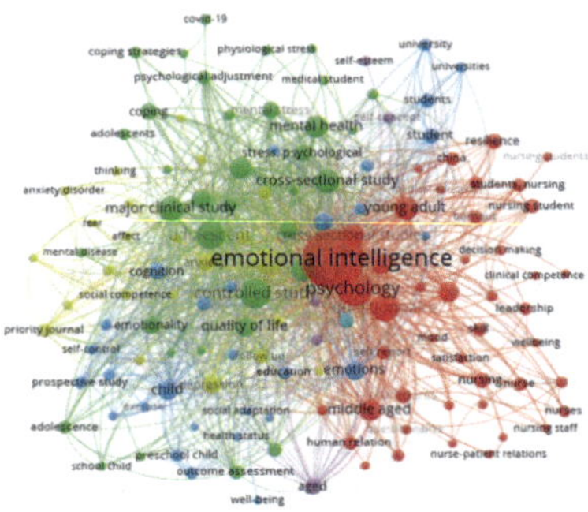

Fig. 1. A heat map showing the development and evolution of concepts over the years is presented. This graphical representation makes it possible to identify the areas of greatest thematic concentration, as well as the periods in which certain terms or ideas had greater relevance within the field of study.

Bibliometric coupling is another analysis that the VOSviewer tool performs. This analysis consists of the reciprocal relationship between scientific documents based on citations, this is visualized through networks of citations, where the nodes represent documents and the connections between them indicate mutual citations, as shown in the following image (Fig. 2).

In this analysis the following documents are highlighted: "The late positive potential: a neurophysiological marker for emotion regulation in children" by Tracy A. Dennis and Greg Hajcak, this article examined the late positive potential (LPP) as a neural marker of emotional regulation in children. Using EEG, it was found that a more modulated LPP after neutral interpretations was associated with less anxiety-depressive symptoms, while a higher LPP indicated worse emotional regulation. The results suggest that LPP could be a key biomarker for identifying mood problems in childhood.

Another featured article is "Associations between emotional intelligence, de-pression and suicide risk in nursing students" by Amor Aradilla-Herrero, Joaquín To-plus-Sábado and Juana Gómez-Benito. This study analyzed suicide risk in nursing students and its relationship to perceived emotional intelligence, depression, anxiety, and self-esteem. Depression and emotional care were found to be key predictors of suicidal ideation,

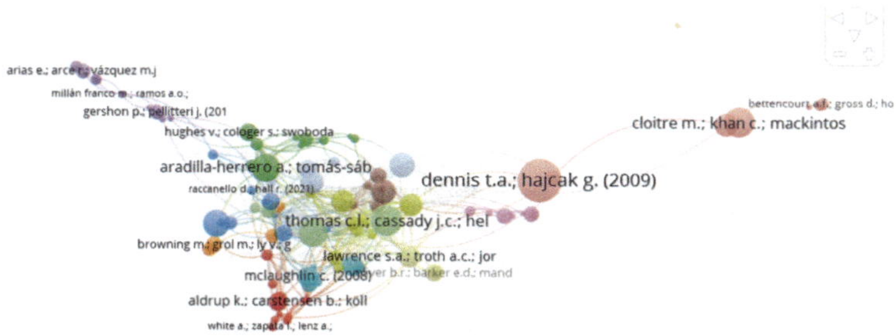

Fig. 2. In the figure presented, the reciprocal relationship between scientific documents based on citations represents an academic link in which two or more researches relate to each other, evidencing a direct dialogue between authors and consolidating a network of knowledge exchange.

while self-esteem, clarity and emotional repair have a protective effect. 14% of the students had a substantial risk of suicide, and women showed higher levels of depression. The findings highlight the importance of detecting mood disorders and strengthening emotional coping skills in preventing suicide in this group.

Finally, the text "Emotional well-being and its relationship to schools and classrooms: a critical reflection" by author Colleen McLaughlin stands out in the bibliometric approach. This article analyses research on emotional well-being in schools and classrooms, examining concepts such as emotional intelligence, resilience, protective factors, mental health in young people and emotional regulation. The interventions and strategies adopted by policy-makers are critically reviewed, pointing out the risks of an exclusively individualistic approach. The author proposes that, in addition to specific programmes, it is essential to strengthen relationships, pedagogy and community building to promote emotional well-being among young people.

In addition, it displays the relevant documents on the topic, taking into account the citations and the relationship of the abstract and title with the subject to investigate, the result of this is evidenced in the following table. (Tables 2 and 3)

Table 2. Relevant documents by year, title, DOI and citations

Año	Paper	Título	DOI	Total Citas
2009	DENNIS TA, 2009, J CHILD PSYCHOL PSYCHIATRY ALLIED DISCIP	The late positive potential: a neurophysiological marker for emotion regulation in children	https://do.org/10.1111/j.1469-7610.2009.02168.x	262
2012	GRANT L, 2012, SOC WORK EDUC	Enhancing Wellbeing in Social Work Students: Building Resilience in the Next Generation	https://do.org/10.1080/02615479.2011.590931	134

(continued)

Table 2. (*continued*)

Año	Paper	Título	DOI	Total Citas
2020	QUINTANA-ORTS C, 2020, J CLIN MED	Untangling the Emotional Intelligence-Suicidal Ideation Connection: The Role of Cognitive Emotion Regulation Strategies in Adolescents	https://do.org/10.3390/jcm9103116	18
2019	ADIBSERESHKI N, 2019, CHILD	The effectiveness of a resilience intervention program on emotional intelligence of adolescent students with hearing loss	https://do.org/10.3390/children6030048	14
2021	ABACIOGLU CS, 2021, CURR PSYCHOL	Teacher interventions to student misbehaviors: The role of ethnicity, emotional intelligence, and multicultural attitudes	https://do.org/10.1007/s12144-019-00498-1	8
2022	BERASTEGUI-MARTÍNEZ J, 2022, HELIYON	Effects of an intervention programme designed to improve emotional intelligence and foster the use of coping strategies among professional female football players	https://do.org/10.1016/j.heliyon.2022.e09860	4
2024	HILUF B, 2024, COGENT EDUC	Exploring the links between emotional intelligence and writing performance among Ethiopian high school learners	https://do.org/10.1080/2331186X.2024.2369970	3
2024	CIOCA IE, 2024, CHILD	Exploring the Links Between Coping Strategies, Emotional Intelligence, and Age in Adolescents with Neuromotor Disabilities	https://do.org/10.3390/children11121466	0

4 Results

Table 3. Relevant documents according to author, title and concomitance

Autores	Título	Concomitancia de la información
DENNIS Y HAJCAK (2009)	The late positive potential (LPP) as a neurophysiological marker of emotional regulation in children.	Cognitive reinterpretation reduces the amplitude of LPP, indicating better emotional regulation. This marker is associated with fewer internalizing symptoms and greater emotional adjustment.
GRANT Y KINMAN (2012)	Resilience, emotional intelligence and well-being in social workers	The research highlights that emotional intelligence, reflection and social skills are essential to develop resilience and well-being in high-stress work contexts, aspects equally applicable in early intervention with children.
QUINTANA-ORTS, C., REY, L., MÉRIDA-LÓPEZ, S., EXTREMERA, N. (2019)	Emotional intelligence, bullying and suicidal risk	Emotional Intelligence acts as a bridge in the relationship between bullying and suicidal risk, suggesting the importance of strengthening it.
ADIBSERESHKI ET AL. (2019)	Effectiveness of a resilience program in the emotional intelligence of adolescents with hearing loss	A 6-week programme significantly improved Emotional Intelligence and its dimensions (emotional regulation, expression and utilization). Highlights the feasibility of short, structured interventions in vulnerable populations.

(*continued*)

Table 3. (*continued*)

Autores	Título	Concomitancia de la información
CEREN SU ABACIOĞLU, MONIQUE VOLMAN Y AGNETA H. FISCHER (2019)	Teaching interventions in response to inappropriate student behaviour: the role of ethnicity, emotional intelligence and multicultural attitudes	The study examines how primary school teachers in the Netherlands intervene in inappropriate behaviour according to the student's ethnicity. It found no significant differences, but positive multicultural attitudes favour dialogical responses over punitive ones. Emotional intelligence was linked to these attitudes, although it did not influence the interventions.
BERASTEGUI-MARTÍNEZ Y LOPEZ-UBIS (2022)	Effects of an intervention program on emotional intelligence and coping strategies in professional footballers.	A 20-session program improved emotional expression and perception, as well as the use of adaptive coping strategies. Confirms that Emotional Intelligence is entrable and applicable in high-performance contexts
BELETE HILUF, MAREW ALEMU, HAILAY TESFAY GEBREMARIAM (2023)	Emotional intelligence and writing performance in Ethiopian students	Emotional Intelligence influences writing skills in students, highlighting the impact of emotional competencies on academic performance.
CIOCA ET AL. (2024)	Relationship between coping strategies and emotional intelligence in adolescents with neuromotor dis-abilities.	Adaptive coping strategies predict 29% of the variance in emotional intelligence. Age influences the development of interpersonal skills and stress management.

The findings support the premise that early intervention in emotional intelligence, through innovative pedagogical approaches, not only strengthens socio-emotional skills but also enhances academic performance. Even-as it has begun to explore how these skills can be of benefit to students with specific needs, such as those with neuromotor disabilities, there is still little knowledge about best practices for teaching and developing emotional intelligence in these populations. This implies the need for studies investigating inclusive strategies that take account of the particularities of these students, as well as continuing teacher training in this field. Addressing these gaps will not only enrich the understanding of emotional intelligence, but also contribute to a more inclusive and equitable educational environment.

Research indicates that pedagogical strategies promoting EI have had a positive impact on the development of socio-emotional competences. Methods such as project-based learning and emotional mediation have been shown to be effective in improving emotional regulation and empathy among students [3, 6]. In addition, collaborative work and the use of digital resources have facilitated emotional self-regulation, allowing learners to manage their emotions more effectively [2, 4]. These practices not only benefit individual learning, but also promote a more cohesive and participatory classroom environment.

Early identification of emotional needs through systematic assessments is crucial to designing personalized interventions. This practice addresses the specific difficulties students face, resulting in decreased levels of anxiety and depression. In addition, there has been an increase in resilience and ability to cope with stressful situations. These findings underscore the need for a proactive and preventive approach to emotional education, where intervention focuses not only on correcting problems but also on strengthening skills that promote the integral well-being of students.

5 Discusión

Emotional intelligence (EI) has been positioned as a crucial element in the integral development of students, especially in the pre-adolescence stage. EI encompasses essential skills such as self-awareness and emotional regulation, which are fundamental to the proper management of emotions and the building of healthy interpersonal relationships. However, authors stress that while these skills are important, their development must be accompanied by an educational context that fosters a safe and positive environment [14]. This means that students must not only be taught to identify and manage their emotions, but also to create an environment where they feel valued and heard, which is particularly relevant in school dynamics.

The pedagogical methodologies that integrate EI have shown to be effective in improving academic performance and school coexistence. Programmes such as RULER and PATHS have shown that the implementation of strategies that foster empathy and self-regulation not only reduces disruptive behaviour, but also improves motivation and performance in various academic areas. However, it is crucial to consider that the effectiveness of these interventions may vary according to the context and characteristics of the students. For example, research has shown that in environments with high levels of stress or violence, interventions must be adapted to address the specific needs of learners [19]. This highlights the importance of a personalised approach to EI teaching, taking into account the particularities of each group.

Despite the obvious benefits of EI in education, there are limitations in the implementation of intervention programmes. The lack of continuing training for teachers in socio-emotional skills may hinder the effectiveness of the proposed strategies [18]. In addition, the identification of emotional needs should be a continuous process and not an isolated event [16]. This means that schools must establish systematic evaluation mechanisms to detect changes in the emotional well-being of students over time, which is essential for tailoring interventions to changing needs. Collaboration between school and family is central to the success of EI strategies. Constant communication between teachers and parents enables a more comprehensive view of the student's emotional development, facilitating the implementation of joint strategies that reinforce emotional learning in different contexts [25]. This synergy not only enhances the impact of interventions, but also contributes to the creation of a supportive environment that promotes emotional well-being for students. In this sense, fostering a school culture that values both academic performance and emotional well-being becomes a key objective for the formation of emotionally competent and resilient citizens.

The results obtained highlight the relevance of emotional intelligence in the educational context, especially in the pre-adolescence stage. Students' ability to manage their emotions and establish healthy interpersonal relationships has been directly linked to their academic success and overall well-being. The literature suggests that emotional intelligence is not only fundamental for personal development, but also acts as a mediator in academic performance, reinforcing the need to integrate it into the school curriculum. The relationship between emotional regulation and empathy is crucial for school coexistence. Programs that promote these skills not only benefit students, but also positively impact the school climate by creating a more inclusive and collaborative environment. Identifying emotional needs and implementing early intervention strategies are essential to addressing emotional and behavioral difficulties. Collaboration between teachers, families and mental health specialists is essential to ensure a holistic approach that respects diversity and promotes the well-being of all students. This collaborative approach not only helps prevent more serious problems in the future, but also strengthens the emotional support network needed for young people's integral development.

6 Conclusions

The present study reaffirms the importance of emotional intelligence as an essential component in the integral development of students, especially at the age of 11 to 12. Early intervention in this area not only promotes psychological well-being, but also enhances social, cognitive and academic skills. The evidence collected shows that integrating emotional education into the school curriculum establishes solid foundations for strengthening self-esteem, emotional self-regulation and building a healthy identity in students.

Research has revealed several knowledge gaps that require urgent attention. One of the most significant is the effectiveness of intervention programmes in various educational contexts. Although positive results have been documented in controlled environments, there is still a lack of evidence on how these interventions are adapted and operate in rural, urban schools or communities with high rates of violence. Variability in socio-economic and cultural conditions suggests that what is effective in one context may not be so in another. It is therefore essential to carry out studies that evaluate and adapt these programmes to the specific realities of each environment in order to maximize their impact. Another important gap is the analysis of the long-term impact of emotional intelligence on students' lives. While many studies focus on the immediate benefits in academic performance and emotional well-being in the school environment, there is a lack of research on how these skills influence mental health, interpersonal relationships and professional success in adult life. Understanding these long-term effects is crucial to justify the integration of emotional intelligence into the school curriculum and to develop strategies that not only benefit students during their time at school, but also to prepare them for the challenges of adult life.

It highlights the urgent need to train teachers in experiential skills and inclusive approaches, enabling them to identify, prevent and effectively address socio-emotional alterations in the classroom. In addition, it is suggested to continue with applied research that validates these strategies in various educational contexts, as well as to promote public policies that recognize emotional intelligence as a transversal axis of the education system. This is the only way to ensure the integral well-being of students and their readiness to face the challenges of the 21st century.

Early identification of emotional needs and the implementation of appropriate support strategies are crucial to preventing emotional and behavioral difficulties in students. Collaboration between school and family, together with an inclusive approach, will ensure that all students have the opportunity to develop fully, thus promoting equitable and quality education.

References

1. Conejo Carrasco, F.: De la inteligencia emocional a la autorregulación del aprendizaje: un camino hacia la autonomía del aprendizaje en el estudiante, pp. 1–164. Corporación Universitaria Minuto de Dios, Bogotá, Colombia (2022)
2. Bechter, B.E., Whipp, P.R., Dimmock, J.A., Jackson, B.: Emotional intelligence and quality of interpersonal relationships as predictors of intrinsic motivation among secondary physical education teachers. Curr. Psychol. **42**, 7457–7465 (2023)
3. Abacioglu, C.S., Volman, M., Fischer, A.H.: Teacher interventions to student misbehaviors: the role of ethnicity, emotional intelligence, and multicultural attitudes. Curr. Psychol. **40**, 5934–5946 (2021). https://doi.org/10.1007/s12144-019-00498-1
4. Berastegui-Martínez, J.: Effects of an intervention programme designed to improve emotional intelligence and foster the use of coping strategies among professional female football players. Heliyon. **8**(10), e09860 (2022). https://doi.org/10.1016/j.heliyon.2022.e09860
5. Hiluf, B., Alemu, M., Gebremariam, H.T.: Exploring the links between emotional intelligence and writing performance among Ethiopian high school learners. Cogent Educ. **11**(1), 2369970 (2024). https://doi.org/10.1080/2331186X.2024.2369970
6. Cioca, I.E., Morcov, M.V., Sporea, C., Apostol, O.A., Pellegrini, A., Bordea, E.-N.: Exploring the links between coping strategies, emotional intelligence, and age in adolescents with neuromotor disabilities. Children. **11**(12), 1466 (2024). https://doi.org/10.3390/children11121466
7. Ryff, C.D., Singer, B.: Know thyself and become what you are: a eudaimonic approach to psychological Well-being. J. Happiness Stud. **9**(1), 13–39 (2017). https://doi.org/10.1007/s10902-006-9019-0
8. Neff, K.D.: The self-compassion scale is a valid and theoretically coherent measure of self-compassion. Mindfulness. **7**(1), 264–274 (2016). https://doi.org/10.1007/s12671-015-0479-3
9. Steger, M.F., Kashdan, T.B.: The meaning in life: a framework for conceptualization, measurement, and growth. Soc. Personal. Psychol. Compass. **11**(10), e12376 (2017). https://doi.org/10.1111/spc3.12376

10. Cohen, S., Janicki-Deverts, D., Miller, G.E.: Psychological stress and disease. JAMA. **298**(14), 1685–1687 (2007). https://doi.org/10.1001/jama.298.14.1685
11. Ryff, C.D.: El bienestar psicológico revisitado: Avances en la ciencia y la práctica de la eudaimonía. Psychother. Psychosom. **87**(1), 10–28 (2018). https://doi.org/10.1159/000477333
12. Best, J.R., Miller, P.H.: Una perspectiva del desarrollo sobre la función ejecutiva. Child Dev. **90**(4), 1081–1099 (2019). https://doi.org/10.1111/cdev.13022
13. Dignath, C., Buettner, G., Langfeldt, H.P.: ¿Cómo pueden los estudiantes de primaria aprender estrategias de aprendizaje autorregulado de manera más efectiva? Un metaanálisis sobre programas de entrenamiento en autorregulación. Educ. Res. Rev. **24**, 40–54 (2018). https://doi.org/10.1016/j.edurev.2018.03.001
14. Blair, C., Raver, C.C.: Cerrando la brecha de logro mediante la modificación de la función neurocognitiva y neuroendocrina: Resultados de un ensayo controlado aleatorio por conglomerados. PLoS One. **13**(10), e0204939 (2018). https://doi.org/10.1371/journal.pone.0204939
15. Brackett, M.A., Rivers, S.E., Salovey, P.: Emotional intelligence: implications for personal, social, academic, and workplace success. Soc. Personal. Psychol. Compass. **15**(2), e12592 (2021). https://doi.org/10.1111/spc3.12592
16. Espinosa, A., Ramos, L., Cueva, M.: Intervención temprana en contextos escolares: estrategias efectivas. Revista Latinoamericana de Psicología Educativa. **10**(1), 55–67 (2020)
17. Jiménez, L., González, M.: Evaluación de necesidades emocionales en el entorno escolar: Herramientas para una intervención efectiva. Revista Latinoamericana de Psicología Educativa. **21**(1), 45–60 (2023). https://doi.org/10.1016/j.rlpede.2023.01.005
18. Jiménez, A., Sánchez, F.: Evaluación e identificación de necesidades emocionales en escolares. Revista Iberoamericana de Psicología y Educación. **18**(2), 124–137 (2022)
19. López, P., Hernández, C.: Intervención escolar y apoyo emocional: estrategias eficaces. Psicología Escolar Contemporánea. **9**(2), 88–102 (2021)
20. Merikangas, K.R., He, J., Burstein, M.: Epidemiology of mental disorders in children and adolescents. Dialogues Clin. Neurosci. **24**(2), 117–128 (2022)
21. OECD: Social and Emotional Skills: Well-being, Connectedness and Success. OECD Publishing, Paris, France (2023)
22. Rivera, N., Guzmán, P., López, C.: Necesidades emocionales no satisfechas en la infancia: una revisión sistemática. Psicología Educativa. **26**(3), 187–196 (2020)
23. Ramírez, L., Torres, A.: Observación docente y detección de necesidades emocionales en estudiantes de primaria. Revista Educación y Desarrollo Emocional. **18**(3), 101–114 (2022). https://doi.org/10.1234/rede.2022.183101
24. Paredes, M., Guzmán, F.: La escucha activa como estrategia de acompañamiento emocional escolar. Psicología Escolar Latinoamericana. **19**(2), 56–70 (2021). https://doi.org/10.5678/psiescolat.2021.192056
25. Castillo, D., Menéndez, J.: Aplicación de instrumentos socioemocionales en educación básica. Estudios en Educación. **25**(4), 67–79 (2020). https://doi.org/10.7890/estedu.2020.254067
26. Herrera, G., Salazar, N.: Alianza escuela-familia en el abordaje de necesidades emocionales. Educación Integral Hoy. **11**(1), 44–59 (2023). https://doi.org/10.1123/edint.2023.111044
27. Martínez, A., Rojas, P., Herrera, D.: Protocolos de identificación temprana de dificultades emocionales en estudiantes de primaria. Psicología Escolar y Educacional. **26**(2), 125–138 (2022). https://doi.org/10.1590/2175-3539/2022.02622
28. Pacheco Olea, L.A.: La inteligencia emocional en estudiantes de Ciencias de la Ingeniería. Revista Ciencia Unemi. **4**(6), 78–87 (2011)
29. Van Eck, N.J., Waltman, L.: Software survey: VOSviewer, a computer program for bibliometric mapping. Scientometrics. **84**(2), 523–538 (2010). https://doi.org/10.1007/s11192-009-0146-3

30. Aria, M., Cuccurullo, C.: Bibliometrix: an R-tool for comprehensive science mapping analysis. J. Informet. **11**(4), 959–975 (2017). https://doi.org/10.1016/j.joi.2017.08.007
31. Grant, L., Kinman, G.: Enhancing wellbeing in social work students: building resilience in the next generation. Social Work Educ. Int. J. **31**(5), 605–621 (2012). https://doi.org/10.1080/02615479.2011.590931
32. Dennis, T.A., Hajcak, G.: The late positive potential: a neurophysiological marker for emotion regulation in children. J. Child Psychol. Psychiatry. **50**(11), 1373–1383 (2009). https://doi.org/10.1111/j.1469-7610.2009.02168.x
33. Quintana-Orts, C., Mérida-López, S., Rey, L., Neto, F., Extremera, N.: Untangling the emotional intelligence–suicidal ideation connection: the role of cognitive emotion regulation strategies in adolescents. J. Clin. Med. **9**(10), 3116 (2020). https://doi.org/10.3390/jcm9103116
34. McLaughlin, C.: Emotional well-being and its relationship to schools and classrooms: a critical reflection. Br. J. Guid. Couns. **36**(4), 353–366 (2008). https://doi.org/10.1080/03069880802364486
35. Aradilla-Herrero, A., Gómez-Benito, J.: Associations between emotional intelligence, depression and suicide risk in nursing students. Nurse Educ. Today. **34**(4), 520–525 (2014). https://doi.org/10.1016/j.nedt.2013.07.001

The Discourse on Political Polarization in the Peruvian Context Through Lima's Digital Media

Hugo Gonzalez Aguilar(✉)

Universidad Autónoma del Perú, Lima, Peru
hugo_augencio@hotmail.com

Abstract. This paper analyzes the discourse on political polarization among some Peruvian officials elected by popular vote, as well as the discourse of some residents regarding their role. The information has been disseminated through newspapers, television, and digital radio stations. The information selected has been disseminated in recent months of this year in various media outlets: "Peru21," "El Comercio," "Infobae," "Expreso," "La República," "Panamericana.pe," "Radio Programas del Perú (RPP)," "Latinanoticias.pe," and "Gestión," totaling 21 documents. The discourse of the President of the Republic of Peru and the current mayor of Metropolitan Lima is analyzed. The speeches present some clashes and disagreements, although controversial or antagonistic discourse is more prevalent. The results have been organized into categories, with controversies over the implementation of transportation projects, the appointment of certain ministers, and problems regarding the lack of social security predominating. It is concluded that the antagonisms occur between elected officials: the president and the mayor of the Lima commune. Thus, like the controversy between the population and the authorities, the former shows its discontent or disapproval.

Keywords: Political polarization · digital media · democracy · people · populism

1 Introduction

Populism has developed in recent years as a complex phenomenon in the political sphere. It conveys diverse ideologies or positions, but shares a common denominator: the distortion of information and the polarization it generates among citizens or between a group and an elite or power group. The information disseminated aligns with the interests of the leader or the group that supports them. It employs a discourse that ostensibly channels the interests of the forgotten and marginalized, but these interests are not actually realized once the leader or group assumes power.

In this context, populism presents homogeneous and antagonistic groups. It is reflected between a group or population that confronts the elite or power group that it represents or practices acts of corruption [1] that affect democracy or democratic processes.

A. Rocha et al. (Eds.): ICITS 2026, LNNS 1966, pp. 44–51, 2027.
https://doi.org/10.1007/978-3-032-27538-7_4

In Latin America, populism, disinformation, and political confrontation have interconnected in recent years, weakening democracy [2]. These problems have generated political instability in the Andean region; in the Peruvian context, this phenomenon has also been reflected in the dismissal of several presidents of the republic.

In Peru, populism has developed antagonistically or through a permanent struggle, starting with the last elections, especially at the beginning of the second round between the left-wing and right-wing candidates. The former represents peasants or people from rural areas, who are generally forgotten by the current government; the latter represents the interests of businesspeople and traditional media outlets that have power or are aligned with those in power and therefore defend them. In this context, a series of reports have been disseminated that denigrate or discredit the left-wing candidate. Even after winning the elections, he was removed from office by the Peruvian Congress on the grounds that he had staged a self-coup. He is currently facing criminal proceedings for rebellion.

One of the main characteristics of populism, whether right-wing or left-wing, is the defense of marginalized people or populations who have no voice or direct participation in the political sphere. However, in practice, representatives never fulfill the promises or commitments they made at the time of their candidacy [2]. For this reason, some have described them as anti-elite, anti-pluralism, and as channeling citizens' concerns through emotional speeches [3]. In this case, they seek one end: to alter political reason and deceive the people without them noticing or suspecting it [4].

On the other hand, populism uses the media to spread its interests. It describes opponents or those in power as corrupt and insensitive to the population's problems [5]. In this case, populism uses the media to spread its position or ideology based on its interests, regardless of the needs of the population.

In this context, populism has allowed certain politicians to rise to power through openly authoritarian discourse on social media or digital media, and in some cases by resorting to fake news or manipulating the media [6]. They spread false information with the intention of influencing the population.

The positions of authorities can vary depending on their goals or interests, and often they only consider their popularity and not the real and urgent needs of the population. In this context, regarding the problem of insecurity, there are indeed common ground between the presidency and the mayor's office of Lima. However, it later becomes clear that the mayor's interest is to put certain trains into operation; but the government, through its representative, does not approve their implementation because there is no project or planning, and it does not have the infrastructure for the cars or locomotives to operate. Indeed, the residents are in favor of the train's existence or operation, but in reality, it is inoperative because the infrastructure for it to operate does not exist.

Regarding political polarization, this consists of divergences or disagreements on ideological aspects, visions or actions of the State, and also includes the perception of the population [7]. It shows an extreme position, which does not achieve consensus between individuals or groups. It also manifests itself as a fragmentation of the population on issues of political-social interest; and as a consequence, no agreement or cooperation is achieved (8). Therefore, polarization is conceived as a division of political and ideological groups that fail to integrate to work together [8, 9] for the benefit of the interests of the population.

Consequently, polarization becomes a problem because it prevents consensus or agreement. It therefore affects democracy and prevents decision-making that benefits society. Consequently, polarization is a threat to the consolidation or development of democracy [10].

Another problem with polarization is the generation or dissemination of false or misleading information [8]. It influences the audience, which changes its beliefs and ends up believing the lie. Therefore, it becomes an anti-democratic element; in this context, disinformation or misinformation is useful for political purposes [11].

2 Method

A qualitative study was conducted based on the analysis of virtual documents disseminated through various media outlets: "Peru21," "El Comercio," "Infobae," "Expreso," "La República," "Panamericana.pe," "Radio Programas del Perú (RPP)," "Latinanoticias.pe," and "Gestión," totaling 21 documents published in recent months. The analysis allowed for the systematization of the following categories (which are later specified or explained in the results): agreements or concordances regarding citizen insecurity; stance on the presidential discourse and its ideological position; antagonism regarding project implementation; changing antagonism regarding citizen insecurity and the executive branch; antagonism regarding the executive branch and other political forces; antagonism regarding salary increases and complaints against the president; stances toward the Attorney General's Office and the judiciary; and public perceptions of the mayor and the president.

3 Results

The relationships between authorities may have certain similarities in some cases, but in others, there are discrepancies.

They have been systematized through the following categories:

3.1 Category 1: Agreements or Concordances Against Citizen Insecurity

Faced with the growing wave of insecurity in Peru, and specifically in Lima, where the Peruvian presidency and metropolitan mayor's office are located, insecurity has increased in recent years. In this context, there are agreements between the president and the mayor to address this problem, which affects many citizens [12].

However, when the interests of the authorities are different or there are no agreements, then various antagonisms arise, which directly affect the population, since projects such as more lines or tracks, railway transport, are no longer carried out or executed.

3.2 Category 2: The Mayor's Position on the Presidential Speech and his Ideological Stance

The mayor (and other authorities) have shown their disagreement with the president's speech since she has not formulated concrete proposals in the short term [13].

The antagonism between the mayor and the president has been developing permanently through various media outlets. The former has called the president a communist since she won the elections with Castillo [14], whom various media outlets or politicians defined as a leader or person who belonged to the radical left. However, once he was removed from office, the current president (Dina Boluarte) took office and aligned herself with the other political forces, forgetting her ideology with which she ran.

3.3 Category 3: Antagonism over Project Execution

Another antagonism between these authorities consists of the opposition or conflict because the representative of the president, who in this case is the Minister of Transport, opposes implementing or putting into operation the trains or the railway project that the mayor is promoting [15]. Faced with this confrontation, the president argues that in order to win over the public, some authorities speak badly of the government, alluding directly to the mayor of Lima [16]. In this same line, the mayor requests or demands that the president remove the Minister of Transport because he opposes or obstructs the execution of the train project that he has acquired through the Municipality of Lima [17].

In this context, the mayor has also threatened to hold a silent protest against the Minister of Transport and Communications for not allowing the implementation of the Lima-Chosica train line [18]. He further suggests that, should the minister continue with the same attitude, he may support a possible impeachment of the president [19].

In other cases, a more aggressive antagonistic discourse is promoted or disseminated, including threats; in this case, the Minister of Transport and Communications has claimed that he has received threats to sign an addendum on the Lima-Chosica railway project [20].

On the other hand, the mayor has shown a defiant attitude toward implementing the train project. To this end, he has announced that he will propose an emergency decree to the executive branch for the implementation or execution of the railway project. In this context, he has also criticized the president's travels or her travel to Asia [21].

3.4 Category 4: The Changing Antagonism Over Citizen Insecurity and the Executive Branch

Initially, the authorities had agreed to combat citizen insecurity. However, the mayor believes the president has no intention of doing anything; in this case, of confronting crime, specifically urban terrorism, which should be prosecuted swiftly or summarily, according to the municipal authority [22]. This demonstrates that populist political discourses are ever-changing, moving from agreements to disagreements or conflicts, which end up harming citizens who live in an environment of permanent social insecurity.

Despite the apparent controversies between the mayor and the Peruvian president, there are also similarities regarding the appointment of some ministers. For example, the mayor is in favor of the appointment of the Minister of Justice and Human Rights, who has once again been appointed by the president, despite having been censured by Congress when he served in the Ministry of the Interior. In this regard, the mayor argues that he agrees with the appointment, given the precedent of having supported him when he served as Minister of the Interior [23].

3.5 Category 5: Antagonism Against the Executive and Other Political Forces

The mayor not only points to the government as an opponent of his projects, but also indicates that there is an alliance between the executive branch and other political forces that are influencing the current government and that this prevents his project on the Lima-Chosica train from materializing because the Ministry of Transportation and Communications opposes it or does not give its approval for its execution [24]. In this regard, various media outlets have reported the mayor's intention: the dismissal of the minister because he does not work or does not support the execution of the railway project [25]. Likewise, in another medium, he has reported his intention that the president really evaluate the permanence of said minister [26].

3.6 Category 6: Antagonism Over Pay Raises and Complaints About the President

In the context of an economic crisis, the president has decided to more than double her salary. The mayor has also opposed the increase, suggesting that her salary should be lowered and she should receive what she earned in her previous job [27].

In other antagonisms or controversies, the mayor in front of the president, are based on the fact that he agrees that the investigations or constitutional complaints against the president continue their course [28].

3.7 Category 7: Positions Regarding the Role of the Prosecutor's Office and the Judiciary

Likewise, the mayor, referring to the role of the Prosecutor's Office and the Judiciary, believes they should be restructured because they are not functioning properly. Specifically, regarding the Prosecutor's Office, he believes it lets criminals go free or frees murderers [29]. In this context, he calls for a restructuring of these branches of government, even though they have autonomy or independence, but in many cases they have become politicized.

3.8 Category 8: Perception of the Population Towards the Mayor and the President

Digital and online media also disseminate the population's true position. It is evident that the majority disapproves of the president's management (96%); however, the mayor has a higher approval rating [30]. Likewise, some experts believe that the president's disapproval is due to the survival or support of Congress and does not depend on the population's approval; the concentration on her travels is more relevant [31].

However, the perception of the leaders of the Aymara and Quechua communities is not one of sympathy towards the mayor, who has given or donated eighty motorcycles to certain mayors of Puno. This attitude of the leaders is because the mayor closed the plazas during the protests carried out by the residents of Puno, against the regime of the president, who murdered several citizens who had protested when she assumed the presidency, after the dismissal of President Castillo [32].

4 Discussion

From the analyzed speech of the two elected authorities, it is evident that they are based on confrontational attitudes that are accompanied by false information.

In this regard, the results show that the antagonisms are heterogeneous, that is, they are changing. For example, in some cases there is agreement on certain issues, such as citizen security, and also on the appointment of certain ministers of state, such as the Minister of the Interior or the Minister of Human Rights. These actions directly affect the population, which is affected by insecurity; this demonstrates a failure to fulfill the commitment offered by the authorities [2]. In this case, there is also clear misinformation intended to deceive without the public noticing [4].

Another relevant aspect is the dissemination or use of the media to convey the purposes or interests that in these cases refer to the implementation or execution of projects such as new roads or railway transport; however, this has not been specified or carried out, which shows that they are outside the interests of the population [5, 6].

The polarization of elected officials reveals that there is no common goal to achieve that favors the interests of the population; this is evidenced by fragmentation regarding issues of political and social interest [8, 9]. Therefore, the antagonistic actions of the authorities prevent the development of a democracy that takes into account the interests of the people [10].

Fragmentation also includes the population's negative perception of the authorities [7]; in this case, there is evidence of a majority disapproval of the president, reaching 96%, and also of the municipal authority. This reflects the population's discontent with the current authorities.

Consequently, there is evidence of polarization among elected officials who lack any consensus or agreement, which affects the interests of society, such as security and the development of a rail transport system. It is also evident that this problem affects the development of democracy by not taking into account the interests of citizens; in this sense, they disapprove of the management of the current authorities because they feel or consider that they are not fulfilling their function or role.

5 Conclusions

After analyzing the documents, it is evident that the antagonisms are shifting and heterogeneous. In the specific case of the issue of insecurity, it is evident that the authorities agree in some cases, but disagree in others. The same occurs in the selection of certain ministers, on which both agree; but in others, they enter into or show a conflict.

Antagonisms arise among elected officials, who disagree on issues related to railway construction. Therefore, they are out of step with the interests of the population and thus hinder the consolidation of democracy and the development of agreements or consensus. On the other hand, there is fragmentation between the authorities and the population; this is evident in the population's perception of the presidency, which is high for both authorities, but with a higher percentage for the presidency, reaching 96%.
There is evidence of antagonism between elected officials on issues of social interest that require urgent resolution; These antagonisms are also evident in citizens' disagreement or disapproval of the authorities' actions.

References

1. Martins, R.S., De Oliveira, A.: Populism and digital bolsonarism: a text mining analysis of Thursday's live broadcasts. Sociedad y. Cultura. **28** (2025). https://doi.org/10.5216/sec.v28.80293
2. Rivera, S., González, G.: Populism, disinformation, and political polarization in the social media communication of Latin American populist presidents. Mex. J. Public Opin. **36**, 79–107 (2024). https://doi.org/10.22201/fcpys.24484911e.2024.36.86828
3. Sánchez, L., Villanueva, S., Faure-Carvallo, A.: Political ideology, populism, information literacy, and critical thinking: challenges for future teachers. Lat. Am. J. Commun. **82** (2024). https://doi.org/10.4185/rlcs-2024-2268
4. Huertas, A., Peres-Neto, L.: Digital culture and populism: vox's use of the X-media (twitter). Athenea Digit. J. Thought Soc. Res. **24**(2) (2024). https://doi.org/10.5565/rev/athenea.3410
5. Yavetz, G.: Populist rhetoric and digital politics: a case study of a Facebook campaign in the 2022 Israeli elections. Cogent Social Sci. **11**, 1 (2025). https://doi.org/10.1080/23311886.2025.2476048
6. Kritsch, R., Da Silva, A., Parra, R.: Democracy, political representation and populism in the age of digital technologies. Lua Nova J. Cult. Polit. **123** (2024). https://doi.org/10.1590/0102-123042rk/123
7. Roca, N., Fernández, D.: TikTok: politics for young people. What are trump and Milei talking about? Polarization, hate, and disinformation in electoral campaigns. Stud. J. Message. **31**(3), 739–751 (2025). https://doi.org/10.5209/emp.102868
8. Berrocal-Gonzalo, S., Waisbord, S., Gómez-García, S.: Political polarization and the media: its impact on democracy and society. Inf. Prof. **32**, 6 (2023). https://doi.org/10.3145/epi.2023.nov.22
9. Pérez, P.: The study of political polarization as academic therapy. Daimon Int. J. Philos. **90**, 163–174 (2023). https://doi.org/10.6018/daimon.560931
10. Pérez, P.: How political polarization threatens democracy and how to confront it. Polit. Anal. **35** (2022). https://doi.org/10.15446/anpol.v35n104.105167
11. Rebollo-Bueno, S., Ferreira, I.: Disinformation and polarization in far-right political advertising in Spain and Portugal. Estudios de la Comunicación. **36** (2023) https://ojs.labcom-ifp.ubi.pt/ec/article/view/1084
12. Perú21. https://n9.cl/giqy3. Last accessed 10 Sept 2025
13. El Comercio. https://n9.cl/21du9. Last accessed 11 Sept 2025
14. Infobae. https://n9.cl/c8z8uc. Last accessed 01 Sept 2025
15. El Comercio. https://n9.cl/vx7qw. Last accessed 12 Sept 2025
16. The republic. https://n9.cl/arsh3. Last accessed 27 Sept 2025
17. Expreso. https://n9.cl/o6030. Last accessed 12 Aug 2025

18. Peru21. https://n9.cl/go36o. Last accessed 02 Aug 2025
19. Expreso. https://n9.cl/cpxn2. Last accessed 20 Aug 2025
20. La República. https://n9.cl/yo699h. Last accessed 22 Aug 2025
21. Infobae. https://n9.cl/sit69h. Last accessed 23 Aug 2025
22. Panamericana.pe. https://n9.cl/39s8i. Last accessed 11 Sept 2025
23. El Comercio. https://n9.cl/ual8ou. Last accessed 29 Aug 2025
24. Panamerican.pe. https://n9.cl/k7hv8. Last accessed 28 Aug 2025
25. Express. https://n9.cl/8ljbh. Last accessed 02 Aug 2025
26. RPP.pe. https://n9.cl/d9bqo. Last accessed 28 Aug 2025
27. Expreso. https://n9.cl/w4tlst. Last accessed 25 Aug 2025
28. RPPP.pe. https://n9.cl/gkux0. Last accessed 22 July 2025
29. Management. https://n9.cl/vb8n5. Last accessed 20 July 2025
30. El Comercio. https://n9.cl/hxjh4l. Last accessed 18 July 2025
31. Latinonoticias.pe. https://n9.cl/zg4dp. Last accessed 12 Sept 2025
32. The República. https://n9.cl/ws55c. Last accessed 02 Sept 2025

The Level of English Language Learning According to the Common European Framework of Reference for Languages in Students of Health-Related Programs at a Private University

Pilar Valeria Zapana-Aguilar, Kevin Mario Laura-De La Cruz(✉), Domingo Nicolás Pérez-Yufra, Miliam Quispe-Vargas, Gabriela Manzur-Vera, and Cynthia Karín Rodriguez-Palumbo

Universidad Nacional Jorge Basadre Grohmann, Av. Miraflores S/N, Tacna, Peru
{pzapanaagu,klaurac,dperezy,mquispev,gmanzurv,crodriguezp}@unjbg.edu.pe

Abstract. The aim of this paper was to assess the proficiency of English language acquisition among students pursuing health occupations at a private institution located in the southern area of Peru, based on worldwide criteria. The research conducted was of a basic nature, specifically descriptive, employing a non-experimental approach as outlined in the study's purpose. The sample comprised 134 students from several study programs, including Human Medicine, Dentistry, Physical Therapy and Rehabilitation, and Clinical and Pathological Laboratory. Essentially, the University of Cambridge employed the B1 level (PET) international exam as a tool to assess proficiency in the four key English language abilities. In addition, a non-random convenience sample was utilized, specifically consisting of students enrolled in the technical English course during the academic year of study. The findings indicate that the most prevalent proficiency level achieved by students was A2, accounting for 40.3% of the total. This was followed by B1, which constituted 29.1% of the sample, and A1, which represented 24.6%. A lower proportion of students, namely 6.0%, attained level B2. This discovery implies that most students are at a low intermediate proficiency level, which may indicate the necessity to enhance English language acquisition in these programs. Nevertheless, a considerable proportion of students possess an upper-intermediate proficiency level, indicating a higher comprehension of both the language and specialized scientific literature within their field of study.

Keywords: Language learning · standards · university students · International Examination · CEFR · health

1 Introduction

English language proficiency has gained crucial importance in both academic and professional contexts, especially in the health sciences. The Common European Framework of Reference for Languages (CEFR) has been established as an international standard for

A. Rocha et al. (Eds.): ICITS 2026, LNNS 1966, pp. 52–61, 2027.
https://doi.org/10.1007/978-3-032-27538-7_5

measuring and describing individuals' linguistic competencies in different languages. According to the Council of Europe, the CEFR provides a detailed framework for assessing language skills, ranging from level A1 (basic) to C2 (full proficiency). This framework has been widely adopted by educational institutions and certification bodies around the world to ensure consistency in foreign language teaching and assessment at the global level.

One of the theories that enriches this research is the Information Processing Theory, proposed by authors such as Atkinson and Shiffrin [1], Baddeley [2], and Anderson [3], among others. This theory explains how humans process the information they receive from the environment and how they store, retrieve, and use it in different situations. It is based on the assumption that the human brain functions as an information-processing system composed of three main components: sensory memory, short-term memory, and long-term memory. The theory also implies that, in order to improve listening and reading comprehension in a foreign language, the development of sensory, short-term, and long-term memory must be stimulated through activities and tasks that involve attention, concentration, repetition, association, elaboration, inference, prediction, reflection, and evaluation, among others.

Similarly, the Sociocultural Theory, developed by psychologist Lev Vygotsky, states that learning is a social and historical process in which students interact with their cultural, linguistic, and material environment and internalize the knowledge, skills, values, and attitudes transmitted by their community. This theory is based on the notion that learning occurs within the Zone of Proximal Development (ZPD) — the distance between the learner's actual developmental level (what they can do independently) and their potential level (what they can achieve with the guidance of an adult or a more capable peer).

On the other hand, several challenges have been identified regarding English language learning levels. Internationally, health science students face multiple difficulties in learning English. Studies conducted in Latin America and Europe have shown that one of the main obstacles is the limited exposure to academic and technical English, which hinders the acquisition of an adequate level for understanding scientific literature and professional communication, as well as restricting their ability to participate in exchange programs and access high-quality bibliographic resources. Similarly, it has been reported that a high percentage of nursing students experience difficulties in understanding scientific texts in English, which negatively impacts their academic training. In this regard, Mayta y Dulanto [4] found that students struggled with medical terminology and the comprehension of complex texts. Furthermore, the lack of time to study English due to heavy academic workloads was identified as a significant barrier. Rodríguez et al. [5] also found that students exhibited varying levels of motivation influenced by their perception of English usefulness in their careers. Lack of confidence and anxiety in communication situations were key barriers, often due to the high stress levels in their specialty courses. Finally, Fernández y Ruiz [6] observed that, despite recognizing the importance of English, health students demonstrated lower performance compared to peers from other fields, which was attributed to academic overload and limited opportunities for real communicative practice.

In the Peruvian context, the challenges are equally significant. A study by Mayta [4] revealed that more than 50% of students in health-related programs had an English level below B1, suggesting insufficient linguistic preparation compared to international standards. Additionally, according to data from the Ministry of Education, there is a lack of pedagogical resources and teachers specialized in English instruction for medical and health contexts, which further exacerbates the problem. This deficiency limits the opportunities of future professionals to participate in international conferences, pursue graduate studies abroad, and engage in collaborative research, as English is the predominant language in these domains. Moreover, there are differences in curricula between public and private universities — public universities tend to allocate fewer hours to English instruction compared to private institutions, which often offer more extensive and personalized language programs.

This research is highly relevant as it addresses a critical issue in the training of health professionals in a context where English has become the lingua franca of science and medicine. The innovation of this study lies in its detailed evaluation of English proficiency according to the CEFR through an international intermediate-level exam among students from various health-related programs at a private university in Tacna, Peru. The findings will contribute to the knowledge society by providing an accurate diagnosis of English proficiency in this specific group, allowing for the design of more effective pedagogical strategies tailored to students' needs.

Furthermore, previous studies such as that of Fernández & Ruiz [6] have emphasized the importance of English mastery for health professionals, arguing that adequate linguistic competence is essential for accessing scientific literature, conducting research, and participating in international professional networks. In summary, this study suggests measuring English language learning among health students from the first to the final academic year, since the technical English course is currently offered only in the fourth and fifth semesters. This would make it possible to track students' progress and encourage them to take international exams to certify their proficiency level.

The purpose of this research is to evaluate the level of English language learning, according to the CEFR, among students of health-related programs at a private university, using the PET international exam (intermediate level B1) as the assessment tool. The study aims to identify strengths and areas for improvement in the linguistic training of these future professionals, in order to propose recommendations that contribute to their academic and professional development in a globalized context.

2 Methodology

The type of research was basic, as it sought to generate and expand knowledge on a specific subject. It is defined as a set of empirically structured processes applied to the study of a phenomenon [7]. Furthermore, it is of a descriptive level, since, as Valderrama [7] points out, it aims to specify the characteristics of the phenomenon under analysis. This means that it involves measuring and collecting information — in this case, measuring the level of English language learning according to international standards.

The research design was non-experimental and descriptive, as it allowed the presentation of information as it was obtained within a specific time and context. In non-experimental designs, there is no manipulation of the independent variable, since the events under study have already occurred prior to the investigation [8].

The population consisted of students enrolled during the 2023-I and 2023-II academic terms in the health-related programs of a private university in Tacna. The sample was composed of students enrolled in the *Technical English* courses. A non-probabilistic convenience sampling method was used, as the total number of participants from each program who were enrolled in these courses was included to obtain more meaningful results.

The technique used was a knowledge test, and the instrument applied was an updated Sample Test from the Cambridge University international exam at the B1 intermediate level, according to the Common European Framework of Reference for Languages (CEFR). This test was used to assess the four language skills of English.

The aforementioned instrument demonstrates and upholds reliability, validity, impact factor, functionality, and quality, as supported by its official [9]. This same instrument is endorsed by the University of Cambridge, which uses it to assess foreign language learning for academic, professional, and/or personal purposes.

3 Results

Table 1. Frequency analysis of English level in the Reading dimension by academic program.

Academic Program		Reading				Total	P-value
		A1	A2	B1	B2		
Medicine	Frequency	10	14	19	21	64	
	Percentage	24,4%	36,8%	67,9%	77,8%	47,8%	
Dentistry	Frequency	15	12	4	1	32	
	Percentage	36,6%	31,6%	14,3%	3,7%	23,9%	
							0,001
Physical Therapy and Rehabilitation	Frequency	11	9	2	5	27	
	Percentage	26,8%	23,7%	7,1%	18,5%	20,1%	
Clinical and Pathological Laboratory	Frequency	5	3	3	0	11	
	Percentage	12,2%	7,9%	10,7%	0,0%	8,2%	
Total	Frequency	41	38	28	27	134	
	Percentage	100,0%	100,0%	100,0%	100,0%	100,0%	

In Tables 1, 2, 3, and 4, it can be observed that across the four assessed dimensions, students from the Medicine program achieved the highest percentages at the upper levels

Table 2. Frequency analysis of English level in the Writing dimension by academic program.

Academic Program		Writing				Total	P-value
		A1	A2	B1	B2		
Medicine	Frequency	6	20	26	12	64	
	Percentage	17,1%	48,8%	56,5%	100,0%	47,8%	
Dentistry	Frequency	19	8	5	0	32	
	Percentage	54,3%	19,5%	10,9%	0,0%	23,9%	
Physical Therapy and Rehabilitation	Frequency	9	9	9	0	27	0,000
	Percentage	25,7%	22,0%	19,6%	0,0%	20,1%	
Clinical and Pathological Laboratory	Frequency	1	4	6	0	11	
	Percentage	2,9%	9,8%	13,0%	0,0%	8,2%	
Total	Frequency	35	41	46	12	134	
	Percentage	100,0%	100,0%	100,0%	100,0%	100,0%	

Table 3. Frequency analysis of English level in the Listening dimension by academic program.

Academic Program		Listening				Total	P-value
		A1	A2	B1	B2		
Medicine	Frequency	12	21	19	12	64	
	Percentage	20,0%	53,8%	90,5%	85,7%	47,8%	
Dentistry	Frequency	22	8	1	1	32	
	Percentage	36,7%	20,5%	4,8%	7,1%	23,9%	
Physical Therapy and Rehabilitation	Frequency	20	7	0	0	27	0,000
	Percentage	33,3%	17,9%	0,0%	0,0%	20,1%	
Clinical and Pathological Laboratory	Frequency	6	3	1	1	11	
	Percentage	10,0%	7,7%	4,8%	7,1%	8,2%	
Total	Frequency	60	39	21	14	134	
	Percentage	100,0%	100,0%	100,0%	100,0%	100,0%	

of proficiency (75% to 100%), with this trend recurring at levels B1 and A2. In contrast, at level A1, the Dentistry program showed the highest percentage (ranging from 36.6% to 60%). Furthermore, statistically significant differences ($p < 0.05$) were found among the programs across the assessed dimensions (Table 5).

Table 4. Frequency analysis of English level in the Speaking dimension by academic program.

Academic Program		Speaking				Total	P-value
		A1	A2	B1	B2		
Medicine	Frequency	3	25	27	9	64	
	Percentage	20,0%	48,1%	49,1%	75,0%	47,8%	
Dentistry	Frequency	9	14	8	1	32	
	Percentage	60,0%	26,9%	14,5%	8,3%	23,9%	
Physical Therapy and Rehabilitation	Frequency	3	11	13	0	27	0,005
	Percentage	20,0%	21,2%	23,6%	0,0%	20,1%	
Clinical and Pathological Laboratory	Frequency	0	2	7	2	11	
	Percentage	0,0%	3,8%	12,7%	16,7%	8,2%	
Total	Frequency	15	52	55	12	134	
	Percentage	100,0%	100,0%	100,0%	100,0%	100,0%	

Table 5. Frequency analysis of English level in the Listening dimension by academic program.

Academic Program		General				Total	P-value
		A1	A2	B1	B2		
Medicine	Frequency	6	22	28	8	64	
	Percentage	18,2%	40,7%	71,8%	100,0%	47,8%	
Dentistry	Frequency	19	9	4	0	32	
	Percentage	57,6%	16,7%	10,3%	0,0%	23,9%	
Physical Therapy and Rehabilitation	Frequency	6	17	4	0	27	0,000
	Percentage	18,2%	31,5%	10,3%	0,0%	20,1%	
Clinical and Pathological Laboratory	Frequency	2	6	3	0	11	
	Percentage	6,1%	11,1%	7,7%	0,0%	8,2%	
Total	Frequency	33	54	39	8	134	
	Percentage	100,0%	100,0%	100,0%	100,0%	100,0%	

Similar to the results observed in the specific dimensions, the overall score shows that the Medicine program reached 100% of students at level B2, 71.8% at level B1, and 40.7% at level A2. At level A1, the Dentistry program had the highest percentage (57.6%). Moreover, when comparing the academic programs, a statistically significant difference ($p < 0.05$) was found.

4 Discussion

The general analysis of English proficiency according to the Common European Framework of Reference for Languages (CEFR) among students in health-related programs at a private university, using the PET exam (B1 level), reveals that the predominant level was A2 (40.3%), followed by B1 (29.1%) and A1 (24.6%), with a smaller proportion of students reaching level B2 (6.0%). This finding suggests that most students are at a lower-intermediate level, indicating a need to strengthen English learning within these programs. Previous studies have reported that the English proficiency level of students in health-related fields is often lower than that of students in other disciplines due to curricular load and the prioritization of specialized content over language learning [10]. This trend is consistent with prior research showing that university students in non-language-related majors tend to have lower linguistic competence. For instance, Medina et al. [11] mention that health science students usually focus more on their technical training, leaving less time for the development of second-language skills.

In the Reading dimension, results indicate that 30.6% of students were at level A1, 28.4% at A2, and 20.9% at B1, while 20.1% achieved level B2. When comparing across programs, Medicine students showed superior performance in Reading, with 77.8% reaching level B2, in contrast to Dentistry students, where level A1 was predominant (36.6%). These results align with previous research highlighting variability in reading skills among different disciplines, as Medicine students are often more exposed to and reliant on reading comprehension of English texts due to the abundance of scientific literature in their field [12, 13, 14, 15].

Regarding Writing, most students were at level B1 (34.3%), with a significant percentage at levels A2 (30.6%) and A1 (26.1%). Only 9.0% reached level B2. Again, Medicine students stood out, with 100% achieving B2, while in Dentistry, level A1 was predominant (54.3%). This pattern is consistent with studies indicating that writing in English is one of the most challenging skills for health students, due to the complexity of academic writing in a second language and the lack of practice in clinical contexts [16–19].

The Listening dimension showed that 44.8% of students were at level A1, followed by 29.1% at A2, 15.7% at B1, and 10.4% at B2. Medicine students presented the highest percentages at B1 (90.5%) and B2 (85.7%), compared to Physical Therapy and Rehabilitation students, who were mostly at A1 (33.3%). These results suggest that listening comprehension is a critical skill requiring a more intensive focus in English teaching for health students, consistent with literature emphasizing the importance of listening for understanding clinical instructions and communicating with patients [20–23].

In the Speaking skill, 41.0% of students achieved level B1, followed by 38.8% at A2, 11.2% at A1, and 9.0% at B2. Medicine students again showed superior performance, with 75.0% at B2, while in Clinical and Pathological Laboratory, level A1 was the most common (60.0%). These findings reflect the need to improve oral English proficiency among health students, as speaking is essential for clinical interaction and medical practice [24, 25, 21].

When analyzing results by gender, women showed better overall performance in Reading (B2, 51.9%), Writing (B2, 66.7%), and Speaking (B2, 75.0%), while men excelled in Listening (B2, 57.1%). Although these differences were not statistically significant, they could be related to sociocultural and motivational factors influencing language learning, as suggested by several studies [26–29].

The general results by gender indicate that women performed better in Reading, Writing, and Speaking compared to men, achieving a higher percentage at level B2. Conversely, men performed better in Listening. However, the differences were not statistically significant, suggesting that both genders have a relatively similar level of competence.

Among the limitations of this study, it should be noted that the sample consisted only of students from a private university, which may limit the generalizability of the results to other institutions. Furthermore, the study focused exclusively on health-related programs, without including other disciplines that might display different English proficiency levels. Additionally, variables such as prior English exposure or personal motivation were not considered, both of which could influence performance in language proficiency tests.

Assessing English proficiency using international standards such as the CEFR is crucial for health students, as it provides an objective measure of their language skills, necessary for accessing scientific literature, participating in academic exchanges, and improving communication with patients in a globalized context. Measuring and improving these competencies is essential to ensure that health professionals are prepared to face the challenges of internationalized medical practice [30–32]. Evaluating the English level of health students according to international standards like the CEFR is fundamental to guarantee that future professionals can access up-to-date scientific research, take part in academic exchanges, and communicate effectively in international clinical settings. Authors such as Fernández and Ruiz [6] emphasize the importance of English proficiency among health professionals, as English is the dominant language in scientific research and global medical collaboration. Moreover, integrating English teaching into health curricula can enhance students' academic, professional, social, and personal opportunities.

5 Conclusions

Most students evaluated in the health-related programs at the private university in Tacna, Peru, demonstrated an English proficiency level corresponding to A2, according to the Common European Framework of Reference for Languages (CEFR). This finding indicates a lower-intermediate level of competence among the majority of students, which may not be sufficient to meet the academic and professional challenges that require understanding and producing scientific texts in English. Significant differences were observed in English proficiency levels among the various programs evaluated. Medical students stood out with a higher percentage at the upper levels (B1 and B2) compared to students in Dentistry, Physical Therapy and Rehabilitation, and Clinical and Pathological Laboratory, who were mostly at levels A1 and A2. This finding suggests the need to design specific reinforcement programs for the disciplines with lower performance.

In terms of specific skills, writing and speaking were found to be the most challenging for students, with the majority performing at levels A1 and A2. This suggests that teaching strategies should focus on developing these skills to improve students' ability to communicate effectively in academic and professional contexts.

Academic overload and limited exposure to English in technical and scientific contexts were identified as factors that may be negatively affecting progress in language learning. Implementing a more integrative pedagogical approach that links English learning with specific health science content could be an effective strategy to improve these outcomes.

The results of this study highlight the need to strengthen technical English programs in health-related fields by increasing instructional hours and enhancing the quality of the content taught. Furthermore, the implementation of international standardized tests from the early academic terms is recommended to monitor and certify students' progress in language proficiency.

References

1. Atkinson, R.C.: Human memory: a proposed system and its control processes. In: disponible, N., Spence, K.W. (eds.) The Psychology of Learning and Motivation: Advances in Research and Theory, pp. 89–195. Academic (1968)
2. Baddeley, A.: Working Memory. Oxford University Press (1986)
3. Anderson, J.R.: The Architecture of Cognition. Harvard University Press (1983)
4. Mayta-Tristán, P., Dulanto-Pizzorni, A.: Prevalencia y factores asociados con la intención de emigración en internos de medicina de una universidad pública, Lima 2007. Revista Peruana de Medicina Experimental y Salud Publica. **25**(3), 274–278 (2008)
5. Rodriguez, A., Alpizar León, N., Pastora, Y., García Hernández, G.: Importancia del idioma Inglés en el campo de la Medicina. Medicentro Electrónica. **24**(2), 413–421. Epub 01 de abril de 2020. Recuperado en 21 de agosto de 2024 (2020). http://scielo.sld.cu/scielo.php?script=sci_arttext&pid=S1029-30432020000200413&lng=es&tlng=es
6. Fernández, C., Ruiz, P.: La importancia del inglés en las ciencias de la salud: Un enfoque global. J. Health Educ. (2018)
7. Valderrama, S.: Pasos para elaborar proyectos de investigación científica: cuantitativa, cualitativa y mixta. (5° ed.). Lima: Editorial San Marcos E.I.R.L. (2015)
8. Hernández, R.F.: Metodología de la investigación. McGraw-Hill Interamericana, México (2014)
9. Cambridge University Press & Assessment. Investigación y Validación. English Language Assessment Cambridge University Press (2024). https://www.cambridgeeng-lish.org/latinamerica/research-and-validation/
10. Richards, J.C.: The changing face of language learning: learning beyond the classroom. RELC J. **46**(1), 5–22 (2015). https://doi.org/10.1177/0033688214561621
11. Medina-Toro, Francys M., Jimenez-Ortiz, Jose L., Aragón-Castillo, Jason, Frutos-Najera, D. G., & Muñoz-Palomeque, Manuel.: Calidad de vida en estudiantes de ciencias de la salud de una universidad privada en Nuevo León, México. Revista de la Facultad de Medicina Humana. **22**(1), 89–94. Epub 31 de diciembre de 2021 (2022). https://doi.org/10.25176/rfmh.v22i1.4119
12. Day, R.R., Bamford, J.: Top ten principles for teaching extensive Reading. Read. Foreign Lang. **14**(2) (2002)
13. Grabe, W., Stoller, F.L.: Teaching and Researching Reading. Routledge (2013)
14. Amini, M., Zahabi, A., Amini, D., Alavi, S.: A review of Reading strategies and models in learning and teaching of English as a foreign language. J. Engl. Lang. Pedagog. **8**, 14–26 (2020)
15. Al-Jamal, D.A.H.: The role of linguistic clues in medical students' reading comprehension. Psychol. Res. Behav. Manag. **11**, 395–401 (2018). https://doi.org/10.2147/PRBM.S174087

16. Leki, I.: Undergraduates in a Second Language: Challenges and Complexities of Academic Literacy Development. Routledge (2017)
17. Hyland, K.: 10 disciplines and discourses: social interactions in the construction of knowledge. Writing in knowledge societies, 193 (2011)
18. Matsuda, P.K., Silva, T.: Second Language Writing Research: perspectives on the Process of Knowledge Construction. Routledge (2014)
19. Ferris, D. R.: Response to Student Writing: Implication for Second Language Students (2003)
20. Field, J.: Listening in the language classroom. ELT J. **64**(3), 331–333 (2010)
21. Goh, C.C., Vandergrift, L.: Teaching and Learning Second Language Listening: Metacognition in Action. Routledge (2021)
22. Rost, M.: Teaching and Researching: Listening. Routledge (2013)
23. Siegel, J.: Exploring L2 listening instruction: examinations of practice. ELT J. **68**(1), 22–30 (2014)
24. Thornbury, S.: How to Teach Speaking. Longman (2005)
25. Burns, A., Seidlhofer, B.: An Introduction to Applied Linguistics: Speaking and Pronunciation. Hodder Education (2010)
26. Nyikos, M., Oxford, R.: A factor analytic study of language-learning strategy use: interpretations from information-processing theory and social psychology. Mod. Lang. J. **77**(1), 11–22 (1993)
27. Ehrman, M.E., Oxford, R.L.: Cognition plus: correlates of language learning success. Mod. Lang. J. **79**(1), 67–89 (1995)
28. Ellis, R.: The Study of Second Language Acquisition. Oxford University Press, New York (1994)
29. Pfenninger, S.E., Singleton, D.: Beyond Age Effects in Instructional L2 Learning: Revisiting the Age Factor, vol. 113. Multilingual Matters (2017)
30. Cook, V., Wei, L.: Introduction: Applied Linguistics and Language Teaching in the Twenty-First Century, pp. 1–9. Language teaching and learning, Contemporary applied linguistics (2009)
31. Crystal, D.: English as a Global Language. Cambridge University Press, Cambridge, Reino Unido (2003)
32. Graddol, D.: English Next: why Global English May Mean the End of 'English as a Foreign Language'. British Council, Londres, Reino Unido (2006)
33. Jenkins, J.: Global Englishes: a Resource Book for Students. Routledge (2014)

Cybersecurity Audit-Based Vulnerability Assessment and Risk Mitigation for the Edutainment Platform

Omar G. Bravo-Quezada[1(✉)], Pedro J. Orellana-Jaramillo[1], Jonathan M. Morán-Velasco[1], and Ledys L. Jiménez-González[2]

[1] GIHP4C, Universidad Politécnica Salesiana, Calle Vieja 12-30 y Elia Liut, 010105 Cuenca, Ecuador
obravo@ups.edu.ec, {porellanaj,jmoranv9}@est.ups.edu.ec
[2] Universidad Bolivariana del Ecuador, Km 5 1/2 Vía Durán - Yaguachi, Durán, Ecuador
lljimenezg@ube.edu.ec

Abstract. The integration of innovative digital platforms in education introduces significant cybersecurity challenges, particularly regarding the management of sensitive data and service continuity. Within this context, the *Loud Computing, Smart Cities & High Performance Computing (GIHP4C)* research group at Universidad Politécnica Salesiana developed the Edutainment platform, conceived as a digital repository designed to centralize, manage, and reuse resources for educational video games and immersive environments. Its design aims to enhance the production of gamified learning experiences and foster pedagogical innovation in digital settings. However, the expansion of its functionalities revealed risks related to unauthorized access, authentication weaknesses, and exposed services. To assess these risks, a comprehensive cybersecurity audit was conducted following ISO/IEC 27001 guidelines and employing penetration testing tools such as Nmap, Metasploit, OpenVAS, and Wireshark. The results enabled the identification of critical vulnerabilities that affected confidentiality, integrity, and availability. The mitigation measures implemented—strengthened authentication policies, continuous monitoring mechanisms, and intrusion detection and prevention systems—resulted in a consistent reduction in unauthorized access attempts across all analyzed address ranges, confirming a substantial decrease in exposure to incidents. These findings reinforce the effectiveness of structured audits in mitigating risks and highlight the need to align educational technologies with international security standards. Future work will replicate the methodology across multiple platforms and explore the integration of artificial intelligence for automated threat detection and adaptive security mechanisms.

Keywords: Cybersecurity · Vulnerabilities · Edutainment Platforms · ISO 27001 · Penetration Testing · Cybersecurity Audit

A. Rocha et al. (Eds.): ICITS 2026, LNNS 1966, pp. 62–73, 2027.
https://doi.org/10.1007/978-3-032-27538-7_6

1 Introduction

The digital transformation of education has increased dependence on technological platforms, exposing institutions to increasingly sophisticated cybersecurity threats [1–3]. Recent reports indicate that more than 60% of Latin American institutions have experienced data breaches or denial-of-service incidents [4], while Ecuador and other Andean countries remain highly targeted by phishing and malware campaigns [5]. Global projections estimate cybercrime costs surpassing 10.5 trillion USD annually by 2025 [6]. In Ecuador, the widespread adoption of LMS platforms such as Moodle, Chamilo, and Ilias has revealed authentication weaknesses, exposure of sensitive data, and permission-management issues, alongside growing malware, ransomware, and RLO-based attacks [1–3].

Within this scenario, the *Edutainment* platform—developed by the GIHP4C research group at Universidad Politécnica Salesiana—functions as a digital repository for educational video games and immersive environments. Its expansion requires systematic cybersecurity audits to ensure integrity and operational continuity. This study applies a structured audit aligned with ISO/IEC 27001 and penetration testing using open-source tools (Nmap, Metasploit, OpenVAS, Wireshark), producing empirical evidence on vulnerability identification and mitigation in educational systems.

Prior literature identifies recurring weaknesses in educational platforms, such as API data exposure, insecure logs, and unauthorized access [7,8], often linked to weak credentials, insufficient updates, and low cybersecurity awareness [9]. Audits and penetration tests remain effective strategies for addressing critical failures [7,10,11]. Complementary studies demonstrate the pedagogical role of gamification for digital-security awareness [9,12], while immersive educational initiatives in Latin America highlight cultural value without addressing systematic auditing [13,14]. NLP-based monitoring systems, such as that proposed by León-Paredes et al., offer relevant insights for educational cybersecurity [15].

From a regulatory perspective, the LOPDP (2021) establishes obligations for managing sensitive educational data, reinforcing ISO/IEC 27001. National agencies such as ARCOTEL and MINTEL additionally promote risk management within the sector.

Recent works on automated vulnerability assessment [16,17] compare SAST tools regarding OWASP Top 10 coverage, particularly in low-resource environments. Unlike code-focused assessments, this study evaluates operational infrastructure, system hardening, and continuous monitoring in near-production conditions, demonstrating measurable reductions in operational risk. The analysis explicitly incorporates constraints typical of the Latin American context—heterogeneous infrastructure, budget limitations, and restricted specialized personnel—integrating ISO/IEC 27001 controls with LOPDP (2021) requirements to produce a practical and context-aware risk model.

To guide the study, the following research question was posed: *To what extent does the structured application of audits based on ISO/IEC 27001, NIST, and OWASP reduce exposure to cybersecurity incidents in an educational environment?* The results show that systematic auditing can remain aligned with international standards while offering a replicable framework for platforms operating under limited resources.

2 Methodology

The methodology adopted was designed to ensure a rigorous and replicable cybersecurity audit process in digital educational environments.

A mixed-method approach was applied, integrating technical analysis, documentary review, and empirical testing, aligned with the **ISO/IEC 27001** standard and the recommendations of **NIST** and **OWASP**. This approach made it possible to identify critical vulnerabilities in the *Edutainment* platform, assess its resilience against simulated attacks, and propose corrective measures aimed at continuous improvement.

2.1 Scope and Description of the Audited Environment

The audited environment corresponds to the technological architecture of the *Edutainment* platform, illustrated in Fig. 1. The topology includes the main entry points, communication routes, and critical servers, such as the web server (Apache), application server, file server, and database server.

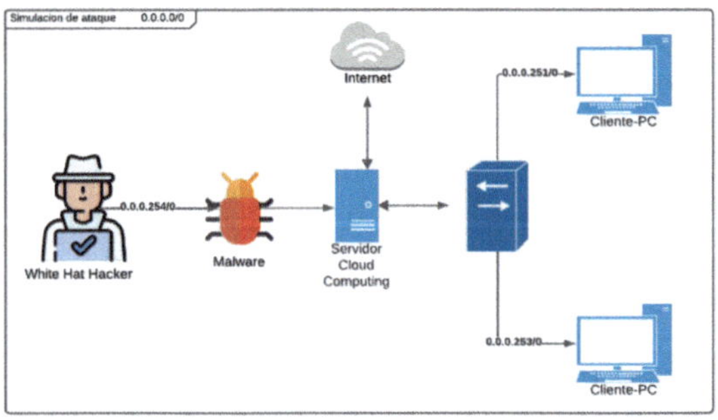

Fig. 1. Attack simulation.

The evaluated architecture comprises web, application, database, and storage servers accessible from both the internal network and the perimeter, with entry points exposed through SSH, HTTP/HTTPS, and auxiliary telemetry ports. The threat model considered external attackers without credentials, students with restricted access, and internal administrators. Assets within scope included personal data, educational resources, intellectual property, and service availability, enabling a focused analysis on plausible exploitation vectors and mitigation priorities aligned with ISO/IEC 27005.

The assessed protocols included HTTPS/TLS for secure web transport, encrypted connections to MySQL/PostgreSQL engines, and authentication mechanisms such as token-based session management, OAuth 2.0, and, when applicable, multifactor authentication (MFA). This definition of the environment clarified the authorized vectors and guided the design of the penetration-testing scenarios.

2.2 Execution Procedure

The cybersecurity audit of the *Edutainment* platform was carried out in five phases aligned with the research objectives.

First, a comprehensive analysis of the system infrastructure, applications, and configurations was conducted to identify vulnerabilities.

In the second phase, security policies and procedures were formulated based on the findings obtained.

The third phase consisted of controlled penetration testing, aimed at simulating ethical attacks and evaluating the robustness of the platform.

In the fourth phase, corrective measures were implemented, focusing on strengthening access control and monitoring mechanisms.

Finally, a process of continuous observation was established to ensure the traceability and long-term sustainability of the improvement actions.

2.3 Methods, Techniques, and Instruments

To guarantee the **traceability of the results** and the **replicability of the study in other digital educational environments**, various methods, techniques, and instruments were applied according to the specific objectives.

The methodological synthesis presented in Table 1 organizes, in a structured way, the relationship between the specific research objectives and the methods, techniques, and instruments employed in each phase.

This representation highlights the study's traceability by linking each objective to verifiable procedures and ensures its replicability in other educational digital contexts.

Likewise, the expected results of each phase reflect the incremental contribution of methodological activities to strengthening the security of the *Edutainment* platform.

3 Results

Network scanning systems made it possible to identify the attack surface and the versions of active services on the Edutainment platform.

However, exploitation attempts did not manage to compromise the system, which suggests a robust configuration and effective controls in the Cloud Computing research group's environment and on the dedicated server. Additionally, security policies were reviewed and adjusted, strengthening the protection posture.

Table 1. Relationship between specific objectives, methods, instruments, and expected results

Objective	Method	Instruments	Expected Results
OE1	Vulnerability analysis and network scanning	Nmap, Kali Linux	Identification of open ports, active services, and insecure configurations
OE2	Documentary review and policy analysis	ISO/IEC 27001, GDPR, checklists	Defined guidelines for access control and data protection
OE3	Controlled penetration testing	Metasploit, OWASP, Burp Suite	Practical validation of vulnerabilities and evidence generation
OE4	Implementation of corrective measures	NIST, OWASP, security patches	Strengthened system with robust authentication, encryption, and *hardening*
OE5	Continuous monitoring and evaluation	System logs, IDS/IPS, scorecards	Active surveillance and periodic validation of security

3.1 Information Gathering and System Vulnerability Search

During the initial exploration, a public application associated with the platform (*Edutainment UPS*), distributed through an official store, was identified. The responsible technical contact and the existence of public repositories linked to the developer ecosystem were also confirmed (links obfuscated).

The initial authentication flow of the augmented reality (*AR*) application was observed to permit navigation without user credentials, indicating the need for additional controls, server-side validations, and session-management policies. Although the unauthenticated access corresponded to non-sensitive demonstration content, it was still considered a potential exposure risk. A preliminary authentication layer and internal-route restrictions were subsequently implemented to reduce the attack surface.

3.2 Comparative Evaluation of Tools

Table 2 summarizes the comparison of tools used or considered.

Nmap stood out for its effectiveness in service discovery; **Metasploit** for its wide catalogue of exploitation modules; **OpenVAS** for its vulnerability coverage (albeit with higher operational cost); and **Wireshark** for focused traffic analysis on internal networks.

3.3 Ports and Services Identified with Nmap

The scan with **Nmap** identified open ports and associated services, and allowed inferring potentially exposed versions.

Table 2. Tools comparison

Tool	Pros	Cons	**Price**
Nmap	Excellent for network scanning and detecting open ports; easy to use	Does not detect specific vulnerabilities; limited to network discovery	Open Source
Metasploit	Powerful for vulnerability exploitation; extensive exploit database	Complex to use without experience; risk if misused	Open Source (Community); paid (Pro)
OpenVAS	Vulnerability scanning with detailed threat analysis	Complex initial configuration; requires resources and tuning	Open Source
Wireshark	Real-time network traffic analysis; ideal for packet inspection	Focused on traffic analysis (not a vulnerability scanner)	Open Source

Table 3 summarizes representative findings and their function in the audited environment. The presence of critical services requires update policies, *hardening* and attack surface reduction.

Table 3. Description of ports and services

Port Number	Protocol	Service	Description
Port 1	TCP	Service X	Critical service for remote administration
Port 2	TCP	Service Y	Provides web functionality
Port 3	TCP	Service Z	Related to database access
Port 4	TCP	Service A	File transfer service
Port 5	TCP	Service B	Enables communication between applications

3.4 Exploitation Verification with Metasploit and Wordlists

Controlled ethical hacking exercises were executed using **Metasploit**, and content reconnaissance was supported by utilities such as `feroxbuster` and `gobuster` using `SecLists` wordlists.

SSH service configurations were evaluated with `sshenum` and credentials were tested using `hydra` with high-entropy lists. No system compromises were obtained, suggesting that current controls mitigate common intrusion vectors.

3.5 Reproducible Commands (Extract) and Traceability

To ensure replicability, the audit documented the tools, versions, execution parameters, and operational conditions used throughout phases OE1–OE3. All servers were evaluated in their production state, recording scan windows, address ranges, and exposed services while preserving security through partial anonymization of IPs and internal routes. Consolidated tools were employed to guarantee traceability: Nmap 7.93 for port scanning, OpenVAS 22.4 for vulnerability detection, Metasploit 6.3.14 for controlled exploitation, Wireshark 4.2.0 for traffic inspection, and Hydra 9.5 for brute-force testing. When applicable, service versions and associated CVEs were logged to correlate findings with up-to-date vulnerability databases.

Below are representative examples extracted from the test protocol; values in brackets (e.g., `[URL]`, `[IP]`) are obfuscated placeholders to be replaced only in controlled environments and with proper authorization.

1. Reconnaissance and web discovery

```
feroxbuster --url [URL]
feroxbuster --url [URL]
    ↪SecLists/Discovery/Web-Content/raft-medium-directories.txt
gobuster dir -u [URL] -w SecLists/Discovery/Web-Content/common.txt
gobuster dir -u [URL] -w
    ↪SecLists/Discovery/Web-Content/raft-medium-files-lowercase.txt
```

2. SSH service enumeration

```
git clone \url{https://github.com/sodomak/sshenum.git}
python3 sshenum.py rhost [IP]:[PUERTO] -w wordlist.txt
```

3. Controlled brute-force attacks (example)

```
hydra -l [USUARIO] -P
    ↪SecLists/Passwords/xato-net-10-million-passwords-100000.txt \
      [IP] -s [PUERTO] -t 4 [SERVICIO]
```

4. Controlled exploitation

```
msfconsole   # Metasploit launch; specific modules were used
    ↪according to identified CVE(s)
```

5. Traffic monitoring and analysis

```
# Selective captures with Wireshark to validate correct TLS
    ↪implementation on HTTPS
tshark -i <iface> -f "tcp port 443" -w captura_tls.pcap
```

3.6 Corrective Measures and Hardening

In this study, the term "exposure" is used in an operational sense to denote the volume of unauthorized access attempts detected by the monitoring system (Fail2Ban) before and after implementing the mitigation measures. The available time windows were not homogeneous, and therefore the data are reported as raw "before" and "after" counts, preserving their integrity and avoiding unsupported temporal inferences.

Based on the findings and aligned with ISO/IEC 27001 and NIST good practices, mitigation actions were prioritized and are summarized in Table 4. Key measures included password hardening via back-end complexity validation, installation of `ClamAV` for handling external content, automated system patching, `ssh` hardening (disabling `PermitRootLogin`, `PasswordAuthentication`, and enforcing key-based access), and enhanced monitoring through `Fail2Ban` to block repeated intrusion attempts.

Table 4. Vulnerabilities, threats and solutions

Vulnerability	Threat	Solution
Weak password	Unauthorized access by brute force	Minimum 8 characters; use of uppercase, lowercase, numbers and symbols (back-end validation); expiration policy and lockout on failed attempts
No antivirus	Malware infection	Install antivirus solution (e.g., `ClamAV`) on all servers and keep signatures/updates; incident response procedures
Lack of updates	Exploitation of known CVEs	Patch policy: `sudo apt update && sudo apt upgrade -y` in controlled environments; patch schedule and testing prior to deployment
Loose SSH access	Credential-based escalation	Disable `root` login, use SSH key pairs, IP allowlisting, and multifactor authentication where possible
Lack of logging	Late detection of intrusions	Enable centralized logs (syslog/ELK/Graylog), retention and alerts; integrate `Fail2Ban` for automatic blocks
Limited monitoring	Undetected persistence	Implement comprehensive monitoring (endpoints, network, processes), alerting rules and periodic reviews; regular audits

3.7 Effect of the Measures on Unauthorized Accesses

To assess the impact of the strengthened controls, the number of IP addresses generating unauthorized access attempts was compared before and after the mitigation actions (Table 5). Figure 2 provides a descriptive visualization of the reduction pattern. The values represent raw counts obtained directly from the monitoring system, reflecting the observed decrease in unauthorized attempts across all evaluated address ranges.

Table 5. Descriptive comparison of unauthorized access attempts before and after mitigation

Address range	Before	After
0.0.0.0/20	320	180
0.0.0.0/21	430	101
0.0.0.0/22	230	60
0.0.0.0/23	497	30
0.0.0.0/24	250	10

Note. The values correspond to raw counts extracted from the system logs (Fail2Ban). They were not normalized by time unit due to heterogeneous recording windows. The results should be interpreted strictly as descriptive comparisons.

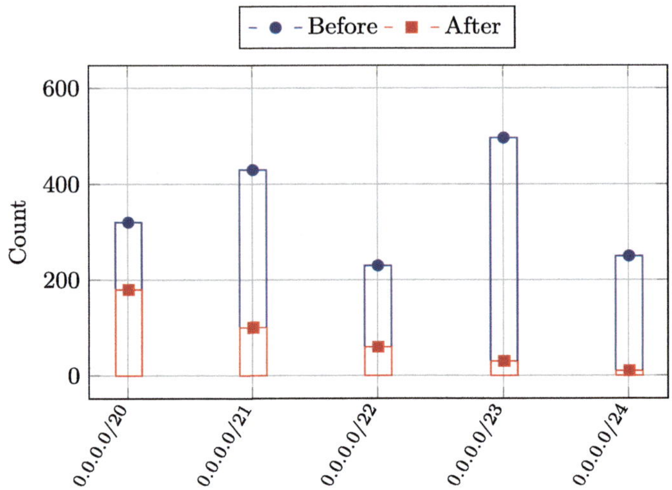

Fig. 2. Comparison of attempts by IP ranges before and after controls.

3.8 Strengthening of Specific Controls

The audit optimized access and monitoring mechanisms through secure password validation, automated updates and reinforcement of remote access. `Fail2Ban` was implemented to block unauthorized attempts and `Prometheus` and `Grafana` were integrated for continuous system monitoring, strengthening overall security.

3.9 Action Plan and Risk Assessment

As part of the action plan and risk assessment, a set of institutional security policies was defined to mitigate critical vulnerabilities and ensure service continuity.

These policies align with ISO/IEC 27001 and the Ley Orgánica de Protección de Datos Personales (Ecuador, 2021), constituting a governance framework that ensures traceability and replicability of the audit in other digital educational environments. Table 6 presents a summary of the main institutional policy axes defined, along with key aspects guiding their implementation.

Table 6. Summary of institutional security policies

Policy axis	Key aspects
Data backup	Location on alternate servers and cloud; frequency according to criticality; defined roles
Technological access	Password vault; key rotation every 90 days; confidentiality agreements
Vulnerability management	White, gray and black box exercises; risk scales; corrective plans
Risk management and classification	ISO/IEC 27005 analysis; information categorization (confidential, restricted, internal, public)
Operational safeguards	Change control and versioning; logs and prior authorizations
Incident management	Incident registry; response protocols; monthly feedback

Severity analysis established differentiated levels of residual risk and corresponding corrective actions. **Critical severity** vulnerabilities present high residual risk and require encryption of sensitive data, role segregation, constant access monitoring, frequent patching, replacement of obsolete equipment and migration to redundant or cloud environments.

For **high-impact** cases where residual risk is medium or high, monitoring of critical processes, periodic updates of antivirus and software, generation of access reports and protection of intellectual property are recommended.

Finally, for vulnerabilities of **medium level**, with low or medium risk, actions include preventive maintenance, power backup, automated monitoring, participation of security committees and change agreements with suppliers. These measures ensure risk management proportional to severity and strengthen the institution's incident response capacity.

3.10 Cost–Benefit Estimation and Financial Planning

The cybersecurity audit required an estimated annual investment of **USD 53,190.00**, combining fixed operational expenses with one-time payments. The largest components corresponded to **network security services** (USD 920.00 per month), **database security** (USD 27,200.00), and ongoing maintenance of **firewall** and **antivirus** systems.

Financial planning—characterized by spending peaks in April, May, June, September, and November—enabled more efficient budget allocation and ensured long-term sustainability of the implemented cybersecurity measures while preserving a favorable cost-benefit balance.

4 Discussion

The cybersecurity audit of the *Edutainment* platform demonstrates the effectiveness of a structured process aligned with ISO/IEC 27001 and supported by open-source penetration-testing tools such as Nmap, Metasploit, OpenVAS, and Wireshark. Critical vulnerabilities were identified and mitigated through corrective policies that strengthened the platform's security while maintaining technical feasibility and operational sustainability. These results are consistent with Sabillon et al. [10] and Ulven and Wangen [11], who emphasize periodic audits as key mechanisms for institutional resilience. Similarly, the risks detected—weak passwords, unauthorized access, and deficiencies in monitoring logs—reflect the patterns described by Alsaadoun [8] and Gasiba et al. [9], underscoring the persistent influence of human and organizational factors on educational cybersecurity.

A central contribution of this work is the contextualized integration of ISO/IEC 27001 and ISO/IEC 27005 with the obligations established by Ecuador's *Organic Law on Personal Data Protection* (2021). This alignment—often absent in regional studies—allows the adaptation of international audit methodologies to environments characterized by heterogeneous infrastructure, institutional limitations, and reduced availability of specialized personnel. As such, the study provides a replicable model for Latin American educational platforms seeking to harmonize global standards with local regulatory demands. Although the monitoring records show a consistent reduction in unauthorized events following mitigation, the analysis remains descriptive due to non-homogeneous recording windows; nevertheless, the observed trend supports the operational effectiveness of the implemented controls in resource-constrained environments.

Future work includes automating audit processes through advanced analytics and continuous monitoring, applying artificial intelligence for anomaly detection and real-time event correlation, and conducting comparative evaluations among Moodle, Chamilo, Ilias, and *Edutainment* to establish an inter-institutional assessment model. Complementary recommendations include strengthening institutional cybersecurity policies, consolidating training programs for teachers, students, and administrators, and promoting national and international partnerships to share best practices and standardize external audits.

References

1. Akacha, S., Awad, A.: Enhancing security and sustainability of e-learning software systems: A comprehensive vulnerability analysis and recommendations for stakeholders. Sustainability **15**(19), 14132 (2023)
2. Mihalache, D.: Assessing web security in e-learning systems. In: Proceedings of the International Conference on Cybersecurity and Cybercrime (IC3) (2024)
3. Rossi, D., Bressan, L., Campos, F., Oliveira, A., Ströele, V.: Educational software and security vulnerabilities: an experimental study. In: Anais do XXXIV Simpósio Brasileiro de Informática na Educação (SBIE 2023), pp. 234–860 (2023)

4. EdTech Hub. State of edtech and cybersecurity in Latin America 2023. Technical report, EdTech Hub, 2023. Consultado en 2023
5. ESET. Ciberseguridad en entornos educativos: Informe latinoamérica 2023. Technical report, ESET Security Research, 2023. Informe regional
6. Cybersecurity Venture. Cybercrime report: 2023–2025 predictions. Technical report, Cybersecurity Ventures, 2023. Proyección del costo global del cibercrimen al 2025
7. Supangat, S., Amna, A.R., Rochman, M.Y.F.: Penetration testing and vulnerability analysis of SINTA platform to strengthen privacy and data protection. J. Inf. Technol. Cyber Secur. **3**(2), 79–83 (2025)
8. Alsaadoun, O., Al-Lawati, B.: Realizing user privacy and security issues in edutainment e-solutions. In: Proceedings of the International Conference on Information Systems Security and Privacy, pp. 278–287 (2019)
9. Gasiba, T., Lechner, U., Pinto-Albuquerque, M.: Sifu - a cybersecurity awareness platform with challenge assessment and intelligent coach. Cybersecurity **3**(1), 1–18 (2020)
10. Sabillon, R., Higuera, J., Cano, J., Higuera, J., Montalvo, J.: Assessing the effectiveness of cyber domain controls when conducting cybersecurity audits: insights from higher education institutions in canada. Electronics **13**(16), 3257 (2024)
11. Ulven, J., Wangen, G.: A systematic review of cybersecurity risks in higher education. Future Internet **13**(2), 39 (2021)
12. Khan, M., Merabet, A., Alkaabi, S., Sayed, H.: Game-based learning platform to enhance cybersecurity education. Educ. Inf. Technol. **27**, 5153–5177 (2022)
13. Bravo-Quezada, O.G., Veletanga-Ordoñez, E.V., Jiménez-González, L.L.: Augmented reality and web services application for the preservation of myths and ancestral legends. In: International Conference on Information Technology & Systems, pp. 314–323, Springer, Heidelberg (2025). https://doi.org/10.1007/978-3-031-93106-2_28
14. Bravo Quezada, O.G., Guallpa-Tenemaza, R.E., Jiménez-González, L.L.: Hachi: development of a gamified video game with artificial intelligence integration for animal protection awareness. In: International Conference on Information Technology & Systems, pp. 95–104. Springer, Heidelberg (2025). https://doi.org/10.1007/978-3-031-93106-2_9
15. León-Paredes, G.A., Bravo-Quezada, O.G., Bermeo-Aguaysa, P.P., Peláez-Currillo, M.J., Jiménez-González, L.L.: Preventing cyberbullying on social networks with Spanish parental control NLP system. Int. J. Adv. Comput. Sci. Appl. **14**(11) (2023)
16. Cano-Crespo, M., García-Ramos, J.C., Guerrero-Contreras, G., Balderas-Díaz, S.: Sast tool evaluation in the age of 5g and iot: Semgrep vs. codacy. Jornadas de Investigación Predoctoral en Ingeniería Informática (JIPII 2024) 51–56 (2024)
17. Alves, A.A., Lamus, A.M., Guerrero-Contreras, G., Balderas-Díaz, S.: Estudio comparativo de fortify sca y horusec en la identificación de las 10 vulnerabilidades principales de owasp. Jornadas de Investigación Predoctoral en Ingeniería Informática (JIPII 2024) 43–49 (2024)

Accounting Education and Entrepreneurship: Keys to Transforming Education in the 21st Century

Liliana Marisol Jumbo Mena, Margoth Iriarte-Solano, Marcelo Juca-Aulestia(✉), and Edwin José Cabrera-Paucar

Universidad Técnica Particular de Loja, Loja, Ecuador
{lmjumbo6,miriarte,jmjuca,jecabrera3}@utpl.edu.ec

Abstract. This article examines the impact of accounting education on promoting entrepreneurship among high school students in Ecuador. A quantitative approach with a descriptive design was used. The specific objectives were: to characterize accounting education as a basis for entrepreneurship, to identify the level of accounting knowledge, and to evaluate the entrepreneurial intent associated with that knowledge. Data collection was carried out through surveys applied to a sample of high school students in Ecuador. The results reveal that accounting education favors the development of entrepreneurial skills such as creativity, innovation, and proactive thinking. With the results obtained from the survey, a data analysis was performed to generate conclusions and recommendations. It was concluded that integrating accounting content into the curriculum strengthens entrepreneurial culture and improves students' preparation to face the challenges of their environment, with positive effects on their academic training and personal development.

Keywords: Accounting education · entrepreneurship · innovation · skills · curriculum · comprehensive training · financial literacy · leadership

1 Introduction

In the current education system, the lack of integration of accounting into the general high school curriculum limits students' preparation for facing both economic and social challenges in a constantly changing world. However, accounting is vitally important for resource management and decision-making, yet its teaching is often limited to technical specialties, leaving many students unprepared in essential skills in accounting education and entrepreneurship.

This is very concerning in educational institutions, as the lack of accounting makes it difficult for students to view entrepreneurship as the basis for their personal and economic development, which highlights the urgent need to implement strategies that link accounting education with entrepreneurship as the basis for a unified general curriculum.

The results of the survey were analyzed to generate conclusions and recommendations.

A. Rocha et al. (Eds.): ICITS 2026, LNNS 1966, pp. 74–84, 2027.
https://doi.org/10.1007/978-3-032-27538-7_7

Accounting education in high school education: a curricular foundation, a tool for financial literacy, and a driver of entrepreneurial culture.

Accounting is considered the language of business, necessary not only in the business world but also in the personal and social development of individuals. Without proper accounting management, businesses could not operate effectively [1]. But why accounting education? Teaching accounting provides students with essential skills to manage their personal finances, make informed decisions, and understand the economic environment around them. In this sense, [2] argue that accounting is not only limited to the professional sphere but also becomes a crucial tool for comprehensive development because it prepares students to face the challenges of the globalized social and financial environment.

Accounting facilitates communication of the financial situation, allowing for informed economic decisions to be made [3]. Therefore that it becomes an essential tool for training critical and responsible citizens who are capable of applying accounting principles in their daily lives [4]. This function is particularly relevant in secondary education, where accounting training can be decisive for the development of entrepreneurial skills [5]. In this context, accounting education is seen as a driver of entrepreneurship, as it provides students with the knowledge and skills necessary to manage limited resources, assess risks, and make strategic decisions, all of which are essential for success in the business world [4].

Without a doubt, the incorporation of accounting content into the high school curriculum not only encourages entrepreneurship, but also develops students' skills in planning, budget management, and evaluating the economic viability of projects [6]. it also allows them to become familiar with technological tools that enhance their analytical and problem-solving skills [7]. By including accounting in their academic training, students not only acquire theoretical knowledge, but also develop practical skills, strengthen their entrepreneurial capacity, and contribute to the formation of responsible and autonomous citizens [8], capable of making informed and strategic decisions that prepare them to face the challenges of the business and economic environment [9, 10],,learning is based on real practice rather than theory [11].

Youth entrepreneurship in secondary education: a focus on personal development, innovation, and social transformation.

Entrepreneurship plays a crucial role in a country's economic development by fostering innovation, generating employment, and promoting self-management. The qualities of an entrepreneur are closely related to their creative capacity [12], and ability to execute new ideas [13], which allows them to transform ideas into both personal and social opportunities. where accounting education plays an essential role in fosterincaneng entrepreneurial spirit [14]. The integration of entrepreneurial skills into the educational curriculum contributes to the economic success and personal development of young people [15].

Entrepreneurship training strengthens students' self-confidence and promotes a positive and proactive vision of their future [16]. This training also stimulates the creation of innovative and sustainable projects, boosting young people's ability to face the challenges of the labor market and society [17, 18]. That is why it is important for young people to acquire these skills, and why education must train critical, observant, and imaginative entrepreneurs with the skills, attitudes, and values that will enable them to take actions that are likely to succeed in the long term. Entrepreneurship means much more than just starting something; it is an opportunity to turn people int. s who are more creative, proactive, and innovative, and everyone should have it in their DNA, regardless of their choice of profession [19].

It is crucial that educational institutions foster the development of entrepreneurial skills through active and innovative methodologies that allow students to apply theoretical knowledge in practical contexts [20]. In this sense, entrepreneurship should be understood as an educational process that goes beyond simply creating businesses, promoting creativity, resilience, and leadership [21]. However, the lack of technological resources and resistance to change in some educational centers represent significant challenges that must be overcome by updating the curriculum and integrating new methodologies [11, 22]. To promote entrepreneurial education among teachers and students, innovation and entrepreneurship must be a fundamental part of the educational process at all levels of education, taking into account attitude and aptitude in order to train successful leaders [23].

Entrepreneurship in high school is an opportunity for students to become resilient leaders with the ability to make a positive impact on their communities. The promotion of entrepreneurial skills should include ethical and social values that enable young people to face challenges with integrity [24]. The link between educational institutions and the business sector is essential to provide students with a richer and more contextualized experience that prepares them for the real world and encourages their creativity [25]. Therefore, integrating entrepreneurship and accounting education into the high school curriculum not only prepares students for professional success but also strengthens their ability to bring about change in society, promoting innovation, leadership, and economic sustainability at the local and global levels [8–10].

2 Methodology

The research has a quantitative approach, which is characterized by its ability to measure and analyze variables numerically, with the aim of establishing relationships and obtaining statistical results [26]. At the same time, it provides the opportunity to work with measurable and achievable data, which is collected using instruments such as surveys, facilitating the obtaining of verified results from the population. In contrast, the qualitative approach focuses on subjectively interpreting phenomena. In addition, it had a descriptive and correlational design. Descriptive because it will focus on describing the level of accounting education and entrepreneurial intent of high school students, analyzing how accounting is taught and its relationship to student entrepreneurship [27]. It is correlational because it seeks to determine the relationship between accounting education and entrepreneurial intent, exploring whether accounting knowledge influences students' entrepreneurial decisions [28].

The population studied includes 100 students who make up the total number of high school students in their first, second, and third years of high school. Therefore, a sample was not used, as the instrument was validated by all students at the aforementioned institution.

On the other hand, the study variables identified are: accounting education as an independent variable and entrepreneurship as a dependent variable.

Likewise, for data collection, a survey was designed structured in two main blocks and one block of general information, in order to meet the following specific objectives: to identify the level of accounting education among general high school students and to identify the degree of entrepreneurial intent with accounting knowledge.

The survey was administered in person during the month of September. Prior to its administration, the instrument was validated by a group of experts in accounting education, entrepreneurship, and technology.

3 Results

The results will be presented according to the objectives set. It begins with a description of the research context and then describes the main findings with the respective discussion of these.

The research was conducted in an educational unit located in the Celica canton, in the province of Loja, Ecuador. This institution is public, located in the urban perimeter, and has 776 students, 376 women and 400 men, 45 teachers, and 4 administrators. Its mission is to guarantee both access and quality in early childhood education, basic general education, and high school, providing comprehensive, holistic, and inclusive education to the community of children, youth, and adults in the canton.

To this end, a survey was used as a research tool and the entire high school student population of the institution, consisting of 100 students, was selected. The survey was designed to measure the level of accounting education with respect to entrepreneurship. In this way, it revealed the students' perception of the impact of accounting education on entrepreneurship.

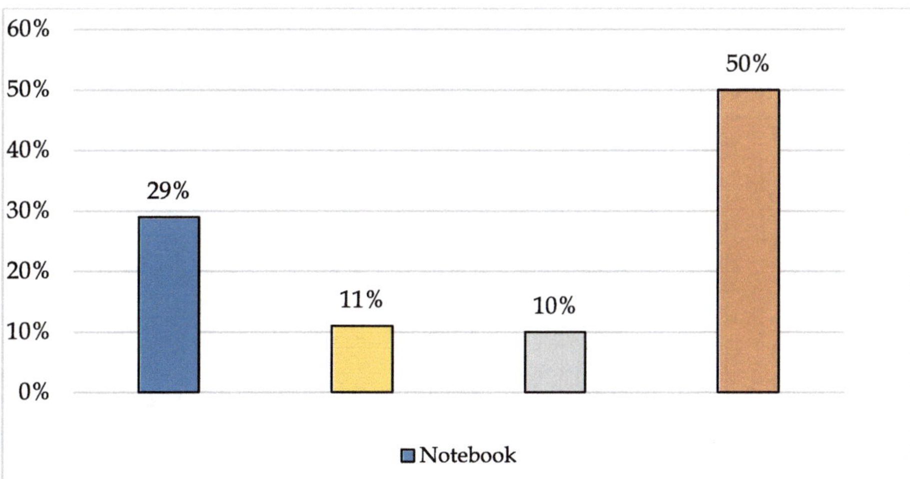

Fig. 1. Tools for recording expenses

Figure 1 analyzes expense tracking, a practice that, according to the Editorial Team (2020), involves the use of an allocated budget, whether by an individual, company, organization, or state entity, for the purpose of acquiring specific goods or services.

In this context, the results reflect that 50% of the students surveyed chose the option "all of the above" as a method for recording their expenses, showing a preference for combining different tools for this purpose.

In this regard, it is considered positive that 50% of students choose to use various methods to record their expenses, such as notebooks, applications, and spreadsheets. This preference shows that using various technological tools increases self-confidence in efficiently managing both personal and business finances. This result coincides with Fercapita (June 10, 2024), who states that combining tools to record expenses facilitates the understanding and efficient management of resources.

This approach not only helps to optimize the budget, but also reduces unnecessary costs, promoting more strategic and sustainable financial planning.

On the other hand, 29% opted exclusively for the use of a notebook, reflecting a preference for traditional methods, possibly due to the simplicity and accessibility they offer. As mentioned in Diario la Hora (October 10, 2016), it is very important to keep track of daily expenses, as this not only allows the user to know where their money is going, but also to better manage their detailed budget. In addition, it points out that the best way to know each person's real financial situation is to keep exhaustive control of their expenses. Although the percentage that prefers the mobile app and spreadsheet is lower (10% and 11% respectively), this should not be overlooked, as it suggests that some students see these digital tools as an opportunity to have more precise control over their finances, even though they may perceive them as less intuitive or more complex.

From the above, it can be deduced that students value expense tracking tools that are versatile and accessible, a key aspect in developing financial applications tailored to the needs of young people. This approach would help them better understand their spending patterns and make financial decisions that positively impact their future.

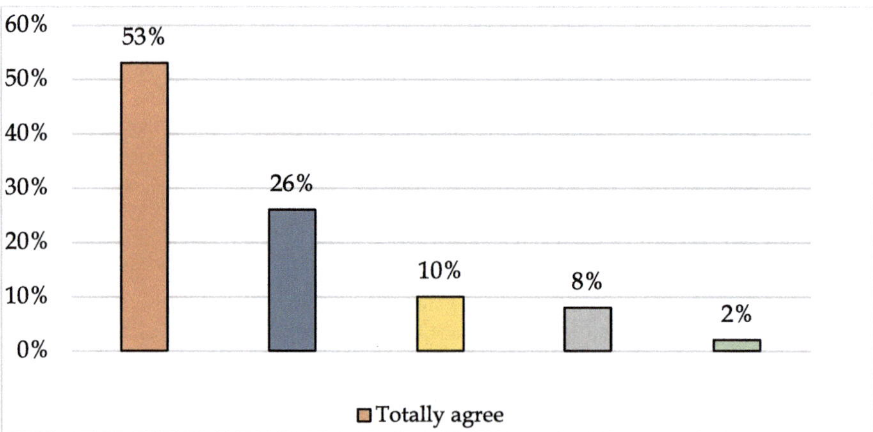

Fig. 2. Does accounting knowledge help generate entrepreneurship?

In Fig. 2, we can see that 53% of students strongly agree that accounting knowledge is crucial for making good decisions in an entrepreneurial venture. In this regard, [29] consider it important because it helps in the management of personal finances or business ideas, through the transmission of knowledge aimed at developing the skills necessary to discern their decisions, an aspect that increases the level of well-being of individuals, families, businesses, and the country.

In this same regard, 26% agree that accounting knowledge allows individuals or entrepreneurs, in this case, to make appropriate decisions regarding the management of their financial resources. According to [29], all of this would have the purpose of positively impacting their quality of life, so that this effect is multiplied in their businesses, their personal lives, and then toward society in general.

A solid understanding of accounting leads to better management of financial resources, which is essential for the long-term success of any business. Having good finances allows risks to be avoided, opportunities to be seized when needed, and the sustainability of any venture to be ensured, which is fundamental for both personal and business growth.

This suggests that a large majority recognizes the importance of accounting in resource management and informed decision-making. As [29] emphasize, accounting knowledge allows entrepreneurs not only to better manage their resources, but also to access more convenient financial services and prepare for financial contingencies, minimizing risks in order to contribute to the balance of the business.

In contrast, only 10% indicate indifference when linking accounting with entrepreneurship. This could be linked to a lack of knowledge and exposure to the subject or, at the same time, a lack of interest in this area of entrepreneurship. The percentage of disagreement represents a minimum of 3%, which must be linked to a lack of knowledge and an underestimation of this science as a driver of personal development.

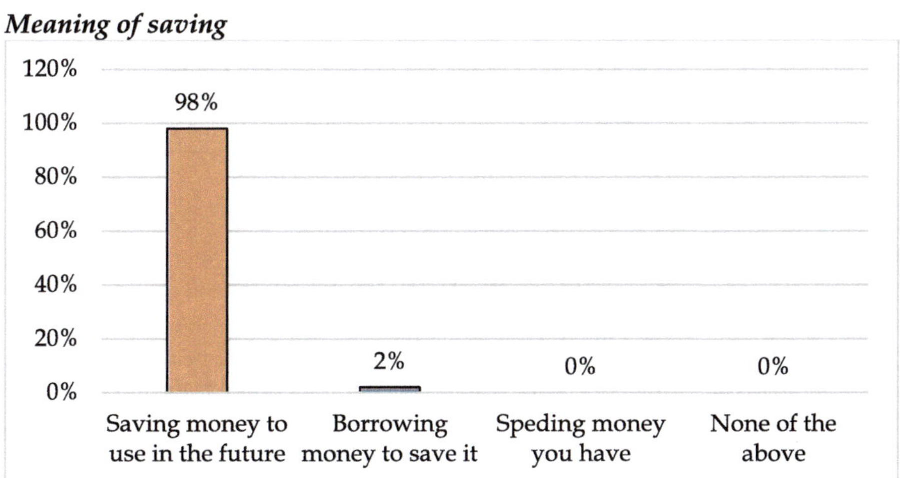

Fig. 3. Meaning of saving

Figure 3 shows that 98% of students understand that saving means putting money aside to use in the future. According to [30], saving is the main activity in personal finance, as it represents the economic planning activity that has the greatest impact on people's development and survival. This high percentage of students who correctly identify the concept suggests a solid understanding of the importance of saving [31], is a set of responsible and intelligent financial habits that must be carried out in a disciplined and consistent manner throughout a person's life. It allows them to develop healthy and responsible financial habits that promote economic stability and the ability to face unforeseen events in the future. It is considered very positive that the majority of students understand the concept of saving, as this habit is fundamental to their financial future. Saving allows them to deal with unforeseen events and at the same time helps them achieve long-term goals, which leads to solid economic stability.

On the other hand, only 2% of respondents answered that saving for them would mean borrowing money and putting it away. Although this is a low percentage, this perception may be associated with confusion about the concept of saving, since saving does not mean borrowing; on the contrary, this would generate debt. This misinterpretation by this minority group underscores the need to adequately reinforce accounting education in the classroom. A lack of understanding of basic financial concepts can have consequences that could be detrimental to financial stability.

From this, it can be deduced that, in general, students do have an adequate understanding of saving, although better management of concepts is required, as they could affect our decisions.

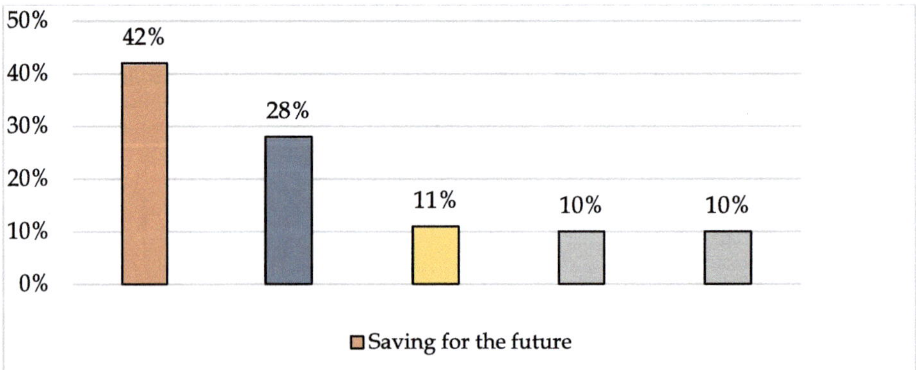

Fig. 4. Saving as the basis for entrepreneurship

Figure 4 shows the results of how students perceive the management of their finances with regard to saving. Thus, 41% believe that saving for the future is the most important thing, which shows positive long-term thinking and gives a sense of financial responsibility for their future. This also indicates a commitment to sustainability, as pointed out by authors [30], who emphasize the importance of saving as a fundamental process that brings happiness. Most students have prioritized saving as a way to manage their finances in the long term. To ensure solid economic stability, it is vitally important to have a savings fund to be able to face future events.

On the other hand, 28% of students have chosen to enjoy the present, but without spending too much, which suggests that they can enjoy themselves, but not in excess, and not exceed spending resources on things that do not bring happiness in the moment, rescuing financial responsibility for the future and promoting a well-founded entrepreneurial mindset.

Furthermore, 11% of students consider that using money to learn or improve is one of the best alternatives, as it allows them to grow through learning, invest in their academic training, prioritize themselves, and focus on their personal development, which will be reflected in the future.

Finally, 10% indicate that planning for the future while continuing to enjoy the present is a relevant option, a balanced view between the present and the future. This preference reveals that students value their present and have the idea of being prepared for the future, which reflects a healthy attitude toward their finances. At the same time, 10% also consider that enjoying the present is more important than waiting or saving for the future, which reflects that they are not very motivated to get ahead and only see the present as an option.

In conclusion, to a greater extent, students demonstrate that they are prepared and have a mature understanding of the importance of balancing the present with the future, highlighting very important data that students highly value saving and investing in themselves, whether in preparation or in other ways. This orientation toward stability and personal growth reflects healthy financial attitudes that, if maintained, can contribute to a more solid personal economy in the future. However, it is also observed that a minority values present enjoyment more, suggesting the importance of continuing to promote financial education to strengthen these positive trends and reduce potential impulsive spending habits.

4 Discussion

The research reveals that accounting education is a key component in the training of high school students, as students value the importance of knowing about accounting in order to make decisions in the field of entrepreneurship. This result is consistent with the findings [1], who assert that proper accounting management is equivalent to the success of a company, as it allows for the management of resources and the sustainability of projects.

In addition, more than half of the students directly associate accounting with entrepreneurship, which is consistent with the postulates [29], who point out that financial training not only improves the individual lives of entrepreneurs but also has a positive impact on overall well-being. In this vein, accounting functions as the language of business and as an engine for entrepreneurial culture [3], enabling risk analysis and strategic decision-making [4].

Similarly, students' understanding of the importance of saving as a foundation for financial stability [30], who link this practice to long-term economic security and happiness. The trend toward responsible saving is evidence of the consolidation of sound financial practices, which are essential for the development of sustainable projects [31]. However, the presence of a small group with misconceptions about saving highlights the need to strengthen early financial education [7].

The research also reveals that, although students appreciate accounting as a tool for starting a business, there is still a fraction that does not consider it indispensable. This finding is consistent with the observations [16], who point out that entrepreneurship training must transcend isolated content and include attitudes, values, and skills that enable young people to see themselves as agents of change. In this case, accounting education faces the challenge of going beyond a technical perspective and adopting a comprehensive educational approach [2].

The connection between entrepreneurship and accounting also highlights what authors such as [12, 13] have stated: that entrepreneurial spirit is linked to creativity and the ability to turn ideas into feasible projects. In this sense, entrepreneurial pedagogy in secondary education promotes leadership and innovation, elements that also emerge from the findings of this research [14].

It is evident that students opt for both conventional methods and digital tools to record their expenses. This finding is consistent with [11], who argues that innovation in education should be based on real practice and the use of affordable technologies. However, the difference between those who use only notebooks and those who choose digital applications highlights the need to strengthen technological integration in accounting education [19], considering that ICTs are currently an essential resource for promoting youth entrepreneurship.

5 Conclusions

Accounting education is essential at the high school level, as it equips students with the practical skills to manage their financial resources and thus make conscious and informed decisions. The results of this study indicate that integrating this discipline into the curriculum is indispensable, as it can transform the educational landscape, providing students with the strategic components and skills to innovate and create a sustainable impact on society.

Accounting education is shown to be a fundamental pillar in the creation of entrepreneurship, providing students with the fundamental basis for managing their resources and making decisions efficiently and informatively. The data obtained contrasts the need for a method that integrates accounting education theory with its practice, allowing students to take their ideas to the highest level in a real-world context.

The accounting education level survey showed that 98% of students have an adequate understanding of basic theory and concepts such as saving and spending, but it should be noted that there is a significant gap in the use of basic financial tools necessary to carry out adequate accounting education. The survey results show that only 11% of students use basic tools to record their expenses. Another important finding is that 29% rely on traditional methodologies, reflecting a need to integrate innovative technologies and methods into the teaching-learning process.

These results confirm that there is a deficit in the use of accounting tools, significantly affecting students and compromising their level of preparation when facing financial challenges. The data obtained on entrepreneurial intent is linked to accounting knowledge. Fifty-three percent of students see accounting as an indispensable tool for promoting a project or venture, while 47% do not link this knowledge to the basis of entrepreneurship.

Acknowledgements. We would like to thank the Master's Degree in Education with a specialization in Educational Innovation and Leadership at the Universidad Técnica Particular de Loja (UTPL) for providing a rigorous, reflective academic environment focused on educational innovation. The teachers, educational resources, and lines of research have been crucial in the development of this work, providing professional, methodological, and theoretical clarity at every stage of the process. I am deeply grateful for the support I have received, as well as for the institutional commitment to excellence and educational transformation.

References

1. Panchi-Mayo, V., Armas-Heredia, I.: Good Accounting Management in the Company as a synonym of success 173–184 (2022)
2. Solis Pereyra, X.G., Casanova Rivera, J.A., Medellin de Dios, E.M.: La educación ética en la formación de contadores en México y el impacto en la práctica: Ethics education in accounting education in Mexico and its impact on practice. LATAM Revista Latinoamericana de Ciencias Sociales y Humanidades. **6**, 1053–1068 (2025)
3. Marcotrigiano Zoppi, L.A.: La comunicación de la información contable: de un leguaje técnico a un lenguaje tecnológico. Revista Visión Contable. **5337**, 9–24 (2018). https://doi.org/10.24142/rvc.n17a1
4. Almachi, C., Saquicela, M.: Planteamiento de un manual de procesos contables para la correcta gestión contable de la empresa Avilés Flor Cia. Ltda. Cantón morona (2015)
5. Arguello, F.L., del Cano Rosas, E., R., Delgado, M.G.: Pedagogical strategy based on simulators for teaching accounting in the first year of high school. **4** (2023)
6. Pazmiño, D.C.: El aprendizaje basado en problemas en el mejoramiento de la enseñanza de la asignatura de Contabilidad de los estudiantes de los primeros cursos de bachillerato del Colegio Fiscomisional "San Jerónimo" de la parroquia Pintag. Cantón Quito durante el año. **2** (2021)
7. Aguaguiña Tirado, E.F., Palacios Morales, M.P., Llerena, S.G., Buenaño Lopez, W.M., del Rocio Paredes Garzón, E.: Aprendizaje Cognitivo en la Era Digital: El Rol de las Plataformas Educativas en la Formación de Competencias Cognitivas en Estudiantes de Bachillerato. Ciencia Latina Revista Científica Multidisciplinar. **8**, 12827–12842 (2024). https://doi.org/10.37811/cl_rcm.v8i5.14732
8. Alva, E.: Plan de formación empresarial para desarrollar las habilidades gerenciales en los estudiantes de contabilidad de un instituto superior tecnológico de Lima. (2024)
9. Hinojosa, C.A., Epiquién, M., Morante, M.A.: Entornos virtuales como herramienta de apoyo al sistema de aprendizaje contable: Un desarrollo necesario. Rev Cienc Soc., 1–8 (2021). https://doi.org/10.31876/rcs.v27i.36489
10. Salinas, E.E.: Uso de las TIC y estrategias de aprendizaje en estudiantes de Contabilidad de un Instituto Público. Villa María del Triunfo. **2019** (2019)
11. Mariño, W.: El proceso de generación de ideas innovadoras para emprendimiento. Revista Internacional de Administración Estudios de la Gestión. **2**, 151–157 (2016)
12. Martínez Canencio, G.: Propuesta de educación significativa para el fortalecimiento del emprendimiento en jóvenes en condición de vulnerabilidad económica del municipio de Cajibío. Cauca. **2**, 1147–1152 (2021)
13. Lira, B., Javier, F.: Entrepreneurship in young people after the covid-19 pandemic Materiales y Métodos, 4–6 (2023)
14. Wernicke, V.: Implementación de la pedagogía emprendedora en el ciclo superior del Bachiller en Economía y Administración de la educación secundaria 1–180 (2021)

15. Caicedo-Quiñónez, Y.N., Guzmán-Hernández, R.: Towards the promotion of an entrepreneurial mentality in higher basic education : A critical and purposeful look Hacia el fomento de una mentalidad emprendedora en la educación básica superior : Una mirada crítica y propositiva Resumen. **9**, 1–27 (2025)
16. González-Tejerina, S., Vieira, M.J.: Entrepreneurship training in primary and secondary education: a systematic review. Revista Complutense de Educacion. **32**, 99–111 (2021). https://doi.org/10.5209/RCED.68073
17. Donoso González, M.: EDUCACIÓN SUPERIOR Y SOSTENIBILIDAD: EMPRENDIMIENTO SOCIAL COMO RESPUESTA (2018)
18. Benavides Nejer, M.A.: PROGRAMA DE CAPACITACION PARA PROMOVER EMPRENDIMIENTOS SUSTENTABLES, DIRIGIDOS A ESTUDIANTES CON ESCOLARIDAD INCONCLUSA EN LA UNIDAD EDUCATIVA ATAHUALPA (2023)
19. Yanza, W.G., Telenchano, G.Y., Sandoval, K.E.: Los emprendimientos productivos: Las Tic como una herramienta de apoyo indispensable para su promoción y difusión. Productive enterprises: ICT as an indispensable support tool for their promotion and dissemination. **5**, 48–61
20. Alegre, M.: Las teorías del aprendizaje y su aplicación en la enseñanza de la contabilidad. Revista Científica Estudios e Investigaciones. **12**, 68–88 (2023). https://doi.org/10.26885/rcei.12.2.68
21. Guachimbosa, V.H., Lavín, J.M., Santiago, N.I.: University for entrepreneurship. Professional training profile and vocation to create companies. Technical University of Ambato, Ecuador. 1–11 (2019)
22. Boldureanu, G., Ionescu, A.M., Bercu, A.M., Bedrule-Grigoruţă, M.V., Boldureanu, D.: Entrepreneurship education through successful entrepreneurial models in higher education institutions. Sustainability (Switzerland). **12**, 1–33 (2020). https://doi.org/10.3390/su12031267
23. Sheyla Acuña-Acuña, B.I., Luzdelia Mendoza-Bravo, K.I.: La educación emprendedora y el aprendizaje-servicio en estudiantes de bachillerato Entrepreneurial education and service-learning in high school students Educação Empreendedora e Aprendizagem em Serviço em Estudantes do Ensino Médio Ciencias de la Educaci. **7**, 1255–1274 (2022). https://doi.org/10.23857/pc.v7i5.4026
24. Hernández Saltos, M., Rodríguez Loor, G.: The learning community in the development of entrepreneurial capacities. Revista Sinapsis. **1** (2022)
25. Mosquera, D. et al.: El Liderazgo como herramienta estratégica en la administración de empresas., 231–242 (2020)
26. Vizcaíno, P., Cedeño, R., Maldonado, I.: Scientific Research Methodology: Practical Guide. (2023)
27. Arias Gonzales, J.L., Gallardo, C.M.: Diseño y metología de la investigación (2021)
28. Ramos Galarza, C.: Los alcances de la investigación clínica. Rev. Fac. Cienc. Med. Cordoba. **78**, 219–220 (2021). https://doi.org/10.31053/1853.0605.v78.n3.32630
29. Beltrán, J.Á., Mulford, E.B., López, M.F., Vega, E.O.: INCIDENCE OF EDUCATION ON COLOMBIAN. **12**, 1–16 (2022)
30. Gómez Hernández, J.F., Gutiérrez Leal, P.: Ahorro y felicidad, vol. 5, pp. 2003–2005. Colegio de estudios superiores de Administración. (2022)
31. Vargas Mateus, F.A., Molina Pelaez, J.P.: ESTUDIO DE EDUCACION EN FINANZAS PERSONALES EN LOS ESTUDIANTES DEL PROGRAMA DE ADMINISTRACIÓN FINANCIERA EN LA CORPORACION UNIVERSITARIA MINUTO DE DIOS VICERRECTORIA LLANOS. Development Studies Research. **3**, 43 (2017)

Accurate and Rapid Time-Synchronization Using Single Camera Image

Masanori Sugimoto(✉)

Hokkaido University, Sapporo, Hokkaido 060-0814, Japan
sugi@ist.hokudai.ac.jp

Abstract. This paper proposes a method for achieving accurate and rapid time-synchronization using a single image captured by a rolling-shutter camera. By leveraging the frequency filtering capability inherent in rolling-shutter cameras, the time difference between the start time of a multi-carrier optical signal transmitted from an LED light source and the shutter opening time of the camera can be estimated. Evaluation experiments conducted under varying SNR and shutter speed conditions of the received signal demonstrated that the proposed method achieved an average time difference estimation error in the range of $1.469 \times 10^{-8} \sim 2.664 \times 10^{-5}$ s, confirming that it could reach or exceed the theoretical time resolution limit of the rolling-shutter camera used in the experiments.

Keywords: Time-synchronization · Rolling-shutter camera · multi-carrier signal

1 Introduction

With the advancement of computer networks and the proliferation of compact, inexpensive, and high-performance computing devices, it has become increasingly feasible for individuals to acquire sensing data from the physical world. For instance, smartphones—ubiquitously used in our daily lives—are equipped with a wide array of sensors, including inertial measurement units (IMUs) such as accelerometers, magnetometers, and gyroscopes.

The application of machine learning algorithms to large-scale datasets obtained from various sensing devices has enabled the development of applications for diverse purposes, such as health monitoring and large-scale infrastructure assessment. As the deployment of sensing devices continues to expand, this trend is expected to further accelerate the digital transformation and "smartification" of our society [2].

In real-world environmental perception, the integrated use of data collected from heterogeneous sensing devices is essential. To achieve this integration, precise time and location information at which the sensing data is acquired is crucial. Consequently, time-synchronization among multiple sensing devices is a

A. Rocha et al. (Eds.): ICITS 2026, LNNS 1966, pp. 85–96, 2027.
https://doi.org/10.1007/978-3-032-27538-7_8

critical issue. Numerous time-synchronization techniques have been proposed in the research fields of distributed systems and wireless sensor networks. Nonetheless, achieving both high accuracy and low-latency synchronization remains a significant challenge and an active research area.

This paper proposes a novel method for accurate and low-latency time-synchronization using a single image captured by a rolling-shutter camera. By leveraging the inherent frequency filtering characteristics of rolling-shutter mechanisms applied to multi-carrier optical signals transmitted from an LED source, the proposed method enables the estimation of the time difference between the signal transmission start time and the camera's shutter opening time with a few tens of nano- to micro-second level precision.

By treating the start time of the signal transmission from the LED as the reference, the proposed method enables accurate and rapid synchronization among multiple camera-equipped sensing devices. To evaluate the proposed approach, the author conducted experiments under various lighting conditions using different shutter speed settings and subjects. The results demonstrated the performance of the method in estimating time differences.

The contributions of this work are summarized as follows:

- The paper proposed a time difference estimation method based on a mathematical model of rolling-shutter cameras, which exploits phase differences among multi-carrier optical signals to achieve high-accuracy and low-latency synchronization.
- The author implemented the proposed method and conducted evaluations under a variety of experimental conditions. Using only a single captured image, the method achieved an average estimation error ranging from 1.469×10^{-8} to 2.664×10^{-5} seconds. These results indicate that the proposed method can achieve the synchronization accuracy which can reach or surpass the theoretical temporal resolution limit (1.570×10^{-5} seconds) of the rolling-shutter camera used in the experiments.

2 Related Work

Many existing time-synchronization techniques and systems use wired or radio-wave communications. The Network Time Protocol (NTP) [16] and Precision Time Protocol (PTP) [8] are used to time-synchronize servers and clients on the Internet to achieve sub-millisecond and sub-microsecond accuracy, respectively. However, it takes from tens of seconds to several hours to achieve their target accuracy.

Many methods for WSNs have been proposed. In reference-broadcast synchronization (RBS) [3], neighboring nodes that receive a reference broadcast compare the arrival times according to their local clocks to estimate their time offset and clock skew. One problem with RBS is that the comparison requires many message exchanges ($(N + 1)/2$ for N nodes). In delay-measurement time-synchronization [18], a one-time broadcast from a master node allows slave nodes to time-synchronize by estimating a time delay at each slave node. In the timing-sync protocol for sensor networks [5], a root node works as a master clock node and extends a hierarchical structure by exchanging timestamps between a parent and its child nodes to identify the clock drift and propagation delay between them.

Radio-wave-based techniques that run on WSNs (such as those proposed by [4,11,15], and [14]) can achieve (sub-) microsecond synchronization but they are implemented using real-time operating systems, which are not employed in current smartphones or tablet PCs. Therefore, it is much more difficult to obtain accurate and precise timestamps using them compared with WSNs. When a target application implements indoor positioning in mobile settings, it is desirable to complete time-synchronization as rapidly as possible. Radio-wave-based techniques conduct repeated message exchanges between nodes to estimate their timestamps accurately, but frequent message exchanges are required within a short period of time, which may induce traffic congestion or lead to longer synchronization times (several seconds to tens of minutes).

An ultrasound time-synchronization technique has been proposed [10]. By calculating the location of the smartphone using TDoA (time difference of arrival) trilateration, the time offset for sending a signal from the speakers time-synchronized with a master clock could be estimated. To implement a microphone array using multiple smartphones to estimate the direction of arrival of sounds, all microphones in the array must be time-synchronized. In Dia [22], two-level synchronization, that is, synchronization between one master smartphone and the other smartphones and synchronization between the CPU clock and the audio I/O clock in each smartphone, was conducted using a least-squares regression method. BeepBeep [17] conducts acoustic ranging for a pair of smartphones by estimating the signal propagation time between the smartphones. These studies do not clearly state how long time-synchronization takes, but completing it instantly appears to be difficult. For example, [22] conducts regression calculations using multiple timestamp data obtained through message exchanges, and [10] and [17] employ specific time intervals between the signal emissions from each smartphone to avoid reverberations, thereby making rapid synchronization difficult.

Regarding time-synchronization techniques using optical signals, FLIGHT [12] is implemented using a fluorescent lamp and a 10 kHz sampling built-in light sensor on a WSN mote called TelesB. The stable nature of the alternating-current frequency enabled time-synchronization within tens of microseconds. [19] uses a single LED for transmitting and receiving optical signals. When an LED is switched to the receiver mode, it can be used to measure the differences in the charged voltages between two slots to achieve time-synchronization. A few techniques using a camera have been proposed for illumination-based time-synchronization. PSync [6] uses a LED illumination as a transmitter transmitting De Bruijn sequences and a photo detector as receiver for rapid and energy efficient time-synchronization. Akiyama et al. [1] and Sugimoto et al. [20] propose techniques using a camera capturing modulated optical signals emitted from

LED lights for acoustic localization. For rapid execution of an algorithm based on a phase-locked loop, a high-speed (1,000 fps) camera is used in [7]. SocialSync [9] can time-synchronize multiple smartphones for 3D construction and scene-depth estimation. The technique estimated the delay between a camera I/O event and a timestamp given by the application software as its time for receiving an image frame with millisecond accuracy.

The proposed technique can be implemented using commercially off-the-shelf camera built-in mobile devices like smartphones without any additional hardware and rapid time-synchronization can be achieved with accuracy in the order of tens of nano- to micro- seconds, which is difficult with existing camera-based systems or techniques.

Fig. 1. An example application that needs time-synchronization among asynchronous camera built-in sensing devices.

3 Proposed Method

3.1 Mathematical Model of Rolling-Shutter Camera

The proposed method involves capturing an object illuminated by modulated light emitted from an LED source using a rolling-shutter camera. From a single image obtained through this process, the time difference between the start time of the signal transmission from the LED illumination and the shutter opening time of the rolling-shutter camera is estimated. When each camera mounted on different asynchronous sensing devices captures the same object illuminated by a common LED light source, the time difference between each camera' shutter opening can be obtained, thereby enabling time-synchronization across the asynchronous devices (Fig. 1).

Compared to global shutter cameras, rolling-shutter cameras consume less power. Consequently, they are typically used in mobile devices to reduce battery consumption and extend operational time. In fact, nearly 100% of the cameras installed in current smartphone models are rolling-shutter cameras.

This section briefly outlines the mathematical model of the rolling-shutter camera, which differs from the global shutter—where all imaging sensors are exposed simultaneously—in that it performs image acquisition by sequentially exposing each line sensor. Details on the mathematical derivation are described in [20][21].

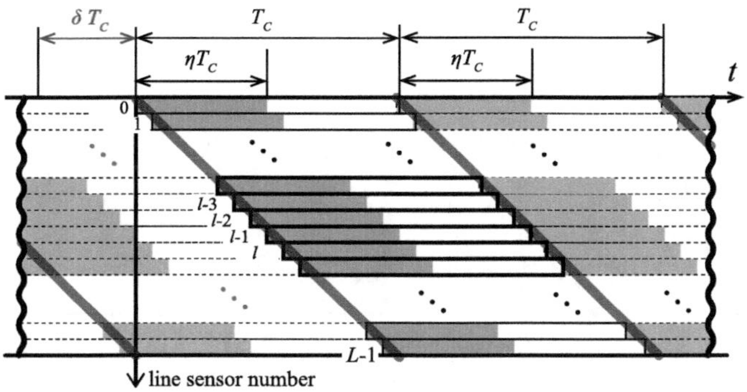

Fig. 2. Timing diagram of rolling-shutter exposure (Note that the readout time discussed in Sect. 4 is not included in the figure).

Figure 2 illustrates how a rolling-shutter camera performs exposure. Let the camera frame period be $T_C = 1/f_C$, the shutter speed be η, the exposure duration be $T_e = \eta T_C$, and the number of line sensors be L. If the fundamental frequency of the transmission signal from the LED light source is $f_C = 1/T_C$ and the frequency order is k $(k = 0, 1, 2, \cdots)$, then the transmission signal consisting of multi carrier waves from the LED illumination can be expressed as:

$$s(t) = \sum_{k=1}^{M} A_k \cos(2\pi f_C k t) \tag{1}$$

Here, M denotes the maximum value of the order k. For simplicity, it is assumed that $A_k = 1$ holds in the following discussion. When the time difference between the transmission start time of the LED signal and the shutter opening time of the rolling-shutter camera is δT_C $(0 \leq \delta < 1)$, the intensity value $r(\delta, l)$ of the l-th line sensor $(l = 0, 1, \cdots, L-1)$ can be expressed by Eq. (2) as follows:

$$r(\delta, l) = \int_0^{\eta T_C} s(t + \delta T_C + \frac{l}{L} T_C) dt \tag{2}$$

By applying a Fourier transform to Eq. (2), Eq. (3) is obtained, from which Eq. (4) can be derived:

$$R_k(l) = \int_0^1 r(\delta, l) e^{-j2\pi k \delta} d\delta \tag{3}$$

$$\frac{R_k(l)}{S_k} = \eta \,\mathrm{sinc}(\pi k \eta) e^{j(\pi k \eta + \frac{2\pi k l}{L})} \tag{4}$$

Figure 3 visualizes the amplitude spectrum ratio of the transmitted and received signals derived from Eq. (4) when η is set to 0.1. From Eq. (4) and Fig. 3, it is confirmed that the amplitude spectrum ratio of the sinusoidal signal with frequency order k is $\eta|\operatorname{sinc}(\eta k\pi)|$ and the phase difference is $k\pi\eta + 2\pi kl/L$, indicating that the rolling-shutter camera functions as a frequency filter.

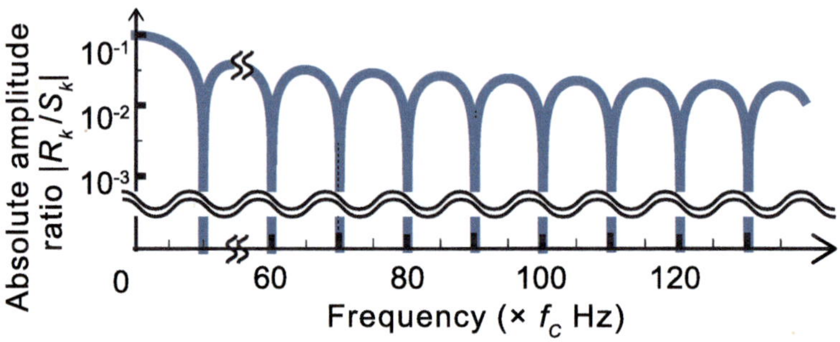

Fig. 3. Frequency filtering characteristic of a rolling-shutter camera.

3.2 Time Difference Estimation

Assume that a multi-carrier signal composed of two frequency orders k_1 and k_2, given by $s(t) = A_1 \cos(2\pi k_1 f_C t) + A_2 \cos(2\pi k_2 f_C t)$, is transmitted from an LED light source and received by a rolling-shutter camera. According to Eq. (4), the phase differences between the transmitted signal and the signal received at the l-th line sensor for each carrier are given by $\pi k_1 \eta + 2\pi k_1 l/L$ and $\pi k_2 \eta + 2\pi k_2 l/L$, respectively. Thus, the received signal phases ϕ_1 and ϕ_2 for each carrier order of $s(x)$ can be expressed by Eq. (5).

$$\begin{aligned} \phi_1 &= \bmod\,(2\pi k_1 \delta t + \pi k_1 \eta + 2\pi k_1 l/L, 2\pi) \\ \phi_2 &= \bmod\,(2\pi k_2 \delta t + \pi k_2 \eta + 2\pi k_2 l/L, 2\pi) \end{aligned} \tag{5}$$

Here, $\bmod\,(x, y)$ denotes the remainder of x divided by y. Using Eq. (5), the time difference δT_C at the shutter opening time ($l = 0$) can be calculated by Eq. (6).

$$\delta T_C = \frac{(\phi_1 - \phi_2) - \pi(k_1 - k_2)\eta}{2\pi(k_1 - k_2)} \tag{6}$$

As shown in Eq. (4), the frequency filtering characteristics of the rolling-shutter depend on the shutter speed η. For example, as illustrated in Fig. 2, when $\eta = 0.1$ holds, sinusoidal signals with orders $k = 10 \times i$ $(i = 1, 2, 3, \ldots)$ are completely blocked and cannot be received by the rolling-shutter camera.

The estimation performance of δT_C using the proposed method is expected to improve with higher signal-to-noise ratio (SNR) of the received signal. Therefore, the values of k_1 and k_2 should be selected such that the amplitude spectrum ratio between transmitted and received signals is maximized. Furthermore, the transmission frequencies of the LED illumination must be higher than the Critical Flicker Frequency (CFF) perceptible to humans and lower than the Nyquist frequency to satisfy the sampling theorem. Thus, the proposed method estimates the time difference using the following algorithm:

1. Given a shutter speed η, let k_{CFF} be the order corresponding to the CFF, L the number of line sensors, and $k_{\text{max}} = L/2$ the maximum order determined by the sampling theorem. From the range of orders satisfying Eq. (7), select k_i $(i = 1, 2)$ that maximizes the amplitude spectrum ratio derived from Eq. (4).
$$k_{\text{CFF}} < k_i \leq k_{\text{max}} \tag{7}$$
2. Let the intensity values of each line sensor obtained by the rolling-shutter camera be represented as an L-dimensional vector $\boldsymbol{I} = \{I_0, I_1, ..., I_{L-1}\}$. Convert $\boldsymbol{I}$ into an analytic signal and compute correlations with the k_1-th and k_2-th order carriers to obtain phases ϕ_1 and ϕ_2 in Eq. (5), respectively.
3. Calculate δT_C using Eq. (6).

4 Experiment

4.1 Experimental Setting

The experimental environment and the detailed setting regarding the experiment are shown in Fig. 4 and Table 1, respectively. In the experiment conducted to evaluate the time difference estimation performance of the proposed method, η was treated as a parameter, and for each value of η, the transmission signal was designed by selecting k_1 and k_2 that maximize the amplitude spectrum ratio. For example, in the case where η was 0.0625, k_1 and k_2 were set to 55 and 56, respectively. To clarify the relationship between the SNR of the received signal and the estimation error, experiments were conducted under different lighting environments (ilm) using multiple reflectors (rfl). The geometric arrangement between the LED light source and the camera, as well as the transmission signal strength, were adjusted such that the resulting SNR was approximately 10, 20, 30, and 40 dB. For example, the $snr40$ setting in Table 1 shows that the SNRs of the received signals obtained using white papers (Wp) photographed under no ambient light environments (Na) ranged from 38.90 to 41.11 (dB) and their average SNR was 39.97 (dB).

As shown in Fig. 4, the transmit signal was generated by a function generator (NF Corporation WF1948), which was connected to an LED source (W-LITE DLFL-001) via a switching power MOS-FET connected to a DC power supply. The purpose of the experiment was to investigate the performance of the proposed method for time difference estimation. The estimation involves the start time of the transmit signal and the shutter opening time of the camera. Being able to release the shutter at an accurate time difference and set camera parameters flexibly is critical to the evaluation of the proposed technique,

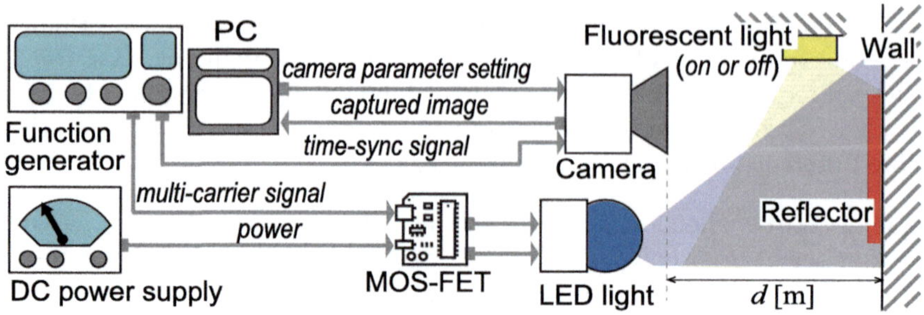

Fig. 4. Experimental environment.

Table 1. Experimental settings

parameter		value
Shutter speed (η)		$0.03175 \times i$ ($0.03175 \leq \eta \leq 0.46875$, $1 \leq i \leq 15$)
Time difference ($\delta^{true}T_C^{RS}$)		$i \times T_C^{RS}/100$ ($i = 0, \cdots, 99$)
Illumination condition (ilm)		No ambient light (Na), Fluorescent light (Fl)
Reflector (rfl)		White paper (Wp), Nature image (Ni)
Signal noise ratio (snr) [dB]	$snr40$: Na + Wp	39.97 (average), 38.90–41.11 (range)
	$snr30$: Fl + Wp	29.95 (average), 28.61–31.05 (range)
	$snr20$: Na + Ni	20.02 (average), 19.02–21.05 (range)
	$snr10$: Fl + Ni	10.03 (average), 8.988–11.16 (range)

and a multi-functional rolling-shutter camera (Point Grey/Teledyne Flir FL3-U3-13S2-C) was therefore used instead of a smartphone built-in camera. The camera received a trigger signal for shutter release directly from the function generator via its general purpose I/O interface. The image of the reflector was captured by the camera and transferred via its USB 3.0 interface to a PC, which conducted the calculations described in Sect. 3.2 to estimate the time difference. A rolling-shutter camera needs readout time T_R, during which its exposure function stops [13]. Thus, T_E must be given as $\eta(T_C - T_R)$, and its maximum is $T_C^{RS} = T_C - T_R$ when $\eta = 1.0$ holds. From the inspection of the camera, T_C^{RS} was found to be 8.028 ms for a frame rate T_C of 50 Hz. Therefore, f_C in this experiment was set to $1/T_C^{RS}$ =124.56 Hz. The proposed technique was evaluated by comparing its estimated time difference with $\delta^{true}T_C$. The resolution of the camera was 1024×1328 pixels and it exposed two neighboring line sensors simultaneously. The total number of line sensors (L) was therefore $1024/2 = 512$. The orders k_{CFF} and k_{max} discussed in Sect. 3.2 were set to 50 and 256, respectively. The time difference estimation was conducted 100 times per each experimental setting and the total number of captured images used in the experiment was 15 (η) $\times 4$ (snr) $\times 100$ ($\delta^{true}T_C^{RC}$) $\times 100$ (no. of measurements) $= 600,000$.

4.2 Experimental Result

The average and standard deviation of the time difference estimation errors achieved by the proposed method for each SNR level, within the range $0.03175 \leq \eta \leq 0.46875$, are shown in Fig. 5. Under the conditions of *snr*40, *snr*30, *snr*20, and *snr*10 as specified in Table 1, the proposed method achieved the following ranges of average errors: $1.469\times10^{-8} \sim 8.257\times10^{-7}$, $2.663\times10^{-8} \sim 8.972\times10^{-7}$, $7.305 \times 10^{-8} \sim 3.995 \times 10^{-6}$, and $7.254 \times 10^{-8} \sim 2.664 \times 10^{-5}$, respectively. The corresponding standard deviations were: $3.333 \times 10^{-7} \sim 5.986 \times 10^{-6}$, $1.170 \times 10^{-6} \sim 1.812 \times 10^{-5}$, $3.235 \times 10^{-6} \sim 6.476 \times 10^{-5}$, and $1.059 \times 10^{-5} \sim 1.821 \times 10^{-4}$, respectively.

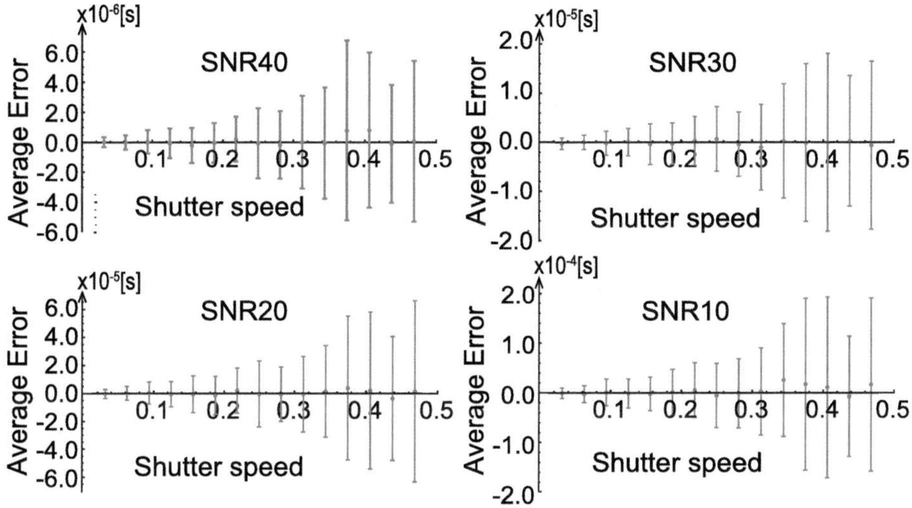

Fig. 5. Average errors and standard deviations of time difference estimation at different shutter speed ($0.03125 \leq \eta \leq 0.46875$) and SNR (*snr*40, *snr*30, *snr*20 and *snr*10) settings. Note that the vertical scales of each figure are different.

Figure 6 presents the cumulative distribution functions (CDFs) for η = 0.03125, 0.25, and 0.46875. Under the conditions of *snr*40, *snr*30, *snr*20, and *snr*10, the 90th percentile errors for these three η values were: $(5.681 \times 10^{-7}, 3.875 \times 10^{-6}, 8.209 \times 10^{-6})$, $(1.887 \times 10^{-6}, 1.140 \times 10^{-5}, 2.743 \times 10^{-5})$, $(5.837\times10^{-6}, 3.783\times10^{-5}, 1.113\times10^{-4})$, and $(1.64\times10^{-5}, 1.122\times10^{-4}, 2.940\times 10^{-4})$, respectively.

5 Discussion

Based on the experimental results, it was found that the proposed method achieved an average time difference estimation error of $1.466 \times 10^{-8} \sim 2.663 \times$

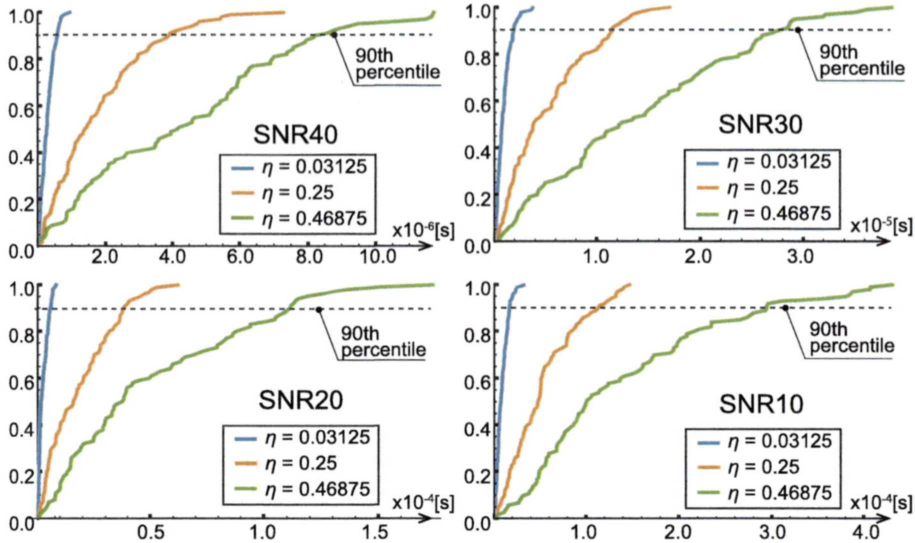

Fig. 6. CDFs at different SNR settings: $snr40$, $snr30$, $snr20$ and $snr10$. Note that the horizontal scales of each figure are different.

10^{-5} s under $snr40$, $snr30$, $snr20$, and $snr10$ conditions, for the range $0.03125 \leq \eta \leq 0.46875$. The rolling-shutter camera used in the experiment had a number of line sensors $L = 512$ and its maximum exposure time $T_e = T_C^{RC} = 8.028$ ms. Therefore, the theoretical time resolution limit of the camera is $T_C^{RC}/L = 1.570 \times 10^{-5}$ s. These results confirmed that the proposed method could estimate time differences with comparable or better accuracy compared to the theoretical limits, using only a single image.

It was also confirmed that the proposed method achieved more accurate time estimation under higher SNR conditions, and that its estimation accuracy deteriorated as η increased in the same SNR conditions. Equation (4) shows that the maximum amplitude spectrum ratio between the transmitted and received signals decreases as η values become smaller, which causes lower SNRs. Further investigations are needed to clarify the relation between shutter speeds, SNRs and the estimation performance of the proposed system not only in laboratory/controlled settings, but also in practical settings.

This paper investigated the time difference estimation performance of the proposed method by treating η and SNR as experimental parameters. Future work should address the following aspects:

- Further improvement of time difference estimation accuracy, for example, by designing, implementing, and evaluating transmission signals with different frequency separations and numbers of carrier components.
- Time-synchronization across multiple devices: As discussed in Sect. 4.1, the proposed method determines the fundamental frequency of the transmission signal based on T_C^{RC}. When T_C^{RC} values of each sensing device are different,

the method cannot correctly estimate the time differences among the devices. One possible approach is to divide the transmission signal period into multiple time slots and assign a specific transmission signal for each sensing device to its corresponding slot.
- Development of practical applications, such as applying the proposed method to cooperative or social sensing using multiple smartphones.

6 Conclusion

This paper proposed a method to accurately estimate the time difference between the signal transmission start time and the shutter opening time from a single image, by capturing an object illuminated by an LED light source using a rolling-shutter camera. Experimental results confirmed that the proposed method achieves time difference estimation performance that can reach or exceed the theoretical limit.

References

1. Akiyama, T., Nakamura, M., Sugimoto, M., Hashizume, H.: SyncSync: time-of-arrival based localization method using light-synchronized acoustic waves for smartphones. In: Proceedings of IPIN 2015, Banff, Canada, pp. 1–9 (2015)
2. Chen, X., et al.: Vehicle as a Service (VaaS): leverage vehicles to build service networks and capabilities for smart cities. IEEE Commun. Surv. Tutor. **26**(3), 2048–2081 (2024)
3. Elson, J., Girod, L., Estrin, D.: Fine-grained network time synchronization using reference broadcasts. In: Proceedings of OSDI'02, Boston, MA, pp. 147–156 (2002)
4. Ferrari, F., Zimmerling, M., Thiele, L., Saukh, O.: Efficient network flooding and time synchronization with glossy. In: Proceedings of IPSN 2011, Chicago, IL, pp. 73–84 (2011)
5. Ganeriwal, S., Kumar, R., Srivastava, M.B.: Timing-sync protocol for sensor networks. In: Proceedings of SenSys'03, Los Angeles, CA, pp. 138–149 (2003)
6. Guo, X., Mohammad, M., Saha, S., Chan, M.C., Gilbert, S., Leong, D.: PSync: visible light-based time synchronization for Internet of Things (IoT). In: Proceedings of INFOCOM 2016, San Francisco, CA, pp. 1–9 (2016)
7. Hou, L., Kagami, S., Hashimoto, K.: An advanced algorithm for illumination-based synchronization of high-speed vision sensors in dynamic scenes. In: Procedings of ICIRA 2010, Shanghai, China, pp. 378–389 (2010)
8. IEEE 1588-2019: IEEE Standard for a Precision Clock Synchronization Protocol for Networked Measurement and Control Systems (2020). https://standards.ieee.org/ieee/1588/6825/. Accessed 15 Apr 2024
9. Latimer, R., Holloway, J., Veeraraghavan, A., Ashutosh, S.: SocialSync: sub-frame synchronization in a smartphone camera network. In: Proceedings of ECCV2014 Workshop on Light Fields for Computer Vision, Zurich, Switzerland, pp. 561–575 (2014)
10. Lazik, P., Rajagopal, N., Sinopoli, B., Rowe, A.: Ultrasonic time synchronization and ranging on smartphones. In: Proceedings of RTAS 2015, Seattle, WA, pp. 108–118 (2015)

11. Lenzen, C., Sommer, P., Wattenhofer, R.P.: Optimal clock synchronization in networks. In: Proceedings of SenSys'09, Berkeley, CA, pp. 225–238 (2009)
12. Li, Z., Chen, W., Li, C., Li, M., Li, X.Y., Liu, Y.: FLIGHT: clock calibration and context recognition using fluoresent lighting. IEEE Trans. Mob. Comput. **13**(7), 1495–1508 (2014)
13. Liang, C.K., Chang, L.W., Chen, H.H.: Analysis and compensation of rolling shutter effect. IEEE Trans. Image Process. **17**(8), 1323–1330 (2008)
14. Lim, R., Maag, B., Thiele, L.: Time-of-Flight aware time synchronization for wireless embedded systems. In: Proceedings of EWSN'16, Austria, pp. 149–158 (2016)
15. Maróti, M., Kusy, B., Simon, G., Lédeczi, A.: The flooding time synchronization protocol. In: Proceedings of SenSys'04, Baltimore, MD, pp. 39–49 (2004)
16. Mills, D.L.: Internet time synchronization: the network time protocol. IEEE Trans. Commun. **39**(10), 1482–1493 (1991)
17. Peng, C., Shen, G., Zhang, Y.: BeepBeep: a high-accuracy acoustic-based system for ranging and localization using COTS devices. ACM Trans. Embed. Comput. Syst. **11**(1), 4:1–4:29 (2012)
18. Ping, S.: Delay Measurement Time Synchronization for Wireless Sensor Networks, Intel Research Berkeley Lab. irb-tr-03-013 Technical report (2003)
19. Schmid, S., Corbellini, G., Mangold, S., Gross, T.R.: Continuous synchronization for LED-to-LED visible light communication networks. In: Proceedings of IWOW, Funchal, Portugal, pp. 45–49 (2014)
20. Sugimoto, M., Kumaki, H., Akiyama, T., Hashizume, H.: Optimally modulated illumination for rapid and accurate time synchronization. IEEE Trans. Signal Process. **65**(2), 505–516 (2017)
21. Sugimoto, M., Kumaki, H., Akiyama, T., Hashizume, H.: High-speed optical camera communication using an optimally modulated signal. In: Proceedings of ICASSP 2018, Calgary, Canada, pp. 3739–3743 (2018)
22. Sur, S., Wei, T., Zhang, X.: Autodirective audio capturing through a synchronized smartphone array. In: Proceedings of MobiSys, Bretton Woods, NH, pp. 28–41 (2014)

Key Competitive Advantage Factors in the Commercialization of Non-timber Forest Products (NTFPs)

Roosa Balk and Jyri Vilko(✉)

Department of Industrial Engineering and Management, Lappeenranta-Lahti University of Technology, Kouvola Unit, Kauppalankatu 13, Kouvola 45100, Finland
jyri.vilko@lut.fi

Abstract. The general competitive advantages of non-wood forest products (NTFPs) classified as food are not well known and the topic has not been researched globally. The purpose of this study is to identify the most important competitive advantages of NTFPs related to the food sector and enable their market value to be improved. The data has been collected as a systematic literature review and has been processed using qualitative coding. Previous literature has shown many functional and structural challenges in the NTFP market. These included, among others, poor information flow, unequal power relations between actors, disorganized markets and challenges in quality control and marketing. The challenges addressed in previous studies allowed the identification of factors that provide more efficient exploitation of the market potential of NTFPs. As a result of the study, the most important generalizable competitive advantages for NTFPs were identified, which included for example innovation, brand and marketing, product quality and standards, and customer orientation and feedback utilization. The study shows that in order to maximize competitive advantage, it is necessary to invest in further research that will help create analytical frameworks for the sector as well as more comprehensive policies and regulations that enable broader integration and comparison of research.

Keywords: Challenges · Commercialization · Competitive Advantage · Food · Natural Product · Non-Timber Forest Product · NTFP

1 Introduction

Non-timber forest products, NTFPs, have a great commercial potential, and their commercialization has a long tradition in human history. They are among the oldest traded goods in the world, whose role and importance in societies have varied from key to secondary goods and back again [12]. Recently, natural products have attracted increasing interest in international markets. The International Tropical Timber Organization (ITTO) defines NTFPs as any natural product that is harvested from the forest, legally or incidentally, and that has no direct connection to timber production [3, 13]. The commercial potential of NTFPs is created by the fact that the products have many usable

A. Rocha et al. (Eds.): ICITS 2026, LNNS 1966, pp. 97–107, 2027.
https://doi.org/10.1007/978-3-032-27538-7_9

components, which enable long-term and versatile use of the products and they support global conservation and community development goals [13, 18]. However, the commercial exploitation of NTFPs is constrained by many factors. The products have low market prices and highly variable productivity, which significantly affects the income of local operators. Difficulties in producing standardized and certified products in line with international market demand limit NTFPs from participating in the market [19].

Previous research has identified the role of NTFPs in supporting livelihoods, nutrition and coping strategies, but the reasons behind the low marketability and structural challenges associated with operations have remained partially unresolved [6, 7, 9, 14]. In order to sustainably exploit the commercial potential of natural products, it is important to identify their key competitive advantages. Analyzing competitive advantage helps to identify the characteristics and conditions that enable the economic exploitation of NTFPs in a sustainable manner, increasing their value and inclusion in the international market. Competitive advantages can ensure that natural products can compete with their competitors in the market.

The purpose of this study is to increase understanding of the competitive advantage and competitive advantage factors of natural products in the food industry. The study takes a sight to identify challenges in the industry that negatively affect the exploitation of the market potential of NTFP products. The challenges aim to identify those factors that can provide competitive advantage for natural products.

2 Methodology

2.1 Systematic Literature Review

A systematic literature review is a scientific research method that extensively analyses existing scientific publications and literature. The goal of a systematic literature review is to assess the quality of existing research data and provide the most objective and unbiased review possible in a selected research area, while also suggesting necessary further research topics. It is built step by step around a selected research question, after which a search strategy is carefully planned. The search strategy defines the desired search terms, databases, delimitations, and possible criteria for searches. Based on these, comprehensive material is narrowed down, the quality and content of which is critically examined. Conclusions and the answer to the research question are created based on a critical review of the material, and the entire process from search processes to conclusions is reported [4].

A systematic literature search was carried out on April 11th 2025 and research materials were retrieved from the Scopus multidisciplinary database. Before the actual data search, several test searches were made with different search terms from the following databases: Scopus, Web of Science and Google Scholar, so that the sampling of materials would be manageable and the materials would contain as much information as possible about the competitive advantage of natural products in the food industry. Finally in the study, was ended up with the search phrase "natural product" OR "non-timber forest product" OR "non-wood forest product" AND "competitive advantage" OR "competitiveness".

The progress of the research material selection process can be seen in Fig. 1. The selected search terms included 81 search results. From these, 10 articles were finally selected for study through various restrictions. In order to make the study sufficiently comprehensive, the sources of the selected articles were reviewed, and 12 more articles were selected for the study using backward snowballing, where the reference list of selected sources is systematically reviewed to identify new papers to include [23].

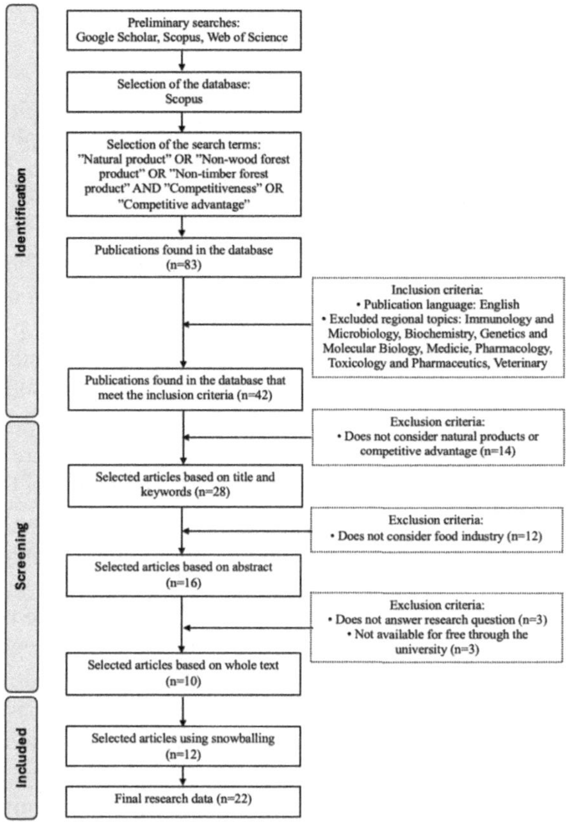

Fig. 1. The progress of the systematic literature review

2.2 Qualitative Coding

Qualitative coding is the systematic classification and interpretation of qualitative data, such as interviews or documents, using different themes, concepts, and phenomena. Qualitative coding helps organize and understand the data, making it easier to answer research questions [22].

The selected research data was analyzed through qualitative coding using NVivo. The topics covered in the selected sources were categorized under different themes to facilitate the analysis of the results. The themes are listed below:

- Factors leading to the topic and increasing general knowledge
- Research methods
- Results of studies that address the current status and use of NTFPs
- Importance of using NTFPs
- Challenges in commercialization of NTFPs
- Development areas in NTFP market use to increase competitive advantage
- Factors improving competitive advantage

3 Analysis and Findings

A systematic literature review highlighted a huge number of challenges that hinder the commercialization of NTFPs. The main challenges identified in previous studies are listed in Table 1. The challenges range widely across structural, logistical and institutional factors in the commercialization of NTFPs. The biggest logistical challenge was perceived to be that many natural products are harvested very far from the marketing and sales point, as forests are located further away from cities. These long distances make transportation difficult [6, 9]. Transportation distances are long, which leads to high transportation costs and makes it difficult to deliver perishable products efficiently to the market. As a solution to product spoilage, processing of products, such as drying, grinding, freezing or preserving, close to the collection points could be utilized. This would simultaneously increase the price of the product and improve the villagers' share of the product value. However, this raises new challenges, which are manifested in the lack of processing facilities and expertise [1, 18]. The phenological characteristics of natural products lead to susceptibility to seasonal and weather variations, which make it difficult to maintain stable production volumes. This, together with the lack of disease control, also negatively affects the marketing of NTFP [15].

One of the most significant challenges identified in this article is the unreliability and lack of transparency of supply chains. There is no common agreement, vision, trust or communication between stakeholders, which weakens competitiveness and slows down market development. Markets are often underdeveloped and based on asymmetric power relations, where intermediaries control prices and quality standards. Information does not reach smaller players and they are easily exploited [5, 6]. It is very difficult for smaller actors to enter the market, as monopolistic positions have emerged in the market, which are controlled by established players [5]. Another challenge is the lack of market information. Prices, quality standards, market positions or product volumes are not known accurately [6]. Price is determined by quality, which in turn varies, as it is not based on the variation in product characteristics. Consequently, quality standards are not uniform and the value of products remains unrecognized and their market price cannot be accurately determined. The lack of quality standards and the arbitrariness of pricing greatly weaken the conditions for commercialization. Price fluctuations can be large and are not fairly reflected on upstream actors [2, 5].

The market potential of NTFPs has not been fully exploited, as the demand potential is greater than the number of current buyers. This may be due to the fact that NTFPs are undervalued compared to wood-based products. When the products are not generally considered important, their markets are not monitored as closely [10]. This is reflected in a lack of funding and insufficient business and marketing expertise [9]. Low technological investments create challenges in monitoring safety and quality standards and disease control measures required in international markets [18]. Inadequate information on the growth stages and regeneration of the products being collected can also hinder the sustainability and development of the market. This, combined with excessively low or high product prices, can lead to overharvesting, which endangers the survival of local species and the ecological state of forests [1, 10]. Sustainability is also undermined by the fact that quantity is often emphasized instead of quality.

The lack of systematic valuation and statistics of NTFPs makes it difficult to assess their importance, which leads to their often being underrepresented or forgotten in forest policy and related decision-making. At EU level, there are no policies or laws regarding NTFPs, which leads to the fact that natural products often have an unclear position in national markets. As a result, they are used locally or sold in unorganized markets, leaving their consumption invisible [10]. In other words, they remain outside of official statistics and economic calculations. In order to develop the successful participation of NTFPs in the market, it is necessary to invest in market analysis and involve them in decision-making [9].

The results of the study show that the export competitiveness of NTFP products is built around productivity, market share, export prices and domestic consumption in particular. Production, marketing and sales expertise, as well as the collection of information related to the target market, therefore prove to be key prerequisites for export success in the NTFP product chain [11, 16]. The greater the market share of actors in international markets, the more competitive they are compared to other actors [16]. The success of commercializing NTFP products can be defined through social justice, improvements in community organization and local culture, and improved economic status. In terms of successful commercialization, it is necessary to understand the structural capabilities of the organization, such as trends in pricing structures, price competitiveness, availability of substitute products, quality requirements, and buyer preferences. In addition to these, the nature of government involvement, the distribution of property rights, market transparency, pressure on resources, and the ability to demand and monitor the realization of market rights also affect the success of products [8, 9].

Many NTFP products have inelastic demand, which allows for the creation and optimization of pricing strategies without significant impacts on export demand [19]. The use of pricing strategies, like other business strategies, is emerging as a significant factor in achieving competitive advantage. The results of the study also show that competitive strategies, supported by systematic performance measurement, improve the ability of companies to respond to market challenges more effectively [20]. The strategic fit between the chosen strategy and the entrepreneur's operating style also acts as a strong driver for success [21]. Especially in SMEs, the proactivity, innovativeness and willingness to take risks associated with entrepreneurship have a positive impact on performance and should be taken into account when creating a strategy [25].

In terms of the competitiveness of NTFP products, product quality, product availability and supply reliability form a key basis for market success. Quality has a direct impact on the value of products, both purchase and sale prices, as it supports the safe

Table 1. The main challenges in commercialization of NTFPs

Challenge	Description	Frequency
Long distances	Long distances cause high transportation costs and cause food to spoil on route	4/22
Lack of funding and marketing	Makes market entry difficult	6/22
Availability of market data	Difficult and expensive to measure, so rough estimates are often made. It is impossible to say the true market value of NTFPs	5/22
Non organized markets	There is no trust or agreements between stakeholders and the flow of information is very poor.	6/22
Overharvesting and scarcity of forests	Can be caused by prices that are too low or exceptionally high. It weakens the condition of forests and can cause the extinction of local products. Insufficient information on how plant species grow and regenerate	7/22
NTFPs underrepresented or forgotten in national policies and legislation	There are no defined operating methods, which leads to different operating methods between markets	5/22
Wide range of different natural products	One marketing strategy cannot be applied to all natural products	4/22
Quality standards are not uniform	The properties of the products do not vary with their quality grades. Prices are determined by the quality grades	5/22
Seasonal variations in crops	Uncertainty and fluctuating production volumes	6/22
Lack of a shared vision of how the market works.	Weakens competitiveness and makes it challenging to develop markets	3/22
Inequality and asymmetric power relations. Exploitation of small producers	More influential producers set prices and standards that are not communicated to small producers. There are large value differences between market actors	7/22
Low and fluctuating prices	Middlemen control pricing and price fluctuations are not passed on to upstream producers	6/22

consumption of products. Compliance with quality standards improves product credibility and supports regulatory compliance and sustainable and ethical practices [6, 24]. Therefore, focusing on quality and adhering to quality standards gives NTFP products a competitive advantage. In addition to quality, companies must be able to meet the needs of the domestic and international markets with sufficient production. When you are able to meet international demand, you can also increase the value of the products [16]. Therefore, competitive advantage is gained when you are able to meet market demand in a timely manner and guarantee good supply reliability for products. The importance of focusing on timely availability and product reliability increases especially when operators operate further away from market centers, as the risk of delivery errors increases as delivery distances increase [6]. Thus, operators located near market centers increase their own competitive advantage over others.

Marketing and sales were seen as the most limiting factors in the overall success of NTFP products in the market, making them important factors in terms of success. Therefore, they can be seen as factors that bring competitive advantage to operators [9]. Building your own product brand is an important part of marketing. You should carefully familiarize yourself with the important elements of international markets and consider which factors increase the attractiveness of products. The role of NTFP products in the market can be improved when factors related to product traceability and certification are combined with branding. The addition of labels, marks and food safety certificates can make customers pay more for products, as they indicate purity, naturalness, healthiness and safety, which are increasingly attracting interest among consumers [10, 24].

Adapting products to customer expectations, requirements and needs is very important in creating customer satisfaction and long-term customer relationships. Providing after-sales services enables a company to accumulate comprehensive information about customer preferences and opinions, which can be used to enable the continuous development of products and operations in the direction demanded by the market. Competitive advantage can be created well when a company has the ability to collect and utilize market information systematically, supporting strategic decision-making in the development of export operations. Without a deep understanding of foreign markets and customer needs, export companies cannot develop as effectively or create closer and deeper relationships with suppliers, vendors or customers. Operators also gain a competitive advantage when they are able to interpret up-to-date, useful information to identify potential customers [11].

Competitive advantage can be built especially through innovation, which serves as a key tool for companies to differentiate themselves in the market. Innovation can be done by utilizing technology and entrepreneurship, providing a skilled and trained workforce, and improving productivity [16]. Financing can play a significant role in implementing innovation, which can also act as a competitive advantage between actors [17]. Strategic action enables a company to organize itself so that the company can utilize the full potential of the resources it possesses to achieve a competitive advantage. Sustainable competitive advantage is created by strategic planning, leadership skills, tacit knowledge, capital, and skilled employment of personnel. However, a company's strategy can only create a significant competitive advantage if its operations differ from those of its competitors and respond to the specific needs of the market in an innovative and efficient manner [21]. Table 2. summarizes the most important competitive advantage factors for natural products that emerged during the study.

Table 2. The most important competitive advantage factors for NTFPs

Competitive advantage factor	Impact	Frequency
Innovation and technological advancement	Enables product differentiation, improves productivity	5/22
Branding, traceability and certification	Increases product credibility, consumer trust and supports market entry and the perceived value of the product.	4/22
Product quality and standards compliance	Ensures safe consumption, complies with regulations	6/22
Marketing and sales expertise	Creates product value	7/22
Supply reliability and product availability	Enables to meet demand and maintains customer trust	4/22
Customer orientation and feedback utilization	Supports product development, customer satisfaction, long-term relationships and continuous improvement.	2/22
Government involvement and institutional support	Affects property rights and market transparency	6/22
Strategic fit and organizational capabilities	Enables efficient use of resources	4/22
Market information and intelligence gathering	Supports strategic decision-making, market goal achievement and demand forecasting	5/22

4 Discussion

The commercialization of NTFPs has aroused more interest as sustainable development principles, poverty reduction, and global environmental awareness have emerged, which aims to better manage and preserve forests [12]. In addition, the commercialization of NTFPs is seen as an important goal for the future because, in addition to income, it can also strengthen the organization of vulnerable communities and improve social justice by increasing the participation of vulnerable members of the community in economic activities [9].

A systematic literature review highlights that the commercialization of natural products is a complex process influenced by socio-economic, ecological and cultural factors. In order to effectively commercialize natural products and to identify NTFPs with high commercialization potential or high risks of failure, careful analysis of ecological, socio-economic and cultural factors is required. Currently, studies have shown that commercialization of NTFPs is often unsuccessful [9]. This may be due to a lack of understanding of the potential markets for NTFPs, trends in pricing structures, availability of substitutes, quality requirements and buyer preferences [6]. In addition, many NTFP markets are based on stakeholders with different priorities, unequal power relations and conflicting views. There is therefore no shared vision for competitiveness and the chain is managed in self-interest, which negatively affects the sustainable management, market development and valuation of NTFPs [5, 6].

5 Conclusions

5.1 Scientifical Implications

This study provides an understanding of NTFP commercialization through a systematic literature review. Previous NTFP research has been limited to case-based analyses without coherent theoretical frameworks, which has made it difficult to draw generalizable conclusions in the field. This literature review helps to develop a general market view of NTFPs, identifying recurring competitive advantages and challenges in the field. The findings of the study are linked to opportunities for developing product market access, competitive advantage and value chain.

The results show that NTFP commercialization is influenced by ecological, socio-economic and institutional factors and their interactions. The study highlights the need for integrative frameworks that link environmental sustainability with business strategy and rural development, highlighting structural challenges in the field such as supply chain transparency, weak marketing and lack of quality standards.

In addition, the study addresses the competitive advantage of food-based NTFPs and its effects on market behavior, which has received less attention in earlier NTFP research. The stronger market position of the products can create new opportunities for future research and at the same time opens doors for more efficient innovation and brand development in the natural products market. The results show how important the flow of information and documentation play in terms of market success and scientific research.

Overall, this study creates a theoretical foundation for combining business and environmental sustainability in NTFP research and consumption, encouraging continuous, sustainable and comparative research in the field and its development.

5.2 Managerial Implications

This study confirms that competitive factors can help to increase the untapped market potential of NTFPs. Competitive advantage arises when a company is able to offer its customers better value than its competitors. Therefore, it is important for NTFP industry actors to understand the factors that allow them to achieve a better position in the market with their products and to utilize these in their strategy. The study offers actors opportunities to improve the value of their products and their own position in the market by identifying general competitive advantages of NTFPs.

First, in order to fully utilize competitive advantage, actors are required to educate themselves and develop better information flow between stakeholders.
Second, they should emphasize the strengths of their products to address existing challenges and improve their position compared to competitors.
Third, NTFP industry also needs clear policies, regulations and financial support to fully utilize its potential.

5.3 Limitations and Future Research

This study has limitations that should be taken into account when analyzing the results of the study. The reviewed literature was collected from a single database, which may have excluded relevant studies. Further research should be conducted on NTFPs and the sector in general to create analytical approaches that help integrate and compare research results related to the sector.

References

1. Andel, T.: Non-timber forest products, the value of wild plants. Agrodok-ser. **39**, 46–48 (2006)
2. Aquilar-Rivera, N., Olvera-Vargas, L.: Innovations for sustainable production of traditional and artisan unrefined non-centrifugal cane sugar in Mexico. In: Leal Filho, W., Krasnov, E., Gaeva, D. (eds.) Innovations and Traditions for Sustainable Development. World Sustainability series., pp. 313–330. Springer (2021)
3. Blaser, J., Frizzo, J., Norgrove, L.: Not Only Timber: the Potential for Managing Non-timber Forest Products in Tropical Production Forests – a Comprehensive Literature Review ITTO Techincal Series. International Tropical Timber Organization, Yokohama, Japan (2020)
4. Carrera-Rivera, A., Ochoa, W., Larrinaga, F., Lasa, G.: How-to conduct a systematic literature review: a quick guide for computer science research. MethodsX. **9**, 103730 (2022)
5. Choudhary, D., Kala, S., Todaria, N., Dasgupta, S., Kollmari, M.: Drivers of exploitation and inequity in non-timber forest product (NTFP) value chains: the case of Indian Bay leaf in Nepal and India. Dev. Policy Rev. **32**(1), 71–87 (2014)
6. Choudhary, D., Kala, S., Todaria, N., Dasgupta, S., Kollmair, M.: Marketing of bay leaf in Nepal and northern India: lessons for improving terms of participation of small farmers in markets. Small-Scale For. **12**, 289–305 (2012)
7. Hodson de Jaramillo, E., Schuler, I.: challenges of sustainable development and productive use of biodiversity: the knowledge based bio-economy approach. In: Environmental Protection and Suistainable Development from Rio to Rio+20, vol. 15, pp. 323–333 (2014)
8. Keča, L., Marčeta, M., Bogojevic, M.: Organization of marketing activities in leader enterprise for purchase, processing and placement of non-wood forest prodicts in Vojvodina. In: CEFood 2012 – Proceedings of 6th Central European Congerr on Food, pp. 1707–1712 (2012)
9. Marshall, E., Newton, A.C., Schreckenberg, K.: Commercialisation of non- timber forest products: first steps in analysing the factors influencing success. Int. For. Rev. **5**(2), 128–137 (2003)
10. Menggala, S.R., Damme, P.V.: Improving Indonesian cinnamon (C. Burmannii (Nees and t. nees) blume) value chains for greater farmers' incomes. IOP Conf. Ser.: Earth Environ. Sci. **129**(1), 012026 (2018)
11. Mohsenzadeha, M., Ahmadia, S.: The mediating role of competitive strategies in the effect of firm competencies and export performance. Procedia Econ. Finance. **36**, 456–466 (2016)
12. Nedeljković, J.: Non-timber forest products (NTFPs). In: Leal Filho, W., et al. (eds.) Life on Land. Springer, Encyclopedia of the UN Sustainable Development Goals (2019)
13. Noweg, T., Keleman, A., Nelson, J.: Bidayuh Jagoi (Bijagoi) community's dependency on non-timber forest products (NTFPS) in Bau District, Sarawak, Malaysia. Malas. For. **87**(2), 182–199 (2024)
14. Pettenella, D., Secco, L., Maso, D.: NWFPS marketing: lessons learned and new development paths from case studies in some European countries. Small-Scale For. **6**, 373–390 (2007)
15. Roland, O., Oyelana, A.: Contribution on non timber forest products to rural household income in Eastern Cape Province, South Africa. Mediterr. J. Soc. Sci. **5**(23), 749 (2014)

16. Sa'diyah, P.F.I., Darwanto, D.H.: Indonesian cinnamon competitiveness and competitor countries in international market. J. Agraris. **6**(2), 123–135 (2020)
17. Sidek, S., Mohamad, M.R., Wan, W.M.N.: Entrepreneurial orientation and SME performance: the serial mediating effects of access to finance and competitive advantage. Int. J. Acad. Res. Bus. Soc. Sci. **9**(9), 81–100 (2019)
18. Takenaka, K., Yamada, M., Shinohara, T., Miyazaki, R.: Role of indigenous horticultural crops for rural smallholders in Mali. Acta Hortic. **1006**, 355–362 (2013)
19. Tanjung, N., Simangunsong, B., Manurung, E.: Export demand of non-timber forest products: a case study of cinnamon products in Indonesia. J. Sylva Lestari. **12**, 181–190 (2024)
20. Teeratansirikool, L., Siengthai, S., Badir, Y., Charoenngam, C.: Competitive strategies and firm performance: the mediating role of performance measurement. Int. J. Product. Perform. Manag. **62**(2), 168–184 (2013)
21. Wekesa, L., Maalu, J., Gathungu, J., Wainaina, G.: Mediating effect of competitive strategy between enterpreuneur characteristics and the performance of non-timber forest product SMEs in Kenya. J. Knowl. Econ. **12**(4), 4293–4317 (2023)
22. Williams, M., Moser, T.: The art of coding and thematic exploration in qualitative research. Int. Manag. Rev. **15**(1), 45–55 (2019)
23. Wohlin, C.: Guidelines for snowballing in systematic literature studies and a replication in software engineering. In: ACM International Conference Proceeding Series (2014)
24. Zamani, S., Fathi, M., Ebadi, M., Máthé, Á.: Global trade of medicinal and aromatic plants. A review. J. Agric. Food Res. **21**, 101910 (2025)
25. Zulfiqar, S., Shah, A., Ahmad, M.: Entrepreneurial orientation and performance of small and medium-sized enterprises: mediating effects of differentiation strategy. Compet. Rev. Int. Bus. J. **29**(5), 551–571 (2019)

Designing a Smart Corporate Governance Prototype for Wellness Tourism Ecosystems: An ICT-Driven Approach

Lady Fernández-Mora[1], Efraín Ortiz-Pabón[2](✉), Daniela Flórez-Villamil[3], Sara Juliana Peña-Gómez[3], Juan David Hernández-Rodríguez[3], María Andrea Méndez-Beltrán[3], and Nestor Armando Nova-Arévalo[4]

[1] Instituto Tecnológico de Costa Rica (TEC), Campus San Carlos, Alajuela, Costa Rica
lfernandez@itcr.ac.cr

[2] Departamento de Ingeniería de Sistemas, Pontificia Universidad Javeriana, Bogotá, Colombia
efrain.ortiz@javeriana.edu.co

[3] Carrera de Ingeniería de Sistemas, Pontificia Universidad Javeriana, Bogotá, Colombia
{fl.daniela,saraj.penag,Juan_hernandezr, mariaa.mendez}@javeriana.edu.co

[4] Departamento de Ciencia de la Información, Pontificia Universidad Javeriana, Bogotá, Colombia
novanestor@javeriana.edu.co

Abstract. The growing importance of Information and Communication Technologies (ICT) in tourism underscores the need for governance models that integrate strategic processes with technological infrastructures. This paper presents the design and evaluation of a smart governance prototype applied to a wellness tourism ecosystem. The study employed a hybrid methodology, combining Design Science Research with the Waterfall model and Scrum, to guide problem identification, requirements analysis, and iterative implementation. The resulting artifact is a microservices-based platform that enables data exchange, coordination, and value co-creation among ecosystem actors. Validation through architectural analysis and technical testing confirmed scalability, reliability, and usability. Beyond the technical contribution, the research expands the literature on smart tourism and governance by showing how methodological hybridity and modular architectures can operationalize governance principles such as transparency, accountability, and collaboration. The findings highlight the importance of aligning digital infrastructures with organizational coordination to strengthen resilience and competitiveness in tourism destinations.

Keywords: Smart Tourism · Smart Governance · Digital Innovation · Shared Value · Microservices Architecture

1 Introduction

The global slowdown experienced by the tourism sector during the pandemic exposed the sector's technological vulnerabilities, particularly in Latin America. This crisis shed light on companies' lack of preparedness to offer tourism value propositions leveraging

A. Rocha et al. (Eds.): ICITS 2026, LNNS 1966, pp. 108–119, 2027.
https://doi.org/10.1007/978-3-032-27538-7_10

information and communications technologies (ICTs) and emerging digital technologies, focusing on smart tourism. This type of tourism refers to the integrated effort of collecting and utilizing large volumes of data from diverse sources to create proposals that enhance the tourist experience, improve efficiency, and promote sustainability [1]. These developments are based on data-generating sensors, large-scale data storage (Big Data), access to complementary open data (e.g., from the web and social media), new forms of connectivity, and the capture and transmission of data through the Internet of Things (IoT) including technologies such as RFID (Radio Frequency Identification), NFC (Near Field Communication), data analytics for value extraction, virtual, augmented, and mixed reality for the design of new experiences, cloud computing for data storage and processing, and artificial intelligence (AI) for increasing value and reducing processing times.

In line with the above, the collaboration, interdependence, coordination, interconnection, synchronization, and alignment required among the numerous actors involved in a tourism ecosystem revealed another major weakness in the sector: the lack of smart governance models useful for bridging the gap between the physical and digital worlds. This is what [2] refers to as the development of digital innovations or innovative technological solutions that integrate digital technologies, allowing companies to share data and join shared value platforms useful for updating and enhancing their business models. Given that the relationships between ecosystem actors are dynamic and complex, a smart tourism ecosystem is the most comprehensive way to describe the conceptual foundations of this new tourism model—smart tourism—which is based on the principle that nothing works in isolation, but rather evolves through interaction within the ecosystem [3]. Smart tourism is characterized by multiple interconnected components and layers, supported by ICTs, which give rise to the development of smart destinations, experiences, and businesses [1].

This article presents the foundations of a design proposal for a platform ecosystem [4, 5] that, using a microservices-based architecture, seeks to operationalize smart governance principles and support shared value creation among the actors in a wellness tourism ecosystem. The prototype acts as a digital infrastructure to coordinate autonomous actors, aligning with the theory of digital ecosystems. This initiative, a consortium collaboration between universities in Colombia and Costa Rica, lays the groundwork for a broader system that could function as a smart governance mechanism for the tourist destination of La Fortuna, Costa Rica. This initiative was conceived as an academic exercise, employing design science research as an integrative methodology, combined with the waterfall methodology for solution development and Scrum during the implementation phase.

The article is structured as follows: Section 2 presents the conceptual foundations, including the definitions of smart tourism, smart governance, shared value, and digital innovation. Section 3 details the methodology and its application, combining Design Science Research (DSR), the waterfall model, and Scrum. Section 4 analyzes the findings and presents the main conclusions, limitations, and future research agenda.

2 Conceptual Foundation

The underlying body of knowledge on which this paper is based involves the linking of four key concepts:

2.1 Smart Tourism

Smart tourism is defined as the integrated effort to collect and use large amounts of data, from diverse sources, to create value propositions that improve the tourist experience, efficiency and sustainability [1]. This concept, which has been frequently used in the literature, is applied to characterize the growing dependence of tourism destinations, their industries, and their tourists on emerging forms of Information and Communication Technologies (ICTs) because they allow the transformation of massive amounts of data into value propositions [1]. Smart tourism is based on the paradigm of high-tech service production, characterized by adopting the principles of interoperability, virtualization, decentralization, service orientation, data analytics, and modularity, which are key aspects of Industry 4.0 [6]. This approach has driven full automation in the production and delivery of tourism goods and services, leading to potential technological disruptions in the sector [7]. The term "smart" refers to the application of a complex amalgamation of technological, economic, and social developments driven by new digital technologies based on the use of data-generating sensors, the storage of large volumes of data (Big Data), access to complementary open data (e.g., originating from the web, social networks), new forms of connectivity, the capture and transmission of data using the Internet of Things - IoT, technologies such as RFID (an acronym for radio-frequency identification), NFC (an acronym for Near Field Communication), associated data analytics to extract value, virtual, augmented, and hybrid reality to design new experiences, cloud computing to store and process large volumes of data, and, more recently, artificial intelligence - AI to add value to them; which in an interconnected, synchronized, and concerted manner allow us to close the gap between the physical world and the digital world. Unlike its predecessor, e-tourism, which focused primarily on the digital world, smart tourism encompasses both the digital and physical worlds. Smart tourism is characterized by the fact that it encompasses multiple components and layers supported by ICTs that give rise to the development of smart destinations, smart experiences, and smart businesses [1].

2.2 Smart Governance

Smart governance is defined as the ability to apply digital technologies and intelligent activities in information processing and decision-making [8]. When applied to tourism, its focus shifts to smart tourism, which sets the requirements for destinations to integrate physical, social, and technological infrastructure in order to enhance the experiences of both tourists and residents, while adapting to the evolving needs of destination stakeholders [3, 9]. The concept of smart governance does not depart from the traditional foundations of corporate governance theory; rather, it builds upon them by incorporating the intelligent use of ICT to improve decision-making processes [3, 10]. Although technological resources are considered a necessary condition within the structure of the smart approach, social capital combined with governance remains a crucial factor for the success of a smart destination. From this perspective, smart governance can be defined through a sociotechnical lens, aligning technological potential with new forms of collaboration between local governments and citizens, with the goal of addressing challenges based on sustainability principles [3, 11, 12]. Smart governance is essential for promoting access to information, transparency, and public participation in decision-making processes involving various stakeholders. This ensures that governance becomes transparent, accountable, collaborative, and participatory—facilitating knowledge sharing and transfer [13–15]. In this context, the tourist is no longer a passive subject but rather becomes an active participant, as smart tourism views the tourist as a co-creator of experiences. To fully understand the scope of smart governance, it is necessary to adopt an ecosystem perspective. This approach encompasses elements such as tourists, governments, natural environments, tourism service providers, information exchange centers, residents, suppliers from other industries, media outlets, destination management organizations, and digital technologies, among others. In other words, an effective smart governance model today must consider most—if not all—of the actors involved in the tourism ecosystem. However, [3] proposes organizing the smart tourism ecosystem around the interaction of four core elements: digital technologies, consumers (tourists and residents), businesses (tourism-related and other sectors), and the tourist destination (governed by destination management organizations and public authorities). Additional actors such as universities, communities, and civil society also play key roles, all of them interconnected through intensive use of technology.

2.3 Digital Innovation

Digital innovation (DI) is defined as the combination of physical and digital products to create new products, services or processes, through the design, development, implementation and exploitation of digital technological artifacts in an organization [16]. From the perspective of [17] DI is conceived as a set of routine and interdependent social interactive activities carried out by humans, mediated by digital technology, and goal-oriented. They can arise from the opportunities that reside in digital ecosystems [17], in the digital transformation processes undertaken by organizations, in the opportunities derived from the application of ICT and new digital technologies. Their results usually incorporate simultaneously product and service characteristics, which makes them identifiable as service innovations or product-service systems [18]. From the point of view of its application to sectors such as tourism, it usually refers to the creation of value through the strategic and transformative use of ICT and digital technologies by making it smarter, because they consider both technological and organizational, contextual and sustainable aspects. In this logic, DI can be seen as a dynamic and multidimensional process that transforms not only products or services, but also the ways in which organizations operate, cooperate, coordinate while competing.

As DI allows proposing IT solutions that integrate digital technologies to support the digitalization of businesses in different sectors such as banking, healthcare, manufacturing, retail, tourism, among others [2], they can be summarized as the process of combining resources, especially digital knowledge and skills to promote the development of organizational capabilities and generate novel solutions [19] it is essential to give body to new strategic management constructs such as smart governance and in this way help improve decision-making, coordination and the evolution of tourism ecosystems. This process involves interactive social activities mediated by digital technology and driven by specific goals [17], where collaboration between government, the private sector, and civil society converges and becomes necessary.

2.4 Shared Value

The concept of shared value refers to operational policies and practices that enhance a company's competitiveness while improving the economic and social conditions of the communities in which it operates. It involves creating economic value in a way that also generates value for society by addressing its needs and challenges. Fundamentally, it represents a reconnection between business success and social progress [20]. This distributed value creation perspective finds its natural embodiment in platform ecosystems, where modular architectures and governance mechanisms enable the coordination of autonomous actors to generate specific complementarities and shared value [4, 5]. Value creation is a long-recognized idea in business, where profit is the revenue earned from customers less the costs incurred. Companies can create economic value by creating social value.

There are three distinct ways to do this: rethinking products and markets, redefining productivity in the value chain, and building supporting industrial clusters at company locations. Each of these is part of the virtuous cycle of shared value; improving value in one area leads to opportunities in others [20]. Precisely, digital platforms are a phenomenon that challenges incumbents by changing the way we consume and provide digital products and services. While traditional businesses create value within the boundaries of a company or supply chain, digital platforms utilize an ecosystem of autonomous agents to co-create value [4]. The tourism sector is no exception; on the contrary, offerings and consumption patterns have transformed as businesses adopt digital transformation processes. With the advent of smart tourism, digital innovations became apparent, and smart tourist destinations and regions began to develop. Tourism governance models also evolved toward smart tourism destination governance models supported by ICTs and new digital technologies. This has led to the consolidation of dynamic tourism ecosystems where collaboration between multiple stakeholders implies a fundamental change in the way tourism value and experiences are created, exchanged, consumed, and shared. The above perspective implies an opportunity for digital innovators because transforming data into meaningful information that supports value creation by the different actors that make up the ecosystem becomes fundamental, therefore, ecosystems can be perceived as a structure, which helps to outline a particular context where multiple actors, dynamically organized around a shared value purpose, co-create value mutually [21] and digital platforms become the indicated mechanism when based on software, extensible code bases that provide basic functionalities, complemented with modular services [4].

3 Methodology

To design the prototype of a smart governance platform for wellness tourism, we adopted a hybrid approach combining Design Science Research (DSR), the Waterfall model, and Scrum. DSR structured the research process, guiding problem identification, objective definition, and the theoretical grounding of the artifact. The Waterfall model ensured rigor in requirements analysis, system design, and documentation, while Scrum introduced flexibility during implementation through iterative cycles and continuous feedback from domain experts. This combination was necessary because the project required both scientific rigor and adaptability to the dynamic needs of tourism ecosystems. The hybrid methodology not only supported the development of the functional prototype but also generated practical insights into how methodological integration can enhance ICT-driven governance models in the tourism sector. Following the iterative DSR research process, the following steps are taken: problem identification, definition of solution objectives, design and development, demonstration, and evaluation and communication of results.

3.1 Problem Identification

The challenge consisted in developing the prototype of a shared-value platform aimed at facilitating better coordination among the actors of the tourism ecosystem through the exchange of data oriented toward the generation of smart governance. Similar approaches have been discussed in the context of challenge-based learning [21], which emphasizes the role of problem-driven design in higher education. However, in this study the focus is placed on applying these principles to the development of a governance prototype for the tourism sector Coordination, from the perspective of smart governance of tourist destinations, involves the use of Information and Communication Technologies (ICTs) and the integration of emerging digital technologies that act as enablers for capturing, storing, processing, and communicating valuable information among the different actors that make up the ecosystem: tourists, tourist attractions, travel agencies, transportation systems, tour guides, chambers of commerce, government, academia, and other stakeholders acting as regulators, trainers, and value creators. Smart governance should enable stakeholders to develop a more holistic understanding of the importance of sustainability and competitiveness for the sector, the region, tourist destinations, and the community.

3.2 Solution Objectives

The objective is to design the functional prototype of the platform to facilitate smart governance and generate shared value among the stakeholders of the tourism ecosystem. The solution should enable coordination among stakeholders. The project must cover the definition of functional and non-functional requirements, propose an architecture, and test and validate the prototype with at least five actors from the ecosystem.

3.3 Design and Development

This part of the process was carried out using the Waterfall methodology, which enabled detailed planning and comprehensive documentation prior to the implementation of the prototype. Agile methodologies fostered flexibility and adaptability through rapid iterations and continuous testing with wellness tourism experts. Before developing the prototype, an analysis of the environment of the selected tourist destination was conducted, and secondary sources such as [22] were consulted to gain a better understanding of the problem context. Interviews were conducted with key stakeholders of the ecosystem, and the results were used to refine the definition of functional and non-functional requirements. A Product Backlog was created to facilitate the management and prioritization of requirements, allowing for greater flexibility in planning and adapting the system to potential changes in the challenge's needs.

3.4 Demonstration

The project included user interfaces, the design of the microservices architectural pattern, and addressed software quality attributes according to ISO 25010. It also covered the justification of the selected technologies and the corresponding design diagrams. The list of functional requirements agreed upon with the client can be found in Table 1. The non-functional requirements can be found in Table 2.

High-Level Architecture: Figure 1 illustrates the overall organization of the system from a high-level architectural perspective. It provides details on the interactions between system actors, the core microservices, the messaging infrastructure, the intermediate processing layer, and the data analysis component. To ensure a modular, scalable, and flexible system, aligned with the principles of evolutionary design, the architecture follows a microservices-based approach. This choice allows each functionality or domain of the system to be developed, deployed, maintained, and scaled independently, encouraging progressive growth and enabling adaptation to new needs within the tourism ecosystem. Each microservice encapsulates a specific responsibility (such as user management, provider services, or visitor preferences) and operates with its own independent database to avoid duplication. SQLite was selected for the prototype due to its simplicity and portability, with the option to migrate to PostgreSQL or MongoDB in later stages. Asynchronous communication is enabled through RabbitMQ under the AMQP protocol, which acts as an intermediary between the microservices and the analytics module. Events are published to the queue and processed by a Python-based consumer, which transforms and stores them in a centralized MySQL database optimized for queries and analytics.

The Data_Analysis component, developed in Python with Flask, accesses this consolidated database to generate metrics, reports, and indicators displayed in an interactive web dashboard. This dashboard provides administrators and providers with insights into service demand, user preferences, and territorial indicators. Despite the additional complexity of managing queues, multiple databases, and different programming languages, the architecture enables horizontal scalability, resilience, technological flexibility, and the future integration of recommendation systems, demand forecasting modules, or artificial intelligence.

Table 1. Functional Requirements Summary

Module	Description
Authentication and Registration	**Tourist Registration:** The system must allow the registration of new tourists, including personal information and access credentials, as long as their email address has not been previously registered.
	Supplier Registration: The system must allow the registration of new suppliers, including personal information and access credentials, as long as their email address has not been previously registered.
	Login: The system must allow registered users to authenticate using a username and password.
	Password Recovery: The system must allow registered users to recover their password using their email address.
Visualization and Metrics	**View Supplier Dashboard:** The system must allow suppliers to view information about their services and tourist behavior.
	View Administrator Dashboard: The system must allow administrators to view information about behavior within the platform.
Geolocation and Recommendations	**Locating Businesses on a Map:** The system must have an interactive map that allows tourists to locate businesses.
	Receiving Recommendations Based on Preferences: The system must provide recommendations to tourists based on their interests.
Service Management and Visualization	**Managing services:** The system must allow administrators and providers to add, read, edit, and delete services.
	Managing Tourists: The system must allow administrators to add, read, edit, and delete tourists.
	Managing Suppliers: The system must allow administrators to add, read, edit, and delete suppliers.
	Service Reading: The system must allow tourists to explore services and view additional information about each one.

3.5 Evaluation

The evaluation of the system architecture was carried out through a validation process using the Architecture Tradeoff Analysis Method (ATAM). This validation involved a

Table 2. Non-functional requirements summary

Quality Attribute	Description
Scalability	The system must be capable of managing a progressive increase in the number of users, data volume, and frequency of interactions without degrading its performance. It should allow the incorporation of new modules and facilitate its expansion into other tourism niches both within and outside of Costa Rica.
Usability	The system must provide a user-friendly and easy-to-navigate interface for users with different levels of technical skills. Interaction should not require prior training, and the dashboards must offer a clean and clear visual design with organized and straightforward information.
Performance	The system must respond efficiently to user requests, maintaining low response times, especially in critical operations such as service registration, metric queries, and event queue processing.
Reliability	The system must ensure continuous and stable operation, guaranteeing data integrity even in the presence of errors or partial failures. Information must be recoverable without loss after restarts or interruptions.
Security	The system must protect personal and sensitive data through the use of secure authentication, role-based access control, and encrypted storage where necessary. Unauthorized access to information must be prevented.
Maintainability	The system must facilitate the incorporation of new features, fixes, or improvements through a clear and well-documented structure. The architecture must allow component updates without interrupting overall operation or compromising system stability.
Availability	The system must remain constantly operational, with backup, monitoring, and fault-tolerance mechanisms that ensure service continuity with minimal interruptions.

detailed analysis of the system's functional and non-functional requirements, identifying key stakeholders and their needs. The provided architectural diagrams, including component, deployment, and data diagrams—were assessed to determine how the architecture meets the requirements and manages various quality attributes. The ATAM process concluded that the proposed architecture demonstrates significant strengths. First, the adoption of microservices provides greater flexibility and scalability, as each system component can be developed, deployed, and scaled independently. Unit tests were conducted to verify the individual functioning of each system component. The tests covered both the service layer and the controller layer. A 90% code coverage was achieved, executing a total of 90 different test cases, all of which were successfully completed. Integration tests were performed to validate the interaction between system components and functional flows. These tests covered key quality scenarios such as authentication (login and registration for tourists, providers, and administrators), service management (creation, editing, deletion, and status changes), and bookings (creation and cancellation). The results showed a 93% success rate, with 14 test cases executed correctly. Load

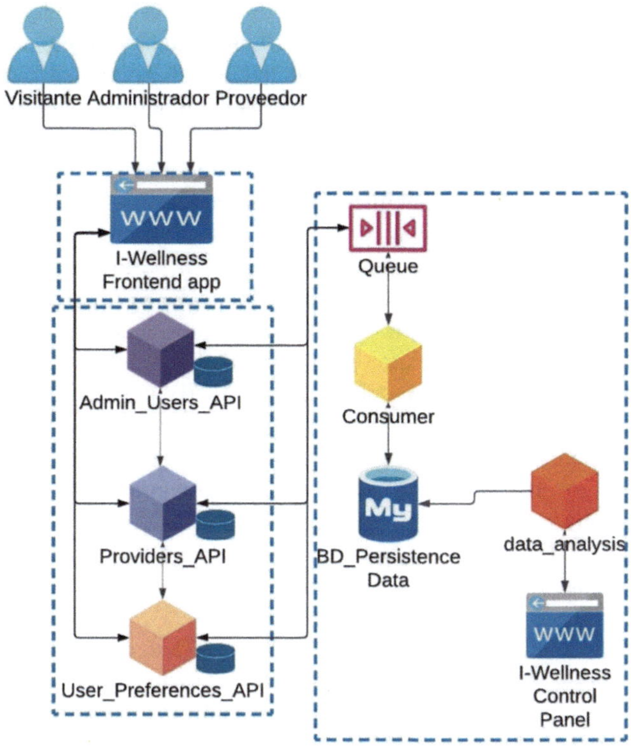

Fig. 1. High Level Diagram

and stress tests were also conducted to evaluate the system's behavior under realistic usage conditions, with the aim of identifying its maximum capacity, degradation points, and stability during periods of high demand. A total of 558 concurrent users were simulated over 3 min and 38 s. Of these, 70% were visitors (approximately 390) and 30% providers. The total test duration was 3 min and 38 s, a period calculated to allow the complete execution of all planned scenarios while maintaining precise control over the load variables.

4 Discussion and Conclusions

The development of the smart governance prototype for wellness tourism ecosystems offers relevant insights at the intersection of information systems design, digital innovation, and governance models. Beyond its technical scope, the study contributes to the debate on how ICT-based solutions can strengthen coordination, transparency, and shared value creation in complex socio-technical ecosystems. The results suggest that integrating DSR with both structured and agile methodologies is particularly effective in contexts where multiple stakeholders demand rigor and adaptability. This aligns with the perspective of Hevner and Gregor [16], who emphasize the capacity of design science to bridge conceptual models and practical artifacts, while also expanding the discussion of methodological hybridity in IS development highlighted by Pereira et al. [8].

The adoption of a microservices-based architecture also demonstrates that architectural choices have direct implications for governance practices. This finding expands the work of Gajdošík [3] on smart tourism ecosystems and aligns with the principles of platform ecosystems [4, 5] by showing how modular and scalable architectures can concretely support governance principles such as transparency, accountability, and participation. In line with Tomor et al. [10], wh41o argue that socio-technical designs are required to operationalize governance principles, our prototype illustrates how ICT infrastructures can be aligned with collaborative processes in multi-actor ecosystems.

Furthermore, this study complements the literature on smart tourism, which has frequently emphasized either the technological dimension of smart destinations [1, 4] or their institutional and organizational challenges [13]. By positioning governance as a mediator between these two spheres, our findings show that effective smart governance depends on their deliberate alignment. The case of La Fortuna, Costa Rica, illustrates how universities, governments, and private actors can work together to design governance solutions tailored to local needs. Ultimately, the findings underscore that smart governance is not merely a technological matter but a socio-technical challenge, in which the sustainable value in tourism emerges from the interplay between governance theory, ICT infrastructures, and collaborative practices.

At the same time, certain limitations must be acknowledged. The evaluation focused primarily on technical validation, without extensive involvement of end-users, which limits insights into adoption, usability, and trust. In addition, the project was constrained by time and resources, which restricted the functionalities that could be included. Future research should validate the platform with stakeholders, expand functionalities with AI-driven analytics and personalization, and test its adaptability across other tourism niches and regions.

References

1. Gretzel, U., Sigala, M., Xiang, Z., Koo, C.: Smart tourism: foundations and developments. Electron Markets. **25**(3), 179–188 (2015). https://doi.org/10.1007/s12525-015-0196-8
2. Khin, S., Ho, T.C.: Digital technology, digital capability and organizational performance: A mediating role of digital innovation. Int. J. Innov. Sci. **11**(2), 177–195 (2019)
3. Gajdošík, T.: Smart tourism destination governance: technology and design-based approach. en smart tourism destination governance. Technol. Des. Based Approach, 160 (2021). https://doi.org/10.4324/9781003269342
4. Hein, A., et al.: Digital platform ecosystems. Electron Markets. **30**(1), 87–98 (2020). https://doi.org/10.1007/s12525-019-00377-4
5. Jacobides, M.G., Cennamo, C., Gawer, A.: Towards a theory of ecosystems. Strateg. Manag. J. **39**(8), 2255–2276 (2018). https://doi.org/10.1002/smj.2904
6. Pencarelli, T.: The digital revolution in the travel and tourism industry. Inf. Technol. Tour. **22**(3), 455–476 (2020). https://doi.org/10.1007/s40558-019-00160-3
7. Ivanov, S.: The impact of automation on tourism and hospitality jobs. Inf. Technol. Tour. **22**(2), 205–215 (2020). https://doi.org/10.1007/s40558-020-00175-1
8. Scholl, H.J., AlAwadhi, S.: Smart governance as key to multi-jurisdictional smart city initiatives: the case of the eCityGov alliance. Soc. Sci. Inf. **55**(2), 255–277 (2016). https://doi.org/10.1177/0539018416629230

9. Marsa-Maestre, I., Lopez-Carmona, M.A., Velasco, J.R., Navarro, A.: Mobile Agents for Service Personalization in Smart Environments. J. Netw. **3**(5), 30–41 (2008)
10. Pereira, G.V., Parycek, P., Falco, E., Kleinhans, R.: Smart governance in the context of smart cities: a literature review. Inf. Polity. **23**(2), 143–162 (2018). https://doi.org/10.3233/IP-170067
11. dos Santos Júnior, A., Mendes Filho, L., Almeida García, F., Manuel Simões, J.: Smart tourism destinations: un estudio basado en la visión de los stakeholders. Rev. Turismo em análise. **28**(3), (2017)
12. Tomor, Z., Meijer, A., Michels, A., Geertman, S.: Smart governance for sustainable cities: findings from a systematic literature review. J. Urban Technol. **26**(4), 3–27 (2019). https://doi.org/10.1080/10630732.2019.1651178
13. Gretzel, U., Scarpino-Johns, M.: Destination resilience and smart tourism destinations. Tour. Rev. Int. **22**(3–4), 263–276 (2018)
14. Antonelli, G., Cappiello, G.: Smart Development in Smart Communities. Routledge, Oxfordshire, UK (2017) Accedido: 21 de mayo de 2025. [En línea]. Disponible en: https://api.taylorfrancis.com/content/books/mono/download?identifierName=doi&identifierValue=10.4324/9781315641850&type=googlepdf
15. Errichiello, L., Micera, R.: A process-based perspective of smart tourism destination governance. Eur. J. Tour. Res. **29** (2021). https://doi.org/10.54055/ejtr.v29i.2436
16. Kohli, R., Melville, N.P.: Digital innovation: a review and synthesis. Inf. Syst. J. **29**(1), 200–223 (2019). https://doi.org/10.1111/isj.12193
17. Kaptelinin, V., Nardi, B.A.: Acting with Technology: Activity Theory and Interaction Design. MIT Press (2006)
18. Hevner, A., Gregor, S.: Envisioning entrepreneurship and digital innovation through a design science research lens: a matrix approach. Inf. Manag. **59**(3), 103350 (2022)
19. Ortiz-Pabón, E., Fernández-Mora, L., Nova-Arévalo, N., González-Rivera, R.: Desafíos Digitales: Análisis de la Madurez Digital de un Clúster Turístico en Seis Cantones de Alajuela, Costa Rica. *Revista Ibérica de Sistemas e Tecnologias de Informação.* **E68**, 57–74 (2024)
20. Kramer, M.R., Porter, M.: Creating Shared Value, vol. 17. FSG, Boston, MA (2011) Accedido: 2 de junio de 2025. [En línea]. Disponible en: https://environz.wordpress.com/wp-content/uploads/2012/11/creating-shared-value-porter-and-kramer-harvard-business-review-article.pdf
21. Burkhalter, M., Betz, C., Auge-Dickhut, S., Jung, R.: Orchestrating value co-creation in business ecosystems. In: Wendt, K. (ed.) Theories of Change: Change Leadership Tools, Models and Applications for Investing in Sustainable Development, pp. 257–291. Springer International Publishing, Cham (2021). https://doi.org/10.1007/978-3-030-52275-9_16
22. Villalobos, M.T., Cardinale-Villalobos, L., Fernández-Mora, L., Víquez-Acuña, L., González-Quirós, R.: Design of an Information System to impulse a process of transformation into a Smart Tourism Destination. Rev. Rosa dos Ventos-Turismo e Hospitalidade. **14**(3), (2022)

Enhancing Sales Forecast Accuracy for New Fashion Items: Leveraging Online Trend Signals and Historical Data

Vera Miguéis[1](✉) and Nelson Alves[2]

[1] Faculdade de Engenharia da Universidade do Porto, INESC TEC, Porto, Portugal
vera.migueis@fe.up.pt
[2] Faculdade de Engenharia da Universidade do Porto, Porto, Portugal

Abstract. Accurately anticipating demand is a critical challenge in fashion retail, where short product life cycles and volatile consumer preferences often lead to costly overstock or missed sales. This preliminary study explores the development of a sales-forecasting model to estimate the sales of new fashion products for the upcoming season. The model jointly leverages (i) sales and product characteristics from a previous season, and (ii) text-mined features from web pages discussing fashion trends relevant to the following season. Using a dataset of more than 1,000 items, we evaluate different machine learning models in this forecasting task. Preliminary results show Random Forest performs best, with $R^2$0.8079 using only historical data and $R^2$0.8127 after adding features referring to fashion trends collected through the web. The incremental gain is modest, likely due to the limited volume of online content, but the findings indicate the promise of combining internal and external information for early demand sensing in fashion retail.

Keywords: sales forecasting · fashion trends machine learning · unstructured data

1 Introduction

Accurately forecasting demand is a critical challenge in the fashion retail sector (Sousa et al., 2025). Short product life cycles, seasonality and volatile consumer preferences often cause companies to either purchase excessive quantities, resulting in storage costs and reduced prices, or under-purchase, resulting in lost sales and dissatisfied customers. For this reason, improving the accuracy of sales forecasts remains a strategic priority in the fashion industry.

In addition to financial and operational risks, inaccurate forecasts also have far-reaching implications for sustainability and responsible production. The fashion sector is increasingly being criticised for its contribution to excessive resource use, unsold inventory and environmentally harmful practices across the supply chain (Dissanayake and Pal,2023). Improving demand anticipation can therefore support leaner operations and a shift towards more sustainable production and consumption models, including a reduced ecological footprint, better alignment with the principles of the circular economy, and enhanced social responsibility.

A. Rocha et al. (Eds.): ICITS 2026, LNNS 1966, pp. 120–130, 2027.
https://doi.org/10.1007/978-3-032-27538-7_11

In recent years, advances in information systems have enabled companies to collect and store large volumes of structured data, including transactional records and product characteristics. However, forecasting demand for new fashion products remains difficult because these items lack a sales history. Researchers and practitioners have therefore explored alternative strategies, such as transferring knowledge from comparable items, applying attribute-based models, or leveraging external signals of consumer interest (Loureiro et al.,2018). Fashion is a particularly challenging area in which to apply these strategies, given the rapid trend cycles, the creative nature of collections, and the speed with which consumer tastes evolve.

Alongside structured internal data, a wealth of unstructured external data is available online. Fashion blogs, specialised websites and social media platforms often reveal information about popular colours, materials or styles early on, which can influence consumer choices in upcoming seasons. Incorporating such information into predictive models enables demand shifts to be anticipated before they are visible in sales records.

In this context, this preliminary study addresses the challenge of forecasting sales for new fashion products in the upcoming season by integrating two complementary sources of information: (i) sales and product characteristics from a previous season, and (ii) text-mined features collected from web content referencing product attributes relevant to the following season. Machine learning techniques are used to forecast the demand using these two types of data. The main contributions of this work are threefold: (i) development of a methodological framework that combines data mining and text mining for sales prediction in fashion retail (ii) empirical evaluation of three machine learning approaches—Random Forest, Support Vector Machine, and Neural Networks—using a real dataset of more than 1,000 products, and (iii) critical assessment of the incremental value provided by online trend information and discussion of directions for future work.

The remainder of this paper is organized as follows. Section 2 reviews related literature on demand forecasting in fashion retail and the use of machine learning and text mining in predictive analytics. Section 3 describes the dataset and presents the methodological framework adopted in this study. Section 4 reports the empirical results obtained from different machine learning models and compares their performance. Finally, Sect. 5 concludes the paper and outlines avenues for future research.

2 Literature Review

Demand forecasting has traditionally relied on statistical time-series methods such as ARIMA, HoltWinters, and regression-based models, which remain widely used for stable markets (Cheong and Ma Nang Laik, 2024). However, these approaches are not suited for the volatility and short life cycles that characterize fashion products, especially new launches for which no historical sales exist. To address these limitations, research has increasingly turned to advanced machine learning techniques. Neural networks, support vector machines, and ensemble methods such as Random Forests have demonstrated superior ability to capture nonlinear relationships among product features and consumer demand (Medina et al., 2022). Random Forests, in particular, have been highlighted for their robustness and interpretability through variable importance measures, while hybrid approaches combining fuzzy logic, clustering, or diffusion models with neural networks have also shown improved accuracy (Wong and Guo, 2010).

A more recent stream of research emphasizes the incorporation of external signals to complement traditional or machine learning models. These external data sources include search engine queries, social media content, online reviews, forums, and blogs, which can act as leading indicators of consumer interest weeks or months before sales are observed. Skenderi et al. (2024), for instance, demonstrated that adding Google Trends data to multimodal forecasting models for fashion improved Weighted Absolute Percentage Error by 1.5%. Similarly, Fu and Fisher (2023) reported substantial gains—between 24% and 57% improvements in mean absolute deviation—by integrating Twitter data and search volume indices into apparel sales forecasts. Tsao et al. (2023) also showed significant reductions in error rates, including a 45.79% drop in mean squared error and a 26.35% decline in root mean squared error when combining social media signals with apparel sales predictions. Steenbergen (2019) further found that incorporating analogous product data reduced forecast error by 19%, underscoring the potential of external and comparable signals.

Evidence from other industries reinforces the value of such data. Kulkarni et al. (2012) showed that search queries improved movie sales forecasting, reducing MAPE from 9.46% to 3.20% when combined with advertising data. Dellarocas et al. (2007) highlighted how online reviews and forums could predict box-office revenues, while Lee et al. (2019) confirmed that IMDb and Amazon reviews enhanced both offline and online DVD sales predictions. In the automotive sector, Kim et al. (2019) reported that incorporating search engine data improved forecast accuracy in over 80% of cases, while Lee et al. (2018) demonstrated the complementary value of patent citations and search traffic for predicting hybrid car adoption. Beyond these contexts, Chumnumpan and Shi (2019) found Google Trends to outperform diffusion models in predicting consumer electronics sales, and Baardman et al. (2018) showed that clustering comparable product data improved demand forecasts for FMCG and fast fashion by 2060%.

The magnitude of performance improvements varies, but most studies report positive effects from integrating external signals. Gains include 1.53.4% reductions in WAPE, 2060% improvements in WMAPE, 2457% reductions in MAD, and up to 45% decreases in RMSE. Search engine data such as Google Trends and Baidu index are particularly effective for short-term forecasting, while patent citations and comparable product data provide longer-term insights. Social media platforms like Twitter and Facebook capture consumer sentiment and short-lived trends, whereas online reviews act as durable proxies of word-of-mouth influence. The timing of these signals also matters: Skenderi et al. (2024) observed predictive power up to 710 months before a fashion collection's release, and Schaer et al. (2019) found that pre-release buzz in video games predicted sales up to 17 weeks before launch.

Taken together, this body of research demonstrates that external signals consistently enhance forecasting accuracy across industries, with the strongest benefits observed in experiential, short-lifecycle, and highly differentiated products such as fashion, entertainment, and consumer electronics. Nonetheless, applications in the fashion sector remain relatively limited. Most studies focus on single external data sources or aggregate-level forecasts, and few attempt to systematically integrate structured sales attributes with unstructured online information within a unified predictive framework. This gap motivates the present study, which investigates whether combining historical sales data with online fashion content can improve demand forecasting for new products in upcoming collections.

3 Methodology and Data

This study develops and evaluates a predictive framework for forecasting the sales of new fashion products by integrating internal sales records with external online fashion content. The methodology combines structured and unstructured data sources, applies systematic preprocessing, and tests multiple machine learning models under comparable conditions. The overall framework is summarized in Fig. 1 and detailed in the following sections.

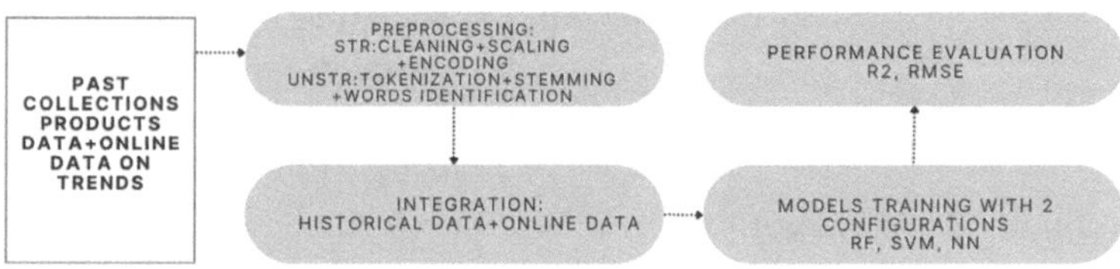

Fig. 1. Overall methodology schema.

3.1 Case Study and Data

To ground the proposed framework in a real-world context, we applied it to a case study of a company operating in the fashion industry. This company operates in the fashion industry, with more than 900 stores worldwide, most of them located in Spain and Portugal. Although the company's product portfolio includes clothing, umbrellas, bags, watches and hats, this study focuses specifically on the sales of 684 types of women's bag, which are this company's core product.

Currently, the company determines the quantity of bags to order without relying on an advanced quantitative model. Instead, decisions are based on three main factors: (i) sales of similar products in previous and comparable seasons, (ii) the experience of staff from the purchasing, development, and marketing departments, and (iii) insights into future trends gathered from various sources, such as fashion blogs and fashion shows.

Internal Data. The dataset provided for this research consists of historical sales records for each product, along with their physical characteristics, logistical and internal organizational aspects, and the insights of domain experts, whose input is considered highly influential on sales performance. These combined factors were used to identify the characteristics that make a product more appealing to customers, thereby assisting the purchasing department in more accurately defining product order quantities. This data forms the structured core of the study and is summarised in Table 2. For each variable, the table specifies its name, type, meaning, and the total number of categories it can assume.

Name	Type	Description	Nr cat.
Family	Categorical	Identifies the type of material and/or pattern used in product production	18
Subfamily	Categorical	Identifies the format of the product	11
Color type	Categorical	Identifies if it is a single-color or multi-color product	2
Color	Categorical	Color of the product	30
Fashion	Categorical	Related to season's trends and to the number of weeks that the product is available in store for sales. The products are classified according to its proximity to the current season's trends	4
International	Categorical	Markets in which the product is sold	5
Matching	Categorical	Identifies if articles that do "Matching" with other gammas	5
Segment	Categorical	Target group	2
Store type	Categorical	Types of store where the product will be available for sales	4
Price	Numerical	Product price	–
Size	Categorical	Product size	4
Expectation level	Categorical	Sales expectation level	4
Sales	Numerical	Total number of units sold	–

It is essential to clarify certain variables that have been adopted. The "Store type" variable. The company classifies its stores into four types (A, B, C, or D), based on size and sales potential. Type "A" stores, with larger display areas and higher potential, receive the full product range, while type "D" stores are limited to items distributed across all locations. This classification is used to define order quantities, with products marked as DCBA available in all stores, and products marked as A restricted to type "A" stores. In what concerns the "Expectation level" variable, it further categorizes products into four groups: SB (high sales expectations), M1 (normal), M2 (average), and M3 (low), as defined by the marketing, development, and purchasing teams. The "Fashion" variable reflects alignment with seasonal trends, ranging from "Trendy" (4 weeks in stores) to "Basic" and "Centralized Distribution" (8 weeks). Finally, the "Segment" variable distinguishes products by target audience: Teen for younger consumers and Woman for adult women.

External Data. A user-generated data extraction program was developed by the author to extract potentially useful information from web pages. The selected web pages contained information relevant to the characteristics of what would be fashionable for the intended season. Accordingly, web pages were chosen that predicted the characteristics of the seasons under study.

For the Spring/Summer season, web pages published prior to that season were selected as sources of information. A total of five web pages were used to extract information for one of the seasons. For the other season, it was more difficult to find web pages with potentially useful content, and therefore ten web pages were used.

The text from each web page was collected and filtered. If the words found in a web page matched those specified in the regular expression, that information was extracted. The filtered words had been previously selected and were related to the products' characteristics stored in the database. For this project, the chosen words referred to the color, family, and subfamily of the products. These selected terms were then counted and analyzed based on their occurrence across the collected web pages.

Data Preprocessing. For the internal data, preprocessing involved cleaning incomplete records, scaling numerical variables (e.g., Z-transformation of price), and encoding categorical attributes into binary indicators.

For the external data, preprocessing followed established text-mining procedures. Raw text was tokenized into words, stopwords were removed, and stemming and lemmatization were applied to reduce words to their canonical forms. Keywords corresponding to product features—such as colors, families, and subfamilies—were identified through regular expressions, and their relative frequencies were computed within each document. These measures provided early indicators of the prominence of different product characteristics in consumer discourse.

Data Integration. The structured and unstructured features were combined into a single dataset. New attributes such as the frequency each color, family and subfamily appeared were created from the text-mined content and added to the original product records. This integration step ensured that each product in the historical dataset was enriched with external signals relevant to its characteristics, thereby linking transactional records with online discourse.

Predictive Models. Three machine learning algorithms were adopted: Random Forests, Support Vector Machines, and Artificial Neural Networks. Random Forests were selected for their robustness, ability to handle heterogeneous features, and provision of variable importance metrics. Support Vector Machines were included for their effectiveness in high-dimensional feature spaces. Neural Networks were tested given their capacity to capture complex nonlinear interactions among product attributes and external signals.

Each model was trained under two configurations: first using only the structured internal dataset, and then using the enriched dataset that incorporated features inferred from external sources. This design allowed the incremental contribution of external data to prediction accuracy to be explicitly evaluated.

Model Evaluation. Model performance was assessed using the coefficient of determination (R^2) as the primary measure of fit. R^2 captures the proportion of variance in sales explained by the model. By comparing performance across models and configurations, i.e. with and without data collected online, the study identifies the approaches that best balance accuracy, robustness, and interpretability.

To assess generalization, the model was trained on data from a given season and then tested on the next one, specifically the Autumn/Winter season.

4 Results

As mentioned before, the models were trained and tested using sales data from a fashion retailer covering two consecutive collections. The first collection was used for training, while the second served as the test set. As mentioned before, three machine learning algorithms, i.e. Random Forest (RF), Support Vector Machine (SVM), and Artificial Neural Networks (ANN) were evaluated under two scenarios: (i) using only structured internal data and (ii) integrating both structured and features obtained from unstructured data.

Table 1 summarizes the performance of the models in terms of the coefficient of determination (R^2).

Table 1. Model performance (R^2) under two data configurations.

Model	Internal Data Only	Internal + External Data
Random Forest	0.8079	0.8127
Support Vector Machine	0.7221	0.7283
Artificial Neural Network	0.6912	0.7025

The results indicate that Random Forest consistently outperformed the other algorithms, achieving an R^2 of 0.8079 when trained exclusively on sales data. The inclusion of online fashion features led to a slight improvement ($R^2 = 0.8127$). Although the gain was modest, it highlights the potential value of external signals. Both SVM and ANN models also showed small but positive increments when enriched with online content, confirming that the additional features contained relevant information.

Overall, the results demonstrate that structured sales data remain the strongest predictor of future demand, but external online sources can complement these signals. The limited performance gain is likely explained by the restricted size of the online dataset used at this stage, which included only a small number of fashion-related documents for each season.

Figure 2 illustrates the products ordered by increasing sales value, which were used for the evaluation of the sales forecast based on unstructured data. The blue points represent the actual sales, arranged in ascending order, while the orange and grey points correspond to the predictions generated by the Random Forest and Support Vector Machine (SVM) models, respectively. This visualization allows for a direct comparison between observed and estimated sales, providing insight into the models' predictive performance across the product range.

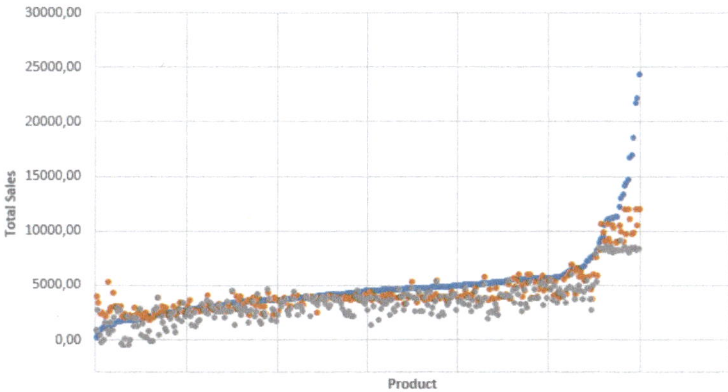

Fig. 2. Overall methodology schema.

An analysis of the graph reveals that the forecasts produced by the Random Forest model are generally closer to the actual sales compared to those of the SVM model. However, it is also observed that for products with higher sales volumes, both models exhibit greater deviations from the real values.

Table 2 presents the importance scores of the variables, highlighting the twenty most influential ones in the forecasting model. Overall, the results show that attributes related to "Store Type" and "Expectation Level" dominate the ranking. This outcome aligns with expectations, as these variables encapsulate expert knowledge about the anticipated sales dynamics for the upcoming season.

Other variables, such as those associated with "Fashion", "Color", and "International", also contribute to the model but with comparatively lower importance scores.

Among the remaining predictors, particular attention should be given to the frequency-related variables. The variable "Subfamily Frequency" stands out with a relatively high importance score (20.63), suggesting that it captures relevant behavioral patterns that enhance the model's forecasting accuracy. In contrast, "Family Frequency" and "Color Frequency" do not appear among the top twenty variables, indicating that their influence on sales forecasting is limited within this modeling context.

Table 2. Score for different categories

	Score
Store Type — ACBA	100
Store Type — ADCBA	68.62
Expectation level — M3	65.56
Store Type — ABA	57.38
Expectation level — M2	54.95
Expectation level — SB	47.39
Fashion — Centralized Distribution	30.22
Color — Camel	28.21
International — Polonia	27.88
Family — Plastic	27.17
Family — Basic PVC	24.52
Family — Beach	24.3
Price	24.01
Family — Embossed PVC	23.48
Family — Bug PVC	22.79
Fashion — Trendy	21.59
Fashion — Basic	21.3
Color — Red	21.12
Submality frequency	20.63
Family — Plain PVC	20.07

5 Conclusion

This study proposed a methodology for forecasting the sales of new fashion products by integrating structured historical sales data with unstructured online fashion content. The empirical results show that Random Forests achieve the highest predictive accuracy, outperforming both Support Vector Machines and Artificial Neural Networks. The incorporation of online fashion features yielded a consistent, though modest, improvement in explanatory power, suggesting that external signals capture elements of consumer preferences that are not fully reflected in past sales.

While preliminary, these findings underscore the potential of hybrid forecasting frameworks that combine transactional data with external sources such as blogs and fashion websites. For practitioners, the results highlight the robustness of Random Forests as a baseline model and point to the value of monitoring online fashion discourse as an early indicator of market trends.

To address the modest marginal gain obtained from web-based features, future research should incorporate formal statistical significance testing to establish whether the improvements in predictive accuracy observed are robust or incidental. Furthermore, a systematic costbenefit analysis is required to evaluate whether the operational effort involved in collecting, cleaning and processing large volumes of online fashion content is justified by the incremental improvement in forecasting performance. These analyses would enable practitioners to assess more rigorously the practical value of integrating unstructured web data into demand prediction pipelines, and identify scenarios in which the additional complexity is warranted.

Future work should also address the limitations of the current study by enlarging the corpus of online data, experimenting with more advanced text representations (e.g., word embeddings, transformer-based models), and testing hybrid architectures that integrate structured and unstructured features more effectively. In addition, extending the approach to other product categories and retailers would strengthen the generalizability of the findings. The use of LLMs to support the identification of fashion trends also represents an interesting venue for future research.

Acknowledgments. This work is funded by national funds through FCT Fundação para a Ciência e a Tecnologia, I.P., under the support UID/50014/2025 (https://doi.org/10.54499/UID/50014/2025).

References

Baardman, L., Levin, I., Perakis, G., Singhvi, D.: Leveraging comparables for new product sales forecasting. Prod. Oper. Manag. **27**(12), 2340–2343 (2018)

Cheong, M.L.F., Laik, M.N.: Demand forecasting. In: Data and Decision Analytics for Business Operations, pp. 11–68. Springer, Cham (2024)

Chumnumpan, P., Shi, X.: Understanding new products' market performance using google trends. Aust. Mark. J. **27**(2), 91–103 (2019)

Dellarocas, C., Zhang, X., Awad, N.F.: Exploring the value of online product reviews in forecasting sales: the case of motion pictures. J. Interact. Mark. **21**(4), 23–45 (2007)

Dissanayake, K., Pal, R.: Sustainability dichotomies of used clothes supply chains: a critical review of key concerns and strategic resources. Int. J. Logist. Manag. **34**(7), 75–97 (2023)

Fu, Y., Fisher, M.: The value of social media data in fashion forecasting. Manuf. Serv. Oper. Manag. **25**(3), 1136–1154 (2023)

Kim, D., Woo, J., Shin, J., Lee, J., Kim, Y.: Can search engine data improve accuracy of demand forecasting for new products? evidence from automotive market. Ind. Manag. Data Syst. **119**(5), 1089–1103 (2019)

Kulkarni, G., Kannan, P.K., Moe, W.: Using online search data to forecast new product sales. Decis. Support Syst. **52**(3), 604–611 (2012)

Lee, C., Xun, X., Lin, C.-C.: Using online user-generated reviews to predict offline box-office sales and online dvd store sales in the o2o era. J. Theor. Appl. Electron. Commer. Res. **14**(1) (2019)

Lee, W.S., Choi, H.S., Sohn, S.Y.: Forecasting new product diffusion using both patent citation and web search traffic. PLoS ONE **13**(4), e0194723 (2018)

Loureiro, A., Miguéis, V., da Silva, L.F.: Exploring the use of deep neural networks for sales forecasting in fashion retail. Decis. Support Syst. **114**, 81–93 (2018)

Medina, H., Peña, M., Siguenza-Guzman, L., Guamán, R.: Demand Forecasting for Textile Products Using Machine Learning Methods. In: Botto-Tobar, M., Montes León, S., Torres-Carrión, P., Zambrano Vizuete, M., Durakovic, B. (eds.) Applied Technologies, pp. 301–315. Springer, Cham (2022)

Schaer, O., Kourentzes, N., Fildes, R.: Estimating the market potential with pre-release buzz. SSRN Electron. J. (2019)

Skenderi, G., Joppi, C., Denitto, M., Cristani, M.: Well googled is half done: multimodal forecasting of new fashion product sales with image-based google trends. J. Forecast. **43**(6), 1982–1997 (2024)

Sousa, M., Loureiro, A., Miguéis, V.: Predicting demand for new products in fashion retailing using censored data. Expert Syst. Appl. **259**, 125313 (2025)

Steenbergen, R.: New product forecasting with analogous products: applying random forest and quantile regression forest to forecasting and inventory management. Thesis/working paper (2019)

Tsao, Y.-C., Liu, Y.-H., Vu, T.-L., Fang, I.-W.: Intelligent design suggestion and sales forecasting for new products in the apparel industry. Fibres Text. East. Eur. **31**(6), 30–38 (2023)

Wong, W.K., Guo, Z.X.: A hybrid intelligent model for medium-term sales forecasting in fashion retail supply chains using extreme learning machine and harmony search algorithm. Int. J. Prod. Econ. **128**(2), 614–624 (2010)

Use of Artificial Intelligence in Clinical Image Analysis to Improve Patient Diagnosis and Treatment

Renato M. Toasa[1(✉)], Karen Stephany Cordóva-Vera[1], Aracelly Núñez-Naranjo[2], and Victor Cusco-V[3]

[1] Universidad Tecnológica Israel, Quito, Ecuador
{rtoasa,kscordova}@uisrael.edu.ec
[2] Facultad de Ciencias de la Educación, Universidad Tecnológica Indoamérica, Ambato 180103, Ecuador
fernandanunez@uti.edu.ec
[3] Universidad de las Fuerzas Armadas ESPE, Sangolquí, Ecuador
vacusco@espe.edu.ec

Abstract. Artificial intelligence (AI) over the years has become an essential technology in certain contexts and areas of knowledge, its capacity has allowed it to be integrated in most of the productive and strategic sectors of society, being the health sciences an area where its intervention is necessary to optimize several processes. Clinical images are resources that health personnel use to detect or rule out any anomaly or disease developed in the human body, which is why knowing how to interpret them in a timely and accurate manner is vital to protect the health of people, in this context it is necessary to resort to technologies such as artificial intelligence to optimize processes and perform the analysis of these graphical representations to have a better diagnosis and treatment. The knowledge and experience of health personnel should prevail in any decision related to the integrity and quality of life of the patient, for this reason, so far artificial intelligence should only be considered as a resource that complements the clinical activity of professionals. In this context, the present study proposes a review of the use of artificial intelligence in the analysis of clinical images to improve the diagnosis and treatment of patients, for which documentary research is carried out to generate knowledge by obtaining relevant and updated information.

Keywords: Artificial intelligence · clinical imaging · diagnostics · treatment

1 Introduction

The modern era has been the protagonist of multiple technological advances that have benefited mankind, at present, artificial intelligence (AI) has become an innovative resource that facilitates the fast and accurate execution of certain activities, this technology has the ability to perform tasks that are normally performed by intelligent beings [1]. Over the years, artificial intelligence has been developed and improved to the point

A. Rocha et al. (Eds.): ICITS 2026, LNNS 1966, pp. 131–140, 2027.
https://doi.org/10.1007/978-3-032-27538-7_12

of integrating it into different strategic sectors of society, among them its application to the field of health; its development is directed to act in safe jobs and complement or replace in certain contexts the work of human beings in order to achieve better results. "Artificial Intelligence (AI) agents are predicted to infiltrate most industries within the next decade, creating a personal, industrial, and social shift towards the new technology" [2].

The different pathologies or diseases are problems that affect the well-being of people, which is why health professionals intervene with their knowledge to correct the anomalies that affect the proper functioning of the human body, and have also incorporated into their activities technologies that help to optimize patient care through timely diagnosis and treatments that help to improve the living conditions of human beings. Artificial intelligence has evolved favorably over the years, its implications are more recurrent in complex areas such as health sciences, this technology quickly and accurately optimizes certain tasks that help to improve decision making through the analysis of data that are available at the time of medical examinations; clinical imaging is one of the resources most used by medical professionals to identify the possible causes that alter the patient's health. Its importance lies in the fact that, through graphic information, it is possible to know the state of organs, tissues, cells, among other anatomical components of the human body, which facilitates an adequate diagnosis and treatment.

The joint work between professionals and technology improves patient diagnoses and treatments; resources such as artificial intelligence make it possible to examine enormous amounts of healthcare data to make decisions based on the knowledge and experience of medical personnel. "The diagnosis and the treatment decisions are referred to as data labels in AI. From healthcare data labeled in this way, AI can learn the decision patterns to apply them to future cases on new data, unseen in the training set" [3]. Due to the continuous technological advances, the medical field needs to integrate the use of artificial intelligence in the analysis of clinical images to improve the diagnosis and treatment of patients, however, it must be defined as a premise that, so far, artificial intelligence is a technology that needs to be supervised by human beings and that in the medical field, the healthcare professional is the one who has the final decision to improve the health conditions of people.

By knowing how artificial intelligence is used in the analysis of clinical images, it is possible to identify safe contexts in which this technology can be used to improve the diagnosis and treatment of diseases that affect the proper functioning of the human body. For this reason, it is necessary to review literature with updated and relevant information so that the proposed study can be properly developed for the benefit of the medical, scientific and technological community.

The document is organized as follows: Sect. 2 describes the methodology; Sect. 3 presents the proposed approach; Sect. 4 discusses the results; and, finally, Sect. 5 outlines the conclusions.

2 Methodology

The work presents a documentary type content that contributes appropriately to the development of the proposed study, the search and analysis of the literature allows to determine researches that support the importance of using artificial intelligence in the analysis of clinical images to improve the diagnosis and treatment of patients. It is important to understand the incidence of artificial intelligence applied to the analysis of clinical images, for this reason, books, magazines or scientific articles that have research that argue the importance of this technology in the field of imaging are consulted; the literature review allows obtaining relevant information from authors who have conducted research to contribute to the development of the field of health and human welfare.

The compilation of outstanding and updated information allows understanding the advances of artificial intelligence applied to the analysis of clinical images. The development of the proposed study supports questions such as: How has the development of artificial intelligence been in the field of imaging? Why is artificial intelligence used in the analysis of clinical images? Is the application of artificial intelligence in imaging safe? How does artificial intelligence affect the diagnosis and treatment of patients? These and other questions will be developed through research that benefits knowledge building.

3 Proposal

The present work reflects a research that shows the need to incorporate innovative technologies to the medical field, in such a way that artificial intelligence should be used more frequently in the analysis of clinical images to achieve better results in terms of diagnosis and treatment of patients.

3.1 Proposal Development

The evolution of artificial intelligence makes it a technology that innovates and optimizes certain tasks, its capacity has even allowed it to become involved in more complex activities, such as medical education and clinical management of patients; with the integration of AI the medical field has improved in certain scenarios, proof of this is that the diagnosis of patients based on pathological, endoscopic, ultrasonographic, radiological and biochemical examinations are more optimal due to its accuracy, and its use generates a lower human workload [4]. Recently, society witnessed the impact of ChatGPT, an artificial intelligence that uses a huge amount of textual data from the Internet, its development and capacity has allowed it to evolve to be integrated in different clinical contexts, such as those shown in Fig. 1, its implications have generated significant interest in medical science, however, caution is required when using this technology [5].

By going to the official site and using the prompt: ChatGPT, as artificial intelligence Is it trained in clinical image analysis to improve patient diagnosis and treatment? relevant information is extracted based on the answer given by ChatGPT: As a language model developed by OpenAI, ChatGPT is not directly designed or trained for clinical image analysis, however, in other clinical contexts, artificial intelligence can be a complementary tool mainly in tasks such as early disease detection, segmentation of anatomical structures, the generation of medical reports, the quantification of biomarkers and the assistance in decision making based on textual evidence, working in conjunction between a model such as ChatGPT and a medical image analysis system, can achieve encouraging results that contribute to the diagnosis and personalized treatment of patients [6].

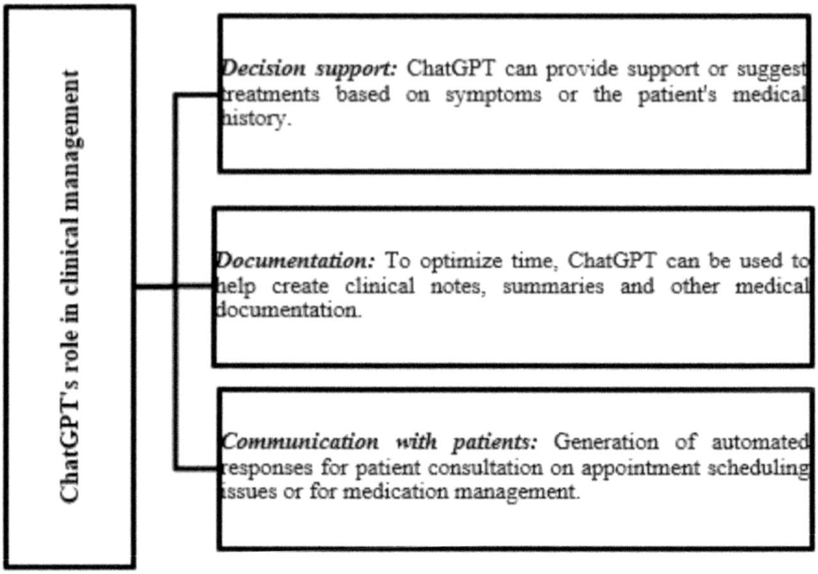

Fig. 1. Clinical contexts for the use of artificial intelligence (adapted from [5]).

3.2 Clinical Image Analysis

The timely detection of diseases allows people to improve their living conditions, for which medical examinations are performed to know the state of the human organism, being the clinical image a resource that reflects graphically the anatomical part that needs to be analyzed. Clinical images have enormous amounts of information, which is why new technologies are used to manage the data contained in these graphic representations [7], the advantage of this technology is that, being a machine that works as an intelligent entity, it can recognize and process data of features omitted by people during the training process of the model to reach and even surpass the accuracy of human processing [8]. In recent years, artificial intelligence has been incorporated to reduce possible medical errors, being a new alternative to improve the diagnosis and treatment of patients.

The use of artificial intelligence in diagnostic imaging can be very promising, since large amounts of data are managed, [9] alludes to the use of artificial intelligence in the imaging of sarcopenia, the analysis of this muscle disease by imaging methods has been laborious and dependent on the skill of the physician, before which the study proposes the integration of AI so that it can help radiologists in repetitive work such as muscle segmentation and analysis of body composition. Artificial intelligence has proven to be a useful resource even in the analysis of histopathological images of tumors and other lesions; the differential diagnosis and prognosis of head and neck tumors have their degree of complication due to their similarities and complexity, which is why technologies such as AI can help in the objective interpretation of histomorphological digital preparations, in this sense, some studies have used digital images such as WSI (Whole Slide Images) for the use of artificial intelligence methods for different approaches in head and neck squamous cell carcinoma (HNSCC) [10].

In this context [11], they examine histological images using WSI to develop an AI training and evaluation method, EfficientNet B0 convolutional neural network model and gradient-weighted class activation mapping (Grad-CAM) were used to determine its validity, which suggested that artificial intelligence focused on the cellular and structural atypia of squamous cell carcinoma and had a tendency to focus on the area surrounding the basal layer as illustrated in Fig. 2.

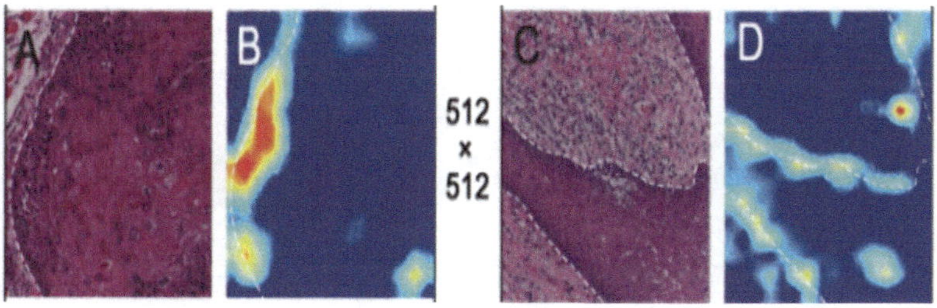

Fig. 2. Artificial intelligence attention to the outer edges of the cancerous parenchyma (A, B) or in the area surrounding the basal layer of normal epithelium (C, D). (adapted from [11])

For artificial intelligence training 90 cases of oral squamous cell carcinoma were used through histological images obtained from a university dental hospital, some of the samples are evidenced in Fig. 3.

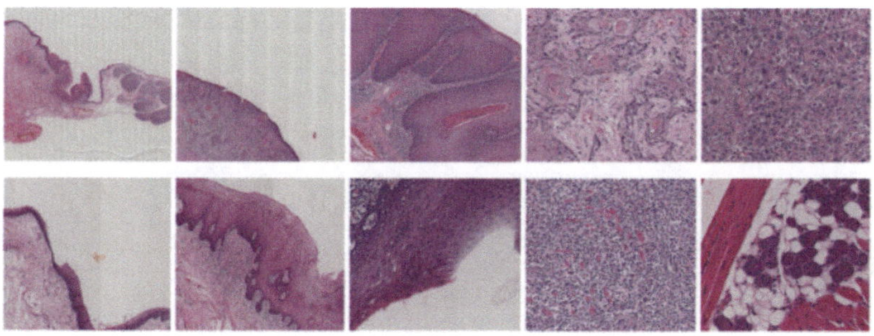

Fig. 3. Histological images used for artificial intelligence training (adapted from [11])

Artificial intelligence when receiving appropriate training based on available and updated information increases its percentage of success in the detection of anomalies that trigger pathologies or diseases, in view of this premise, [12] provide through Fig. 4 a sequence of the training that was performed in the study entitled "Artificial intelligence-based image analysis can predict outcome in high-grade serous carcinoma via histology alone" during this process refers to deep learning in image analysis, which can predict the outcome or identify underlying molecular alterations in certain types of tumors, such as lung carcinoma, melanoma, colorectal carcinoma, among others. The aforementioned study took into consideration Aiforia artificial intelligence, as it equips the health sciences field with powerful deep learning AI and cloud technology to increase the speed, consistency, and accuracy of medical image analysis, especially in pathology [13].

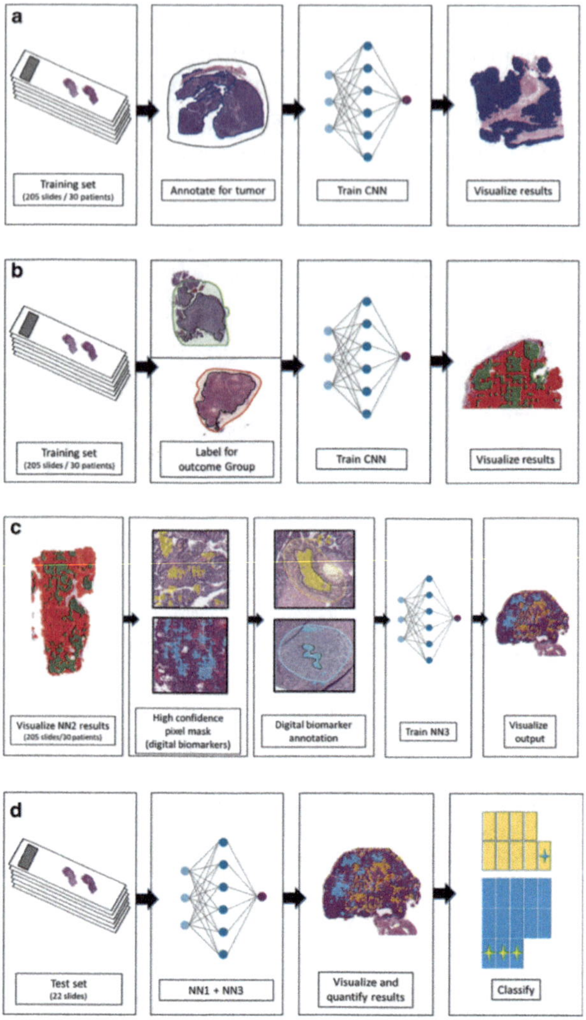

Fig. 4. Training process. Neural network 1 (NN1), level a, shows learning under supervision of tumor segmentation, manual annotations of semantic segmentation and convolutional neural network (CNN) performs learning. Neural network 2 (NN2), level b, shows weakly supervised learning, information from neural network 1 and the neural network performs tumor feature learning of the two sets of results. Neural network 3 (NN3), level c, shows supervised learning based on digital biomarker annotations, visualization and filtering of results from neural network 2, digital biomarker features are revised to have new annotations, neural network performs tumor feature learning related to the result group. Combined inference pipeline, level d, shows the combination of neural networks 1 and 3 into a single inference pipeline, neural networks applied to the validation test set, visualization and classification of the results, blue star symbolizes misclassified WSI; yellow stars reflect misclassified slides. (adapted from [12]).

Artificial intelligence is increasingly being considered as a tool that can be used in typical medical image analysis tasks such as diagnosis, segmentation or classification, its features are of great help in specialties such as radiology, oncology and pathology, where image analysis is critical [14]; the safe application of this technology lies with trained and knowledgeable professionals.

The appropriate use of artificial intelligence in the analysis of clinical images favors the diagnosis and treatment of patients, this is due to the training and learning of this technology to detect in certain cases anomalies that may go unnoticed by the health professional, similarly, speed and accuracy are attributes that make AI a resource that complements the medical performance, since, despite technological advances, health personnel is the one who makes the final decision regarding the welfare and quality of life of people suffering from any disease.

3.3 Responsibility and Professional Ethics

At present, healthcare personnel are solely responsible for decisions related to human health; knowledge, empathy and experience are attributes that artificial intelligence cannot replicate in their entirety at the moment. This technology is a complementary resource that in certain contexts can help in the activities of the medical field.

Professional ethics within the health sciences should prevail in any context that involves the life and integrity of the person; for this reason, in the face of technological innovation, it is necessary to know when and in what situations it is correct to resort to technology. The responsible use of artificial intelligence should be considered as a premise of medical activity in order to align technological advances with the central values of medical ethics [15].

4 Discussion and Results

Artificial intelligence is a modern technology used in certain activities where speed and accuracy are mainly needed; the healthcare field is a strategic sector that has integrated AI to optimize and improve patient care, with clinical image analysis being the area where work has been done to develop intelligent machines that are capable of complementing clinical practice.

By means of artificial intelligence training, this type of technology acquires a learning process that allows it to detect anomalies that trigger different diseases. By optimally analyzing clinical images through AI, the work of healthcare personnel can be complemented, since the diagnosis and treatment of patients can be improved. The results achieved with the use of artificial intelligence are reflected in the speed and capacity that this technology acquires when going through the learning process of convolutional neural networks.

Although the literature used in the proposed work reflects great advances and positive results, it is still a period in which the intervention of health personnel is irreplaceable, since it must be considered that so far there is no technology capable of simulating or surpassing all human intellect and capacity, there is no doubt that, in certain tasks such as clinical reports and information management, AI has a clear advantage, especially in aspects such as speed, but the human being is the one who decides and supervises that the work of the technology is carried out safely and correctly, AI has a clear advantage, especially in aspects such as speed, but it is the human being who decides and supervises that the work of the technology is carried out safely and correctly, and in the clinical context, technology cannot develop or simulate empathy, a human quality necessary to understand the emotions that arise in a health area.

The integration of innovative technologies in the medical field is necessary to improve the conditions and quality of life of patients. In order to control and detect possible anomalies that affect the functioning of the human body, it is necessary to resort to medical examinations which reflect information that health personnel analyze for an adequate diagnosis; clinical images are a valuable resource to identify the state of an affected area. As there is an abundance of graphic data, including microorganisms, the integration of artificial intelligence in certain clinical cases is an appropriate complement that facilitates the detection of structures that the human eye may ignore.

Health related information and technology needs to be updated on a regular basis in order to have data available to assist in the analysis of clinical images through artificial intelligence, as these technological resources help to improve the diagnosis and treatment of patients.

In order to develop the present study, a search for relevant information was carried out using websites or platforms that host important data. PubMed, SciELO, ScienceDirect, Elsevier among other prominent media in health and technology contributed to the research; the selected literature shows that artificial intelligence has developed favorably in the analysis of clinical images, the training of convolutional neural networks has allowed preparing this technology to detect anomalies or diseases such as carcinoma.

The different positions and contributions made by the selected authors contributed to the development of this study, in which several questions revolving around artificial intelligence in image analysis are clarified, its application in certain clinical cases has made it possible to detect anomalies that interfere with the proper functioning of the organism, This technology has shown a great impact on the medical community, since with appropriate training, AI is capable of improving its performance to reach high levels of accuracy. The authors are also aware that its participation in the health field will gradually improve, which is why further research is needed, mainly to achieve clinical validation.

The selected literature shows the progress and development of artificial intelligence over time, which is why it is necessary to carry out periodic reviews to help understand how it is applied and the level of safety of AI when intervening in the diagnosis and treatment of patients.

The advances and capacity of AI evolve in short periods of time, which is why it is necessary to investigate, update and generate new information that facilitates the dissemination and construction of knowledge to apply modern and safe technologies that help to complement the activity of healthcare professionals. It should be taken into consideration that, despite the promising results of artificial intelligence in the analysis of clinical images, the role that the healthcare professional should play over the technology should be considered, i.e., the final decision is made by the human being and technologies such as AI complement and help to improve the diagnosis and treatment of patients.

5 Conclusions

The evolution and results obtained through artificial intelligence have turned it into a technology capable of intervening in different human activities, being health sciences one of the areas in which it has caused the greatest impact, since its integration has allowed health personnel to have at their disposal an innovative and complementary resource that optimizes clinical activities, improves diagnosis and helps to find alternatives for the benefit of human health.

Artificial intelligence training is a necessary process for learning convolutional neural networks to acquire knowledge that allows them to identify or rule out human anomalies and diseases.

The analysis of medical images through artificial intelligence is promising, since the diagnosis of certain clinical cases through this technology reaches a considerable percentage of security, however, its action is a valuable complement for the health personnel to make the decision and seek the best procedure for the benefit of the patient's well being.

The use of artificial intelligence in the analysis of clinical images should be considered for the moment as a complementary resource for healthcare personnel with their experience and knowledge to take advantage of the features offered by this technological innovation.

References

1. Chiu, T.K.F., Ahmad, Z., Ismailov, M., Sanusi, I.T.: What are artificial intelligence literacy and competency? A comprehensive framework to support them. Comput. Educ.Open. **6** (2024). https://doi.org/10.1016/j.caeo.2024.100171
2. Kelly, S., Kaye, S.A., Oviedo-Trespalacios, O.: What factors contribute to the acceptance of artificial intelligence? A systematic review. Telematics Inform. **77** (2023). https://doi.org/10.1016/j.tele.2022.101925
3. Kundu, S.: How will artificial intelligence change medical training? Commun. Med. **1** (2021). https://doi.org/10.1038/s43856-021-00003-5
4. Liu, P.R., Lu, L., Zhang, J.Y., Huo, T.T., Liu, S.X., Ye, Z.W.: Application of artificial intelligence in medicine: an overview. Curr Med. Sci. **41**(6), 1105–1115 (2021). https://doi.org/10.1007/s11596-021-2474-3
5. Khan, R.A., Jawaid, M., Khan, A.R., Sajjad, M.: ChatGPT - Reshaping medical education and clinical management. Pak J Med Sci. **39**(2), 605–607 (2023). https://doi.org/10.1007/978-3-030-49186-4_31
6. OpenAI, ChatGPT: Respuesta a consulta sobre análisis de imágenes clínicas, (2025). [En línea]. Disponible en: https://chat.openai.com/ [Accedido: 22-may-2025]
7. Diaz, O., et al.: Data preparation for artificial intelligence in medical imaging: a comprehensive guide to open-access platforms and tools. Phys. Med. **83**, 25–37 (2021). https://doi.org/10.1016/j.ejmp.2021.02.007
8. Ma, D., Dang, B., Li, S., Zang, H., Dong, X.: Implementation of computer vision technology based on artificial intelligence for medical image analysis. Int. J. Comput. Sci. Inf. Technol. **1**(1), 69–76 (2023). https://doi.org/10.62051/ijcsit.v1n1.10
9. Rozynek, M., Kucybała, I., Urbanik, A., Wojciechowski, W.: Use of artificial intelligence in the imaging of sarcopenia: a narrative review of current status and perspectives. Nutrition. **89** (2021). https://doi.org/10.1016/j.nut.2021.111227

10. Pereira-Prado, V., et al.: Artificial intelligence for image analysis in oral squamous cell carcinoma: a review. Diagnostics. **13**(14) (2023). https://doi.org/10.3390/diagnostics13142416
11. Oya, K., Kokomoto, K., Nozaki, K., Toyosawa, S.: Oral squamous cell carcinoma diagnosis in digitized histological images using convolutional neural network. J Dent Sci. **18**(1), 322–329 (2023). https://doi.org/10.1016/j.jds.2022.08.017
12. Laury, A.R., Blom, S., Ropponen, T., Virtanen, A., Carpén, O.M.: Artificial intelligence-based image analysis can predict outcome in high-grade serous carcinoma via histology alone. Sci. Rep. **11** (2021). https://doi.org/10.1038/s41598-021-98480-0
13. Aiforia: "About Us", Aiforia. [En línea]. Disponible en: https://www.aiforia.com/about-us. Accedido 26 May 2025
14. Barragán-Montero, A., et al.: Artificial intelligence and machine learning for medical imaging: a technology review. Phys. Med. **83**, 242–256 (2021). https://doi.org/10.1016/j.ejmp.2021.04.016
15. Weidener, L., Fischer, M.: Role of ethics in developing ai-based applications in medicine: insights from expert interviews and discussion of implications. JMIR AI. **3** (2024). https://doi.org/10.2196/51204

Use of ICT in Education: Descriptive Analysis Based on Secondary Data from Ecuador

María José Carrera Yánez(✉)

Universidad Tecnológica Israel, E4-142, Marieta De Veintimilla y Fco. Pizarro, Quito, Ecuador
mjcarrera@uisrael.edu.ec

Abstract. Information and Communication Technologies (ICT) are essential in modern education, expanding access to learning resources, encouraging innovative pedagogical practices, and strengthening digital literacy across academic levels. Their integration promotes participatory, equitable, and student-centered learning aligned with 21st-century competencies. This study aims to describe the access, use, and distribution of ICT in Ecuadorian educational institutions through the analysis of secondary data, identifying inequalities and institutional challenges in the national context. A descriptive approach was applied using official sources, including the National Institute of Statistics and Census (INEC), the Ministry of Education of Ecuador (MINEDUC), the United Nations Educational, Scientific and Cultural Organization (UNESCO), and the World Bank. Findings reveal persistent gaps between urban and rural areas, limited technological infrastructure in public schools, and unequal distribution of digital resources. These results emphasize the urgency of strengthening digital inclusion policies, improving technological infrastructure and internet connectivity, and promoting continuous teacher training to ensure equitable and high-quality ICT-based learning for all students in Ecuador.

Keywords: Education · ICTs · Equity · Digital Divide

1 Introduction

Digitization in education has become an unavoidable priority for all education systems worldwide, particularly in light of the challenges posed by the global COVID-19 health crisis, which I see as an urgent need to introduce digitization into the education system and promote resilience in the face of in-person shutdowns (UNESCO, 2021). According to the Organization for Economic Cooperation and Development (OECD, 2020), "The effective integration of ICT in educational settings contributes to optimizing learning effectiveness and the development of essential digital skills" (p. 45). In this sense, ICTs transcend their role as a means of communication and also promote the acquisition of digital skills that are essential for contemporary society. Its proper integration allows for the creation of a solid environment in the teaching and learning process, forging more dynamic, personalized educational areas that respond to students' demands.

In Latin America, the implementation of ICTs presents significant challenges related to infrastructure, gaps in equitable access, and poor teacher training. These limitations

A. Rocha et al. (Eds.): ICITS 2026, LNNS 1966, pp. 141–155, 2027.
https://doi.org/10.1007/978-3-032-27538-7_13

are reflected in the so-called "digital divide," evidenced by the inequality that exists in terms of connectivity, availability of technological devices and development of digital skills among students and teachers from different social, economic, and territorial sectors (UNICEF, 2021).

Adel 2024 analyzed financial inclusion in 56 developing countries, showing that comparative secondary data allow the identification of common patterns, regional differences, and contextual factors that shape inequality. This perspective is relevant to the present study, as the use of official secondary data in Ecuador enables the detection of specific gaps and challenges in ICT access within the education system. As in international research, evidence-based analysis supports understanding structural digital inequalities and provides key insights for strengthening public policy and educational inclusion.

Recent studies show that, despite global progress in digital access, significant usage gaps persist in developing regions (Caragliu & Del Bo, 2025). Evidence highlights that digitalization can effectively reduce social and educational inequalities when supported by equitable public policies and capacity-building initiatives (Plaza de la Hoz et al., 2024). In Latin America, ICT integration is essential for learning and social development. In this context, the present study analyzes ICT access and use in Ecuador's education system, finding improvements in connectivity but persistent disparities between urban and rural areas, socioeconomic groups, and public and private schools (Ministry of Education, 2022). These inequalities limit learning opportunities, underscoring the need for more inclusive digital policies.

Far beyond the availability of equipment and access to the internet, the lack of teacher training poses a significant barrier to the proper integration of ICT, therefore, having technology at hand but without accompanying training does not guarantee progress in education, so the lack of digital skills among teachers restricts the integration of technology into the classroom. In this sense, teacher training emerges as an indispensable process for turning the use of ICT into a tangible benefit in learning.

In this regard, all educational efforts must focus on providing electronic devices and expanding connectivity, but also on prioritizing ongoing teacher training. The experience that many went through during the COVID-19 health pandemic, which highlighted both the potential and weaknesses of digitization in education. In Ecuador, the shift to virtual teaching methods revealed the importance of having adequate technological infrastructure and trained teachers and students, as well as highlighting the urgent need to implement educational strategies that ensure equity and inclusion (UNESCO, 2021; OECD, 2020).

Beyond the provision of technological devices and access to connectivity, it also involves the application of transformative teaching methodologies that promote autonomy in learning, collaborative work, and the development of higher cognitive functions. Pérez and Torres 2023states that: "Educational technology alone is not enough to transform education; it is necessary to design dynamic learning environments where teachers are competent in integrating ICT into pedagogical practice and enhancing high-level skills" (p. 55).

Caragliu and Del Bo (2025) note that the global expansion of the Internet and mobile technologies has shifted academic debate toward explaining why certain countries and social groups adopt digital tools at a slower pace. These differences have deepened the digital divide, as shown by international comparative research, which emphasizes that this gap is not only technological or economic but also has significant educational implications. Limited access to ICT restricts learning opportunities and reinforces social inequalities. Therefore, analyzing ICT use in Ecuadorian educational institutions is essential to understand how these disparities manifest in the national context and to evaluate whether current policies are effectively promoting equitable and high-quality digital inclusion.

This study analyzes secondary data to describe the current state of ICT use in Ecuadorian education, highlighting persistent disparities and opportunities for improvement. Its purpose is to inform decision-making aimed at reducing educational inequality, ensuring universal digital access, and fostering innovation, even in contexts with limited infrastructure. While ICT integration offers a strategic path to improving educational quality, significant territorial, social, and economic gaps persist, requiring structural policies, teacher training, and sustainable inclusion strategies. The findings show that meaningful transformation depends on a comprehensive approach that combines infrastructure, digital competencies, and updated pedagogical practices.

1.1 ICT in Education

Information and Communication Technologies (ICT) are defined as the set of technological resources, devices, software, and digital interfaces that facilitate the collection, storage, processing, and transmission of data, as well as collaboration and interaction among users across different contexts (UNESCO, 2021). In the educational field, ICT encompasses computers, smartphones, tablets, and virtual learning environments that provide interactive access to digital content and supplementary materials. Their main contribution lies in their dynamic and flexible capacity to enable learning regardless of temporal or geographical constraints. This aspect proved particularly relevant during the global health crisis, when many education systems were forced to transition to remote learning. In Ecuador, this shift posed major challenges due to the limited availability of digital infrastructure and resources necessary for effective virtual teaching and learning.

In the field of education, ICTs are taking a step beyond merely serving as a tool for accessing information, as they are becoming guides for promoting transformative educational strategies. Proper technological integration promotes the creation of activities aimed at providing students with dynamic, collaborative, and individualized learning, which strengthens student motivation and promotes the development of advanced skills, including critical thinking, problem solving, digital literacy, and socio-emotional skills related to ethical information management. Likewise, educational digitization also provides hybrid learning spaces where the face-to-face and virtual contexts complement each other to cater to the diversity of learning styles and rhythms (Cabero-Almenara, & Palacios-Rodríguez, 2020).

In Ecuador, the integration of ICT into the education system can be observed through the use of Ministry of Education platforms such as EducarEcuador, the Open Educational Resources (OER) portal, and institutional management systems like SIMESE. Additionally, national indicators—such as the proportion of schools with internet access, availability of computer labs, student-to-device ratios, and teacher training in digital competencies—serve to evaluate the country's level of digital development in education. These platforms and indicators make it possible to assess not only infrastructure access but also the degree of pedagogical integration of ICT and the persistent disparities between urban and rural educational settings.

ICT can improve education, but its effective implementation requires more than just equipment and connectivity: a comprehensive plan is needed that includes ongoing training for teachers, technical support, and the creation of appropriate digital content. Without these elements, its impact is limited or misguided. Therefore, technological integration must be a holistic process where infrastructure, training, and educational management work together to ensure quality learning environments (Fig. 1).

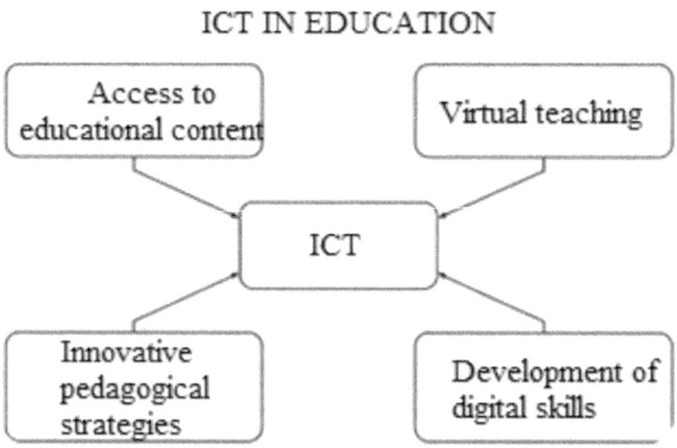

Fig. 1. Diagram Education and ICT
Note: This diagram shows how ICT facilitates teaching and learning processes in addition to digital skills. Taken from (Pérez and Torres 2023)

The use of information and communication technologies promotes inclusion in education, which is why the systemic interface and tools are so important in facilitating access to topics and resources that are tailored to students with cultural and functional educational needs. That is why responding to this digital access has become a fundamental part of education today, enabling all students to be part of this digitization process (UNICEF, 2021). Therefore, information technologies must be understood not only as physical devices that contain technology, but also as a means of ensuring equitable and equal processes in the development of education, responding to a fully planned landscape and being part of the diversification of the student body.

1.2 The Digital Divide and Equity

The concept of the digital divide refers to inequalities in access to and use of technologies, visible at the regional, institutional, and socioeconomic levels (World Bank, 2022). In education, this divide particularly affects low-income students, who lack both devices and connectivity and the skills necessary to use them properly. Therefore, the digital divide is a multidimensional phenomenon that includes infrastructure, digital skills, and conditions for effective learning (Fig. 2).

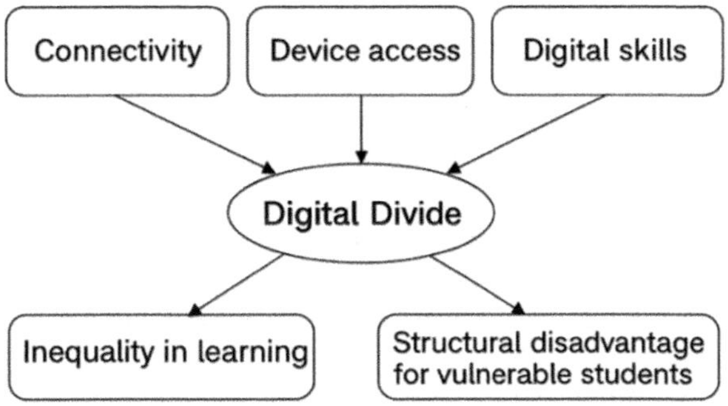

Fig. 2. Causal relationship diagram
Note: The diagram shows the relationship between the digital divide and educational equity, allowing us to identify the factors related to digital access and skills that produce disparities in educational outcomes. Taken from (World Bank, 2022)

The digital divide in Latin America is particularly pronounced in rural areas and among socially vulnerable groups. Students living in remote areas have difficulty accessing the Internet and owning electronic devices, while trained human resources to support digital education are often limited.

These circumstances exacerbate educational inequality, which has been persistent over time and has been part of the problem of low student performance in terms of grades. This is further aggravated by the fact that students drop out of school. However, studies show that in order to minimize these discrepancies, it is necessary to implement public policies that include investment in technological infrastructure, continuing education for teachers, and ICT-based curriculum planning.

In order to speak of digital equivalence, aspects such as cultural diversification must be taken into account. Although it is true that the digital divide has narrowed recently, albeit not significantly, differences in the appropriate use of ICTs are evident, at least among students in rural areas where economic resources are quite limited. Similarly, it is important that digital compendiums take into account linguistic, social, and cultural diversity so that all students, without exception, have the necessary access to these technologies for an optimal learning process. Therefore, digital equity considers not only the material object of technology, but also the pedagogical use of these tools, ensuring inclusion and quality.

1.3 Public ICT Policies in the Ecuadorian Context

In Ecuador, the government has implemented a variety of policies aimed at promoting the integration of ICTs in education at all levels, including initiatives such as "Internet for All" and the "National Plan for Educational Connectivity." Each of these strategies has the primary goal of extending internet coverage and providing devices to schools, colleges, and residences, as well as ensuring adequate digital training for both teachers and students (Ministry of Education, 2022). The main idea is to minimize this gap and offer equal access to digital resources, ensuring that geographical, social, or economic factors do not hinder the learning process.

The results of this study not only support the formulation of policies to reduce the digital divide, but also guide the design of ICT-based education systems in Ecuador. The inequalities detected between urban and rural areas reveal the need to create platforms adapted to low-connectivity contexts, accessible educational tools, and hybrid school management systems. Likewise, the findings can guide the development of mobile applications, interoperable infrastructures, and mechanisms for monitoring the pedagogical use of technology. Taken together, the evidence provides criteria for strategically investing in innovative technological solutions that respond to the real needs of students and teachers in the country.

Aunque Although progress has been made in the creation of government policies focused on ICT, there are still certain obstacles to their proper implementation. According to reports provided by UNESCO and the Ministry of Education, there are currently shortcomings in technological infrastructure, connectivity, and teacher training, specifically in rural areas and vulnerable communities (UNESCO, 2021; INEC, 2023). These gaps hinder the equitable distribution of technological advantages and lead to learning gaps, as shown in Fig. 3:

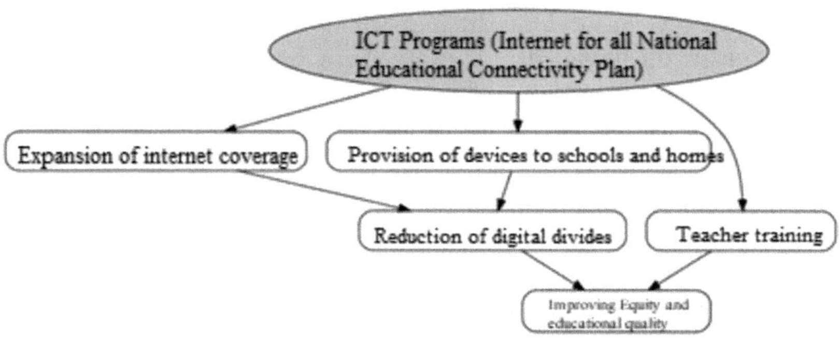

Fig. 3. Diagram of government policies
Note: This chart shows government policies and their desired effects, transparently representing how these ICT programs translate into educational objectives. Taken from (INEC, 2022).

To address these barriers, education experts recommend implementing holistic and sustainable tactics that not only provide training programs for teachers but also involve processes such as technical support, appropriate digital learning compendiums, and methodological processes tailored to each context. It is clear that it is essential to monitor and evaluate everything related to ICTs to ensure the effective use of these tools (Ramírez and López, 2022).

In this regard, ICT in Ecuador is guided by policies that promote pedagogical innovation and its proper implementation, fostering inclusion. It is important to note that these policies are aligned with the Sustainable Development Goals (SDGs) and the Education 2030 Agenda. Thus, studies accurately mention that only through a comprehensive approach combined with active participation, programming, and monitoring will it be possible to reduce the digital divide and ensure that ICT has a positive impact at the national level in the area of education.

It has been proven that information technologies comprise a series of essential tools for the modern educational context. However, their effectiveness depends on a process of appropriate pedagogical unification, the minimization of this digital limitation, and the ability to integrate inclusive policies. For Ecuadorian society, a series of public policies and programs have achieved significant progress. However, challenges still require monitoring to ensure that ICT in education is equitable and of great pedagogical value.

The results of this analysis show patterns consistent with recent research on digital divides in Latin America, especially in relation to urban-rural inequalities in internet access, devices, and teacher digital training (Plaza de la Hoz et al., 2024; UNESCO, 2023). Similar to the findings reported by Ortiz et al. (2022) and Navarro & Álvarez (2021), this study confirms that socioeconomic and territorial context continues to be a critical determinant in the effective adoption of educational technologies. However, the secondary data used here provide more up-to-date and specific evidence for the Ecuadorian case, allowing us to observe not only the persistence of the digital divide, but also a slow reduction in unequal access, particularly in domestic and school connectivity in urban areas.

Unlike studies focused on technological availability or teacher perceptions (Ramírez & López, 2020; Mena et al., 2021), this study integrates access, device type, and digital training, showing that the expansion of smartphone use does not guarantee equitable educational opportunities, partially coinciding with Benavides & Chávez (2023) in Peru.

The study shows that the digital divide exists both at school and home, emphasizing the need for coordinated educational policies and local technological development. Despite improvements in school connectivity, challenges remain in rural teacher training and equipment, highlighting that achieving digital equity requires multidimensional and sustained interventions.

Finally, the study's main contribution is its updated, descriptive-comparative analysis using recent official data, providing guidance for educational policy and the design of contextually adapted, accessible ICT systems in Ecuador. It highlights that inclusive digital education depends not only on infrastructure but also on developing pedagogically relevant and sustainable technological tools.

2 Methodology

This study is based on a descriptive approach, as its ultimate goal is to analyze and characterize the current usefulness, access, and availability of these information and communication technology tools in the educational field in Ecuadorian society. This type of approach is appropriate because it enables the identification of relevant trends, patterns, and gaps that affect educational equity, without assuming the establishment of causal relationships between variables.

This study adopted a descriptive-documentary design based on the analysis of secondary data from official national and international sources. The period analyzed covered the years 2020 to 2023, which allowed us to capture recent developments in access to and use of information and communication technologies (ICT) in the Ecuadorian education system, in a context marked by the post-pandemic transition and the consolidation of educational digitization strategies.

Sources were selected based on criteria of relevance, reliability, timeliness, and public availability. Only recognized, open-access institutional databases were included, including those of the National Institute of Statistics and Census (INEC), the Ecuadorian Ministry of Education (MINEDUC), the United Nations Educational, Scientific and Cultural Organization (UNESCO), and the World Bank. In addition, statistical reports, educational yearbooks, and official surveys such as the National Survey of Employment, Unemployment, and Underemployment (ENEMDU) and the MINEDUC Educational Yearbook were considered.

Table 1 Study variables and operational definitions

Study Variable	Operational Definition	Indicators
Households with connectivity	Level of Internet access in Ecuadorian households where students reside.	% of households with Internet access; urban–rural differences; type of connection available.
Devices available in households with students	Availability of technological devices in households with students for educational activities.	% of households with a computer; percentage of households with a smartphone; number of devices per household.
Connectivity in schools	Degree of Internet connection availability in educational institutions in the country.	% of educational institutions connected to the Internet; type and quality of connection.
Ecuador vs. Latin America – Connectivity in Schools	Comparison of the level of school connectivity between Ecuador and the average for Latin American countries.	% of schools connected in Ecuador vs. other countries in the region; relative gaps
Access to ICT in Urban vs. Rural Schools	Territorial disparities in access to educational technological infrastructure between urban and rural contexts.	% of urban schools with connectivity; percentage of rural schools with connectivity; availability of ICT laboratories; teacher training

This analysis is based on secondary data obtained from official sources and international organizations, including: the National Institute of Statistics and Census (INEC), through the National Survey of Employment, Unemployment, and Underemployment (ENEMDU 2023); the Ministry of Education of Ecuador (MINEDUC) through the 2022 Education Yearbook; UNESCO with its Global Education Monitoring Report; and the

World Bank through the World Development Indicators. Using these sources of information provides reliable and up-to-date data to enable a comprehensive, well-founded, and rigorous analysis of the phenomenon under study.

The criteria for including data were: Information published between 2020 and 2023, indicators directly related to access, use, and availability of ICT in education, data with national representativeness or urban–rural disaggregation, verifiable statistics reported by official agencies or multilateral entities.

Documents without institutional support, unverifiable data or data from periods prior to 2020, as well as opinion studies without statistical support, were excluded. The information collected was systematized in comparative matrices and analyzed using descriptive techniques, prioritizing the analysis of trends, urban-rural gaps, and the availability of infrastructure and digital training in educational institutions.

The study examines key variables related to the digital divide and educational equity, including home connectivity, access to devices, internet in schools, ICT integration in teaching, and regional differences. Analyzing these factors helps identify persistent gaps and their impact on education quality.

With regard to the techniques used for the analysis, a document review was chosen as the main tool, which allowed for the analysis of statistical sources, technical reports, and relevant regulatory frameworks, with the aim of comparing results and establishing trends. This methodology provided a comprehensive overview of internet connectivity and ICT use in education, including a comparative analysis that revealed digital disparities linked to social and territorial factors.

Thus, the methodology used is not limited to presenting an overview of ICT access and use in the country, but also provides evidence for the formulation and strengthening of public policies that seek to reduce existing gaps and guarantee fair and quality education.

3 Results and Discussion

The integration of Information and Communication Technologies into the Ecuadorian education system has seen significant progress in recent times, driven both by government initiatives and increased access to digitalization in homes. However, there continue to be marked inequalities related to connectivity, the provision of devices, and the development of digital skills, which translate into gaps in learning quality and educational inequalities. In view of this, it is essential to examine the available statistical and documentary data in order to identify trends, difficulties, and possible opportunities that reveal the real scope of the policies implemented and their orientation towards new actions aimed at digital equity.

3.1 Households with Connectivity

The key benchmarks used to assess digital inclusion and equity are based on the situation of individual households in Ecuador. In Ecuador, this indicator highlights the marked influence of the territorial divide, as there are significant differences in connectivity levels between urban and rural areas. As can be seen in Table 1, this percentage difference is clearly shown, with internet access in rural households being considerably lower than in urban areas, reflecting structural disparities in access to technology and repercussions on learning in terms of educational equity (Fig. 4).

Table 2 Internet Access in Ecuadorian Households

Area	% of households with internet access
Urban	72,4%
Rural	38,6%
National	59,1%

Note: Taken from the National Survey on Employment, Unemployment, and Underemployment (ENEMDU 2023), INEC.

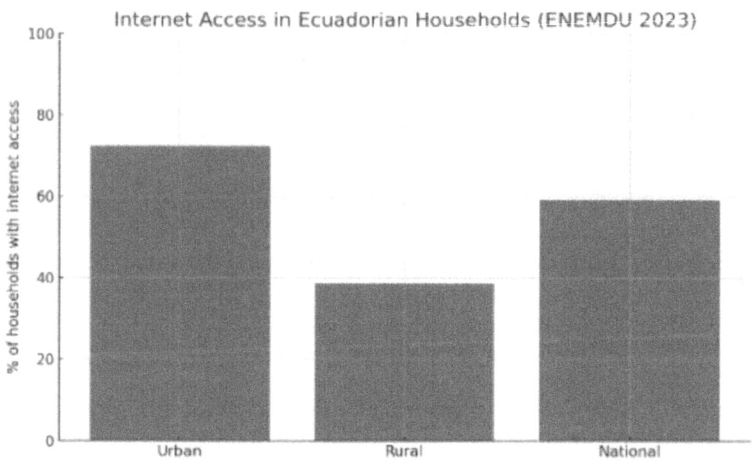

Fig. 4. Internet Access in Ecuadorian Households

The difference between urban and rural areas highlights a structural inequality that goes beyond the merely technological, as it also reflects the country's socioeconomic and territorial divide. The lack of connectivity in rural areas limits access to digital educational resources, conditions participation in online learning environments, and widens the gap with urban students. This can be validated as the urban-rural difference is 33.8 percentage points, showing structural inequality in internet access.

3.2 Devices Available in Households with Students

The availability of technological devices in households with students is a key factor in ensuring the continuity and quality of digital learning. In this regard, it is important to analyze what types of devices predominate and how they are distributed according to geographical location. Table 3 shows that although smartphones are widespread, computers (more useful for academic tasks) are much less common in rural areas, as can be seen in Table 2.

Table 3 Usable devices

Device	National	Urban	Rural
Computer	45,2%	60,1%	28,4%
Smartphone	89,5%	93,0%	84,2%
Tablet	23,7%	30,8%	15,4%

Note: Taken from the National Survey of Employment, Unemployment, and Underemployment (ENEMDU 2023), INEC.

The concentration of access on smartphones, compared to the lower availability of computers and tablets, shows a pattern of technology consumption that is more oriented toward personal use than academic use. Although smartphones facilitate communication, their functionality for school activities is limited, which creates a deficit in the quality of digital learning in the most vulnerable households. Although smartphones are widely used in households with students (89.5% nationwide), computers, which are more suitable for academic activities, are unevenly distributed: 60.1% in urban households compared to 28.4% in rural households, reinforcing the digital divide in educational contexts.

3.3 Connectivity in Schools

Internet access in educational institutions reflects the institutional capacity to integrate ICT into teaching and learning processes. In Ecuador, the differences between public and private schools in terms of connectivity reveal structural gaps that affect educational quality, as shown in Table 4.

Table 4 Internet access in educational institutions

Type of School	With Internet	Without Internet
Public	54,7%	45,3%
Private	92,8%	7,2%
National Average	63,1%	36,9%

Note: Taken from the 2022 Education Yearbook, Ecuadorian Ministry of Education.

While 92.8% of private schools have internet access, only 54.7% of public institutions have this service. This difference reflects a critical deficiency in the digital infrastructure of the public sector, with direct implications for the quality and equity of learning. Internet access in educational centers reflects the institutional capacity to integrate ICTs into teaching and learning processes. In Ecuador, the differences between public and private schools in terms of connectivity reveal structural gaps that affect educational quality. Public schools face critical connectivity deficiencies, which reinforce educational inequalities.

3.4 Ecuador vs. Latin America – Connectivity in Schools

Ecuador's position relative to other Latin American countries in terms of school connectivity allows us to gauge its relative progress in the digital education agenda. The regional comparison provides a benchmark for assessing the degree of competitiveness and educational equity in the country, as shown in Table 5.

Table 5 Connectivity in institutions Ecuador and Latin America

Country	% of schools with Internet access
Chile	95%
Argentina	84%
Colombia	73%
Ecuador	63%
Regional Average	78%

Note: Taken from the Global Education Monitoring Report, UNESCO (2023).

Ecuador's lag behind the regional average reflects the need for more aggressive and sustainable public policies on technological investment in education. Lagging behind neighboring countries limits the country's competitiveness and makes it difficult to achieve regional standards of digital inclusion, affecting both equity and preparedness for a globalized economy. Ecuador is below the regional average, limiting its competitiveness in digital education. Ecuador, with 63% of schools connected, is below the regional average (78%) and lags behind countries such as Chile (95%) and Argentina (84%). This regional gap limits the country's competitiveness in digital education.

3.5 Access to ICT in Urban vs. Rural Schools

Geographic location is a determining factor in the provision of technological infrastructure. When comparing conditions in urban and rural schools in Ecuador, inequalities are identified that directly impact digital learning opportunities, as shown in Table 6.

Differences in infrastructure and teacher training between urban and rural areas reinforce inequality in educational opportunities. The lower availability of resources in rural areas not only limits students' access to digital content, but also conditions the professional development of the teaching staff, creating a cycle of exclusion that

Table 6 Access to ICT

Indicator	Urban (%)	Rural (%)
Schools with Internet access	85%	43%
Schools with computer labs	76%	38%
Teachers trained in ICT	61%	29%

Note: Taken from the 2022 Education Yearbook Report, Ecuadorian Ministry of Education.

perpetuates the digital divide. The urban-rural divide is significant, limiting the quality of digital learning in rural areas. The urban-rural divide is marked: 85% of urban schools have internet access, compared to only 43% in rural areas. Similarly, access to computer labs and teacher training in ICT is significantly lower in rural areas, limiting digital learning opportunities for the most vulnerable students.

Descriptive data show that Ecuador has made progress in integrating ICT into education, but significant inequalities remain. Urban areas enjoy connectivity and technological resources comparable to the Latin American average, while rural and Amazonian provinces face severe limitations. Although gender differences exist, they are minor compared to territorial and socioeconomic inequalities. Teacher training in ICT is an additional challenge; many teachers lack the skills to effectively integrate technology into teaching.

The results highlight significant gaps in ICT access and use at both household and educational levels, reflecting challenges in achieving SDG 4 (Quality Education) and SDG 9 (Industry, Innovation, and Infrastructure). The findings underscore the need to strengthen Ecuador's digital policies, including the National Digital Transformation Plan and educational connectivity programs. Addressing urban-rural inequalities requires enhanced digital infrastructure, teacher training, and device access to ensure effective ICT integration in education and promote a more equitable, knowledge-based society.

The study uses reliable secondary data but faces limitations due to database availability, lack of longitudinal continuity, and exclusion of qualitative insights on ICT use in education. National data may underrepresent local contexts, especially remote rural and Amazonian areas. These constraints suggest the need for future research using mixed methods to explore educational experiences in greater depth.

4 Conclusions

The findings show that, although Ecuador has made significant progress in terms of incorporating and using Information and Communication Technologies (ICT), structural gaps remain that limit the possibility of guaranteeing equitable education. These gaps are mainly evident in three areas: territorial, with marked inequality between urban and rural areas; institutional, with significant differences between public and private schools; and socioeconomic, where lower-income households have lower levels of connectivity and access to adequate learning devices.

The analysis shows that access to ICT is positively associated with better educational indicators, such as higher enrollment and completion rates for basic education. These results show that digital inclusion transcends the technological mechanism and is a decisive factor in ensuring educational equity, school continuity, and improved learning. In this context, government strategies that prioritize infrastructure provision and teacher training are fundamental and essential for reducing existing inequality and thus moving toward a more equitable education system.

At the regional level, comparative data show that Ecuador has a lower level of school connectivity than other Latin American countries, ranking below the average of other countries that are leaders in this field. This gap hinders the potential of the national system, as it must respond to the demands of a global environment undergoing digital transformation, which is essential for contributing to innovation and sustainable progress.

In short, the results of the analysis show that the main challenge for Ecuador is not limited to increasing technological infrastructure, but rather to providing equitable and high-quality access, especially for vulnerable groups and across all social sectors. Achieving this goal requires a comprehensive approach that combines infrastructure transformation, solid institutional management, and digital inclusion strategies. In this way, the country will be able to close persistent gaps and consolidate a more resilient and professional education system in the regional and international context.

This study contributes theoretically to the analysis of the educational digital divide by offering recent evidence on ICT access and use in Ecuador, highlighting urban-rural differences and their impact on educational equity. On a practical level, the results provide key information for strengthening policies and technological systems aimed at improving school infrastructure, access to devices, and teacher training, supporting actions to move toward a more inclusive and digitally integrated education.

References

Adel, N.: Financial inclusion and digital finance: evidence from developing countries. Heliyon. **10**(24) (2024). https://doi.org/10.1016/j.heliyon.2024.e40951

Andrade, M., Cueva, M.: ICT in teaching, learning, and inclusion: benefits and difficulties in Ecuador. Memorias Y Boletines De La Universidad Del Azuay. **1**(1), 84–103 (2023). https://doi.org/10.33324/memorias.vi1.491

Mundial, B.: World Development Indicators (2022). https://databank.worldbank.org/reports.aspx?source=2&country=WLD

Barros, M., Barros, J.: Digital Literacy and Ict in Learning and Inclusion - Ecuador, pp. 97–118 (2020). https://doi.org/10.24917/9788395373732.5

Cabero-Almenara, J., Palacios-Rodríguez, A.: Digital competence of educators: identification of levels and training needs. Educ. Sci. **10**(3), 86 (2020). https://doi.org/10.3390/educsci10030086

Caragliu, A., Del Bo, C.: Regional institutions and the urban digital divide. Pap. Reg. Sci. **104**(6) (2025). https://doi.org/10.1016/j.pirs.2025.100118

Erazo-Luzuriaga, A.F.: Integración de las TICs en el aula: Un análisis de su impacto en el rendimiento académico. Revista Científica Zambos. **3**(1), 56–72 (2024). https://doi.org/10.69484/rcz/v3/n1/12

Fondo de las Naciones Unidas para la Infancia: La brecha digital impacta en la educación (2021). https://www.oecd.org/en/publications/digital-strategies-in-education-across-oecd-countries_33dd4c26-en.html

Guamaní, S., Reyes, V., et al.: Is the use of ICTs related to academic performance in technological education? A case study in the rural sector. Prohominum. Revista de Ciencias Sociales y Humanas. **6**(4), 192–203 (2024). https://doi.org/10.47606/acven/ph0295

Instituto Nacional de Estadística y Censos: Encuesta Nacional de Empleo, Desempleo y Subempleo (ENEMDU). INEC (2023a). https://www.ecuadorencifras.gob.ec/enemdu-historico-empleo-2023/

Instituto Nacional de Estadística y Censos: Encuesta nacional de empleo, desempleo y subempleo (ENEMDU) (2023b). https://www.ecuadorencifras.gob.ec/documentos/web-inec/EMPLEO/2023/Trimestre_I/2023_I_Trimestre_Mercado_Laboral.pdf

Ministerio de Educación del Ecuador: Estadística Educativa Vol. 3-Datos Abiertos (2022). https://educacion.gob.ec/wp-content/uploads/downloads/2022/12/Estadistica-Educativa-Vol3.pdf

Organización de las Naciones Unidas para la Educación: la Ciencia y la Cultura: Education in the time of COVID-19 (2020). https://unesdoc.unesco.org/ark:/48223/pf0000374075_eng

Organización de las Naciones Unidas para la Educación: la Ciencia y la Cultura: Informe de seguimiento de la educación en el mundo, 2023: tecnología en la educación: ¿una herramienta en los términos de quién? (2023). https://unesdoc.unesco.org/ark:/48223/pf0000388894?posInSet=1&queryId=c32b4ce5-c9db-4e0e-b49f-f514238bc551

Pérez, J., Torres, L.: Proposal of a model for the analysis of the state of the use of ICT in education in Ecuador. **11**(7), 112 (2023). https://doi.org/10.3390/computers11070112

Plaza de la Hoz, J., Espinosa Zárate, Z., Camilli Trujillo, C.: Digitalisation and poverty in Latin America: a theoretical review with a focus on education. Humanit. Soc. Sci. Commun. **11**, 1194 (2024). https://doi.org/10.1057/s41599-024-03692-0

Ramírez, M., López, F.: Brechas digitales y equidad educativa en Ecuador. Universidad Central del Ecuador, Quito (2020) https://repositorio.uce.edu.ec/handle/25000/21015

Van der Vlies, R.: Digital strategies in education across OECD countries: exploring education policies on digital technologies. OECD Educ. Work. Pap. **226** (2020). https://doi.org/10.1787/33dd4c26-en

Patterns of Technology Utilization and Academic Dishonesty in Secondary Education

Juan Carlos Torres-Diaz[1], Pablo Vicente Torres-Carrión[1], Mónica Hinojosa Becerra[2], and Diana Rivera-Rogel[3](✉)

[1] Faculty of Engineering and Architecture, Universidad Técnica Particular de Loja, Loja CP 11-01-608, Ecuador

[2] Department of Audiovisual Communication and Advertising, Universidad de Málaga, Málaga, España

[3] Faculty of Social Sciences, Education and Humanities, Universidad Técnica Particular de Loja, Loja CP 11-01-608, Ecuador

derivera@utpl.edu.ec

Abstract. Plagiarism requires educational strategies that promote comprehensive and ethical practices. This study examines the correlation between technology usage and plagiarism among a sample of 6,396 students, aged 14 to 18, across 52 schools in Ecuador. The participants were divided into two cohorts: first-year high school students, consisting of 3,340 respondents, and last-year high school students, comprising 3,056 respondents. Technology usage was assessed by categorizing students into clusters based on their preferences, and these clusters were then analyzed in relation to levels of plagiarism using a logistic regression model. The findings reveal that first-year high school students exhibit a higher tendency to engage in plagiarism when they dedicate more time to activities such as social networking, academics, and entertainment. However, this association diminishes as students' progress to their senior year. The only consistent correlation across both groups pertains to the level of trust students place in internet-based information and tools.

Keywords: Plagiarism · Internet · Secondary School · Education · Technology

1 Introduction

Academic integrity constitutes the ethical foundation that guides teaching and research practices, grounded in values such as honesty, fairness, and responsibility (ICAI, 2021; Eaton, 2023). These principles foster trust within educational communities and ensure the credibility of knowledge creation and use. Plagiarism, understood as the improper appropriation of ideas or expressions without acknowledgment, represents a transgression against academic integrity; however, its perception and characterization vary across educational contexts and individual interpretations (Krou et al., 2021).

Traditionally, most studies on academic plagiarism have focused on higher education and university settings, including research from Latin America (Torres-Díaz, Duart &

A. Rocha et al. (Eds.): ICITS 2026, LNNS 1966, pp. 156–168, 2027.
https://doi.org/10.1007/978-3-032-27538-7_14

Hinojosa, 2018; Amida, Appianing & Marafa, 2022; Waltzer & Dahl, 2022). These works have identified factors such as academic experience, access to digital resources, and institutional policies as determinants of dishonest practices. Nevertheless, the analysis of behaviors and perceptions of plagiarism among secondary school students remains limited, despite this stage being crucial for developing ethical and technological habits that shape future university conduct (ENAI, 2018).

In Ecuador, internet access has expanded steadily yet unevenly, transforming learning environments by intertwining educational, recreational, and social uses of technology (Newton, 2018). This situation generates new ethical tensions, especially among adolescents, for whom distinguishing the boundaries between entertainment and learning can be challenging.

To interpret behaviors related to self-reported plagiarism, this study employs the Theory of Planned Behavior (TPB), which posits that attitudes, subjective norms, and perceived behavioral control determine the intention to perform an action (Uzun & Kilis, 2020). Within this framework, time spent online is conceived as an indicator of opportunity and control, while trust in digital information reflects attitudes toward the validity and ethical use of sources.

The aim of this study is to analyze patterns of technological use and their relationship with self-perceived plagiarism among Ecuadorian secondary school students. The research questions are:

P1. Are there significant differences in self-reported plagiarism behaviors between first and third-year high school students?

P2. How are time spent online and trust in digital information associated with the probability of engaging in self-perceived plagiarism?

2 Theoretical Framework: Academic Integrity and Technology

2.1 Academic Integrity and Student Behavior

Academic integrity represents a set of principles and values that guide intellectual work in education. Its analysis extends beyond the formal compliance with institutional norms, encompassing the cognitive, social, and technological processes that influence students' ethical decision-making. Recent research suggests that student plagiarism often does not stem from malicious intent but rather from misconceptions about authorship boundaries, academic collaboration, and legitimate use of digital resources (Dhillon & Murray, 2021; Stone, 2023). These perceptions vary significantly depending on the level of digital literacy, individual motivation, and external performance pressures.

In this context, self-perceived plagiarism gains relevance as an indicator of the beliefs, attitudes, and justifications that students attribute to their own practices. According to Vera-San-Miguel, Núñez-Lezcano, and Sanabria-Báez (2023), students' perceptions of plagiarism are closely linked to prior ethical formation and experience in digital environments. Self-reports—unlike digital detection records—allow for the identification of attitudinal and cognitive patterns underlying dishonest academic behavior, thus providing a deeper and more nuanced understanding of the phenomenon.

This behavioral analysis can be enriched through explanatory theories such as the TPB (Ajzen, 1991). The TPB asserts that intentional action results from the interaction of three essential components: attitudes toward the behavior, perceived social norms, and perceived behavioral control. In the digital domain, trust placed in online information can be understood as a favorable attitude toward technology use, while time spent navigating the internet reflects both opportunity and perceived control (Uzun & Kilis, 2020).

In parallel, the Technology Acceptance Model TAM (Davis, 1989) emphasizes that the perceived usefulness and ease of use of technologies directly influence their adoption. When the internet is perceived as a tool that facilitates task completion and enhances academic performance, students may develop more permissive attitudes toward the unethical use of digital content—a phenomenon corroborated by studies in higher education (Torres-Díaz, Duart & Hinojosa, 2018; Waltzer & Dahl, 2022).

Integrating TPB and TAM proves particularly useful in analyzing plagiarism as the outcome of interactions among personal motivations, social influences, and technological competencies.

2.2 Technology and Plagiarism Detection

The rapid development of digital technologies has transformed academic oversight processes, enabling increasingly sophisticated methods for detecting plagiarism (Rejeb et al., 2024; Gill et al., 2024). Current tools, supported by artificial intelligence and machine learning, efficiently compare texts and minimize detection errors using strategies such as semantic cleaning and lexical or semantic similarity analysis (Alromema & Hezzam, 2020; Hu, Zhang & Zhang, 2020; Chang et al., 2021). In addition, systems such as BPlag have been designed to detect behavioral similarity in source code, while deep learning-based solutions have achieved high precision levels (Cheers, Lin & Smith, 2021; Hambi & Benabbou, 2020).

However, technology also has limitations: automated detection of textual matches does not always distinguish between intentional plagiarism and legitimate information use. Several authors emphasize the need for human interpretation and supervision to ensure the ethical validity of the results produced by these systems (Foltýnek et al., 2020). Therefore, this study focuses not only on algorithmic aspects but also on exploring patterns of internet use—such as time spent online and trust in digital sources—in relation to self-perceived plagiarism.

In summary, this theoretical framework articulates two complementary dimensions: the behavioral analysis of the attitudes and beliefs that predispose individuals to plagiarism, and the technological dimension concerning the digital tools used for its detection. Integrating both perspectives provides a robust foundation for empirical research and data interpretation, fostering a comprehensive understanding of the relationship between ethics, technology, and learning in contemporary educational contexts.

3 Materials and Methods

The primary objective of the study is twofold: firstly, to identify the underlying structures explaining students' technology usage and to elucidate them by describing profiles or user types; secondly, to ascertain whether disparities exist between first-year and final-year high school students concerning the impact of internet usage on plagiarism.

Data were collected from 52 schools nationwide. The first group comprised first-year high school students, totaling 3,340 surveys, while the second group consisted of final-year high school students, with a total of 3,056 surveys. Categories were established in both groups through principal component analysis, followed by cluster analysis applied to these results (factors).

Thirteen variables were utilized for principal component analysis (refer to Table 2), with resulting scores for each factor treated as variables (factors) used as input for subsequent cluster analysis. In both groups, five categories were identified based on perceptions regarding information and internet tools, internet usage duration, and years of internet experience. For the first-year high school group, resulting categories included: dedicated, entertainment-focused, careless, trusting, and distrustful. Similarly, the final-year high school group exhibited categories such as dedicated, trusting, distrustful, careless, and entertainment-focused.

To determine existing relationships, chi-squared tests were conducted between plagiarism and variables such as student type (cluster) and gender, revealing significant associations. These findings were further verified through a binary logistic regression model, where the dependent variable was plagiarism (yes, no), and the independent variables were the student's cluster membership (five groups) and gender (male, female).

The study was conducted in accordance with the ethical principles of educational research. Participation was voluntary and anonymous, and informed consent was obtained from students, their legal guardians, and the participating educational institutions. Data were collected and analyzed in aggregate form, and no personally identifiable information was requested from any participant. All data were treated with strict confidentiality, and their use was limited exclusively to academic and scientific purposes.

4 Results and Discussion

4.1 Description of the Sample

The sample comprises 6396 students aged 14 to 18, with a gender distribution of 50.7% male and 49.3% female. A notable imbalance exists between groups: in the first year, males predominate with 54.3% (1813) versus 45.7% females (1527), while in the final year this reverse, with 46.7% males (1427) and 53.3% females (1629), making males the minority.

Socioeconomic data show that 60% of students come from the two lowest income levels, highlighting concentration in disadvantaged households. Regarding technology access, 16.3% lack a mobile phone—a notable finding given age and income—while 75.6% have internet at home, indicating substantial but incomplete connectivity.

The dependent variable, plagiarism, is detailed in Table 1. Only 11% of students in both groups report no plagiarism, with many showing higher levels, underscoring its prevalence in the sample.

Table 1. Distribution of plagiarism levels.

Plagiarism level	Frequency	Percentage
0	703	11
1	372	5.8
2	490	7.7
3	493	7.7
4	426	6.7
5	631	9.9
6	479	7.5
7	697	10.9
8	607	9.5
9	1498	23.4
Total	6396	100

4.2 First Year High School Group

We constructed a classification for first-year high school students in two phases. First, principal component analysis was applied to condense the data, producing factors later used as input for cluster analysis. The original variables, consistent for both first and third-year groups, appear in the first column of Table 2. Three factors were identified, explaining 51.85% of the variance.

Table 2. Scores of the variables associated with each factor.

Variables	Factors		
	Factor 1 Trust	Factor 2 Time on Line	Factor 3 Entertainment
Iinternet_facilitatesLearning	0.782		
Internet_doesWithoutLibrary	0.749		
Internet_reliesOnInformation	0.741		
Internet_improveScores	0.694		
Internet_jobsFast	0.666		
hours_social_network_phone		0.792	
hours_connect		0.782	
fun_phone_hours		0.773	
internet_days		0.579	
subjects_phone_hours		0.435	

(*continued*)

Table 2. *(continued)*

Variables	Factors		
	Factor 1 Trust	Factor 2 Time on Line	Factor 3 Entertainment
hours_games_phone_hours			0.831
video_phone_hours			0.706
connect_years			0.440

Factor 1 encompasses the variables listed in column 1 (Refer to Table 3); Factor 2 comprises the variables presented in column 2, and so forth. The inherent characteristics of the variables within each factor enable the assignment of descriptive labels to them:

- Factor 1: Trust
- Factor 2: Time on Line
- Factor 3: Entertainment

Each factor was treated as a variable, resulting in three variables utilized as input data for cluster analysis. This analysis generated five distinct groups, which are depicted in the following figure (Fig. 1):

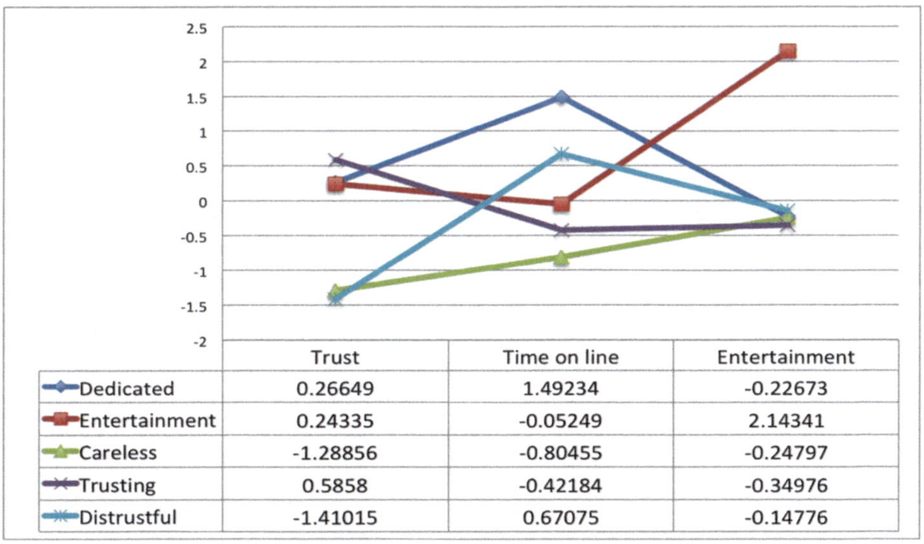

Fig. 1. Cluster's graph of the high school first year students.

Group one (Dedicated) is characterized by spending more time on the phone for leisure, social networking, and studying subjects, with longer connection durations.

Group two (Entertainment) is distinguished by a greater inclination towards watching videos and playing video games, and it boasts the longest tenure as users.

Group three (Careless) is the cluster that allocates the least time on the phone for leisure, social networking, and studying subjects; it can also be noted as the group with the shortest online presence.

Group four (Trusting) exhibits the highest level of trust in internet-based information and tools overall.

Group five (Distrustful) displays the lowest confidence in internet-derived information.

Each of these clusters was encoded as a dichotomous dummy variable, where a value of 1 denotes membership in the cluster, and 0 denotes non-membership.

A variable was created for each student category.

To ascertain the relationships between these categories and plagiarism, the chi-square test was employed, revealing significant associations. Gender of the student was also included in the analysis (Refer to Table 3).

Table 3. X2 test of the plagiarism variable with the independent variables.

Dependent variables	Chi square Value	Sig
Dedicated	6,921	0,009
Entertainment	9,47	0,002
Less Dedication	43,2	0,00
Confident	10,2	0,01
Distrustful	15,7	0,00
Gender	8,9	0,003

To examine these relationships, a statistical model was developed using binomial logistic regression. In this model, plagiarism serves as the dependent variable (either does not plagiarize or does plagiarize), while the independent variables consist of the five categories and the student's gender. The outcomes of this analysis are presented in Table 4.

Table 4. Logistic regression first year students.

		B	E.T	Wald	gl	Sig.	Exp(B)
Variables	Dedicated	−.835	.199	17.678	1	.000	.434
	Entertainment	−.983	.234	17.717	1	.000	.374
	Trusted	−.672	.161	17.418	1	.000	.511
	Gender	.261	.106	6.111	1	.013	1.298
	Consistent	3.642	.503	52.383	1	.000	38.167

The outcomes of the model reveal that the Odds Ratio (OR) of being part of the group that does not engage in plagiarism decreases by 0.434 if the student belongs to the dedicated group. This suggests that students who devote more time to leisure activities, social networks, and online engagement are less inclined to belong to the non-plagiarizing group.

Similarly, the OR of belonging to the non-plagiarizing group decreases by 0.374 for students in the entertainment group, indicating that those who spend more time watching online videos, playing video games, and have a longer online history are less likely to be part of the group that refrains from plagiarism.

Furthermore, the OR of belonging to the non-plagiarizing group decreases by 0.511 for students in the trusting group, suggesting that individuals who place trust in internet-based information are less likely to belong to the non-plagiarizing group.

Lastly, the likelihood of belonging to the group that abstains from plagiarism increases by 1.298 times for female students.

These results suggest that students who dedicate more time to online activities particularly social networking, entertainment, and academic searching tend to exhibit higher levels of self-perceived plagiarism. From the perspective of the Theory of Planned Behavior (Ajzen, 1991; Uzun & Kilis, 2020), these behaviors reflect a combination of positive attitudes toward digital resources and low perceived behavioral control regarding the proper use of information. The more students trust online content and perceive it as a legitimate shortcut to accomplish school tasks, the more likely they are to rationalize plagiarism as acceptable. This aligns with findings by Dhillon and Murray (2021) and Stone (2023), who emphasize that digital autonomy without ethical guidance can foster unintentional academic misconduct.

4.3 Third Year of High School Group

To contrast these findings, a second analysis was conducted with third-year high school students, following the same methodological procedure. This comparison allows evaluating whether the observed associations between technology use and plagiarism persist as students advance in age and educational experience.

The second group we analyzed consists of students in the third year of high school. The methodology applied to this group mirrors that of the first group. Specifically, principal component analysis was conducted, resulting in the identification of three factors that collectively explain 53% of the variance (Refer to Table 5).

Table 5. Table captions should be placed above the tables.

Variables	Factors		
	Trust	Time on Line	Entertainment
Internet_facilitatesLearning	0.799		
Internet_trustOnInfomation	0.768		
Internet_doesWithoutLibrary	0.763		
Internet_improveSkills	0.69		
Internet_WorksFast	0.686		

(continued)

Table 5. (*continued*)

Variables	Factors		
	Trust	Time on Line	Entertainment
hours_connect		0.813	
hours_social_network_phone		0.796	
fun_phone_hours		0.754	
phone_hours_subjects		0.516	
internet_days		0.482	
connect_years		0.408	
game_phone_hours			0.849
video_phone_hours			0.724

Each factor aggregates the variables listed in Table 5, following the same methodology used for first-year students, and their descriptions remain consistent: Factor 1 (Trust), Factor 2 (Time Online), and Factor 3 (Entertainment). Cluster analysis was also applied, with results shown in Fig. 2.

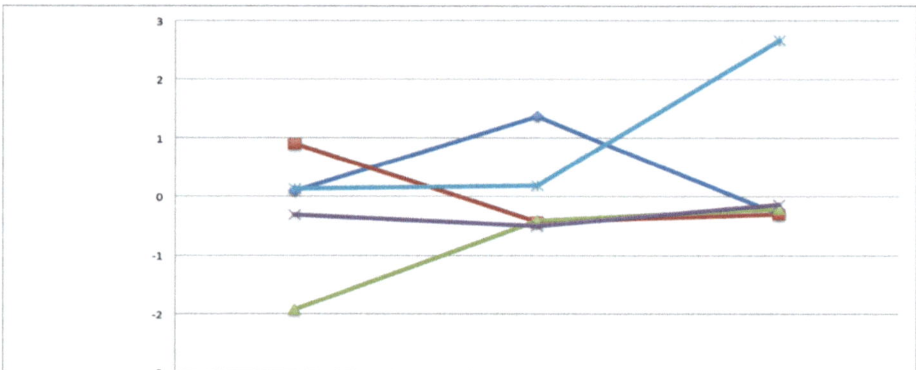

	Trust	Time on line	Entertainment
Dedicated	0.09511	1.37813	-0.29676
Trusting	0.90665	-0.42956	-0.30151
Distrustful	-1.92959	-0.41186	-0.21174
Careless	-0.31012	-0.49967	-0.13224
Entertainment	0.1326	0.19073	2.67241

Fig. 2. Cluster's graph of the high school last year students.

Group 1(Dedicated) is characterized by a high level of online time.
Group 2 (Entertainment) is characterized by the highest level of time spent on entertainment.
Group 3 (Careless) is characterized by having the lowest level of time online
Group 4 (Trusting) is characterized by the highest level of trust.

Group 5 (Distrustful) is characterized by having the lowest level of confidence.

"Dummy" variables were generated to represent the presence or absence of each student within the respective category, thereby depicting the student cluster. Subsequently, these variables, along with gender, were examined in relation to the level of plagiarism, initially utilizing the X2 statistic for this analysis (Refer to Table 6).

Table 6. X2 of the independent variables and the level of plagiarism.

Dependent variables	Chi square Value	Sig
Dedicated	2,89	0,089
Trust	8,9	0,003
Distrustful	20,4	0,000
Careless	1,79	0,180
Entertained	0,350	0,554
Gender	0,088	0,767

A relationship is only evident within the profiles associated with trust. To provide a more comprehensive explanation of the relationships between variables, a regression model was constructed (Table 7).

Table 7. Logistic regression last year students.

	B	E.t	Wald	gl	Sig.	Exp(B)
Distrustful	.760	.277	7.507	1	.006	2.138
Constant	1.569	.748	4.397	1	.036	4.802

The results reveal that the Odds Ratio (OR) of belonging to the group that abstains from plagiarism increases by 2.138 times when the student belongs to the Distrustful group. Belonging to other groups or the student's gender shows no significant effect.

Compared with first-year students, third-year participants display a weaker relation between technological behavior and self-perceived plagiarism, suggesting ethical and digital maturity, where older students develop stronger self-regulation and awareness of academic norms. According to the TAM (Davis, 1989), experience reduces the novelty and perceived utility of digital tools as shortcuts, leading to more selective use. Students who remain distrustful of online information consistently report lower plagiarism levels, indicating that critical digital literacy protects against academic dishonesty (Torres-Díaz et al., 2018; Vera-San-Miguel et al., 2023).

Among first-year students, plagiarism correlates with academic technology use, those using it longer tend not to plagiarize, though this pattern vanishes by the third year. Findings align with Torres-Díaz et al. (2018) and others (Amida et al., 2022; Waltzer & Dahl, 2022), where e-book use predicts plagiarism. First-year women plagiarize less than men, but the gap narrows by the third year, echoing Torres-Díaz et al. (2018), who found age did not significantly affect plagiarism. The cause of early-year differences remains unclear.

Relationships evident in the first year time spent on academic, social, or leisure activities via mobile devices, diminish by the third year, and the reasons for their disappearance remain unknown. A similar pattern occurs among students spending more time on videos or games: they plagiarize more initially, but less by the third year, consistent with Waltzer & Dahl (2022), who found information literacy predicts plagiarism.

Taken together, these findings highlight a developmental transition in students' technological and ethical competences. During early high school years, plagiarism appears linked to exploratory or instrumental uses of technology, where digital engagement is driven by curiosity rather than ethical awareness. By the final year, such associations fade as learners internalize academic norms and distinguish between legitimate and dishonest practices. This evolution reinforces the importance of embedding integrity education and digital ethics training from early stages of secondary schooling, ensuring that technological proficiency is accompanied by moral reasoning and responsible information use (Eaton, 2023).

A pattern observed in first year high school students and that persists with third year students has to do with the relationship between distrust of web information and the tendency not to plagiarize; that is, distrustful students tend to plagiarize less, this being the only relationship that does not change over time.

5 Conclusion

The impact of gender on plagiarism levels seems to diminish over time. Although this influence is notable among first-year high school students, as females tend to plagiarize less, it disappears when students reach the final year of high school.

Regarding technological behavior, students who devote longer time to online activities—especially for entertainment or academic purposes—show a higher likelihood of plagiarism in early secondary education. This pattern can be interpreted, according to the TPB (Ajzen, 1991; Uzun & Kilis, 2020), as the result of positive attitudes and weak perceived control toward the use of online resources. However, as students mature, these associations weaken, consistent with the TAM (Davis, 1989), which posits that experience reduces the perceived utility of technology as a shortcut for academic tasks.

Across both groups, trust in digital information emerges as the most consistent predictor: students who critically question the reliability of online sources are less likely to engage in plagiarism. This finding emphasizes the protective role of critical digital literacy in preventing academic dishonesty (2018, Vera-San-Miguel et al., 2023).

These results underscore the need to strengthen digital ethics education in secondary schools. Integrating explicit instruction on information evaluation, citation practices, and the responsible use of technology may foster more sustainable academic integrity behaviors. Future research could explore the influence of contextual variables—such as teacher modeling, institutional culture, and socioeconomic background—on the development of ethical digital competences.

References

Ajzen, I.: The theory of planned behavior. Organ. Behav. Hum. Decis. Process. **50**(2), 179–211 (1991. [Online]. Available:). https://doi.org/10.1016/0749-5978(91)90020-T

Alromema, W., Hezzam, E.A.: Towards the enhancement of text plagiarism detection effectiveness: experimental study. Int. J. Adv. Trends Comput. Sci. Eng. **9**(4), 5555–5559 (2020). https://doi.org/10.30534/ijatcse/2020/201942020

Amida, A., Appianing, J., Marafa, Y.A.: Testing the predictors of college students' attitudes toward plagiarism. J Acad Ethics. **20**, 85–99 (2022). https://doi.org/10.1007/s10805-021-09401-9

Chang, C.-Y., Lee, S.-J., Wu, C.-H., Liu, C.-F., Liu, C.-K.: Using word semantic concepts for plagiarism detection in text documents. Inf. Retr. J. **24**(4), 298–321 (2021). https://doi.org/10.1007/s10791-021-09394-4

Cheers, H., Lin, Y., Smith, S.P.: Academic source code plagiarism detection by measuring program behavioral similarity. IEEE Access. **9**, 50391–50412 (2021). https://doi.org/10.1109/ACCESS.2021.3069367

Davis, F.D.: Perceived usefulness, perceived ease of use, and user acceptance of information technology. MIS Q. **13**(3), 319–340 (1989). https://doi.org/10.2307/249008

Dhillon, S., Murray, N.: An investigation of EAP teachers' views and experiences of E-learning technology. Educ. Sci. **11**(2), 1–16 (2021). https://doi.org/10.3390/educsci11020054

Eaton, S.E.: Postplagiarism: transdisciplinary ethics and integrity in the age of artificial intelligence and neurotechnology. Int. J. Educ. Integr. **19**(1) (2023). https://doi.org/10.1007/s40979-023-00144-1

ENAI: Glosario de Integridad Académica. Proyecto Erasmus+. (2018) https://www.academicintegrity.eu/wp/wp-content/uploads/2023/02/ES-Glosario-de-Integridad-Academica.pdf

Foltýnek, T., et al.: Testing of support tools for plagiarism detection. Int. J. Educ. Technol. High. Educ. **17**(1) (2020). https://doi.org/10.1186/s41239-020-00192-4

Gill, S.S. et al.: Transformative effects of ChatGPT on modern education: emerging era of AI Chatbots. Internet Things Cyber-Phys. Syst. **4**, 19–23 (2024). https://doi.org/10.1016/j.iotcps.2023.06.002

Hambi, E. M., & Benabbou, F.: A new online plagiarism detection system based on deep learning. (IJACSA) Int. J. Adv. Comput. Sci. Appl., 11(9), 470–478. (2020). doi:https://doi.org/10.14569/IJACSA.2020.0110956

Hu, B., Zhang, J., Zhang, L.: A plagiarism detection approach for Chinese documents based on semantic textual similarity. J. Shenzhen Univ. Sci. Eng. **37**, 107–111 (2020). https://doi.org/10.3724/SP.J.1249.2020.99107

ICAI.: The Fundamental Values of Academic Integrity, 3rd ed., [Online]. Available: https://academicintegrity.org/wp-content/uploads/2022/05/ICAI-Fundamental-Values-2021.pdf (2021)

Krou, M.R., Fong, C.J., Hoff, M.A.: Achievement motivation and academic dishonesty: a meta-analytic investigation. Educ. Psychol. Rev. **33**, 427–458 (2021). https://doi.org/10.1007/s10648-020-09557-7

Newton, P.M.: How common is commercial contract cheating in higher education and is it increasing? a systematic review. Front. Educ. **3**, 67 (2018). https://doi.org/10.3389/feduc.2018.00067

Rejeb, A., Rejeb, K., Appolloni, A., Treiblmaier, H., Iranmanesh, M.: Exploring the impact of ChatGPT on education: a web mining and machine learning approach. Int. J. Manag. Educ. **22**(1), 100932 (2024). https://doi.org/10.1016/j.ijme.2024.100932

Stone, A.: Student perceptions of academic integrity: a qualitative study of understanding, consequences, and impact. J. Acad. Ethics. **21**(3), 357–375 (2023). https://doi.org/10.1007/s10805-022-09461-5

Torres Diaz, J.C., Duart, J.M., Hinojosa Becerra, M.: Plagiarism, internet and academic success at the university. NAER. **7**(2), 98–104 (2018). https://doi.org/10.7821/naer.2018.7.324
Uzun, A.M., Kilis, S.: Investigating antecedents of plagiarism using extended theory of planned behavior. Comput. Educ. **144** (2020). https://doi.org/10.1016/j.compedu.2019.103700
Vera-San-Miguel, A., Núñez-Lezcano, F., Sanabria-Báez, G.: Actitudes frente al plagio académico en estudiantes de medicina de universidades privadas y públicas. Medicina Clínica y Social. **7**(2), 70–76 (2023). https://doi.org/10.52379/mcs.v7i2.287
Waltzer, T., Dahl, A.: Why do students cheat? Perceptions, evaluations, and motivations. Ethics Behav. **33**(2), 130–150 (2022). https://doi.org/10.1080/10508422.2022.2026775

Applying a Quantitative Method to Analyze the Cybersecurity Risk a Portuguese Hospital

Ana Luísa Teixeira[1], Ana Patrícia Borges[1], Dário Carreira[2], Victor Costa[3], João Pavão[4], and Nelson Pacheco Rocha[5](✉)

[1] Science and Technology School, University of Trás-os-Montes and Alto Douro, Vila Real, Portugal

[2] Instituto Superior da Maia, Avenida Carlos Campos Maia, Maia 4475, Portugal
dariocarreira@netcabo.pt

[3] Centro Hospitalar de Trás-os-Montes e Alto Douro, Av. da Noruega, Vila Real 5000-508, Portugal
vcosta@chtmad.min-saude.pt

[4] Science and Technology School, University of Trás-os-Montes and Alto Douro INESC-TEC, Vila Real, Portugal
jpavao@utad.pt

[5] Department of Medical Sciences, University of Aveiro IEETA, Aveiro, Portugal
npr@ua.pt

Abstract. Cyber resilience models aim to mitigate the security risks of information systems due to a wide range of security threats. This article reports on the application of a quantitative model, that is the Factor Analysis of Information Risk, to analyze the cybersecurity risks of two departments of a Portuguese hospital, where information systems are essential: the medical imaging and clinical pathology departments. The results show that FAIR model allows quantifying cybersecurity risks and their impacts, to prioritize measures to improve hospital's cyber resilience.

Keywords: Healthcare · Medical imaging · Clinical pathology · Cyber resilience · Cybersecurity · Factor Analysis of Information Risk · FAIR

1 Introduction

The healthcare sector, due to the adoption of digital technologies [1–6], is a particularly vulnerable sector and a constant target for cybersecurity-related events (e.g., phishing, denial of service, malware, or destructive software) [7–11]. It is known that, in 2019, 24% of cyberattacks targeted healthcare services [12]. In particular, the increased connectivity of medical devices is considered one of the greatest vulnerabilities of the healthcare sector, as direct attacks can put at risk patients' safety and, additionally, can have a very significant impact on organizations in terms of financial and reputational losses [1].

Cybersecurity comprises a collection of tools, policies, concepts, protection agents, guidelines, approaches, actions, training, practices, controls and technologies that might

A. Rocha et al. (Eds.): ICITS 2026, LNNS 1966, pp. 169–180, 2027.
https://doi.org/10.1007/978-3-032-27538-7_15

be used to protect cyberspace and their assets [13]. Because it is impossible to prevent all cyberattacks, whether due to their diversity and sophistication or a lack of adequate resources, it is widely accepted that implementing cybersecurity measures alone is not sufficient to address the problems caused by successful attacks. Organizations must also develop integrated plans to achieve cyber resilience, meaning the ability to fully or at least partially recover from cyberattacks. This has led to growing interest in cyber resilience across various sectors [14, 15], including the healthcare sector [16–18].

Cyber resilience is defined by the National Institute of Standards and Technology (NIST) as the ability to anticipate, resist, recover from, and adapt to adverse conditions, stresses, attacks, or compromises [19]. To help organizations improve their cyber resilience, NIST proposed the NIST Cybersecurity Framework (NIST-CSF) [19], which is based on recognized and effective practices from other models and references and provides phased processes for identifying, analyzing, evaluating, and prioritizing risks to minimize the impact of potential threats.

In Portugal, based on the NIST-CSF, the National Cybersecurity Reference Framework was created to enable Portuguese organizations to reduce the risk associated with cyberthreats. Moreover, the Centro Nacional de Cibersegurança (National Cybersecurity Center) released the Capability Assessment Framework, which defines three capability levels for different cybersecurity measures, enabling organizations to implement them according to their contexts and size [20].

Since risk analysis is considered a fundamental component of cyber resilience, the ISO 27005 standard [21] should be taken into account, since it systematizes best practices for information security risk management through a systematic and modular approach.

Certain approaches to risk analysis present risk matrices representing results obtained through qualitative analysis [22], such as risk matrices based on color maps (e.g., green, yellow, red), descriptions, or scales (e.g., 1 to 5). Unlike qualitative risk assessment, quantitative approaches provide results in value ranges that allow for more objective risk quantification, a better understanding of the level of risk and its impact, and, consequently, more assertive decision-making [23]. The Factor Analysis of Information Risk (FAIR) model [24] is an example of quantitative risk analysis, providing a comprehensive framework for the perception, analysis, and quantification of risk in terms of financial impacts, which allows organizations to aggregate, compare, and prioritize the measures to be implemented and ensure that decisions are based on objective parameters [23].

In this context, the main objective of this study was to evaluate the applicability of quantitative risk analysis techniques based on the FAIR model in a Portuguese public hospital within the National Health Service, focusing specifically on two departments where information systems are essential: the medical imaging and clinical pathology departments.

2 Methods

The FAIR model is based on an analysis of objective data, considering essential risk factors, in order to avoid unclear matrices and scales. It contemplates several steps to determine the probable frequency and magnitude of future losses (i.e., Annualized Loss Exposure and Loss Exceedance Curve) from the estimation of Loss Event Frequency (i.e., the probable frequency, during a given time interval, for the loss to materialize as a result of an action by a threat agent) and Loss Magnitude (i.e., probable magnitude of losses that result from the occurrence of a threat) [24].

Loss Event Frequency and Loss Magnitude can be estimated from lower-level risk factors such as Threat Event Frequency (i.e., the likely frequency, within a given period, that threat actors will act in a manner that could result in loss) and Vulnerability (i.e., the probability of an action taken by a threat actor that results in loss), considering specific risk scenarios.

Thus, to determine potential Loss Event Frequency and Loss Magnitude values for the medical imaging and clinical pathology departments of a Portuguese public hospital, four possible threat scenarios were defined, two for each department. The Loss Event Frequency for each scenario was estimated directly, considering the types of threats and the specific vulnerabilities associated with each scenario.

In turn, to determine the Loss Magnitude, it was necessary to define the types of losses, which, according to the FAIR model, are primary and secondary losses. Primary losses correspond to direct losses that materialize as a result of a threat event affecting primary stakeholders, while secondary losses correspond to losses that involve the negative reactions of secondary stakeholders to the primary losses.

To calculate primary losses, the following factors were considered:

- Productivity. Costs associated with the non-operational nature of assets affected by a successful attack.
- Replacement. Costs associated with replacing assets affected by a realized by a successful attack.
- Response. Costs associated with managing the loss event, including logistical expenses to the person-hours required to handle the event. These expenses should include the costs of internal human resources required to manage an incident (e.g., representatives from management of the hospital and the affected departments and representatives from the information technologies and human resources departments) and investigation costs by an external specialized team to collect as much evidence as possible to eventually identify those responsible and recover compromised information.

In turn, in terms of secondary losses, the following factors were considered:

- Response. Secondary response costs include notification costs (to patients or suppliers), costs associated with conducting a response strategy, legal costs, and public relations costs. The costs of personnel involved in secondary response are calculated similarly to those used to determine the costs of internal personnel involved in primary response.
- Competitive advantage. Losses resulting from the impossibility of using assets that differentiate the organization from its potential competitors.
- Legal actions' costs. Expenses arising from legal or regulatory actions faced by the organization. These include fines resulting from personal data breaches under the General Data Protection Regulation (GDPR), as well as potential claims for damages and litigation-related costs.

- Reputation. Losses associated with the external perception that an organization's value proposition has been diminished, potentially leading, for example, to employee departures.

Once the Loss Event Frequency and Loss Magnitude were determined for each scenario, the FAIR-U application [25] provided by the FAIR Institute was used to conduct Monte Carlo simulations considering random values for the unknown variables [24]. This application allows determining the Annualized Loss Exposure, that is, the potential loss that an organization faces each time it makes a decision that implies monetary expenditures (e.g., implementing a new process within the organization or acquiring additional hardware), and the Loss Exceedance Curve, that is, the probability of a loss event exceeding its expected value.

3 Results

3.1 Scenarios

Based on the characteristics of the medical imaging and clinical pathology departments, four scenarios were developed (i.e., two scenarios for each department).

Medical Imaging Department. The medical imaging department that was the subject of this study has a total of 82 employees and, annually, performs an average of approximately 760,000 medical imaging studies, using a variety of diagnostic equipment in terms of radiology, radioscopy, mammography, ultrasound, computerized axial tomography and magnetic resonance imaging.

It is important to mention that much of this equipment is not physically in the department, following a provisional strategy for carrying out imaging studies in other clinical departments where it is necessary to provide patient care.

In addition to all the medical imaging equipment mentioned above, the department has also several computers running the Windows operating system, a set of proprietary applications required for producing and viewing medical images, and a set of printers. All these devices are interconnected through the hospital's internal network, and their maintenance is the responsibility of the information technologies department. The same does not apply to diagnostic equipment, whose maintenance, whether on-site or remote, is the responsibility of its suppliers.

Based on the potential threats and vulnerabilities that were identified, two scenarios (i.e., S1 and S2) were developed relating to phishing (S1) and ransomware (S2).

In the phishing scenario (S1), an individual or group of individuals not belonging to the hospital organization, with an exclusive financial motive, carries out the threat. The attacker acts as a member of the organization, such as an IT department employee, and proceeds to send an email to the employees asking them to access a specific link to change their authentication credentials. Some employees, unaware of the sender's authenticity, innocently act as requested, allowing the attacker access to the information systems of the medical imaging department and, consequently, patient records. In the ransomware scenario (S2), the threat intends to deny access to the device that receives images produced by the computed tomography equipment. This medical imaging modality was selected because it was considered the most critical, as it must be constantly available. It was assumed that the malicious attack is carried out by a group of criminals not belonging to the organization, taking advantage of vulnerabilities of a software application for the visualization of computed tomography studies. Therefore, in this scenario, an attack is expected to deny the visualization of imaging studies until a ransom is paid, preventing the organization from performing imaging studies on the computed tomography equipment.

Clinical Pathology Department. The clinical pathology department selected for this study has 32 employees, including physicians, diagnostic and therapeutic technicians, and technical assistants. The technical assistants are responsible for receiving analyses from various hospital services, namely inpatient, outpatient, emergency, and outpatient departments.

The clinical pathology department operates 24 h a day and consists of three laboratories: i) biochemistry and hematology; ii) microbiology; and iii) immunology. Each laboratory is equipped with information systems and various medical diagnostic equipment, all connected to the hospital's communications network. For biological sample collection, the department has an area with collection stations. After collection, the samples are transported to the service's laboratories, where they are analyzed.

Two different scenarios were considered (i.e., S3 and S4), one in which the threat is related to the appropriation of sensitive patient information (i.e., S3) and another focused on the threat to physical assets, more specifically an attack on a diagnostic medical device (i.e., S4).

In scenario S3 (information access) it was considered improper access of information from the clinical pathology department, specifically patient information, for the purpose of undue profit. It was assumed that an employee could copy one hundred different records daily for six months before being detected. This number is possible if the employee is regularly alone and the record retrieval process is simple. In fact, despite the existence of different levels of access to clinical information, all employees have access to sensitive patient information. The data storage security policy prohibits the use of any removable data storage device, as USB device connections are blocked on all computers. However, information can be obtained through photographs, for example, using a smartphone. In scenario S4 (attack on a medical device), specifically a diagnostic medical device, a cybercriminal external to the department exploits a vulnerability in a diagnostic medical device, forcing it to shut down and, consequently, preventing exams from being performed. Sensitive information may also be accessed through the computer connected to the diagnostic device.

3.2 Loss Event Frequency

Table 1 presents the probable value and the minimum and maximum values for Loss Event Frequency values for the four scenarios. Additionally, we assumed a high secondary Loss Event Frequency (i.e., a high probability of primary loss events resulting in secondary stakeholder losses).

To determine the values in Table 1, we assumed that for scenario S1 (phishing), that the probability of occurring this type of event is high. Thus, we assume that this event occurs at least three times in four years (i.e., a frequency of 0.75) and at most four

Table 1. Loss Event Frequency values estimated for the four scenarios.

	Medical imaging		Clinical pathology	
	S1	S2	S3	S4
Probable value	1.50	0.25	0.75	0.20
Minimum	0.75	0.13	0.20	0.10
Maximum	4.00	2.00	2.00	3.00

times in a year (4.00). Adjusting for a more likely value, we obtained a frequency of 1.50 (corresponding to three times in two years). In turn, for scenario S2 (ransomware), we assumed that the probability of the threat occurring is once every eight years (i.e., approximately 0.13). The probable value is 0.25 (i.e., once every four years) and that the event can occur at most twice in a year (2.00). For scenario S3 (information access), we assume the event occurs at least once every five years (0.20), which is very likely to occur three times every four years (0.75) and has a maximum frequency twice a year (2.00). Finally, for scenario S4 (attack on a medical device), we assume the event occurs only once every five years (0.20). Since this is our most likely value, we assumed the minimum is once every ten years (0.10) and the maximum is three times a year (3.00).

3.3 Loss Magnitude

Table 2 presents the primary Loss Magnitude values for the different scenarios, while Table 3 presents the respective secondary Loss Magnitude values.

Since there is no equipment replacement, replacement losses are not applicable and the same applies to productivity losses in scenarios S1 (phishing) and S3 (information access), as these threats do not affect the activities of the medical imaging and clinical pathology departments or the operation of their assets.

In turn, in scenario S2 (ransomware), productivity losses occur due to the unavailability of medical imaging studies. To determine productivity losses in scenario S2 (ransomware), we assumed that access to computed tomography studies would be unavailable for one day. We estimated that 65 exams could be performed per day, a value that could vary from a minimum of 48 exams per day to 96 exams per day, and that the average cost of each computed tomography study is 143 €.

Finally, for scenario S4 (attack on a medical device), given that the clinical pathology department has a second system of the same model with the same configuration, productivity losses would be associated with the process of transferring clinical samples from the attacked system to the second system. However, if this is not possible, we assume the costs of lost revenue due to the equipment's inability to operate for a week.

Primary response costs are associated with the costs of internal and external human resources. For the different scenarios, the probable investigation time was subdivided into internal and external teams (i.e., 50% for each team) and the most likely value was considered to be 120 k€, with a potential range from 110 k€ to 130 k€.

Table 2. Primary Loss Magnitude.

	Medical imaging		Clinical pathology	
	S1	S2	S3	S4
Productivity				
Probable value	–	9 k€	–	14 k€
Min.–max.	–	7–14 k€		7 k€–4.8 M€
Replacement	–	–	–	–
Response				
Probable value	120 k€	120 k€	120 k€	120 k€
Min.–max.	110–130 k€	110–130 k€	110–130 k€	119–124 k€

Table 3. Secondary Loss Magnitude

	Medical imaging		Clinical pathology	
	S1	S2	S3	S4
Patients' notification	–	–	–	–
Suppliers' notification	–	–	–	–
Support calls	–	–	–	–
People in meetings				
Probable value	40 k€	40 k€	40 k€	40 k€
Min.–max.	30–50 k€	30–50 k€	30–50 k€	30–50 k€
Public relations'				
Probable value	0 €	–	0 €	0 €
Min.–max.	0–179 k€	–	0–179 k€	0–179 k€
Legal costs				
Probable value	240 k€	–	240 k€	–
Min. – max.	30–480 k€	–	30–480 k€	–
Competitive advantage	–	–	–	–
Legal actions' costs				
Probable value	10 M€	–	10 M€	–
Min.–max.	0–20 M€	–	0–20 M€	–
Reputation	–	–	–	–

Secondary responses include i) patients' notification, ii) suppliers' notification, iii) support calls, iv) people in meetings, v) public relations' costs, vi) legal actions' costs, and vii) reputation costs.

In this case, since we're talking about a state-owned institution with no significant direct competition, calculating how many patients would leave the hospital isn't particularly relevant. While some patients may leave the hospital and decide to use private hospitals, this minority won't have an impact on the hospital's revenue. Therefore, reputational costs weren't considered. Similarly, since there's no loss of reputation, it also doesn't make sense to calculate the costs associated with the loss of competitive advantage, even considering expectations regarding a potential reduction in central government funding.

For all scenarios, considering that if public communication is necessary, this can be done through the institution's website, we can therefore consider the costs of notification to users to be insignificant, the same happening with the costs to suppliers and, consequently, with support calls.

For the costs of the internal event management team, the most likely value was considered to be 40 k€, with a potential range from 30 k€ to 50 k€.

In terms of the most likely value for public relations costs, the value of 179 k€ was considered, which corresponds to the amount spent on public relations by the hospital unit in 2022.

For scenarios S1 (phishing) and S3 (information access), given that there may be third parties seeking compensation in court, it is necessary to consider the court fee (i.e., the amount to be paid to initiate legal proceedings), the costs of charges (i.e., extra expenses involved in the process, such as costs for expert assessments, postage, transport or telecommunications) and the party costs (i.e., the total amount of the costs of the process that is reimbursed by the losing party to the winning party).

For scenarios S2 (ransomware) and S4 (attack on a medical device), there are no further secondary losses to consider. Legal costs are not applicable, as the medical imaging department has alternative patients' redirection processes, which avoids filing complaints. Reputation and competitive advantage are also not affected. The same applies to legal actions' costs, as no breach of patients' personal data occurred.

In turn, regarding legal actions' costs, in scenarios S1 (phishing) and S4 (attack on a medical device), given that there was improper access to sensitive information, according to the GDPR, the institution may be committing a serious crime (which corresponds to a fine ranging from 2.5 k€ to 10 M€) or a very serious crime (which corresponds to a fine ranging from 5 k€ to 20 M€). Therefore, we assume the average fine amount as the likely value, and a minimum and maximum value between 0 € (if no fine is applied) and 20 M€.

Once the primary and secondary losses have been determined, the Loss Magnitude for each scenario can be calculated, as shown in Table 4.

3.4 Annualized Loss Exposure

Once the Loss Event Frequency and Loss Magnitude were estimated, it was possible to determine the average, minimum, and maximum Annualized Loss Exposure values for the four scenarios using the FAIR-U application (Table 5).

Table 4. Loss Magnitude values for the four scenarios.

	Medical imaging		Clinical pathology	
	S1	S2	S3	S4
Primary Loss Magnitude				
Probable value	120 k€	129 k€	120 k€	134 k€
Min.–max.	110–130 k€	117–144 k€	110–130 k€	117 k€–4.9 M€
Secondary Loss Magnitude				
Probable value	10.3 M€	40 k€	10.3 M€	40 k€
Min.–max.	60 k€–20.7 M€	30–40 k€	60 k€–20.7 M€	30–218 k€
Total				
Probable value	10.4 M€	169 k€	10.4 M€	174 k€
Min.–max.	170 k€–20.8 M€	147–373 k€	170 k€ -20.8 M€	147 k€–5.2 M€

Table 5. Annualized Loss Exposure values for the four scenarios.

	Medical imaging		Clinical pathology	
	S1	S2	S3	S4
Average	18.1 M€	37.6 k€	4.7 M€	649 k€
Minimum	0 €	0 €	0 €	0 €
Maximum	74.4 M€	232.2 k€	38.4 M€	8.3 M€

3.5 Loss Exceedance Curve

Also based on the use of the FAIR-U application, Fig. 1 presents the Loss Exceedance Curves that resulted from the Loss Event Frequency and Loss Magnitude that were calculated for the four scenarios considered.

4 Discussion

The FAIR model emerged to address the challenges associated with qualitative risk analysis, relying on a quantitative approach that allows for objective decision-making regarding the risks facing an organization. The starting point for applying the FAIR model is creating scenarios. It is necessary to define the asset at risk, the threat community, the type of threat and its impact (e.g., in terms of data confidentiality), and the existing controls. The scenarios are used to estimate the Loss Event Frequency and Loss

Magnitude, values that are necessary for performing simulations based on Monte Carlo statistical techniques, using the FAIR-U tool. These values allow to determine the range of possible Annualized Loss Exposure values and their probabilities of realization (i.e., Loss Exceedance Curve).

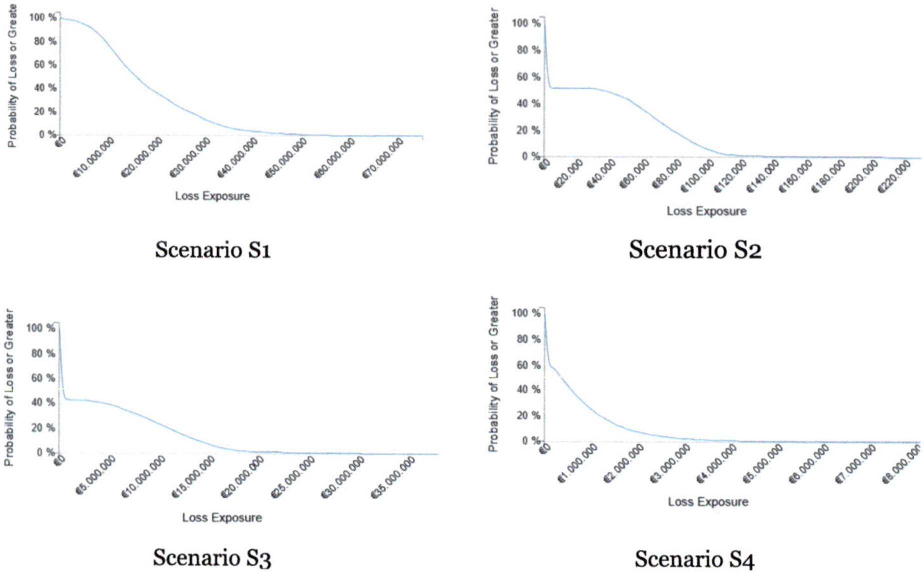

Fig. 1. Loss Exceedance Curves for the four scenarios.

Annualized Loss Exposure is the key metric in the simplest way to communicate risk. It represents the potential loss a company faces each time it makes a decision that involves monetary expenditure. This expenditure ranges from implementing a new process within the institution to purchasing new hardware. In this context, Annualized Loss Exposure allows us to compare distinct risk issues that may have different frequencies and impacts, enabling us to identify the potential impact and thus decide which risk management strategies should be implemented to avoid financial losses.

The minimum and maximum values for Annualized Loss Exposure correspond, respectively, to the lowest and highest results generated by Monte Carlo simulations. The fact that the minimum values obtained for the different scenarios are close to 0 € means that, in at least one simulation of the model, the loss event did not occur. On the other hand, the maximum value results from at least one simulation, but its occurrence is unlikely, as can be inferred from the Loss Exceedance Curve for each scenario.

Considering the four scenarios, the average values vary from 37.6 k€ for scenario 2 to 18.1 M€ for scenario 1. Regarding the other two scenarios, values of 4.7 M€ and 649 k€ were obtained for scenarios S3 and S4. From these values, it is possible to verify that the scenarios that imply greater losses are those related to undue access to patient information, taking into account the fines provided for in the GDPR.

Based on the values obtained, it is possible to prioritize the interventions to be implemented. Clearly, the scenario that deserves the most attention is scenario S1, followed by scenario S3. Thus, the hospital can implement actions and controls that reduce its vulnerability to the respective types of events and, consequently, reduce the financial impact it may have on the institution. Examples of measures to be implemented are: i) access privilege management that reports access to critical hospital infrastructure and systems; ii) surveillance controls on access to the medical imaging department assets; and iii) controls on the use of mobile devices within medical imaging department during working hours; or iv) promoting employees' training.

Given that the free access FAIR-U Tool was used in this test work, as future work, it will be interesting to carry out the same simulations in the paid version, where it is possible, in a single graph, to compare the different LEC curves with each other, as well as their relationship with the organization's risk tolerance curve.

5 Conclusion

The methods that allowed the realization of this work are based on the FAIR model, using the FAIR-U tool to perform Monte Carlo statistical techniques that allowed thousands of simulations of possible scenarios for a given uncertainty event.

This study showed that the FAIR model and its tools can be adopted by healthcare organizations to deal with uncertain events and, consequently, have a better understanding of the impact of certain risks and, consequently, enhance sustained decision-making to mitigate them.

References

1. Triplett, W.J.: Cybersecurity vulnerabilities in healthcare: a threat to patient security. Cybersecurity and Innov. Technol. J. **2**(1), 15–25 (2024)
2. Paccoud, I., Leist, A.K., Schwaninger, I., van Kessel, R., Klucken, J.: Socio-ethical challenges and opportunities for advancing diversity, equity, and inclusion in digital medicine. Digital Health. **10**, 20552076241277705 (2024)
3. Galazzi, A., et al.: Recommendations to promote the digital healthcare transformation in the clinical practice: findings from an international consensus development method. BMC Health Serv. Res. **25**(1), 929 (2025)
4. Aytekin, A., et al.: Digital health technologies in patient experience literature: a scoping review and future outlook for sustainable digital health interventions. Sustainability. **17**(2), 456 (2025)
5. Coelho, R.R., Bonato, S.V., Pacheco, D.A.D.J., Brito, P.D.S.: Moving health 4.0 forward: a systematic literature review, challenges and future research agenda. TQM J., 1–34 (2025)
6. Dalingwater, L.: Digital health in a post-COVID age. Int. J. Health Wellness Soc. **15**(1), 1 (2025)
7. Bhosale, K.S., Nenova, M., Iliev, G.: A study of cyber attacks: in the healthcare sector. In: Sixth Junior Conference on Lighting (Lighting), pp. 1–6. IEEE, Piscataway (2021)
8. Arafa, A., Sheerah, H.A., Alsalamah, S.: Emerging digital technologies in healthcare with a spotlight on cybersecurity: a narrative review. Information. **14**(12), 640 (2023)
9. Cartwright, A.J.: The elephant in the room: cybersecurity in healthcare. J. Clin. Monit. Comput. **37**(5), 1123–1132 (2023)

10. Benmalek, M.: Ransomware on cyber-physical systems: taxonomies, case studies, security gaps, and open challenges. Internet Things Cyber-Phys. Syst. **4**, 186–202 (2024)
11. Kirubavathi, G., Regis Anne, W., Sridevi, U.K.: A recent review of ransomware attacks on healthcare industries. Int. J. Syst. Assur. Eng. Manag. **15**(11), 5078–5096 (2024)
12. Martignani, C.: Cybersecurity in cardiac implantable electronic devices. Expert Rev. Med. Devices. **16**(6), 437–444 (2019)
13. Craigen, D., Diakun-Thibault, N., Purse, R.: Defining cybersecurity. Technol. Innov. Manag. Rev. **4**(10), 13–21 (2014)
14. Ahmadi-Assalemi, G., Al-Khateeb, H., Epiphaniou, G., Maple, C.: Cyber resilience and incident response in smart cities: a systematic literature review. Smart Cities. **3**(3), 894–927 (2020)
15. Verma, P., Newe, T., O'Mahony, G.D., Brennan, D., O'Shea, D.: Towards a unified understanding of cyber resilience: a comprehensive review of concepts, strategies, and future directions. IEEE Access. **13**, 49945–49965 (2025)
16. Garcia-Perez, A., Cegarra-Navarro, J.G., Sallos, M.P., Martinez-Caro, E., Chinnaswamy, A.: Resilience in healthcare systems: cyber security and digital transformation. Technovation. **121**, 102583 (2023)
17. Iyer, V.R., Babu, K.: Resilience at the core: enhancing cyber-resilience and cybersecurity practices in healthcare sector. J. Conting. Crisis Manag. **33**(3), e70061 (2025)
18. Ulloa-Zamora, M., Barría-Huidobro, C., Sánchez-Rubio, M., Galeazzi, L.: Integral security pillars for medical devices: a comprehensive analysis. Appl. Sci. **15**(12), 6634 (2025)
19. NIST: Framework for Improving Critical Infrastructure Cybersecurity. National Institute of Standards and Technology, Gaithersburg (2018)
20. CNCS: Quadro Nacional de Referência de Cibersegurança. Centro Nacional de Cibersegurança. CNCS, Lisboa (2019)
21. ISO/IEC.: ISO/IEC 27005:2022Information Security, Cybersecurity and Privacy Protection Guidance on Managing Information Security Risks. ISO/IEC, Geneva (2022)
22. Akinrolabu, O., Nurse, J.R., Martin, A., New, S.: Cyber risk assessment in cloud provider environments: current models and future needs. Comput. Secur. **87**, 101600 (2019)
23. Hubbard, D.W., Seiersen, R.: How to Measure Anything in Cybersecurity Risk. John Wiley & Sons, Hoboken (2023)
24. Freund, J., Jones, J.: Measuring and Managing Information Risk: A Fair Approach. Butterworth-Heinemann, Oxford (2014)
25. Dreyling, R., Jackson, E., Pappel, I.: Cyber security risk analysis for a virtual assistant G2C digital service using FAIR model. In: 2021 Eighth International Conference on eDemocracy & eGovernment (ICEDEG), pp. 33–40. IEEE, Piscataway (2021)

Artificial Intelligence in Teacher Training: Technological, Ethical, and Pedagogical Challenges for Critical Education

Elza Mesquita, Bruno F. Gonçalves, and Maria Raquel Patrício(✉)

Transdisciplinary Research Center in Education and Development (CITeD), Instituto Politécnico de Bragança, Bragança, Portugal
{elza,bruno.goncalves,raquel}@ipb.pt

Abstract. The integration of Artificial Intelligence (AI) into teacher training presents significant technological, ethical, and pedagogical challenges. This article explores the need for teacher training that goes beyond technical proficiency to encompass ethics and innovation. We analyze technological challenges, such as infrastructure and digital skills; ethical challenges, including privacy, algorithmic biases, and academic integrity; and pedagogical challenges, such as curriculum integration and the risk of technological dependence. Based on this analysis, we propose a training model structured around three axes: ethical training, technological and pedagogical training, and innovation and experimentation. We argue that a holistic approach, balancing technological, ethical, and pedagogical dimensions, is crucial to empowering future teachers to use AI critically, responsibly, and creatively, ensuring that this technology becomes an ally in promoting equitable, quality education. The article concludes with suggestions for future research, aiming to deepen the understanding and implementation of AI in the educational context by future teachers.

Keywords: Artificial Intelligence · Teacher Training · Pedagogical Innovation · Technological Challenges

1 Introduction

The growing integration of Artificial Intelligence (AI) in various sectors of society has driven a profound transformation in economic, political, and social dynamics. Education is no exception to this wave of innovation, with AI appearing as a powerful tool capable of (re) defining teaching-learning processes and teaching practice itself. The presence of AI in the educational context has intensified in recent decades (we are referring to the 2020s), reflecting a significant change in the way information is produced, accessed, and used. This technology offers considerable potential to enhance teaching and learning processes, promoting innovative solutions to pedagogical, organizational, administrative, and evaluative challenges.

However, the integration of AI into the educational environment is not limited to its technical capabilities, as it also raises in-depth reflection on its ethical, pedagogical,

A. Rocha et al. (Eds.): ICITS 2026, LNNS 1966, pp. 181–191, 2027.
https://doi.org/10.1007/978-3-032-27538-7_16

and social implications [1, 2]. Academic debate has focused on understanding how these technologies influence educational practices in higher education, particularly in the debate on teacher-student relationships and the meanings attributed to teaching and learning at this level of education [2]. There is an urgent need to rethink the traditional roles of educational agents, since AI stimulates new forms of knowledge mediation, in which pedagogical decisions are, to a certain extent, guided by data and algorithms. This movement, often associated with the so-called "datafication of education," has profound implications for school culture, curriculum planning, and conceptions of learning [2]. Given this scenario, whether typical or atypical, it is imperative to outline paths for teacher training that are not only technically proficient in the use of AI but also based on an ethical and innovative perspective.

This article aims to identify and analyze the primary challenges associated with implementing AI in initial teacher training. Based on this analysis, it proposes guidelines and strategies for training that better integrate AI in an ethical, innovative, and effective manner. The central question guiding this research is: How can we outline paths for teacher training that integrate AI in an ethical, innovative, and effective way?

2 Artificial Intelligence in the Reconfiguration of Teacher Training

The growing integration of Artificial Intelligence (AI) in the education sector is not limited to a mere technological tool but requires a profound reconfiguration of initial and continuing teacher training. AI acts as a catalyst for change, requiring future educators to develop new skills and a critical attitude towards technology [3]. This theoretical framework aims to explore the tensions and opportunities that AI introduces in teacher training, analyzing the technological, ethical, and pedagogical challenges, proposing paths for training that are simultaneously critical, ethical, and innovative.

Artificial Intelligence (AI) has emerged as a transformative force in multiple sectors, and education is no exception. Its growing integration redefines teaching-learning processes and pedagogical practice itself, positioning AI not only as a technological tool, but as a "pedagogical agent" capable of mediating knowledge, personalizing learning, and optimizing administrative tasks [4, 5]. However, this technological evolution is not without its complexities, presenting a dual movement between the innovative potential of AI and the ethical challenges that emerge from its application [4].

The discussion about AI in education can be anchored in several theoretical perspectives that help to understand its implications.

Vygotsky's Social Constructivism Theory, for example, emphasizes the importance of social interaction and cultural mediation in the learning process. AI, in this context, can be seen as a new mediator, altering the dynamics of interaction between children, teachers, and knowledge itself.

Activity Theory (Leontiev) offers a lens through which to analyze how AI tools become an integral part of pedagogical activities, modifying the objectives, actions, and operations of educational agents.

Critical Theory of Technology (Feenberg) is fundamental to analyzing AI in education, arguing that technology is not neutral but rather shaped by social and political values. This perspective allows us to question the dominant narratives about AI as a universal solution to educational problems, exposing its potential risks and biases.

As Kenski points out, teacher training should promote an understanding of technologies as social and political constructs, not merely as neutral instruments [4].

The integration of AI into teacher training is a complex process, marked by interrelated challenges that require a holistic approach.

2.1 Technological, Ethical, and Pedagogical Challenges

The integration of AI into teacher training is a complex process, marked by interrelated challenges that require a holistic approach. These challenges can be categorised into three main areas, which serve as fundamental constructs for assessing the perceptions of future teachers: Technological Challenges; Ethical Challenges; and Pedagogical Challenges.

Technological Challenges

Technological challenges are often the most obvious, ranging from inadequate infrastructure to a significant digital skills gap. The lack of access to high-quality internet and sufficient technological devices in schools creates a "digital divide" that AI can exacerbate rather than mitigate.

The technological challenges focus on infrastructure, digital literacy, and equitable access to AI resources [5]. The lack of adequate infrastructure in educational institutions, such as high-speed connectivity and compatible devices, is a primary barrier to the effective adoption of AI [6]. In addition, teachers' digital literacy must be deepened, going beyond basic software use to include computational thinking and an understanding of the principles underlying AI algorithms [7]. Training should mitigate the risk of AI exacerbating digital inequalities by ensuring that all future teachers, regardless of their context, have access to and training in the use of these tools.

On this issue, Guimarães warns that introducing artificial intelligence into schools without connectivity, training, or digital inclusion reinforces inequalities under the guise of innovation [2]. He also stresses the need for robust public policies that ensure equity in access to and use of AI, preventing it from becoming yet another factor of social exclusion.

Ethical Challenges

Ethical challenges are central to the discussion on AI in education. Ethical challenges require in-depth reflection on data privacy, algorithmic transparency, and the risk of dehumanisation in educational practices [8, 9]. The General Data Protection Regulation (GDPR) in Europe shows an unavoidable legal framework, making it imperative that future teachers understand the implications of collecting and using student data by AI systems [10]. The United Nations Educational, Scientific, and Cultural Organization (UNESCO) emphasises that the ethical incorporation of AI must be based on principles of transparency, explainability and accountability, ensuring that technology is a tool to support human development and not a factor of opacity or exclusion [11]. Training should prepare teachers to find and mitigate the risk of algorithmic bias, which can perpetuate or merge social prejudices [12].

The collection and analysis of sensitive data from students and teachers raises serious concerns about privacy and security. Additionally, algorithmic biases inherent in training data can perpetuate and amplify social prejudices, thereby affecting equity and inclusion.

O'Neil describes these algorithms as automated decision-making models and systems that function as veritable *weapons of mathematical destruction*, i.e., instruments that, under the guise of objectivity and technical precision, perpetuate and intensify social inequalities. These mechanisms often disproportionately penalize the most vulnerable populations. They combine existing privileges and hide their harmful consequences behind a cloak of supposed scientific neutrality, making it difficult to challenge their effects and hold their creators accountable [9].

Eubanks contributes to this debate with a warning about the so-called *automation of exclusion*, which is seen as a process by which deeply rooted and structural social inequalities are translated into parameters and calculation variables within computer systems. In this way, historical injustices are no longer perceived as political constructs and become naturalized within algorithmic models, converting into seemingly objective criteria for automated decision-making. The result is the technological legitimization of discriminatory practices, which do not reproduce, but rather crystallize and amplify social vulnerabilities [10].

Academic integrity and teacher autonomy are also called into question. The threat of plagiarism generated by AI, along with the potential dehumanization of teaching, needs a deep reflection on the delicate balance between technological innovation and the irreplaceable role of the human element in education.

UNESCO emphasizes that the ethical incorporation of AI in the field of education must be based on fundamental principles, such as transparency of processes, explainability of automated decisions, accountability of the agents involved, and effective participation of school communities. These criteria aim to ensure that technology is used as a tool to support human and social development, and not as a factor of opacity or exclusion [11].

Pedagogical Challenges

The pedagogical challenges focus on how AI can be effectively and meaningfully integrated into teaching and learning processes. The curricular integration of AI requires a reconceptualization of traditional approaches, promoting digital literacy that goes beyond the technical use of tools, including computational thinking and understanding the principles underlying AI.

AI has the potential to transform learning paradigms, requiring educators/teachers to develop skills to use AI to promote accessibility, personalization, and engagement for children. However, excessive technological dependence is a risk and may compromise the development of students' autonomy and critical thinking.

Selwyn argues that the introduction of artificial intelligence in the school context cannot be conceived as something neutral. On the contrary, it is a force that reconfigures the role of teachers, fragments pedagogical knowledge into technical procedures, and subordinates' educational practice to a logic of algorithmic monitoring and control [1]. This critical reading highlights the urgency of professional training that prepares future teachers to act not only as executors of automated routines, but also as curators of meaningful experiences [1], capable of critically mediating educational processes and promoting emancipatory learning [12].

In this context, the presence of AI in education requires a reflective and active stance on the part of teaching staff, so that technology can be transformed into a resource that supports human development, rather than a mechanism of subordination [13].

The integration of Artificial Intelligence into teacher training is a contemporary imperative that requires a balanced and critical approach.

By equipping teachers with a thorough understanding of the technological, ethical, and pedagogical aspects of AI, it is possible to transform this technology into a powerful ally for a more equitable, inclusive, and quality education [2, 14].

Far from being diminished, the role of the teacher is reconfigured and valued as an essential mediator in a constantly evolving educational scenario, given the experiences of children in plural contexts that should characterize educational practices.

3 Methodology

This research adopted an exploratory and mixed-methods approach, combining qualitative and comparative analysis, to critically examine the functionality, potential, and results.

This research adopts a quantitative approach, outlined through a survey based on a questionnaire, to collect data from a representative group of higher education students. This approach is particularly suitable for exploring beliefs, attitudes, and behaviors of a population, allowing the generalization of results to a wider universe [13]. As Coutinho [15] and Vilelas [16] highlight, the role of quantitative research in the systematic measurement of variables, a fundamental pillar for understanding social and educational phenomena.

The participants in this study were higher education students. Participants were selected through convenience sampling. Participation was voluntary and anonymous, ensuring the confidentiality of the data collected, following the ethical principles of research [17].

The study included a sample of 72 participants, sourced from a higher education institution located in the Northern region of Portugal, all enrolled in the Basic Education Degree and the Professional Master's Degree in Teaching programs.

Data collection was carried out through a questionnaire survey. The questionnaire is a widely used tool in quantitative research, allowing for the efficient collection of standardized information from many participants [18]. Its structure was designed to address beliefs of AI, its challenges, and training proposals.

The questionnaire consisted of an unspecified number of questions, predominantly closed-ended (Likert scale, multiple choice), which eases statistical analysis of the data [19]. Before its final application, the questionnaire was pre-tested with a small group of students to verify the clarity of the questions and the appropriateness of the format, ensuring its validity and reliability [19]. It was also sent to an ethics committee, whose case number is 492842.

The evaluation, through the instrument, contemplated three thematic axes: Technological Challenges (infrastructure, digital literacy, and access to resources), Pedagogical Challenges (integration of AI into teaching strategies and redefinition of the teaching role), and Ethical Challenges (data privacy, algorithmic transparency, and the potential dehumanization of practices).

The questionnaire was administered online. Participants were informed about the study aims, the voluntary nature of their participation, and the confidential treatment of data. Informed consent was obtained from all participants before data collection.

The data collected through the questionnaire were subjected to descriptive and inferential statistical analysis. Frequencies, percentages, means, and standard deviations were calculated to characterize the participants' responses. The results of this analysis are presented in the following section.

4 Results

4.1 Sociodemographic Characterization of the Sample

The sociodemographic analysis of the sample reveals an asymmetrical distribution by gender, characterized by a female predominance (88.9%), in contrast with a male minority (9.7%), and a residual percentage of participants who did not specify their gender (1.4%). This distribution reflects the phenomenon of feminization typically associated with teacher training programs in Portugal.

Age Profile and Nationality

Regarding the age variable, a significant concentration was observed in the 17 to 23 age group (81.9%), followed by smaller representations in the 24–28 age range (9.7%), over 35 years (6.9%), and 29–34 years (1.4%). This pattern indicates a predominantly young profile, consistent with the typical entry and academic progression in study cycles leading to teaching qualifications.

Concerning nationality, the sample is mainly composed of students of Portuguese nationality (84.7%), complemented by a significant percentage of students in international mobility/Erasmus programs (15.3%). This data suggests a notable degree of internationalization in the institution's educational offerings.

Educational Level and Curricular Distribution

In terms of educational level, students enrolled in Bachelor's Degrees predominate (84.7%), with a minority consisting of Master's Degree students (15.3%).

The analysis of the distribution by curricular year in the Bachelor's Degree in Basic Education points to a marked concentration in the first year (75.4%), followed by the third year (18%), and a smaller representation in the second year (6.6%).

Regarding the Professional Master's Degrees in Teaching, the distribution by specialization is as follows:

- Teaching in the 1st Cycle of Basic Education and of Portuguese and History and Geography of Portugal in the 2nd Cycle of Basic Education: 5 students (distributed between the 1st and 2nd years).
- Pre-school Education and Teaching in the 1st Cycle of Basic Education: 4 students (equally distributed between the two curricular years).
- Teaching in the 1st Cycle of Basic Education and of Mathematics and Natural Sciences in the 2nd Cycle of Basic Education: 2 students (both in the 1st year).

4.2 Challenges of Integrating Artificial Intelligence into Teacher Training

The integration of Artificial Intelligence (AI) into initial teacher training, while promising, is accompanied by a complex set of challenges that require attention and coordinated strategies. These challenges can be categorized into three main areas: technological, ethical, and pedagogical, each with its own particularities and implications for teaching practice and, concomitantly, for improving the quality of education.

Technological Challenges

Technological challenges are often the most visible and immediate when discussing the implementation of AI in educational contexts. Infrastructure and access to adequate technological resources are significant barriers. The quality of the internet available in schools and the availability of a sufficient number of technological devices for all children and teachers are limiting factors.

The study reveals that 52.8% of participants strongly agree and 25.0% partially agree that internet quality is a limiting factor, and that device availability is a concern, with an average Likert scale of 3.40, showing that resources may not fully meet the needs of AI integration.

Besides infrastructure, the lack of digital skills and technical knowledge about AI among future teachers constitutes a substantial barrier. Many students (future teachers) feel unprepared to effectively use AI tools, which manifests in the difficulty of using these tools.

The survey shows that the lack of technical knowledge about AI among future teachers is a significant barrier, with an average score of 4.32. The complexity of the tools and the need for ongoing training are evident. Teachers' resistance to adopting AI, often due to a lack of training or fear of being replaced, is also a pedagogical challenge with technological roots.

Another crucial aspect is maintenance and technical support. AI needs frequent updates and ongoing maintenance, which poses a challenge for educational institutions. A lack of ongoing technical support can hinder the implementation and regular use of AI, as indicated by a mean of 4.06. The compatibility of AI tools with technology systems already implemented in educational institutions is also a concern, with 40.3% of respondents partially agreeing and 29.2% strongly agreeing that AI tools are not always compatible.

Ethical Challenges

Ethical challenges are paramount in integrating AI into education, as they concern fundamental issues of privacy, equity, and accountability. Privacy and data protection are central concerns, as the collection of sensitive data from students and teachers raises questions about the security and ethical use of information.

The General Data Protection Regulation (GDPR) is a crucial benchmark, and legal compliance in the adoption of AI in the educational context is widely recognized as important, with 84.3% of respondents agreeing on the need for regulation.

Algorithmic biases and equity in the effective use of AI pose another significant ethical challenge. Algorithms can reproduce social biases or discriminate against vulnerable groups if the training data is not representative. Concern about the perpetuation of existing biases in society and the impact on diversity and inclusion in the classroom is clear, with a mean of 3.46, showing that a significant proportion of participants recognize that AI can perpetuate social biases and affect diversity and inclusion in the educational context.

Academic integrity and teacher autonomy are also questioned. The threat of plagiarism, associated with the use of AI-generated content, and the need to ensure the integrity of assessments are valid concerns among the students who took part. Furthermore, the dehumanization of teaching and the possible reduction of teacher autonomy are risks highlighted, reinforcing the need to balance technological innovation with the human role in education. Most respondents (54.4%) oppose the idea that AI should make educational decisions without human supervision, and 71.9% agree that AI systems should help children and not replace the role of educators/teachers.

Pedagogical Challenges

The pedagogical challenges of AI in teacher education center on how this technology can be effectively and meaningfully integrated into teaching and learning processes. Curricular integration is one of the main challenges, as harmonizing AI with existing methodologies and curricula requires a reconceptualization of traditional approaches.

The need to develop digital literacy skills in future educators/teachers that go beyond the technical use of tools, including computational thinking and an understanding of the principles underlying AI-related technologies, is crucial.

Reconceptualizing practices is imperative. AI has the potential to transform pre-established learning paradigms, requiring educators/teachers to develop skills to use AI to promote accessibility, personalization, and children's engagement. However, excessive reliance on technology poses a risk.

According to many participants (99%), the excessive use of AI can compromise the development of autonomy and critical thinking among higher education students. There is also the imminent danger that AI will diminish the role of educators/teachers at other levels of education, which are essential for children's development.

The integration of AI into education still faces several challenges, from technical and pedagogical limitations to ethical, social, and economic issues. These challenges range from a lack of infrastructure and adequate, certified training to concerns about privacy, unequal access, and financial sustainability, requiring a more thoughtful and collaborative approach among teachers, administrators, and that responsible for/mentoring technology development.

4.3 Proposals for Ethical and Innovative Training

To address the identified challenges and promote the effective and responsible integration of Artificial Intelligence (AI) into teaching, it is essential to develop educator/teacher training that is both ethical and innovative. The proposals presented below aim to empower future educators/teachers to use AI critically, consciously, and in a manner grounded in pedagogical and ethical principles.

Ethical Training Axis

Ethical training is a central pillar for the integration of AI in education. It is imperative to incorporate mandatory modules addressing AI ethics into initial teacher training curricula. These modules should cover topics such as privacy and data protection, algorithmic biases, transparency, human accountability, and the social implications of AI. The inclusion of these topics is considered significantly important by most participants, with 87.7% attributing great importance to integrating these topics into educator/teacher training.

Beyond theory, it is crucial to promote critical analysis of AI tools. Future educators/teachers must be trained to identify and mitigate ethical risks associated with the use of AI, as well as the potential for perpetuating bias or compromising children's privacy. This involves developing the ability to question, evaluate, and select AI tools that are transparent, fair, and safe. The discussion about the need for specific regulation for the use of AI in education is also pertinent, as a lack of regulation can lead to insecurity and uncertainty about its uses and implications.

Technological and Pedagogical Training Axis

Technological and pedagogical training must go beyond the instrumental use of AI. It is essential to develop AI literacy that focuses on understanding its underlying principles, computational thinking, and its pedagogical implications. This means that all teachers (at all levels of education) must not only know how to use an AI tool, but also understand how it works, its limits, and how it can be adapted to different learning contexts.

Training should, therefore, promote a deep understanding of AI and view it as an ally in the educational process, not as a substitute for the role of the teacher. In addition to formal training, it is essential to foster the creation of communities of practice. These collaborative spaces allow for the sharing of experiences, the joint development of innovative teaching approaches with AI, and joint problem-solving.

Peer-to-peer knowledge exchange and practice-based learning are crucial for building a culture of innovation and continuous adaptation. Continuous training and investment in proper technological infrastructure are also essential to support this capacity building.

Innovation and Experimentation Axis

The innovation and experimentation axis is vital for future educators/teachers to explore the transformative potential of AI. It is crucial to encourage pedagogical experimentation with AI in controlled environments, allowing for the exploration of new methodologies and the adaptation of existing tools to the specific needs of children. This practical approach allows future educators/teachers to develop confidence and creativity in the use of AI, transforming it into a tool for pedagogical innovation.

Fostering partnerships between training institutions, schools, and technology mentors is equally important. These collaborations can facilitate the creation of AI tools that respond to real pedagogical needs, ensuring that the technology is developed with an educational perspective and is tested and validated in classroom settings. Collaboration between different stakeholders is essential to overcome the challenges and maximize the benefits of AI in education.

5 Conclusions

The integration of Artificial Intelligence into the initial training of educators/teachers stands for a multifaceted challenge, but also a unique opportunity to improve the quality and relevance of education.

This article analyzed the main technological, ethical, and pedagogical challenges associated with this integration and based on this analysis, proposed a set of guidelines for training that is both ethical and innovative.

In the technological domain, the findings revealed that inadequate infrastructure, insufficient digital skills, and the lack of continuous technical support appear as critical barriers. Many prospective teachers reported difficulties using AI tools due to unfamiliarity with underlying concepts and irregular access to equipment and the Internet.

On the ethical front, there was widespread concern about privacy and data protection, as well as the need to follow relevant regulations. Participants also acknowledged the risk of algorithmic bias and emphasized the importance of supporting human oversight in pedagogical decisions to safeguard fairness, transparency, and academic integrity.

From a pedagogical perspective, the results showed that integrating AI into the traditional curriculum requires a reconceptualization of teaching practices. Future teachers highlighted the need to develop not only digital literacy but also computational thinking and the capacity to foster students' autonomy and critical thinking. They also warned that excessive reliance on AI could undermine these very competencies.

Based on these findings, we propose a training model structured around three interdependent pillars. The first pillar, ethical training, should include mandatory modules on data privacy, algorithmic transparency, human accountability, and the social implications of AI. The second pillar, technological and pedagogical training, aims to foster a deep understanding of AI principles, supported by the creation of communities of practice for experience sharing and action research. The third pillar, innovation and experimentation, encourages the implementation of practical projects in controlled environments and the establishment of partnerships among training institutions, schools, and technology mentors.

In conclusion, the findings confirm that a holistic approach—combining investment in infrastructure, ethical training, and pedagogical reinvention—is essential for AI to serve as a true ally in promoting fair, high-quality education.

For future research, we suggest conducting longitudinal studies to evaluate the effectiveness of the proposed guidelines; carrying out intervention studies to explore new methodologies for integrating AI into teaching practice; and undertaking investigations that deepen our understanding of the impact of AI on pedagogical practice and on teacher professional development.

Acknowledgments. This work has been supported by FCT – Fundação para a Ciência e Tecnologia within the Project Scope: UID/05777/2025.

References

1. Selwyn, N.: Should Robots Replace Teachers? AI and the Future of Education, 1st edn. Polity Press (2019)
2. Guimarães, A.M.: Inteligência artificial e educação: entre a inovação pedagógica e os desafios éticos. Revista Acadêmica da Lusofonia. **2**(9), 1–13 (2025) https://revistaacademicadalusofonia.com/index.php/lusofonia/article/view/152
3. Guimarães Junior, J.C., et al.: Desafios Éticos e Pedagógicos da Inteligência Artificial na Educação. Interference: J. Audio Cult. **11**(2), 976–992 (2025). https://doi.org/10.36557/2009-3578.2025v11n2p976-992
4. Caitano, T.F., et al.: A inteligência artificial como agente pedagógico: Potencialidades e riscos éticos na formação docente. Revista Políticas Públicas & Cidades. **14**(5), 1–16 (2025) https://journalppc.com/RPPC/article/view/2245
5. Reis, M.R., Mota, J.L.B.: Inteligência artificial na educação: Personalização e desafios éticos em perspectiva. Revista Ibero-Americana de Humanidades, Ciências E Educação. **11**(6), 5399–5413 (2025). https://doi.org/10.51891/rease.v11i6.20095
6. Silva, M.R., et al.: Inteligência artificial na educação básica: possibilidades e limites éticos. Cuadernos De Educación Y Desarrollo-QUALIS A4. **17**(4), e8081 (2025). https://doi.org/10.55905/cuadv17n4-106
7. Portilho, M.M.A., Lima, A.F., Delgado, V.B.S.: Inteligência artificial na educação: Potencialidades e desafios éticos. Revista Missioneira. **27**(6), 145–157 (2025). https://doi.org/10.46550/gxty8y14
8. Carvalho, F., Oliveira, D., Lima, M., Teles, L.: Inteligência artificial e educação: criatividade na formação docente. Cadernos RCC#35. **10**(4), 142–153 (2023) https://www.researchgate.net/publication/378798643_Inteligencia_Artificial_e_Educacao_criatividade_na_formacao_docente
9. O'Neil, C.: Weapons of Math Destruction: How Big Data Increases Inequality and Threatens Democracy. Crown (2016)
10. Eubanks, V.: Automating Inequality: How High-Tech Tools Profile, Police, and Punish the Poor. St. Martin's Press (2018)
11. UNESCO.: Recomendação sobre a Ética da Inteligência Artificial. Unesco (2021)
12. de Duque, R.C.S., et al.: Formação de professores e a Inteligência Artificial: Desafios e perspectivas. Contribuciones a las Ciencias Sociales. **16**(7), 6864–6878 (2023). https://doi.org/10.55905/revconv.16n.7-158
13. Creswell, J.W.: Research Design: Qualitative, Quantitative, and Mixed Methods Approaches, 4th edn. Sage Publications (2014)
14. Holmes, W., Miao, F.: Guia para a IA generativa na educação e na pesquisa. UNESCO Publishing (2023) https://www.unesco.org/pt/articles/guia-para-ia-generativa-na-educacao-e-na-pesquisa
15. Coutinho, C.P.: Reimpressão 2024. In: Metodologia de investigação em ciências sociais e humanas: Teoria e prática, 2nd edn. Almedina (2023)
16. Vilelas, J.: Investigação: O processo de construção do conhecimento, 4th edn. Edições Sílabo (2025)
17. American Psychological Association: Ethical Principles of Psychologists and Code of Conduct. (2017). https://www.apa.org/ethics/code
18. Malhotra, N.K.: Marketing Research: An Applied Orientation, 7th edn. Pearson (2019)
19. Fink, A.: How to Conduct Surveys: A Step-by-Step Guide, 7th edn. Sage Publications, Inc (2024)

Word Embedding to Analyze Innovation Projects in Higher Education Institutions

Carlos Tandazo-Malla and Ruth Reátegui(✉)

Universidad Técnica Particular de Loja, Loja, Ecuador
{catandazo3,rmreategui}@utpl.edu.ec

Abstract. Teaching innovation projects store information about methodologies, techniques, and technologies that teachers use in their classrooms to innovate the teaching-learning process. Based on artificial intelligence and natural language processing techniques, these types of document can be analyzed to extract knowledge that improves the decision-making process. In this work, word embedding techniques such as word2vec and doc2vec have been used to represent information in a vector format. In this way, it has been possible to identify sematic relationships between words with word2vec and between documents with doc2vec. The use of word embedding models has revealed patterns and connections that might have gone unnoticed with conventional methods, highlighting their utility in extracting valuable information, such as analogies where the use of arithmetic operations allows to infer and understand more complex semantic relationships.

Keywords: word embedding · word2vec · doc2vec · university

1 Introduction

University professors engage in various activities, including the development of innovative teaching projects. These projects often focus on specific areas of knowledge or transversal aspects of the educational process, such as assessment, technology use, inclusion of students with special needs, and more. These projects are tools for improving and modernizing education, potentially benefiting both teachers and students by fostering creativity, collaboration, and continuous learning. Most universities have information systems that store and manage information generated by professors. This information grows continuously, making its analysis and knowledge discovery challenging, especially when dealing with unstructured or narrative-formatted information.

Although Word2Vec and Doc2Vec are well-established models in the field of natural language processing, their application in the analysis of teaching innovation projects remains limited. No previous studies have been identified that apply these Word Embedding techniques to explore large institutional repositories of educational projects, especially in Spanish. That is why this work provides

A. Rocha et al. (Eds.): ICITS 2026, LNNS 1966, pp. 192–207, 2027.
https://doi.org/10.1007/978-3-032-27538-7_17

a novel applied approach that demonstrates how these techniques can reveal patterns, semantic relationships, and institutional trends that are not visible using traditional methods of document analysis.

Artificial intelligence and natural language processing techniques can handle unstructured information, and one key method is word embedding. This technique transforms words into a continuous vector space, allowing words with similar meanings to have similar representations. Word embeddings can capture context of a word within a document but also its semantic and syntactic relationships with other words [1–3].

In recent years, word embedding has become a widely used text analysis tool in various fields, including academia. In [4], the authors show the relevance of word order in text processing in spoken CALL shared-task systems, and text processing systems are improved using machine learning techniques. They worked with three algorithms to generate word embeddings: GoogleNews, word2vec, ST word2Vec, and ST doc2vec

Another study [5] proposes a unified approach for recommending academic collaborators based on the similarity of their research interests and the collaboration network structure. To achieve this goal, word2vec and node2vec models are used to extract and learn relevant features. Also, in [6] authors propose a Semantic Academic Profiler (SAP) approach for thematic analysis of research documents and evaluation of researchers, institutions, and publications using semantic techniques. The SAP framework employs advanced topic modeling methods and an innovative visual interface to highlight key themes associated with articles, researchers, and institutions.

Furthermore, in [7] authors propose a framework called FTSci that leverages the power of word embedding techniques to incorporate semantics into scientific topic clustering and dynamic field analysis. This framework is used to examine the evolution of publications on "Artificial Intelligence" over time, identifying subfields and their dynamics. Additionally, the study aims to identify research trends and visualize how the AI field has evolved over time.

This work proposes the application of word embedding in a corpus of innovative teaching projects. The development of studies focusing on innovative teaching projects through the lens of AI and NLP is crucial for advancing education. Word2vec and doc2vec models that use neural network algorithms to learn word associations in large documents were used. These models represent coordinates in a specific vector space. Word2vec allows for identifying nearby or analogous words based on the distance between their vectors, thus identifying similar terms (e.g., methodologies, innovation). On the other hand, doc2vec helps identify similar documents.

Furthermore, the analysis carried out adds value to institutional knowledge management processes, as it allows real patterns in educational documentation to be identified, highlights the importance of preprocessing in administrative corpora, and shows how word embedding techniques can support the understanding of large volumes of information. This approach helps to address common problems in educational data management, such as lack of standardization, document heterogeneity, and difficulty in extracting meaningful information.

2 Methodology

Currently, there is no established methodology for projects implementing word embedding. Therefore, through related works, a pattern in the steps for developing and applying word embedding techniques has been identified. The steps followed using the Python language are detailed below.

The methodological approach adopted does not seek to propose new algorithms, but rather to evaluate the real usefulness of word embedding models already validated in the literature when applied to a type of corpus that has been little explored, such as teaching innovation projects. The main contribution lies in adapting these techniques to an institutional context, building a replicable workflow that allows universities to interpret their internal production and facilitate data-driven decision-making.

This study does not incorporate reference methods such as Bag-of-Words, TF-IDF, or LDA, given that its main purpose is to evaluate the institutional applicability of word embedding models. However, these approaches constitute relevant baselines, and we propose to integrate them into future work to compare their performance against embedding models in tasks such as document retrieval, thematic analysis, and classification.

2.1 Data Collection

For this work, innovative teaching projects from an Ecuadorian university were considered. These projects aim to improve educational quality by integrating innovative plans, ideas, and methodologies to strengthen and prepare students. A total of 527 projects presented between 2018 and 2022 were collected in an Excel file, all written in Spanish. To work with Word embedding techniques, the introduction, description, general objective, specific objectives, and results described in the projects were merged to create an Excel file with 525 records and a single field called "text". Two files with incomplete information were removed.

2.2 Tools

The present study was developed with Pyhton using Google Colab. Some of libraries used are pandas, numpy, spaCy and gensim.

2.3 Data Preprocessing

Data preprocessing aims to clean the data to work with relevant and concise information. The following activities were performed:

- Data Cleaning:

- Missing values: Rows with null values were discarded to ensure the dataset was homogeneous and free of irregularities.
- Duplicate removal: Removing duplicate rows helped maintain consistency and avoid redundancies, eliminating 52 records, resulting in 473 records for analysis.
- Conversion to lowercase: Uniform capitalization facilitated analysis by reducing complexity.
- Specific word removal: Irrelevant or problematic words were removed to improve semantic analysis quality, removing words like 'pueden', 'través', 'vez', 'tal', 'así', and 'además'.
- Leading whitespace removal: Unnecessary leading spaces were removed to improve presentation.
- Numeric value removal: Numbers were removed to reduce noise in semantic analysis.
- Special characters and punctuation removal: Simplified the text for better interpretation and reduced analysis complexity.

- Stop words removal: spaCy was used to remove stop words, reducing dimensionality and improving analysis quality.

In certain cases, words such as "can," "sometimes," or "in addition" do not contribute relevant semantic content to the analysis and were therefore removed. This decision responds to the need to reduce noise and focus the model on conceptual terms linked to the teaching innovation process. Likewise, the dataset presents biases inherent to the institutional record, such as differences in style between faculties, variation in the length of projects, and the predominance of certain thematic lines. These aspects should be considered when interpreting the results.

2.4 Word Embedding Model Creation

Once the data was processed, the word embedding model creation began to learn vector representations of words based on their context. Word2vec models, mainly using the Skip-gram algorithm to predict word context and learn high-quality vector representations, were used. Furthermore, doc2vec and the Distributed Memory (DM) algorithm was used to learn document vector representations.

Word2Vec is a technique used in natural language processing (NLP) to create vector representations of words. It uses a neural network model trained on a large corpus of text to produce word embeddings, where each word is represented as a vector in a continuous vector space [8,9]. The vectors capture semantic relationships between words, meaning that words with similar meanings are located close to each other in the vector space. This is achieved by considering the context in which words appear, typically using methods like Skip-gram and Continuous Bag of Words (CBOW) [9,10]. Furthermore Word2Vec is widely used in various NLP tasks such as sentiment analysis, named entity recognition, and machine translation [9]. It has been shown to perform better than traditional bag-of-words models in many cases [11].

Doc2Vec is an extension of Word2Vec designed to create vector representations for larger text units such as sentences, paragraphs, or entire documents [12]. It aims to capture the semantic meaning of a document as a whole. Also Doc2Vec works by embedding documents into a continuous vector space, similar to how Word2Vec embeds words. It can be used to represent documents in a way that allow for efficient comparison and classification [12,13]. It is used in tasks such as document classification, information retrieval, and sentiment analysis. It has been shown to produce high-quality document representations that can outperform other methods in certain tasks [14,15].

Word Embedding Model Creation with Word2Vec: for this model the steps developed are described below:

- Data Reading: The first stage involved loading and reading data on innovative teaching projects from an .xlsx file using the Pandas library. The preprocessed dataset provided the model with relevant information, including project-related text.
- Bigram and Trigram Generation: Natural language processing techniques were applied to generate bigrams and trigrams from tokenized sentences, capturing richer semantic relationships between words in the projects.
- Word2Vec Model Training: Using the Gensim library, a word2vec model was trained with the Skip-gram algorithm to capture semantic and contextual relationships between words in the projects. Various parameters, including vector size, context window, minimum word count, parallel processing workers, and automatically adjusted final vocabulary, were configured. Figure 1 show the Skip-gram model.
- Word Vector Visualization: A function called plot_word_vectors_zoomable was created to visualize word vectors effectively. This function used Principal Component Analysis (PCA) to reduce vector dimensionality to two dimensions, and the Plotly library created an interactive graph showing words and their relationships in a two-dimensional space. Figure 2 shows an example of this visualization.
- A WordCloud was generated to provide an additional graphical representation of the most relevant words in the projects, enriching the thesis analysis with insights into the key words and semantic relationships within the projects. Figure 3 provides an example of some of these significant words: "proyecto" (project), "aprendizaje" (learning), "estudiante" (student), and "docente" (professor).

Word Embedding Model Creation with Doc2vec: for this model the steps developed are described below:

- Data Reading: Innovative teaching projects stored in an .xlsx file were accessed via the Pandas library for effective data loading and reading.

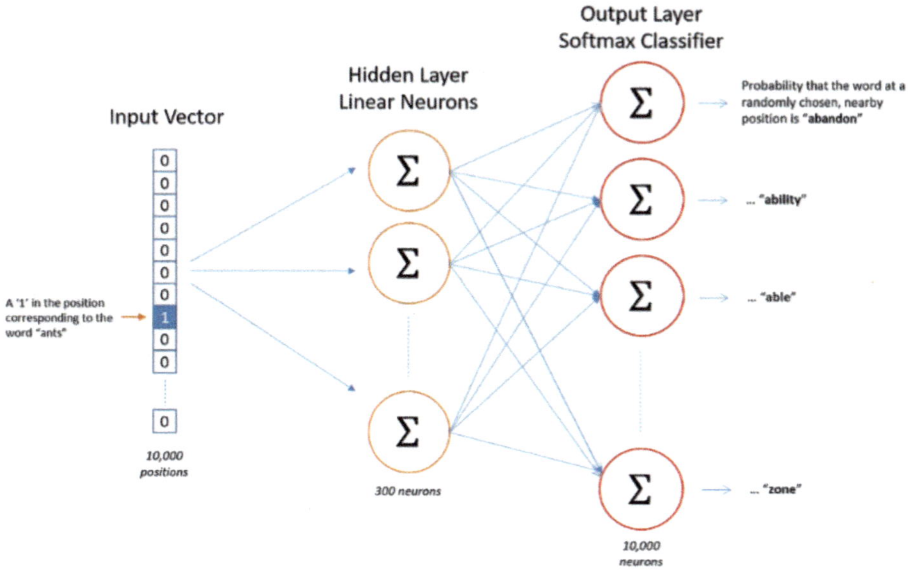

Fig. 1. A simple Skip-Gram Model. Source: [16].

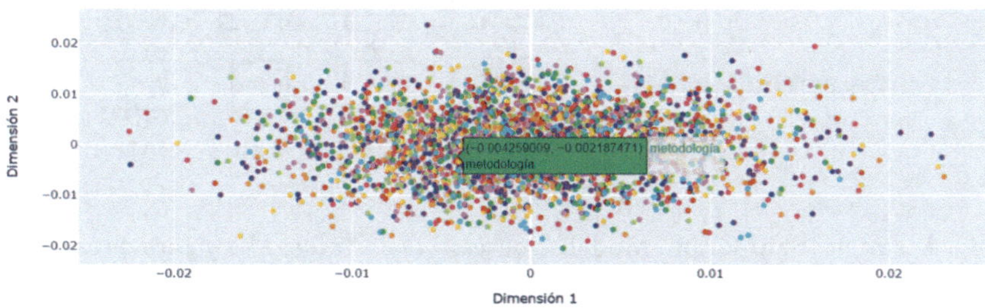

Fig. 2. Word vectors from the word2vec model.

Fig. 3. Example of word cloud.

- Tagged Document Creation: TaggedDocument objects associating each prepro-cessed project with a unique label were created for clear and concise project information affiliation with unique identifiers.
- Doc2vec Model Definition and Training: A doc2vec model was defined with parameters such as vector size, context window, minimum word count, model type (dm= 1 for Distributed Memory), and training epochs. Figure 4 shows the algorithm.
- Similar Document Search: A function called search_similar was created to search for documents similar to a query text. The query text was preprocessed similarly to the projects and used to infer a document vector. The most similar documents were searched in the doc2vec vector space, and the results were represented in a horizontal bar graph showing the most similar documents with their similarity levels. Figure 5 shows the results.

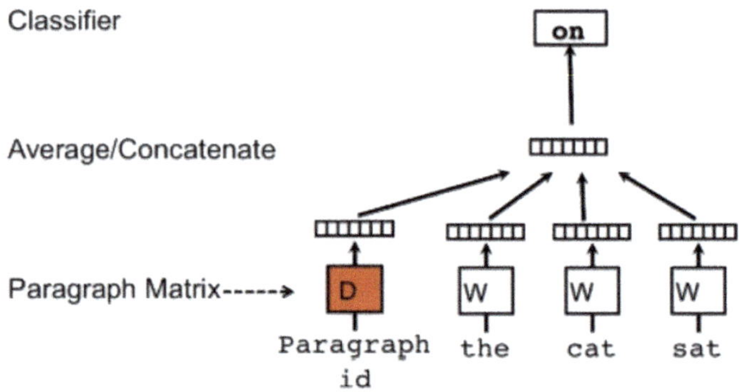

Fig. 4. A simple Skip-Gram Model. Source: [17]. *Note: The figure presented is a schematic representation of the Skip-Gram model. It is included for illustrative purposes only, given that the conceptual architecture of Doc2Vec, particularly in its Distributed Memory (DM) variant, is based on similar principles of context prediction. It should not be interpreted as an exact representation of the Doc2Vec algorithm.*

3 Results

3.1 Results with Word2vec

After training the word embedding model, the resulting vectors are analyzed using vector operations such as addition and subtraction to examine semantic similarities and word relationships.

This study is conceived as an exploratory and descriptive analysis of the potential of word embeddings applied to teaching innovation projects. Therefore, formal quantitative metrics that would be necessary in a study aimed at comprehensive validation of the model were not incorporated. Future research

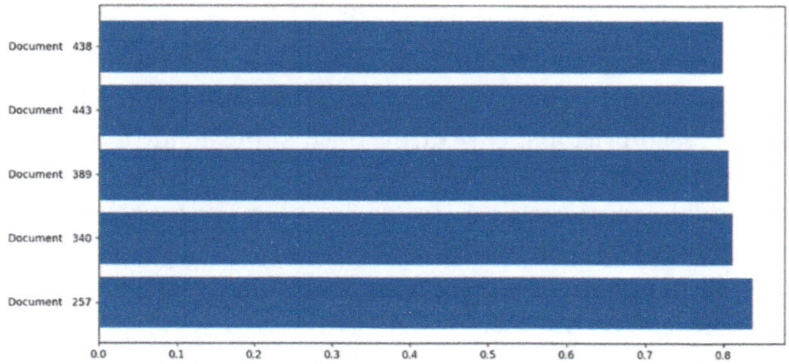

Fig. 5. Results of doc2vec search.

may incorporate these metrics to complement the qualitative perspective presented here.

Semantic Similarity measures the interrelation between two words or terms. This analysis highlights the specific evaluation of semantic similarity, revealing how the Word2vec model captures and reflects these relationships. The diverse set of words and phrases in the projects were subjected to these Word embedding techniques, with results showing that Word2vec is useful to represent contextual similarities. As an example, the word "docente" is introduced and the desired number of words that the model should present is specified; in this case, 10 additional words are requested. The model displays the results, validating on a scale from 0 to 1, where values closer to 1 indicate a stronger semantic relationship. Figure 6 shows the results.

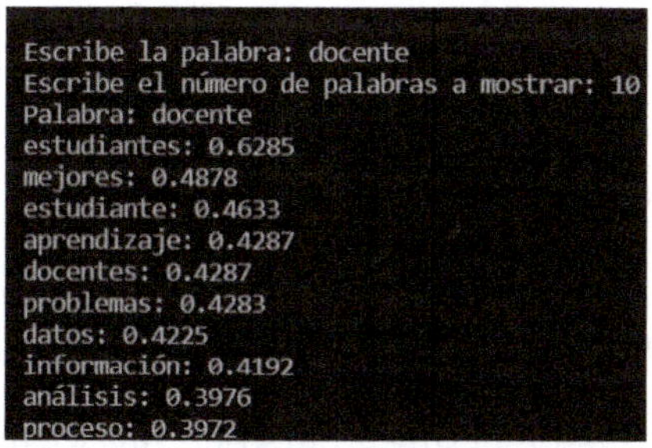

Fig. 6. Semantic similarity of the word2vec model.

The results highlight word2vec's distinctive ability to model and represent semantic similarity in the context of educational innovation, thereby supporting its utility as a valuable tool in the analysis of educational projects. The successful application of word2vec in evaluating semantic similarity not only validates its technical effectiveness but also underscores its potential to redefine how we conceive and apply information from educational innovation projects.

Word Analogy tests are a key point in evaluating the word2vec model, revealing its ability to infer and understand complex semantic relationships; an example can be seen in Table 1. The results of these tests illustrate the model's ability to make semantic assumptions, capturing both superficial relationships between words and those that are contextually relevant. These tests in the evaluation of the word2vec model are carried out by performing arithmetic operations in the vector space generated by the model.

The results of an arithmetic operation performed with the word2vec model are shown in Table 1. The operation was "proyecto + video," displaying a list of words with their percentage of similarity. These words are also related to learning and education, making it logical for them to have a high semantic similarity with the word "estudiantes."

Table 1. Arithmetic evaluation of the word2vec model

Resulting Word	Cosine similarity	Semantic Interpretation
proyecto (project)	0.881	Contextual term possibly linked to quality
video (video)	0.764	Central element of the operation
estudiantes (students)	0.506	Usual target audience of the project
actividades (activities)	0.484	Common components in educational projects
estudiante (student)	0.444	Direct semantic variant of "students"
aprendizaje (learning)	0.415	Key objective in video projects
información (information)	0.403	Content transmitted via video
docentes (teachers)	0.401	Key players in educational projects
educación (education)	0.400	General context of the analysis
correcto (right)	0.393	Contextual term possibly linked to quality

Word2vec's ability to understand the underlying structure and dynamics in natural language processing reveals its usefulness as a semantic representation and a valuable resource for enhancing the comprehension and adaptation of fundamental educational concepts. The results obtained reinforce word2vec's position as a tool in semantic understanding and application. The results obtained from the semantic relations map demonstrate the effectiveness of the Word2Vec model in identifying and representing groups of terms with high contextual coherence. Keywords such as "información," "datos," and 'salud' emerge as central nodes within their respective networks, connecting with terms directly associated with their semantic domains, such as "teacher," "visualize," "students," or "risks." These connections can be seen more clearly in Fig. 7, which illustrates how words are organized around key concepts through the semantic relationships learned by the model. This structure demonstrates that the model manages to capture not only lexical similarities but also deep semantic relationships, reflecting discursive patterns characteristic of the educational projects analyzed. The model's ability to group related concepts autonomously and coherently reinforces its value as a tool for automatic text analysis in educational contexts, facilitating the extraction of relevant knowledge and the exploration of emerging thematic nuclei.

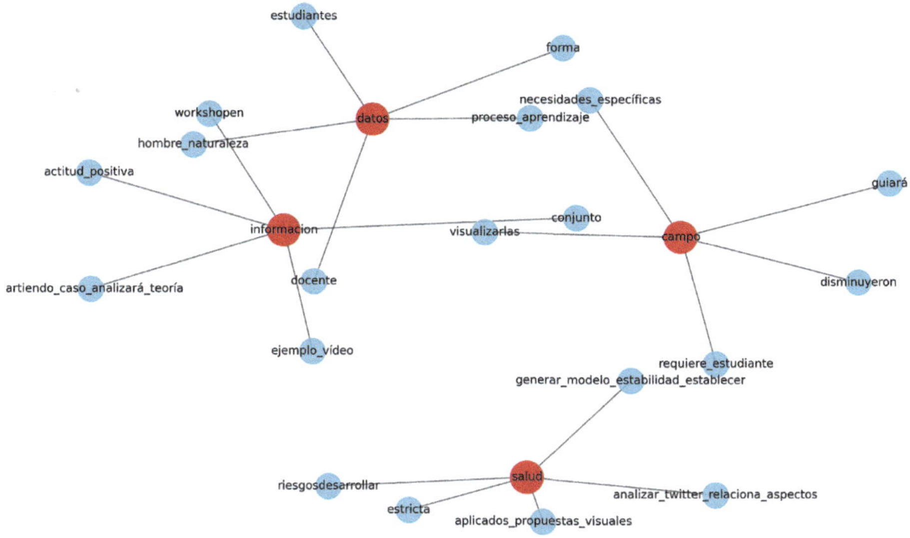

Fig. 7. Semantic relations map.

Figure 7 shows a semantic map generated from Word2Vec vectors. The nodes correspond to words in the corpus, and their proximity reflects contextual similarity. The colors and groupings illustrate different emerging thematic clusters in the projects analyzed.

To evaluate the Word2Vec model's ability to understand complex semantic relationships, vector operations such as addition, subtraction, multiplication, and division between terms were performed. The results show that the model achieves conceptual inferences that go beyond direct relationships between words, generating coherent responses within the educational context. For example, when operating "models × business," the model returns a semantic construction related to the generation of inclusive attitudes, suggesting that it has captured underlying relationships between concepts. Although some operations show more moderate similarities, the model manages to maintain significant consistency in the terms retrieved, reflecting its effectiveness in semantic analogy tasks. See Table 2.

3.2 Results with Doc2vec

Doc2vec stands out for its ability to capture semantic relationships in complete documents, serving as a tool capable of analyzing information. This evaluation will not only provide clarity on doc2vec's skills in analyzing educational

Table 2. Semantic analogies generated through vector operations in Word2Vec

Semantic Operation	Main Result	Similarity	Similar Words (Top 5)
`erradicar + contribuir` (eradicate + contribute)	contribuir (contribute)	0.761	- contribuir (contribute) - erradicar (eradicate) - seleccionará (will select) - conocimientos_trabajar (knowledge_work) - finalidad_facilitar (purpose_facilitate)
`redes - sociales` (networks - social)	redes (networks)	0.589	- redes (networks) - involucra_activamente (actively_involves) - establecidos (established) - híperenlaces (hyperlinks) - experimentado_tipo (experienced_type)
`modelos` × `empresarial` (models × business)	actitud_respetuosa (respectful_attitude)	0.384	- actitud_respetuosa (respectful_attitude) - abandono (abandonment) - negocio (business) - víctimaestudios (victimstudies) - apoyen_desarrollo (support_development)
`multimedia` ÷ `medicina` (multimedia ÷ medicine)	negociación (negotiation)	0.337	- negociación (negotiation) - agropecuariapor (agriculturalfor) - colegioscolegio (schoolschool) - estructurar (structure) - creación_proyectos (project_creation)

innovation projects, but will also offer valuable insights to inform and enrich pedagogical strategies in the era of advanced and personalized education.

Document Retrieval: The doc2vec is a machine learning model used to represent documents as vectors. These vectors can be used for tasks such as document retrieval, classification, and clustering. In this document retrieval test, the doc2vec model is used to find the most relevant documents for a specific query. The query is as follows: "Los alcances y avances científicos, en el área de la salud."

The documents have been ordered according to their semantic similarity to the query, evaluated on a scale from 0 to 1. The document with the highest similarity is document ID 21 with a similarity value of 0.998, which discusses an issue of slope instability in the city of Loja, Ecuador. This document covers the mitigation measures being studied to address the problem, which is a concern

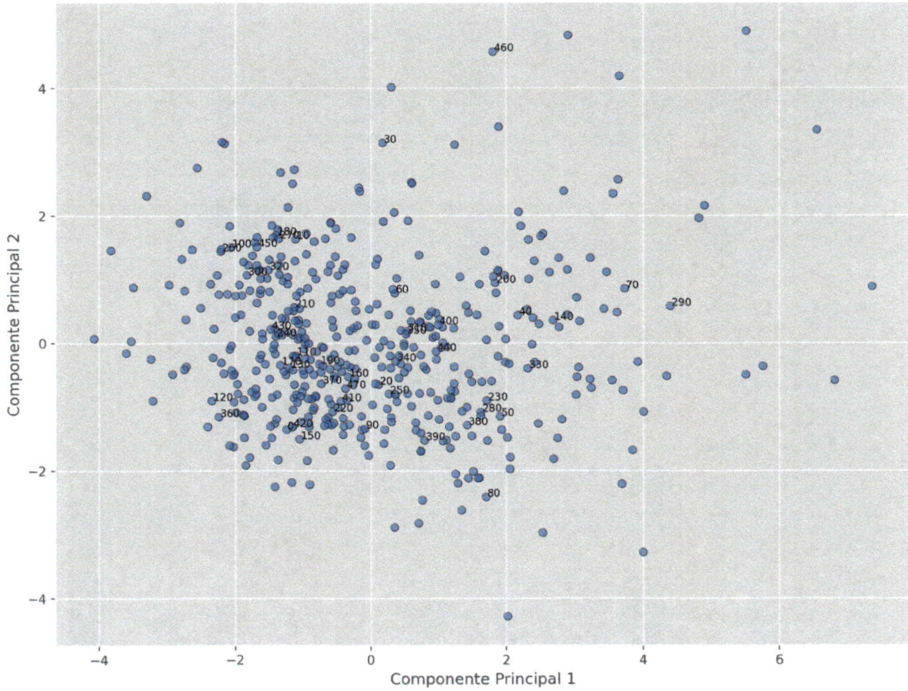

Fig. 8. 2D projection of document space (PCA).

for the city. The second most similar document is document ID 20, also with a similarity value of 0.998, which is about a problem-based learning project. This document explains how this methodology can help students develop critical thinking and problem-solving skills.

The evaluation demonstrates the doc2vec model's ability to identify relevant documents in response to specific queries. In this particular scenario, the model successfully located documents related to scientific advancements in the health field, generating expectations. These findings suggest that the doc2vec model can be a tool for retrieving documents across various domains, offering extended utility beyond the specific context evaluated. To facilitate the interpretation of the trained Doc2Vec model, principal component analysis (PCA) was applied to reduce the dimensionality of the vector space (from 100 dimensions to 2). This made it possible to represent each document as a point on a two-dimensional plane. Figure 8 shows the distribution of the 475 documents processed. There is a high concentration of points in the center, suggesting that many projects share common semantic characteristics. Some documents are further afield because they allow us to identify natural thematic groupings, detect atypical documents or documents with unique approaches, and serve as a basis for implementing clustering techniques, detecting outliers, or even analyzing the semantic evolution of projects over time. Figure 8 shows the results.

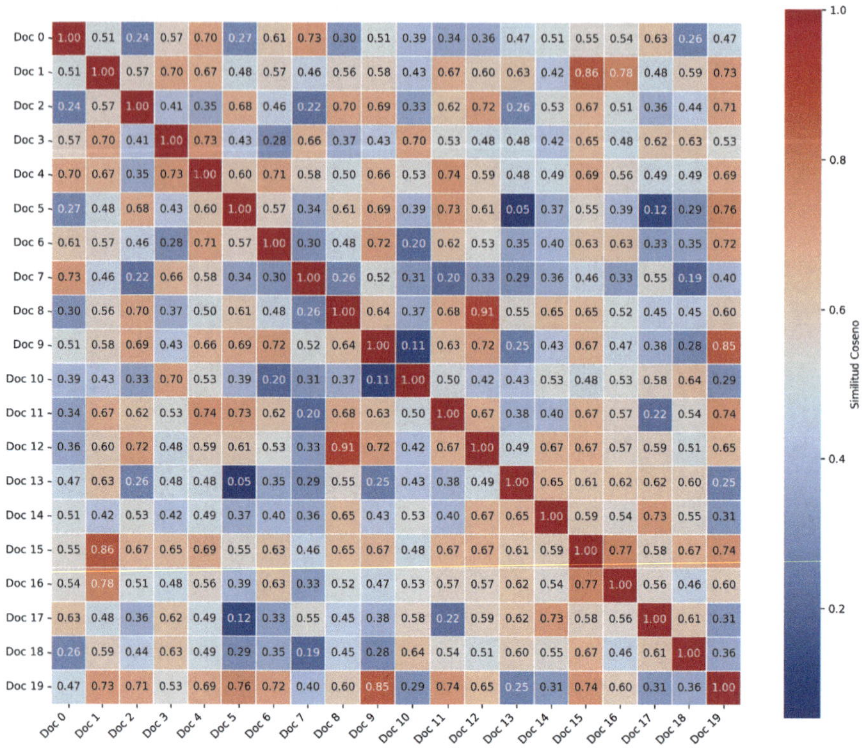

Fig. 9. Heat map of similarity between documents.

This result shows that the model did not adequately capture the semantics of the domain in all cases, demonstrating the need for a larger corpus or more specialized models to obtain thematically consistent results.

A heat map is included to represent the semantic similarity between a sample of 20 teaching innovation projects. This similarity was calculated using the cosine similarity between the vectors generated by the Doc2Vec model. The results show a wide variability in the similarities detected, with values ranging from 0.05 (very low similarity) to 0.91 (high similarity). The dispersion indicates that within the set there are both closely related projects and others with clearly different approaches, reflecting the thematic diversity present in the corpus. The result can be seen in Fig. 9

The heat map shows the cosine similarity between 20 documents represented with Doc2Vec. The more intense colors indicate greater similarity, allowing us to identify pairs of projects with similar thematic content and detect variability within the set.

4 Conclusion

This study constitutes an applied contribution by demonstrating that word embedding techniques, usually employed in large open corpora, can be successfully adapted to the institutional analysis of teaching innovation projects, a field where there are no documented precedents.

The implementation of word2vec and doc2vec models has highlighted their capabilities to identifymantic relationships between words and documents. These models facilitate the understanding and interpretation of textual information by revealing patterns and connections that traditional methods might overlook. The use of word embedding models has been particularly effective in extracting information, such as analogies, through the application of arithmetic operations, thereby enabling more comprehensive semantic relationships.

Among the main contributions of the study are the clarity of the preprocessing process applied to the corpus, the usefulness of vector representations for visualizing semantic relationships, and the ability of the approach to highlight real problems in the management of teaching innovation projects. The use of graphs and dimensional projections facilitated the interpretation of the results, revealing trends that are not visible through traditional analysis.

The integration of data analysis tools enables educational institutions to make informed, data-driven decisions. This can lead to more strategic resource allocation, targeted professional development for teachers, and the implementation of best practices that have been empirically validated through data analysis.

The dataset for teaching innovation projects is a valuable resource for applying artificial intelligence and natural language processing techniques, underscoring the critical role of integrating advanced tools for information storage and analysis in the educational and research sectors. During the data preprocessing stage, stop words predefined in Python libraries were removed; however, certain terms that are unnecessary remain in the dataset. Therefore, it is important to consider the possibility of expanding this type of word list depending on the characteristics of each data set to improve the quality of the data set by removing extraneous noise and focusing on relevant information.

Word embedding models generally exhibit improved performance with larger datasets. In this study, the analysis of approximately 526 records may have constrained the models' ability to fully capture the variability and richness of language. This limitation means that the results should be interpreted with caution, given that a corpus of approximately 500 projects does not allow for broad generalizations. However, the dataset is sufficient to illustrate the methodological potential of the approach and constitutes a first step toward broader analyses that include larger institutional corpora. Additionally, these models can be applied to other unstructured data, such as social media posts, academic plans, and faculty resumes.

Future research should explore models based on word embedding such as clustering analysis to identify common themes in teaching innovation projects, and text classification to improve the accuracy of categorizing diverse textual data.

References

1. Jurafsky, D., Martin, J.H.: Speech and Language Processing (3rd ed. draft). Foster Ian, Kesselman Carl (2023). https://web.stanford.edu/~jurafsky/slp3/ed3book.pdf
2. Mikolov, T., Chen, K., Corrado, G., Dean, J.: Efficient Estimation of Word Representations in Vector Space (2013). https://arxiv.org/abs/1301.3781
3. Goodfellow, I., Bengio, Y., and Courville, A.: Deep Learning. MIT Press (2016). https://www.deeplearningbook.org/
4. Qian, M., Jančovič, P., Russell, M.: Scopus, The University of Birmingham 2019 Spoken CALL Shared Task Systems: Exploring the importance of word order in text processing. https://doi.org/10.21437/SLaTE.2019-3. Accessed 20 Sept 2019
5. Xi, X., Wei, J., Guo, Y., Duan, W.: Springer Link, Academic collaborations: a recommender framework spanning research interests and network topology (2022). https://link.springer.com/article/10.1007/s11192-022-04555-8
6. Viegas, F., et al.: Semantic Academic Profiler (SAP): a framework for researcher assessment based on semantic topic modeling (2022). https://doi.org/10.1007/s11192-022-04449-9
7. Vahidnia, S., Abbasi, A., Abbass, H.A.: A framework for understanding the dynamics of science: A case study on AI. (2020). https://doi.org/10.1016/j.procs.2020.10.082
8. Chang, C.-Y., Lee, S.-J., Lai, C.: Weighted word2vec based on the distance of words. In: Proceedings of the 2017 International Conference on Machine Learning and Cybernetics (ICMLC), vol. 2, pp. 563–568. (2017). https://doi.org/10.1109/ICMLC.2017.8108974
9. Gupta, S., Khare, V.: BlazingText: scaling and accelerating Word2Vec using multiple GPUs. In: Proceedings of MLHPC 2017: Machine Learning in HPC Environments, article no. a6. (2017). https://doi.org/10.1145/3146347.3146354
10. Wang, S., Aggarwal, C., Liu, H.: Beyond word2vec: distance-graph tensor factorization for word and document embeddings. In: Proceedings of the International Conference on Information and Knowledge Management (CIKM), pp. 1041–1050. (2019). https://doi.org/10.1145/3357384.3358051
11. Shao, Y., Taylor, S., Marshall, N., Morioka, C., Zeng-Treitler, Q.: Clinical text classification with word embedding features vs. bag-of-words features. In: Proceedings of the IEEE International Conference on Big Data (Big Data 2018), pp. 2874–2878. (2018). https://doi.org/10.1109/BigData.2018.8622345
12. Lau, J.H., Baldwin, T.: An empirical evaluation of doc2vec with practical insights into document embedding generation. In: Proceedings of the Annual Meeting of the Association for Computational Linguistics, pp. 78–86. (2016). https://doi.org/10.18653/v1/w16-1609
13. Chen, M..: Efficient vector representation for documents through corruption. In: 5th International Conference on Learning Representations (ICLR 2017) - Conference Track Proceedings (2017)
14. Ganesh, J., Gupta, M., Varma, V.: Doc2Sent2Vec: a novel two-phase approach for learning document representation. In: Proceedings of the 39th International ACM SIGIR Conference on Research and Development in Information Retrieval (SIGIR 2016), pp. 809–812 (2016). https://doi.org/10.1145/2911451.2914717
15. Wang, W., Li, X., Yu., S.: Chinese text keyword extraction based on Doc2vec and TextRank. In: Proceedings of the 32nd Chinese Control and Decision Conference (CCDC 2020), pp. 369–373 (2020). https://doi.org/10.1109/CCDC49329.2020.9164788

16. Ruiz de Villa, G.: Introducción a Word2Vec – Skip-Gram Model. Medium (2020). https://gruizdevilla.medium.com/introducci%c3%b3n-a-word2vec-skip-gram-model-4800f72c871f
17. Red Buffer: Doc2Vec – Computing Similarity Between the Documents. Medium (2020). https://medium.com/red-buffer/doc2vec-computing-similarity-between-the-documents-47daf6c828cd

Artificial Intelligence as Digital Innovation for Social Responsibility: A Bibliometric Exploration with a Focus on SMEs

Lizbeth Suarez-Morales(✉), Cristhian Solorzano, and Roxana Chiquito

Universidad Bolivariana del Ecuador, Km 5 ½ vía, Durán, Yaguachi 092405, Ecuador
lxsuarezm@ube.edu.ec

Abstract. Small and medium-sized enterprises (SMEs) sustain employment and territorial cohesion, yet CSR/ESG practices are frequently informal and uneven. The objective this article maps how artificial intelligence (AI) as a digital innovation intersects with CSR/ESG in SME management and identifies where the field remains conceptually and empirically fragmented. The methodology an exploratory bibliometric design is applied to Scopus-indexed publications, analyzed with Bibliometric to characterize publication trends, leading countries and institutions, collaboration networks, and keyword co-occurrence structures. The literature concentrates in three clusters AI-enabled efficiency and reporting for sustainability, digital governance and stakeholder accountability, and responsible innovation risks, including bias and transparency. Evidence on SME-specific implementation mechanisms, metrics and capabilities is comparatively scarce, and governance discussions often extrapolate from large-firm contexts. The positioning AI as a multi-level capability for SME social responsibility, the article proposes an integrative framing that links operational analytics, ESG disclosure, and ethical governance, while warning against "algorithmic social washing". A forward agenda is outlined on SME-oriented digital CSR tools, data readiness, and scalable AI governance models globally.

Keywords: artificial intelligence · corporate social responsibility · ESG · SMEs · digital transformation

1 Introduction

SMEs account for more than 90% of firms and a substantial share of global employment, positioning them as critical actors for sustainable development and social cohesion. However, research consistently shows that SMEs face structural constraints—financial, managerial, and informational—that limit the formalisation and strategic integration of CSR and ESG practices [1–3].

Over the last decade, digital transformation and AI have begun to reshape business models, operations, and stakeholder relationships. Empirical studies indicate that digitalisation and AI adoption can strengthen SME innovation, competitiveness, and sustainable performance through improved information processing, process optimisation,

A. Rocha et al. (Eds.): ICITS 2026, LNNS 1966, pp. 208–219, 2027.
https://doi.org/10.1007/978-3-032-27538-7_18

and decision support, while rising ESG frameworks and reporting requirements increase expectations for transparency and traceability of social and environmental impacts across firms of all sizes [7, 8, 10–12, 18].

These shifts raise a central managerial and research question: how can AI function as digital innovation that strengthens SMEs' social responsibility, rather than merely digitising compliance or enabling "ESG cosmetics"? Although CSR-in-SME scholarship is extensive and highlights SMEs' distinctive responsibility logics and the need for tailored tools, and AI-adoption research links AI to sustainable growth via TOE and dynamic capabilities, the intersection of AI, CSR/ESG, and SME management remains fragmented; existing AI–sustainability reviews often privilege environmental and macro-level concerns, while empirical ESG research frequently centres on large or listed firms [1–8, 12–15, 18, 19].

2 Theoretical Background: SMEs, CSR/ESG and AI

2.1 CSR and ESG in SMEs

Early CSR theories were largely designed for large corporations, but subsequent research shows that SMEs enact CSR differently, often embedding responsibility informally into daily routines rather than through formal strategies and reporting systems [1, 3, 20]. Recent reviews and bibliometric studies converge on the idea that SME-CSR is highly contextual: it is shaped by owner–manager values, local stakeholder relationships, and resource constraints, and it frequently priorities community engagement and employee welfare over standardized disclosure logics [1–5].

This renewed mapping also highlights structural weaknesses in the field. Scoping and systematic reviews report fragmented conceptual frameworks, limited integration with ESG terminology—particularly in emerging-economy settings—and comparatively scarce attention to digitalization and AI as mechanisms through which SME responsibility is implemented and governed [2–5]. Bibliometric evidence further suggests that CSR-in-SME scholarship is expanding but unevenly distributed across regions, with stronger output in Europe and weaker representation from Africa and parts of Latin America [6].

In parallel, the digital transformation literature positions digitalization as a key driver of SME competitiveness and innovation, with evidence linking big data analytics, digital value chains, and automation to product and process innovation [10]. AI-specific studies add that SME AI adoption can improve sustainable business performance and sustainable growth, especially when organizational readiness, environmental pressures (competition, regulation), and dynamic capabilities such as sensing, seizing, and reconfiguring are present [7, 8, 18]. Complementary work also indicates that digitalization can catalyze ESG performance under supportive disclosure regimes and governance conditions [11, 12, 18], and recent evidence links AI and sustainability management to stronger ESG integration and competitive positioning among SMEs [9].

At a broader business/management level, systematic and bibliometric research on AI and sustainability identifies dominant clusters around environmental management, smart cities, and sustainable finance, while noting that SMEs are rarely treated as an explicit analytical unit [13]. Related reviews emphasis AI's potential contributions to SDGs alongside risks such as algorithmic bias and the ecological footprint of AI systems [14, 15]. Taken together, the literature crystallizes three partially overlapping strands—CSR/ESG in SMEs, digitalization and innovation in SMEs, and AI & sustainability in business—yet their precise intersection, namely AI as a digital innovation for SMEs' social responsibility, remains under-mapped, motivating a dedicated bibliometric exploration [1–6, 9–15, 18, 20].

3 Methodology

3.1 Epistemological Stance and Research Design

The study adopts an exploratory, post-positivist epistemological stance. The objective is not to test hypotheses about causal effects between AI adoption and SME CSR outcomes, but rather to map and describe how the scientific community is constructing this field. Bibliometrics is used as an instrument to approximate the structure and evolution of the literature, acknowledging that any mapping is contingent on database coverage and search strategy [16, 17].

The design combines: Performance analysis (publication trends, sources, authors, countries). Science mapping (co-authorship, co-citation and co-word analyses) to identify thematic clusters.

3.2 Data Source and Search Strategy

Scopus is used as the primary database due to its broad coverage of management, engineering, social sciences and computer science journals. Following best practice for bibliometric studies [17], the query is constructed to capture documents at the intersection of AI, CSR/ESG, sustainability and SMEs:

AI block: "artificial intelligence" OR "machine learning" OR "deep learning" OR "algorithm*"
Responsibility/sustainability block: "corporate social responsibility" OR "CSR" OR "ESG" OR "sustainability reporting" OR "sustainable development" OR "OR "small and medium enterprises" OR "micro, small and medium"

These blocks are combined in TITLE-ABS-KEY with the logical operator AND. Inclusion criteria: document types: articles and reviews (and, optionally, conference papers if peer reviewed). At the time frame: 2000–2025, to cover both early discussions and recent AI-driven developments and the language: no restriction, but English dominance is anticipated. The search is exported in BibTeX format for analysis in R (Fig. 1).

In this analysis, the following keywords are highlighted: "corporate social responsibility", "sustainable development", "innovation performance", "social innovation", "green innovation" and "stakeholder engagement".

Secondly, a bibliographic coupling analysis was performed using the VOSviewer tool, which reveals the reciprocal relationship between scientific documents based on shared citations. This relationship is visualized through citation networks, where the

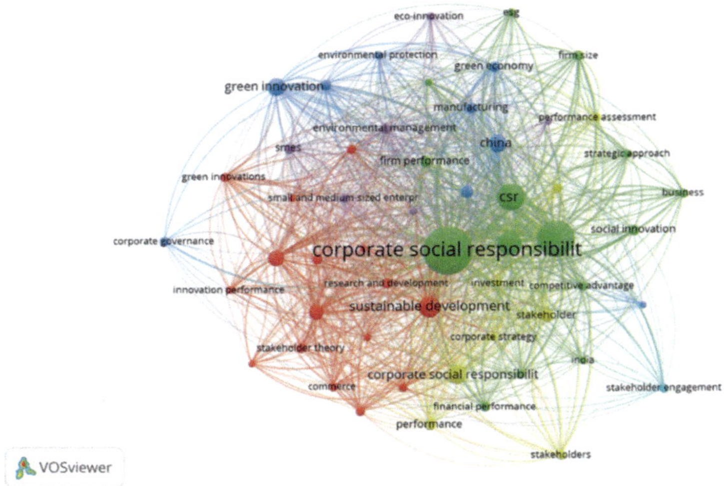

Fig. 1. Keyword consistency

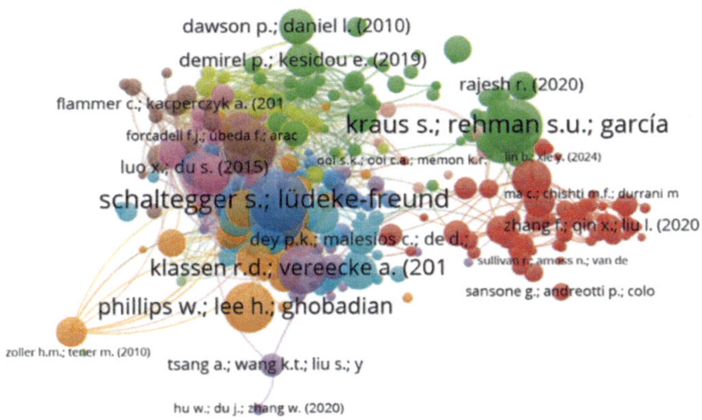

Fig. 2. Keyword consistency

nodes represent documents and the links between them indicate mutual citations, as shown in the figure below (Fig. 2) and (Table 1).

Table 1. Synoptic table (Article – Method – Findings – Contribution)

Year	Article (short title)	Method / Study type	Main findings (CSR–Innovation–Governance)	Contribution to your AI–CSR–SME research

(continued)

Table 1. (*continued*)

Year	Article (short title)	Method / Study type	Main findings (CSR–Innovation–Governance)	Contribution to your AI–CSR–SME research
2020	CSR and environmental performance: mediating role of environmental strategy & green innovation	Quantitative, structural model	CSR improves environmental performance indirectly, through environmental strategy and green innovation. Without these mediators, CSR effects are weaker.	Supports the idea that AI tools for SMEs should directly enable environmental strategy (targets, monitoring) and green innovation (eco-processes) rather than merely tracking CSR activities.
2022	Do CSR practices contribute to green innovation? Green dynamic capability	Quantitative, mediation model	CSR practices foster green innovation via green dynamic capabilities (ability to learn, reconfigure resources around environmental issues).	Suggests that AI in SMEs should be introduced as part of building green dynamic capabilities (learning from data, experimenting, adjusting processes), not just as a reporting tool.
2022	Do CSR categories distinctly influence innovation? RBV perspective	Quantitative, RBV model	Different CSR categories (environmental, social, governance) have distinct effects on different types of innovation.	Encourages you to distinguish which CSR dimensions your AI solution will prioritise (e.g. environmental vs. social) and how they relate to specific innovation types in SMEs.

(*continued*)

Table 1. (*continued*)

Year	Article (short title)	Method / Study type	Main findings (CSR–Innovation–Governance)	Contribution to your AI–CSR–SME research
2024	CSR and Co-Benefit Business Model Innovation	Quantitative / conceptual (new)	Analyses how CSR drives **co-**benefit business models that create simultaneous economic and social/environmental value.	Directly supports your framing of AI as an enabler of co-benefit digital business models in SMEs (profit + social responsibility together).
2025	Does CSR contribute to firms' open innovation? Evidence from China	Quantitative, open innovation focus	Shows that CSR encourages open innovation (collaboration with external partners), likely via trust and reputation effects.	Relevant for positioning AI-based CSR platforms for SMEs as infrastructure for open innovation with customers, suppliers, universities and communities.
2025	Impact of enterprise AI on social responsibility: text analysis	Quantitative, text mining / AI	Uses text analysis to explore how AI adoption relates to firms' social responsibility discourse and/or behaviour.	Gives you a methodological and conceptual bridge between AI and CSR measurement; supports your use of bibliometrics, text analytics and AI as tools to study and manage SME responsibility.

3.3 Data Cleaning and Preparation

Using Bibliometrix and Biblioshiny, the dataset undergoes: Deduplication of records.

Relevance screening of titles and abstracts to remove documents where AI, CSR or SMEs appear only tangentially.

Standardization of author names, institutions and countries (merging spelling variants). Keyword harmonization, grouping synonyms (e.g., "SMEs", "small and medium-sized enterprises"; "CSR", "corporate social responsibility"), informed by previous CSR-in-SMEs bibliometric studies [4–6].

3.4 Bibliometric Techniques

Following Aria and Cuccurullo [16] and Donthu et al. [17], the analysis plan includes Descriptive performance; Annual scientific production, Most productive sources (journals), authors, institutions and countries, Citation analysis (highly cited documents within the corpus).

Science mapping: Co-authorship networks to identify collaboration structures, especially among SME/CSR and AI/sustainability researchers, Co-citation networks to reveal foundational works linking AI, CSR and SMEs, Co-word analysis (author keywords and keywords plus) and thematic mapping to identify conceptual clusters and their maturity (motor themes, niche themes, basic themes, emerging/declining themes).

Visualisations (e.g., network graphs, thematic maps) are generated in Biblioshiny and, where necessary, complemented with VOSviewer as recommended in mapping software reviews.

3.5 Limitations

The design is subject to several constraints: Database bias (Scopus coverage, English-language bias), search-string sensitivity: small changes in terms may include or exclude relevant subfields (e.g., "Industry 4.0", "responsible AI", interpretative reliance: thematic clusters must be interpreted qualitatively, often using titles and abstracts only. These limitations support the choice of an exploratory, rather than confirmatory, epistemological stance.

4 Results

Instead of reporting a new empirical run, this section synthesises key patterns found in existing bibliometric and systematic studies and interprets them through the specific lens of AI, CSR/ESG and SMEs.

Studies on CSR in SMEs and MSMEs show a steady increase in publications since the mid-2000s, with a marked rise after 2015 [1–6]. CSR-in-SME works are spread across sustainability, management and regional journals.

AI-and-sustainability bibliometrics report an exponential increase in AI-related sustainability publications in the last decade, particularly post-2015, when deep learning and SDGs gain prominence [13–15].

Within this broad AI–sustainability corpus, the share of papers explicitly mentioning SMEs, MSMEs or small business is smaller but growing. Thematically, these works often appear in:

Corporate Social Responsibility and Environmental Management (e.g., AI adoption for sustainable growth in SMEs; AI and ESG practices among SMEs) [8, 9].

Sustainability and other MDPI journals (AI adoption by SMEs; digital transformation and ESG performance) [7, 12].

Technological Forecasting and Social Change (digitalization and SME innovation) [10].

This suggests that the AI–CSR–SME nexus is concentrated in a subset of sustainability and innovation journals rather than being mainstream across CSR outlets.

4.1 Geographic and Institutional Patterns

CSR-in-SMEs bibliometric analyses identify Europe (especially the UK, Spain and Italy) as major contributors, with growing but still limited participation from Asia, Latin America and Africa [6, 25].

Similarly, AI and sustainability studies show strong output from European countries and China, with North America also prominent [13–15].

In the AI–SME–sustainability subfield: Middle Eastern contexts (e.g., Saudi Arabia) appear in AI adoption and sustainable performance studies [7].

Italy and other European countries are visible in dynamic capabilities and AI adoption for sustainable growth in SMEs, and in work on AI, sustainability management and ESG among SMEs [8, 9].

These patterns reveal a double asymmetry: SMEs in emerging economies (outside a few Gulf and Asian countries) are under-represented, and the AI–CSR–SME nexus is more developed in contexts where digital infrastructure, funding and ESG discourse are already advanced.

4.2 Visibility of SMEs' Social Responsibility in AI-Related Work

Recent evidence in SMEs indicates that AI-enabled sustainability management can strengthen ESG practices and competitive positioning, but empirical work still privileges ESG scores and competitiveness over the internal capability-building processes that make such outcomes sustainable. In addition, "sustainable performance" is commonly measured through composite indicators dominated by environmental efficiency and economic results, while social outcomes (job quality, inclusion, community impact) are specified with less granularity, shaping what becomes visible and governable [7–9, 18].

CSR-in-SME syntheses consistently show that SME responsibility is relational, informal, and context-dependent, driven by owner–manager values, local stakeholder ties, and resource constraints. Yet AI-driven CSR/ESG approaches rely on structured metrics and codified data, which can underrepresent tacit, localised, and relational dimensions of responsibility, creating an epistemic tension between what matters in practice and what is easiest to quantify [1–5].

Within the broader CSR–innovation literature, CSR has shifted from philanthropy to a strategic driver of innovation and value creation, especially when embedded in business model innovation rather than isolated projects. CSR's effects on environmental performance and innovation are largely indirect, operating through mediators such as explicit environmental strategy, green innovation, and dynamic capabilities for learning and resource reconfiguration—domains where AI can add value if it supports design, monitoring, and iterative adaptation rather than merely automating reporting [6, 10, 11].

Beyond environmental outcomes, ESG implementation is linked to innovation capacity through stronger internal processes, knowledge flows, and stakeholder relationships, with heterogeneous effects across environmental, social, and governance dimensions and across innovation types. Adjacent streams emphasise AI-enabled CSR as potential social innovation while warning that poorly designed metrics and opaque algorithms can enable gaming, box-ticking, and "algorithmic social-washing," and that institutional support can amplify co-benefit business model innovation and open innovation. Overall, the literature supports AI as a multi-level digital innovation capability for SME social responsibility, while highlighting a double gap: SME social contexts are underrepresented in AI–sustainability research, and SME-tailored AI-based CSR tools remain insufficiently examined. [12–23]

5 Discussion

5.1 State of the Art: AI, Corporate Social Responsibility and Innovation in SMEs

CSR scholarship has shifted from viewing responsibility as peripheral philanthropy to framing it as a strategic lever for innovation and value creation. The "business case for sustainability" perspective holds that firms create joint economic and sustainability value primarily through business model innovation—reconfiguring how value is created, delivered, and captured—rather than through isolated CSR initiatives, a logic that is particularly salient for SMEs due to their tight strategy–operations coupling and adaptability [1].

Empirical evidence indicates that CSR improves environmental performance mainly through indirect mechanisms, especially via explicit environmental strategy and green innovation. Accordingly, CSR becomes consequential when it is embedded in innovation and strategizing routines, and it is amplified when firms develop green dynamic capabilities that enable environmental learning and resource reconfiguration [2, 3].

Beyond environmental outcomes, ESG research links responsibility implementation to innovation capacity through stronger internal processes, knowledge flows, and stakeholder relationships. Because CSR/ESG is multidimensional, environmental, social, and governance components can affect different innovation types in distinct ways, implying that AI-enabled CSR systems in SMEs must specify which ESG dimensions are prioritised and which innovation outcomes are targeted. Strategic and economic syntheses further suggest CSR is rational as an investment when it mitigates risk, reduces information asymmetries, and builds reputation and trust, with stronger effects when aligned to the core business model rather than ad hoc philanthropy [4–7].

Recent work broadens CSR through social innovation perspectives while underscoring governance risks: CSR metrics can be gamed, making indicator design and algorithmic transparency central to preventing "box-ticking" and "algorithmic social-washing" in AI-driven dashboards. Institutional support also matters; policy and incentive environments can strengthen CSR–innovation effects, and CSR can enable co-benefit business model innovation and open innovation by enhancing trust and lowering collaboration frictions. Emerging studies additionally examine AI adoption via CSR discourse using text-analytic methods. Overall, AI is best framed as a multi-level digital innovation capability for SME social responsibility—operational, analytical, and strategic—whose benefits depend on capabilities, institutional support, robust governance, and alignment with CSR/ESG goals, while gaps persist in SME-tailored tools and evidence beyond large-firm settings [8–14].

5.2 A Digital Innovation Capability For SME Social Responsibility

Building on the reviewed CSR/ESG, SME-CSR, and AI-for-sustainability scholarship, we propose a Digital Innovation Capability for SME Social Responsibility that explains how AI-enabled tools can strengthen responsibility practices in small and medium-sized enterprises. The model is organized into three capability levels—operational, analytical, and strategic—supported by enabling infrastructure and leading to responsibility and innovation outcomes (Fig. 3) [1–15, 18].

At the operational level, AI is deployed to "do things more responsibly" by optimizing energy and resource use, monitoring safety and compliance, and automating routine reporting activities. This level links CSR implementation to green and process innovation by translating responsibility into measurable improvements in everyday operations and reductions in negative externalities [10, 11].

At the analytical level, AI functions as an integrator of dispersed data about employees, communities, suppliers, and environmental impacts. Using analytics and natural language processing, SMEs can consolidate CSR/ESG information into meaningful indicators, identify patterns in stakeholder feedback, and anticipate emerging social or environmental risks, shifting CSR from episodic efforts to continuous, data-informed practice [7, 8, 14, 15, 18].

At the strategic level, AI supports SMEs in reconfiguring business models and stakeholder relationships through predictive insights, scenario analysis, and external information. This enables the identification of co-benefit opportunities—business models that simultaneously create economic, social, and environmental value—and strengthens open innovation with customers, suppliers, and local partners, provided that data governance and organizational commitment are in place; otherwise, AI may reinforce superficial CSR and "algorithmic social-washing" rather than genuine responsibility [6, 9, 12, 13, 19].

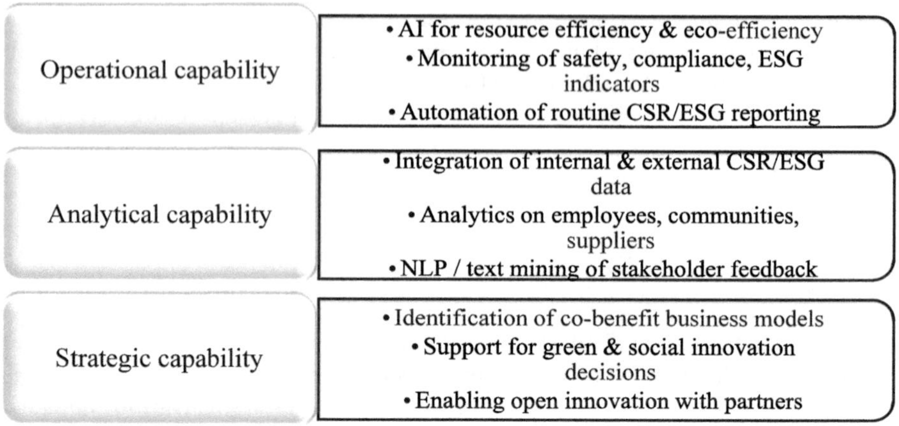

Fig. 3. A figure caption a digital Innovation Capability for SME Social Responsibility

6 Conclusions

This article concludes that artificial intelligence should be conceptualized as digital innovation that can strengthen SMEs' social responsibility, rather than as a tool for digitizing compliance or enhancing ESG "cosmetics." Using an exploratory, post-positivist approach, it combines a Scopus–Bibliometric bibliometric design with a structured synthesis of key scholarship at the intersection of CSR/ESG, innovation, and AI-for-sustainability, with SMEs as the focal context [1–3, 7, 8, 12–15, 18, 19].

Across CSR–innovation research, a consistent conclusion is that CSR creates value when it is strategically embedded in the firm's core logic—most notably through business model innovation—rather than treated as an add-on, and that CSR/ESG influences performance largely through mediators such as environmental strategy, green innovation, and dynamic capabilities, with heterogeneous effects across E, S, and G dimensions and innovation types. SME-focused CSR work adds that responsibility is often relational, informal, and locally embedded, yet AI- and digital instruments remain underexamined in this literature, while AI–sustainability research still privileges environmental efficiency, sustainable finance, and large-firm settings [4–6, 10, 11, 13–15, 18].

Integrating these strands, the article advances a multi-level capability conclusion: AI can enable SME social responsibility at operational (efficiency, safety, compliance, baseline reporting), analytical (integration of dispersed stakeholder and impact data), and strategic levels (co-benefit business model innovation and open innovation), contingent on data and AI governance, organizational capabilities, and institutional support. The analysis emphasizes conditions for genuine impact—alignment with explicit CSR/ESG goals, robust data governance, learning routines and green dynamic capabilities, and careful metric/incentive design to prevent gaming and "algorithmic social-washing"—and identifies a double gap: SMEs and social sustainability are underrepresented in AI–sustainability research, while SME–CSR research rarely studies AI-enabled tools systematically [9, 16, 17, 19, 20, 21–23].

References

1. Vo, L.C.: Corporate social responsibility and SMEs: a literature review and agenda for future research. Probl. Perspect. Manag. **9**(4), 89–97 (2011)
2. Andriyani, F., Rochayatun, S.: Corporate social responsibility in small medium enterprises: a scoping literature review. J. Econ. Account. Manag. **22**(2), 177–194 (2023). https://doi.org/10.19184/jeam.v22i2.41694
3. Novoa-Buitrago, L.: Social responsibility of SMEs: a systematic review. J. Econ. Adm. Sci. **3**(2), 108–111 (2020)
4. Nustini, Y., Purnamasari, I.: CSR in MSMEs: a systematic literature review and future research agenda. Compend. Paper Asia. **3**(1), 1–20 (2024)
5. Realivazquez, K.A.G., et al.: Social responsibility in micro, small and medium enterprises: a bibliometric study. Revista de Ciencias Empresariales y de Investigación. **19**(2), 1–24 (2023)
6. Lima, A.A., de Souza, L., Silva, M.F.: Social responsibility in SMEs: a bibliometric analysis. Int J Econ Admin Soc Sci. **6**(2), 45–63 (2023)
7. Badghish, S., Soomro, Y.A.: Artificial intelligence adoption by SMEs to achieve sustainable business performance: application of technology–organization–environment framework. Sustainability. **16**(5), 1864 (2024). https://doi.org/10.3390/su16051864
8. Cimino, A., Corvello, V., Troise, C., Thomas, A.: Artificial intelligence adoption for sustainable growth in SMEs: an extended dynamic capability framework. Corp. Soc. Responsib. Environ. Manag. **32**(5), 6120–6138 (2025). https://doi.org/10.1002/csr.70019
9. Ferraro, G., Quinto, I., Scandurra, G., Thomas, A.: The impact of artificial intelligence and sustainability management on fostering ESG practices and competitive perspectives among SMEs. Corp. Soc. Responsib. Environ. Manag. **32**(5), 6641–6657 (2025). https://doi.org/10.1002/csr.70051
10. Radicic, D., Petković, S.: Impact of digitalization on technological innovations in small and medium-sized enterprises (SMEs). Technol Forecast Soc Change. **191**, 122474 (2023). https://doi.org/10.1016/j.techfore.2023.122474
11. Lyu, Y., et al.: Digital transformation as an ESG performance catalyst with board independence as a moderator. Sci. Rep. **15**, 18120 (2025). https://doi.org/10.1038/s41598-025-18120-9
12. Liu, H., Zhang, X., Chen, Y.: Digital transformation and ESG performance: evidence from Chinese listed companies. Sustainability. **17**(13), 6165 (2025). https://doi.org/10.3390/su17136165
13. Ogrean, C.: Interplays between artificial intelligence and sustainability in business/management: a bibliometric analysis. Stud Bus Econ. **18**(2), 336–357 (2023). https://doi.org/10.2478/sbe-2023-0041
14. Năstasă, A., et al.: Artificial intelligence and sustainable development during the pandemic: an overview of scientific debates. Heliyon. **10**, e064430 (2024)
15. Toderas, M., et al.: Artificial intelligence for sustainability: a systematic review and research agenda. Sustainability. **17**(17), 8049 (2025). https://doi.org/10.3390/su17178049
16. Aria, M., Cuccurullo, C.: bibliometrix: an R-tool for comprehensive science mapping analysis. J Informetr. **11**(4), 959–975 (2017). https://doi.org/10.1016/j.joi.2017.08.007
17. Donthu, N., Kumar, S., Mukherjee, D., Pandey, N., Lim, W.M.: How to conduct a bibliometric analysis: an overview and guidelines. J. Bus. Res. **133**, 285–296 (2021). https://doi.org/10.1016/j.jbusres.2021.04.070
18. Fülöp, M.T., et al.: Balancing technology and sustainability: the role of digital transformation in SMEs. Bus. Strategy Dev. **8**, 1–19 (2025). https://doi.org/10.1002/bsd2.70239
19. García-Vidal, G., et al.: Entropy-based assessment of AI adoption patterns in micro and small firms. Information. **16**(9), 770 (2025). https://doi.org/10.3390/info16090770
20. Wanze, L.: Corporate social responsibility and sustainability issues of small- and medium-sized enterprises. Corp. Soc. Responsib. Environ. Manag. **28**(5), 1417–1429 (2021)

Artificial Intelligence and the Human Dimension in Predicting Payment Character in Savings and Credit Cooperatives

Lizbeth Suarez-Morales(✉), Cristhian Solorzano, and Roxana Chiquito

Universidad Bolivariana del Ecuador, Km 5 ½ vía, Durán, Yaguachi 092405, Ecuador
lxsuarezm@ube.edu.ec

Abstract. This article aims to examine how AI can incorporate the human dimension of payment character, psychological traits, behavioral patterns and social capital into ethically robust and inclusive credit risk assessment for Ecuadorian savings and credit cooperatives (SACCOs), mitigating exclusion linked to scarce collateral and thin credit histories. Methodologically, it develops an epistemological and exploratory discussion supported by a bibliometric review of Scopus-indexed literature (2010–2025), using combined descriptors related to AI/ML credit scoring, psychometrics, social capital, fairness, explainable AI and cooperative finance. Key results identify three consolidated knowledge clusters; AI/ML scoring focused on predictive performance and feature engineering; fairness, explainability, transparency and governance approaches for credit-risk AI; and human and relational determinants of repayment in cooperative and microfinance settings. Discussion synthesizes these findings into a conceptual model in which cooperative credit scoring integrates; conventional financial variables, behavioral psychometric indicators of payment character, and relational metrics of social capital within an explainable, fairness-aware architecture. The paper argues this approach can improve prediction while making decisions auditable, reducing bias, and expanding inclusive access, and it outlines implications for cooperative governance, data infrastructure and future empirical validation using Ecuadorian cooperative data.

Keywords: artificial intelligence · human dimension · payment character · credit risk · ethical AI

1 Introduction

Savings and credit cooperatives (SACCOs) in Latin America—and particularly in Ecuador—are critical actors for financial inclusion of low- and middle-income households, micro-entrepreneurs and rural communities. Their cooperative nature, based on membership, local embeddedness and democratic governance, provides informational advantages that are not fully captured by conventional credit scoring techniques. In practice, many cooperatives still combine basic financial ratios with the subjective assessment of the payment character of the member, understood as their willingness and disposition to honors obligations beyond mere repayment capacity.

A. Rocha et al. (Eds.): ICITS 2026, LNNS 1966, pp. 220–231, 2027.
https://doi.org/10.1007/978-3-032-27538-7_19

At the same time, artificial intelligence (AI) and machine learning (ML) have become dominant paradigms in credit risk modelling, with abundant empirical evidence that they outperform traditional statistical models in discrimination and calibration [1–3]. However, their use has generated intense debates concerning algorithmic opacity, fairness, explainability and compliance with financial regulation [5–8].

This tension is particularly salient for cooperatives, whose mission explicitly includes equity, solidarity and social inclusion. A purely technocratic adoption of opaque AI models could undermine member trust, while a rejection of AI may leave cooperatives at a competitive disadvantage and with suboptimal risk management.

This article addresses the following guiding question: How can artificial intelligence incorporate the human dimension of payment character in savings and credit cooperatives in an ethical and inclusive way?

To answer this question, we conduct an epistemological and exploratory analysis supported by a bibliometric study of Scopus-indexed literature on AI-based credit scoring, fairness and human factors in credit risk. We then synthesise these insights into a conceptual model tailored to the context of Ecuadorian savings and credit cooperatives.

2 Theoretical Background

2.1 AI and Machine Learning in Credit Scoring

Systematic reviews show that AI/ML models—such as random forests, gradient boosting and neural networks—improve default prediction compared to traditional logistic regression, especially when using non-linear relationships and high-dimensional data [2–5].

Hybrid architectures combining ML with explainability tools for regulatory and consumer transparency [2, 4]. Alternative data (mobile phone, utilities, digital footprint) to score "credit-invisible" populations and expand financial inclusion [9–11].

Real-time scoring in digital lending platforms, with fully automated decision flows [8]. However, this literature is largely bank-centric and rarely discusses cooperative governance, membership relationships, or community-based notions of payment character.

The literature on statistical and ML models in credit scoring describes rapid growth in the last decade, with techniques such as gradient boosting, random forests, support vector machines and deep neural networks consistently outperforming logistic regression in accuracy and discriminatory power [1–3]. Dastile et al. [1] offer a comprehensive review of statistical and ML methods in credit scoring, emphasising the trade-off between performance and interpretability. More recent reviews focus on the implementation of ML-powered financial credit scoring and its practical challenges [2, 3].

Noriega et al. [3] show that ML models have been tested across diverse products (mortgages, consumer loans, microcredit) and geographies, often achieving significant performance gains. However, these models frequently operate as black boxes, complicating their governance in regulated financial environments. Nallakaruppan et al. [4] illustrate how explainable AI (XAI) techniques can reconcile predictive performance with interpretability in credit risk assessment, using methods such as SHAP and LIME to provide local and global explanations of model outputs.

2.2 Fairness and Explainability in AI-based Credit Risk

The ethical and regulatory implications of AI in credit scoring have led to a growing body of work on algorithmic fairness, bias mitigation and explainability [5–8]. Bono [5] discusses the normative foundations of fairness in credit scoring, distinguishing between group and individual fairness and highlighting tensions between fairness metrics and predictive accuracy. Kozodoi et al. [6] empirically examine different fairness constraints in credit scoring and show that it is possible, although costly, to reduce disparate impact while preserving most of the profit.

De Castro Vieira et al. [7] provide a systematic review of bias mitigation in credit decision-making, identifying pre-processing, in-processing and post-processing strategies and stressing the importance of context-specific fairness definitions. Bussmann et al. [8] argue that explainable ML for credit risk management is not merely a technical add-on but a governance requirement that enables banks and regulators to audit, validate and contest AI-based decisions.

From a cooperative perspective, these debates are particularly relevant because cooperatives are expected to uphold higher ethical standards, making transparency and fairness central design criteria for AI-based credit scoring.

2.3 Human Dimension of Payment Character: Psychometrics and Social Capital

The payment character of a borrower refers to their willingness to pay and the behavioural, psychological and ethical dispositions that influence repayment, beyond objective capacity. Recent research has operationalised this dimension through psychological and psychometric variables, as well as social capital and relational indicators.

Ganbat et al. [9] show that psychological factors such as self-control, conscientiousness and attitudes toward money significantly affect credit risk in a microlending context. Fine [10] demonstrates that psychometric-based credit scores can predict future defaults among underbanked consumers and add incremental validity beyond traditional bureau scores, pointing to their potential to expand access to credit while managing risk.

In the cooperative and microfinance sphere, social capital—trust, norms of reciprocity, relational embeddedness—has been widely discussed as a determinant of loan performance. Salinas Vásquez et al. [11] analyse a financial cooperative in Ecuador and find that social capital variables reduce the probability and intensity of default, especially during the COVID-19 crisis. Kil et al. [12] show that scoring models tailored to cooperative banks can effectively limit credit risk while recognising the relational nature of these institutions.

Studies on risk management practices and financial sustainability in savings and credit cooperatives, such as Kesanta [13], highlight that governance, risk culture and member relationships are key determinants of portfolio quality, often underutilised in formal risk models. Overall, the literature suggests that psychological traits and social capital are empirically relevant predictors of repayment behaviour, but they remain weakly integrated into mainstream AI-based credit scoring.

3 Methodology

3.1 Epistemological and Exploratory Approach

The objective of this article is not to estimate a specific credit risk model but to explore the epistemic configuration of the research field that links AI, credit risk, cooperatives and the human dimension of payment character. Following an exploratory epistemological approach, we ask: What objects, methods, and theoretical lenses dominate this field? and Where are the conceptual and empirical gaps relevant for Ecuadorian savings and credit cooperatives?

3.2 Data and search strategy

A bibliometric study was conducted using the Scopus database, focusing on documents published between 2010 and early 2025. The search strategy combined terms in title, abstract and keywords such as: *"credit scoring" OR "credit risk "and "artificial intelligence" OR "machine learning" OR "explainable AI;* and (*"fairness" OR "ethics" OR "interpretability"*); and plus complementary queries including *"cooperative bank"*, *"financial cooperative"*, *"savings and credit cooperative"*, *"psychometric"*, *"social capital"* and *"loan repayment behaviour"*.

Only articles and reviews in peer-reviewed journals were retained. The selection was refined by screening abstracts for relevance to credit risk and AI or to the human dimension of credit behaviour. This produced a corpus dominated by publications in journals such as *Applied Soft Computing*, *Risks*, *Journal of Risk and Financial Management*, *European Journal of Operational Research*, *Behavioral Sciences*, *Journal of Co-operative Organization and Management*, and *Cogent Business & Management* [1–4, 6, 7, 9–13].

3.3 Bibliometric Techniques

Following bibliometric practice [14–16], the exploratory analysis used; descriptive indicators (annual publication trends, leading journals and countries) to characterise the evolution of the field, co-occurrence analysis of keywords, using tools such as VOSviewer [14] to identify thematic clusters and their interconnections and citation patterns to highlight influential works in algorithmic credit scoring, fairness and human factors. For transparency and replicability, the article adopts the main reporting principles for literature reviews [15], adjusting them to an exploratory bibliometric purpose. Although specific numerical values depend on the exact query, the bibliometric exploration reveals three robust clusters.

Once the data had been retrieved from the Scopus database, they were processed using the VOSviewer software. This enabled a co-occurrence analysis, through which the joint appearances of keywords across the articles under review were identified. The results of this analysis are shown in the following figure (Fig. 1).

Subsequently, a bibliometric coupling analysis was performed, using another functionality provided by VOSviewer. This technique examines the relationships between scientific documents based on the references they share: two documents are said to be bibliometric ally coupled when they cite, in common, a substantial portion of the same

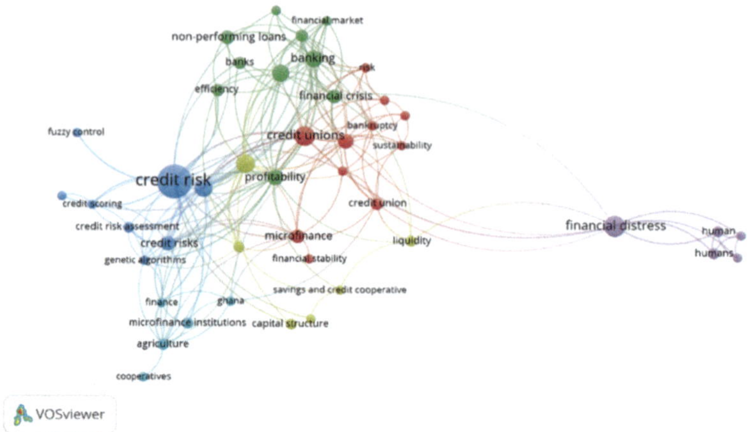

Fig. 1. co-occurrence analysis

underlying literature. The results are represented as citation networks, where the nodes correspond to the documents included in the analysis and the links between them reflect the strength of their connection (the more references they share, the stronger the coupling). The structure of this network, as well as the clusters formed by documents with similar citation patterns, can be observed in the following figure (Fig. 2).

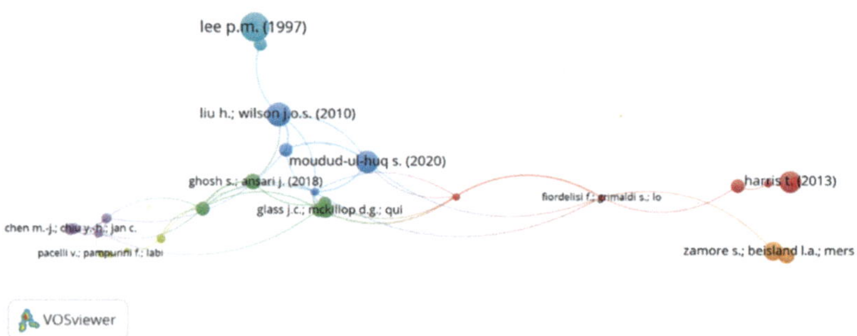

Fig. 2. Bibliometric coupling

4 Results

The bibliometric analysis of Scopus-indexed literature situates this research at the intersection of three main streams: (i) AI and machine-learning credit scoring, (ii) algorithmic fairness and explainability in credit decisions, and (iii) human and relational factors in repayment behaviour, particularly in cooperative and microfinance contexts.

From a technical perspective, the first group of studies provides a consolidated baseline for AI/ML credit scoring. Dastile et al. [1] show that machine-learning models—such as gradient boosting, random forests and neural networks—generally outperform traditional statistical approaches in terms of discriminatory power and accuracy. However, they also emphasise that these gains often come at the cost of interpretability, which can create challenges in regulated financial environments. Building on this foundation, Ayari et al. [2] and Noriega et al. [3] update and refine the picture, documenting how ML-based scoring has moved from experimental prototypes to real implementations in financial institutions, while mapping the most frequently used algorithms, features and performance metrics. Taken together, these articles confirm that AI-based credit scoring is a mature and expanding field, but one still dominated by performance-oriented designs.

The contribution by Nallakaruppan et al. [4] begins to bridge this performance focus with concerns about transparency. Their work demonstrates that explainable AI (XAI) techniques such as SHAP and LIME can be successfully applied to credit risk models, enabling practitioners to understand which variables drive risk predictions at both global and individual levels. This is particularly relevant for savings and credit cooperatives, which must not only comply with regulatory expectations but also explain credit decisions to their member–owners to maintain trust and legitimacy.

A second block of literature focuses explicitly on fairness, bias and governance in AI-based credit decisions. Bono [5] provides a conceptual and normative framework for understanding algorithmic fairness in credit scoring, clarifying the differences between individual and group fairness and discussing the associated policy implications. Kozodoi et al. [6] complement this with an empirical investigation, showing that the imposition of fairness constraints can effectively reduce disparate impact against protected groups while preserving most of the profitability of a scoring model. De Castro Vieira et al. [7] systematize bias mitigation strategies across the AI pipeline—from data pre-processing to in-processing regularization and post-processing adjustments—highlighting that the choice of technique must be context-dependent. Bussmann et al. [8] extend the discussion to governance, arguing that explainability is not optional but central to model risk management and regulatory compliance in credit risk.

For the present research, these studies collectively establish that any attempt to introduce AI-based scoring in cooperatives cannot be purely technocratic. Instead, fairness, bias control and explainable decision-making must be treated as first-order design criteria, fully aligned with cooperative principles of equity, transparency and member participation.

The third cluster centers on the human and relational dimensions of repayment behaviour, which are at the core of the "payment character" concept. Ganbat et al. [9] provide empirical evidence that psychological traits—such as self-control, risk attitudes and financial behaviour patterns—significantly influence credit risk in a microlending context. Fine [10] goes further by validating psychometric-based credit scoring for underbanked consumers and demonstrating that these scores add predictive power beyond traditional bureau data. These findings are crucial for the present research, as they show that the "human dimension" of payment character is not merely anecdotal but can be measured and integrated into predictive models.

At the relational level, Salinas Vásquez et al. [11] analyses an Ecuadorian financial cooperative and show that social capital—captured through variables such as trust, networks and cooperative involvement—reduces both the probability and the severity of default. Their study provides a direct empirical anchor for the Ecuadorian context, confirming that cooperative-specific relational factors are relevant to credit risk and should not be ignored in model design. In a broader cooperative banking setting, Kil et al. [12] demonstrate that scoring models adapted to cooperative banks can effectively limit credit risk, suggesting that it is feasible to tailor quantitative models to the institutional logic of member-based organizations. Kesanta [13] complements this by linking risk management practices to financial sustainability in savings and credit cooperative societies, reinforcing the idea that robust and well-governed risk models are essential to the long-term viability of cooperative finance.

When these three streams are read together, several key insights emerge for the design of a human-centric AI model for predicting payment character in savings and credit cooperatives:

Technical Feasibility and Maturity
The AI/ML literature shows that there is a mature toolkit—algorithms, feature engineering strategies, evaluation metrics—that cooperatives can draw upon to improve the predictive accuracy of their credit risk models [1–4]. This addresses the purely technical question of whether AI can be applied in cooperative credit risk management.

Ethical and Governance Constraints
The fairness and XAI literature makes clear that deploying AI in credit decision-making without explicit fairness and transparency safeguards risks reproducing or amplifying existing inequalities [5–8]. For cooperatives, which are explicitly value-driven and member-owned, this implies that any AI-based scoring system must include fairness constraints, bias mitigation strategies and explainability features from the outset.

Centrality of the Human and Relational Dimension
Evidence from psychometrics and social capital research demonstrates that individual psychological traits and relational embeddedness significantly affect default risk [9–11]. For cooperatives, where long-term relationships and community ties are central, these factors are not peripheral; they are fundamental. Ignoring them would mean discarding a distinctive informational advantage that cooperatives possess over other financial institutions.

Cooperative-Specific Adaptation and Sustainability
Studies on cooperative banks and SACCOs show that risk models can be adapted to cooperative logic and that improved risk management is directly linked to financial sustainability [12, 13]. This suggests that introducing AI-based scoring in cooperatives is not just a technical innovation; it is a strategic lever to strengthen resilience and inclusion if aligned with cooperative values.

In summary, the state of the art reveals a fragmented landscape where performance-oriented AI/ML models, fairness and explainability debates, and research on human and relational factors have developed largely in parallel. The contribution of the present research is to bring these strands together in the specific context of Ecuadorian savings and credit cooperatives, arguing that an AI-based model of payment character must simultaneously: exploit the predictive power of ML, respect fairness and transparency requirements, and explicitly encode the human and relational dimensions that define cooperative credit relationships.

This integrated perspective lays the conceptual foundation for the empirical development and validation of a human-centric AI model for ethical and inclusive credit risk in cooperatives.

Although specific numerical values depend on the exact query, the bibliometric exploration reveals three robust clusters.

Cluster 1: AI/ML Credit Scoring and Technical Performance

The first and largest cluster brings together works focused on AI and ML methods for credit scoring, emphasizing predictive performance and model comparison [1–4]. Keywords such as *"machine learning"*, *"credit scoring"*, *"default prediction"*, *"deep learning"* and *"feature selection"* dominate this cluster.

Articles by Dastile et al. [1], Ayari et al. [2] and Noriega et al. [3] are highly cited, providing systematic reviews of statistical and ML models, performance metrics and application domains. Recent contributions such as Nallakaruppan et al. [4] incorporate explainability into this performance-driven paradigm, showing that interpretable ML models can achieve competitive accuracy.

Cluster 2: Fairness, Bias and Explainable AI in Credit Decisions

A second cluster focuses on fairness, ethics, regulatory compliance and explainability in AI-based credit risk models [5–8]. Typical keywords are *"algorithmic fairness"*, *"discrimination"*, *"interpretability"*, *"XAI"* and *"regulation"*.

Bono [5] frames fairness in credit scoring from a policy perspective, while Kozodoi et al. [6] empirically show the trade-offs between fairness constraints and profitability. De Castro Vieira et al. [7] classify bias mitigation techniques and argue for hybrid, context-sensitive approaches. Bussmann et al. [8] stress that explainable ML is a prerequisite for trustworthy AI in finance, enabling model risk management and auditability.

This cluster is conceptually close to the first one but shifts the focus from pure accuracy to governance, accountability and human oversight, concerns that are central to cooperative principles.

Cluster 3: Human and Relational Factors in Loan Repayment

A third, more interdisciplinary cluster articulates psychology, behavioural economics, social capital and cooperative/microfinance studies [9–13]. Here the dominant keywords include *"psychological factors"*, *"psychometrics"*, *"social capital"*, *"microfinance"*, *"financial cooperatives"* and *"loan repayment behaviour"*.

Ganbat et al. [9] show that psychological traits significantly predict credit risk in microlending. Fine [10] demonstrates the practical value of psychometric credit scoring for underbanked consumers. Salinas Vásquez et al. [11] provide evidence from an Ecuadorian financial cooperative that social capital reduces both the probability and intensity of default. Kil et al. [12] and Kesanta [13] extend the analysis to cooperative banks and savings and credit cooperatives, highlighting how relational nature, risk culture and governance shape portfolio quality.

This cluster offers conceptual and empirical tools to operationalize the human dimension of payment character, but it remains only weakly connected to the more technical AI/ML literature.

5 Discussion

The bibliometric evidence suggests that AI/ML models dominate the credit scoring literature, with a growing but still secondary stream addressing fairness and explainability, and a relatively separate stream on psychological factors and social capital. For savings and credit cooperatives, this fragmentation is problematic because their mission requires simultaneous attention to risk, inclusion and member-centred values.

Drawing on the three clusters, an AI-based credit scoring framework for cooperatives should integrate at least three types of variables:

Financial and transactional variables (income, instalment ratios, arrears histories) aligned with the mainstream ML literature [1–4].
Psychometric and behavioural indicators (self-control, conscientiousness, financial attitudes) derived from validated instruments, as in Ganbat et al. [9] and Fine [10].
Relational and social capital metrics (participation in assemblies, co-signed loans, peer references, community involvement), inspired by Salinas Vásquez et al. [11] and cooperative governance studies [12, 13].

Properly designed, AI can combine these dimensions to predict default probabilities while providing individual-level explanations of how payment character—understood as a configuration of personal traits and social ties—contributes to the risk assessment.

5.1 Epistemological Tensions: Risk, Inclusion and Ethics

From an epistemological standpoint, the field reveals tensions between; a technocratic-objectivist vision, where risk is a purely statistical phenomenon optimized by ML algorithms; and a relational-humanist vision, where repayment behaviour is embedded in psychological and social structures.

Cooperatives, especially in a context like Ecuador, cannot fully subscribe to either extreme. They must manage risk rigorously while acknowledging the moral and relational character of credit relationships. AI systems that ignore the human dimension risk perpetuating or amplifying exclusionary patterns; systems that rely solely on subjective judgement risk arbitrariness and bias.

The literature on fairness and XAI [5–8] offers a bridge: by making models transparent and by explicitly tracking the distributional impact of decisions across groups, cooperatives can align their use of AI with cooperative principles of equity and solidarity.

5.2 Conceptual Model Proposal for Cooperatives

Based on the literature reviewed, we propose a conceptual AI-based model for predicting payment character in savings and credit cooperatives, structured around four modules:

- **Data Layer**

  Financial data: income, debt ratios, product mix, payment histories [1–4].

  Psychometric data: scores on validated scales of self-control, conscientiousness, financial attitudes and trust [9, 10].

  Social capital data: indicators of member participation, length of membership, co-guarantees, network centrality in cooperative relations, derived from cooperative MIS [11–13].

- **Modelling Layer**

  A hybrid ML architecture (e.g., gradient boosting plus interpretable models such as generalized additive models) that jointly estimates probability of default and behavioral risk profiles.

  Incorporation of fairness-aware constraints as discussed by Kozodoi et al. [6] and de Castro Vieira et al. [7], ensuring that sensitive attributes (e.g., gender, ethnicity, rurality) are not used directly and that outcome disparities are monitored.

- **Explainability and Governance Layer**

  Use of XAI tools (SHAP, LIME, rule-based surrogates) to provide member- and portfolio-level explanations [4, 8].

  Documentation and audit trails compatible with regulatory expectations and cooperative governance practices.

- **Decision and Feedback Layer**

  Integration with credit committees, allowing human experts to validate or override automated recommendations with documented justifications.

  Continuous learning from realized defaults and repayments to recalibrate both the financial and human-factor components of the model [1–4, 9–11].

  For Ecuadorian cooperatives, such a model could be piloted using anonymized data from selected institutions, combining internal cooperative datasets with psychometric and social capital instruments adapted to local cultural contexts (Fig. 3).

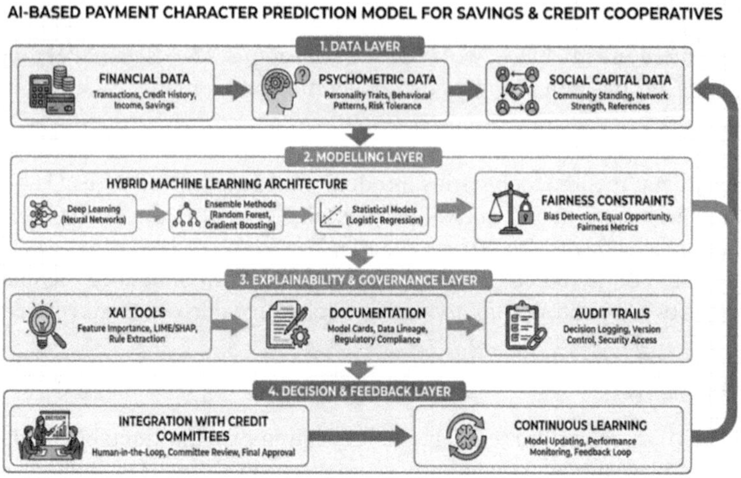

Fig. 3. Models that Capture Payment Character in Cooperatives

6 Conclusions

This article explored, from an epistemological and bibliometric standpoint, how artificial intelligence can incorporate the human dimension of payment character in credit risk models for savings and credit cooperatives. The review of Scopus-indexed literature shows that:

AI and ML credit scoring research has achieved substantial advances in predictive performance but tends to under-represent human and relational factors [1–4].
A growing body of work on fairness and explainable AI offers tools to align credit risk models with ethical and regulatory standards [5–8].
Empirical evidence from psychometrics, microfinance and cooperative studies demonstrates that psychological traits and social capital significantly influence credit risk and repayment behaviour, including in Ecuadorian financial cooperatives [9–13].

For Ecuadorian savings and credit cooperatives, the proposed conceptual model suggests that AI does not have to be opposed to cooperative values. On the contrary, if designed with explicit ethical, fairness and human-centric considerations, AI-based credit scoring can help:

Make visible and measurable the contribution of payment character to credit risk;
Provide more nuanced, personalised decisions for members lacking conventional collateral or thick credit histories; Strengthen transparency and accountability through explainable models; Support financial inclusion while controlling portfolio risk.

Future research should include (i) empirical implementation and validation of the proposed model with Ecuadorian cooperative data, (ii) qualitative studies on member perceptions of AI-assisted credit decisions, and (iii) comparative analyses across countries and cooperative systems.

References

1. Dastile, X., Celik, T., Potsane, M.: Statistical and machine learning models in credit scoring: a systematic literature survey. Appl. Soft Comput. **91**, 106263 (2020)
2. Ayari, H., Yousfi, S., Ghnimi, S.: Machine learning powered financial credit scoring: a systematic literature review. Artif. Intell. Rev. **59**, 13 (2025). https://doi.org/10.1007/s10462-025-11416-2
3. Noriega, J.P., Chamba, G., Rojas, E., García, L.: Machine learning for credit risk prediction: a systematic literature review. Data. **8**(11), 169 (2023)
4. Nallakaruppan, M.K., et al.: Credit risk assessment and financial decision support using explainable artificial intelligence. Risks. **12**(10), 164 (2024)
5. Bono, T.: Algorithmic fairness in credit scoring. Oxf. Rev. Econ. Policy. **37**(3), 585–617 (2021)
6. Kozodoi, N., Jacob, A., Lessmann, S.: Fairness in credit scoring: assessment, implementation and profit implications. Eur. J. Oper. Res. **297**(3), 632–648 (2022)
7. de Castro Vieira, J.R., Barboza, F., Cajueiro, D., Kimura, H.: Towards fair AI: Mitigating bias in credit decisions—a systematic literature review. J. Risk Financial Manag. **18**(5), 228 (2025)
8. Bussmann, N., Papakonstantinou, K., Thoma, G., Wozabal, D.: Explainable machine learning in credit risk management. Comput. Econ. **57**, 203–227 (2021)
9. Ganbat, M., et al.: Effect of psychological factors on credit risk: a case study of the microlending service in Mongolia. Behav. Sci. **11**(4), 47 (2021)
10. Fine, S.: Character counts: Psychometric-based credit scoring for underbanked consumers. J. Risk Financial Manag. **17**(9), 423 (2024)
11. Salinas Vásquez, J., Sarmiento Jara, J.P., Urgilés Salinas, M.P., Oña Avendaño, D.F.: Social capital and credit risk in a financial cooperative of Ecuador. J. Co-op. Organ. Manag. **12**(2), 100247 (2024)
12. Kil, K., Ciukaj, R., Chrzanowska, J.: Scoring models and credit risk: The case of cooperative banks in Poland. Risks. **9**(7), 132 (2021)
13. Kesanta, M.N.: Risk management practices and financial sustainability: Evidence from savings and credit cooperative societies in Tanzania. Cogent Bus. Manag. **12**, 2563034 (2025)
14. Van Eck, N.J., Waltman, L.: Software survey: VOSviewer, a computer program for bibliometric mapping. Scientometrics. **84**(2), 523–538 (2010)
15. Page, M.J., McKenzie, J.E., Bossuyt, P.M., Boutron, I., Hoffmann, T.C., Mulrow, C.D., et al.: The PRISMA 2020 statement: An updated guideline for reporting systematic reviews. BMJ. **372**, n71 (2021)
16. Liu, J., Wang, H., Zhang, Q.: A bibliometric analysis of financial risk (2015–2024). Int. J. Financial Stud. **13**(3), 132 (2025)
17. Kedia, P., Mishra, L.: Credit risk management: a systematic literature review and bibliometric analysis. J. Credit Risk. **20**(1), 51–76 (2024)
18. Zhou, Z., et al.: How do online emotions support credit risk decision-making? J. Technol. Manag. Innov. **16**(3), 47–56 (2025)

Reading Assistant for Visually Impaired People Implemented in Raspberry Pi with 3D Casing Made from Recycled Materials

Elizabeth Salazar-Jácome[1](✉), Rolando Salazar-Paredes[2], Gustavo Guaigua-Albarracín[1], and Javier De la Torre-Guzmán[1]

[1] Department of Engineering Sciences, Universidad Tecnológica Israel, Quito, Ecuador
{msalazar,gguaigua,jdelatorre}@uisrael.edu.ec

[2] Department of Exact Sciences, Universidad de las Fuerzas Armadas ESPE, Sangolquí, Ecuador
rxsalazar@espe.edu.ec

Abstract. Today, artificial vision represents a significant contribution to the advancement of systems oriented towards technological accessibility; The reading assistant, in particular, seeks to facilitate the interpretation of the environment for people with visual disabilities. Although there are a variety of commercially available devices, their high price makes them inaccessible, which creates an inequality in digital inclusion. This study proposes the design and implementation of a low-cost prototype, based on Raspberry Pi, which in-corporates a camera to take images, processing using computer vision algorithms and the conversion of the results into audio using the gTTS library, in addition to the creation of the case with recycled materials. The methodology used was based on the creation of an automated process flow that allows managing everything from image taking to speech synthesis, including physical controls for navigation and parameter adjustment. The results reveal that the system can efficiently identify images and convert them to audio, providing an accessible and functional experience. This research represents an economical and feasible option compared to commercial products, helping to decrease technological barriers, foster sustainability, and strengthen digital inclusion.

Keywords: Raspberry Pi · sustainability · artificial vision · visual impairment

Abbreviations

gTTS	Google Text-to-Speech
OCR	Optical Character Recognition
TTS	Text-to-Speech
PSM	Page Segmentation Mode
GPIO	General Purpose Input / Output
MS	Milliseconds
CPU	Central Processing Unit

A. Rocha et al. (Eds.): ICITS 2026, LNNS 1966, pp. 232–244, 2027.
https://doi.org/10.1007/978-3-032-27538-7_20

1 Introduction

The most important challenge for the inclusion of visually impaired people is the technological field, especially what has to do with access to printed texts [1]. Although there are several studies developed on systems that combine OCR (Optical Character Recognition) and TTS (Text-to-Speech) to convert physical documents into audio, the cost of acquiring one of them makes it impossible or limited to adopt it by low-income users [2]. For example, commercial products such as Kurzweil 1000 or specialized professional OCR systems can exceed a thousand dollars, which restricts their accessibility [3].

In the academic field, there are numerous projects that seek to reduce costs through the use of general-purpose hardware, free software and low-cost components [4]. An ex-ample is the article "A Novel Real-Time Text-to-Speech System Using Raspberry Pi for Assisting the Visually Impaired" where a system embedded with Raspberry Pi 4 is proposed that integrates dynamic image preprocessing stages to improve OCR accuracy [5]. An-other relevant study is "Reader for Blind Using the Raspberry Pi", published in 2023, which implements a reader for blind people using accessible components along with OCR and speech synthesis [6].

Despite these advances, many similar products still fail to achieve sufficiently robust accuracy [7], or have limitations in terms of languages, special fonts, lighting conditions, or physical accessibility of the device (e.g., size, button size, simplified interface) [8]. In "Text Reader for Visually Impaired Person" [9], for example, a mobile application is presented that recognizes printed text by camera and vocalizes it [10], but the accuracy varies depending on whether the text is in English, Thai, or a mixture of the two languages, which denotes that general systems still face technical barriers for all linguistic contexts [11].

In addition, recent research has addressed more specialized variants, such as newspaper readers ("Newspaper Reader Application") that allow images of newspapers to be captured and converted to audio using OCR + TTS [12], but again many of these systems use smartphones or software/licensing infrastructure that can increase the final cost to the user [13]. Also, high-performance commercial OCR devices often rely on expensive li-censes, proprietary hardware, or cloud services that involve recurring payments, which increases operational costs [14].

Faced with these limitations, the need for solutions that combine low cost, technical robustness and sustainability is growing, especially in contexts with economic constraints [15]. The use of free hardware such as Raspberry Pi [16], simple cameras, 3D printing for the cases (ideally with recycled materials) and open source software (OpenCV, Pytesseract, free or low-cost TTS libraries) [17] appears as a promising alternative. Some of the works cited already go down this route, but often without incorporating all the components [18] (for example, some versions omit the accessible physical interface, or the custom case, or the ability to operate without an internet connection) [19].

This work aims to contribute to that space by developing and implementing a reading assistant for blind people [20] that incorporates: (1) low-cost hardware based on Raspberry Pi [21], (2) 3D printed housing with recycled materials oriented towards sustainability [22], (3) OCR and TTS integrated with open source libraries [23],and (4) physical interface accessible via buttons [24]. This solution is expected to not only significantly reduce the cost compared to high-performance commercial devices, but also be viable in rural or resource-poor contexts, contributing to both social inclusion and the sustainable development goals [25].

The main contribution of this work lies in the integration of a low-cost, sustainable reading assistant that combines open-source OCR and TTS technologies with a 3D-printed casing made from recycled materials. Unlike similar implementations, our approach emphasizes circular design principles, off-grid usability, and full local processing without internet dependency, thus ensuring accessibility in low-resource environments.

2 Materials and Methods

Using an organized methodology, it was possible to examine the results of the reading assisted prototype, which involved clearly establishing the goals of the project, obtaining data in a structured way in different test scenarios, through a thorough analysis that allowed the effectiveness, accuracy and ease of use of the system to be thoroughly assessed. Indicators such as accuracy in character identification, the speed of reaction of the system and the percentage of success when reading printed texts were considered, facilitating the direction of the tests of the system and a rigorous and measurable study of its operation.

The selection of performance indicators was based on prior research on OCR and assistive systems. Accuracy, response time, and user satisfaction were chosen because they represent the most relevant dimensions of effectiveness, efficiency, and usability in assistive reading devices. Each indicator was quantitatively assessed through repeated measures across different lighting and text-type scenarios.

2.1 Material Selection

With the careful selection of materials, electronic components and peripherals, necessary to assemble the assisted reading prototype, it was possible to ensure that the system works well and is efficient. With the XL4015 module, the issue of power supply and voltage regulation that the prototype needed was solved, for the audio output a PAM8610 amplifier was used that had a potentiometer, connected to a 4580 rectangular speaker of 4 Ω and 3 W, with the use of the Raspberry Pi 4, a fluid central processing of the system was obtained, allowing the OCR engine and TTS module to run smoothly.

An 8-megapixel V2 camera module with 1080p resolution was used for image capture, ensuring optimal quality for character recognition. To make the usability of the product the most suitable, a four-button membrane keypad and a 1x4 pushbutton matrix were used, allowing simple and intuitive control of the main functions. The combination of these elements ensured a balance between performance, accessibility and energy efficiency, constituting the technical basis of the prototype (Fig. 1).

The prototype housing was designed in 3D using SolidWorks software, resulting in an accurate and functional model. Recycled materials were used for construction, especially recovered plastics, which made it possible to reduce costs, minimize environmental impact and demonstrate that it is possible to develop accessible and sustainable technological solutions for visually impaired people with very little money (Fig. 2).

Fig. 1. Prototype Connections

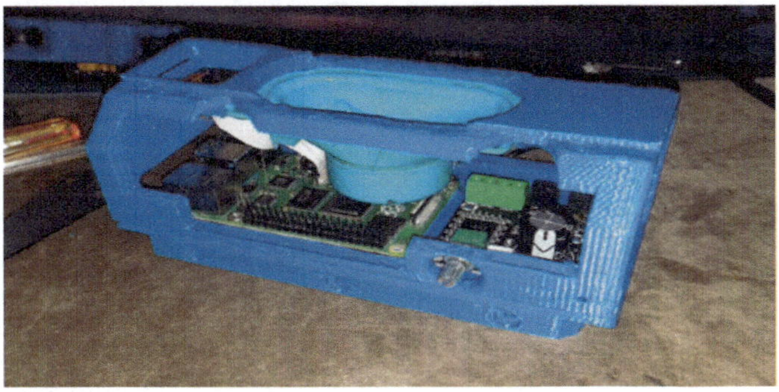

Fig. 2. Prototype Assembly

For data collection, the prototype was executed in multiple test scenarios, it was necessary to have several characteristics of the text, font type, size and format, as well as lighting conditions, since we worked with spaces with intense natural light to low light situations. All relevant results were recorded one by one, including correct and incorrect readings, processing times and any errors or failures detected in the system, all this al-lowed the creation of a robust body of data, facilitating an in-depth study of the performance of the prototype.

3 Results

Once the prototype was implemented, it was necessary to activate and use it, a scheme was devised with three approaches that complement each other. To begin with, how well the OCR worked was measured, bluntly comparing what the system transcribed with the source text, using similarity scales to know with numbers how faithful the recognition was. Then, it was put under the magnifying glass how long it took the system to react according to the tests, detecting if there were strange changes or points where the process got stuck. To close, a thorough inspection was made of the faults that the OCR and TTS modules released, putting them in order by how often they occurred, what kind they were and why they could happen, thus finding the key points to polish. This comprehensive approach provided a deep understanding of the prototype's performance and formed a solid foundation for its optimization and validation.

3.1 Electrical Calculations

The XL4015 module, configured as a Step-Down regulator, made it possible to efficiently reduce an input voltage from 12 V to 5 V, suitable for powering the Raspberry Pi, with an efficiency of 96% and a maximum current of 4.5 A, close to the module's 5 A limit, ensuring a stable and safe supply of power to the system (Fig. 3).

$$P - out = 5V \bullet 3A = 15W \tag{1}$$

$$P - in = \frac{15W}{0.96} = 15.625 \tag{2}$$

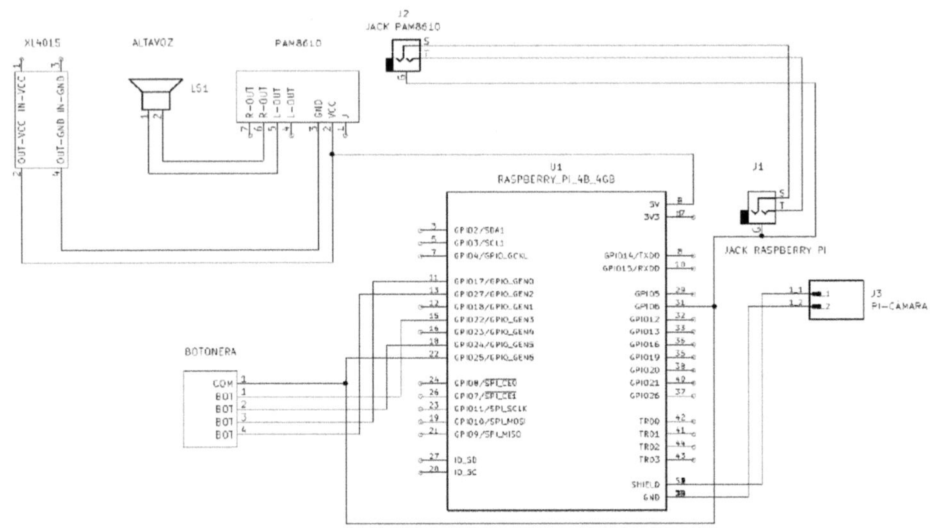

Fig. 3. Electronic schematic of the prototype in Ki Cad software

The PAM8610 amplifier, powered by 12 V and 3 A (36 W maximum), allowed a correct amplification of the audio signal for the 4 Ω and 3 W speaker, ensuring adequate performance without exceeding the power limits of the module and ensuring clear and efficient sound reproduction for the prototype.

$$P = \frac{V^2}{R} = \frac{12V^2}{4\Omega} = 36W \tag{3}$$

$$I = \sqrt{\frac{P}{R}} = \sqrt{\frac{15W}{4\Omega}} = 1.94A \tag{4}$$

The Raspberry Pi, rated at 5 V and up to 3 A (15 W), showed typical power consumption between 1.2 and 2.5 A (6–12.5 W), ensuring efficient operation within its power limits. The associated camera module, with an additional 250 mA consumption, allowed the capture of high-resolution images necessary for OCR processing, integrating stably into the system without compromising the energy efficiency of the prototype.

$$Prasp - cam = 5V \bullet 3.25A = 16.25W \tag{5}$$

3.2 Interpretation of Results

For the interpretation of the results, the use of data visualization tools was considered. Planned to generate graphs to show OCR accuracy, response times, and other relevant data. Additionally, the creation of comparative tables was contemplated to contrast the results with existing systems or previous versions of the prototype. Once all the reading assistant tests have been carried out, the quantitative and qualitative findings derived from them are presented. The performance of the OCR system is sensitive to lighting conditions. Table 1 summarizes the results obtained.

Table 1. Environmental Testing (OCR System).

Condition of Lighting	Types of Text	OCR Accuracy (%)	Bugs Caught (out of 40)	Response Time (Sec)	PSM Mode	Confidence threshold
Neutral (Día)	Books/ Magazines	92.5	3	0.8	6	65%
Artificial (LED)	Documents	88.3	8	1.1	7	70%
Low Lighting	Tags	71.2	18	1.8	3	80%
Mixed	Documents	85.7	6	1.0	6	65%
Indirect shadow	Newspapers	89.1	5	0.9	7	70%
Directional Light	Manuals	90.8	4	0.7	6	60%

According to the results, lighting turns out to be the key element that most affects the performance of the OCR system, impacting both accuracy and speed of response. The settings that automatically adapt to Page Segmentation (PSM) modes proved to be useful, indicating that PSM 6 works wonderfully in conditions where the light does not change, while PSM 3 optimizes performance in places where the light varies, so it is essential to adjust the system parameters depending on the environment

As shown in Table 2, the error analysis identified specific patterns: recognition failures are mainly due to small text and complex fonts, capture failures are mainly due to insufficient lighting and blurring, and TTS failures are due to connectivity and long texts, while high resolution rates indicate that most are solvable through configuration adjustments and technical improvements.

Table 2. OCR System Error Analysis (Classification, Frequency and Implemented Solutions).

Type of Error	Frequency (%)	PSM Most Affected	Main causes	Impact severity	Implemented Solutions	Resolution Rate (%)
Incorrect recognition	35.2	11	Small text, complex fonts	High	Threshold setting, confidence, pre-processing	78
Capture error	22.8	8	Insufficient lighting, blur	Middle	Improves lighting, stabilization	92
TTS Processing	18.5	All	Internet connections, long text	Middle	Local Cache, Audio Optimization	85
GPIO Failure	8.7	N/A	Loose connections, debounce	Low	Connection review, improved debounce	95
Time Exceeded	9.3	3	Intensive processing, memory	Middle	Optimization Code, memory	82
Configuration Error	5.5	13	Incorrect parameters, calibration	Low	Parameter validation, user guides	96

Error frequencies were obtained by analyzing 40 test samples under six lighting conditions. Each OCR output was manually compared with the ground-truth text, and discrepancies were classified according to their origin: recognition, capture, or processing errors. The percentage frequency was computed as the number of errors per category divided by the total number of errors detected across all tests.

In Table 3 it is possible to show that the performance data indicates that the Raspberry Pi 4 is the critical component of the system, the overall efficiency is high although Tesseract is less efficient, the thermal management is kept within safe ranges and the hardware components have a high mean time between failures, which reflects the robustness and stability of the prototype.

Several measurements were taken in order to gauge how user-friendly and accessible the system turned out to be (view Table 4).

Usability data were collected through a structured questionnaire applied to ten participants (five visually impaired users and five sighted users). The evaluation used a

Table 3. Hardware/Software Performance Comparison.

Component/Module	Power Consumption (W)	Processing Time (MS)	Operational Efficiency (%)	Operating Temperature (°C)	Resource utilization (%)	Level Criticality	MTBF Estimated (hours)
Raspberry Pi 4 CPU	3.8	250	85.2	52	65.3	Critical	8760
Camera V2 CSI	0.78	100	92.7	38	15.7	High	17520
PAM8610 Amplifier	2.1	50	91.3	45	8.9	Middle	26280
Tesseract Module	1.2	800	78.5	N/A	45.2	Critical	N/A
gTTS Module	0.3	600	82.1	N/A	12.4	High	N/A
CPIO/Keypad	0.1	10	98.5	32	2.1	Low	43800

Table 4. Usability and accessibility metrics

Functionality	Ease of use (1 to 10)	Learning Time (min)	User Satisfaction (%)	Novice User Errors	Motor Accessibility	Cognitive accessibility	Tactile feedback	System Response Time
Image capture	8.5	5	87.3	2/10	Excellent	Good	Yes	Immediate
Menu navigation	7.2	12	74.6	4/10	Good	Regular	Yes	< 1 s
Configuration Adjustment	6.8	18	68.9	6/10	Regular	Difficult	Partial	< 2 s
Audio playback	9.1	3	92.1	1/10	Excellent	Excellent	No	Immediate
Volume Control	8.7	4	89.4	1/10	Good	Good	Yes	Immediate
Preview mode	7.5	8	76.8	3/10	Good	Regular	Yes	< 1 s

10-point Likert scale to rate ease of use, learning time, and satisfaction. Accessibility aspects such as tactile feedback and menu navigation were observed during controlled testing sessions.

Usability analysis indicates that the audio playback and image capture functions are the most accessible to users, while the system configuration and menu navigation require improvement. The time needed to learn how to use it ranges from three to eighteen minutes; The most difficult thing is the configuration. The overall satisfaction, with an average of 81.6%, indicates that the prototype has been received quite well.

To contextualize the performance of the proposed prototype, Table 5 summarizes a quantitative comparison with similar systems reported in recent studies. The proposed solution achieves an OCR accuracy of 92.5% under optimal lighting, outperforming other low-cost Raspberry Pi-based systems that typically range from 85% to 90%. In addition, the overall cost of our system (approximately USD 75) is substantially lower than that of comparable commercial readers exceeding USD 1000, while maintaining similar usability levels

4 Discussion

Both qualitative and quantitative evaluation show that the accuracy of optical character recognition is closely linked to light conditions; it reaches 92.5% in good lighting (such as natural daylight on texts in books or magazines) and drops to 71.2% in low light (such as on labels). This difference of 21.3% shows that the quality of light is critical. The system requires an average response time of 0.8 s under optimal conditions, extending to 1.8 s in low light.

Among the initial functional tests that were carried out on the prototype was to verify fonts and sizes, where each sample was placed at a certain distance from the camera and thus some favorable results were obtained and others not so much.

It was used in different types of settings for the determination of printed text, modifying the confidence and similarity parameters, and at different sound levels in the speaker (Fig. 4).

Fig. 4. Functional tests and parameter configuration.

Error analysis identified that character recognition issues (35.2%) and image capture errors (22.8%) are the most prevalent. However, the system is robust, as between 78% and 96% of detected errors can be solved through configuration adjustments or technical improvements. The most critical hardware component is the Raspberry Pi 4's CPU, with a resource utilization of 65.3%, mainly due to OCR processing, which consumes 45% of the total system processing time (Fig. 5).

The usability of the system by users was high, 81.6% consider it easy to use, the audio playback and image capture functions are particularly intuitive. However, there are potential improvements related to system settings and menu navigation for ease of use.

Beyond the lighting dependency, additional limitations include the processing delay inherent to the Raspberry Pi 4 during OCR execution and the restricted multilingual

Fig. 5. Text reading and playback in the audio system.

capability of the Tesseract engine. Nevertheless, the use of recycled plastics for the 3D-printed casing proved structurally adequate and reduced material costs by approximately 40%, demonstrating the feasibility of integrating sustainability within assistive technologies

5 Conclusions and Future Work

This project led to the development of an effective reading assistant for visually im-paired people, which allows printed text to be captured, processed and reproduced aloud. The implementation on a Raspberry Pi showed that it is possible to create accessible, portable and low-cost solutions that directly contribute to technological and social inclusion for visually impaired users.

By merging essential technologies such as computer vision, character recognition (OCR), and text-to-speech (TTS) for this prototype, it was possible to offer users a seamless, independent, and trustworthy experience, enabling the use of hardware components accessible as open-source software when creating devices designed for people with special abilities.

The design and creation of the 3D model of the structure, produced by three-dimensional printing, made it possible to improve ergonomics, ease of transport and physical defense of the internal elements. This configuration lays the foundation for future individualization updates, adjusting the device to various scenarios and user requirements.

The Raspberry Pi showed an efficient performance in the execution of all the processes necessary for the production of assisted reading, such as image capture or voice production, confirming the expectations that were had regarding its high potential to be used as a training platform or to make a prototype quickly; In addition, given its versatility, this device can be positioned as a fundamental resource within the development of projects that deal with technological accessibility. The assisted reading system developed had a quality of 92.5% of the OCR Tesseract engine under optimal lighting conditions, confirming that when reading texts, its quality must be high, well printed and well lit; But the lack of light has an effect on performance, being the most determining factor since variations in light levels affect the accuracy of the recognition and the response time of the system.

The most significant bottleneck is OCR processing in terms of time, although most of the errors identified, such as character recognition failures and capture errors, can be re-solved through configuration adjustments, parameter optimization, and technical improvements in image preprocessing. In the operational field, the prototype operates at temperature tangos that ensure a minimum of energy consumption, guaranteeing stability and reliability when operating continuously. The user evaluates the experience as positive, highlighting the function of vocalization of status messages and results that significantly improves accessibility especially for blind people, it is necessary to improve the interface of the system, especially in the configuration and navigation through the menus, where the improvement should focus on simplicity and clarity to reduce learning times and improve general satisfaction.

As a future work, we propose to address the possibility of mixing artificial intelligence algorithms not only for text recognition, but also to apply the automatic description of scenes and objects, which would make the device more useful in the daily lives of users. We must include multilingual support, improve character recognition (OCR) in complex documents and consider the possibility of incorporating rechargeable batteries with a long life to have more autonomy; as well as scaling this prototype to achieve a product with a greater social reach, with design optimization processes and alliances with educational institutions, health institutions and NGOs dedicated to inclusion, which would generate a positive impact on the reduction of technological barriers to improve the quality of life of blind people.

Future developments will focus on integrating lightweight deep learning models for real-time OCR enhancement, adding automatic language detection, and incorporating on-device scene description through convolutional neural networks. Additionally, optimizing power management and implementing wireless connectivity (Bluetooth or Wi-Fi) will increase system autonomy and user interaction capabilities.

References

1. Okolo, G.I., Althobaiti, T., Ramzan, N.: Assistive systems for visually impaired persons: challenges and opportunities for navigation assistance. Sensors. **24**(11), 3572 (2024). https://doi.org/10.3390/s24113572
2. Lavric, A., Beguni, C., Zadobrischi, E., Căilean, A.-M., Avătămăniței, S.-A.: A Comprehensive survey on emerging assistive technologies for visually impaired persons: lighting the path with visible light communications and artificial intelligence innovations. Sensors. **24**(15), 4834 (2024). https://doi.org/10.3390/s24154834
3. Preece, A.: Fine reader professional version 11 by ABBYY and text cloner pro version 11.5 by premier literacy: an evaluation and comparison of two optical character recognition products. AccessWorld. **14**(10), 15752 (2025) [Online]. Available: https://afb.org/aw/14/10/15752
4. Lancioni, G.E., et al.: Low-cost technology-aided programs for supporting people with motor, visual, and intellectual disabilities in functional forms of occupation and communication: proof-of-concept study. JMIR Rehabil. Assist. Technol. **10**, e44239 (2023). https://doi.org/10.2196/44239

5. Atitallah, A.B., et al.: A novel real-time text-to-speech system using raspberry Pi for assisting the visually impaired. Trait. Signal. **41**(6), 3183–3192 (2024). https://doi.org/10.18280/ts.410634
6. Reddy, Y.P.K., Hemadri, G., Nithya, U.J., Basha, D.S.H., Krishna, Y.H.: Reader for blind using the raspberry Pi. In: Kumar, A., Senatore, S., Gunjan, V.K. (eds.) Proceedings of the 3rd International Conference on Data Science, Machine Learning and Applications (ICDSMLA 2021) Lecture Notes in Electrical Engineering, vol. 947, pp. 741–748, Singapore: Springer Nature Singapore (2023). https://doi.org/10.1007/978-981-19-5936-3_69
7. Skulimowski, P.: Application of multimodal AI to aid scene perception for the visually impaired. Appl. Sci. **15**(12), 6442 (2025). https://doi.org/10.3390/app15126442
8. Han, Y., Beheshti, M., Jones, B., Hudson, T.E., Seiple, W.H., Rizzo, J.-R.: Wearables for persons with blindness and low vision: form factor matters. Assist. Technol. **36**(1), 60–63 (2024). https://doi.org/10.1080/10400435.2023.2205490
9. Patel, I., Kulkarni, M., Mehendale, N.: Review of sensor-driven assistive device technologies for enhancing navigation for the visually impaired. Multimed. Tools Appl. **83**(17), 52171–52195 (2023). https://doi.org/10.1007/s11042-023-17552-7
10. Zhang, X., Huang, X., Ding, Y., Long, L., Li, W., Xu, X.: Advancements in smart wearable mobility aids for visual impairments: a bibliometric narrative review. Sensors. **24**(24), 7986 (2024). https://doi.org/10.3390/s24247986
11. Sundar, C., Abinaya, O., Ajithkumar, V.: An optical character recognition framework based newspaper reader application for blind. In: Proceedings of the First International Conference on Combinatorial and Optimization, ICCAP 2021, pp. 1–7 (2021). https://doi.org/10.4108/eai.7-12-2021.2314975
12. Mashiata, M., et al.: Towards assisting visually impaired individuals: a review on current status and future prospects. Biosens. Bioelectron. X. **12**, 100265 (2022). https://doi.org/10.1016/j.biosx.2022.100265
13. Reddy, K.K., Badam, R., Alam, S., Shuaib, M.: IoT-driven accessibility: a refreshable OCR-braille solution for visually impaired and deaf-blind users through WSN. J. Econ. Technol. **2**, 128–137 (2024). https://doi.org/10.1016/j.ject.2024.04.007
14. Thiyagarajan, S., Saravana Kumar, D.G., Praveen Kumar, E., Sakana, G.: Implementation of optical character recognition using Raspberry Pi for visually challenged person. Int. J. Eng. Technol. **7**(3.34), 65–68 (2018). https://doi.org/10.14419/ijet.v7i3.34.18718
15. Fernando, S., Ndukwe, C., Virdee, B., Djemai, R.: Image recognition tools for blind and visually impaired users: an emphasis on the design considerations. ACM Trans. Access. Comput. **18**(1), 1–21 (2025). https://doi.org/10.1145/3702208
16. Kavitha, D., Radha, V.: Raspberry Pi based OCR implementation in multi-line refreshable braille display device. J. Intell. Fuzzy Syst. **46**(3), 6235–6248 (2024). https://doi.org/10.3233/JIFS-236527
17. Chopde, A., Bhole, G., Tyagi, A., Tyagi, A.: Blind text reader and currency detector using Raspberry Pi. In: 2023 7th International Conference On Computing, Communication, Control And Automation (ICCUBEA), pp. 1–5. IEEE, Pune, India (2023). https://doi.org/10.1109/ICCUBEA58933.2023.10392241
18. Kathiria, P., Mankad, S.H., Patel, J., Kapadia, M., Lakdawala, N.: Assistive systems for visually impaired people: a survey on current requirements and advancements. Neurocomputing. **606**, 128284 (2024). https://doi.org/10.1016/j.neucom.2024.128284
19. Islam, M.R., Sheikh Sadi, M.: Generative AI-assisted standalone wearable reading glasses for the visually impaired. In: 2024 IEEE 3rd International Conference on Robotics, Automation, Artificial-Intelligence and Internet-of-Things (RAAICON), pp. 173–176. IEEE, Dhaka, Bangladesh (2024). https://doi.org/10.1109/RAAICON64172.2024.10928631

20. Khan, W., Hussain, A., Khan, B.M., Crockett, K.: Outdoor mobility aid for people with visual impairment: Obstacle detection and responsive framework for the scene perception during the outdoor mobility of people with visual impairment. Expert Syst. Appl. **228**, 120464 (2023). https://doi.org/10.1016/j.eswa.2023.120464
21. Mehta, A., Parikh, P., Shah, P.: Raspberry-Pi based physical media to audio conversion device for visually impaired individuals. Int. J. Sci. Res. Sci. Eng. Technol. **11**(4), 249–260 (2024). https://doi.org/10.32628/IJSRSET24114127
22. Jasmine, G.S., Marry, D.M., Lakshmi, S.S., Rishiwanth, R., Sreehariprasath, K., Surendhar, J.: Camera based text and product lable reading for blind people. In: 2021 7th International Conference on Advanced Computing and Communication Systems (ICACCS), Coimbatore, pp. 1122–1126. IEEE, India (2021). https://doi.org/10.1109/ICACCS51430.2021.9441860
23. Wallace, S.E., Hux, K., Knollman-Porter, K., Brown, J.A., Parisi, E., Cain, R.: Reading behaviors and text-to-speech technology perceptions of people with aphasia. Assist. Technol. **34**(5), 599–610 (2022). https://doi.org/10.1080/10400435.2021.1904306
24. Xu, J., et al.: Multimodal navigation and virtual companion system: a wearable device assisting blind people in independent travel. Sensors. **25**(13), 4223 (2025). https://doi.org/10.3390/s25134223
25. Raffoul, S., Jaber, L.: Text-to-speech software and reading comprehension: The impact for students with learning disabilities. CJLT. **49**(2), 1–18 (2023). https://doi.org/10.21432/cjlt28296

Exploring Digital Metaliteracy Through a Bibliometric Lens: Intergenerational Perspectives in Education

Eric Le Fur(✉)

INSEEC Grande Ecole, Bordeaux, France
elefur@inseec.com

Abstract. The concept of metaliteracy—encompassing critical thinking, ethical awareness, online collaboration, and content creation—has emerged as a key competence in the digital age. While initially developed within higher education contexts, its scope extends far beyond universities, particularly in a world increasingly marked by misinformation, digitalized public services, and intergenerational interactions online. This paper presents a bibliometric analysis of research on digital metaliteracy, with a special focus on its intergenerational dimension. Data are from Web of Science databases covering the period 2015–2025, using keywords such as metaliteracy, digital literacy, and intergenerational learning. We use VOSviewer to identify clusters and thematic developments. Preliminary results suggest that the majority of research remains focused on university students, primarily on online courses, misinformation, and critical thinking skills. Intergenerational approaches, especially those involving adults in lifelong learning or senior populations, are largely absent. This gap highlights a paradox: despite its universal ambition, the academic study of metaliteracy continues to privilege formal educational contexts and younger learners. By mapping the existing literature, this paper identifies both strengths and omissions in current research and calls for a broader integration of intergenerational perspectives. Such a shift is essential to fostering inclusive digital education and ensuring that metaliteracy contributes to social participation and equity across generations.

Keywords: Bibliometric Analysis · Digital Education · Digital Metaliteracy · Intergenerational Learning

1 Introduction

The rapid digital transformation of societies has significantly reshaped the skills necessary for participation in education, the labor market, and civic life. This transformation has brought digital literacy to the forefront as a foundational competence, supplementing traditional forms of literacy and encompassing the critical and ethical use of information and technologies [2, 12]. In addition to digital literacy, the concept of metaliteracy, introduced by Jacobson and Mackey [19, 24], underscores the learners' ability not only to consume but also to produce, curate, and critically evaluate digital content. Concerns

A. Rocha et al. (Eds.): ICITS 2026, LNNS 1966, pp. 245–256, 2027.
https://doi.org/10.1007/978-3-032-27538-7_21

have emerged about intergenerational differences in digital skills, often discussed in terms of 'digital natives' [5, 17, 20] and 'digital immigrants' [1, 3, 9]. These generational disparities raise crucial questions about equity, inclusion, and the intergenerational transmission of competencies.

Despite the acknowledged importance of these literacies, the research landscape remains fragmented. While digital literacy is a widely explored area, metaliteracy is less frequently addressed in empirical studies, and intergenerational literacy is vastly underexplored. In this context, the need for a systematic approach, such as bibliometric analysis, becomes evident. It can help map the field, identify dominant trends, reveal research gaps, and propose future directions. To reduce this gap, this article presents a bibliometric study of Web of Science-indexed articles published between January 2015 and September 2025 in the field of education, with a focus on digital literacy, metaliteracy, and intergenerational literacy.

Our analysis suggests that metaliteracy is a promising concept, but its application is often limited to specific contexts like libraries, curation, and higher education, rather than broad population studies. The intergenerational approach, when present, tends to be more descriptive (comparisons of use) than prescriptive (targeted programs for intergenerational transmission). This lack of targeted programs is a significant limitation in the field. Moreover, the literature on informal practices (digital practices outside of class, migrant practices, personal curation) is a rich and growing field. This bibliometric study is a crucial first step towards a more detailed understanding of the dynamics of research on digital metaliteracy. The mapping of publications highlights areas for strengthening, particularly the empirical study of seniors and adults, as well as the comparison of practices for transmitting metaliteracy between generations. Ultimately, this work aims to contribute to a renewal of thinking on digital education by placing intergenerational equity at the heart of academic and political debates. The rest of this article is as follows. Section 2 presents the methodology. Section 3 highlights the results discussed in Sect. 4. Finally, Sect. 5 concludes.

2 Methodology

The bibliometric study was conducted using the Web of Science Core Collection. The keywords' metaliteracy', 'digital literacy', and 'intergenerational learning' were meticulously chosen to ensure the academic quality and comparability of documents. The search was limited to records classified under 'Education Educational Research', to the period January 2015–August 2025, and to peer-reviewed journal articles written in English. Conference proceedings, book chapters, and monographs were excluded. After extraction, the data underwent a rigorous cleaning process to eliminate duplicates and harmonize metadata, ensuring the integrity and reliability of the dataset. The final database includes 493 articles.

The analysis, conducted with the robust and comprehensive tool VOSviewer, identifies general trends, thematic clusters, and underrepresented areas in the literature [6, 33]. The retrieved records were exported in plain text and processed to extract key bibliographic fields: title, authors, journal, year, abstract, and cited keywords. Data cleaning was performed to unify author names and keywords. The analysis proceeded in several steps, covering descriptive statistics (annual output, distribution by journals, authors, and countries), keyword co-occurrence analysis to identify thematic clusters, and citation analysis to highlight the most influential works. This comprehensive approach, coupled with the use of VOSviewer, ensures that the study covers a wide range of aspects, providing a thorough understanding of the literature. Limitations of the method include the limited coverage of the Web of Science, which may underrepresent contributions published in regional journals or in grey literature [7, 26, 31, 34].

3 Results

3.1 Temporal Evolution

Figure 1 illustrates the number of academic articles published on digital literacy between January 2015 and August 2025. The field has experienced a clear upward trend, although growth has not been linear. Between 2015 and 2017, research output remained modest, ranging from 21 to 32 articles per year. This period represents the early consolidation of digital literacy as a research field. In 2017, a dip to 23 articles suggests that the theme was still emerging and had not yet reached critical mass. From 2018 to 2019, publication volumes stabilized at around 33–34 articles per year, indicating that digital literacy has gained recognition as a legitimate research theme, attracting consistent scholarly attention. A sharp increase occurred between 2020 and 2022, with 53 to 55 articles per year. This surge, coinciding with the COVID-19 pandemic, which accelerated the integration of digital tools in education and heightened the need for empirical studies on students' and teachers' digital skills, underscores the urgency and relevance of digital literacy research. The pandemic acted as a catalyst, bringing digital literacy to the forefront of educational research. The peak occurred in 2023 and 2024 (68 and 71 articles, respectively). This period marks the field's maturity. Indeed, from this period, digital literacy was studied not only as a practical skill but also in relation to critical thinking, inclusion, and broader socio-educational debates.

3.2 Leading Journals, Disciplinary Fields and Most Cited Articles

Out of the 130 journals that publish on digital literacy, Table 1 highlights the significant role of leading academic journals in shaping the landscape of digital literacy research. This data emphasizes the influence of a select group of technology-focused educational journals, followed by a more widespread distribution across various literacy and language learning outlets. The concentration of publications not only signifies the field's maturity but also its interdisciplinary nature, providing a comprehensive view for our academic audience.

Education and Information Technologies, with its 65 articles (13.2% of total publications), emerges as the most influential outlet, underscoring the central role of techno-pedagogical approaches in framing digital literacy research. Following closely is Computers & Education, with 27 articles (5.5%), serving as a flagship journal in the field of educational technology.

Another group of journals, forming an intermediate research core, plays a crucial role in broadening the scope of digital literacy research. These include Interactive Learning

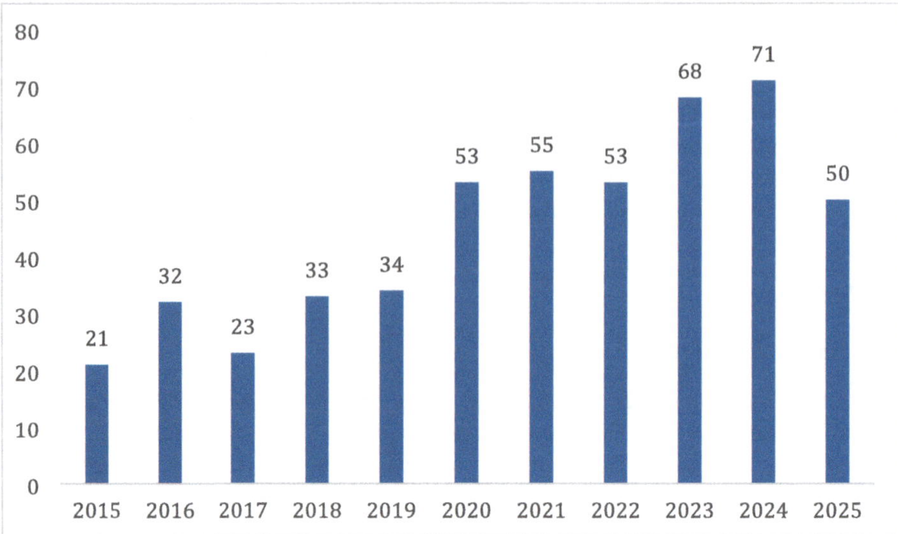

Fig. 1. Number of annual publications.

Environments (18 articles), Journal of Adolescent & Adult Literacy (13 articles), British Journal of Educational Technology (11 articles), and Educational Technology Research and Development (11 articles). They extend the boundaries of digital literacy research by linking it to classroom practices (such as using digital tools for collaborative learning) and broader issues of technology integration (like the impact of digital resources on curriculum design), thereby enriching the field.

The remaining journals in the list serve as a testament to the field's interdisciplinary nature. These publications bridge digital literacy with literacy studies, language learning, and the pedagogy of reading and writing. Their presence underscores that digital literacy research is not just about technological skills but also about communication, language, and critical thinking, making the field diverse and complex. This diversity is something our academic audience can appreciate and learn from. This distribution indicates several main insights. There is a clear polarization, with two leading journals dominating the publication landscape. Interdisciplinarity is evident, as research spans across educational technology, literacy, and language domains. The balance between generalist educational technology journals and specialized literacy journals reveals the field's dual nature, providing our academic audience with a comprehensive coverage of the field. For academics, successful dissemination strategies should target both high-impact technology journals and more specialized outlets to reach diverse academic communities.

The article by Falloon [13], "From digital literacy competence: the teacher digital competency (TDC) framework," published in ETR&D-Educational Technology Research and Development, is the most cited. Following this, the paper of Prior, Mazanov, Meacheam, Heaslip and Hanson [27] focuses on attitude, digital literacy, and self-efficacy. Lastly, Mohammadyari and Singh [25] provide valuable insights into the practical applications of digital literacy in their publication in Computers & Education. They

Table 1. Leading journals.

Education and Information Technologies	65	Technology, Pedagogy and Education	10
Computers & Education	27	Communicar	9
Interactive Learning Environments	18	Computer Assisted Language Learning	9
Journal of Adolescent Adult Literacy	13	Journal of Literacy Research	9
British Journal of Educational Technology	11	Literacy	9
ETR&D-Educational Technology Research and Development	11	Language Learning Literacy	7
Journal of Research on Technology in Education	10	Reading and Writing	7

specifically discuss its role in understanding the effect of e-learning on individual performance, such as how digital literacy can enhance student engagement and learning outcomes in online courses.

3.3 Geographic Distribution and Institutions

Out of the 659 institutions across 66 countries in the research field, more than a third of the publications originate from American universities. China and Australia follow, with Israel, South Korea, Turkey, and Taiwan emerging as major publishing countries in Asia. In Europe, Spain, England, and Norway lead the pack. The strong presence of education departments, centers for educational technology, and applied linguistics in the affiliations underscores their significant role in the research field (Table 2).

Figure 2 presents the country co-occurrences. The USA and China emerge as global hubs, followed by several regional clusters. In Europe (Spain, England, Italy, the Netherlands, and Scandinavia), there are highly interconnected relationships, reflecting a tradition of research in critical literacy and comparative pedagogy. In the Asia-Pacific region (Australia, South Korea, Malaysia, Indonesia, and Taiwan), regional clusters are closely connected to China and focus intensely on digital education and academic performance. Finally, emerging countries (Iran, Turkey, and South Africa) appear on the periphery but are connected via collaborative bridges. The geography of research reflects differentiated national priorities. In the USA, the focus is on formal education and social inequality. In China and the Asia-Pacific region, the priority is integrating technologies into mass education. In Europe, the critical approach focuses on citizenship and digital culture. Intergenerational issues are implicit in these perspectives, particularly in the USA, where social divides often intersect with age divides. In Europe, critical thinking focuses on young citizens versus adult citizens, while in Asia, the emphasis is on younger generations and performance, leaving seniors behind.

Table 2. Leading institutions.

Institution	Country	Nb. of articles
University System of Georgia	USA	11
State University System of Florida	USA	10
University of Oslo	Norway	9
Education University of Hog Kong	China	8
Open University Israel	Israel	8
University of California System	USA	8
Monash University	Australia	7
University of Hong Kong	China	7
Georgia State University	USA	6
Pennsylvania Commonwealth System of Higher Education	USA	6
University of British Columbia	Canada	6
University of London	England	6
University of Texas System	USA	6
Chinese University of Hong Kong	China	5
Pennsylvania State University	USA	5
Simon Fraser University	USA	5
Stanford University	USA	5
Texas AM University	USA	5
University of Haifa	Israel	5
University of Jyvaskyla	Finland	5
University of Valencia	Spain	5
University of Washington	USA	5
Abant Izzet Baysal University	Israel	5

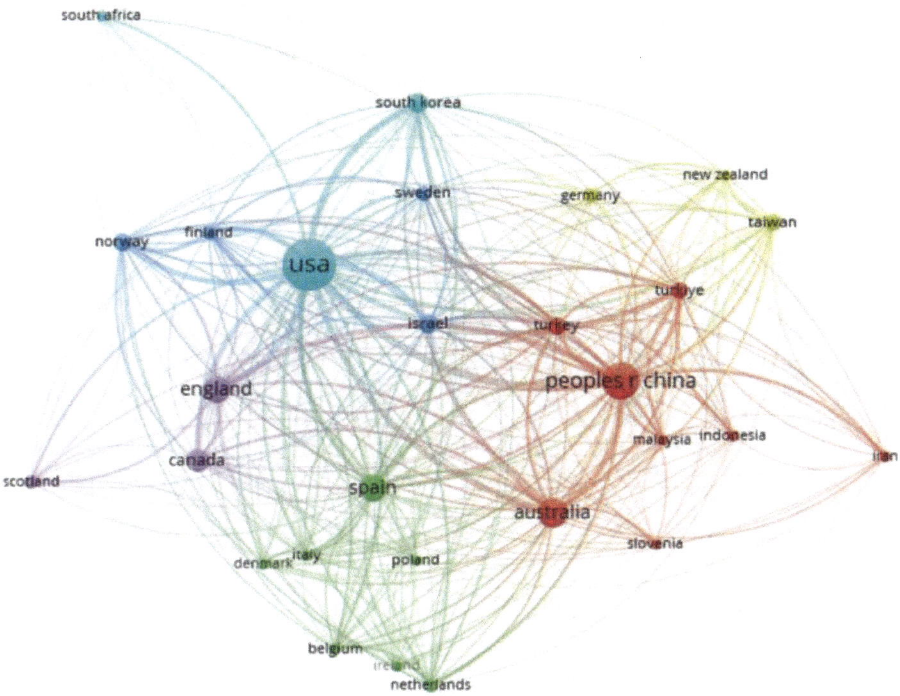

Fig. 2. Country co-occurrences.

4 Keyword Analysis and Potential Future Ways of Investigation

Table 3 presents the frequency of top keywords. Keyword analysis reveals digital literacy as the dominant term, followed by digital competence, information literacy, critical thinking, self-efficacy, teacher training, adult migrants, digital curation, and online learning engagement. Two major thematic clusters can be distinguished: (1) academic skills, focusing on students, teachers, libraries, and pedagogical approaches; (2) social inclusion, addressing migrants, adults, workplace practices, and citizenship. Metaliteracy is rarely mentioned, and intergenerational literacy is virtually absent, confirming their marginal status in the empirical literature. Research explicitly addressing intergenerational literacy is minimal. A small number of studies adopt the digital natives versus digital immigrants' framework, often with descriptive comparisons of technology use across age groups. There are a few comparative studies that directly address the significant issue of generational differences and categories, such as digital native/digital immigrant/pre-digital immigrant, as seen in an intergenerational analysis of ICT use at work in Spain. However, the vast majority of publications focus on higher education and students, or on specific populations (children, migrants, teachers) rather than on systematic comparisons between generations. This emphasis on specific groups highlights the significance of generational differences in ICT use. However, very few works propose systematic programs or models for the intergenerational transmission of digital skills. This absence represents a clear research gap, particularly in the context of lifelong learning, workplace adaptation, and family-based digital practices.

Table 3. Frequency of top keywords.

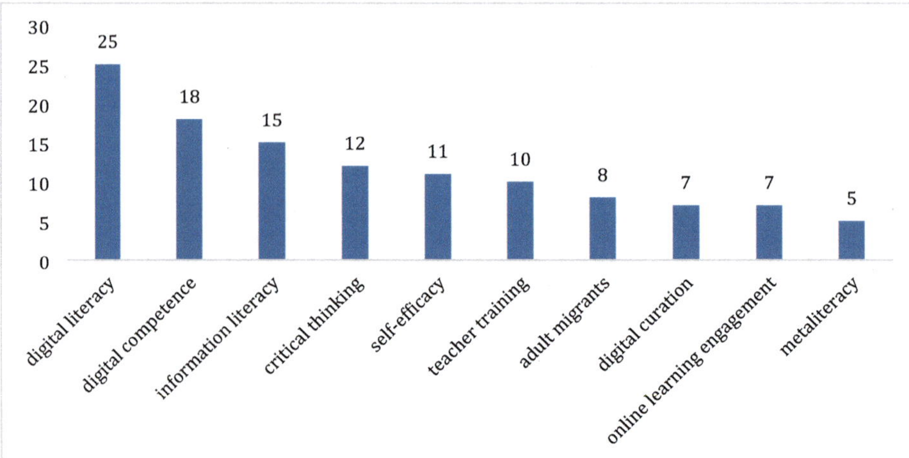

Figure 3 highlights five clusters of keywords, each represented by a distinct color. As mentioned previously, digital literacy is the key term in all articles and serves as the central node. Around this latter node, several thematic clusters gravitate. In green, the axis of teacher education includes notably information literacy, classroom competence, pre-service teachers, and technology integration. In this cluster, we find the concept of formal transmission, or intergenerational transmission between teachers and students. For example, Ballarino and Bernardi [4] compare the intergenerational transmission of inequality and education in fourteen countries. They suggest that including both occupation-based and income-based measures yields a richer, detailed picture of the intergenerational socio-economic transmission.

In red, the axis critics and social focuses on critical literacy, identity, media literacy, and pedagogy agency. In other words, this cluster highlights the issues of power, identity, and practice relevant to generation inequalities in access and mastery of digital technologies. For example, Fang et al. [14] explore privilege in the digital divide. From 55 articles published between 2006 and 2016, they highlight the role of age, education, income, gender, and generational status. Many articles examine digital inequalities across generations [e.g., 16, 28–30].

In yellow, the axis reading/understanding includes comprehension, readers, adolescents, children, and online reading. This cluster focuses on the intergenerational transition between young digital natives, a term used to describe individuals who have grown up in the digital age, and older generations. For example, despite using different reading measures, Liman Kaban & Karadeniz [22] observed no significant difference in reading comprehension levels among the young generation. In addition, Duncan, McGeown, Griffiths, Stothard and Dobai [11] show that word identification emerges as a strong predictor of reading comprehension across adolescence and text genres. Meta-analyses [e.g., 8, 10, 21] and mode-effect study [32] show that reading comprehension on paper is better than on screen among young adults.

Finally, in blue, the axes include self-efficacy, achievement, online learning, equity, and COVID-19. That serves as a reminder that the pandemic has accelerated the digital divide between generations. The literature on the impact of COVID-19 in the education sector is predominantly descriptive and focuses on the difficulties teachers face in transitioning to online teaching [e.g., 15, 18, 23].

This map shows that the notion of intergenerational does not appear directly as a keyword, but rather that it implicitly crosses the debates: education (formal transmission), social practices (age divides and identities), and crises (COVID-19). Metaliteracy, therefore, remains an underutilized conceptual framework for addressing these tensions between generations.

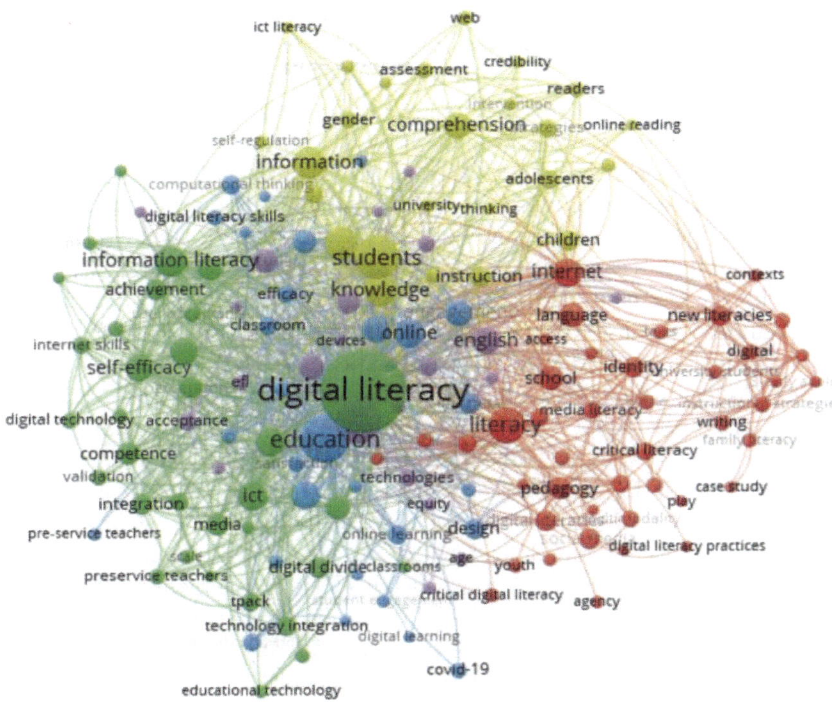

Fig. 3. Clusters of keywords.

Our analysis highlights several exploitable gaps for academic research, and your expertise is crucial in addressing these. First, a systematic mapping of intergenerational comparative studies, with your insights, would help address the lack of aggregation and synthesis in this field. Second, we could implement a standardized measurement of the transmission mechanisms (family, school, workplace) of metaliteracy, with your guidance. Third, future research could conduct longitudinal studies to evaluate the effects of programs on the sustainability of digital skills among adults and seniors, with your participation. Finally, investigations of transmission inequalities will allow us to analyze particularly how socialization circuits promote or hinder the acquisition of metaliteracy between generations, and your contributions to this research will be invaluable.

5 Conclusion

The surge in academic publications on digital literacy, more than tripling between 2015 and the 2024 peak, is a significant indicator of the field's growth. The pandemic period (2020–2022) was not just a pivotal moment, but a critical one that underscored the urgent and relevant role of digital skills in crisis management. The subsequent post-pandemic peak signifies the field's maturation. These statistics, beyond being mere numbers, mirror the broader shifts in education and society: the digitalization of classrooms, the ascent of online learning, and the recent emergence of generative artificial intelligence tools. These developments reignite discussions on the essence of digital literacy in the 2020s. The future challenge lies in incorporating critical and intergenerational dimensions into digital literacy research, ensuring it promotes equity and democratic participation.

This bibliometric study of digital literacy, metaliteracy, and intergenerational literacy between 2015 and 2025 uncovers a consolidated research domain, an emerging but underdeveloped conceptual framework, and a particularly intriguing yet neglected dimension of intergenerational transmission. By mapping current trends, the analysis provides a clearer picture of where the field stands and where it should go next. The potential of future studies to prioritize longitudinal and comparative approaches, expand metaliteracy into applied contexts, and integrate intergenerational perspectives to address digital inequalities is a cause for optimism in the field's progress.

Several methodological patterns are discernible. Surveys and structural equation models are prevalent in quantitative studies, while interviews and case studies are common in qualitative work. Bibliometric and text-mining approaches are gaining traction. However, the scarcity of longitudinal designs is a significant gap, hindering our understanding of how digital skills evolve. There is a pressing need for more diverse research methods to ensure the field remains innovative and adaptable.

The findings call for a research agenda that expands beyond measuring individual competencies to exploring contexts of transmission, including intergenerational learning, workplace adaptation, and community-based initiatives. Moreover, metaliteracy could be more fully integrated into educational practice, fostering critical engagement and content creation alongside the consumption of information.

References

1. Agárdi, I., Alt, M.A.: Do digital natives use mobile payment differently than digital immigrants? A comparative study between generation X and Z. Electron. Commer. Res. **24**(3), 1463–1490 (2024)
2. Audrin, C., Audrin, B.: Key factors in digital literacy in learning and education: a systematic literature review using text mining. Educ. Inf. Technol. **27**(6), 7395–7419 (2022)
3. Autry Jr., A.J., Berge, Z.: Digital natives and digital immigrants: getting to know each other. Ind. Commer. Train. **43**(7), 460–466 (2011)

4. Ballarino, G., Bernardi, F.: The intergenerational transmission of inequality and education in fourteen countries: a comparison. In: Education, Occupation and Social Origin, pp. 255–282. Edward Elgar Publishing (2016)
5. Bennett, S., Maton, K., Kervin, L.: The ‘digital natives’ debate: a critical review of the evidence. Br. J. Educ. Technol. **39**(5), 775–786 (2008)
6. Bukar, U.A., Sayeed, M.S., Razak, S.F.A., Yogarayan, S., Amodu, O.A., Mahmood, R.A.R.: A method for analyzing text using VOSviewer. MethodsX. **11**, 102339 (2023)
7. Chadegani, A.A., et al.: A comparison between two main academic literature collections: Web of Science and Scopus databases. arXiv preprint arXiv, 1305.0377 (2013)
8. Clinton, V.: Reading from paper compared to screens: a systematic review and meta-analysis. J. Res. Read. **42**(2), 288–325 (2019)
9. Creighton, T.B.: Digital natives, digital immigrants, Digital Learners: an international empirical integrative review of the literature. Educ. Leadersh. Rev. **19**(1), 132–140 (2018)
10. Delgado, P., Vargas, C., Ackerman, R., Salmerón, L.: Don’t throw away your printed books: a meta-analysis on the effects of reading media on reading comprehension. Educ. Res. Rev. **25**, 23–38 (2018)
11. Duncan, L.G., McGeown, S.P., Griffiths, Y.M., Stothard, S.E., Dobai, A.: Adolescent reading skill and engagement with digital and traditional literacies as predictors of reading comprehension. Br. J. Psychol. **107**(2), 209–238 (2016)
12. Estrada, F.J.R., George-Reyes, C.E., Glasserman-Morales, L.D.: Security as an emerging dimension of Digital Literacy for education: a systematic literature review. J. E-Lear. Knowl. Soc. **18**(2), 22–33 (2022)
13. Falloon, G.: From digital literacy to digital competence: the teacher digital competency (TDC) framework. Educ. Technol. Res. Dev. **68**(5), 2449–2472 (2020)
14. Fang, M.L., Canham, S.L., Battersby, L., Sixsmith, J., Wada, M., Sixsmith, A.: Exploring privilege in the digital divide: implications for theory, policy, and practice. The Gerontologist. **59**(1), e1–e15 (2019)
15. Hamdan, K.M., Al-Bashaireh, A.M., Zahran, Z., Al-Daghestani, A., Al-Habashneh, S., Shaheen, A.M.: University students’ interaction, Internet self-efficacy, self-regulation and satisfaction with online education during pandemic crises of COVID-19 (SARS-CoV-2). Int. J. Educ. Manag. **35**(3), 713–725 (2021)
16. Helsper, E.: The digital disconnect: The social causes and consequences of digital inequalities, pp. 1–232. SAGE Publications Ltd (2021)
17. Helsper, E.J., Eynon, R.: Digital natives: where is the evidence? Br. Educ. Res. J. **36**(3), 503–520 (2010)
18. Hong, J.C., Liu, X., Cao, W., Tai, K.H., Zhao, L.: Effects of self-efficacy and online learning mind states on learning ineffectiveness during the COVID-19 lockdown. Educ. Technol. Soc. **25**(1), 142–154 (2022)
19. Jacobson, T.E., Mackey, T.P.: Proposing a metaliteracy model to redefine information literacy. Commun. Inf. Lit. **7**(2), 84–91 (2013)
20. Jones, C., Ramanau, R., Cross, S., Healing, G.: Net generation or digital natives: is there a distinct new generation entering university? Comput. Educ. **54**(3), 722–732 (2010)
21. Kong, Y., Seo, Y.S., Zhai, L.: Comparison of reading performance on screen and on paper: a meta-analysis. Comput. Educ. **123**, 138–149 (2018)
22. Liman Kaban, A., Karadeniz, S.: Children’s reading comprehension and motivation on screen versus on paper. SAGE Open. **11**(1), 2158244020988849 (2021)
23. Ma, K., Chutiyami, M., Zhang, Y., Nicoll, S.: Online teaching self-efficacy during COVID-19: Changes, its associated factors and moderators. Educ. Inf. Technol. **26**(6), 6675–6697 (2021)
24. Mackey, T.P., Jacobson, T.E.: Reframing information literacy as a metaliteracy. C&RL. **72**(1), 62–78 (2011)

25. Mohammadyari, S., Singh, H.: Understanding the effect of e-learning on individual performance: the role of digital literacy. Comput. Educ. **82**, 11–25 (2015)
26. Norris, M., Oppenheim, C.: Comparing alternatives to the Web of Science for coverage of the social sciences' literature. J. Informet. **1**(2), 161–169 (2007)
27. Prior, D.D., Mazanov, J., Meacheam, D., Heaslip, G., Hanson, J.: Attitude, digital literacy and self-efficacy: Flow-on effects for online learning behavior. Internet High. Educ. **29**, 91–97 (2016)
28. Robinson, L., et al.: Digital inequalities and why they matter. Inf. Commun. Soc. **18**(5), 569–582 (2015)
29. Robinson, L., et al.: Digital inequalities 2.0: Legacy inequalities in the information age. First Monday. **25**(7), 1–27 (2020a)
30. Robinson, L., et al.: Digital inequalities 3.0: Emergent inequalities in the information age. First Monday. **25**(7), 1–21 (2020b)
31. Singh, V.K., Singh, P., Karmakar, M., Leta, J., Mayr, P.: The journal coverage of Web of Science, Scopus and Dimensions: a comparative analysis. Scientometrics. **126**(6), 5113–5142 (2021)
32. Støle, H., Mangen, A., Schwippert, K.: Assessing children's reading comprehension on paper and screen: a mode-effect study. Comput. Educ. **151**, 103861 (2020)
33. Yan, L., Zhiping, W.: Mapping the literature on academic publishing: a bibliometric analysis on WOS. SAGE Open. **13**(1), 21582440231158562 (2023)
34. Zhu, J., Liu, W.: A tale of two databases: the use of Web of Science and Scopus in academic papers. Scientometrics. **123**(1), 321–335 (2020)

Unraveling the Macroeconomic Determinants of Economic Growth

Ferdina Paththamesthrige, Gayani Hawpe Liyanage, Uthpala Punchihewa, Danish Khokhar, Daniyal Jawad, and Jyri Vilko(✉)

Department of Industrial Engineering and Management, Lappeenranta-Lahti University of Technology, Kouvola Unit, Kauppalankatu 13, 45100 Kouvola, Finland
jyri.vilko@lut.fi

Abstract. This study aims to investigate the macroeconomic factors responsible for Finland's economic growth from 2004 to 2023, considering unemployment, inflation (CPI), exports, and the volume of retail sales. Utilizing secondary data from the World Bank data sets and multiple linear regression, results demonstrate that GDP growth is largely driven by consumer spending, exports, and moderate inflation. Although unemployment correlates negatively with GDP, its impact on Finland's GDP is statistically insignificant. These findings challenge the conventional interpretations of Okun's Law, providing evidence for the claim that strong welfare and labor policies in advanced economies in advanced economies can offset the typical growth employment trade-off. This study is naval as it highlights the underexplored but crucial role of retail sales as a predictor of Finnish GDP and the positive role of moderate inflation in sustaining consumption-led growth. Further, the findings underscore the importance of domestic demand, trade diversification, and inflation management for strengthening Finland's long-term economic resilience, providing theoretical contributions and policy insights.

Keywords: Macroeconomic Indicators · Gross Domestic Product (GDP) · Economic Resilience

1 Introduction

The economic Growth of Finland during the period of 2004 to 2023 has been influenced by a combination of internal and external factors. Finland's economy which is highly industrialized and openness to trade has faced both significant, challenges and opportunities during this period. The global financial crisis that took place during 2008 and 2009 has influenced profoundly, leading to a deflation in GDP and an increase in unemployment. However, country's strong public sector policies and a robust welfare state model supported for Finland's economic recovery [1]. Another major challenge was posted due to the COVID 19 pandemic situation in 2020, causing a transitory decline in the economic growth of the country. Regardless of these obstacles, Finland's economy demonstrated a resilience through the recovery driven by expanded individual consumption and export growth [1]. Further, structural reforms and investment in technology and education have significantly affected sustaining long term economic growth [2]. In spite

A. Rocha et al. (Eds.): ICITS 2026, LNNS 1966, pp. 257–268, 2027.
https://doi.org/10.1007/978-3-032-27538-7_22

of extensive research on the determinants of GDP, a comprehensive analysis of how multiple macroeconomic indicators affect GDP is still needed. Hence, the purpose of this study is to fill this gap by examining the relationships between GDP and its possible predictors. Further, understanding and analyzing these determinants provide crucial insights into the dynamics of Finland's economic development over the past two decades. In doing this, the study aims to answer the following research questions: *What are the key determinants of Economic Growth and their trends in last two decades in Finland? What is the relationship of each variable with Economic growth in Finland? and How do these key macroeconomic determinants impact on Economic growth in Finland?*

2 Literature Review

Gross Domestic Product (GDP) is the most common metric of economic growth, which is the value of goods and services produced by a country during a set time [3]. Its development is affected by labor, accumulated capital and natural resources, as well as indirectly by the efficiency of government, its fiscal policies and the structure of institutions [4]. Despite the view that GDP does not give enough attention to well-being, human capital, non-monetary transactions and environmental concerns, many continue to rely on it to measure economic activities and make policies [5]. It is important to understand macroeconomic factors that affect GDP for analyzing the dynamics of economic development.

The correlation between GDP and unemployment has been historically studied under the theory of Okun, which is a negative relationship. It is hypothesized that an increment in unemployment by 1 % decreases GDP by 3 % [6]. The negative correlation between GDP and unemployment is backed by a variety of studies. Abdul-Khaliq and others [7] found it in Arab countries considering period 1994–2010, Xesibe and Nyasha [8] confirmed there was a negative impact by unemployment to GDP in South Africa considering data from 1994–2017, and Kukaj [9] confirmed similar results in the Western Balkans considering the period 2001–2015. Yet, this relationship is not consistent in all contexts. In an example of what can happen under rising unemployment, it was found that in Jordan for period 1970–2008, there was no direct relationship between unemployment and low growth, with structural and institutional issues influencing the relationship between the two variables [6].

Another important determinant of GDP, which affects consumption, investment, and price stability, is inflation, which is measured via the Consumer Price Index (CPI). It is an outcome of the increase of prices over time, and it influences how consumers shop and how the economy functions [10]. Most of the literature presents a negative correlation, because long term inflation shrinks the purchasing power and discourages investment. In cross-country analysis, Barro [11] covering 100 countries for period 1960–1990, found inflation growth-reducing and Saungweme and Odhiambo [12], confirmed this inverse relationship in a study of Kenya covering the period 1970–2019. However, other research shows that good results are possible where Aboobucker and Jahufer [13] indicated a very high positive correlation in Sri Lanka in 1960–2016 period. Song [10] also added that moderate inflation may stimulate spending and investment, yet too much inflation is damaging. These findings imply that the inflation-growth relationship is largely sensitive to context, therefore the results vary according to the level of inflation, the economic structure, and the efficacy of the monetary policies.

In connection with globalization and integration, exporting plays a crucial role in economic development. It is considered a necessary driver for domestic production that utilizes natural and other resources in the best possible ways [14]. The relationship between export and economy growth has been studied and analyzed by numerous scholars from around the world. Hagemejer and Mućk [15] found a powerful relation between exports and economic growth for Central and Eastern Europe during 1995–2014. Nguyen [14] pointed out the positive impact of exports on Vietnam's industrial and modernization processes. But some research results differ from each other. Researching Mexico in 1960–2003, Waithe [16] found that exports initially raised GDP, but this relationship later became negative. The study explained that one probable cause is the fact that exports depend more on imported goods and less on locally manufactured content, with fewer linkages to local suppliers. This decreases the possibility of spreading economic gains within the country. It seems that export promotion can help the economy grow but doing this requires additional policies to tackle long-term problems.

Retail sales is another factor, which includes consumption and household demand, in which the macroeconomic factors influence GDP. It shows changes in how people consume and develop the country, its cultural and material norms, the income of people, and the retail sector [17]. The Retail Sales Volume Index (RSVI) which is a proxy, has been popularly used as a short-run measure of commercial activities [18]. The empirical evidence indicates a close relationship between retail activity and GDP and especially in an economy which relies on domestic consumption. Zhang [17], for example, notes that the increasing sales in Chinese retail contributed to GDP growth in long term, between 1992 and 2019. Further, proving the correlation between the Retail Sales Volume and Economic Growth, Anagboso & McLare [18] found out when these is sustainable economic Growth, the Retail Sales volumes tend to increase. These demonstrate why domestic spending or consumption is necessary for the economy to grow, especially considering the uncertainties in global trade.

3 Methodology

This research examines the observable, measurable phenomena and by using that data aims to identify patterns, trends, and relationships. For that purpose, the study carried out as a quantitative study focuses on objective analysis of macroeconomic data and trends over a long period (from year 2004 to 2023). The data on macroeconomic indicators from 2004 to 2023 were collected using the Global Economic Monitor data in the World Bank Group [19]. Further, studies developed hypotheses using existing theories and validating them through empirical data to test hypotheses about the macroeconomic determinants and economic growth relationship. Hence, deductive approach is utilized for finding the determinants of Economic Growth of Finland for past two decades [20]. Several statistical techniques are being used to analyze the data and test the hypothesis developed. In the data analysis, Inferential statistical methods were applied using the SPSS statistics software. Further, Microsoft excel has been used for visualizing the data graphs. As statistical tests for data analysis, test for normality, correlation analysis and multiple linear regression analysis were conducted, and the below hypothesis were tested.

Hypothesis for Normality Test
Normality tests are carried out to measure the central tendency and dispersion of quantitative data set which tests the normality of the data. This is a prior requirement to be satisfied for various tests including correlation analysis and regression analysis [21]. The following hypothesis were tested through Kolmogorov-Smirnov and Shapiro-Wilk Test calculating the significance level which is called p value. ($p > 0.05$)

H0: The Variables (Dependent and Independent) are normally distributed.
Ha: The Variables (Dependent and Independent) are not normally distributed.

Hypothesis for Correlation Between Dependent and Independent Variables
Pearson correlation is used to measure the correlation between the dependent and independent variables, which measures the direction and strength of a linear relationship between two variables. The measurement lies between -1 to $+1$ where -1 indicates highly negative correlation, 0 represents there is no correlation between variables and $+1$ indicates highly positive correlation with the variables [20].

H0: There is no linear relationship between the dependent and i^{th} Variable (The two Variables are Independent)
Ha: There is a linear relationship between the dependent and i^{th} Variable (The two Variables are dependent)

At last, Multiple Linear Regression Analysis is conducted to analyze how do these key macroeconomic determinants Impact on Economic growth in Finland. Further, the hypotheses below were tested during the regression analysis and used to test the below conceptual framework.

H0: All independent variables do not explain the dependent variable
Ha: At least one independent variable explains the dependent variable.

The above hypothesis was tested in regression ANOVA while below has tested the impact of each individual variable on dependent variable.

H0: There is no impact on Economic Growth (GDP) by I^{th} Variable
Ha: There is an impact on Economic Growth (GDP) by I^{th} Variable
Note: The rejection criteria of the null hypothesis for the above analysis was $P < 0.05$
Where, I_1 – Unemployment Rate, I_2 – Consumer Price Index, I_3 – Exports, I_4 – Retail Sales Volume Index

The below conceptual framework in Fig. 1 was tested through these hypotheses.

4 Results and Discussion

In research, the chapter on Analysis and Findings is significant in the sense that it explains how each research question is evaluated and how the results are formally interpreted.

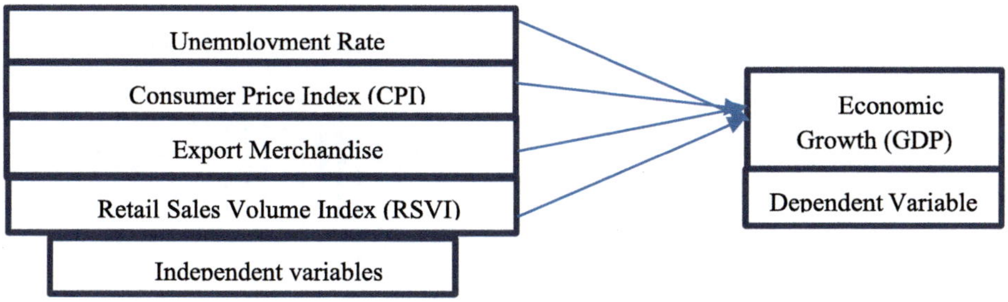

Fig. 1. Conceptual Framework

Key Determinants of Economic Growth and Their Trends in Last Two Decades

According to the Literature review four Key determinants have been identified that drives the Economic Growth of a country. They are *Unemployment Rate*, *Inflation* (measure using the Consumer Price Index), *Exports* and the *Retails Sales Volume* (measured using the Retail Sales Volume Index). The trends of each of these variables changes over 2004 to 2023 and Microsoft Excel used to visualize the data.

Finland's GDP growth over two decades from 2004 to 2023 has a steady increase until 2008 followed by a dip for 2 years. After 2015, there's a gradual increase till 2023. Overall, the long-term trend is upward with some exceptions as shown in Fig. 2.

Fig. 2. GDP at Market Prices of Finland (2004–2023)

The unemployment rate from 2004 to 2023 has been reducing over the years. From 2004 to 2008 there is a gradual decrease, followed by an increase till 2015. From 2015 onward, the rate gradually decreases with some exceptions as shown in Fig. 3. The consumer price index in Finland gradually increased represents in Fig. 4 indicating the Inflation rate has been increasing over the years. The growth rate is highest for the last 2 years in 2022 and 2023.

Finland Export merchandise from 2004 to 2023 rose steadily with a peak in 2008 and dropped in 2009. From 2009 to 2011, there was an increase, followed by fluctuations up to 2020. A notable dip occurred in 2016 and 2021 onwards the export merchandise rate increased and almost remains the same till 2023 as seen in Fig. 5. The Retail Sales Volume Index graph from 2004 to 2023 shows in Fig. 6 with an upward trend until 2008, with a peak in 2008. From 2009 onwards till 2014 the growth rate remains almost same.

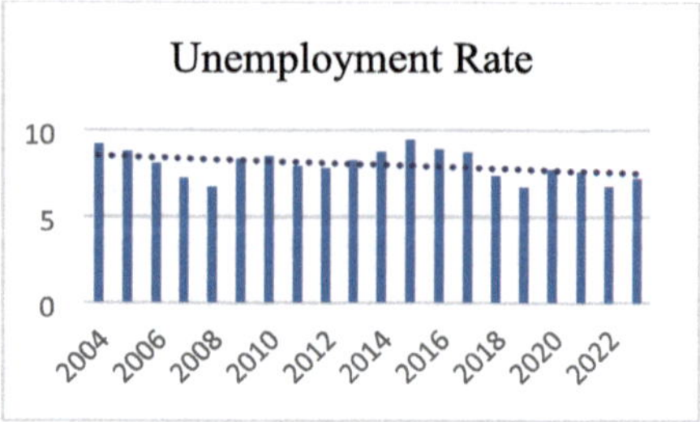

Fig. 3. Unemployment Rate of Finland (2004–2023)

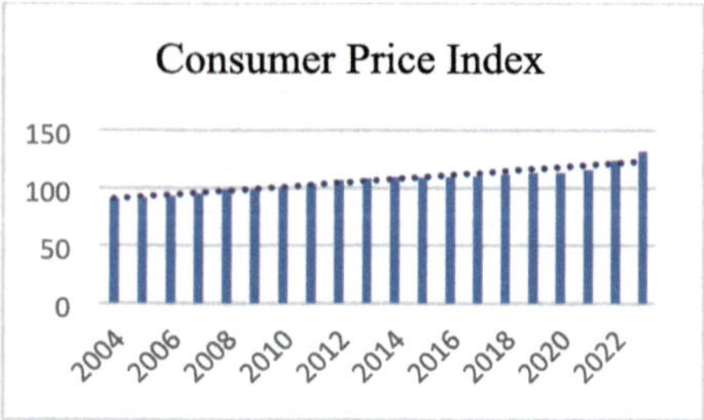

Fig. 4. Consumer Price Index of Finland (2004–2023)

In 2015 there is a sudden decrease, and the retails sale volume remains almost the same till 2023, with the peak in 2021.

Fig. 5. Export Merchandise of Finland (2004–2023)

Fig. 6. Retal Sales Volume Index (2004–2023)

The Relationship Between Macroeconomic Determinants with Economic Growth
After the analyzing the trends, the relationship between the independent variables and the Dependent Variable was carried out to answer the research question 2. However, before that the normality of the variables was tested using the Kolmogorov-Smirnov and Shapiro-Wilk Test [21].

According to the test results from both Kolmogorov-Smirnov and Shapiro-Wilk Test, the P value is greater than 0.05 as shown in Table 1. Hence, the null hypothesis is accepted and concluded the variables are normally distributed. As the Variables are normally distributed, the Pearson Correlation Analysis was conducted to discover the relationship between each independent variable with the dependent variable.

According to the Pearson test results in Table 2, all considered macroeconomic determinants were correlated with the GDP as the P value is less 0.05 and rejected the hypothesis. The unemployment rate has a − 0.678 moderate negative correlation with GDP in Finland, while the Consumer Price Index with 0.721, Export Merchandise with 0.619 and Retail Sales Volumes with 0.709 have a strong positive correlation with the Finland's Economic growth.

Table 1. Kolmogorov-Smirnov and Shapiro-Wilk tests statistics for Normality

Tests of Normality						
	Kolmogorov-Smirnov[a]			Shapiro-Wilk		
	Statistic	df	Sig.	Statistic	df	Sig.
GDP at Market Prices	.139	20	.200*	.932	20	.170
Unemployment Rate	.109	20	.200*	.961	20	.554
Export Merchandise	.089	20	.200*	.974	20	.837
Consumer Price Index	.123	20	.200*	.954	20	.435
Retail Sales Volume Index	.154	20	.200*	.965	20	.652

* This is a lower bound of the true significance, [a] Lilliefors Significance Correction
Note: The rejection criteria of the null hypothesis for the above analysis were $P < 0.05$

Table 2. Pearson Correlation Test with 20 samples

Correlations		GDP at Market Prices, Current US $, Millions
Unemployment Rate	Pearson Correlation	−.678**
	Sig. (2-tailed)	.001
Export Merchandise	Pearson Correlation	.619**
	Sig. (2-tailed)	.004
Consumer Price Index	Pearson Correlation	.721**
	Sig. (2-tailed)	.000
Retail Sales Volume Index	Pearson Correlation	.709**
	Sig. (2-tailed)	.000

** Correlation is significant at the 0.01 level (2-tailed), * Correlation is significant at the 0.05 level (2-tailed).
Note: The rejection criteria of the null hypothesis for the above analysis were $P < 0.05$

The Impact of Key Macroeconomic Determinants on GDP Growth in Finland

In order to find out the impact of each key determinant on GDP growth of Finland, a Multiple Linear Regression analysis was conducted. However, prior to running a regression analysis, there are key assumptions to be satisfied where there should be a linear relationship between dependent and independent variables, the residuals should be normally distributed, the residuals should be random, the variance of the residuals has a constant value, no Multicollinearity should be existed [20]. Accordingly, the normality of the residuals was further confirmed based on the P-P plot. As per the results P-P plot in Fig. 7, the residuals are distributed close to the line and indicate the normality of the residuals. Further, to test the randomness of the and constant variance of the residuals were tested using the residual scatter. In the scatter plot illustrated in Fig. 8, standardized residuals have been presented in relation to standardized predicted values. They are randomly distributed without having a predicted pattern. The variance of residuals is constant. Thus, there is no heteroscedasticity problem in the regression model and results are highly valid.

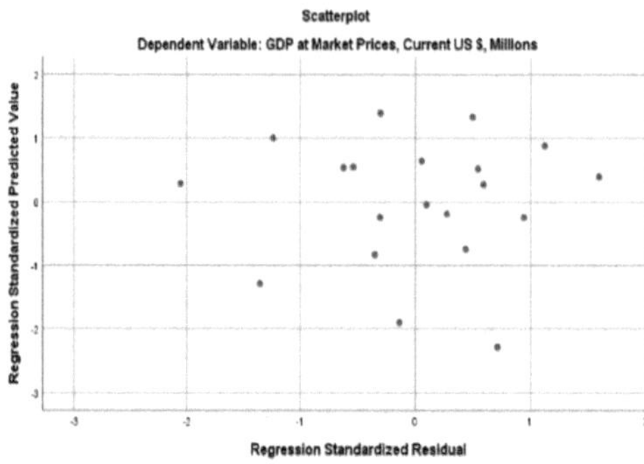

Fig. 7. P-P Plot for Standardize residuals

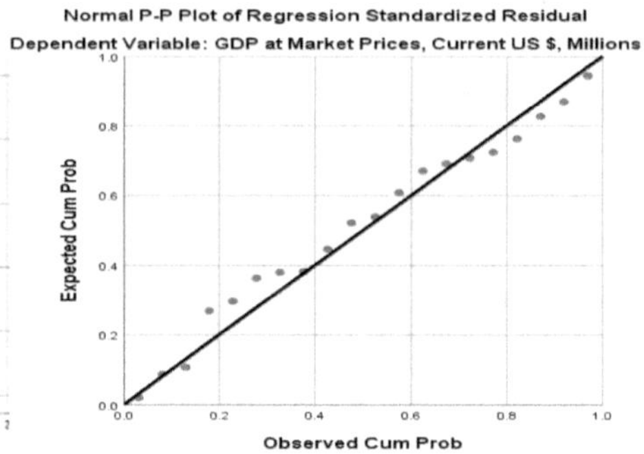

Fig. 8. Scatter Plot

After that, test has been carried out to find out whether there is any multicollinearity problem. Multicollinearity refers to the correlation with two or more independent variables that are measured with the variance inflation factor (VIF). If the VIF value lies between 1–10, it is considered that there is no multicollinearity within the independent variables. According to the results all the values lie between the appropriate levels as seen in Table 3. Hence, no Multicollinearity problem in the model.

Table 3. Collinearity Statistics for Independent Variables

Model		Collinearity Statistics	
		Tolerance	VIF
1	(Constant)		
	Unemployment Rate	.306	3.273
	Consumer Price Index	.792	1.263
	Export Merchandise	.302	3.316
	Retail Sales Volume Index	.642	1.557

Since all the prior assumptions being satisfied, regression analysis was conducted, and the results are as below.

Table 4. Regression Model Summary

Model Summary[b]					
Model	R	R Square	Adjusted R Square	Std. Error of the Estimate	Durbin-Watson
1	.993[a]	.986	.982	3682.4853	1.542

[a] Predictors: (Constant), Retail Sales Volume Index, CPI, Unemployment Rate, Export Merchandise
[b] Dependent Variable: GDP at Market Prices, Current US $, Millions

According to the model summary in Table 4, Retail Sale Volume Index (RSVI), Consumer Price Index (CPI), Unemployment Rate and Exports merchandise are strongly correlated with the GDP as the R value is 99.3%. Further, the regression model has explained 98.6% dependent variable (GDP) which indicated by R Square values in model summary. Moreover, the adjusted coefficient of determination says that 98.2% of GDP has been explained by the regression model. Hence, all these independent variables represent the regression model. Durbin-Watson test statistics are 1.542 which lies between the limits of 1.5 and 2.5. Therefore, residuals are independent, and the model is appropriate. Further, the regression ANOVA results as below. An ANOVA table suggests relationships between all independent variables considered in the study on the selected dependent variable chosen.

Table 5. Regression ANOVA

ANOVA[a]						
Model		Sum of Squares	df	Mean Square	F	Sig.
1	Regression	14186174143.126	4	3546543535.781	261.531	.000[b]
	Residual	203410464.396	15	13560697.626		
	Total	14389584607.522	19			

[a] Dependent Variable: GDP at Market Prices, Current US $, Millions
[b] Predictors: (Constant), Retail Sales Volume Index, CPI, Unemployment Rate, Export Merchandise

In the regression ANOVA in Table 5, probability of F test statistics is 0.000. This is highly significant, as the p value is less than 0.05. The model is appropriate and jointly significant, where Unemployment Rate, Consumer Price Index, Export merchandise and Retail Sales Volume Index jointly influence GDP. The individual effect of these factors on GDP is provided by the individual coefficients of the following Table 6.

According to the coefficient table (see Table 6), Consumer Price Index has a positive coefficient (1696.505) and highly significant with 0.000, suggesting that Consumer Price Index strongly impacts the GDP. Further, Exports Merchandise and Retail Sales Volume Index have positive coefficient values 0.344 and 1836.66 respectively and both variables are highly significant as the p values are less than 0.05. Therefore, both variables have a

Table 6. Regression Coefficient Table

Unstandardized Coefficients		Standardized Coefficients	t	Sig.	
B	Std. Error	Beta			
(Constant)	−115632.732	30023.049		−3.851	.002
Unemployment Rate	−1431.923	1841.655	−.043	−.778	.449
CPI	1694.505	90.008	.650	18.826	.000
Export Merchandise	.344	.147	.131	2.335	.034
Retail Sales Volume Index	1836.666	122.404	.575	15.005	.000

[a] Dependent Variable: GDP at Market Prices, Current US $, Millions

highly positive impact on GDP growth. However, the Unemployment Rate has a negative coefficient which is −1431.923 but is it not significant as the P value is greater than 0.05. Thus, the null hypothesis is being accepted. Hence, Unemployment rate isn't significantly impacted on the GDP individually with the considered predictors but jointly influence the GDP. Considering above, the below equation for Multiple regression can be constructed.

$$GDP = 1694.505(Consumer\ Price\ Index) + 0.344(Export\ Merchandise) \\ + 1836.666(Retail\ Sales\ Volume\ Index) - 115632.732$$

The empirical analysis shows that Finland's GDP growth from 2004 to 2023 mainly driven from domestic consumption, exports, and moderate inflation while unemployment is statistically insignificant and did not have a meaningful individual impact. However, there was a negative correlation between unemployment and GDP, its insignificance in the regression model indicates that Finland's strong labor policies, social safety nets, and active labor market programs lessen the typical trade-off between growth and employment. Further the findings challenge the conventional application of Okun's Law in advanced welfare economies. The importance of demand-side factors in fostering economic growth has demonstrated by the significant role of the Consumer Price Index (CPI) and retail sales volume. In Finland, moderate increases in prices appear to stimulate both consumption and investment, contrasting with the perspective on inflation in many other economies. Further, findings indicate favorable effects of retail sales volume suggest that indicators of household consumption could be crucial predictors of overall economic performance, an area that has not been thoroughly investigated. Collectively these findings point to the aspect of domestic demand, external trade and inflation management as the central mechanisms influencing the economic future of Finland.

5 Conclusions

The study's conclusions emphasize the importance of demand-side strategies and imply that boosting household spending with focused incentives can successfully promote economic expansion. The importance of preserving robust trade policy and boosting sectoral competitiveness is shown by the substantial role that exports play. Although unemployment and GDP have a negative relationship, its statistical insignificance indicates the value of Finland's extensive labor market policies and social safety nets that focused on job matching, retraining, and job security. The future studies need to explore the disaggregated household consumption components, sector-wise export contributions, and the interplay of labor market dynamics with economic growth. Further probe into the structural characteristics of the Finnish labor market, such as social protections and participation dynamics, as well as the nonlinear impacts of inflation and comparative analysis across welfare economies, could offer wider perspectives on growth determinants.

This paper illustrates that Finland's economic growth from 2004 to 2023 has been primarily driven by domestic demand, trade performance and the stabilizing effects of moderate inflation. While unemployment customarily has been a central indicator in growth models, the Finnish examples reveal that robust labor and social policies can weaken the impact and stimulate conventional macroeconomic theories, offering new perspectives on Okun's Law. Retail sales as a GDP predictor, on the other hand, provided further evidence that consumer behavior continues to support growth in advanced economies. Further, the positive export contribution shows that the requirement for diversified trade strategies to enhance resilience to global uncertainty. When combined, these findings illustrate that policy, consumption and external competitiveness in creating a sustainable economic performance. This study fills the gap in knowledge by considering retail activity and inflation management as underexplored but critical variables, this study brings forth new ideas to the macroeconomic growth literature. In the final analysis, the Finnish experience demonstrates how a balanced macroeconomic policy can maintain long-term stability from which strengthening adaptive capacity in a changing world economy.

References

1. European Commission: Country Report – Finland 2023. https://economy-finance.ec.europa.eu/system/files/2023-06/ip250_en.pdf. Last accessed 8 Dec 2024
2. Bank of Finland: Finland's economy is gradually moving out of recession. https://www.bofbulletin.fi/en/2024/3/finland-s-economy-is-gradually-moving-out-of-recession/. Last accessed 8 Dec 2024
3. Bătuşaru, C.M., Baboş, A., Pîrvuţ, V.: Tracking Economic Growth: a correlation analyse on the determinants of gross domestic product. Int. Conf. Knowl. Based Organ. **29**, 24–29 (2023)
4. Boldeanu, F.T., Constantinescu, L.: The main determinants affecting economic growth. Bull. Transilvania Univ. Braşov, Ser. V Econ. Sci. **8**, 329–338 (2015)
5. Giannetti, B.F., Agostinho, F., Almeida, C.M.V.B., Huisingh, D.: A review of limitations of GDP and alternative indices to monitor human wellbeing and to manage eco-system functionality. J. Clean. Prod. **87**, 11–25 (2014)
6. Kreishan, N.: Economic growth and unemployment: an empirical analysis. J. Soc. Sci. **7**, 228–231 (2011)
7. Abdul-Khaliq, S., Soufan, T., Shihab, R.A.: The relationship between unemployment and economic growth rate in Arab country. Dev. Country Stud. **4**, 62–66 (2014)
8. Xesibe, Z., Nyasha, S.: Unemployment and economic growth in South Africa: a re-examination. Int. J. Sustain. Econ. **12**, 101–116 (2020)

9. Kukaj, D.: Impact of unemployment on economic growth: evidence from western balkans. Eur. J. Mark. Econ. **1**, 9–17 (2018)
10. Song, Y.: The relationship between CPI and employment and national GDP: empirical research based on French Data. In: Proceedings of the 2024 2nd International Conference on Finance, Trade and Business Management (FTBM 2024), pp. 236–245 (2024)
11. Barro, R.: Inflation and Economic Growth. (1995). https://doi.org/10.3386/w5326
12. Saungweme, T., Odhiambo, N.M.: Inflation and economic growth in Kenya: an empirical examination. Adv. Decis. Sci. **25**, 1–25 (2021)
13. Aboobucker, H., Jahufer, A.: Impact of consumer price index and gross domestic saving on economic growth in Sri Lanka: an econometric analysis using Johansen Co-Integration Approach. Int. J. Innov. Res. Dev. **7**, 184–191 (2018)
14. Nguyen, T.H.: Impact of export on economic growth in Vietnam: empirical research and recommendations. Int. Bus. Manag. **13**, 45–52 (2016)
15. Hagemejer, J., Mućk, J.: Export-led growth and its determinants: evidence from Central and Eastern European countries. World Econ. **42**, 1994–2025 (2019)
16. Waithe, K., Lorde, T., Francis, B.: Export-led growth: a case study of Mexico. Int. J. Bus. Humanit. Technol. **1**, 33–44 (2011)
17. Zhang, J.: The relationship between total retail sales of consumer goods and economic growth: based on the time series analysis of 1992–2019. In: 2020 International Symposium on Frontiers of Economics and Management Science (FEMS 2020), pp. 199–204. Clausius Scientific Press, Beijing (2020)
18. Anagboso, M., McLaren, C.: The impact of the recession on retail sales volumes. Econ. Labour Mark. Rev. **3**, 22–28 (2009). https://doi.org/10.1057/elmr.2009.140
19. World Bank Group: Global Economic Monitor. https://datacatalog.worldbank.org/search/dataset/0037798/Global-Economic-Monitor. Last access 2 Dec 2024
20. Saunders, M., Lewis, P., Thornhill, A.: Research Methods for Business Students, 9th edn. Pearson Education (2023)
21. Mishra, P., Pandey, C.M., Singh, U., Gupta, A., Sahu, C., Keshri, A.: Descriptive statistics and normality tests for statistical data. Ann. Card. Anaesth. **22**, 67–72 (2019)

Digital Literacy through Robotics: A Fablab-Based Pathway to Strengthen Programming Skills and Academic Achievement in Secondary Education

Estefany Nayeli Quispe Tica, Bryan Antony Quinta Ccosi, and Luis Daniel Haro Laque(✉)

Fabrication Laboratory, Continental University, Arequipa, Perú
{72936807,bquinta,dharo}@continental.edu.pe

Abstract. This study evaluates the impact of a structured Digital Literacy through Robotics intervention delivered in a university FabLab environment on secondary-school students' academic performance in mathematics, the arts, and communication. Recognizing that conventional in-school robotics classes may lack the resources and collaborative spaces needed for deep experiential learning, the objective of this research is to demonstrate that access to a fully equipped FabLab yields greater cross-disciplinary gains than standard instruction alone. Two independent cohorts of nine students each were tracked longitudinally across seven hands-on FabLab sessions. The experimental group attended weekly sessions covering hardware assembly, programming logic, and autonomous control while the control group received standard classroom instruction without robotics integration. End-of-term scores in mathematics, arts, and communication were collected for both cohorts. We applied independent samples t-tests and repeated-measures ANOVAs to compare within and between-group differences over time. Results show that students in the FabLab- based robotics pathway achieved statistically significant improvements in mathematics ($p < 0.01$) and communication ($p < 0.05$) and demonstrated a positive trend in the arts ($p = 0.08$); the control group exhibited no comparable gains. These findings underscore the value of FabLab environments for enhancing digital literacy and fostering measurable academic benefits across disciplines.

Keywords: Digital Literacy · Robotics Education · Academic Performance · fabrication laboratory

1 Introduction

Rapid advancements in technology have significantly reshaped educational paradigms, placing digital literacy at the core of 21st-century competencies required by students globally. Digital literacy encompasses more than mere operational skills with digital devices; it includes critical understanding, manipulation, and creative usage of technological resources [1]. Particularly important within this broader definition is the integration of robotics, which has emerged as a powerful interdisciplinary educational tool

A. Rocha et al. (Eds.): ICITS 2026, LNNS 1966, pp. 269–277, 2027.
https://doi.org/10.1007/978-3-032-27538-7_23

to foster problem-solving, critical thinking, and algorithmic reasoning skills in students [2].

Educational robotics aligns closely with the STEM (Science, Technology, Engineering, Mathematics) educational framework, emphasizing inquiry-based learning through hands-on activities that integrate theory and practice [3]. Research consistently indicates that robotics education not only improves technical proficiency in hardware assembly and programming but also contributes positively to cognitive and non-cognitive outcomes such as student motivation, collaboration, and self-efficacy [4]. However, while robotics education has been widely explored, most traditional school settings, particularly in developing countries, face significant resource constraints limiting students' exposure to comprehensive robotics programs [5].

This resource gap is particularly pronounced in Latin America, where educational assessments such as the Programme for International Student Assessment (PISA) consistently highlight significant deficits in science, mathematics, and reading skills among students [6]. For instance, recent PISA scores indicate that Latin American countries frequently rank below the OECD average, emphasizing the urgent need for innovative educational strategies that can enhance student learning and engagement in these critical areas.

Fabrication Laboratories, commonly known as FabLabs, represent a significant innovation in educational spaces by providing students with access to sophisticated technological tools, including 3D printers, CNC machines, and electronic workstations [7]. FabLabs function not only as technological resource hubs but also as collaborative environments fostering creativity, innovation, and community-driven learning [8]. This holistic approach supports students' development of a deeper understanding of technology through direct, practical experience, enabling experimentation, prototyping, and iterative refinement of ideas and projects [9].

Despite the demonstrated benefits of FabLab environments, empirical studies investigating their direct impact on traditional academic outcomes, especially in Latin American educational contexts, remain relatively sparse. This gap in literature is particularly significant given that many secondary schools in these regions lack dedicated fabrication facilities, limiting student exposure to interdisciplinary learning opportunities afforded by digital fabrication and robotics [10]. The current study directly addresses this research gap by examining the impact of a structured robotics intervention delivered within a FabLab environment at a university in Arequipa, Peru. The research aims to determine whether participation in FabLab-based robotics sessions significantly improves students' digital literacy specifically in hardware, programming, and autonomous logic development and to assess whether these skills translate into measurable academic improvements across other core subject areas, such as mathematics, arts, and communication.

The specific objectives of this research are:

a) To quantitatively assess the effect of a seven-session FabLab robotics curriculum on students' competence in robotics-specific digital literacy.
b) To evaluate the transferability of robotics-related skills to broader academic performance improvements, thus validating the interdisciplinary benefits of FabLab- based learning environments.

c) To provide empirical evidence supporting the integration of FabLab and robotics-based curricula into secondary education, thereby guiding educational policy and curricular decisions in resource-constrained regions.

Employing a controlled experimental design with a pre-test/post-test measurement approach, the study involved two groups of secondary-school students: an experimental group receiving robotics instruction in a university FabLab setting and a control group following standard classroom instruction without FabLab exposure. Inferential statistical analyses, including independent-samples t-tests, repeated-measures ANOVAs, and Pearson correlation analyses, were applied to rigorously evaluate the intervention's effectiveness and its broader academic implications.

By systematically exploring these research questions, this study contributes critical insights into the value of FabLab-based robotics education, providing a robust empirical foundation for integrating digital fabrication facilities within secondary education frameworks. Such integration promises not only enhanced technical proficiency and digital literacy among students but also meaningful academic benefits across core curricular areas, reinforcing the importance of comprehensive STEM education pathways.

2 Methodology

2.1 Experimental Protocol

In recognition of the study's modest sample size ($n = 9$ per group), we conducted an a priori power analysis using G*Power, which indicated that detecting a large effect ($d = 1.2$) at $\alpha = .05$ and $1-\beta = .80$ would require at least 16 participants per condition. Although our observed effect sizes ($\eta^2 = .65$–$.77$ for Digital Literacy and Mathematics) exceeded this threshold, future work should recruit larger cohorts to confirm these results with adequate statistical power. Eighteen secondary-school students (ages 14–16) from a local school in Arequipa, Peru, They were randomly assigned to two independent groups ($n = 9$ each):

- Experimental (FabLab-Robotics) group: These nine students were invited to our university FabLab for seven 2-h sessions. Before Session 1, they completed digital literacy through robotics pre-test designed to assess their baseline competence in hardware assembly, programming logic, and autonomous control. After Session 7, they took the same test to quantify learning gains and the robot's teaching impact.
- Control group: Nine matched peers continued their normal in-school coursework, without any robotics or FabLab exposure. They completed equivalent pre- and post-tests in their classrooms.

2.2 Intervention (FabLab Sessions)

- Electronics fundamentals (motors, H-bridge, sensors)
- Hardware assembly & wiring safety
- Programming basics (if/while/for control structures)
- Debugging techniques & simple algorithm design
- Sensor integration & data acquisition
- Motor control routines (speed/position)
- Autonomous logic development and robot-led demonstrations

2.3 Prototype

Figure 1 presents both an isometric and exploded view of the tracked robot developed for the "Digital Literacy through Robotics" program. The isometric rendering shows the fully assembled robotic platform, designed with a compact, tank-like form and equipped with dual continuous tracks supported by a symmetrical wheel and sprocket system on each side, allowing for all-terrain navigation. An ultrasonic sensor module is mounted at the front of the chassis, enabling distance-based obstacle detection and autonomous behavior development. The chassis features a low-profile central housing optimized for stability and ease of access, containing the primary electronic components such as the microcontroller, motor drivers, battery pack, and wiring harness. The accompanying exploded view reveals the modular architecture of the system, consisting of 16 distinct components. These include: (1) a top shell enclosure for mechanical protection and design aesthetics; (2–4) a sensor mounting bracket with an ultrasonic rangefinder; (5–8) a tread and wheel system with rubberized tracks, support wheels, and connecting axles (11); (9–10) a structural mounting plate for the internal electronics; (12–13) motor coupling interfaces and a programmable microcontroller board (e.g., Arduino); (14–15) power supply infrastructure including battery holder and wiring components; and (16) two DC motors with encoders for differential drive and position feedback. This design not only supports autonomous logic routines and motor-sensor integration but also serves as a pedagogical tool, allowing students to engage directly with each subsystem during assembly, debugging, and performance testing within a FabLab environment.

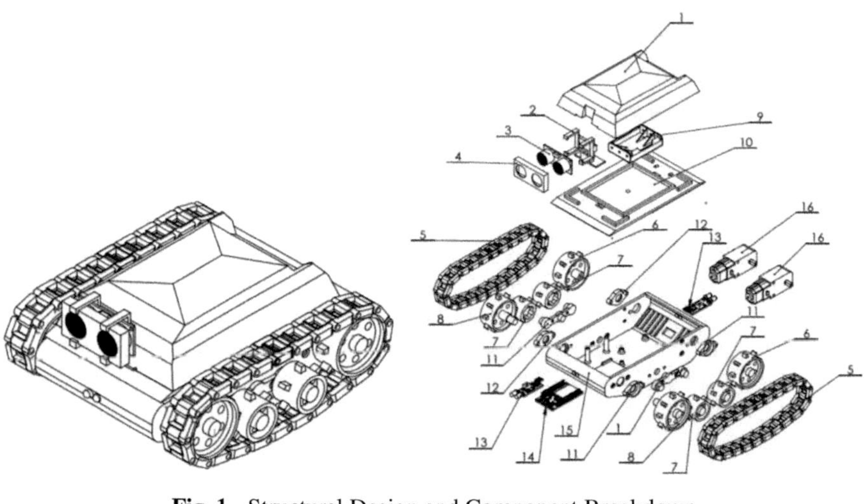

Fig. 1. Structural Design and Component Breakdown

2.4 Academic Measures

- End-of-term grades (0–20) in Mathematics, Arts, and Communication, collected one week before Session 1 (pre-test) and one week after Session 7 (post-test).

- Digital Literacy Test Scores (0–20): Same scale, based on a custom robotics-literacy instrument.

3 Results

Table 1 summarizes the pre- and post-test performance (on a 0–20 grading scale) for both the Experimental (FabLab-Robotics) and Control groups across four measures: Digital Literacy, Mathematics, Arts, and Communication. For each measure, the mean (M) and standard deviation (SD) are reported at baseline (Pre) and after the seven-session intervention (Post), along with the raw gain score ($\Delta = \text{Post} - \text{Pre}$). Notably, the Experimental group began with scores comparable to the Control group (e.g., Digital Literacy Pre: M = 12.0 vs. 12.3), but exhibited substantially higher gains across all domains (e.g., Mathematics $\Delta = +3.1$ vs. $+0.2$). These descriptive results suggest that participation in the FabLab-based robotics pathway was associated with larger improvements in both domain-specific and cross-disciplinary academic performance.

Table 1. Descriptive statistics (0–20 Scale)

Measure	Group	Pre M ± SD	Post M ± SD	Gain (Δ)
Digital Literacy	Experimental (n = 9)	12.0 ± 1.0	16.0 ± 0.9	+4.0
	Control (n = 9)	12.3 ± 1.3	12.5 ± 1.1	+0.2
Mathematics	Experimental (n = 9)	12.0 ± 1.0	15.1 ± 0.9	+3.1
	Control (n = 9)	12.4 ± 1.3	12.6 ± 1.1	+0.2
Arts	Experimental (n = 9)	13.7 ± 1.0	14.8 ± 1.0	+1.1
	Control (n = 9)	13.9 ± 1.1	14.0 ± 1.0	+0.1
Communication	Experimental (n = 9)	13.1 ± 1.1	14.6 ± 0.9	+1.5
	Control (n = 9)	13.2 ± 1.2	13.3 ± 1.2	+0.1

Table 2 presents the results of independent-samples t-tests on gain scores and two-way repeated-measures ANOVAs (Group × time interaction) for each of the four measures. For the t-tests, the gain in each domain was compared between Experimental and Control groups, yielding highly significant differences in Digital Literacy ($t(16) = 8.94$, $p < .001$), Mathematics ($t(16) = 6.03$, $p < .001$), and Communication ($t(16) = 4.47$, $p = .0004$), as well as a significant effect in Arts ($t(16) = 2.58$, $p = .02$). The ANOVA interaction terms confirm that the pattern of pre-to-post change differed by group: $F(1, 16) = 53.7$ ($p < .001$, $\eta^2 = .77$) for Digital Literacy, $F(1, 16) = 29.1$ ($p < .001$, $\eta^2 = .65$) for Mathematics, $F(1, 16) = 5.23$ ($p = .036$, $\eta^2 = .25$) for Arts, and $F(1, 16) = 17.3$ ($p = .0006$, $\eta^2 = .52$) for Communication. These inferential statistics corroborate that the FabLab intervention produced significantly greater learning gains than standard instruction alone.

Table 3 shows that there is a very strong, positive correlation between FabLab participation and gains in Digital Literacy ($r = .88$, $p < .001$) and Mathematics ($r = .85$, $p < .001$), indicating that students who attended the seven-session robotics pathway

Table 2. Inferential statistics

Measure	Test	Statistics	df	*p*	η^2 (partial)
Digital literacy	*t*(16) ANOVA G × T interaction F	8.94 53.7	16 1, 16	$< .001$ $< .001$	.77
Mathematics	*t*(16) ANOVA G × T interaction F	6.03 29.1	16 1, 16	$< .001$ $< .001$	.65
Arts	*t*(16) ANOVA G × T interaction F	2.58 5.23	16 1, 16	.02 .036	.25
Communication	*t*(16) ANOVA G × T interaction F	4.47 17.3	16 1, 16	.0004 .0006	.52

t-tests compare gain scores (Post – Pre) between groups, ANOVA G × T = interaction of Group and Time, η^2 = partial eta-squared effect size.

tended to exhibit larger score improvements. The correlation with Communication gains is strong (r = .72, p = .002), and the correlation with Arts gains is moderate (r = .51, p = .033). These results reinforce that assignment to the FabLab intervention is significantly associated with greater academic gains across all measured domains.

Table 3. Pearson correlation between group membership (0 = Control, 1 = FabLab) and Gain Scores

Measure	*r*	df	*p*
Digital Literacy	0.88	16	$< .001$
Mathematics	0.85	16	$< .001$
Arts	0.51	16	.033
Communication	0.72	16	.002

4 Conclusions

This study provides compelling evidence that a structured, seven-session Digital Literacy through Robotics intervention delivered in a fully equipped university FabLab can produce significant cross-disciplinary learning gains among secondary- school students. Compared to a matched control group, participants who engaged in hands-on robotics activities demonstrated:

- Substantial improvements in domain-specific digital literacy, with an average gain of 4.0 points on a 0–20 scale ($\eta^2 = .77$).
- Marked academic benefits in mathematics, showing a mean gain of 3.1 points ($\eta^2 = .65$).
- Notable enhancements in communication skills, with a 1.5-point increase ($\eta^2 = .52$).
- Moderate but meaningful gains in the arts, reflected by a 1.1-point rise ($\eta^2 = .25$).

Pearson correlations further underscored the strength of these associations, with group membership strongly linked to gains in digital literacy (r = .88) and mathematics (r = .85), and moderately to the arts (r = .51) and communication (r = .72).

These findings suggest that relocating robotics instruction to a FabLab setting where students have access to specialized tools, collaborative workspaces, and iterative prototyping amplifies both technical competencies and broader academic outcomes. Embedding robotics within a maker-centered environment appears to reinforce problem-solving, systems thinking, and creative expression, which in turn transfer to traditional school subjects.

With a modest sample size (n = 9 per group) and a short, seven-session intervention, the study's generalizability is constrained. Future research should replicate these results with larger cohorts, diverse educational contexts, and extended follow-up to assess long-term retention. Additionally, qualitative investigations into students' motivation, self-efficacy, and team-based learning processes would enrich understanding of the mechanisms driving academic transfer. We are planning a multi-site replication involving at least 30 students per group, stratified by school and socioeconomic background, and powered to detect medium effects ($d \approx 0.5$) across all domains. This scaled study will also incorporate random-block assignments and blinded grading to further strengthen causal inference and generalizability.

By demonstrating statistically and practically significant gains across multiple disciplines, this work advocates for the integration of FabLab-based robotics curricula into secondary education. Such initiatives not only cultivate essential digital-age skills but also bolster performance in mathematics, the arts, and communication thereby fostering a more holistic, interdisciplinary approach to learning, being our proposal the following.

Session	FabLab Activity	STEM Subjects	Curriculum Linkage	Key Learning Outcomes
1	Electronics Fundamentals	Science, Engineering	Physics: circuit theory; technology: component identifica-tion	Understand voltage/current Identify motors, sensors
2	Hardware Assembly & Wiring Safety	Engineering, Technology	Engineering Design: mechanical assembly; Safety & lab practices	Apply mechanical tolerancing Enforce safety norms
3	Programming Basics (if/while/for)	Mathematics, Technology	Math: algorithmic thinking; ICT: coding fundamentals	Translate logic into code Control program flow
4	Debugging & Algorithm Design	Engineering, Technology	Design Thinking: iterative testing; Computational Thin-king	Systematically isolate faults Optimize algorithms

(*continued*)

(*continued*)

Session	FabLab Activity	STEM Subjects	Curriculum Linkage	Key Learning Outcomes
5	Sensor Integration & Data Acquisition	Science, Engineering	Scientific method: data collection; STEM Lab: measurement techniques	Interface sensors Record & interpret sensor data
6	Motor Control Routines (speed/position)	Mathematics, Engineering	Math: PID control concepts; Engineering Systems: feedback loops	Implement closed-loop control Model system response
7	Autonomous Logic Development & Demonstration	Mathematics, Technology	Computer Science: autonomy; Robotics: system integration	Design decision-making logic Validate end-to-end system

Cross-disciplinary projects: Link each session to existing units (e.g., Physics in Session 1, Algebra in Session 3).

- Assessment integration: Use the gain-score framework (pre/post tests) within regular grading to measure impact.
- STEM portfolio: Students document each session's design-build-test cycle as part of a cumulative STEAM portfolio.
- This table can be inserted into your curriculum guide under a "FabLab Robotics Integration" module, highlighting both the practical FabLab sessions and their clear alignment with standard STEM learning objectives.

References

1. Diseiye, O., Ukubeyinje, S.E., Oladokun, B.D., Kakwagh, V.: Emerging technologies: leveraging digital literacy for self-sufficiency among library professionals. Metaverse Basic Appl. Res. **3**, 2 (2024)
2. Adorni, G., et al.: Development of algorithmic thinking skills in K-12 education: a comparative study of unplugged and digital assessment instruments. Comput. Hum. Behav. Rep. **15**, 100466 (2024)
3. Aguilera, D., Lupiáñez, J.L., Perales-Palacios, F.J., Vílchez-González, J.M.: IDEARR model for STEM education—a framework proposal. Educ Sci (Basel). **14**(6), 638 (2024)
4. Alam, A., Mohanty, A.: Integrated constructive robotics in education (ICRE) model: a paradigmatic framework for transformative learning in educational ecosystem. Cogent Educ. **11**(1), 2324487 (2024)
5. De los Rios Rosado, J., Salcedo, J.D.C., Lucano, J.J.L., Condori, K.V.: Enhancing technological maturity in school-level robotics: a case study of integrating LunaBot V2.3 in educational environments in remote areas of Arequipa, Peru. In: *Lecture Notes in Educational Technology*, vol. Part F424, pp. 121–132 (2025). https://doi.org/10.1007/978-981-96-3698-3_9

6. Adkins, D., Brendler, B.M.: Libraries and reading motivation: a review of the Programme for International Student Assessment reading results. IFLA J. **41**(2), 129–139 (2015)
7. Khalifa, S.: FabLab as an innovative educational approach in higher education. In: Innovative Educational Frameworks for Future Skills and Competencies, pp. 81–134. IGI Global Scientific Publishing (2025)
8. García-Ruiz, M.-E., Lena-Acebo, F.-J., Rocha Blanco, R.: Early stages of the FabLab movement: a new path for an open innovation model. Sustainability. **15**(11), 8907 (2023)
9. Gil, L.: A fablab at the periphery: decentering innovation from São Paulo. Am. Anthropol. **124**(4), 721–733 (2022)
10. Salcedo, J.D.C., Lucano, J.J.L., Hinostroza, A.J.T., Quispe, A.R.R., Ramos, J.A.D.A.: Factors affecting the use of 'Maker Route' E-learning management systems for the development of digital manufacturing skills in higher education-engineering. In: Proceedings of the LACCEI international Multi-Conference for Engineering, Education and Technology (2024). https://doi.org/10.18687/LACCEI2024.1.1.247

Innovation in Neuromuscular Electrostimulation: An Integrative Review and Case Study

Manuel Au-Yong-Oliveira[1,2](✉), Suzete Carvalho[3], and Ana Palma-Moreira[4]

[1] Institute for Systems and Computer Engineering, Technology and Science (INESC TEC), Porto 4200-465, Portugal
mao@ua.pt

[2] Research Unit on Governance, Competitiveness and Public Policies (GOVCOPP), Department of Economics, Management, Industrial Engineering and Tourism (DEGEIT), University of Aveiro, Aveiro 3810-193, Portugal

[3] Department of Economics, Management, Industrial Engineering and Tourism (DEGEIT), University of Aveiro, Aveiro 3810-193, Portugal

[4] Faculdade de Ciências Sociais e Tecnologia, Universidade Europeia, Quinta do Bom Nome, Estr. da Correia 53, Lisboa 1500-210, Portugal
ana.moreira@universidadeeuropeia.pt

Abstract. This study aims to analyze the effects of neuromuscular electrostimulation (NMES) (Aussie current, in particular) on muscle strength and functional rehabilitation. Peer-reviewed articles published between 2015 and 2025 were selected using databases such as PubMed, Scopus, and Web of Science. A clinical case was also presented. The results indicate that NMES significantly contributes to the preservation and enhancement of muscle strength, particularly in individuals with limited mobility or during post-operative recovery. Improvements in functional performance and reduced muscle atrophy were reported. NMES is a valuable therapeutic tool in clinical and sports contexts. Its integration into rehabilitation protocols can accelerate recovery and optimize muscle performance, although further research is needed to standardize application parameters and evaluate long-term results. In the clinical case a prominent senior amateur golfer who was not able to walk more than 3 km per day due to injuries to his feet (and ankles) can now walk over 15 km per day due to his treatment with NMES. The recovery has been far superior to applying ice (cold) packs directly to the feet soles (twice daily - morning and night – to decrease the pain and inflammation) and self-massaging of the ankles. Treatment with NMES twice a week, on an on-going basis, is seen to be ideal.

Keywords: NMES · physiotherapy · muscle strength · rehabilitation · functional recovery · Aussie current

A. Rocha et al. (Eds.): ICITS 2026, LNNS 1966, pp. 278–289, 2027.
https://doi.org/10.1007/978-3-032-27538-7_24

1 Introduction

Neuromuscular electrostimulation (NMES) has gained prominence as a therapeutic and physical conditioning tool, used both in-patient rehabilitation and in optimizing athletic performance. By applying low-frequency electrical currents, NMES promotes involuntary muscle contractions that can help preserve muscle mass, increase strength, and support functional recovery.

Several studies have demonstrated the benefits of electrostimulation in populations with reduced mobility, such as post-operative patients, individuals with neuromuscular injuries, and the elderly. In addition, its use has expanded into sports contexts, being incorporated into training programs for high-performance athletes.

This article aims to analyze the effects of neuromuscular electrostimulation on muscle strength and functional rehabilitation through an integrative review of current scientific literature.

The Aussie current, in particular, is examined, an innovative treatment (using a machine and technology) invented by Dr. Alex Ward, who lived in Melbourne, Australia at that time. The aim was to help rehabilitate patients with muscular lesions. Dr. Alex Ward determined the best, most adequate and comfortable value as regards NMES and for each patient.

The article proceeds with a literature review, followed by a methodology section, then a clinical case study. After the discussion the authors conclude with some final remarks.

2 Literature Review

Neuromuscular electrostimulation (NMES) has been widely studied as an effective strategy for preserving muscle strength, especially in critically ill patients, athletes, and individuals undergoing rehabilitation. Several studies report that the application of low-frequency electrical currents can promote involuntary muscle contractions, contributing to the maintenance of muscle mass and improved functionality.

According to Cecílio et al. (2025), NMES has demonstrated positive effects in patients admitted to intensive care units (ICUs), such as reduced mechanical ventilation time, prevention of muscle atrophy, and increased functional independence after hospital discharge. An application for seven consecutive days was considered necessary to obtain significant results.

In the sports context, Espíndola and Ribeiro (2023) highlight the benefits of whole-body electrostimulation (WB-EMS) training for athletes, especially football players, including increased strength, muscle mass, and maintenance of physical conditioning. Short sessions, held twice a week, proved effective even during periods of low training load.

In addition, Silva et al. (2016) conducted a systematic review that demonstrated the effectiveness of NMES in reducing weaning time from mechanical ventilation and improving muscle strength in critically ill patients. The authors emphasize that the technique may be a viable alternative to conventional treatment (such as ice, heat, and massage), especially in cases of prolonged immobility.

Despite the promising results, there are still gaps in the standardization of application protocols, such as the frequency, intensity, and duration of sessions. The heterogeneity of the studies makes it difficult to directly compare the results, reinforcing the need for additional research with greater methodological rigor.

2.1 Aussie Current – The Aid of Technology

The Aussie Current, developed in Australia, is a type of medium-frequency electrostimulation (1000 Hz) with short pulses (2–4 ms) designed to provide greater patient comfort during application (Ward & Chuen, 2009; Ward & Lucas-Toumbourou, 2007). Unlike Russian Current, which uses longer pulses and can cause discomfort, Aussie Current is recognized for its ability to generate effective muscle contractions with less pain perception (Lima & Silva, 2020).

Comparative studies show that Aussie Current can be as effective as other forms of electrostimulation for muscle activation, with significant advantages in tolerability. Gobbo et al. (2014) observed that electrode position directly influences muscle response and that Aussie Current provides more localized and comfortable activation.

In addition, the current has been used in post-operative rehabilitation, muscle strengthening in the elderly, and even in aesthetic programs for toning and reducing flaccidity. Its application is recommended for 20 to 30 min per session, weekly, and can be combined with active exercises to enhance results.

Despite the reported benefits, there are still few randomized clinical studies evaluating the long-term effectiveness of Aussie Current or in specific populations, such as neurological patients or high-performance athletes. Standardization of protocols and expansion of scientific evidence are necessary to consolidate its use in clinical practice.

Aussie Current has the following characteristics:

- Base frequency: 1,000 Hz (motor) or 4,000 Hz (sensory)
- Modulation: Short pulses of 1 ms, with a controlled burst
- Current type: Alternating, biphasic, symmetrical
- Advantages: Greater comfort, lower skin resistance, better tolerance to stimulation

2.2 Applications for the Aussie Current in Sports Physiotherapy

Aussie current has proven effective in various rehabilitation and muscle performance protocols:

- Muscle strengthening: Recommended for athletes in recovery or for preventing atrophy due to disuse. The stimulation promotes tetanic contractions with less discomfort, encouraging treatment adherence.
- Analgesic effect: Used to control acute or chronic pain, especially in musculoskeletal injuries. A 4,000 Hz frequency, modulated between 50 and 100 Hz, stimulates Aβ sensory fibers and promotes the release of endorphins.
- Motor re-education: Applied in cases of muscle imbalance or neuromotor deficit, assisting in the selective activation of muscle groups.
- Injury prevention: By improving muscle tone and resistance, the Aussie current contributes to joint stability and reduces functional overload.

- Sports aesthetics: Used in muscle toning and flaccidity reduction programs, with good tolerance by patients.

In the application of the Aussie current:

- Silicone conductive electrodes with neutral gel are used to improve conductivity, or with free electrodes that already come with gel.
- The electrodes are positioned on the muscle insertions and secured with elastic straps.
- Sessions last 15 to 20 min, with a recommendation of 10 to 15 sessions, twice per week, until improvements are felt; then once a week should be sufficient.

2.3 Therapeutic Objectives of the Aussie Current

The Aussie current has the following therapeutic objectives:

- Relief of muscle and joint pain: Acts as an analgesic through nerve stimulation.
- Post-injury or post-operative recovery: Helps with muscle rehabilitation and prevents atrophy.
- Muscle strengthening in cases of immobilization: Ideal for patients who cannot perform conventional exercises (attend gyms or have a personal trainer or teacher).
- Improvement of neuromuscular function: Stimulates motor nerves and promotes muscle re-education.
- Reduction of oedema: Activation of lymphatic circulation promotes the drainage of accumulated fluids.

2.4 Muscle Recovery with Aussie Current

Muscle recovery sessions with Aussie current aim to rehabilitate and strengthen weakened or injured muscles, improve neuromuscular function, and relieve pain.

The recommended parameters are as follows:

- Frequency: 1,000 Hz (ideal for deep motor stimulation).
- Mode of application: Visible contractions, with rest between stimuli.
- Intensity: Progressive, always respecting the comfort limit.

For Aussie current therapy to be most effective, patients should combine sessions with light active exercises, maintain good hydration and a protein-rich diet, and avoid muscle overload on application days.

The main benefits of muscle recovery with Aussie current are:

- Strengthening of weakened muscles
- Post-injury or post-operative rehabilitation
- Relief of muscle and joint pain
- Prevention of atrophy due to immobilization
- Improvement of neuromuscular function

2.5 Comparison Between Aussie Current and Russian Current

As mentioned above, there are differences between the Russian and Aussie currents. Russian current uses longer pulses, which can cause discomfort, while Aussie current is recognized for its ability to elicit effective muscle contractions with less pain perception (Lima & Silva, 2020; Mariano et al., 2020). This is why we are comparing the characteristics of these current types.

Table 1 compares the main characteristics of the Aussie current and the Russian current (Lima & Silva, 2020).

Table 1. Comparison between Aussie current and Russian current (adapted from Lima & Silva, 2020)

Characteristic	Aussie current	Russian current
Frequency	1000 Hz	2500 Hz
Pulse type	Short pulses (2–4 ms)	Modulated pulses (bursts)
Main objective	Muscle strengthening with comfort	Intense muscle strengthening
Patient comfort	High	Moderate below
Clinical indications	Rehabilitation, strengthening	Rehabilitation, strength gain
Muscle contractions	Efficient and tolerable	Intense, but less comfortable
Patient reaction feedback	More comfortable throughout - including during adjustment and preparation of treatment session phase – by the therapist	Considerable pain may be felt especially during adjustment of current immediately prior to treatment session – despite care by therapist

3 Methodology

This article seeks to provide a literature review on the topic and a practical solution for individuals who suffer from lesions - including in the feet and ankles. For the literature review, peer-reviewed articles published between 2015 and 2025 were selected using databases such as PubMed, Scopus, and Web of Science. The search terms were "Physiotherapy", "Sport", "Aussie current" and "Russian current". A clinical case is also presented herein, regarding a patient one of the authors has been treating (a former professional golfer who is still a prominent senior amateur playing at the national level). The results have been remarkable as the patient had been considering surgical intervention to a spur in his right foot - but has now put that idea aside. The case is relevant as spurs are more prevalent than supposed and even people who have not been linked to sports careers may feel similar symptoms to those described in this case study. The treatment also involves lesser lesions in the ankles, legs, hands, wrists, elbows, and shoulders where similar significant response and gains have been registered (Fig. 2).

4 Clinical Case Study

Below, we present a clinical case study treated with Aussie currents.

The patient in question was 55 years old, currently a non-professional golfer, and had injuries to both feet (and ankles), which led to significant pain after walking and playing golf. Treatment began on 1 August 2025. Let it be noted that walking and movement at one stage was limited to less than 2 km per day (at that time the individual also played padel – but following two doctors' indications he ceased to play that sport).

H.P.I. (History of Previous Illness): The patient, a former professional golfer, reports feeling pain after walking about 5 km, even on flat ground and grass (while playing golf). However, the walking and the excruciating pain worsened without proper treatment. After 3 km, he already felt considerable pain and was unable to continue, so he sat down to rest his feet (the main focus of pain – though the ankles were also a problem – due to excessive sport as a youth and during his life). After playing golf the patient applied ice to the soles of his feet in the morning and at night, but he caught a cold from the contact with cold ice, and the pain relief was much less intense and shorter-lasting, as was the anti-inflammatory action, as compared to that felt with the Aussie current treatment. Additionally, self-massage of the ankles was not sufficient also.

4.1 Medical Reports

The patient had been to the hospital over a year prior to the treatment described herein. This was due to pain felt in the knees and feet and to a general increasing lack of mobility (pain felt when walking). X-rays were thus performed, and the doctor's report may be found in Fig. 1. For anonymity purposes, hospital and doctor information have been removed - as well as the name of the patient.

The medical report of the radiographic examination of the knees (front and side view under load) (Fig. 1) indicated the following (translated from the report from Portuguese to English using deepl.com Pro):

> "Normal trabecular bone texture in the segments examined, with regularity of the femoro-tibial-patellar joint surfaces and normal interline width.
>
> Evidence of enthesopathic calcification in the patellar insertion of the quadriceps, in both the right and left knees.
>
> No relevant changes in the periarticular tissue planes, although there is slight irregularity of the cortical contour of the posterior aspect of the tibia and fibula, in both the right and left knees, related to 'tug lesions', i.e., microavulsions of the cortical bone induced by muscle insertions, without particular significance."

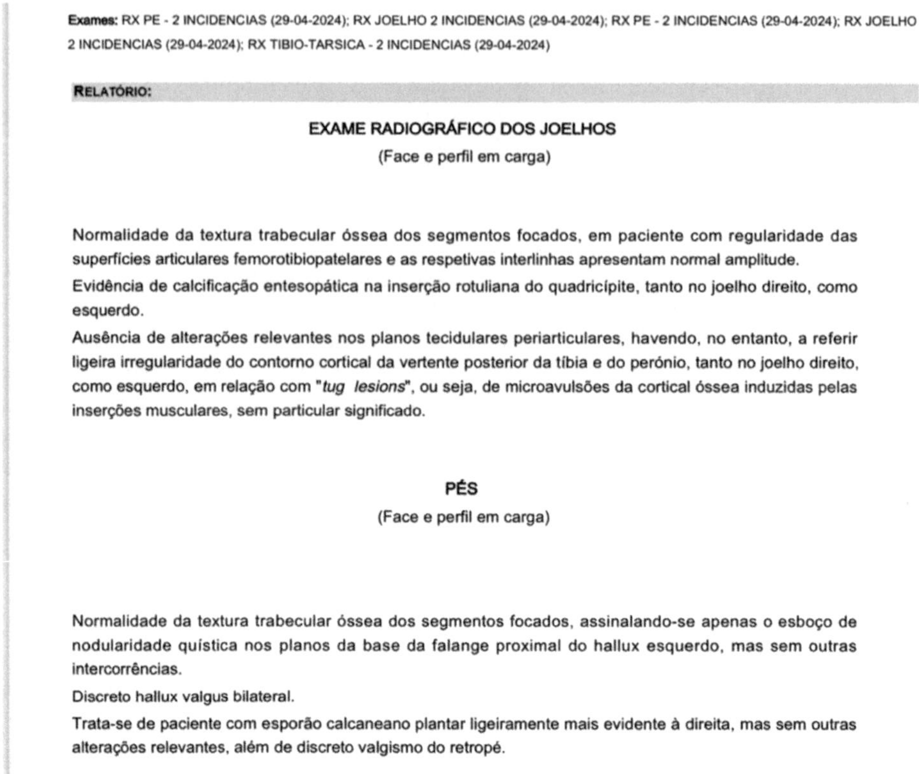

Exames: RX PE - 2 INCIDENCIAS (29-04-2024); RX JOELHO 2 INCIDENCIAS (29-04-2024); RX PE - 2 INCIDENCIAS (29-04-2024); RX JOELHO 2 INCIDENCIAS (29-04-2024); RX TIBIO-TARSICA - 2 INCIDENCIAS (29-04-2024)

RELATÓRIO:

EXAME RADIOGRÁFICO DOS JOELHOS

(Face e perfil em carga)

Normalidade da textura trabecular óssea dos segmentos focados, em paciente com regularidade das superfícies articulares femorotibiopatelares e as respetivas interlinhas apresentam normal amplitude.

Evidência de calcificação entesopática na inserção rotuliana do quadricípite, tanto no joelho direito, como esquerdo.

Ausência de alterações relevantes nos planos tecidulares periarticulares, havendo, no entanto, a referir ligeira irregularidade do contorno cortical da vertente posterior da tíbia e do perónio, tanto no joelho direito, como esquerdo, em relação com "*tug lesions*", ou seja, de microavulsões da cortical óssea induzidas pelas inserções musculares, sem particular significado.

PÉS

(Face e perfil em carga)

Normalidade da textura trabecular óssea dos segmentos focados, assinalando-se apenas o esboço de nodularidade quística nos planos da base da falange proximal do hallux esquerdo, mas sem outras intercorrências.

Discreto hallux valgus bilateral.

Trata-se de paciente com esporão calcaneano plantar ligeiramente mais evidente à direita, mas sem outras alterações relevantes, além de discreto valgismo do retropé.

Fig. 1. Medical report of the patient – knees and feet - hospital and doctor information removed as well as the name of the patient

The medical report of the radiographic examination of the feet (front and side view under load) indicated the following (translated from the report from Portuguese to English using deepl.com Pro):

> "Normal trabecular bone texture in the segments examined, with only slight cystic nodularity noted in the planes of the base of the proximal phalanx of the left hallux, but no other complications.
>
> Mild bilateral hallux valgus.
>
> This is a patient with a plantar calcaneal spur slightly more evident on the right, but with no other relevant changes, apart from mild hindfoot valgus."

4.2 Treatment with Aussie Current

As mentioned above, this patient began treatment on 1 August 2025. Treatment has been carried out twice a week with two electrostimulation sessions (Fig. 2).

The rhythm of each session is as follows and in the following order:

1. Aussie current for 15 min.

2. Ultrasound on each foot for 6 min.
3. Manipulation of the affected area.
4. Infrared to relax the entire treated area and increase blood flow.

Current Medical History (CMH): The patient is regularly treated and reports that he almost forgets he has limitations due to his heel spur. The inflammation is much less. The pain now appears after 10 km, but it is still bearable.

The patient believes that after about two months of treatment, he may improve even more. The treatment should be performed twice a week.

The patient has been reporting progressive improvements, so much so that he can now walk 15 km (in November 2025) on the golf course and has won three more championships, proving that he is no longer in as much pain as before.

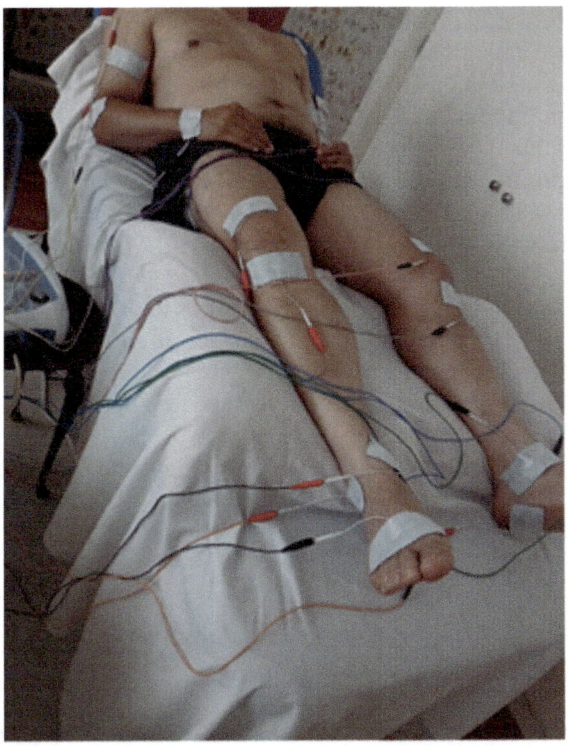

Fig. 2. A patient undergoing treatment with Aussie currents with self-adhesive fixtures with gel (duly authorized picture) – shoulders, elbows, wrists, hands, legs, knees, ankles and feet are targeted with the treatment

5 Discussion

This study aimed to analyze the effects of neuromuscular electrostimulation on muscle strength and functional rehabilitation. The patient analyzed in this clinical case, treated with Aussie currents, has reported progressive improvements, claiming that he can now walk 15 km or more on the golf course and has won three more championships, as he no longer experiences as much pain (feet and ankles) as before. Lesions felt in the shoulders, elbows, arms, wrists and knees have also benefitted from similar treatment. Feeling fit and apt in the named zones has also led to important gains in suppleness and subsequently in technique – strength and speed of the golf movement (the golf swing). When younger (at 20 years of age, for example) these feelings of being healthy are often taken for granted (not even thought about by the athlete). The results obtained in this clinical case are consistent with the literature (Silva & Andrade, 2022). According to Espíndola and Ribeiro (2023), the use of electrostimulation increased muscle mass and maintained physical conditioning in athletes, even during periods of high training load. These results indicate that NMES can be used not only in rehabilitation but also as a complement to conventional training.

Let it be noted that the patient was also advised by the therapist to change shoes to a new and more comfortable pair (made by a premium brand of footwear). This resulted in less pain felt after walking, according to the patient. Shoes should be changed regularly and especially before they are completely worn out (e.g., after 300 km or about 6 months of use).

Despite the observed benefits, the literature points to the need for standardization of application protocols. The variability in the parameters used, such as frequency, intensity, and duration, makes direct comparison between studies difficult and may influence the results obtained. In addition, few studies address the long-term effects of NMES, which limits understanding of its therapeutic sustainability.

Continued research, with greater methodological rigor and larger samples, is essential to consolidate its use in different therapeutic contexts.

5.1 Suggestions for Future Research

It is suggested that research be conducted in the following areas:

1. Neurological Applications
   - Functional electrical stimulation (FES) is used in neurodegenerative diseases such as Parkinson's and multiple sclerosis.
   - Transcranial electrical stimulation (tDCS) for cognitive modulation and treatment of depression or anxiety.
   - FES in post-stroke rehabilitation: comparison between different protocols and application parameters.
2. Muscle Rehabilitation and Performance
   - Comparison between different types of electrical current (Russian, faradic, interferential) in muscle hypertrophy.

- Electrostimulation is associated with physical exercise: synergistic effects on muscle strength and endurance.
- Use of electrostimulation in athletes: post-training recovery and injury prevention.

3. Biomarkers and Personalization

- Studies on individual variability in response to electrostimulation (genetics, age, gender).
- Development of personalized protocols based on artificial intelligence and machine learning.

4. Technology and Innovation

- Integration with wearable devices for real-time monitoring.
- Wireless electrostimulation is controlled by mobile applications: feasibility and effectiveness.
- Nanotechnology applied to electrostimulation: innovative electrodes and biofeedback.

5. Electrostimulation in critical patients

- Investigation of the effectiveness of neuromuscular electrostimulation (NMES) in ICU patients to prevent acquired muscle weakness.
- Longitudinal studies on functional recovery after hospital discharge using NMES.

6. Integration with wearable technologies

- Development of smart devices that automatically adjust stimulation parameters based on biometric sensors.
- Evaluation of the adherence and efficacy of electrostimulation vests in home or gym training.

7. Functional electrostimulation (FES) in rehabilitation

- Comparison between FES and conventional physiotherapy in patients with neurological or orthopedic injuries.
- Studies on safety, perception and practice of physiotherapists in hospital and outpatient settings.

8. Impact on sports performance

- Meta-analyses on the use of NMES (neuromuscular electrical stimulation) in high-performance athletes to improve strength and muscle mass.
- Studies on ideal frequency, intensity and duration protocols for different sports.

9. Psychosocial aspects and treatment adherence

- Research on patients' perceptions of comfort, pain and motivation during electrostimulation sessions.
- Studies on factors influencing adherence to continuous technology use in prolonged treatments.

6 Conclusions

Aussie current stands out as an effective and versatile electrotherapy technique, with both aesthetic and therapeutic applications. In the context of muscle recovery, its specific parameters enable deep, comfortable stimulation, promoting strengthening, functional rehabilitation, and pain relief (Costa & Menezes, 2020). The results observed throughout the sessions demonstrate its potential as a complementary resource in physiotherapy and rehabilitation protocols.

Self-treatment – with ice blocks/cold gel and self-massage are not seen to be nearly as effective as Aussie currents. Patients are advised to seek professional help.

Despite advances, there are still gaps in the scientific literature that warrant further research. Comparative studies with other currents, applications in specific populations, and personalized protocols can increase understanding of its effectiveness and safety. Thus, Aussie current represents not only a promising clinical tool but also a fertile field for new research in electrotherapy (Mariano et al., 2020).

Aussie current is an effective form of electrostimulation, with relevant applications in both the aesthetic and therapeutic fields (Ward & Chuen, 2009; Ward & Lucas-Toumbourou, 2007). In the context of muscle recovery, its specific parameters favor deep, comfortable stimulation of muscle fibers, contributing to strengthening, functional rehabilitation, and pain relief. The results obtained throughout the sessions reinforce its potential as a complementary resource in physiotherapy protocols.

However, there is a recognized need for further studies to explore its effectiveness across different populations, to compare it with other electrotherapeutic currents, and customize clinical protocols. Future research could increase the scientific basis and consolidate the use of Aussie current as a safe tool.

Acknowledgements. The authors thank the patient for his participation and authorization and the ICITS 2026 organization for making this publication possible.

References

Cecílio, M.B., Oliveira, A.L., Silva, R.T.: Efeitos da eletroestimulação neuromuscular em pacientes críticos: uma revisão sistemática. Rev. Bras. Fisioter. **29**(1), 45–52 (2025). https://doi.org/10.1016/j.rbfis.2025.01.005

Costa, L.F., Menezes, D.R.: Tecnologias emergentes na fisioterapia esportiva: uma análise crítica. Revista Saúde em Movimento. **18**(3), 225–239 (2020). https://doi.org/10.5935/rsm.v18i3.2020.225

Espíndola, F.M., Ribeiro, J.C.: Whole-body EMS training in professional soccer players: effects on strength and performance. J. Sports Sci. Med. **22**(3), 210–218 (2023). https://doi.org/10.1123/jssm.2023-0032

Gobbo, M., Gaffurini, P., Bissolotti, L., Esposito, F., Orizio, C.: Transcutaneous neuromuscular electrical stimulation: influence of electrode positioning on muscle activation. Eur. J. Appl. Physiol. **114**(4), 805–812 (2014). https://doi.org/10.1007/s00421-013-2814-5

Lima, M.O., Silva, A.P.: Comparação entre correntes Aussie e Russa na reabilitação muscular. Fisioterapia em Movimento. **33**(2), 1–9 (2020). https://doi.org/10.1590/fm.2020.3302-

Mariano, M.R., de Oliveira, R.A., de Souza, M.A., da Silva, R.A.: Efeito agudo das correntes Russa e Aussie na força muscular de flexores de punho e dedos. Fisioterapia Brasil. **21**(5), 438–445 (2020) https://pesquisa.bvsalud.org/portal/resource/pt/biblio-1283433

Silva, M.T., Andrade, R.F.: Aplicações clínicas da eletroestimulação na fisioterapia esportiva: uma abordagem baseada em evidências. Rev. Bras. Fisioter. **26**(4), 310–318 (2022). https://doi.org/10.1016/j.rbfis.2022.04.005

Silva, A.P., Costa, M.L., Andrade, F.M.: Neuromuscular electrical stimulation in intensive care: a systematic review. Physiother. Res. Int. **21**(4), 234–242 (2016). https://doi.org/10.1002/pri.1652

Ward, A.R., Chuen, W.L.: Lowering of sensory, motor, and pain-tolerance thresholds withburst duration using kilohertz-frequency alternating current electric stimulation: part II. Arch. Phys. Med. Rehabil. **90**(9), 1619–1627 (2009). https://doi.org/10.1016/j.apmr.2009.02.022

Ward, A.R., Lucas-Toumbourou, S.: Lowering of sensory, motor, and pain-tolerance thresholds with burst duration using kilohertz-frequency alternating current electric stimulation. Arch. Phys. Med. Rehabil. **88**(8), 1036–1041 (2007). https://doi.org/10.1016/j.apmr.2007.04.009

Artificial Intelligence in Management: Transforming Decision-Making, Operations, and Organizational Efficiency

Ricardo Marcão[1(✉)], Vasco Santos[2,3], Filipa Martinho[3], Rodrigo Nascimento[3], and Bablu Kumar Dhar[4]

[1] NECE-UBI and ISLA Santarém – Polytechnic University, Santarém, Portugal
ricardo.marcao@islasantarem.pt
[2] Polytechnic University of Tomar, NECE-UBI, GOVCOPP, CiTUR, Tomar, Portugal
vasco.santos@islasantarem.pt
[3] ISLA Santarém – Polytechnic University, Santarém, Portugal
filipa.martinho@islasantarem.pt
[4] Mahidol University International College, Salaya, Thailand
bablu.kum@mahidol.ac.th

Abstract. Artificial Intelligence (AI) is revolutionizing the way organizations manage their resources and make strategic decisions. Its growing adoption represents a significant transformation in the automation of processes and the analysis of large volumes of data. This article reviews the scientific literature focused on the main applications of AI in management, with special attention to the areas of operations, human resources, marketing and finance. The analysis reveals that AI is contributing to a notable increase in organizational efficiency, by improving processes, personalizing services and supporting more informed decision-making. However, major obstacles remain. Among the main challenges are ethical issues, a shortage of professionals with the right skills and resistance to change within organizations. These factors can limit the full potential of AI, requiring a strategic approach.

This review provides an up-to-date and comprehensive understanding of the role of AI in management. In addition, it offers relevant clues for future research, pointing to the need to further study the social and organizational impacts of AI, as well as strategies for more effective adoption.

Keywords: Artificial Intelligence · Management · Automation · Decision-making · Organizational efficiency

1 Introduction

During the last decade, digital transformation has become a central force behind organizational change. Among the enabling technologies, artificial intelligence (AI) stands out because of its ability to process large volumes of data, detect complex patterns, and generate recommendations for action in real time. Organizations are using AI to

A. Rocha et al. (Eds.): ICITS 2026, LNNS 1966, pp. 290–296, 2027.
https://doi.org/10.1007/978-3-032-27538-7_25

optimize supply chains, customize marketing campaigns, increase the productivity of human resource management, support financial decisions, and design new data-driven business models [1–3].

AI is, therefore, no longer a purely technical subject. It is a management subject. Managers must understand how AI creates value, how it alters decision processes, which risks it introduces, and what competencies are required to implement and govern it. At the same time, AI adoption still raises questions concerning algorithmic bias, transparency, data protection, workforce upskilling, and employees' resistance to automation [4–6].

The purpose of this paper is to systematize recent contributions on AI in management and to highlight (i) the main functional areas where AI is already delivering value; (ii) the main organizational and ethical challenges reported in the literature; and (iii) the directions that future research and managerial practice should follow.

The paper is organized as follows. Section 2 describes the methodology adopted for selecting the literature. Section 3 provides the conceptual background on AI and its main branches relevant to management. Section 4 discusses AI applications in five management domains: decision-making; operations and logistics; human resource management; marketing and consumer behavior; and financial management. Section 5 presents the main challenges and limitations to AI adoption. Section 6 outlines future trends and research opportunities. Section 7 concludes.

2 Methodology

This paper follows a narrative literature review approach. Searches were carried out in Scopus, Web of Science, SpringerLink, and Google Scholar between January 2022 and October 2025, using the keywords: "artificial intelligence" AND management, AI AND decision-making, AI AND operations, AI AND human resource management, AI AND marketing, and AI AND finance. Preference was given to peer-reviewed journal articles and book chapters from recognized publishers (Springer, Elsevier, Taylor & Francis, IEEE, ACM), as well as to recent policy documents on responsible AI.

Inclusion criteria were: (i) the study explicitly examined AI in an organizational/managerial context; (ii) the study reported an application, a framework, or an adoption challenge; (iii) the publication language was English. Conceptual duplicates and papers focused exclusively on technical AI model optimization without managerial implications were excluded. The final corpus provided the basis for the thematic structure used in Sects. 4, 5, 6.

3 Conceptual Background

AI is commonly defined as the capability of computer systems to perform tasks that normally require human intelligence, such as perception, learning, reasoning, and problem-solving [1]. Classic definitions make reference to learning (acquiring information and rules), reasoning (using the rules to reach conclusions), and self-correction. The Turing Test, originally proposed by Alan Turing, is still used as a conceptual benchmark to evaluate machine behavior that is indistinguishable from human behavior [7].

In management, the most relevant distinction is between narrow AI (designed for specific tasks, such as demand forecasting or churn prediction) and general AI (a theoretical form of AI able to perform any intellectual task a human can do). Current business applications rely almost exclusively on narrow AI.

Machine Learning (ML): Algorithms that learn from data to make predictions or classifications. Widely used for predictive analytics, customer segmentation, credit scoring, fraud detection, and inventory optimization [2, 8].

Natural Language Processing (NLP): Enables machines to understand and generate human language. Applied in chatbots, virtual assistants, sentiment analysis, and HR chat interfaces, improving customer and employee experience [9].

Robotics and Intelligent Process Automation: AI-powered robots and software bots automate repetitive and rule-based tasks, increasing speed and accuracy in operations, logistics, and shared service centers [10].

Expert and Decision Support Systems: AI systems that emulate the reasoning of human experts to support strategic, tactical, or operational decisions, especially when large data sets need to be integrated [11].

Early AI adoption was experimental and confined to R&D or to very specific industrial automation. From 2015 onward, three drivers accelerated adoption: (i) the availability of cloud-based AI services; (ii) the drop in data storage and processing costs; and (iii) pressure for real-time analytics in competitive markets [3, 12]. More recently, research highlights that AI is moving from process-level optimization to strategy-level enablement, where AI informs resource allocation, innovation portfolios, and even business model redesign [5, 13].

4 AI Applications in Management

A central contribution of AI to management is the enhancement of decision quality. By processing large, heterogeneous data sets and delivering insights faster than human analysts, AI helps managers identify patterns, trade-offs, and opportunities that would otherwise be missed [2, 14]. Predictive analytics and scenario planning tools simulate market, demand, or supply conditions and recommend the best course of action, reducing uncertainty and improving resource allocation [15].

Large firms such as Walmart or Amazon have demonstrated how AI-based demand forecasting and pricing can support strategic and operational decisions simultaneously, leading to improved competitiveness and agility [16]. The literature, however, warns that overreliance on algorithmic recommendations without human oversight can lead to suboptimal or ethically problematic outcomes [4].

AI is transforming operations and supply chains through automation, prediction, and real-time optimization. In routine activities (order processing, stock monitoring, document checking), AI and robotic process automation (RPA) reduce human error and shorten cycle times. In inventory management and logistics, ML models predict demand, optimize safety stocks, and generate optimal delivery routes, which results in lower costs and higher service levels [10, 17, 18].

AI also improves end-to-end visibility in supply chains by integrating data from IoT sensors, ERP systems, and transportation platforms, which supports proactive risk management and disruption response [19]. Future supply chains are expected to become more autonomous, with AI coordinating human workers, robots, and external partners in real time.

In HRM, AI supports three main areas:

i) Talent acquisition and selection: AI tools screen résumés, rank candidates according to predefined job profiles, and interact with applicants through conversational agents, reducing time-to-hire and improving the candidate experience [20];
ii) People analytics and retention: Predictive models detect early signs of turnover, absenteeism, or disengagement, allowing HR to design targeted retention and development actions [21];
iii) Learning and performance: Adaptive learning platforms recommend individualized training paths, and AI-enabled performance systems provide real-time feedback to employees.

At the same time, HRM is one of the domains where ethical risks are more visible: biased datasets can lead to discriminatory hiring decisions, lack of transparency can reduce employees' trust, and excessive monitoring can harm organizational climate. Consequently, authors argue that AI in HRM must be accompanied by explicit governance, human-in-the-loop arrangements, and clear communication with employees [6, 22].

Marketing is one of the earliest and most successful fields for AI adoption. Firms use AI to segment audiences, personalize content, predict purchase intentions, and optimize campaign timing [23]. Recommender systems in e-commerce platforms are a well-known example: they analyze users' past behavior and similar users' preferences to suggest products in real time, increasing conversion rates.

AI is also changing consumer behavior dynamics. Interactive, AI-mediated channels (chatbots, virtual shopping assistants, conversational commerce) shape consumers' information search, evaluation, and post-purchase experiences, making the decision process shorter and more data-driven [24]. This, however, raises privacy concerns, and marketers need to ensure compliance with data protection rules and avoid manipulative practices [25].

In finance, AI enables faster, more accurate, and more automated processes. Predictive models support budgeting and forecasting, AI-driven anomaly detection improves internal controls, and intelligent automation accelerates reconciliation and reporting processes [26]. In the capital markets and investment context, AI helps identify market patterns and optimize portfolio allocation.

A second important contribution is to risk management and fraud detection. AI systems can analyze large transaction volumes and flag unusual behaviors that may indicate fraud, money laundering, or credit deterioration, allowing earlier intervention and better compliance [27]. Nevertheless, financial AI systems must comply with strict auditability and data protection requirements.

5 Challenges and Limitations

Even though the potential benefits are clear, organizations report a relatively stable set of obstacles.

AI systems depend on large, reliable, and well-structured datasets. Many organizations still operate with fragmented, outdated, or inconsistent data, which undermines model accuracy and trustworthiness [28]. Robust data governance, standardized data models, and secure data-sharing mechanisms are prerequisites for AI at scale.

Integrating AI into existing ERPs, CRMs, or production systems is often complex and costly. Legacy architectures are not always designed for real-time data flows or for the computational requirements of AI workloads, so companies need transitional solutions (APIs, data lakes, microservices) and careful change management [29].

Most firms report a lack of AI, data science, and MLOps talent, as well as a shortage of managers who can translate business needs into AI requirements. On the user side, employees may resist automation because they fear job loss or do not understand how AI reaches its decisions. This calls for continuous training, transparent communication, and participation mechanisms during AI implementation [5, 30].

Algorithmic bias, explainability, and personal data protection are now core topics in AI management. Biased data can generate unfair outcomes; black-box models can make it impossible to justify a decision to regulators or customers; and sensitive data can be exposed if security is weak. Companies are therefore encouraged to adopt responsible AI frameworks, perform model audits, and align with international principles such as the OECD AI Principles or the EU AI Act [4, 6, 25].

6 Future Trends and Research Opportunities

Three lines of development emerge from the reviewed literature:

i) Advanced and autonomous automation: AI will increasingly orchestrate entire processes across departments (procure-to-pay, order-to-cash, hire-to-retire), deciding when to involve humans and when to act autonomously;
ii) Human–AI collaboration: Future research needs to look at how managers and employees work with AI systems — not just at task automation. Issues such as trust in AI, shared decision-making, and role redefinition will become central [13];
iii) Education and upskilling for AI management: Business and management programs must integrate AI, data analytics, and digital ethics into their curricula so that future managers can interpret AI outputs, assess risks, and design AI-enabled business models [30].

7 Conclusion

AI has moved from a technological novelty to a strategic capability for management. It improves decision-making, increases operational efficiency, personalizes customer and employee experiences, and strengthens financial and risk management. Yet, the benefits do not come automatically. They depend on data quality, integration capacity, digital skills, ethical governance, and – above all – leadership that frames AI as part of organizational strategy.

For practitioners, the main implication is that AI initiatives should be led from the top, aligned with business objectives, accompanied by change management, and measured through value indicators (cost, time, quality, customer satisfaction). For researchers, there is room to deepen empirical studies on real AI implementations, to investigate human–AI collaboration in decision-making, and to develop governance models that balance innovation and accountability.

References

1. Russell, S., Norvig, P.: Artificial Intelligence: A Modern Approach, 4th edn. Pearson, Boston (2021)
2. Davenport, T.H., Ronanki, R.: Artificial intelligence for the real world. Harv. Bus. Rev. **96**(1), 108–116 (2018)
3. Brynjolfsson, E., McAfee, A.: The Second Machine Age: Work, Progress, and Prosperity in a Time of Brilliant Technologies. Norton, New York (2014)
4. Cath, C.: Governing artificial intelligence: ethical, legal and technical opportunities and challenges. Philos. Trans. R. Soc. A. **376**, 20180080 (2018)
5. Shrestha, Y.R., Ben-Menahem, S.M., von Krogh, G.: Organizational decision-making structures in the age of AI. Calif. Manag. Rev. **61**(4), 66–83 (2019)
6. Lepri, B., Oliver, N., Letouzé, E., Pentland, A., Vinck, P.: Fair, transparent, and accountable AI for an inclusive society. Computer. **51**(7), 34–44 (2018)
7. Kilani, Y.A., Alshammari, R.: Artificial intelligence and the Turing test: towards a better evaluation method. Int. J. Adv. Comput. Sci. Appl. **9**(6), 161–165 (2018)
8. Jeble, S., Dubey, R., Childe, S.J., Papadopoulos, T., Roubaud, D.: Impact of big data and predictive analytics capability on supply chain sustainability. Int J Log Res Appl. **21**(3), 1–17 (2018)
9. Adamopoulou, E., Moussiades, L.: An overview of chatbot technology. In: IFIP International Conference on Artificial Intelligence Applications and Innovations, pp. 373–383. Springer, Cham (2020)
10. Ivanov, D., Dolgui, A.: Artificial intelligence in the supply chain: joint design and management. Ann. Oper. Res. **308**, 1–20 (2022)
11. Cui, L., Wang, H., Li, J.: Expert systems for decision support in the era of big data and artificial intelligence: a review. Expert Syst. Appl. **204**, 117597 (2022)
12. Arsenyan, J., Piepenbrink, A.: Drivers of AI investment and adoption across industries. In: Proceedings of the 56th Hawaii International Conference on System Sciences (2023)
13. Raisch, S., Krakowski, S.: Artificial intelligence and management: the automation–augmentation paradox. Acad. Manag. Rev. **46**(1), 192–210 (2021)
14. Sultana, R.: AI-powered analytics and strategic decision-making in modern enterprises. Int. J. Bus. Intell. Data Min. **19**(2), 101–120 (2024)
15. Yesufu, M., Alajlani, S.: Predictive analytics and scenario planning: leveraging AI for strategic resource optimization. J. Artif. Intell. Bus. Strategy. **3**(1), 51–67 (2025)
16. Huang, M.-H., Rust, R.T.: Artificial intelligence in service. J. Serv. Res. **21**(2), 155–172 (2018)
17. Zhang, H.: Predictive analytics in inventory management: reducing operational costs and improving customer satisfaction. J. Supply Chain Anal. **12**(1), 45–59 (2024)
18. Thenmozhi, R., Krishnakumari, S.: AI-driven automation in supply chain operations: enhancing efficiency and reliability. Int. J. Oper. Supply Chain Manag. **17**(1), 88–111 (2024)
19. Lei, Z., Hui, C.: AI integration in smart logistics infrastructure: a roadmap for future supply chain innovation. Logistics Transp. Rev. **6**(2), 1–19 (2024)
20. Kadirov, D., Shukurov, A.: AI in talent management: predictive analytics and automation in HR practices. Hum. Resour. Manag. Rev. **34**(1), 100987 (2024)

21. Min, J.: Automation in auditing: enhancing accuracy and operational efficiency through AI. Int. J. Auditing Technol. **4**(1), 55–68 (2024)
22. Bharadwaj, R.: Data privacy concerns in AI-driven HR systems: risks and regulatory frameworks. Int. J. Inf. Manag. **75**, 102012 (2024)
23. Oğuz, A.: Artificial intelligence in predictive marketing: understanding consumer behavior through data analytics. J. Digit. Mark. Res. **5**(3), 201–219 (2024)
24. Farooq, R., Yen, D.C.: Consumer preference prediction using AI in digital platforms: a behavioral perspective. J. Consum. Anal. **4**(1), 33–50 (2024)
25. Patil, R.: Personalized advertising using AI: the future of targeted campaigns. Int. J. Mark. Technol. **12**(2), 77–93 (2025)
26. Rohatgi, A.: Artificial intelligence for market trend analysis and financial performance evaluation. J. Financ. Innov. **10**(2), 1–15 (2023)
27. Deshmukh, A.P., Kamat, S., Rao, P.: Predictive modeling for financial risk management: scenario simulation and policy adjustment. J. Risk Anal. **16**(1), 88–104 (2024)
28. Bérubé, M., Tremblay, A.: Data quality challenges in AI systems: impact on reliability and outcomes. J. Data Sci. Technol. **9**(2), 215–229 (2025)
29. Madanchian, M., Taherdoost, H.: Challenges in integrating AI solutions with legacy systems: technological and operational perspectives. Int. J. Inf. Syst. Manag. **14**(1), 45–61 (2025)
30. Chigwada, J.: Building ethical and technical capacity for digital transformation: the role of education and training. Int. J. Digit. Educ. Innov. **6**(1), 87–111 (2024)

The Impact of Transformational Leadership on Organizational Performance

Tiago Costa[1], Ricardo Marcão[2(✉)], Vasco Santos[3,1], Filipa Martinho[1], and Isabel Reis[4]

[1] ISLA Santarém – Polytechnic University, Santarém, Portugal
filipa.martinho@islasantarem.pt

[2] NECE-UBI and ISLA Santarém – Polytechnic University, Santarém, Portugal
ricardo.marcao@islasantarem.pt

[3] Polytechnic University of Tomar, NECE-UBI, GOVCOPP, CiTUR, Tomar, Portugal
vasco.santos@islasantarem.pt

[4] NECE-UBI, CEFAGE, and ISLA Santarém – Polytechnic University, Santarém, Portugal
isabel.reis@islasantarem.pt

Abstract. Transformational leadership is often linked to better organizational results, but the pathways are not always clear. This literature review brings together credible evidence on how transformational behaviors support performance across settings. We focus on mechanisms that consistently show up in the research: stronger intrinsic motivation, higher psychological safety, more effective knowledge sharing, and faster learning that feeds innovation. We also place transformational leadership alongside complementary practices such as clear goal setting and fair reinforcement, since most successful leaders blend styles in real life. The review highlights three cautions that matter for practice: do not romanticize the leader at the expense of systems, mind the context in which teams operate, and be careful with measurement tools that blur boundaries between leadership constructs. Finally, we point to trends that are reshaping day to day leadership, including digital collaboration, inclusion, and adaptability. Taken together, the evidence suggests that transformational leadership helps most when inspiration is paired with good design: strategy that people can use, feedback loops that support learning, and cultures that reward fairness, voice, and accountability.

Keywords: Transformational leadership · Organizational performance · Motivation · Psychological safety · Knowledge sharing · Organizational learning · Innovation · Organizational culture · Ethical leadership · Inclusive leadership

1 Introduction

Organizations are working through rapid change, tighter competition, and higher expectations from employees and stakeholders. In this setting, leadership affects not only what gets done but how people commit to the work. Transformational leadership stands out because it aims to elevate purpose, align values, and develop capability rather than

A. Rocha et al. (Eds.): ICITS 2026, LNNS 1966, pp. 297–305, 2027.
https://doi.org/10.1007/978-3-032-27538-7_26

relying only on rules and incentives [1]. The basic claim is straightforward: leaders who communicate a clear and compelling direction, challenge people to think in new ways, and coach them as individuals can spark effort and learning that compound into stronger results.

This review examines what the literature says about how transformational leadership relates to performance. We look at mechanisms that link leadership to outcomes, including intrinsic motivation, psychological safety, knowledge sharing, and innovation. We also situate transformational leadership alongside transactional practices that clarify goals and reinforce performance, since most effective leaders blend styles in real settings [2]. Because culture shapes how any behavior is understood, we note the leader's role in creating and evolving the assumptions and routines that guide everyday work [3].

Our aim is practical. We synthesize consistent findings, call out limits and common critiques, and highlight trends that are changing day to day leadership, including digital collaboration, inclusion, and adaptability [4]. The picture that emerges is a style that works best when inspiration is matched with sound design: usable strategy, feedback loops that enable learning, and cultures that reward fairness, voice, and accountability.

2 Leadership Concepts

Leadership is best understood as a social process in which people influence one another toward shared outcomes, shaped by the interaction of leaders, followers, and context rather than by traits alone [4]. Over time, research moved from searching for universal leader traits to studying observable behaviors and the conditions under which they work. In practice, two families of behavior matter most for results. Transactional leadership creates clarity through goals, roles, and fair reinforcement. Transformational leadership adds a different layer by articulating a compelling direction, inviting people to think in new ways, and coaching individuals so they can grow and contribute at a higher level [1, 2]. Most effective managers blend both: they set expectations and hold people accountable, while also building purpose and capability.

Culture explains much of what people take from a leader's actions, which is why shaping culture is part of the job. Through the systems they design, the routines they sponsor, the stories they tell, and the behaviors they reward, leaders help create and evolve the assumptions that guide everyday work, and those assumptions channel performance over time [3]. Alongside transformational and transactional ideas, several positive approaches are often combined in organizations. Authentic leadership stresses self-awareness, values, and transparent relationships that build durable trust and credibility [5]. Ethical leadership focuses on fairness and moral guidance, modeling and reinforcing standards that support integrity and accountability [6]. In knowledge-intensive and agile settings, empowering and shared leadership distribute influence and give teams real decision latitude, which helps with speed and ownership [10].

Motivation is the mechanism that connects these ideas to outcomes. A Self-Determination Theory lens shows why transformational behaviors are effective: people thrive when work supports autonomy, competence, and relatedness. Leaders who set meaningful goals, provide coaching that builds mastery, and foster a climate of respect and belonging make it easier for employees to internalize objectives and sustain effort, which accelerates learning and innovation [8, 9]. Put together, transactional practices provide clarity and fairness, transformational practices generate energy and growth, culture determines whether stretch feels like opportunity or risk, and motivation turns good intentions into consistent performance gains [1–3].

3 Types of Leadership

In practice, most organizations see a blend of leadership types rather than a single pure style. Transactional leadership focuses on clarity: setting goals, defining roles, monitoring progress, and using fair reinforcement to align effort with expectations. Transformational leadership adds a different engine for performance by articulating a compelling direction, encouraging people to think in new ways, and coaching individuals so they can grow and contribute at a higher level. The evidence suggests that the strongest results come from a thoughtful mix of both, where clear structures support inspired, capability-building work [1, 4]. Alongside these two, several positive approaches often operate together. Authentic leadership emphasizes self-awareness, values, and transparent relationships that build durable trust and credibility [2]. Ethical leadership centers on fairness and moral guidance, modeling and reinforcing standards that support integrity and accountability [6]. Servant leadership prioritizes developing people and serving stakeholder needs, and research shows it can add explanatory power beyond transformational effects in some outcomes [7]. In knowledge-intensive and agile settings, empowering and shared leadership distribute influence and give teams real decision latitude, which tends to improve ownership and speed [9]. Inclusion has become a core operating principle rather than a side initiative, with leaders intentionally creating conditions where diverse voices are invited, respected, and used in decisions [10]. Finally, adaptive leadership capabilities help organizations sense change, experiment, and scale what works, which is increasingly crucial in dynamic environments [8]. Taken together, these types are best viewed as complementary tools. Effective leaders choose and combine them to fit the task, the team, and the culture, using transactional clarity to reduce friction while drawing on transformational, authentic, ethical, empowering, inclusive, and adaptive practices to unlock motivation, learning, and sustained performance [1, 2, 4, 6–10].

4 Transformational Leadership: Definition, Characteristics, and Applications

Transformational leadership is a way of leading that elevates purpose, strengthens values, and develops people so they can contribute at a higher level. Classic work defines it through four mutually reinforcing behaviors: idealized influence, where leaders model credibility and shared values; inspirational motivation, where they describe a compelling direction that people can believe in; intellectual stimulation, where they invite people to question assumptions and solve problems creatively; and individualized consideration, where they coach and support each person's growth [1]. These behaviors are not theatrical. They work by shaping day to day experience so that employees feel their work matters, see how it connects to strategy, and gain the skills and confidence to deliver.

Several features explain why this style is effective. First, it tends to activate intrinsic motivation. When people have real latitude to decide how to meet clear goals, receive feedback that helps them build mastery, and work in respectful relationships, effort becomes self driven and more resilient [3, 14]. Second, it improves learning and innovation. Leaders who encourage questions, welcome ideas, and protect smart experimentation create the routines that move organizations from insight to better execution [5]. Third, it travels well with transactional clarity. Clear expectations and fair reinforcement are still essential, and meta analytic evidence shows that combining them with transformational behaviors strengthens satisfaction, extra effort, and effectiveness across settings [4].

The practical applications are broad. During change programs, transformational behaviors help people make sense of the why, reduce anxiety about the how, and keep attention on learning rather than blame, which improves adoption and performance [1, 5]. In innovation work, intellectual stimulation and psychological safety support idea generation and iteration, while individualized coaching builds the skills needed to scale what works [5]. In knowledge intensive or agile teams, empowering practices that distribute influence fit naturally with transformational leadership and are associated with higher engagement and better performance [9, 11]. In remote or hybrid environments, inspirational clarity and reliable coaching are substitutes for constant presence, and leaders must design simple rituals and feedback loops that keep trust and momentum high [8]. Culture remains the moderator. In organizations where systems, symbols, and rewards align with espoused values, people interpret stretch goals as opportunity; where norms are rigid or status heavy, the same messages may feel risky, which is why leaders must actively shape assumptions through what they measure, recognize, and consistently model [12].

Transformational leadership is also complementary to authentic and ethical leadership. Authenticity helps leaders act consistently with their values and build durable trust, which makes purpose credible [2]. Ethical leadership sets the guardrails of fairness and responsibility, which protects the long term and enables voice and accountability [6]. Across these contexts, the through line is simple. When leaders pair clear goals and fair reinforcement with vision, learning, and individualized growth, organizations are more likely to see sustained improvements in engagement, innovation, and results [1, 4, 5].

5 Critiques and Limitations of the Transformational Leadership Model

Transformational leadership has strong appeal, but several cautions matter for research and practice. First, there is the risk of leader romanticism. Attributing outcomes mainly to a single leader can hide the role of systems, resources, and culture, and it can discourage healthy dissent when followers idealize the person in charge [1, 12]. Second, effects are context dependent. In highly regulated, tightly coupled, or resource-constrained settings, frequent experimentation or broad discretion may be impractical, so transactional clarity can dominate results even when leaders communicate an inspiring vision [4, 8]. Third, measurement is a recurring issue. Common tools capture overlapping constructs and rely heavily on follower self-reports collected at one time, which raises concerns about common method variance, endogeneity, and inflated correlations [13]. Related to this, publication bias and the tendency to study convenience samples can produce optimistic estimates of impact.

Another critique centers on incremental validity. Meta-analytic work shows that transformational leadership predicts important outcomes, yet authentic, ethical, and servant leadership can explain additional variance in attitudes and performance once transformational effects are accounted for [7]. This suggests that organizations may benefit more from an integrated approach that combines clarity, purpose, and ethical guardrails than from a single style. Ethical concerns also appear on the dark side. Charisma without integrity can drift into pseudo-transformational leadership that manipulates values or suppresses voice. Clear norms, transparent decisions, and fair reinforcement are needed to keep influence aligned with responsibility [6]. Finally, sustainability is a practical limit. Constant inspiration can fatigue teams if not balanced with workload design, skill development, and reliable routines that convert ambition into steady execution. In short, the model is most credible when treated as a toolkit coordinated with culture and systems, not as a stand-alone remedy for complex organizational problems [1, 4, 6–8, 12, 13].

6 Current and Future Trends in Organizational Leadership

Leadership is shifting toward practices that make organizations more adaptive, inclusive, and learning centered. Digital and hybrid work require clear purpose, lightweight routines, and reliable feedback loops so that trust and momentum do not depend on physical presence. Leaders who design simple cadences for goals, check-ins, and learning reviews help distributed teams stay aligned and innovative [5, 8]. Inclusion has moved from a side initiative to a core operating principle. Evidence links inclusive climates with stronger engagement and decision quality, which means leaders must invite voice, remove friction in processes, and hold themselves accountable for fair outcomes [10]. Empowering and shared leadership are also becoming standard in knowledge intensive and agile settings. By distributing influence and giving teams real decision latitude within clear constraints, organizations gain ownership, speed, and better problem solving [9, 11].

Adaptability is now a strategic capability. Leaders who help their organizations sense change, run disciplined experiments, and scale what works build resilience and performance in dynamic markets [8]. Motivation science remains central to this shift. Practices that support autonomy, competence, and relatedness make effort more self driven and sustain learning over time, which is essential when priorities change quickly [3, 14]. Looking ahead, the rise of data and AI will increase the premium on judgment and ethics. Leaders will need to pair analytics with human centered design, protect privacy, and ensure that automated decisions meet fairness standards. Finally, sustainability and stakeholder expectations are expanding what performance means. Leaders who translate environmental and social goals into practical metrics and routines align purpose with execution and reduce the risk of values washing. Across these trends, the strongest results appear when inspiration is paired with systems. Clear expectations and fair reinforcement remain foundational, while transformational, inclusive, and empowering practices provide the energy and learning that keep organizations improving [1, 4, 5, 8–11, 14].

7 Organizational Performance

Performance is more than financial results at quarter end. Healthy organizations track a balanced set of outcomes that include process quality, learning and innovation, customer and stakeholder experience, and responsible conduct. Transformational leadership contributes by making strategy usable in everyday work and by building the conditions that turn insight into execution. Leaders who communicate a clear direction, link roles to purpose, and keep attention on learning routines help teams translate effort into measurable improvements in throughput, reliability, and service quality [1, 5]. Meta-analytic evidence shows that when these behaviors sit alongside transactional clarity, organizations see higher effectiveness, satisfaction, and discretionary effort, which are precursors to durable results [4]. Culture is the amplifier. When systems, symbols, and rewards align with espoused values, people experience stretch goals as genuine opportunity rather than risk, and the day to day choices they make tend to reinforce strategy rather than drift from it [12]. Motivation science explains why this matters for performance. Work that supports autonomy, competence, and relatedness invites people to internalize goals, sustain effort, and improve faster, which shows up in cycle times, defect rates, and customer outcomes [3]. Inclusion strengthens the picture by widening the set of ideas considered and by improving the fairness of processes that allocate opportunities and make decisions [10]. In dynamic markets, adaptability ties it all together. Leaders who help their organizations sense change, run disciplined experiments, and scale what works are more likely to convert learning into consistent financial and operational gains [8].

8 Impact of Transformational Leadership on Organizational Performance

Transformational leadership improves performance by activating motivation, strengthening learning, and aligning day to day work with strategy. At the individual level, clear vision, intellectual stimulation, and personalized coaching help employees internalize goals and sustain effort beyond minimum requirements, which shows up in quality, speed, and service outcomes [1, 3]. Meta-analytic evidence links these behaviors to higher effectiveness, satisfaction, and discretionary effort, especially when paired with transactional clarity that sets expectations and reinforces results fairly [4]. At the system level, transformational leadership accelerates organizational learning and innovation. Leaders who invite questions, protect smart experiments, and connect teams across boundaries create routines that turn insight into better processes and products, which in turn support financial and operational results [5]. Culture amplifies or dampens these effects. Where systems, symbols, and rewards align with espoused values, people experience stretch as opportunity and coordinate more reliably; where norms are rigid or status heavy, the same messages can feel risky, limiting gains [12]. The mechanism that ties these pieces together is motivation quality. Work that supports autonomy, competence, and relatedness tends to produce self driven effort and faster skill growth, and transformational behaviors are effective partly because they create those conditions in everyday practice [3, 14]. In knowledge intensive and agile settings, empowering and shared leadership complement the model by distributing influence and giving teams real decision latitude within clear guardrails, which improves ownership and problem solving [9, 11]. Finally, adaptability has become a performance differentiator. Leaders who help their organizations sense change, run disciplined tests, and scale what works are better at converting learning into durable financial and operational gains, and transformational behaviors provide much of that engine when they are matched with simple, reliable execution routines [8].

9 Organizational Culture: Relevance and Impact on Transformational Leadership and Performance

Organizational culture shapes how people interpret a leader's words and actions, and it channels day to day choices toward or away from strategy. Schein's classic view helps make this practical: culture operates at three levels, from visible artifacts to espoused values to the deeper assumptions that quietly guide behavior [12]. Transformational leadership is effective when these layers line up. If systems, symbols, and rewards reinforce the stated direction, employees tend to experience stretch as a real opportunity. If they conflict, even a compelling vision can feel performative, and discretionary effort fades [1, 12].

Culture also explains part of the link between leadership and results. Transformational behaviors encourage learning, experimentation, and cross boundary collaboration, but those routines become durable only when the culture protects voice and treats disciplined trials as normal work. Studies show that learning rich climates help convert insight into better processes and products, which in turn support financial and operational outcomes [5]. Ethical norms are another pillar. When people believe decisions are fair and responsibility is shared, trust rises and the organization can move faster without sacrificing accountability [6]. Inclusion strengthens this effect by widening the range of ideas considered and improving the quality of everyday decisions, especially in complex or uncertain contexts [10].

As environments change, adaptable cultures give transformational leadership more room to work. Leaders who design simple cadences for goals, feedback, and reflection help teams sense shifts, test responses, and scale what works, which is the practical engine of adaptability [8]. In short, culture is the amplifier. Align systems with values, protect fairness and inclusion, and use learning routines to translate vision into execution. Under those conditions, transformational leadership is more likely to produce sustained gains in engagement, innovation, and performance [1, 5, 6, 8, 10, 12].

10 Conclusion

Transformational leadership helps organizations perform when it is paired with clarity, fairness, and routines that turn intent into execution. The evidence is consistent: leaders who communicate a credible direction, stimulate new thinking, and coach people as individuals raise motivation quality, strengthen learning, and improve results, especially when transactional basics such as clear goals and fair reinforcement are in place [1, 3–5]. Culture is the amplifier. When systems, symbols, and rewards align with espoused values, employees experience stretch as opportunity and coordinate more reliably; when they do not, even strong messaging loses power [12]. Ethical guardrails and inclusive practices protect trust and decision quality, widening the set of ideas considered while maintaining accountability [6, 10].

Three practical implications follow. First, design the environment. Make strategy usable, establish simple cadences for goals and feedback, and measure what matters so learning compounds into better processes and products [5, 8]. Second, build people. Treat coaching and development as core work, not extras, so autonomy and mastery can grow in step with rising expectations [3]. Third, match style to context. In dynamic settings, combine transformational behaviors with adaptive routines that sense change, test responses, and scale what works; in tightly regulated or highly coupled systems, keep experimentation within clear guardrails while preserving voice and fairness [4, 8]. Taken together, these choices move leadership from inspiration to sustained performance: a balanced blend of vision, design, and disciplined follow through that helps organizations improve and adapt over time [1, 4, 5, 8, 12, 14].

References

1. Bass, B.M., Riggio, R.E.: Transformational Leadership, 2nd edn. Lawrence Erlbaum Associates, Mahwah (2006). https://doi.org/10.4324/9781410617095
2. Judge, T.A., Piccolo, R.F.: Transformational and transactional leadership: a meta-analytic test of their relative validity. J. Appl. Psychol. **89**(5), 755–768 (2004). https://doi.org/10.1037/0021-9010.89.5.755
3. Schein, E.H.: Organizational Culture and Leadership. Jossey-Bass, San Francisco (1985)
4. Avolio, B.J., Gardner, W.L.: Authentic leadership development: getting to the root of positive forms of leadership. Leadersh. Q. **16**(3), 315–338 (2005). https://doi.org/10.1016/j.leaqua.2005.03.001
5. Deci, E.L., Connell, J.P., Ryan, R.M.: Self-determination in a work organization. J. Appl. Psychol. **74**(4), 580–590 (1989). https://doi.org/10.1037/0021-9010.74.4.580
6. Brown, M.E., Treviño, L.K.: Ethical leadership: a review and future directions. Leadersh. Q. **17**(6), 595–616 (2006). https://doi.org/10.1016/j.leaqua.2006.10.004
7. Shore, L.M., Cleveland, J.N., Sanchez, D.: Inclusive workplaces: a review and model. Hum. Resour. Manag. Rev. **28**(2), 176–189 (2018). https://doi.org/10.1016/j.hrmr.2017.07.003
8. Uhl-Bien, M., Arena, M.: Leadership for organizational adaptability: a theoretical synthesis and integrative framework. Leadersh. Q. **29**(1), 89–104 (2018). https://doi.org/10.1016/j.leaqua.2017.12.009
9. Monje-Amor, A., Abeal Vázquez, J.P., Faíña, J.A.: Transformational leadership and work engagement: exploring the mediating role of structural empowerment. Eur. Manag. J. **38**(1), 169–178 (2020). https://doi.org/10.1016/j.emj.2019.06.007
10. Pourbarkhordari, A., Zhou, E.H.I., Pourkarimi, J.: How individual-focused transformational leadership enhances its influence on job performance through employee work engagement. Int. J. Bus. Manag. **11**(2), 249–268 (2016). https://doi.org/10.5539/ijbm.v11n2p249
11. Baard, P.P., Deci, E.L., Ryan, R.M.: Intrinsic need satisfaction: a motivational basis of performance and well-being in two work settings. J. Appl. Soc. Psychol. **34**(10), 2045–2068 (2004). https://doi.org/10.1111/j.1559-1816.2004.tb02690.x
12. García-Morales, V.J., Jiménez-Barrionuevo, M.M., Gutiérrez-Gutiérrez, L.: Transformational leadership influence on organizational performance through organizational learning and innovation. J. Bus. Res. **65**(7), 1040–1050 (2012). https://doi.org/10.1016/j.jbusres.2011.03.005

13. Hoch, J.E., Bommer, W.H., Dulebohn, J.H., Wu, D.: Do ethical, authentic, and servant leadership explain variance above and beyond transformational leadership? A meta-analysis. J. Manag. **44**(2), 501–529 (2018). https://doi.org/10.1177/0149206316665461
14. Antonakis, J., Day, D.V. (eds.): The Nature of Leadership, 3rd edn. Sage, Thousand Oaks (2018)

Consumer Preferences on Natural Food Products: Perspectives on NTFP Demand

Vihtori Hänninen and Jyri Vilko(✉)

Department of Industrial Engineering and Management, Lappeenranta-Lahti University of Technology, Kouvola Unit, Kauppalankatu 13, 45100 Kouvola, Finland
jyri.vilko@lut.fi

Abstract. This systematic literature review comprehensively examines consumer demand for non-timber forest products (NTFPs), with a particular focus on food-related products such as wild berries, mushrooms, and culinary herbs. Although global markets for NTFPs are expanding as a result of growing consumer interest in health, sustainability, and natural lifestyles, the academic literature addressing the factors influencing purchasing behavior remains fragmented and lacks systematic consolidation. This review is guided by two primary research questions, namely the extent of consumer interest in NTFPs and their willingness to pay (WTP) for such products. The analysis is based on a sample of 17 peer-reviewed journal articles published between 2005 and 2025. Through thematic analysis, the study identifies and categorizes key drivers of consumer demand, including price sensitivity, demographic factors, access to reliable information, and marketing and certification mechanisms. Perceived health benefits and the natural origin of products emerge as the most significant consumer motivations, whereas economic limitations and informational barriers constrain the broader development of NTFP markets. While certification schemes have been found to enhance consumer trust, skepticism regarding label credibility continues to persist among certain consumer groups.

Keywords: Non-timber forest products (NTFPs) · Consumer behavior · Willingness to pay (WTP) · Health benefits · Certification · Sustainability

1 Introduction

Recent global developments—such as climate change, socio-economic transformations, and increased environmental awareness—have shifted forest management from a timber-focused view to recognizing forests as multifunctional ecosystems offering goods and services, including non-timber forest products (NTFPs) such as wild berries, mushrooms, herbs, and honey [1, 2]. This transition supports sustainability and bioeconomy goals. NTFPs represent an emerging market responding to consumer demand for natural, healthy, and ecologically sound products [3], while holding socio-economic value for rural livelihoods, particularly in the Global South, where up to 80% of rural households depend on wild forest resources [2]. According to the FAO, their global market value

A. Rocha et al. (Eds.): ICITS 2026, LNNS 1966, pp. 306–316, 2027.
https://doi.org/10.1007/978-3-032-27538-7_27

is about USD 88 billion, projected to reach USD 5 trillion by 2050 [4]. In developed countries, interest is also increasing; for example, one-quarter of European households collect NTFPs, worth an estimated €23.3 billion annually [5]. In central Europe, demand has grown especially for wild food and natural health products [6].

Despite this relevance, the drivers of consumer behavior toward NTFPs remain poorly understood. Literature indicates influences such as health benefits, sustainability awareness, price sensitivity, trust in certification, and cultural preferences [3, 7], yet findings are fragmented and lack synthesis. This review consolidates current knowledge on consumer demand for NTFPs, focusing on food-related products.

2 Methodology

This review was conducted as a focused systematic literature review (SLR), drawing on selected principles from the guidelines presented by [1] and PRISMA 2020 [2]. Although no formal review protocol was registered (e.g., in PROSPERO), the process followed four basic stages: (1) scoping and criteria definition, (2) database searching, (3) screening and eligibility assessment, and (4) synthesis. Process illustrated in the Prisma flow diagram.

2.1 Review Scope and Objectives

The aim of this review was to examine consumer demand and preferences for NTFPs, particularly those related to food. Search strategy and selection searches were performed on April 17, 2024, in Scopus and Web of Science databases (Table 1). Articles from irrelevant scientific domains were excluded using database filters. Inclusion criteria specified journal articles addressing consumer-related aspects of NTFPs in tended for human consumption.

Table 1. Databases searched, search queries applied, and corresponding search dates.

Databases	Query string	No of articles	Date of acquisition
Scopus	TITLE-ABS-KEY (“non-timber forest products” OR “non-wood forest products” OR “NTFP” OR “NWFP”) AND TITLE-ABS-KEY (“consumer” OR “demand” AND “food”) AND PUBYEAR >2004 AND PUBYEAR <2026 AND (EXCLUDE (SUBJAREA, “CHEM” OR “CENG” OR “COMP” OR “BIOC” OR “MEDI” OR “MATH” OR “PHAR” OR “IMMU” OR “PHYS” OR “VETE” OR “EART” OR “NURS” OR “MULT” OR “ARTS” OR “PSYC” OR “ENER” OR “MATE”)) AND (LIMIT-TO (LANGUAGE, “English”))	172	17/4/2025

(*continued*)

Table 1. (*continued*)

Databases	Query string	No of articles	Date of acquisition
Web of science	TS = ("non-timber forest products" OR "non-wood forest products" OR "NTFP" OR "NWFP") AND TS = ("consumer" OR "demand") AND PY = (2005–2025) NOT WC = (Materials Science OR Energy & Fuels OR Psychology OR Art OR Multidisciplinary Sciences OR Nursing OR Physics OR Immunology OR Mathematics OR Pharmacology & Pharmacy OR Chemistry OR Veterinary Sciences)	72	17/4/2025

2.2 Screening and Inclusion Process

The initial search yielded 234 records. After removing 27 duplicates, 207 records were screened by title and abstract. Of these, 154 were excluded due to lack of relevance to consumer themes. The remaining 53 full-text articles were assessed for eligibility. 39 were excluded for not meeting the inclusion criteria, resulting in a final sample of 14 studies. Three additional studies were identified via reference search, resulting in 17 studies included in total (Table 2), comprising 17 peer-reviewed journal articles.

Table 2. Studies included

Author	Year	Journal	Topic	Methodology
Kilchling, P.; Hansmann, R.; Seeland, K.	2009	Forest Policy and Economics, 11(4)	The study explores the role of NTFPs in multifunctional forestry in Switzerland, focusing on their income potential, consumer preferences, and the value of quality and environmental attributes.	Mixed
Okunlola, M.I.; Onyekwelu, J.C.; Oluwajuwon, T.V.; Olasuyi, K.E.; Ojo, D.O.; Israel, R.	2023	Environment, Development and Sustainability, 27	The paper examines NTFP trading in Akure, Nigeria, assessing its potential and challenges for enhancing value chains and socioeconomic impact.	Mixed

(*continued*)

Table 2. (*continued*)

Author	Year	Journal	Topic	Methodology
Singer, H.; Özsahin, S.	2024	Journal of Retailing and Consumer Services, 78	Consumer decision-making criteria for purchasing non-wood forest products using fuzzy AHP modeling and rural livelihoods in Jammu and Kashmir.	Quantitative
Peerzada, I.A.; Chamberlain, J.; Reddy, M.; Dhyani, S.; Saha, S.	2021	Sustainability, 13(21)	The study explores the valuation and economic potential of NTFPs in supporting a sustainable bioeconomy	Mixed
Meinhold, K.; Dumenu, W.K.; Darr, D.	2022	Forest Policy and Economics, 134	The paper examines the commercialization of baobab (Adansonia digitata L.) in global markets, emphasizing sustainability and stakeholder roles in the supply chain.	Qualitative
Schunko, C.; Lechthaler, S.; Vogl, C.R.	2019	Sustainability, 11(7)	Commercialisation of wild plants by European organic farmers and its rural development potential through NTFPs.	Qualitative
Paudel, B.; Zhou, M.	2024	Forest Policy and Economics, 168	Willingness to pay for eco-labelled maple syrup from sustainably managed forests.	Quantitative
Sheppard, J.P.; Chamberlain, J.; Agúndez, D.; et al.	2020	Current Forestry Reports, 6	The study examines the transition to sustainable forest management, highlighting the integration of NWFPs and their economic importance for trade and livelihoods.	Mixed

(*continued*)

Table 2. (*continued*)

Author	Year	Journal	Topic	Methodology
Diaz-Balteiro, L.; Alfranca, O.; Voces, R.; Soliño, M.	2023	Forest Policy and Economics, 152	The study examines the rising demand for quality wild fungi, especially porcini mushrooms, the challenges in demand modeling due to limited data, and the impact of internet search trends on prices.	Quantitative
Weiss, G.; Emery, M.R.; Corradini, G.; et al.	2020	Forests, 11:165	The role of NWFPs, specifically chestnuts, their historical significance, and potential for economic opportunities in rural areas.	Qualitative
Maisharou, A.; Larwanou, M.	2015	International Forestry Review, 17	Focuses on non-wood forest products (NWFPs) in Sahelian countries.	Qualitative
Alfaro, A.V.; Schunko, C.; Callo-Concha, D.	2024	Small-Scale Forestry, 23(2)	Assesses the sustainable commercialization potential and economic viability of selected NTFPs in Pando, Bolivia.	Mixed
Vaughan, R.C.; Munsell, J.F.; Chamberlain, J.L.	2013	Journal of Forestry, 111(1)	Explores NTFP management in the U.S.	Mixed
Bardule, A.; Jurmalis, E.; Libiete, Z.; et al.	2020	Silva Fennica, 54	Discusses increasing interest in NWFPs in Europe, highlighting economic, cultural, and ecological significance, and the impact of meteorological conditions on availability and pricing.	Quantitative

(*continued*)

Table 2. (*continued*)

Author	Year	Journal	Topic	Methodology
Stryamets, N.; Elbakidze, M.; Angelstam, P.	2012	Scandinavian Journal of Forest Research, 27	Role of NWFPs for local livelihoods in countries with transition and market economies: case studies in Ukraine and Sweden.	Mixed
Stoyanova, M.; Stoyanov, N.	2013	Forest Systems, 22(2)	Examines the economic roles and categories of NWFPs in Bulgaria, focusing on exportable products like herbs and mushrooms.	Quantitative
Chauhan, K.V.S.; Sharma, A.K.; Kumar, R.	2008	International Forestry Review, 10(2)	Holistic management of NTFPs: ecological value, measurement challenges, and socio-economic impacts.	Quantitative

3 Demand of NTFPs

The material reviewed in this study consists of 17 peer-reviewed journal articles that examines NTFPs from a consumer perspective. The literature includes both quantitative and qualitative approaches. Overall, the methodological landscape is diverse, yet most studies remain confined within a single disciplinary or analytical frame. Broad, interdisciplinary analyses are still relatively rare. Furthermore, the studies are highly context-specific: while the dataset includes European, African, and Asian cases, many of the findings are difficult to generalize due to localized market structures and cultural factors. This heterogeneity justifies the thematic structure adopted in the following sections: drivers of demand (3.1), demographic influences (3.2), the role of information access, awareness and certification (3.3), price and willingness to pay (3.4), and marketing and value creation (3.5).

3.1 Demand of NTFPs

Demand for wild foods and plant-based cosmetics fluctuates with consumer trends, unlike the relatively stable demand for medicinal raw materials. In recent years, wild foods have gained popularity in several European countries [6]. Some NTFP markets experience boom-and-bust cycles from resource scarcity or overharvesting, which can be mitigated through domestication strategies such as truffle-inoculated oaks or shade-grown American ginseng [8]. Health benefits remain the most important purchase driver, followed by economic and quality-related factors, while appearance, marketing, and environmental values are less critical [3]. Studies on willingness to pay (WTP) report 11–44% premiums for eco-labeled products, such as coffee and maple syrup [7], yet a gap persists between stated intentions and actual purchases due to accessibility and market constraints [1]. In West Africa, demand for products like wild mango seeds and locust beans reflects seasonal consumption and economic reliance, with over 60% of traders citing them as their main income source [2]. Similarly, baobab market growth shows that strategic marketing and awareness campaigns can significantly boost demand [5]. Digital tools have further improved market understanding; in Spain, online search trends for wild fungi have been effective in forecasting shifts in consumer interest and complementing conventional market analysis [9].

3.2 Demographic Influences

Consumer decision-making related to NTFPs is shaped by multiple factors, including health, price, and origin of products. A study conducted in Turkey employed an interval-valued intuitionistic fuzzy AHP framework and purposive sampling of individuals over 18 years of age with knowledge or experience in purchasing NTFPs. This approach enabled focused analysis of how these key attributes influence consumer choices. Although the study did not include demographic segmentation, related literature suggests that variables such as age and education level can significantly affect consumer priorities [3].

Gender and age distributions can provide further insight into purchasing behaviors for NTFPs. A study from Indiana, United States, revealed that 62.6% of respondents were women, with the most prominent age groups being 25–34 and 45–54 years. The same study underscored the role of the household as a unit of decision-making, with an average household size of 3.17 people, indicating that NTFP purchases are often influenced by broader family needs rather than individual preferences alone [4].

Demographic patterns also reveal generational and economic divides in consumer interest in NTFPs. A nationally representative Swiss survey found that 56% of respondents were female and 38% were between 15 and 29 years of age, while individuals over 60 were underrepresented. Furthermore, 83% of respondents disclosed their income level, with most situated in the middle- to lower-income categories. These findings imply that both age and economic status may shape consumers' ability and willingness to invest in sustainably sourced NTFPs [5].

Differences in demographics are equally evident on the production side of the NTFP market. In Finland and Sweden, commercial collectors are often older, less educated, and retired individuals. To address seasonal labor shortages during peak harvest times, the sector relies heavily on migrant workers from Eastern Europe and Southeast Asia. This reliance highlights the structural role of demographic composition in sustaining the availability of wild-collected products in Northern European markets [6].

3.3 Information Access, Awareness and Certification

Access to information and awareness levels strongly influence consumer behavior toward NTFPs. Demographic and psychological factors, such as age and education, are closely linked to preferences for sustainably certified products, with older and more educated consumers showing stronger inclinations to choose conservation-certified goods. Interest in sustainability further increases willingness to pay (WTP) for eco-labeled products [7].

Perceived credibility of certification varies. A U.S. study on eco-labeled maple syrup found that "green" labels were more positively received when self-declared or linked to conservation programs, while third-party certifications sometimes faced skepticism due to perceived greenwashing [7]. From the producer's perspective, limited consumer knowledge can weaken market potential, as shown in an Italian study where experiential knowledge was vital, yet many buyers were unaware of certification benefits [6].

The baobab market illustrates the role of targeted awareness efforts, including trade fairs, media campaigns, in-store marketing, and consistent branding, which strengthened demand through producer–distributor collaboration [5]. In Switzerland, surveys reveal strong interest in domestic NTFPs, suggesting opportunities for targeted marketing and ecological education programs such as SILVIVA [1]. However, integrating NTFPs into schemes like FSC or PEFC remains difficult due to a lack of standardized protocols; regional cooperatives have been proposed to address this gap [1].

3.4 Price and Willingness to Pay

Economic factors remain a fundamental component of consumer decision-making related to NTFPs. A comprehensive study employing an interval-valued intuitionistic fuzzy AHP model identified "product price" as the most influential economic criterion, ranking above considerations such as product origin, payment flexibility, customization, and supply-demand dynamics. The five highest-rated sub criteria across the model were health benefits, additive-free composition, price, quality assurance, and nutritional value [3].

Parallel research examining consumer preferences for eco-labelled NTFPs—such as sustainably harvested maple syrup—demonstrates a consistent willingness to pay a premium for products associated with positive environmental attributes. Across multiple label formats—including self-declared, bird-friendly, conservation program-verified, and third-party certified—consumers favored "green" products over conventional alternatives. However, willingness to pay varied considerably by certification type. Products verified through conservation programs received the highest consumer preference, whereas third-party certifications elicited the lowest premium. This lower valuation was attributed to consumer skepticism rooted in concerns about greenwashing, lack of transparency, and limited understanding of certification systems [4].

Demographic characteristics such as education level and age were also found to influence label preferences. Individuals with higher educational attainment expressed greater support for conservation-linked labels, yet exhibited a relatively lower preference for third-party certifications. Furthermore, familiarity with sustainability topics and active interest in environmental issues were positively associated with stronger preferences for eco-labelled products. These findings underscore the importance of consumer education and transparent certification communication in mitigating skepticism and fostering more informed purchasing behavior. They also suggest that market communication strategies should be tailored to reflect demographic differences in environmental attitudes [4].

Women and urban residents emerged as key consumer segments in the NTFP market. Individuals who valued sustainability and product traceability showed heightened receptiveness to environmentally certified products. Nevertheless, despite strong stated preferences—particularly among environmentally conscious consumers—actual purchasing behavior was often constrained by financial considerations. This disparity indicates that while the willingness to pay exists in principle, it may not consistently translate into practice, especially within lower-income groups [5].

3.5 Marketing and Value Creation

Marketing is a fundamental component in NTFP commercialization, enabling consumer access and income generation [2]. In Akure, Nigeria, it secures market participation and livelihood stability, underscoring the role of value addition in expanding reach and profitability for small enterprises. Beyond quality and sustainability, marketing also conveys symbolic and experiential value, with products linked to regional identity positioned as artisanal, healthy, and ecologically grounded [10].

Structural barriers include poor transport infrastructure, price volatility, seasonal scarcity, distant harvest sites, limited storage, and lack of capital [2]. For products like Moringa oil and Shea butter, commercialization is further constrained by fragmented research, low processing capacity, and high domestic taxes [11]. In Bolivia, açaí demand has grown, while other native palms remain underutilized [12]. Globally, low wages and limited value-added activities constrain income, though social networks can improve access to niche markets [13].

In high-income countries such as Switzerland, high labor costs require premium positioning. Quality assurance and cooperation with certification bodies have improved competitiveness [1]. Swiss urban markets show interest in locally sourced foods and health goods, but pricing and information barriers remain [1]. In Latvia, extreme weather caused sharp price fluctuations in mushrooms and berries, revealing the climate sensitivity of these markets [14].

Case studies from Ukraine and Sweden illustrate contrasting uses: in Roztochya, Ukraine, commercial collection provides vital seasonal income, while in Småland, Sweden, gathering is mainly for personal use, limited by forest degradation [15]. Internationally, NTFPs support natural cosmetics, herbal remedies, and botanical extracts, contributing to livelihoods and local development [3]. Higher market values encourage trade, with low-income countries exporting raw products and high-income ones processed goods [16]. In India, branding and market training have expanded reach and doubled incomes for some producers [17]. Finally, while the commercialization of wild plants offers potential for supporting rural livelihoods, market competition may threaten traditional micro-enterprises. Short value chains and direct marketing strategies, however, have shown promise in reaching niche consumer segments effectively [6].

4 Conclusions

4.1 Scientific Implications

This review synthesizes existing consumer-focused research on NTFPs, showing that health-related motivations and product affordability are primary drivers of purchasing decisions, while sustainability often remains secondary unless supported by credible certification or labeling. Current research is limited, context-specific, and shaped by regional and cultural factors. Incorporating empirical data on actual purchasing alongside survey-based data could better capture real choices and long-term patterns.

It provides a structured overview of consumer demand studies, consolidating fragmented and case-specific findings to identify underexplored areas. Lifestyle values—such as health orientation, sustainability, and naturalness—strongly influence preferences and should inform value creation, marketing, and the positioning of NTFPs within food systems and sustainable consumption frameworks.

Methodological gaps remain, as economic methods dominate; combining them with cultural, sensory, and behavioral approaches could yield deeper insights. Future research should address information asymmetries, cultural framing, and the long-term effects of commercialization on producers and ecosystems.

4.2 Managerial Implications

Producers and marketers should emphasize health and naturalness, ensure transparency in sourcing and pricing, and use certified labeling, regional branding, and storytelling to build trust. Pricing must reflect sensitivity in emerging economies, and marketing should target urban, young, and health-conscious consumers.

4.3 Limitations and Future Research

This review is limited by the small number of studies (n = 17) and the absence of a formal quality assessment. Many did not directly address consumer preferences or long-term behavioral patterns, and cross-cultural willingness-to-pay data are scarce. Future research should adopt stronger empirical designs, examine the effects of digital marketing, and enhance transparency while strengthening producer–consumer linkages.

References

1. Kilchling, P., Hansmann, R., Seeland, K.: Demand for non-timber forest products: surveys of urban consumers and sellers in Switzerland. Forest Policy Econ. **11**, 294–300 (2009). https://doi.org/10.1016/j.forpol.2009.05.003
2. Okunlola, M., Onyekwelu, J., Oluwajuwon, T., Olasuyi, K., Ojo, D.O., Israel, R.: Marketing of non-timber forest products in Nigeria: a case of Akure metropolis. Environ. Dev. Sustain. **27**, 7595–7614 (2023). https://doi.org/10.1007/s10668-023-04208-7
3. Singer, H., Özsahin, S.: An interval-valued intuitionistic fuzzy analytic hierarchy process model for understanding consumer decision-making in non-wood forest product purchases. J. Retail. Consum. Serv. **78**, 103734 (2024). https://doi.org/10.1016/j.jretconser.2024.103734

4. Peerzada, I., Chamberlain, J., Reddy, M., Dhyani, S., Saha, S.: Policy and governance implications for transition to NTFP-based bioeconomy in Kashmir Himalayas. Sustainability. **13**, 11811 (2021). https://doi.org/10.3390/su132111811
5. Meinhold, K., Dumenu, W.K., Darr, D.: Connecting rural non-timber forest product collectors to global markets: the case of baobab (Adansonia digitata L.). Forest Policy Econ. **134**, 102628 (2022). https://doi.org/10.1016/j.forpol.2021.102628
6. Schunko, C., Lechthaler, S., Vogl, C.R.: Conceptualising the factors that influence the commercialisation of non-timber forest products: the case of wild plant gathering by organic herb farmers in South Tyrol (Italy). Sustainability. **11**, 2028 (2019). https://doi.org/10.3390/su11072028
7. Paudel, B., Zhou, M.: Are consumers "green" enthusiasts or skeptics? evidence from non-timber forest products. Forest Policy Econ. **168**, 103302 (2024). https://doi.org/10.1016/j.forpol.2024.103302
8. Sheppard, J. et al.: Sustainable forest management beyond the timber-oriented status quo: transitioning to co-production of timber and non-wood forest products—a global perspective. Curr. For. Rep. **6**, 26–40 (2020). https://doi.org/10.1007/s40725-019-00107-1
9. Diaz-Balteiro, L., Alfranca, O., Voces, R., Soliño, M.: Using Google scholar search patterns to explain the demand for wild edible mushrooms. Forest Policy Econ. **152**, 102993 (2023). https://doi.org/10.1016/j.forpol.2023.102993
10. Weiss, G., Emery, M.R., Corradini, G., Živojinović, I.: New values of non-wood forest products. Forests. **11**, 165 (2020). https://doi.org/10.3390/f11020165
11. Maisharou, A., Larwanou, M.: Market potential of non-wood forest products in the Sahelian countries. Int. For. Rev. **17**, 125–135 (2015). https://doi.org/10.1505/146554815816007018
12. Alfaro, A., Schunko, C., Callo-Concha, D.: Uncovering the potential for the sustainable commercialization of non-timber forest products: palm fruits in Pando, Bolivia. Small-Scale For. **23**, 313–350 (2024). https://doi.org/10.1007/s11842-024-09562-6
13. Vaughan, R., Munsell, J., Chamberlain, J.: Opportunities for enhancing nontimber forest products management in the United States. J. For. **111**, 26–33 (2013). https://doi.org/10.5849/jof.10-106
14. Bardule, A., Jurmalis, E., Libiete, Z., Paulina, I., Donis, J., Treimane, A.: Use of retail market data to assess prices and flows of non-wood forest products in Latvia. Silva Fenn. **54**(3), 10341 (2020). https://doi.org/10.14214/sf.10341
15. Stryamets, N., Elbakidze, M., Angelstam, P.: Role of non-wood forest products for local livelihoods in countries with transition and market economies: case studies in Ukraine and Sweden. Scand. J. For. Res. **27**, 74–87 (2012). https://doi.org/10.1080/02827581.2011.629622
16. Stoyanova, M., Stoyanov, N.: Non-wood forest products in Bulgaria. Silva. **14**, 5–21 (2013)
17. Chauhan, K., Sharma, A., Kumar, R.: Non-timber forest products subsistence and commercial uses: trends and future demands. Int. For. Rev. **10**, 201–216 (2008). https://doi.org/10.1505/ifor.10.2.201

Flipped Classroom and Gamification for Training Engineers

Aracelly Núñez-Naranjo[1](✉), María-Stefanie Vasquez-Peñafiel[2], Mauricio Carvache-Franco[3], Orly Carvache-Franco[4], Aleksandra Minchenkova[5], and Wilmer Carvache-Franco[6]

[1] Centro de Investigación en Ciencias Humanas y de la Educación-CICHE, Facultad de Ciencias de la Educación, Universidad Tecnológica Indoamérica, Ambato 180103, Ecuador
fernandanunez@uti.edu.ec
[2] Universidad Internacional del Ecuador, Quito, Ecuador
mavasquezpe@uide.edu.ec
[3] Departamento de Ciencias de la Ingeniería, Universidad Internacional del Ecuador, Quito, Ecuador
smcarvachef@ube.edu.ec
[4] Universidad Espíritu Santo, Km. 2.5 Vía a Samborondón, Samborondón, Ecuador
[5] Institutional Research Center, Bernardo O'Higgins University, Avenida Viel 1497, Santiago 6513491, Chile
aleksandra.minchenko@ubo.cl
[6] Facultad de Ciencias Sociales y Humanísticas, Escuela Superior Politécnica del Litoral (ESPOL), Campus Gustavo Galindo Km 30.5 Vía Perimetral, Guayaquil 090902, Ecuador
limavera@espol.edu.ec

Abstract. This study analyzes the impact of implementing active learning strategies, specifically flipped classroom in Mathematics and gamification in Physics, among first-year engineering students. The research followed a quantitative approach with a quasi-experimental design, involving 71 participants divided into a control group and an experimental group. Results show that the experimental group achieved a significant improvement in academic performance, with an average increase of 2.76 points compared to 0.62 points in the control group. The results were statistically significant ($p < 0.001$), with both Mathematics and Physics analyzed independently. The use of digital platforms such as Google Classroom and Educaplay enhanced student engagement, facilitated problem-solving, and fostered intrinsic motivation. These findings suggest that active digital-supported methodologies can inform teacher training and curriculum design in early engineering education. In conclusion, the integration of active methodologies supported by technology constitutes an effective strategy to overcome learning difficulties in critical areas such as Mathematics and Physics during the initial stages of engineering education.

Keywords: Active learning · flipped classroom · gamification · educational technology · academic performance · engineering education

A. Rocha et al. (Eds.): ICITS 2026, LNNS 1966, pp. 317–330, 2027.
https://doi.org/10.1007/978-3-032-27538-7_28

1 Introduction

1.1 A Subsection Sample

In higher education, students' academic performance is a decisive factor for their persistence and success in university programs. This performance reflects both content mastery and the students' ability to adapt to the cognitive, social, and emotional demands of university life [1]. Several studies indicate that academic performance is influenced by multiple variables, including the pedagogical methodologies employed in the classroom, student motivation, study habits, and the availability of instructional resources. However, engaging engineering students in active participation with course content remains a challenge for instructors, particularly in critical areas such as Mathematics and Physics, where historically higher rates of dropout, failure, and low achievement are recorded [2, 3]. These difficulties directly affect academic continuity, since foundational science courses often act as a "filter" in the early stages of engineering education.

Meaningful learning, understood as the integration of new information with students' prior knowledge, has proven to be a key strategy for enhancing deep and lasting understanding of course content [4, 5]. Unlike traditional rote approaches, this pedagogical model emphasizes the active construction of cognitive schemas, enabling students to relate concepts, transfer them to different contexts, and apply them to solve more complex problems [6, 7]. Within this framework, active methodologies have gained prominence, as they promote student-centered teaching and stimulate not only academic performance but also motivation and self-regulated learning.

In the context of engineering education, the implementation of active strategies represents an effective approach to overcoming the inherent limitations of traditional expository teaching. While conventional methods often focus on one-way transmission of knowledge, active methodologies encourage students to assume a central role in their learning process, fostering autonomy, responsibility, and commitment to their professional development [8, 9]. This pedagogical shift enhances knowledge assimilation and promotes the construction of transferable learning outcomes applicable to practical and professional contexts.

In particular, these methodologies contribute to strengthening transversal competencies essential for the education of 21st-century engineers: problem-solving, enhanced through the application of knowledge to real-world scenarios; critical thinking, indispensable for analyzing, evaluating, and proposing innovative solutions; and the proficient use of digital tools, a requirement in academic and professional environments increasingly mediated by technology [10, 11]

The integration of technological resources into pedagogical practice, such as Google Classroom and Educaplay, further supports the construction of meaningful learning and adapts to the demands of dynamic educational environments [12, 13]. At the same time, these tools foster more inclusive and collaborative settings, generating positive impacts on both academic performance and student motivation [14, 15].

Despite these advances, there is still a gap in the literature regarding the combined and differentiated application of active methodologies in critical areas of engineering, specifically Mathematics and Physics, during the first stages of university education. Most prior studies focus on isolated experiences of flipped classroom or gamification, but few have contrasted their comparative and quantitative impact on the initial academic performance of engineering students.

Based on this context, the present study seeks to analyze the impact of applying active strategies—flipped classroom and gamification—among first-year engineering students, with emphasis on two critical areas of knowledge: Mathematics and Physics.

However, few studies have simultaneously compared the differentiated effect of flipped classroom and gamification on the early academic performance of engineering students in Mathematics and Physics. This study addresses this gap.

2 Background

Academic performance in higher education has been widely studied as a multifactorial phenomenon influenced by cognitive, socio-emotional, pedagogical, and contextual variables [16, 17]. In engineering programs, Mathematics and Physics have been identified as critical subjects, since they present the highest failure and dropout rates during the first years of study [18, 19].

2.1 Active Learning Strategies

Active learning strategies are designed to engage students in their own learning process, fostering autonomy, critical thinking, and meaningful knowledge construction [20, 21]. These methodologies break away from the traditional passive teaching model, offering experiences that stimulate reflection, interaction, and problem-solving. Among the most widely used active strategies in higher education are:

Project-Based Learning (PBL). This approach orients instruction toward solving real and contextualized problems, fostering 21st-century skills such as collaboration, creativity, and innovation [22]. Through PBL, students not only acquire technical knowledge but also develop the capacity to integrate interdisciplinary knowledge to address complex situations, simulating professional challenges.

Cooperative Learning. This strategy promotes small group work with the goal of achieving shared objectives, where each member assumes responsibility for the collective outcome. Cooperative learning not only strengthens academic performance but also enhances social skills such as communication, empathy, and leadership [23].

Academic Debates. Debates constitute an effective tool for developing critical thinking and logical argumentation, which are essential in university education. By confronting ideas and defending positions based on evidence, students strengthen their analytical capacity and learn to build sound reasoning, skills that are indispensable in engineering and research [24].

2.2 Flipped Classroom

The flipped classroom is one of the most relevant methodologies in contemporary education. In this model, students access theoretical content in advance through digital resources (videos, readings, simulations) and use classroom time for practical activities, discussions, and problem-solving [25]. Recent studies show that this methodology increases motivation and participation, particularly in complex subjects such as Mathematics [26]. This approach is especially relevant to the current study, as Mathematics requires prior conceptual access and class-time optimization.

2.3 Gamificación

Gamification involves the incorporation of game elements—such as points, badges, levels, and rewards—into educational environments. This strategy enhances intrinsic motivation, engagement, and content retention, making it an effective tool in higher education [27, 28]. In engineering education, gamification has yielded positive outcomes by stimulating logical-mathematical reasoning and problem-solving skills. Gamification is aligned with the challenges of teaching Physics, a subject where low motivation and conceptual abstraction are recurrent issues.

2.4 Tecnologías Digitales en el Aprendizaje

The use of information and communication technologies (ICT) in higher education has become a cornerstone for educational innovation. Platforms such as Google Classroom, Educaplay, and virtual simulators make learning processes more flexible and adaptable to different cognitive styles [29, 30]. Likewise, these tools enhance teacher–student interaction and foster collaborative environments that positively impact academic performance.

Flipped classroom and gamification are active pedagogical strategies which, when integrated with digital resources, can significantly improve the academic performance of engineering students, particularly in critical areas such as Mathematics and Physics.

3 Methodology

3.1 Research Approach and Design

This research followed a quantitative approach, aimed at objectively measuring the impact of implementing active learning strategies on the academic performance of first-year engineering students. A quasi-experimental design was used, with one control group and one experimental group, both evaluated through diagnostic tests and pre/post assessments. The scope of the study was descriptive–explanatory, as it not only identified critical learning areas but also explained the effect of the applied strategies on students' performance.

3.2 Population and Sample

Participants were selected through non-probabilistic convenience sampling based on class enrollment. No inclusion or exclusion criteria beyond course registration were applied. The study population consisted of 71 first-year engineering students, divided into two groups:

Group A (control): 36 students who continued under traditional teaching methodologies.
Group B (experimental): 35 students in whom active strategies were applied, specifically the flipped classroom in Mathematics and gamification in Physics.

The selection of these groups was based on the need to identify differences in academic performance in two core areas that historically present greater difficulty in the early stages of engineering studies: Mathematics and Physics.

3.3 Pedagogical Strategies Applied

Flipped Classroom (Mathematics). Students accessed theoretical content in advance through videos and interactive materials on Google Classroom and Educaplay. Classroom time was dedicated to solving problems, participating in discussions, and applying concepts to practical situations.

Gamification (Physics). Game dynamics such as competitions, rewards, and challenges were incorporated through digital platforms. This methodology aimed to increase motivation, foster logical reasoning, and improve problem-solving skills.

3.4 Data Collection Instruments

The instruments were reviewed by three subject-matter experts to ensure content validity. Reliability was verified using Cronbach's alpha, yielding acceptable coefficients ($\alpha = 0.82$ for Mathematics and $\alpha = 0.79$ for Physics). The following instruments were applied:

Diagnostic tests in Mathematics and Physics to identify the initial level of students.
Pretest and posttest, administered to both groups, to measure the evolution of academic performance.
Perception surveys, applied to the experimental group, to assess their learning experience with the active strategies.

3.5 Statistical Analysis

Descriptive statistics included means, standard deviations, variances, and standard errors. The results were processed using descriptive statistics (means, variances) and Student's *t*-test for independent samples to determine significant differences between the control and experimental groups before and after the intervention.

4 Results

The results obtained allow us to contrast the academic performance between the control group and the experimental group in Mathematics (Area 1) and Physics (Area 2). The main findings are presented in tables followed by their respective analysis. Mathematics and Physics results were analyzed independently in all statistical procedures to avoid aggregation bias.

4.1 Initial Diagnostic

Table 1 presents the results of the diagnostic test applied to the control group, composed of 36 students, in the areas of Mathematics (Area 1) and Physics (Area 2). The average score in Mathematics was 8.04, reflecting an acceptable initial performance and a moderate mastery of basic concepts in this subject. In contrast, the average in Physics was 6.53, indicating significantly lower performance and confirming that Physics represents one of the main difficulties for first-year engineering students.

Table 1. Control Group (A) – Initial Diagnostic Results

Student	Area 1 (Mathematics)	Area 2 (Physics)
1	9.33	3.25
2	5.04	5.55
3	8.67	3.35
...	...	...
36	8.19	7.95
Average	8.04	6.53

*a** The control group showed an acceptable performance in Mathematics but significantly lower performance in Physics.

The data analysis shows that although some students achieved high grades in Mathematics (e.g., 9.33), their results in Physics were considerably lower (e.g., 3.25). This disparity highlights the performance gap between the two areas. Overall, the control group demonstrated greater consistency in Mathematics, while clear limitations were observed in Physics, particularly in the comprehension and application of fundamental concepts.

This initial diagnostic supports the choice of Mathematics and Physics as critical areas for intervention, as they represent different levels of difficulty for students and have a direct influence on academic persistence and success during the first semesters of engineering programs.

4.2 Experimental Group – Initial Diagnostic

Table 2 presents the results of the diagnostic test applied to the experimental group, composed of 35 students, in Mathematics (Area 1) and Physics (Area 2). The average score in Mathematics was 8.30, while in Physics the mean was 6.63. These results are similar to those of the control group, although with a slightly better performance in both areas.

Table 2. Experimental Group (B) – Initial Diagnostic Results

Student	Area 1 (Mathematics)	Area 2 (Physics)
1	8.38	5.20
2	8.92	7.00
3	9.79	6.90
…	…	…
35	9.15	7.00
Average	8.30	6.63

*a** The experimental group presented results similar to those of the control group but with slightly better performance in both areas.

As in the control group, the experimental students showed greater mastery in Mathematics compared with Physics. However, the slightly higher averages in both subjects suggest a more favorable starting point for this group. For instance, some students achieved high scores in Mathematics (e.g., 9.79) while maintaining only moderate performance in Physics (e.g., 6.90), reaffirming the gap between the two areas of knowledge.

This initial diagnostic confirms that although the two groups started from relatively homogeneous conditions, the experimental group displayed marginally higher potential. This similarity between groups ensures the validity of subsequent comparisons after applying the active strategies of flipped classroom and gamification.

4.3 Pretest–Posttest Comparison – Control Group

Table 3 shows the results of the control group (36 students) in the tests applied before (pretest) and after (posttest) the pedagogical intervention, in both Mathematics (Area 1) and Physics (Area 2).

The average score in the pretest was 6.18, while in the posttest it increased to 6.80, representing a gain of only 0.62 points. Although this difference reflects a minor improvement, the change in academic performance is not significant in pedagogical terms.

Despite maintaining traditional teaching methodologies, students achieved a slight progression in both areas, likely attributable to continuity in the learning process and exposure to content throughout the semester. However, the improvement was not sufficient to overcome structural learning difficulties, especially in Physics, where lower results persisted compared with Mathematics.

These findings indicate that traditional lecture-based methods have a limited impact on improving academic performance, underscoring the need for innovative strategies that can generate substantial progress in critical engineering subjects.

Table 3. Control Group – Pretest and Posttest Results

Student	Area 1 Pretest	Area 2 Pretest	Area 1 Posttest	Area 2 Posttest
1	5.00	6.00	6.80	7.50
2	6.20	7.10	6.50	6.90
…	…	…	…	…
36	6.00	6.20	6.70	7.00
	Average Pretest: 6.18		Average Posttest: 6.80	
Mean	6.18	–	6.80	–
SD	0.56	–	0.16	–
SE	0.093	–	0.027	–
n	36	–	36	–
Mean	6.18	–	6.80	–

*a**: The control group showed a slight improvement (+0.62), but without meaningful pedagogical impact.

4.4 Statistical Analysis – Control Group

Table 4 presents the results of the Student's *t*-test for paired samples applied to the control group (36 students). The objective was to determine whether significant differences existed between pretest and posttest scores.

Table 4. Paired t-Test – Control Group

Variable	Mean	Variance	SD	SE	Observations
Pretest	6.175	0.319	0.56	0.093	36
Posttest	6.830	0.025	0.16	0.027	36
t-statistic	6.72	–	–	–	–
Critical t (two-tailed)	2.02	–	–	–	–
p-value	4.53E-08	–	–	–	–

The average increased from 6.175 in the pretest to 6.83 in the posttest, with reduced variance in both cases, indicating greater homogeneity in performance. The calculated t-statistic (6.72) exceeded the two-tailed critical value (2.02), with $p < 0.05$, confirming a statistically significant difference.

However, despite this statistical significance, the actual gain was only 0.62 points, which indicates that the pedagogical impact of traditional methodologies in the control group was limited and practically irrelevant. This analysis reinforces the conclusion that although learning occurs naturally as the semester progresses, lecture-based approaches do not generate substantial improvements in engineering students' performance in critical areas such as Mathematics and Physics.

4.5 Pretest–Posttest Comparison – Experimental Group

Table 5 presents the results obtained by the experimental group (35 students) in the tests conducted before (pretest) and after (posttest) the pedagogical intervention in Mathematics (Area 1) and Physics (Area 2).

Table 5. Experimental Group – Pretest and Posttest Results (Selected Students)

Student	Area 1 Pretest (Math)	Area 2 Pretest (Physics)	Area 1 Posttest (Math)	Area 2 Posttest (Physics)
1	5.50	5.80	8.00	8.70
2	5.20	5.40	7.20	8.80
…	…	…	…	…
35	5.60	5.90	8.30	8.80
Mean	5.74	–	8.50	–
SD	0.214	–	0.293	–
SE	0.036	–	0.049	–
n	35	–	35	–
Gain	+2.76	–	–	–

*a**: The experimental group showed a substantial improvement in both areas following the pedagogical intervention.

The average pretest score of the experimental group was 5.74, which increased to 8.50 in the posttest, reflecting an overall gain of +2.76 points. This improvement is considerably greater than that of the control group (+0.62), demonstrating the positive impact of the active strategies applied—flipped classroom in Mathematics and gamification in Physics.

Individual results also show consistent progress. For example, one student who started with 5.50 in Mathematics and 5.80 in Physics reached 8.00 and 8.70, respectively, after the intervention. These results indicate that the strategy benefited both overall averages and individual performance.

The application of active methodologies produced substantial and sustained improvements in both areas, overcoming the limitations of traditional models and validating their effectiveness in strengthening learning in the initial stages of engineering education.

4.6 Statistical Analysis – Experimental Group

Table 6 presents the results of the Student's *t*-test applied to the experimental group (35 students), with the aim of evaluating the difference between pretest and posttest scores.

The initial average was 5.73, while after the intervention it increased to 8.48, reflecting a substantial improvement of +2.75 points. Although variance remained low in both cases (0.046 and 0.086), indicating homogeneity within the group, the change in means was highly significant.

Table 6. Paired t-Test – Experimental Group

Variable	Pretest	Posttest
Mean	5.73	8.48
Variance	0.046	0.086
Observations	35	35
t-statistic	44.83	–
Critical t (two-tailed)	1.99	–
p-value	5.85E-49	–

The calculated t-statistic (44.83) far exceeded the critical value (1.99), and the p-value (5.85E-49) confirms that the difference was not due to chance. These results validate that the implementation of the flipped classroom in Mathematics and gamification in Physics had a positive and statistically significant impact on academic performance. Moreover, they reinforce the hypothesis that active methodologies supported by digital resources constitute an effective alternative to the limitations of traditional teaching in the early stages of engineering education.

4.7 Comparative Synthesis – Control vs. Experimental

Table 7 summarizes the overall results of both groups, allowing a direct comparison of the evolution in academic performance following the application of pedagogical strategies.

Table 7. Global Comparison Between Groups

Group	Pretest Average	Posttest Average	Gain	SD / SE Availability
Control (A)	6.18	6.80	+0.62	SD/SE calculated from Table 4
Experimental (B)	5.74	8.50	+2.76	SD/SE calculated from Table 6

*a** While the control group improved by only +0.62 points, the experimental group achieved a gain of +2.76 points, validating the hypothesis that active strategies enhance academic performance.

The control group (A), taught with traditional methodologies, progressed from a pretest average of 6.18 to 6.80 in the posttest, achieving a limited gain of +0.62 points. Although statistically significant, this improvement was pedagogically modest, confirming the constraints of lecture-based methods.

In contrast, the experimental group (B), exposed to active methodologies (flipped classroom in Mathematics and gamification in Physics), achieved a much greater advance: from 5.74 in the pretest to 8.50 in the posttest, with a gain of +2.76 points.

These results demonstrate that the impact of active methodologies was significantly greater compared to the learning achieved through traditional approaches. The comparative analysis validates the research hypothesis and confirms that the integration of innovative, technology-supported strategies boosts academic performance in critical areas of engineering education.

The comparison between both groups shows that the adoption of active methodologies generates statistically significant improvements and produces real, sustained pedagogical progress, consolidating their effectiveness as an alternative for early stages of higher education.

5 Discussions

The analysis of the results obtained makes it possible to affirm that the implementation of active strategies had a positive and significant impact on the academic performance of first-year engineering students, particularly in the areas of Mathematics and Physics.

First, it was evident that the experimental group substantially improved its grades compared to the control group. While the latter increased its average by only 0.62 points, the experimental group achieved an improvement of 2.76 points. This finding confirms the hypothesis that active methodologies foster meaningful learning and enhance the cognitive skills required in highly complex subjects [5].

The most notable difference was observed in Mathematics (Area 1), where the application of the flipped classroom allowed students to access digital materials prior to class. This optimized face-to-face time for problem solving and case discussions. This methodology is consistent with the findings of [6], who emphasize that the flipped classroom increases student autonomy and motivation—essential factors for improving performance in abstract areas.

In Physics (Area 2), the application of gamification promoted active participation, increased motivation, and facilitated the understanding of complex concepts through interactive dynamics. These results align with the evidence reported in [15], which shows that gamification in educational settings enhances engagement and knowledge retention. Furthermore, the findings of this study confirm previous research highlighting the potential of ICT in higher education. Platforms such as Google Classroom and Educaplay energized the teaching–learning process, creating an inclusive and collaborative environment in which students felt like protagonists of their own training [31].

In contrast, the control group, which maintained traditional methodologies based on lectures, showed minimal improvements. This highlights the limitations of passive approaches and the need to promote a shift toward student-centered pedagogies. As pointed out by [28, 29, 31], active methods improve academic outcomes while also strengthening critical thinking, collaboration, and problem-solving skills.

The results of this study validate the effectiveness of active strategies as pedagogical tools in engineering programs. However, they also reveal the importance of training faculty in the use of digital technologies and innovative methodologies, as well as the need to design flexible curricula that enable their sustainable implementation over time.

These findings have important implications for teacher training, as engineering instructors require competencies in digital pedagogy and active methodology design. Institutions should incorporate structured professional development and revise curricula to integrate sustainable technology-supported strategies. This study is limited by its sample size, non-randomized design, and reliance on a single institution. Future research should replicate these results across diverse engineering programs and compare different active methodologies. Future studies could explore the long-term retention effects of flipped classroom and gamification, evaluate hybrid combinations of active methodologies, or analyze differential impacts by gender, prior knowledge, or learning styles."

6 Conclusions

The present study demonstrated that the implementation of active strategies in the teaching–learning process has a positive impact on the academic performance of first-level engineering students, especially in the critical areas of Mathematics and Physics.

First, it was shown that the experimental group, in which the flipped classroom was applied in Mathematics and gamification in Physics, achieved significant improvements compared to the control group, which continued with traditional methodologies. The results of statistical tests confirm highly significant differences, with an increase of 2.76 points in the experimental group versus only 0.62 points in the control group.

The application of the flipped classroom facilitated a more efficient use of class time, promoting problem-solving and active interaction among students, which translated into better performance in Mathematics. For its part, gamification fostered motivation, participation, and logical thinking, contributing to a better understanding of Physics content.

Second, it was verified that the use of digital technologies such as Google Classroom and Educaplay played a key role in the implementation of these strategies, fostering a collaborative and dynamic environment that motivated students to actively engage in their learning.

It is concluded that active methodologies are effective in improving academic performance and student motivation in highly complex areas. The flipped classroom and gamification are especially useful strategies in Mathematics and Physics, respectively. The integration of technological resources enhances the effectiveness of these strategies, although it requires continuous teacher training to optimize their use. It is necessary to strengthen educational policies that promote methodological innovation in higher education in order to overcome the limitations of the traditional expository model. This study validates that active learning is an effective way to improve academic performance in the early stages of engineering education.

Acknowledgments. We would like to thank Indoamérica Technological University for its support in conducting this research as part of project IIDI-001-23.

References

1. Beck, D.: Be prepared: online school experience and student achievement during the pandemic. Front. Educ. **8**, 1161003 (2023). https://doi.org/10.3389/feduc.2023.1161003
2. Spady, W.G.: Dropouts from higher education: an interdisciplinary review and synthesis. Interchange. **1**(1), 64–85 (1970). https://doi.org/10.1007/BF02214313

3. Núñez-Naranjo, A.F.: Analysis of the determinant factors in university dropout: a case study of Ecuador. Front. Educ. **9**, 144534 (2024). https://doi.org/10.3389/feduc.2024.1444534
4. Ausubel: Vacation ideas, field trips, awards are stressed by Minneapolis committee. J. Am. Oil Chem. Soc. **31**(8), a23–a25 (1954). https://doi.org/10.1007/BF02632177
5. Hattie, J.: Visible Learning. Routledge, London (2008). https://doi.org/10.4324/9780203887332
6. Bergmann, J., Sams, A.: Flip your classroom: reach every student in every class every day. International Society for Technology in Education (ISTE), (2012)
7. Visarrea, B.T., Del Salto, E.M.: Application of the STEAM methodology for the design of a scaled-up hydroprinting robotic arm using LEGO Mindstorms EV3. In: 2024 IEEE 4th International Conference on Advanced Learning Technologies on Education & Research (ICALTER), pp. 1–4. IEEE, Tarma, Peru (2024). https://doi.org/10.1109/ICALTER65499.2024.10819227
8. Miryam, A., Ramos, E.A., Vito, N., Arnaiz, Q.: Estrategia neuroeducativa para optimizar el aprendizaje matemático de los estudiantes de educación básica elemental. Mikarimin. Rev. Cient. Multidiscip. **8**(3), 85–104 (2021) https://orcid.org/0000-0001-6220-2319
9. Oliveira, A., Behnagh, R.F., Ni, L., Mohsinah, A.A., Burgess, K.J., Guo, L.: Emerging technologies as pedagogical tools for teaching and learning science: a literature review. Hum. Behav. Emerg. Technol. **1**(2), 149–160 (2019). https://doi.org/10.1002/hbe2.141
10. Jian, M.J.K.O.: Personalized learning through AI. Adv. Eng. Innov. **5**(1), 16–19 (2023). https://doi.org/10.54254/2977-3903/5/2023039
11. Guerrero-Rodríguez, O.A., Herrera-Navas, O.D., Guzmán-Hernández, R., Baque-Arteaga, M.E.: Metodología B-learning en el desarrollo de competencias en programación y bases de datos en estudiantes de primero bachillerato informática. MQRInvestigar. **9**(1), e265 (2025). https://doi.org/10.56048/MQR20225.9.1.2025.e265
12. Bolaños-Caisapanta, B., Gómez-Trigueros, I.: Motivación y aprendizaje a través de B-Learning para estudiantes de bachillerato a través de un estudio experimental intragrupo. 593 Digit. Publ. CEIT. **9**(3), 19–33 (2024). https://doi.org/10.33386/593dp.2024.3.2349
13. Getenet, S., Cantle, R., Redmond, P., Albion, P.: Students' digital technology attitude, literacy and self-efficacy and their effect on online learning engagement. Int. J. Educ. Technol. High. Educ. **21**(1), 3 (2024). https://doi.org/10.1186/s41239-023-00437-y
14. Koivisto, J., Hamari, J.: The rise of motivational information systems: a review of gamification research. Int. J. Inf. Manag. **45**, 191–210 (2019). https://doi.org/10.1016/j.ijinfomgt.2018.10.013
15. Wild, S., Rahn, S., Meyer, T.: Dropout predictors in the academic fields of economics and engineering in cooperative education: an observation of the first academic year using cox regression. Empir. Res. Vocat. Educ. Train. **15**(1), 13 (2023). https://doi.org/10.1186/s40461-023-00152-y
16. Ibda, H., Wulandari, T.S., Abdillah, A., Hastuti, A.P., Mahsun, M.: Student academic stress during the COVID-19 pandemic: a systematic literature review. Int. J. Public Heal. Sci. **12**(1), 286 (2023). https://doi.org/10.11591/ijphs.v12i1.21983
17. Albarracín-Villamizar, C.Z., Hernández-Suárez, C.A., Prada-Núñez, R.: Learning objects and number thinking skills development: analysis using a quasi-experimental design. Aibi, Rev. Investig. Adm. e Ing. **8**(3), 131–137 (2020). https://doi.org/10.15649/2346030X.725
18. Peña, M., Aponte, M.: Internacionalización Conectiva: El Currículo en Mundo en Red. **11**(1) (2019) http://scioteca.caf.com/bitstream/handle/123456789/1091/RED2017-Eng-8ene.pdf
19. González-Pérez, L.I., Ramírez-Montoya, M.S.: Components of education 4.0 in 21st century skills frameworks: systematic review. Sustainability. **14**(3), 1493 (2022). https://doi.org/10.3390/su14031493

20. da Silva, M.L., et al.: Using active learning strategies during a quality improvement collaborative: exploring educational games to enhance learning among healthcare professionals. BMJ Open Qual. **13**(2), e002427 (2024). https://doi.org/10.1136/bmjoq-2023-002427
21. Rienties, B., et al.: Education 4.0 in higher education and computer science: a systematic review. Comput. Appl. Eng. Educ. **31**(5), 1339–1357 (2023). https://doi.org/10.1002/cae.22643
22. Wang, L.: The Application of Cooperative Learning Theory in College English Multimedia Teaching, pp. 102–104 (2016). https://doi.org/10.2991/ssehr-16.2016.131
23. Alén, E., Domínguez, T., de Carlos, P.: University students' perceptions of the use of academic debates as a teaching methodology. J. Hosp. Leis. Sport Tour. Educ. **16**, 15–21 (2015). https://doi.org/10.1016/j.jhlste.2014.11.001
24. Akçayır, G., Akçayır, M.: The flipped classroom: a review of its advantages and challenges. Comput. Educ. **126**, 334–345 (2018). https://doi.org/10.1016/j.compedu.2018.07.021
25. Galindo-Dominguez, H.: Flipped classroom in the educational system. Educ. Technol. Soc. **24**(3), 44–60 (2021) https://www.jstor.org/stable/27032855
26. Abdullah, M.I., Inayati, D., Karyawati, N.N.: Nearpod use as a learning platform to improve student learning motivation in an elementary school. J. Educ. Learn. **16**(1), 121–129 (2022). https://doi.org/10.11591/edulearn.v16i1.20421
27. Núñez-Naranjo, A., Sinailin-Peralta, J., Morales-Urrutia, E.: Gamification: From Motivation and Challenges to Improving Academic Performance in Learning Mathematics, pp. 106–113 (2024). https://doi.org/10.1007/978-3-031-44131-8_11
28. Núñez-Naranjo, A., Cumbicus, F.C., Ocaña, J.M.: TIC as a Didactic Tool for the Development of Reading Comprehension, pp. 144–154 (2024). https://doi.org/10.1007/978-3-031-44131-8_15
29. Núñez-Naranjo, A.F., Ocaña, J.M., Martinez, V.L.: Gamification for psychomotor development: an experience with genially in pre-school education. In: 2024 IEEE Eighth Ecuador Technical Chapters Meeting (ETCM), pp. 1–6. IEEE (2024). https://doi.org/10.1109/ETCM63562.2024.10746034
30. Morales-Urrutia, E., Ocaña, J., Yánez-Rueda, H., Núñez-Naranjo, A.F.: Methodological innovation for the teaching of TIC in higher education. RISTI - Rev. Iber. Sist. e Tecnol. Inf. **2021**(E46), 507–517 (2021) https://www.scopus.com/record/display.uri?eid=2-s2.0-85128733045&origin=resultslist
31. Lee, E.: Implementation of an academic software for children from 6 to 8 years old for easy learning in basic mathematics and reading comprehension. J. Adv. Res. Appl. Sci. Eng. Technol. **45**(2), 38–55 (2024). https://doi.org/10.37934/araset.45.2.3855

Usability Evaluation of Cross-Platform Mobile Development Frameworks

Jairo Pozo, William Montalvo(✉), and Rodrigo Tufiño

Salesian Polytechnic University, Quito 170146, Ecuador
jpozoc6@est.ups.edu.ec, {wmontalvo,rtufino}@ups.edu.ec

Abstract. The evolution of mobile application development has fostered the adoption of cross-platform frameworks, which allow for the creation of applications for multiple platforms from a single codebase, optimizing both cost and development time. This study presents a technical comparison of the Flutter, React Native, and .NET MAUI frameworks through functional prototypes developed using Clean Architecture and SOLID principles, incorporating key native features such as camera access, geolocation, and push notifications. The tests, conducted on physical devices, employed tools like Firebase Performance and Xcode Instruments. Flutter demonstrated the best overall performance, with low UI latency (<50 ms), lower energy consumption (145 mAh), and smaller executable sizes. React Native stood out in usability and adoption but showed higher resource consumption due to its bridge-based architecture, .NET MAUI effectively integrated the Microsoft ecosystem, albeit with higher resource demands. Therefore, Flutter is best suited for fluid UI projects, React Native for agile development, and MAUI for enterprise environments.

Keywords: Usability · Performance · Cross-Platform Frameworks · Mobile Development · Prototypes

1 Introduction

The continued growth of mobile environments has intensified the demand for development solutions that deliver high-quality applications with reduced time to market and controlled costs. Although native development ensures optimal performance and full access to device capabilities, it requires maintaining separate code bases, which negatively affects maintainability, scalability, and resource allocation [1–3]. Cross-platform frameworks attempt to address these limitations by allowing a single code base for multiple operating systems, but this advantage is often accompanied by trade-offs related to usability, performance efficiency, and integration with hardware-dependent features.

From the point of view of software quality, standards such as **ISO/IEC-25010** highlight efficiency, ease of maintenance, and usability as fundamental attributes that influence the overall quality of mobile applications. Likewise,

A. Rocha et al. (Eds.): ICITS 2026, LNNS 1966, pp. 331–340, 2027.
https://doi.org/10.1007/978-3-032-27538-7_29

usability models, including Nielsen's heuristics and the System Usability Scale, emphasize ease of learning, error prevention, and user satisfaction as key factors that determine the developer's experience when adopting new technologies in programming.

Previous studies have analyzed specific aspects of **Flutter**, **React Native** y **.NET-MAUI**, but most studies only analyze them in isolation or through descriptive comparisons that do not adhere to established principles of software quality and usability engineering, creating a gap in integrated evaluations that jointly assess developer usability, runtime performance, and the interoperability of native features within a unified architectural and experimental framework.

In order to address this need, this study contains a rigorous comparative evaluation of three multi-platform frameworks using functional prototypes developed using the principles of Clean Architecture and SOLID, to ensure structural homogeneity, given that this methodology combines quantitative performance measurements, usability tests based on criteria aligned with the SUS, and a systematic examination of native integrations such as the camera, geolocation, and notifications, offering a more comprehensive and theoretically based evaluation to support decision-making in cross-platform mobile development.

2 Methodology

The study followed a comparative experimental methodology grounded in principles of usability engineering and software quality evaluation [4,5]. The procedure (Fig. 1) was executed sequentially to ensure internal validity, reproducibility, and equivalence between the three evaluated frameworks.

First, twelve developers from the National Police of Ecuador (DNTICS) were selected, all participants had verified experience in web development but no prior exposure to Flutter, React Native, or .NET MAUI. This profile was intentionally defined to isolate factors related to learnability and developer usability without the confounding influence of previous framework familiarity, as recommended in controlled usability evaluations [4]. Their homogeneous technical background contributed to reducing variability in the execution of tasks.This selection strategy follows methodological recommendations for software usability benchmarking, ensuring neutrality and minimizing prior technology bias [4,6].

Second, development tasks were designed to reflect real-world mobile scenarios, including interface creation, navigation flows, REST API consumption, and the implementation of native functionalities such as camera access, geolocation, and push notifications. The tasks were adapted from prior comparative studies [5] and validated through expert review by two senior mobile developers who verified equivalence in complexity and relevance across the three frameworks. A small pilot test with additional developers ensured clarity and proportionality of the tasks before the main experiment.

Third, functional prototypes were built for each framework under identical architectural constraints. To guarantee structural comparability, all prototypes were developed following Clean Architecture and SOLID principles, consistent with methodological recommendations for cross-platform evaluation [7]. The use of Clean Architecture is consistent with recommendations from recent comparative evaluations, which highlight the importance of structural equivalence when testing cross-platform frameworks [7,8]. The prototypes were deployed and tested on the same hardware configuration (Google Pixel 8 with Android 14 and iPhone 13 with iOS 17), controlling for performance variability due to device differences.

Fourth, task execution was performed individually in a controlled environment, all sessions followed a standardized briefing and identical workstation setup. The think-aloud method was applied to capture participants' reasoning and difficulties, while screen recordings and device logs were used to register completion times, configuration issues, runtime warnings, and crashes. Error analysis followed an established classification scheme distinguishing syntax errors, API misuse, configuration failures, and interoperability issues, allowing for a systematic assessment of each framework's robustness.

Fifth, quantitative performance data—load times, UI latency, memory and CPU usage, battery impact, and metrics for GPS, camera, and notifications—were processed through descriptive statistics and comparative tests. Qualitative data were collected through a modified System Usability Scale (SUS) and a Likert-based questionnaire focused on documentation quality, ease of use, and developer satisfaction.

Finally, quantitative and qualitative findings were integrated through methodological triangulation [4], increasing the credibility of interpretations. The consolidated evidence supported the formulation of conclusions and practical recommendations regarding the suitability of Flutter, React Native, and .NET MAUI for different development contexts.

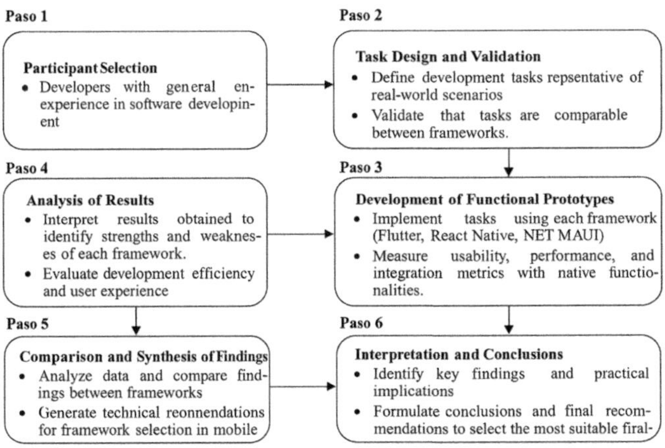

Fig. 1. Procedural workflow used in the study.

The Fig. 1 illustrates in detail the methodological flow used in the study, integrating experimental design, measurement tools, and validation phases.

3 Analysis and Results

The results are presented in three analytical dimensions: ease of use for developers, integration of native functions, and application performance, in accordance with the experimental design described above. Rather than focusing on raw numerical comparisons, this section emphasizes the practical implications of the observed differences for development, maintenance, and real-world user experience.

3.1 Developer Usability

Developers completed equivalent programming tasks in all three frameworks, followed by a SUS-based evaluation. React Native achieved the highest usability scores, supported by its familiar JavaScript syntax and modular UI components, which reduced onboarding time (Fig. 2). Flutter ranked closely behind, with participants highlighting its strong documentation and predictable widget-based architecture, although its reactive model required an initial adaptation period. .NET MAUI demonstrated the steepest learning curve, reflecting the complexity inherent to XAML and the broader Microsoft ecosystem.

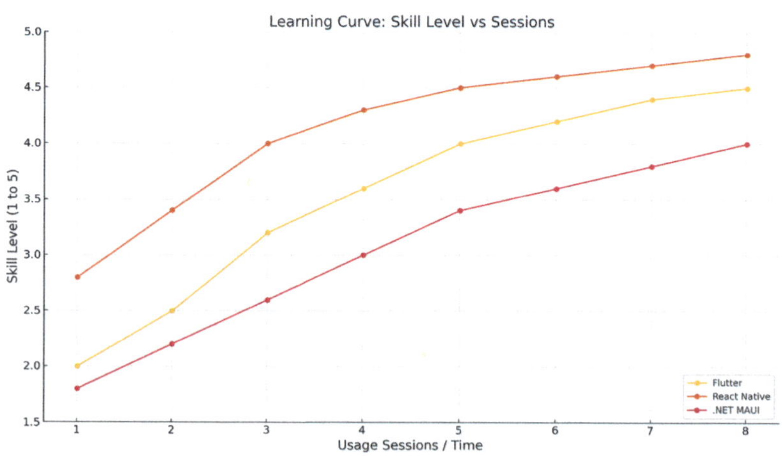

Fig. 2. Learning curve progression across sessions. React Native shows the fastest onboarding, followed by Flutter and .NET MAUI.

These findings reinforce prior studies [4,5], which argue that familiarity with JavaScript ecosystems significantly reduces learning barriers. Additional works confirm that JavaScript-based frameworks provide faster early progress [2,9], whereas frameworks like MAUI require deeper architectural understanding [8].

The Fig. 3 provides an overall view of developers' perceived usability across the three evaluated frameworks. React Native shows the highest average ratings, reflecting its accessible JavaScript-based ecosystem and strong community support reported in prior studies [2,5]. Flutter presents slightly lower but still stable scores, consistent with research that highlights its robust documentation and coherent development model despite a steeper initial learning curve [3,4]. Meanwhile, .NET MAUI exhibits the lowest evaluations, which aligns with findings describing its relative immaturity and the ongoing evolution of its ecosystem [7,8]. Overall, the figure reinforces trends observed in the literature: React Native is generally perceived as the most developer-friendly option, Flutter offers a solid and consistent experience, and .NET MAUI appears technically capable but less usable in current practice.

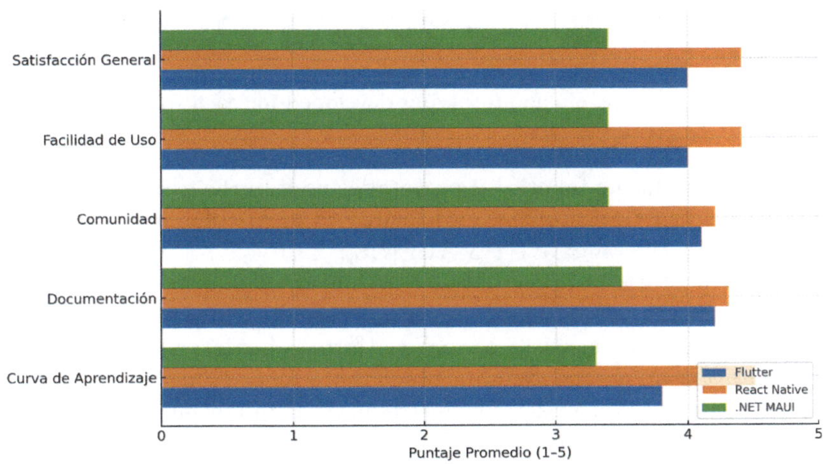

Fig. 3. Subjective usability evaluation (SUS dimensions).

3.2 Native Feature Integration

The functional prototypes developed for Flutter, React Native, and .NET MAUI (Fig. 4) served as the foundation for evaluating native feature behavior, ensuring that all comparisons were performed under equivalent architectural constraints. Each prototype implements the same interface structure, navigation flow, and native capabilities—camera access, geolocation, and push notifications—following Clean Architecture and SOLID principles as recommended in prior comparative studies [7,8]. This uniformity guarantees fairness in the analysis of integration complexity, performance, and cross-platform consistency.

Fig. 4. Screens of the prototypes developed for each framework.

Figure 4 presents the functional prototypes used in the evaluation, each built with identical structure and native features to ensure fair comparison across Flutter, React Native, and .NET MAUI [7,8].

The implementation of core native capabilities—GPS access, camera usage, and push notifications—revealed significant architectural differences. Flutter demonstrated the most consistent cross-platform behavior, requiring fewer manual configuration steps and offering predictable interactions with native modules. These results align with previous research highlighting Flutter's stability in hardware-dependent features [2,10].

React Native showed good performance but required platform-specific adjustments, particularly on iOS. This reflects its hybrid bridge architecture [7], which introduces additional points of failure. .NET MAUI provided intuitive APIs via .NET Essentials but exhibited inconsistencies on iOS, consistent with recent reports describing its early-stage maturity [8].

Figure 5 illustrates the quantitative metrics recorded for GPS compatibility, camera latency, and notification delivery times.

In practical terms, Flutter minimizes integration risks, React Native requires additional validation across platforms, and MAUI is currently more suitable for Microsoft-centered enterprise environments where its tooling advantages outweigh performance uncertainty.

3.3 Application Performance

Performance testing focused on load times, UI latency, resource usage, and energy consumption, following methodologies used in previous performance anal-

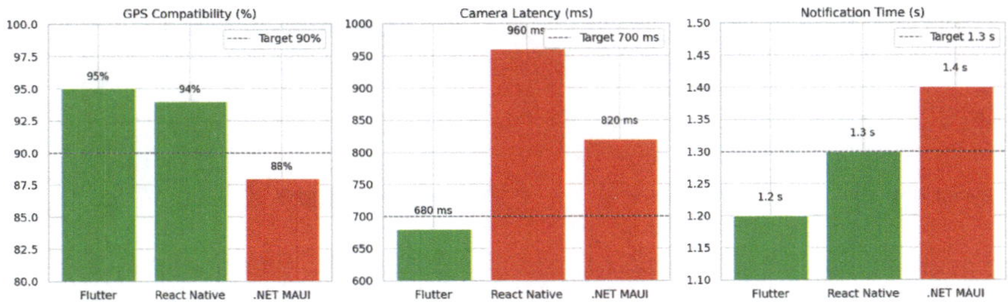

Fig. 5. Native feature metrics.

yses [11–13]. The comparative results reflect each framework's rendering model and runtime architecture.

Flutter showed the lowest startup times and UI latency, which is consistent with earlier findings attributing its performance advantages to the Skia rendering engine [5]. React Native produced moderate performance, affected by the JavaScript bridge. .NET MAUI exhibited higher RAM and CPU consumption, which aligns with previous evaluations noting overhead from .NET runtime abstractions [3].

Figure 6 presents the load-time bullet chart, revealing Flutter's consistent lead.

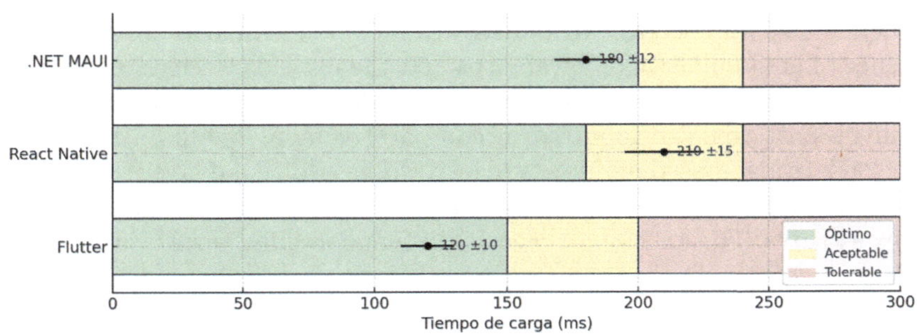

Fig. 6. Load-time bullet chart comparing frameworks.

Additionally, UI latency ranges (Fig. 7) show Flutter maintaining optimal responsiveness, while React Native and MAUI frequently surpass expected thresholds for high-fluidity interfaces.

These patterns suggest that Flutter is preferred for animation-intensive applications, React Native balances performance and productivity, and MAUI prioritizes maintainability and integration within enterprise ecosystems.

Discussion

The findings of this study show that the differences between Flutter, React Native, and .NET MAUI extend beyond performance metrics and reflect dis-

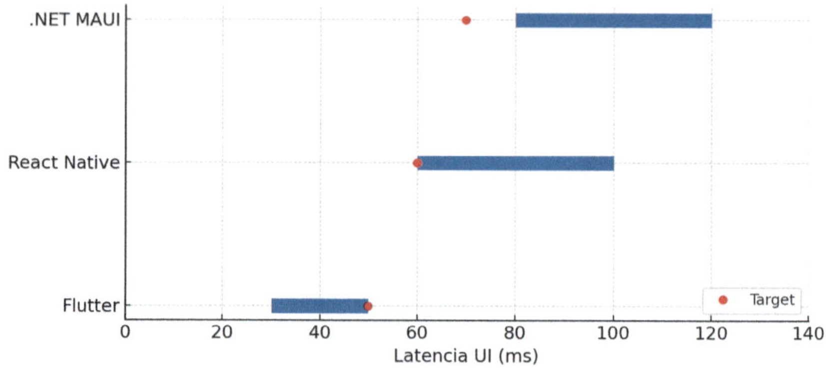

Fig. 7. UI latency comparison across frameworks.

tinct implications for developer workflows, long-term maintainability, and end-user experience. Flutter's stable rendering model and consistent cross-platform behavior contribute to smoother interfaces and reduced energy usage, aspects that directly enhance user perception and align with prior evidence highlighting its advantage in UI-intensive applications [zahra2022systematic]. These characteristics suggest that Flutter is particularly suitable for scenarios where visual fluidity, animation stability, and predictable performance are central to user satisfaction.

React Native, in contrast, demonstrates its value primarily in the developer experience. The familiarity of JavaScript and the extensive ecosystem surrounding it facilitate rapid onboarding, shorter development cycles, and easier team scalability—patterns also reported in comparative analyses [2,5]. However, the reliance on a bridge-based architecture introduces additional layers during native interactions, which may increase debugging effort and impact responsiveness on complex features. This tension between agility and architectural overhead implies that React Native is best suited for teams prioritizing delivery speed and iterative development.

.NET MAUI presents a different profile driven by enterprise-oriented design. Its integration with the Microsoft ecosystem, MVVM structure, and tooling support provide a strong foundation for maintainable and standardized development processes, especially in corporate environments [8]. Although its runtime resource demands are higher, these trade-offs may be acceptable for organizations where long-term maintainability, security policies, and integration with existing .NET infrastructure outweigh strict performance requirements.

Taken together, the results highlight that the choice of framework should not be determined solely by performance metrics but by a strategic alignment between technical characteristics and organizational needs. Flutter delivers user-centric performance benefits, React Native maximizes developer productivity, and .NET MAUI reinforces architectural consistency within enterprise ecosystems. These distinctions emphasize that cross-platform development involves balancing efficiency, maintainability, and experiential quality for both users and developers.

4 Conclusions

This study provides an integrated technical comparison of Flutter, React Native, and .NET MAUI through functional prototypes developed under identical architectural constraints, the analysis combined usability assessments, native feature evaluations, and performance measurements to provide a holistic understanding of how each framework behaves in real development scenarios.

- **Flutter** stands out for its technical consistency and efficient rendering model, offering predictable performance and reliable integration of native features, these strengths make it suitable for applications where fluid interfaces and stable cross-platform behavior are critical.
- **React Native** excels in developer usability and onboarding, supported by the familiarity of JavaScript and a mature ecosystem, its agility makes it ideal for rapid development cycles, although teams must be prepared to manage occasional complexities in native interoperability.
- **.NET MAUI** is positioned as an enterprise-focused solution, benefiting projects that require alignment with the Microsoft ecosystem, strong architectural structure, and long-term maintainability. Although its performance profile is less favorable, its tooling and integration capabilities remain valuable in corporate contexts.

Overall, the findings indicate that each framework offers meaningful advantages depending on project priorities. Rather than identifying a single "best" technology, this study highlights the importance of matching framework characteristics with organizational needs, development expertise, and the expected user experience.

References

1. Stanojevic, J., et al.: An overview of modern cross-platform mobile development frameworks. In: Proceedings of FOI TSE, pp. 489–497 (2022)
2. Gulcuoglu, E., Ustun, A.B., Seyhan, N.: Comparison of flutter and react native platforms. J. Internet Appl. Manage. **12**(2), 129–143 (2021). https://doi.org/10.34231/iuyd.888243
3. Nawrocki, P.: A comparison of native and cross-platform frameworks for mobile applications. Computer **54**(3), 18–27 (2021). https://doi.org/10.1109/MC.2020.2983893
4. Paz, F., Paz, F.A., Sánchez, M., Moquillaza, A., Collantes, L.: Quantifying the usability through a variant of the traditional heuristic evaluation process. In: Marcus, A., Wang, W. (eds.) DUXU 2018. LNCS, vol. 10918, pp. 496–508. Springer, Cham (2018). https://doi.org/10.1007/978-3-319-91797-9_36

5. Markowski, M., Smolka, J.: A comparative analysis of the flutter and react native frameworks. J. Comput. Sci. Inst. **29**, 346–351 (2023). https://doi.org/10.35784/jcsi.3794
6. Anastacio, C.J., Vega, A., Morales, D.: Análisis comparativo de entornos de desarrollo móvil en entornos multiplataforma. *Ciencia Latina Revista Científica Multidisciplinar*, **6**(4), 1–15 (2022). https://doi.org/10.37811/cl_rcm.v6i4.2950
7. Jangassiyev, R., et al.: Comparative analysis of cross-platform development methodologies: a comprehensive study. TELKOMNIKA **23**(1), 108–118 (2024). https://doi.org/10.12928/telkomnika.v23i1.26331
8. Ponomarev, I.: Features of the .net maui framework for creating cross-platform applications. Syst. Technol. **1**, 7–12 (2023). https://doi.org/10.34185/1562-9945-1-144-2023-07
9. Javeed, A.: Performance optimization techniques for reactjs. In: IEEE International Conference on Electrical, Computer and Communication Technologies (ICECCT), pp. 1–5 (2019). https://doi.org/10.1109/ICECCT.2019.8869134
10. Barros, L.P., et al.: Analyzing the performance of apps developed by using cross-platform and native technologies. In: Proceedings of SEKE (2020). https://doi.org/10.18293/SEKE2020-122
11. Mushtaq, F., Azam, F., Anwar, M.W.: Performance comparison of single code base development tools: flutter, react native, and xamarin. In: 14th International Conference on Software Technology and Engineering (ICSTE), pp. 17–23 (2024). https://doi.org/10.1109/ICSTE63875.2024.00011
12. Drzewiecki, A.: Comparison of the cross-platform mobile applications performance in 2D graphics processing. J. Comput. Sci. Inst. **25**, 362–365 (2022). https://doi.org/10.35784/jcsi.3024
13. Kishore, K., et al.: Performance and stability comparison of react and flutter: cross-platform application development. In: International Conference on Cyber Resilience (ICCR), pp. 1–4 (2022). https://doi.org/10.1109/ICCR56254.2022.9996039
14. Jost, G., Taneski, V.: State-of-the-art cross-platform mobile application development frameworks: a comparative study of market and developer trends. Informatics **12**(2), 45 (2025). https://doi.org/10.3390/informatics12020045
15. Sinaga, A.M., et al.: Performance of automation testing tools for android applications. In: 10th International Conference on Information Technology and Electrical Engineering (ICITEE), pp. 534–539 (2018). https://doi.org/10.1109/ICITEED.2018.8534756
16. Palmieri, M., Singh, I., Cicchetti, A.: Comparison of cross-platform mobile development tools. In: 16th International Conference on Intelligence in Next Generation Networks (ICIN), pp. 179–186 (2012). https://doi.org/10.1109/ICIN.2012.6376023
17. Ohrt, J., Turau, V.: Cross-platform development tools for smartphone applications. Computer **45**(9), 72–79 (2012). https://doi.org/10.1109/MC.2012.121

Smart Irrigation in Miniature Roses: A Case Study with IoT and Agriculture 4.0

Blanca Topon-Visarrea(✉), Adriana Guevara-Arteaga, and Miryam Johana Salcedo-Quishpe

Facultad de Ingenierías, Ingeniería Industrial, Universidad Tecnológica Indoamérica, Quito 170103, Ecuador
blancatopon@uti.edu.ec

Abstract. Inefficient water use remains one of the main challenges in floriculture, especially in high-value crops such as miniature roses, where manual irrigation generates significant losses, high costs, and sanitary risks. In response to this problem, an automatic irrigation system was designed based on IoT technologies, integrating soil moisture sensors, flowmeters, a PLC, a centrifugal pump, and a monitoring screen to improve water management and crop productivity.

The system incorporates a control interface that allows the operator to monitor and adjust irrigation parameters in a simple way, complemented by dashboards in ThingSpeak that display real-time soil moisture and water flow, accessible from any device connected to the internet. This technological integration enables faster, more reliable monitoring adapted to real production conditions.

Simulations showed a reduction of 51.43% in water consumption and an increase of 14% in productivity, from 4.33 to 4.92 stems per dollar invested. With an estimated investment of 6,624.33 USD and a payback period of less than one year, the system proved to be technically and economically feasible while improving crop health and reducing labor dependence.

These results confirm that the incorporation of IoT technologies in agriculture can transform production processes, optimize resource use, and strengthen the sustainability of floriculture within the framework of Agriculture 4.0.

Keywords: Smart irrigation · IoT · control interface · sustainable agriculture · Agriculture 4.0

1 Introduction

Water is an essential resource for agriculture and, at the same time, one of the most vulnerable in the face of population growth, climate change, and the pressure to ensure food security. In this scenario, efficient water management is no longer just a good practice but an urgent need to sustain agricultural productivity at a global level [3].

A. Rocha et al. (Eds.): ICITS 2026, LNNS 1966, pp. 341–349, 2027.
https://doi.org/10.1007/978-3-032-27538-7_30

Several studies confirm that optimizing irrigation reduces losses, ensures a more stable supply, and contributes to equitable food access. In contemporary agriculture, where efficiency and sustainability are priorities, the use of soil moisture sensors and automated irrigation systems has marked a significant change. These tools not only improve precision in water application but also help farmers address current challenges with an environmentally responsible approach [6,7,11,12].

Advances in automation have opened the possibility of integrating technologies such as programmable logic controllers, monitoring screens, cloud-based platforms, and various sensors (moisture, ultrasonic, rain), allowing irrigation systems to adapt to the conditions of each crop. Furthermore, the potential of the Internet of Things (IoT) applied to agriculture goes beyond measuring soil moisture, enabling the monitoring of essential nutrients such as nitrogen, phosphorus, and potassium (NPK). This technological integration drives precision agriculture, reduces resource waste, and strengthens environmental sustainability [5,6,11,12].

In Ecuador, floriculture is one of the most representative sectors of the economy and one of the most demanding in terms of quality. However, many production units still rely on manual irrigation, which generates water waste, increases operating costs, and affects crop performance [3,4]. These practices limit sustainability and reduce competitiveness in increasingly demanding international markets.

This study presents a case applied to an Ecuadorian floriculture company, where an intelligent automated irrigation system was designed and validated. The objective was to demonstrate that integrating sensors, controllers, and IoT platforms can reduce losses, optimize water use, and establish a more efficient and sustainable production model within the framework of Agriculture 4.0 [11,12].

2 Methodology

2.1 Study Approach and Initial Diagnosis

This study was conducted with an applied and experimental approach aimed at solving a specific problem: the inefficient use of water in miniature rose cultivation. The initial diagnosis identified two main limitations of manual irrigation: uneven water distribution and over-irrigation, both associated with productivity losses and an increased presence of fungal diseases.

To characterize the situation, production was compared with losses across different varieties. The Yellow Babe variety showed the highest percentage of waste, while Snowflake presented better utilization (Fig. 1). In addition, irrigation deficiencies favored the spread of diseases such as *Botrytis cinerea*, *Peronospora sparsa*, and *Podosphaera pannosa* (Fig. 2) [8–10].

2.2 Technical Design of the Automated System

The proposed system included soil moisture sensors, flowmeters, a centrifugal pump, a PLC (Programmable Logic Controller), and an HMI (Human-Machine

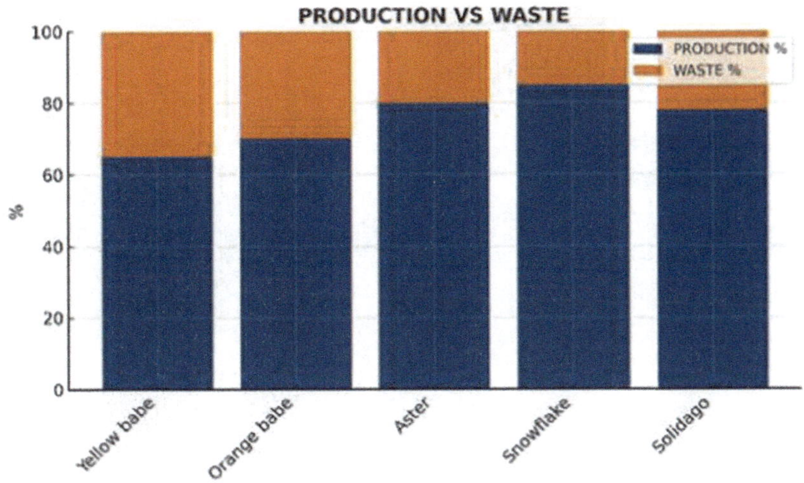

Fig. 1. Production vs. waste in miniature rose varieties.

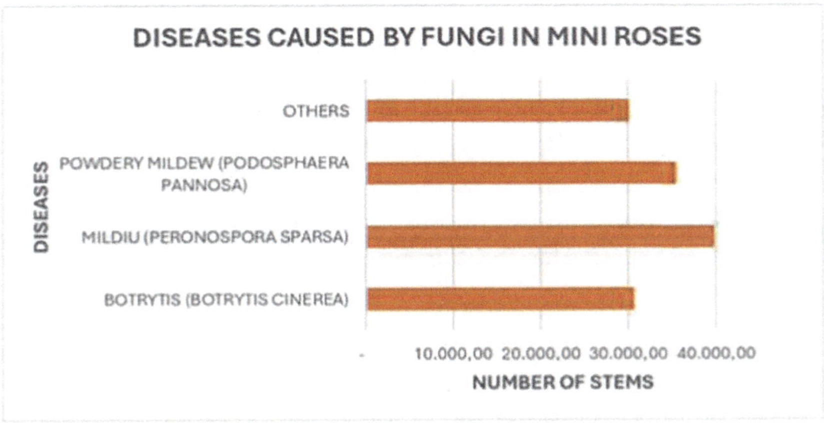

Fig. 2. Fungal diseases in miniature roses by number of affected stems.

Interface). The selection of these elements was based on flow, pressure, and terrain requirements. Programming was carried out in TIA Portal, where irrigation activation conditions were defined based on sensor data [6,7].

Before integration, the soil moisture sensors were calibrated using the direct comparison method with a reference digital hygrometer, performing 28 measurements at different substrate points to verify stability. The average error recorded was less than ±2%, considered adequate for agricultural applications.

For the flowmeter, volumetric verification tests were conducted by comparing measured flow with a standard volume, yielding an accuracy of ±3%. These margins remained constant during simulations and validation tests (Fig. 3).

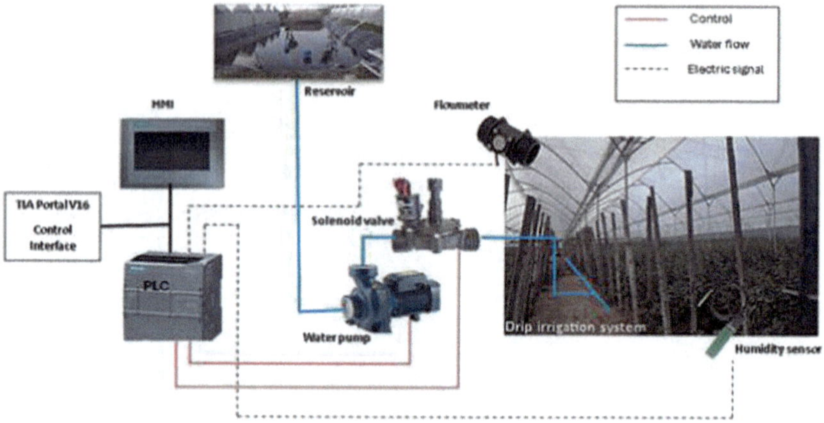

Fig. 3. Schematic diagram of the automated irrigation system.

2.3 IoT Integration and Remote Monitoring

To strengthen autonomy and control, the IoT was integrated into the system. The ThingSpeak platform was used as a data management environment, configuring channels for variables such as soil moisture and flow (Fig. 4). Dashboard visualization allowed users to monitor system behavior in real time (Fig. 5).

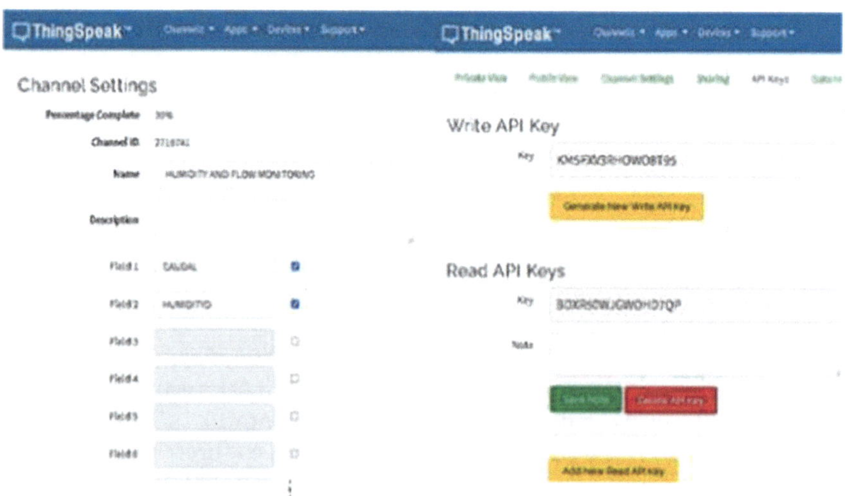

Fig. 4. ThingSpeak channel configuration for monitoring moisture and flow.

The sensors transmitted data to ThingSpeak at a 30-s interval, allowing near-real-time monitoring without overloading the network. This interval was selected after preliminary tests showed that frequencies below 15 s caused unnecessary server spikes, while intervals above 60 s reduced system responsiveness.

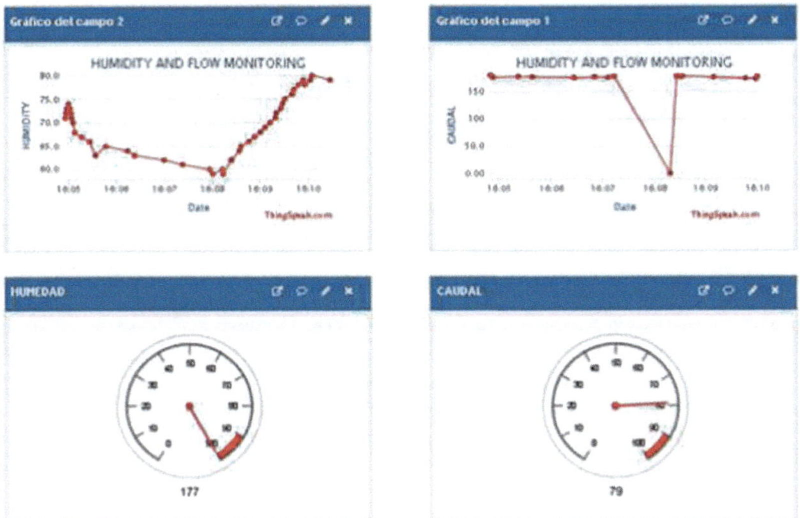

Fig. 5. Real-time visualization of soil moisture and flow in customized dashboards.

Node-RED was used to link the PLC, HMI, and the cloud, facilitating data flow management through graphical logic that reduces human error [6,7] (Fig. 6).

2.4 Electrical Design of the System

The electrical layout was organized into two circuits: a control circuit consisting of sensors, PLC, HMI, and relays, and a power circuit that supplied the actuators (solenoid valves and pump). Separating both circuits improved system safety and maintenance. Figure 7 shows the complete electrical diagram, which includes a power supply converting 110 V AC to 24 V DC to protect sensitive electronic equipment.

2.5 Validation and Performance Tests

Tests were conducted in TIA Portal and ThingSpeak, configuring soil moisture sensors so that irrigation was activated when values dropped below the optimal 70–80% range. The validation methodology verified the system's ability to respond in real time, collect data, and maintain soil moisture within appropriate limits.

Simulations were complemented with field measurements considering ambient temperature (16–22 °C) and greenhouse relative humidity (60–85%). These

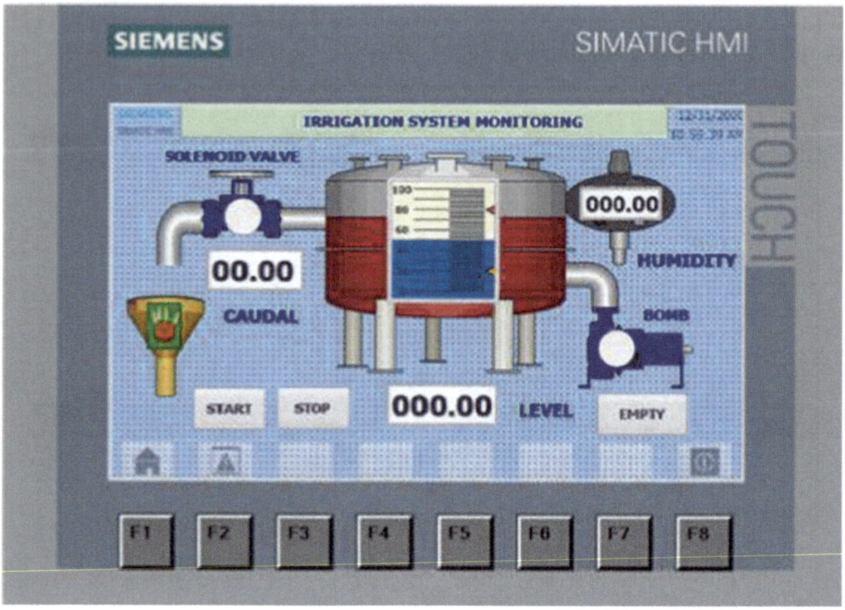

Fig. 6. HMI interface for irrigation system supervision.

variables influence evapotranspiration; therefore, recording them was essential to validate the actual performance of the system. Under these conditions, soil moisture remained within the recommended range.

To evaluate the system beyond its operational performance, technical and economic indicators were defined:

- **Water efficiency:** determined from flow records, comparing manual versus automated irrigation.
- **Agricultural productivity:** measured as the number of marketable stems per dollar invested [1,2].
- **Return on investment (ROI):** estimated based on reduced water consumption and labor in contrast to system cost.

3 Results

3.1 System Performance in Irrigation Control

The data confirmed the effectiveness of automation. Figure 8 shows soil moisture fluctuating between 60% and 80%, with irrigation triggered only when the lower threshold was reached. This prevented unnecessary irrigation and ensured more stable conditions.

Figure 9 compares performance before and after implementation: manual irrigation showed pronounced oscillations, while the automated system maintained stability near the recommended range.

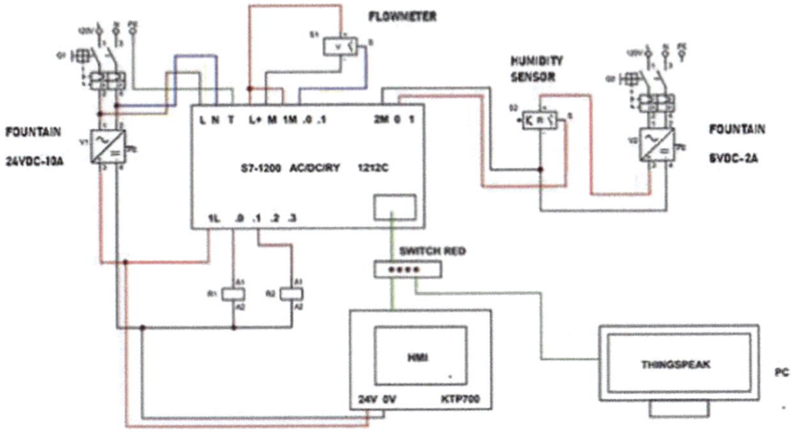

Fig. 7. Electrical diagram of the automated irrigation system.

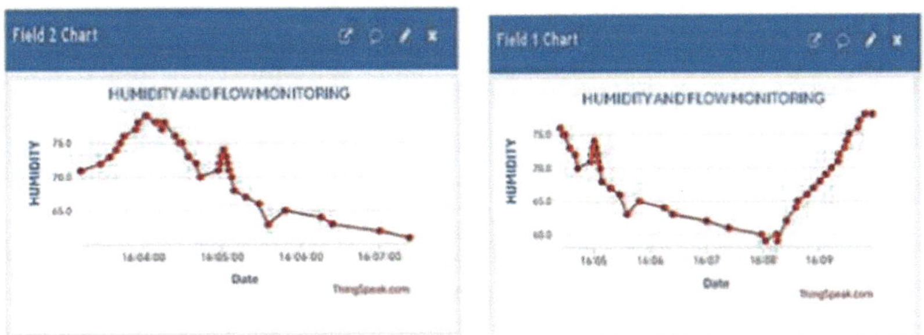

Fig. 8. Soil moisture and flow records in ThingSpeak during simulation.

3.2 Impact on Efficiency and Productivity

The results confirmed that irrigation automation significantly improved miniature rose cultivation:

- **Water efficiency:** a 51.43% reduction in water consumption was recorded.
- **Productivity:** increased by 14%, from 4.33 to 4.92 stems per dollar invested.
- **Crop health:** fungal disease incidence decreased due to more stable moisture.
- **Operating costs:** four workers previously required for manual irrigation were no longer needed.
- **Economic feasibility:** the system achieved a payback period of less than one year.
- **Scalability:** the system architecture allows expansion through additional sensor nodes and segmented irrigation.

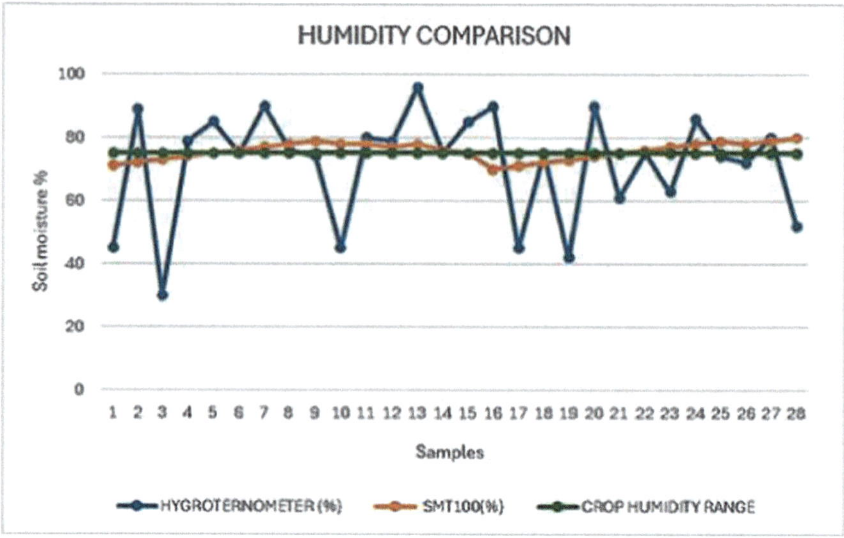

Fig. 9. Comparison of soil moisture levels before and after the automated system.

4 Conclusions

The study confirmed that manual irrigation in miniature rose cultivation generates significant inefficiencies, expressed in high waste levels, excessive water consumption, and strong dependence on labor for repetitive tasks. These findings highlight the need to adopt more sustainable solutions.

The implementation of an automatic irrigation system with IoT technology demonstrated that transforming water resource management is achievable. The new design reduced water consumption by 51.43% and increased productivity by 14%, from 4.33 to 4.92 stems per dollar invested. This not only improved resource efficiency but also reduced fungal disease incidence, strengthening crop quality.

The system proved technically and economically viable. Its integration with sensors, flowmeters, a centrifugal pump, a PLC, an HMI, and remote monitoring platforms such as ThingSpeak ensures reliable and scalable operation. With an initial investment of 6,624.33 USD and a payback period of less than one year, the proposal is an accessible alternative for producers seeking to optimize their processes.

This research supports a more efficient, competitive, and sustainable floriculture sector, enabling intuitive control interfaces and IoT platforms that facilitate producers' work and improve daily agricultural management.

5 Future Work

Future improvements could include predictive irrigation models based on machine learning (such as LSTM or ARIMA) to anticipate water requirements using historical humidity, temperature, and flow data.

Integrating weather APIs (OpenWeather, NOAA) would enable automatic adjustments based on rainfall or solar radiation forecasts.

Additionally, future implementations could migrate to distributed architectures with multiple networked PLCs, supporting scalability for large plantations and centralized monitoring on more robust IoT platforms.

References

1. Carro, J., González, P.: Application of the Craig Harris method to analyze productivity in agricultural systems. J. Agric. Stud. **15**(2), 45–58 (2021)
2. Chavarro, M.: Flower exports and international market competitiveness. Agribus. Rev. **18**(3), 200–212 (2022)
3. LatinAmericanPost: The role of floriculture in Ecuador's employment and economy. LatinAmericanPost J. (2024). https://latinamericanpost.com
4. Vidal, R.: Ecuador leads flower exports with fertigation innovations, (2023). Expoflores Report
5. Oksen, T.: Precision agriculture and robotics in smart farming. Agric. Technol. Rev. (2023)
6. Rosero Montalvo, C.: IoT-based monitoring in Brazilian crops: an approach to precision agriculture. Int. J. Smart Agric. **9**(1), 12–25 (2023)
7. Olmos, L.: Tempus air: IoT smart irrigation system for sustainable farming. J. IoT Agric. **6**(4), 67–75 (2023)
8. Horticultura: fungal diseases in floriculture and their relationship with soil humidity. Horticultura Technical Bull. (2020)
9. Sela, M.: Environmental factors influencing fungal disease propagation in greenhouse crops. Plant Pathol. J. **40**(5), 321–333 (2024)
10. Yong, D.: Soil moisture and plant physiology: impacts on water management. Agric. Sci. Rep. **29**(3), 115–128 (2004)
11. Valencia-Aragón, K., Zapata, M., Toapanta, C., Arias-Flores, H.: Optimizing agriculture with LoRaWAN and HCI: a smart approach to sustainable farming. In: Stephanidis, C., Antona, M., Ntoa, S., Salvendy, G. (eds.) HCI International 2025 Posters. CCIS, vol. 2527, Springer, Cham (2025). https://doi.org/10.1007/978-3-031-94165-8_42
12. Topón Visarrea, B.L., Muñoz del Salto, E., Pozo Espín, I., Ruano Cañar, L.: Smart agriculture: designing a cocoa pulping machine with IoT and Industry 4.0. In: Stephanidis, C., Antona, M., Ntoa, S., Salvendy, G. (eds.) HCI International 2025 Posters. CCIS, vol. 2527, Springer, Cham (2025). https://doi.org/10.1007/978-3-031-94165-8_18

Personalization of Educational Assessments: An Artificial Intelligence Perspective

Alexander Puentes-Castro(✉) and Diana Janeth Lancheros-Cuesta

Maestría en Inteligencia Artificial, Universidad de La Salle, Bogotá, Colombia
{apuentes19,dilancheros}@unisalle.edu.co

Abstract. Designing assessments that adapt to each student's level, progress, and individual needs remains a significant challenge across all educational levels, from K–12 to higher education. This article presents a descriptive analysis of research on the application of artificial intelligence (AI) and machine learning (ML) to the personalization of educational assessments following the PRISMA methodology. The study identifies three main categories: large language model (LLM) systems, machine learning techniques for education, and AI-driven adaptation of academic content. The results show a balanced distribution across these categories, with a predominance of studies published in high-impact journals (Q1 and Q2). The global research landscape is led by Asia, Europe, and the Americas. The most frequent machine learning techniques are natural language processing, neural networks, and deep learning. The review highlights critical challenges in multimodal data integration, algorithmic fairness, and the sustainability of implementation. A conceptual framework is proposed for ethical adaptive systems, emphasizing the importance of human oversight and transparency in personalized assessment tools.

Keywords: Personalized assessment · Artificial intelligence · Machine learning · Large language models · Educational technology · Ethics in AI · Adaptive learning

1 Introduction

The lack of personalization in educational assessment poses a critical challenge to students' academic development. Traditional assessment systems often rely on generic instruments that overlook individual characteristics, leading to results that are not fully representative, reduced motivation, and inefficiencies in learning processes [3,13]. In this context, the integration of artificial intelligence and machine learning emerges as a solution, enabling the dynamic adaptation of assessments to each student's academic, social, emotional, and contextual needs.

Adaptive assessment is a systematic and continuous process aimed at determining the degree to which learning objectives are achieved, and it requires consideration of students' diverse styles, paces, and contexts [7]. Likewise, personalized learning, facilitated by artificial intelligence, seeks to tailor the educational

A. Rocha et al. (Eds.): ICITS 2026, LNNS 1966, pp. 350–359, 2027.
https://doi.org/10.1007/978-3-032-27538-7_31

experience to individual needs, interests, and paces, providing more precise and meaningful assessments [1].

These adaptive assessment systems dynamically adjust the difficulty and type of questions based on students' prior responses, overcoming the limitations of traditional standardized tests [15]. Moreover, personalized feedback generated by advanced algorithms helps students understand their strengths and areas for improvement in an individualized manner, taking into account their context and learning style [14].

Learning analytics, through the analysis of large volumes of data, identifies patterns and trends that can inform educational decision-making and optimize learning environments [15].

Recent research underscores the transformative potential of artificial intelligence for personalized learning by automating assessment processes, providing instant feedback, and adapting educational resources to individual needs [6]. Adaptive recommendation systems, based on machine learning algorithms, enable the dynamic adjustment of learning pathways and content according to user behavior, preferences, and context, thereby enhancing the relevance and effectiveness of educational interventions [14]. Likewise, artificial intelligence facilitates the identification of behavioral and performance patterns, contributing to more objective assessment and the personalization of pedagogical strategies [10].

Next, Sect. 2 details the methods, the search strategy, and the inclusion and exclusion criteria; subsequently, Sect. 3 presents the results, the identified trends, and discusses the findings, their practical implications, and limitations; and, finally, Sect. 4 offers the conclusions and directions for future research.

2 Methods

This article addresses the following research questions: What artificial intelligence and machine learning techniques predominate in the personalization of educational assessments? How are individual and contextual variables integrated into current models? What technical and ethical challenges persist in the implementation of artificial intelligence in educational assessments?

2.1 Search Model and Strategy

The systematic literature review was conducted following the PRISMA model. Searches were performed in the Scopus, Web of Science, and IEEE Xplore databases, using terms such as "adaptive assessment," "personalized assessment," "artificial intelligence," "machine learning," and "LLM." Articles published between 2022 and 2024 in indexed journals (Q1–Q4) that addressed the application of artificial intelligence to the personalization of educational assessments were included. The procedure is detailed in Fig. 1.

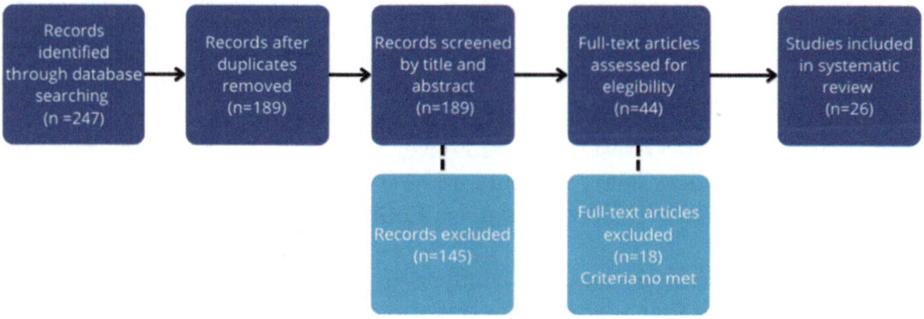

Fig. 1. PRISMA flow diagram for the identification, screening, and inclusion of studies.

2.2 Inclusion and Exclusion Criteria

The following inclusion and exclusion criteria were established for the selection of studies. Eligible publications included articles published between 2022 and 2024, as well as papers in indexed journals or peer-reviewed conference proceedings that addressed LLM systems, machine learning applied to education, or AI-driven adaptation of educational content. Mixed-methods research, both quantitative and qualitative, was also included, provided that it reported verifiable results. Articles in English or Spanish were accepted. Publications prior to 2022, studies not directly related to the application of artificial intelligence in education, research lacking a clear methodology or verifiable results, as well as grey literature and non-peer-reviewed publications, were excluded.

2.3 Data Processing and Analysis

An analysis matrix was developed that considered dimensions such as the artificial intelligence techniques employed, the contextual variables considered, the validation metrics, and ethical considerations. The analysis also included categorizing the studies, identifying trends, and synthesizing key findings.

2.4 Correlation Analysis

Based on the analysis matrix, variables from each reviewed study were identified, and correlation coefficients were calculated between these variables to explore potential relationships. This analysis was performed using NVivo statistical software, with a significance level of 5%.

3 Results

The descriptive analysis of the studies enabled identification of key contributions and emerging trends in applying machine learning to personalize educational assessments. It begins with a recent systematic review showing how artificial intelligence has transformed assessment methods in primary and secondary education, highlighting innovations such as natural language processing for essay scoring, the use of educational robots, and data mining systems to predict academic performance [3], while underscoring the importance of high-quality data and the need to balance automation with human judgment. On the other hand, various studies focused on K–12 education have examined the role of student competencies in AI-based learning, showing that teamwork and collaboration between students and technological tools are fundamental and that there are no significant differences by gender or school type [12].

Likewise, recent research has identified the predominance of logistic regression in evaluation systems, as well as the relevance of gender as a sensitive variable, revealing gaps in multivariable analyses and the need for more comprehensive evaluation frameworks [10]. Similarly, successful experiences are evident with adaptive educational platforms based on machine learning, which have demonstrated significant improvements in academic performance, student satisfaction, and diversification of learning strategies in higher education [14]. In addition, qualitative studies have highlighted the reduction of bias in assessment, the provision of timely feedback, and the adaptation of pedagogical strategies based on real-time data [11].

Regarding advanced techniques, research combines neural networks with fuzzy logic to enhance adaptation to learning styles, as well as the use of factor analysis to validate assessment instruments [2,15]. Additionally, the integration of computer vision and speech recognition systems has enabled real-time competency assessment, expanding the reach of educational personalization across contexts and needs. In the realm of LLM systems, the literature identifies four primary areas of application: adaptive learning, intelligent tutoring, intelligent assessment, and emerging products, reflecting the field's multidisciplinary nature and the need to expand research into new educational contexts [9]. It is essential to recognize that generative AI is viewed as a complementary tool rather than a replacement for teachers, underscoring the importance of critical thinking, empathy, and socioemotional competencies [5]. The review also shows improvements in learning efficiency and the quality of automated feedback, although challenges related to ethical and legal aspects persist [8]. Furthermore, robust instruments have been validated to measure technological self-efficacy among university students [7]. In the area of educational content adaptation, hybrid frameworks are proposed that balance human and automated decisions, emphasizing the importance of continuous monitoring [4]. There is also an emphasis on the need for deeper, bidirectional analyses of human–AI interactions, as well as the importance of human oversight throughout the AI lifecycle [1]. Finally, AI-based decision-support systems are reported to facilitate personalized and effective interventions, highlighting the mitigation of the impact of learning disabilities through study strategies [13], and demonstrating high accuracy and adaptability of advanced speech recognition technologies for personalized learning [6]. The reviewed literature presents a constantly evolving landscape in which the integration of advanced AI technologies and consideration of contextual factors are fostering personalized assessment. However, it should be noted that, despite these advances, relevant challenges remain related to equity, ethics, and the ability to adapt to diverse educational environments. The incorporation of artificial intelligence into educational assessment has shown considerable progress in recent years, as highlighted by recent reviews [15]. For this review, 26 articles were analyzed; the thematic distribution revealed that LLM systems accounted for 32% of the studies (8 articles), while machine learning applied to education and AI-driven adaptation of educational content represented 35% and 33%, respectively (9 articles in each category). Regarding academic quality, there was a marked concentration of publications in high-impact journals, with 45% of studies in Q1 journals, 38% in Q2, 7% in Q3, and 10% in conferences or other sources. Finally, the geographic distribution of research showed global coverage, led by Asia (35%), followed by Europe (27%), the Americas (23%), and other regions (15%) (Figs. 2, 3 and 4).

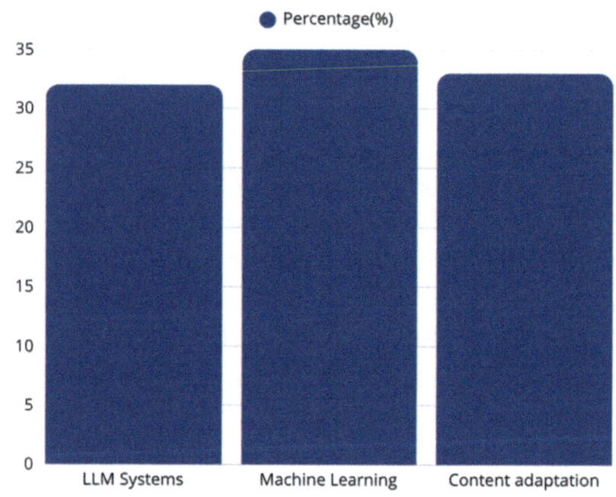

Fig. 2. Distribution of articles by analysis category.

3.1 Predominant Machine Learning Techniques

Analysis of the most commonly used machine learning techniques revealed that natural language processing was the most frequent (38%), followed by neural networks (31%), deep learning (23%), supervised learning (19%), and reinforcement learning (12%). The predominance of natural language processing and neural networks aligns with prior reviews that highlight the potential of these techniques for learning personalization [15]. Table 1 summarizes these frequencies, while Fig. 5 illustrates the diversity of approaches identified.

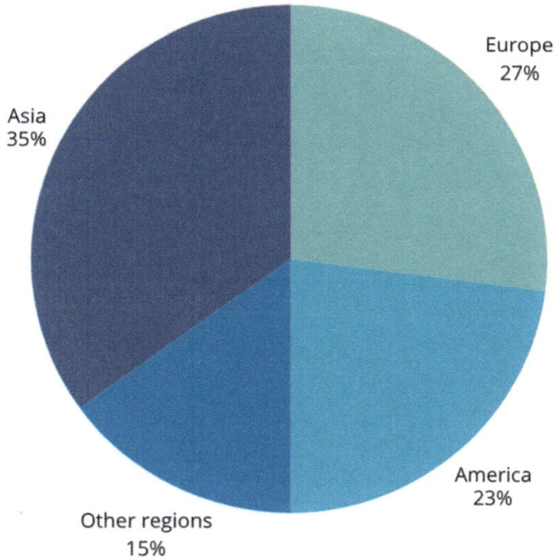

Fig. 3. Geographic distribution of publications.

3.2 Predominant Machine Learning Approaches

To clarify the different levels of classification, Table 1 organizes the approaches into learning paradigms (categories), core techniques, specialized subareas, and application fields.

Table 1. Classification of Machine Learning Approaches

Level	Approach	Frequency (%)
Learning paradigm (category)	Supervised learning	19
	Reinforcement learning	12
Core technique	Neural networks	31
Specialized subarea	Deep learning	23
Application field	Natural language processing (NLP)	38

Notes:

- *Learning paradigms* group algorithms by training strategy (e.g., supervised vs. reinforcement learning).
- *Core techniques* denote individual algorithm families (e.g., neural networks).
- *Specialized subareas* refer to focused domains within core techniques (e.g., deep learning).
- *Application fields* specify the domain of use (e.g., NLP for text analysis).

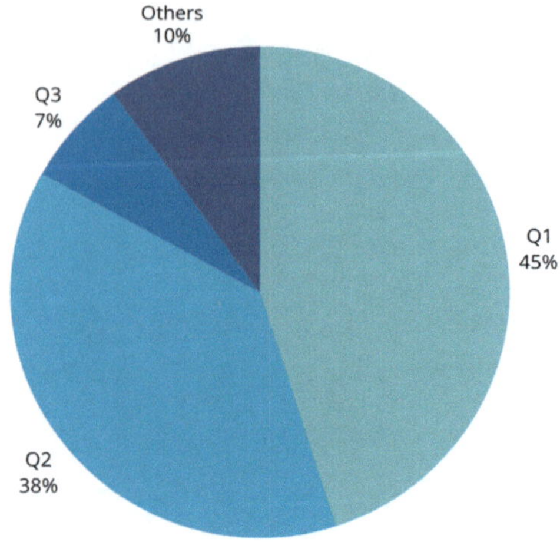

Fig. 4. Distribution of articles by journal type.

Beyond the techniques reported in the table, the review identified studies employing methods such as random forests; support vector machines (SVM); fuzzy logic; clustering; topic modeling; factor analysis; and hybrid models (e.g., neural networks combined with fuzzy logic or LSTM). This variety reflects AI's adaptability to different educational assessment needs [15]. Likewise, some articles use advanced statistical techniques and qualitative analyses, broadening the methodological spectrum of research in this field. The analysis also revealed that the reviewed studies employ a diverse range of inputs, including behavioral data (such as platform interactions and response times), physiological data (facial expressions and heart rate), images, audio, surveys, academic history, and demographic data. Model outputs span performance prediction and diagnosis, personalized recommendations, learning pathways, adaptive feedback, and psychometric evaluation.

3.3 Relationships Among Key Variables

To explore potential relationships among the main variables of analysis, co-occurrence was examined between AI techniques, educational level addressed, and contextual variables in the reviewed studies. The qualitative analysis of the data matrix revealed that studies employing advanced machine learning techniques—such as natural language processing and neural networks—are more prevalent in higher education contexts and, to a lesser extent, in secondary education [15]. It was also observed that the explicit integration of contextual variables (emotional, socioeconomic, or cultural) is more frequent in studies addressing assessment personalization and multimodal data analysis.

By contrast, studies focused on basic education and those prioritizing traditional assessment tend to use more basic data analysis techniques and show less consideration of non-academic variables. This qualitative analysis suggests a

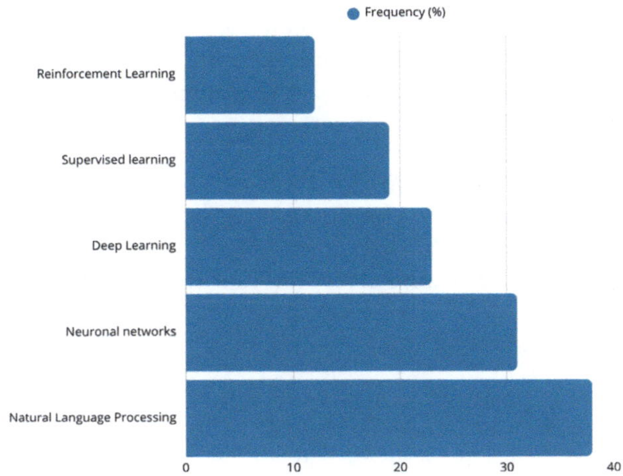

Fig. 5. Frequency of machine learning techniques in the reviewed studies.

clear tendency associating the adoption of innovative AI approaches with greater consideration of the diverse needs and contexts of students [15].

These findings reinforce the importance of integrating contextual variables and advanced machine learning techniques to achieve truly personalized educational assessment, especially in higher education and in contexts that require more complex adaptation to individual student characteristics [15].

The literature indicates that most studies incorporate academic data (92%), while the inclusion of emotional indicators (47%), socioeconomic context (23%), and cultural variables (11%) is significantly lower [10,15]. It is essential to note that these percentages are not mutually exclusive categories, as a single study may consider multiple types of variables simultaneously. The predominance of academic data suggests a prioritization of aspects related to school performance to the detriment of other relevant dimensions of student development. This limited integration of emotional, socioeconomic, and cultural variables may constrain systems' capacity to adapt to the diversity of students' needs and contexts, negatively affecting the potential of personalized assessments [10].

4 Conclusion

The personalization of educational assessment through artificial intelligence and machine learning emerges as a promising pathway to transform teaching and learning processes. The findings of this systematic review suggest that adaptive systems can enhance accuracy, fairness, and motivation in educational assessment, provided that contextual variables are integrated, human oversight is ensured, and robust ethical protocols are implemented. Future research should focus on the empirical validation of these systems, their adaptability to diverse educational contexts, and the development of more comprehensive and transparent evaluation frameworks.

The diversity of techniques, educational contexts, and measurement objectives evidenced in the literature highlights the importance of designing adaptive, ethical, and flexible solutions capable of responding to both individual and collective student needs across different settings.

Acknowledgment. The support of the Artificial Intelligence Master's Program at Universidad de La Salle is gratefully acknowledged.

Disclosure of Interests. The authors declare that they have no conflicts of interest.

References

1. Bressane, A., et al.: Understanding the role of study strategies and learning disabilities on student academic performance to enhance educational approaches: a proposal using artificial intelligence. Comput. Educ. Artif. Intell. **6**, 100196 (2024). https://doi.org/10.1016/j.caeai.2023.100196
2. Chan, C.K.Y., Tsi, L.H.Y.: Will generative AI replace teachers in higher education? A study of teacher and student perceptions. Stud. Educ. Eval. **83**, 101395 (2024). https://doi.org/10.1016/j.stueduc.2024.101395. https://www.sciencedirect.com/science/article/pii/S0191491X24000749
3. Martínez-Comesaña, M., Rigueira-Díaz, X., Larrañaga-Janeiro, A., Martínez-Torres, J., Ocarranza-Prado, I., Kreibel, D.: Impact of artificial intelligence on assessment methods in primary and secondary education: systematic literature review. Revista de Psicodidáctica (English ed.) **28**(2), 93–103 (2023). https://doi.org/10.1016/j.psicoe.2023.06.002. https://www.sciencedirect.com/science/article/pii/S2530380523000072
4. Memarian, B., Doleck, T.: Human-in-the-loop in artificial intelligence in education: a review and entity-relationship (ER) analysis. Comput. Hum. Behav. Artif. Hum. **2**(1), 100053 (2024). https://doi.org/10.1016/j.chbah.2024.100053. https://www.sciencedirect.com/science/article/pii/S2949882124000136
5. Morales-García, W.C., Sairitupa-Sanchez, L.Z., Morales-García, S.B., Morales-García, M.: Adaptation and psychometric properties of a brief version of the general self-efficacy scale for use with artificial intelligence (GSE-6AI) among university students. Front. Educ. **9**, 1293437 (2024). https://doi.org/10.3389/feduc.2024.1293437. https://www.frontiersin.org/journals/education/articles/10.3389/feduc.2024.1293437/full
6. Munir, H., Vogel, B., Jacobsson, A.: Artificial intelligence and machine learning approaches in digital education: a systematic revision. Information **13**(4), 203 (2022). https://doi.org/10.3390/info13040203. https://www.mdpi.com/2078-2489/13/4/203
7. Ninaus, M., Sailer, M.: Closing the loop – the human role in artificial intelligence for education. Front. Psychol. **13**, 956798 (2022). https://doi.org/10.3389/fpsyg.2022.956798. https://www.frontiersin.org/journals/psychology/articles/10.3389/fpsyg.2022.956798/full
8. Orosoo, M., et al.: Transforming English language learning: advanced speech recognition with MLP-LSTM for personalized education. Alexandria Eng. J. **111**, 21–32 (2025). https://doi.org/10.1016/j.aej.2024.10.065. https://www.sciencedirect.com/science/article/pii/S1110016824012195

9. Peláez-Sánchez, I.C., Velarde-Camaqui, D., Glasserman-Morales, L.D.: The impact of large language models on higher education: exploring the connection between AI and Education 4.0. Front. Educ. **9** (2024). https://doi.org/10.3389/feduc.2024.1392091. https://www.frontiersin.org/journals/education/articles/10.3389/feduc.2024.1392091/full
10. Pham, N., Pham Ngoc, H., Nguyen-Duc, A.: Fairness for machine learning software in education: a systematic mapping study. J. Syst. Softw. **219**, 112244 (2025). https://doi.org/10.1016/j.jss.2024.112244. https://www.sciencedirect.com/science/article/pii/S0164121224002887
11. Ravuri, A., Lourens, M., Aswini, S., Nijhawan, G., Zabibah, R.S., Chandrashekar, R.: Improving personalized education: a machine learning method for flexible learning environments. In: 2023 10th IEEE Uttar Pradesh Section International Conference on Electrical, Electronics and Computer Engineering (UPCON), pp. 1715–1720 (2023). https://doi.org/10.1109/UPCON59197.2023.10434888. https://ieeexplore.ieee.org/document/10434888
12. Sanusi, I.T., Olaleye, S.A., Agbo, F.J., Chiu, T.K.F.: The role of learners' competencies in artificial intelligence education. Comput. Educ. Artif. Intell. **3**, 100098 (2022). https://doi.org/10.1016/j.caeai.2022.100098. https://www.sciencedirect.com/science/article/pii/S2666920X22000534
13. Sarker, S., et al.: Personalized learning and assessment in digital education: a review of AI-driven approaches. Comput. Educ. **189**, 104592 (2024). https://doi.org/10.1016/j.compedu.2023.104592
14. Tang, X., Chen, Y.: Adaptive education platform based on machine learning: a new way to improve the quality of higher education. In: 2024 International Conference on Interactive Intelligent Systems and Techniques (IIST), pp. 310–316 (2024). https://doi.org/10.1109/IIST62526.2024.00054. https://ieeexplore.ieee.org/document/10734474
15. Wang, S., Wang, F., Zhu, Z., Wang, J., Tran, T., Du, Z.: Artificial intelligence in education: a systematic literature review. Expert Syst. Appl. **252**, 124167 (2024). https://doi.org/10.1016/j.eswa.2024.124167. https://www.sciencedirect.com/science/article/pii/S0957417424010339

Breaking Barriers: Technology as a Tool for the Educational Inclusion of Students with Visual Impairments

Víctor Manuel Zamudio García and Glendamira Serrano Franco(✉)

Metropolitan Polytechnic University of Hidalgo, Boulevard access to Tolcayuca, 1009, Tolcayuca, Hidalgo, Mexico
{vzamudio,gfranco}@upmh.edu.mx

Abstract. Ensuring truly inclusive higher education with equitable opportunities in Mexico and Latin America involves addressing aspects such as visual functional diversity as one of the main barriers. This work presents an alternative that reinforces the conditions for genuine educational inclusion by using technological tools with artificial intelligence, designed under the principles of the Universal Design for Learning (UDL) concept, with the purpose of eliminating obstacles and promoting the full participation and autonomy of students with visual disabilities in the university setting.

The research was based on a person-centered approach, implementing deep learning models (AlexNet and MobileNetV2) through transfer learning. The process was enriched with collaborative usability tests conducted with students, whose feedback was essential for the design. The results demonstrated that MobileNetV2 not only offers greater accuracy and efficiency, but its integration into an accessible application significantly improves the users' perception of inclusion and self-confidence.

In conclusion, artificial intelligence establishes itself as a strategic ally for building learning environments that adapt to the diverse needs of students, paving the way for real educational inclusion. It is recommended to expand the research by incorporating intersectional approaches and co-creation with the community of people with disabilities, ensuring that technological development is always at the service of equity and educational justice.

Keywords: Intelligent systems · cognitive technological assistance · lightweight neural network architectures · deep learning models

1 Introduction

Educational inclusion constitutes one of the main challenges for educational systems. In the case of Mexico, INEGI (2022) indicates that approximately 5% of the population has some type of disability, with visual disability being one of the most prevalent.

Although in recent years policies aimed at inclusion have been promoted, significant limitations persist regarding technological resources adapted to the needs of the higher

A. Rocha et al. (Eds.): ICITS 2026, LNNS 1966, pp. 360–368, 2027.
https://doi.org/10.1007/978-3-032-27538-7_32

education sector in Mexico and Latin America. According to the World Health Organization (2023), more than 2.2 billion people live with some degree of visual impairment, which directly impacts their educational and employment opportunities.

Here, Artificial Intelligence (AI) technology emerges as a key tool for the development of accessible applications. However, there is still a considerable gap between technological advances and their effective application in inclusive educational environments. Therefore, this work aims to generate an inclusive educational environment with an AI-based application to support students with visual disabilities, which will allow comparing the performance of deep learning models in identifying and describing objects. The central hypothesis is that the principles of Universal Design for Learning (UDL) combined with the implementation of artificial intelligence will help eliminate attitudinal and instructional barriers to favor the full participation and autonomous learning of students with visual disabilities in higher education environments, thereby promoting greater inclusion, favoring accessibility, and promoting student autonomy in the field of higher education.

2 Problem Statement

Inclusive education at the higher level faces the permanent challenge of strengthening its infrastructure and expanding research spaces. These should be oriented towards the co-design of technological and pedagogical solutions that effectively promote the participation and learning of the entire student diversity. Progress in this direction is fundamental to harness the diverse talent of educational communities and fully value their academic and cultural contributions, transcending the focus on institutional competitiveness to prioritize the construction of fairer educational systems.

Inclusion is consolidated when the university community recognizes and draws from the skills and perspectives of all the groups that comprise it. This principle is the basis for generating safe environments, sustained by diversity, and for fostering practices that favor an ecosystem of integral well-being. This approach, as Gupta and Zhao (2020) point out, enables the elimination of stigmas, drives innovation in educational processes, and strengthens constant and equitable development for all.

In this framework, the application of artificial intelligence techniques through accessible tools, such as a web application for mobile devices, emerges as a promising strategy to eliminate barriers and facilitate the full access and participation of people with visual disabilities in higher education. The ultimate goal is to improve their educational experience and guarantee their right to a quality education, a fundamental principle of true educational inclusion.

3 Research Question

How can an artificial intelligence algorithm, based on the principles of Universal Design for Learning (UDL), eliminate attitudinal and instructional barriers to favor the full participation and autonomous learning of students with visual disabilities in higher education environments, thereby promoting more diverse and equitable educational communities?

4 Objective

To determine how an artificial intelligence algorithm, based on the principles of Universal Design for Learning (UDL), can eliminate attitudinal and instructional barriers to favor the full participation and autonomous learning of students with visual disabilities in higher education environments, thereby promoting more diverse and equitable educational communities.

5 Hypothesis

The principles of Universal Design for Learning (UDL) combined with the implementation of an artificial intelligence algorithm will help eliminate attitudinal and instructional barriers to favor the full participation and autonomous learning of students with visual disabilities in higher education environments, thereby promoting more diverse and equitable educational communities.

6 State of the Art

Various recent studies have highlighted the potential of AI in the field of inclusive education. For example, convolutional neural networks (CNNs) have demonstrated effectiveness in recognizing visual patterns, which facilitates the design of applications for people with visual disabilities (Zhang et al., 2021; Kumar and Mapoli, 2020).

In Latin America, digital accessibility initiatives are still incipient. Projects such as the use of virtual assistants or screen readers based on AI have shown promising results, but their integration into universities is limited (Díaz-Parra et al., 2023). Research on MobileNet and AlexNet applied to computer vision problems shows that MobileNetV2 offers a balance between precision and low computational cost, ideal for mobile devices (Dong et al., 2019).

However, few studies have explored these applications in real educational contexts in Mexico, which opens an opportunity for scientific and practical contribution.

7 Methodology

This research was developed under an applied approach and with a quasi-experimental research design, oriented towards the creation and validation of a support tool for educational inclusion. The core of the study consisted of the co-design of an accessible prototype application, supported by artificial intelligence models, specifically convolutional neural networks, for the identification and description of elements in the academic environment. The final purpose of this tool is to eliminate barriers to access to information and promote the autonomy of students with visual functional diversity, facilitating their full participation in the educational environment.

To evaluate the impact on the inclusive learning experience, the methodology integrated a mixed approach. A pre-test and post-test design will be applied to a control group, in order to measure significant changes in the perception of self-efficacy, autonomy, and sense of belonging among the participants after using the tool. The sample selection was carried out using random probabilistic sampling to ensure representativeness and minimize bias, starting from a total population of 180 students. This methodological choice ensures that the results reflect the voices of a diverse community and are generalizable to inform broader inclusive practices.

8 Important Note on Training Data

While the initial technical development used the public dataset "cat_vs_dogs" to validate the model architecture and transfer learning techniques, it is crucial to highlight that the final crucial phase of prototype training and validation will be carried out with contextually relevant data. The model will be re-trained with a custom image set, captured in real educational environments (classrooms, laboratories, libraries), representing the most common objects, texts, and spaces that students need to access. This step is indispensable to ensure that the technology responds to real educational needs and genuinely aligns with the principles of Universal Design for Learning (UDL).

The technical development process (Flatten, Dense layers, ReLU activation functions, monitoring with TensorBoard) was guided by the principle of accessibility from the design (Design for Accessibility), prioritizing computational efficiency to ensure that the final application was usable on mid and low-range devices, thus promoting digital equity. The tool was implemented as a responsive web application, ensuring its cross-platform accessibility from any internet-connected device, which democratizes its potential use in diverse educational contexts.

9 Results

The tool validation was carried out in a simulated educational environment, replicating real pedagogical scenarios in the area of Information Technologies, with the primary objective of evaluating its usefulness for promoting equitable participation. Users evaluated the application positively, identifying certain intermittencies in object identification as an area for improvement. However, its effectiveness in recognizing the academic elements for which it was trained was corroborated, demonstrating its potential as a support tool for autonomy within the classroom. The findings reflect that, although the educational community perceives incipient efforts in matters of inclusion, such as improvements in physical accessibility and promotion of respect for diversity, these are considered insufficient and lacking systemic implementation. The scarce preparation of the teaching staff and the ambiguity of institutional policies generate uncertainty, which evidences the urgent need to strengthen an inclusive culture through continuous training, clear communication, and concrete and visible actions.

The surveyed individuals seem to notice some efforts by the institution, both in improving physical accessibility and in promoting respect for functional diversity. However, these advances are not perceived as sufficient or decisive. The preparation of the teaching staff and the clarity of institutional policies are also aspects that generate doubts or neutral perceptions. This suggests that, although there is a foundation to build upon, it is necessary to reinforce communication, training, and visible actions to achieve a more positive and solid perception of inclusion within the university environment.

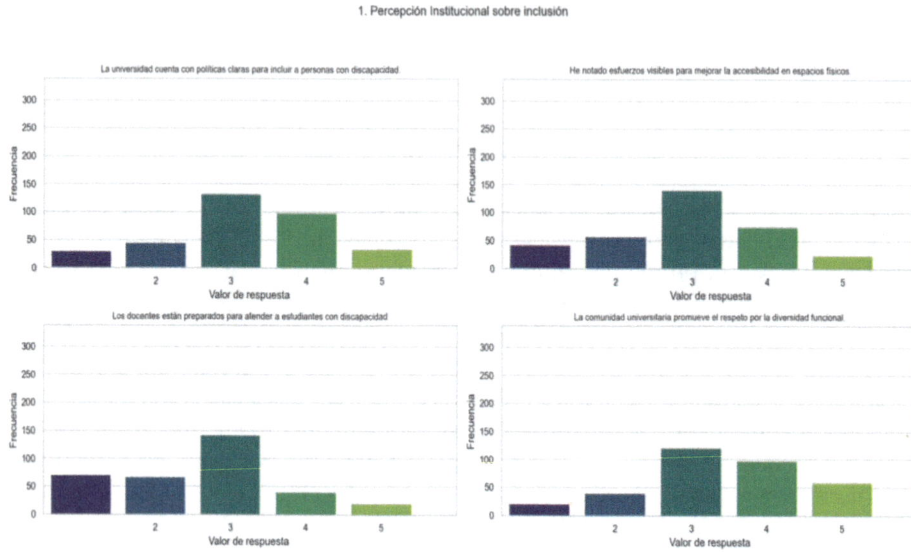

A notably receptive attitude and genuine interest on the part of the student community towards technological accessibility was also identified. There is a positive assessment of the potential of artificial intelligence as an enabler of fairer educational environments. However, this willingness contrasts with the evidently limited curricular integration of accessibility and universal design topics in the IT curriculum. This gap points to a strategic opportunity to mainstream inclusion in the training of future engineers, aligning the curriculum with the principles of education for all.

The most significant impact was observed in the transformation of student perceptions and attitudes. The experience served as a catalyst for empathy and social awareness, motivating many to get involved in projects with social impact and to reflect critically on their own privileges. This subjective change is as crucial as the technical one, as it lays the foundation for a more cohesive and supportive university community.

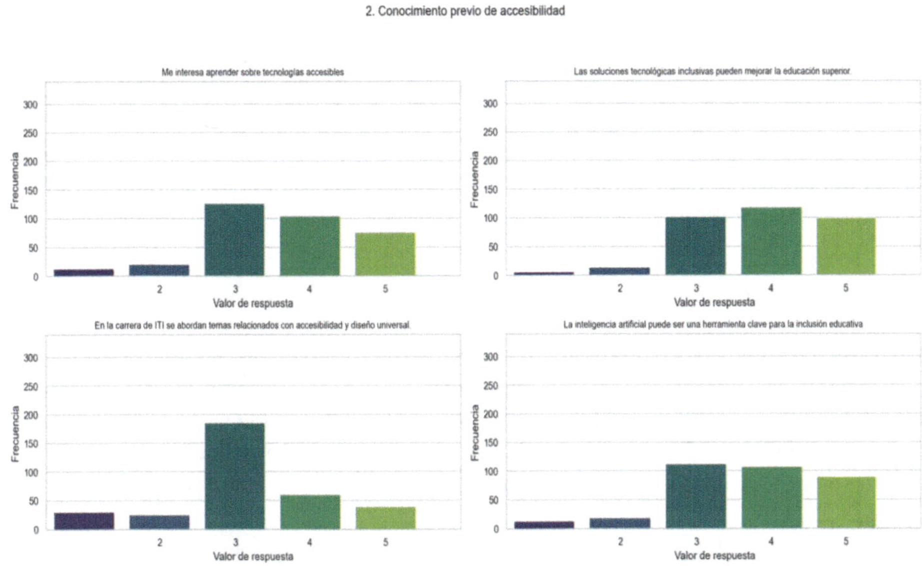

Finally, the notion that accessibility must be a transversal principle and not an isolated topic is consolidated. The comprehensive analysis of the results, both qualitative and quantitative, empirically validates the theoretical framework and highlights the profound impact that co- designed technologies with an inclusive focus can have on building a true culture of inclusion within higher education, where equity ceases to be an aspiration and becomes a tangible practice.

3. Cambio de percepción personal

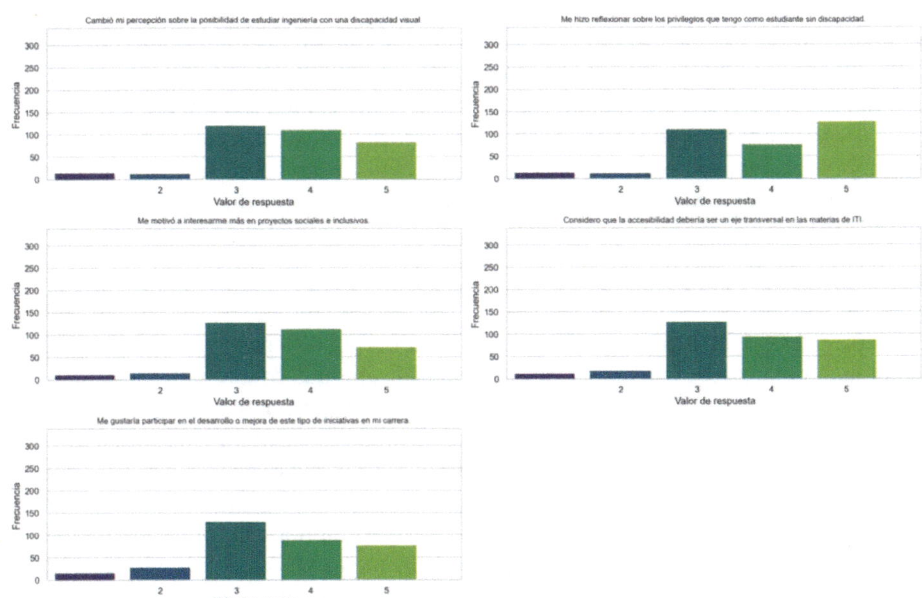

The analysis of the results obtained from the statistical models, as well as the performance metrics of the classification algorithms, allows for establishing significant links with the approaches of the Theoretical Framework and the findings identified in the State of the Art. This contrast not only reinforces the empirical validity of the research but also deepens the understanding of the impact that inclusive technologies can have on the perception, attitude, and disposition of the student community towards visual disability in the university setting.

10 Discussion

The results of this study go beyond the technical performance of an algorithm to position themselves in the realm of the ethics of educational inclusion. The suitability of MobileNetV2 for low-cost devices is not just a technical advantage, but a digital equity imperative: ensuring that support tools are accessible to all institutions, regardless of their resources, is fundamental to not perpetuating inequality gaps (Dong et al., 2019; Li et al., 2021).

However, the identified limitations – the dataset size, the need for extended validation – underscore a key principle of inclusion: co-creation with the user community. A technically sound model but developed without the active participation of students with visual functional diversity runs the risk of becoming a well-intentioned solution disconnected from the real barriers and lived experiences in the classroom. This adds to the indispensable ethical reflection on data privacy and algorithmic justice, aspects that must be pillars of technological development in inclusive educational environments.

In summary, this research goes beyond the validation of an AI model. Its true value lies in its capacity to catalyze an attitudinal and cultural change. The results demonstrate that when technology is designed with a vocation of service towards diversity, it becomes a powerful agent of educational transformation. Its impact is not measured only in accuracy, but in its potential to foster empathy, deconstruct stigmas, and consolidate university spaces where equity and dignity are lived experiences and not just aspirations.

This contrast evidences that the value of AI for inclusion is not inherent to its code, but emerges from its ethical design, its pedagogical contextualization, and its capacity to amplify the agency and voice of students. Its final success depends on it being at the service of building a deeply human and consciously inclusive educational culture.

11 Conclusions

Universal Design for Learning (UDL) proposes offering different ways of teaching and learning, so that every student, regardless of their abilities or circumstances, can participate and progress. Therefore, technologies like AI, leveraging their use in various devices, function as integrating elements that revolutionize and motivate students, containing educational materials adapted culturally and linguistically, since it is not just about including more students in school, but about ensuring that what they learn is meaningful, relevant, and useful for their lives. The combination of Universal Design for Learning and technology like AI can foster inclusive physical spaces with a participatory and replicable approach, seeking to achieve the permanence and well-being of diverse students, fostering a culture of inclusion, and generating dissemination products that can be shared with other educational institutions and society in general.

Artificial Intelligence, implemented with ethical and pedagogical rigor, is not an end in itself, but a powerful means to make effective the human right to a quality higher education for all, without exclusions. This path demands a constant alliance between technology, pedagogy, and advocacy for inclusion.

AI is a strategic lever for educational justice when subordinated to the principles of Universal Design for Learning (UDL). The prototype application, supported by MobileNetV2, proved to be a viable tool for reducing barriers to access to information and strengthening the autonomy of students with visual disabilities, critical factors for their full participation in higher education.

To transcend the prototype state and achieve systemic impact, it is recommended:

- Scale the research under participatory design methodologies, protagónically incorporating students with visual functional diversity in all phases of development.
- Integrate the solution into an ecosystem of accessibility that combines computer vision with text readers, audio description, and voice assistants, avoiding isolated solutions. Prioritize teacher training for the critical appropriation of these technologies, ensuring that their implementation in the classroom is pedagogically sound and student-centered.

References

Díaz-Parra, O., et al.: Smart white cane for visually impaired people for hearing recreation in tourist environment. Int. J. Comb. Optim. Probl. Informatics. **14**(2), 3–17 (2023). https://doi.org/10.61467/2007.1558.2023.v14i2.348

Dong, K., Ruan, Y., Zhou, C., Li, Y.: MobileNetV2 model for image classification. In: International Conference of Information Technology and Computer Application ITCA. (2019) https://conferences.computer.org/ictapub/pdfs/ITCA2020-6EIiKprXTS23UiQ2usLpR0/114100a476/114100a476.pdf

Gupta, A., Zhao, Y.: Advancing accesibility: the transformative role of technology in empowering individuals with visual impairment. J. Artif. Intell. Mach. Learn. Data Sci. 2020. **1**(1), 431–430 (2020). https://doi.org/10.51219/JAIMLD/yijia/119

Hernández Sampieri, R., Fernández Collado, C., Baptista Lucio, P.: Metodología de la investigación, 6th edn. McGraw-Hill (2014)

INEGI. Estadísticas sobre discapacidad en México. Instituto Nacional de Estadística y Geografía (2022)

Kumar, K., Mapoli, P.V.: An intelligent assistant for the visually impaired & blind people using machine learning. Int. J. Imag. Robot. 20(3) (2020). ISSN 2231–525X. https://www.researchgate.net/publication/342561852

Li, F., Zhang, Y., Cosman, P.: MMMNet: an end-to-end multi-task deep convolution neural network with multi-scale and multi-hierarchy fusion for blind image quality assessment. Trans. Circuits Syst. Video Technol. (2021). https://doi.org/10.1109/TCSVT.2021.3055197

Organización Mundial de la Salud (OMS): World Report on Vision. World Health Organization (2023)

UNICEF. AI and children: a parent's guide to AI. United Nations Children's Fund (UNICEF) (2021). Website: https://www.unicef.org/innocenti/media/1371/file/UNICEF-Global-Insight-AI-guide-for-parents-2021_ES.pdf

Urbán, D., García, J., E inglés, C.: Social anxiety risk profiles for interpersonal difficulties in a sample of Spanish adolescents. J. Psychodidact. **29**(1), 9–18 (2024). https://doi.org/10.1016/j.psicoe.2023.11.001

Yanes, B., Coca, Y.: Artificial intelligence in preschool. In: 4th International Scientific Conference, Havana, Cuba (2021) Website: https://repositorio.uci.cu/bitstream/123456789/9646/1/UCIENCIA_2021_paper_209.pdf

Yuquilema, M., Arízaga, F., Aguirre, M.Y., García, A.: Impact of artificial intelligence on education: challenges and opportunities. Sci. J. World Res. Knowl. **8**(2), 24–34 (2024). https://doi.org/10.26820/recimundo/8.(2).abril.2024.24-34

Semantic Resizing of Neutral Emotions in Multi-label Classification Systems with SBERT and Cascade Models

Christian Salamea-Palacios(✉), Juan Vasquez-Avila, Javier Viñanzaca-Figueroa, and Jairo Sacoto-Cabrera

Universidad Politécnica Salesiana, Calle Vieja 12-30, Cuenca, Ecuador
{csalamea,esacoto}@ups.edu.ec, {jvasqueza9, fvinanzaca}@est.ups.edu.ec

Abstract. A hierarchical emotion classification system for text is presented, based on pre-trained natural language models, with the goal of improving the interpretation of ambiguous sentences, especially those initially labeled as "neutral." In the first phase, a synthetic Spanish dataset inspired by the GoEmotions corpus was generated, with sentences labeled across 27 micro-emotions. This corpus was used to train DeBERTa models specialized by macro-emotional class. Subsequently, the classes were semantically reorganized into five balanced macro categories, enabling the training of a BETO model for primary classification. During inference, the system first predicts the macro-emotion using BETO; if the result is ambiguous or requires refinement, the corresponding DeBERTa model is activated to retrieve the specific micro-emotion. The approach was evaluated using an independent set of unseen sentences generated with GPT-3.5-Turbo, employing accuracy, F1-score, and confusion matrices as evaluation metrics for both stages of the system.

Keywords: motion Classification · Hierarchical Models · BERT · DeBERTa · GoEmotions · GPT-3.5 · Sentiment Analysis · Natural Language Processing (NLP)

1 Introduction

Automatic emotion analysis in text is one of the principal challenges in Natural Language Processing (NLP), due to semantic complexity, expressive ambiguity, and the necessity for contextual precision [1]. This challenge is particularly relevant to applications such as digital emotional support, mental health monitoring, and human computer interaction. Previous studies indicate that structuring the emotion detection task into well-defined stages enhances the system's ability to capture subtle emotional cues, especially in the case of the "neutral" category, which is often misclassified [2], In addition, previous work on a sentiment analysis system for chatbot responses using fuzzy logic shows what can be achieved in this area [3].

A. Rocha et al. (Eds.): ICITS 2026, LNNS 1966, pp. 369–380, 2027.
https://doi.org/10.1007/978-3-032-27538-7_33

In this context, [4] emphasizes the significance of integrating psychological theories with NLP techniques that go beyond basic text classification. Structured approaches such as role labeling and stimulus detection are considered beneficial, and it is equally important to preserve emotional sensitivity from both ethical and social perspectives.

This paper presents a hierarchical emotion classification system based on BERT models. At the first level, the system employs the BETO model to classify input into five macro-emotions. At the second level, specialized DeBERTa models are used to refine the classification into specific micro-emotions. The training datasets were synthetically generated using GPT-3.5-Turbo, inspired by the GoEmotions corpus [5]. Initially, 27 micro-emotions were used to train the DeBERTa models; these were subsequently reorganized into five balanced macro categories to train the BETO model. During inference, the system first predicts a macro-emotion using BETO. In cases of ambiguity, particularly those labeled as "neutral" the corresponding DeBERTa model is activated to refine the prediction.

The hierarchical structure draws inspiration from models such as CMC-HF [6], where feedback mechanisms increase robustness and reduce dependency on preprocessing techniques that may remove emotional nuances [7]. Previous work by Akhtar et al. [8] also demonstrates that affective tasks benefit from multitask learning, which aligns with the macro–micro configuration of the proposed system. The NRC Affect Intensity Lexicon (AIL) [9] supports the conceptualization of emotions as continuous rather than discrete categories.

To train the models, GPT-3.5-Turbo was employed along with data augmentation and balancing strategies based on [10], aligned with recent research on empathetic text generation. Finally, the system adopts a modular architecture inspired by the SemEval-2024 task structure [11], where dividing the process into stages, classification, refinement, and evaluation states, enhances overall system performance. The proposed approach seeks to provide a precise, interpretable, and flexible solution for emotion classification in Spanish-language texts.

2 Description of the Technique

2.1 Architecture of the System

Figure 1 illustrates the architecture of the proposed emotion classification system. The process begins with the input of a text sentence in Spanish, which is then preprocessed and tokenized to prepare it for model inference. The preprocessed text first passes through a primary classifier based on the BETO model, which predicts the overarching or macro-emotion category. If the primary prediction is ambiguous or requires further refinement, the corresponding specialized DeBERTa model is selected to perform a secondary classification, producing a more granular micro-emotion prediction. The system ultimately outputs the final sentiment of the input text, combining both macro- and micro-emotion information to provide a detailed understanding of the expressed emotion. This hierarchical approach allows for both efficiency and precision, ensuring that subtle emotional nuances are captured while maintaining robust overall classification performance.

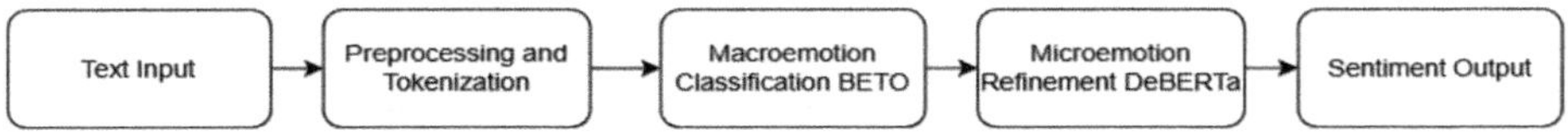

Fig. 1. Block diagram of the classification of emotions system

2.2 Generation and Translation of the Balanced Dataset

The initial process consisted of constructing a synthetic and balanced dataset derived from the GoEmotions corpus, with the objective of equitably representing the 27 micro-emotions. To achieve this, the original English sentences were grouped according to their assigned emotional labels, generating a multivalued dictionary that mapped each emotion to a corresponding list of textual instances. The following pseudocode summarizes this stage (Table 1).

Table 1. Grouping of Sentences by Emotion

Algorithm 1

```
for each row r in df_train do
    for each emotion e in r.emotion_names do
        emotion_to_texts[e] ← emotion_to_texts[e] ∪
        {r.text}
    end for
end for
```

Subsequently, random sampling was applied to select up to 100 sentences per emotion, ensuring a homogeneous distribution across classes (Table 2).

Table 2. Balanced Selection by Emotional Class

Algorithm 2

```
for each emotion e in emotion_to_texts do
    S ← random selection of up to 100 sentences from emotion_to_texts[e]
    Add S to the final balanced list
end for
```

The resulting corpus was stored as a new DataFrame, containing the original English sentences along with their corresponding micro-emotional labels. For use in models trained in Spanish, automatic translation was carried out using the multilingual model*facebook/m2m100_418M*, which allows direct translation between language pairs without requiring a pivot language (Table 3).

It is important to emphasize that M2M100 performs inference at the full-sequence level, preserving both the semantic and affective context of the message. Unlike word-by-word translators, this model maintains the overall communicative intent, which is crucial

Table 3. Translation of Sentences Using the M2M100 Model

Algorithm 3

```
for cada lote B de tamaño fijo en balanced_texts do
        Codificar B con el tokenizador M2M100
        Generar traducción hacia idioma objetivo .es"
        Decodificar secuencias y almacenar en text_es
end for
```

in emotion analysis tasks where the affective load depends on the integral meaning of the sentence.

Finally, the balanced and translated dataset was consolidated, generating a CSV file with approximately 1,000 examples, evenly distributed across classes. This corpus was used as the basis for training the hierarchical classifiers described in the following sections.

2.3 Semantic Grouping of Micro-emotions into Macro Categories

To hierarchically structure the emotional space, a semantic reorganization process was implemented, grouping the 27 specific emotions (micro-emotions) into five general categories (macro-emotions): very happy, happy, neutral, sad, and very sad.

This grouping was directly inspired by the organization proposed by the Google Research team in the original GoEmotions dataset design, where positive, negative, and ambiguous patterns are identified through the semantic analysis of each emotion. Figure 2 illustrates this reference structure.

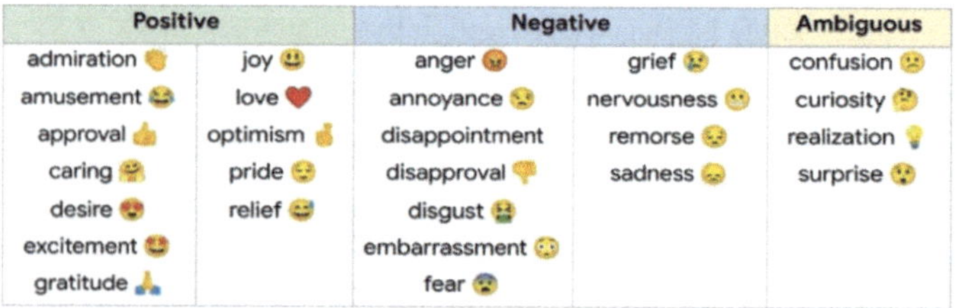

Positive		Negative		Ambiguous
admiration	joy	anger	grief	confusion
amusement	love	annoyance	nervousness	curiosity
approval	optimism	disappointment	remorse	realization
caring	pride	disapproval	sadness	surprise
desire	relief	disgust		
excitement		embarrassment		
gratitude		fear		

Fig. 2. Semantic classification of emotions in GoEmotions. Source: Google AI Blog

Based on this foundation, a mapping dictionary from micro_emotion → macro_emotion was defined using emotional affinities and balanced class distribution. The process was automated by iterating over the synthetic dataset, labeling each sentence with its corresponding macro category (Table 4).

Table 4. Grouping used for macro-emotional resizing

Macroemoción	Example
Very Happy	joy, love, excitement, gratitude, pride, approval
Happy	admiration, amusement, optimism, relief, desire, caring
Neutral	confusion, curiosity, realization, surprise, neutral
Sad	disappointment, disapproval, nervousness, remorse, sadness
Very Sad	anger, annoyance, disgust, embarrassment, fear, grief

2.4 Generation and Evaluation of Synthetic Data

To expand the semantic coverage of the corpus and strengthen the hierarchical classification stages, synthetic data were generated using the GPT-3.5-Turbo language model. Starting from a balanced base set inspired by the GoEmotions corpus, an initial synthetic dataset was constructed containing emotional sentences in Spanish that represented the 27 micro-emotions.

Subsequently, a second, fully independent corpus was generated, also using GPT-3.5-Turbo, composed of examples distinct from those in the training set. This evaluation dataset consisted of previously unseen sentences, one for each micro-emotion. Each sentence was processed hierarchically: first, the corresponding macro-emotion was inferred using the BETO model; then, based on the predicted category, the specialized DeBERTa model was applied to refine the classification toward the original micro-emotion.

This dual-corpus scheme, one for training and the other for evaluation, allowed for an empirical assessment of the hierarchical architecture's generalization capability on novel emotional texts, ensuring a clear separation between training and testing phases.

2.5 Hierarchical Evaluation Using Novel Synthetic Data

To validate the system's performance on unseen data, the synthetic evaluation corpus generated with GPT-3.5-Turbo was employed. This set consisted of previously unseen sentences, evenly distributed across the 27 micro-emotions, and completely independent from the corpus used for model training. Its purpose was to assess the hierarchical system's generalization capability under controlled yet realistic conditions.

This hierarchical procedure allowed for evaluating the system's performance both at the first stage (macro-classification) and at the second stage (refinement toward the micro-emotion), validating its accuracy and consistency across the complete processing pipeline.

Figure 3 illustrates the proposed hierarchical architecture for automatic emotion classification in text. The system is structured into three main stages: training, inference, and evaluation.

In the training stage, the process begins with the co-generation of a synthetic dataset using GPT-3.5, designed to cover 27 micro-emotions. These micro-emotions are semantically grouped into five macro-emotions to ensure a balanced dataset for model training.

From this data, the BETO model is trained for macro-emotion classification, along with five specialized DeBERTa models each dedicated to refining predictions toward specific micro-emotions.

During the inference stage, the system receives an input sentence, which is tokenized and preprocessed. The primary classification is first performed by BETO to determine the corresponding macro-emotion. In cases where the prediction is ambiguous particularly within the neutral category the secondary classification stage is activated, invoking the appropriate DeBERTa model to generate a refined micro-emotion prediction, thus enhancing accuracy.

The evaluation stage involves generating an independent synthetic corpus with GPT-3.5-Turbo, followed by a hierarchical assessment of model performance. Quantitative metrics such as accuracy, precision, and F1-score are computed at both macro- and micro-emotion levels to ensure consistency and validate the system's overall effectiveness.

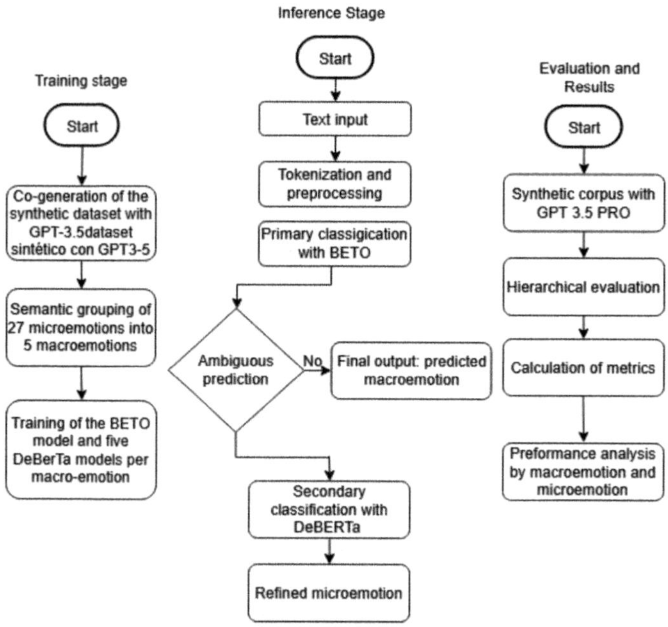

Fig. 3. Flowchart of the hierarchical emotion classification process

3 Results

The BETO model was evaluated on the task of classifying five macro-emotions: very happy, happy, neutral, sad, and very sad. Figure 4 shows the confusion matrices for the five macro-emotion classes and for the micro-emotions associated with the Very Sad class. For the macro-emotions, the best-performing classes were Neutral and Very

Sad, while Sad exhibited greater confusion with other categories. Regarding the micro-emotions, Anger, Disgust, Embarrassment, and Grief were classified with high precision, whereas Annoyance and Fear showed confusion with other classes.

Table 5. Classification Report Table for Microemotions (Very Sad).

Class	Precision	Recall	F1-score
Anger	0.94	1.00	0.97
Annoyance	1.00	0.26	0.41
Disgust	0.57	1.00	0.73
Embarrassment	1.00	0.98	0.99
Fear	0.97	0.73	0.84
Grief	0.81	1.00	0.89
Accuracy		0.82	
Macro avg	0.88	0.83	0.81
Weighted avg	0.88	0.82	0.80

Table 6. Classification Report Table for Microemotions (Sad).

Class	Precision	Recall	F1-score
Disappointment	1.00	0.92	0.96
Disapproval	1.00	1.00	1.00
Nervousness	1.00	0.82	0.90
Remorse	1.00	1.00	1.00
Sadness	0.72	1.00	0.84
Accuracy		0.95	
Macro avg	0.94	0.95	0.94
Weighted avg	0.96	0.95	0.95

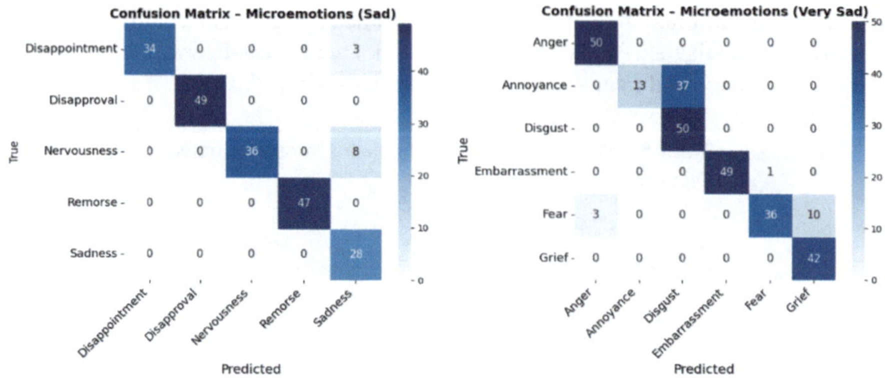

Fig. 4. Confusion matrix of the BETO model for macro-emotions and of the DeBERTa model for micro-emotions (Sad and Very sad).

Tables 5 and 6 summarize the performance of the specialized DeBERTa models for the Very Sad and Sad macro-emotions, respectively. The Very Sad category achieved strong precision in emotions such as Anger, Embarrassment, and Grief, all exceeding 0.85 in F1-score, while Annoyance showed lower recall due to overlaps with semantically similar emotions. In contrast, the Sad category displayed more balanced and consistent results, with Disapproval and Remorse achieving perfect F1-scores (1.00), confirming the reliability of the hierarchical approach in differentiating negative emotional nuances (Fig. 5).

Table 7. Classification Report Table for Microemotions (Happy).

Class	Precision	Recall	F1-score
Admiration	1.00	0.94	0.97
Amusement	1.00	0.94	0.97
Caring	1.00	1.00	1.00
Desire	1.00	1.00	1.00
Optimism	0.94	1.00	0.97
Relief	0.94	1.00	0.97
Accuracy		0.98	
Macro avg	0.98	0.98	0.98
Weighted avg	0.98	0.98	0.98

Table 8. Classification Report Table for Microemotions (Very Happy).

Class	Precision	Recall	F1-score
Approval	1.00	0.90	0.95
Excitement	1.00	0.95	0.97
Gratitude	1.00	0.64	0.78
Joy	0.98	1.00	0.99
Love	0.71	0.98	0.82
Pride	0.89	1.00	0.94
Accuracy		0.91	
Macro avg	0.93	0.91	0.91
Weighted avg	0.93	0.91	0.91

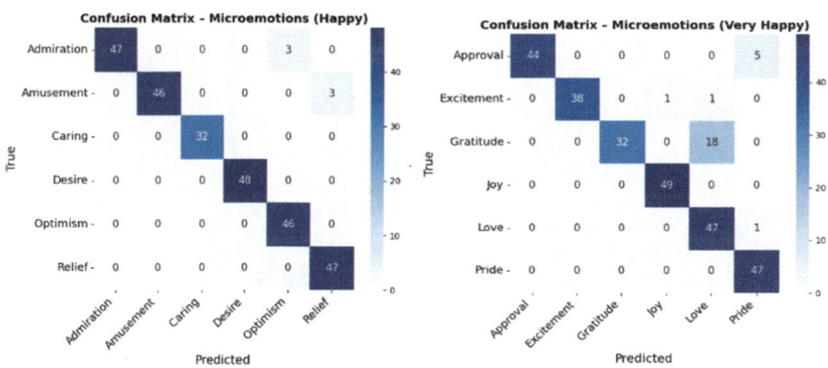

Fig. 5. Confusion matrix of the BETO model for macro-emotions and of the DeBERTa model for micro-emotions (Happy and Very Happy).

Tables 7 and 8 present the results obtained for the Happy and Very Happy macro-emotions. The Happy group reached an outstanding overall accuracy of 0.98, with all classes such as Admiration, Caring, and Desire showing balanced precision and recall close to 1.00. Similarly, the Very Happy category achieved a macro-average F1-score of 0.91, demonstrating the system's ability to generalize across diverse positive emotions. However, minor confusion was observed between Gratitude and Love, whose semantic proximity slightly affected their recall (Fig. 6).

Table 9. Classification Report Table for Microemotions (Neutral).

Class	Precision	Recall	F1-score
Confusion	0.96	1.00	0.98
Curiosity	1.00	1.00	1.00
Neutral	0.98	0.96	0.97
Realization	1.00	0.97	0.98
Surprise	1.00	1.00	1.00
Confusion	0.96	1.00	0.98
Accuracy		0.99	
Macro avg	0.99	0.98	0.99
Weighted avg	0.99	0.99	0.99

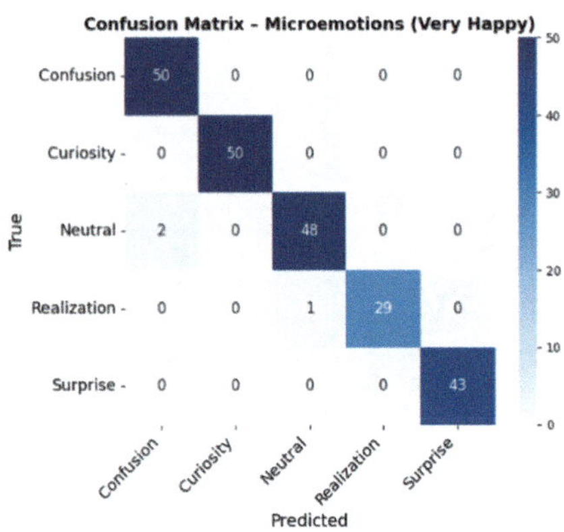

Fig. 6. Confusion matrix of the BETO model for macro-emotions and of the DeBERTa model for micro-emotions (Neutral).

Table 9 reports the classification metrics for Neutral macro-emotion, which achieved the highest overall accuracy (0.99) among all categories. Emotions such as Curiosity, Surprise, and Realization obtained perfect or near-perfect scores across all metrics, reflecting the model's capacity to preserve semantic neutrality while maintaining emotional sensitivity. The high precision and F1-scores confirm that the hierarchical combination of BETO and DeBERTa effectively mitigates the ambiguity often associated with neutral or context-dependent expressions, establishing this category as one of the most reliable within the proposed framework.

4 Conclusions

The proposed hierarchical system, composed of a BETO model for macro-emotion classification and specialized DeBERTa classifiers for micro-emotions, demonstrated strong and consistent performance across all evaluation stages. The two-level architecture enhanced discrimination among closely related emotional categories, as reflected in the overall metrics and the reduction of classification errors.

At the macro-emotion stage, the BETO model achieved an overall accuracy of 91% and a macro F1-score of 0.91. Outstanding performance was observed in the Neutral and Happy classes, whereas the Sad class showed some confusion with Very Sad, reducing its recall to 0.82. Nevertheless, the weighted average of the metrics remained above 90%, validating the model's effectiveness at this stage.

At the micro-emotion stage, the specialized DeBERTa models achieved remarkable metrics, with emotions such as Joy, Disapproval, Remorse, Caring, Desire, Confusion, Curiosity, Amusement, and Excitement reaching precision and F1-score values above 95%. However, challenges were identified for emotions such as Love (F1-score of 0.82) and Gratitude (F1-score of 0.78), which tend to be confused with semantically similar classes. This finding suggests the need to explore data augmentation or class-balancing techniques to improve performance.

The Neutral macro-emotion was one of the most consistent, achieving a macro precision of 0.988 and a macro F1-score of 0.987. This confirms that the proposed hierarchical architecture can effectively differentiate ambiguous emotions, a common challenge in affective analysis systems.

Overall, the BETO-DeBERTa pipeline demonstrated substantial improvement compared to direct classification schemes, achieving an average performance above 90% across all macro-emotions. These results also validate the strategy of generating synthetic data using GPT-3.5-Turbo and the semantic mapping of emotions, which was key for ensuring class coverage and balance.

References

1. Poria, S., Majumder, N., Mihalcea, R., Hovy, E.: Emotion recognition in conversation: research challenges, datasets, and recent advances. arXiv preprint arXiv:1905.02947. (2019) https://arxiv.org/abs/1905.02947
2. Zhang, S., Zhang, H., Zhang, J., Zhang, X., Zhuang, Y., Wu, J., et al.: arXiv preprint arXiv:2404.16905. (2024) https://arxiv.org/abs/2404.16905
3. Arcos-Salamea, E., Salamea-Palacios, C., Martel-Sócala, W.: Development of a sentiment analysis system for chatbot responses enhanced with fuzzy logic. In: Rocha, A., Ferrás, C., Calvo, H. (eds.) Information Technology and Systems: ICITS 2025, vol. 1448, pp. 71–80. Springer, Cham (2025)
4. Štajner, S., Klinger, R.: Emotion analysis from texts. In: Zanzotto, F.M., Pradhan, S. (eds.) Proceedings of the 17th Conference of the European Chapter of the Association for Computational Linguistics: Tutorial Abstracts, pp. 7–12. Association for Computational Linguistics, Dubrovnik, Croatia (2023) https://aclanthology.org/2023.eacl-tutorials.2/
5. Demszky, D., Movshovitz-Attias, D., Ko, J., Cowen, A., Nemade, G., Ravi, S.: GoEmotions: a dataset of fine-grained emotions. In: Jurafsky, D., Chai, J., Schluter, N., Tetreault, J. (eds.) Proceedings of the 58th Annual Meeting of the Association for Computational Linguistics,

pp. 4040–4054. Association for Computational Linguistics, Online (2020) https://aclanthology.org/2020.acl-main.372/
6. Liu, X., Xu, Z., Huang, K.: Multimodal emotion recognition based on cascaded multichannel and hierarchical fusion. Comput. Intell. Neurosci. **2023**(1), 9645611 (2023) https://onlinelibrary.wiley.com/doi/abs/10.1155/2023/9645611
7. Rezapour, M.: Emotion detection with transformers: a comparative study. arXiv preprint arXiv:2403.15454. (2024) https://arxiv.org/abs/2403.15454
8. Akhtar, M.S., Chauhan, D.S., Ghosal, D., Poria, S., Ekbal, A., Bhattacharyya, P.: Multi-task learning for multi-modal emotion recognition and sentiment analysis. arXiv preprint arXiv:1905.05812. (2019) https://arxiv.org/abs/1905.05812
9. Mohammad, S.: Word affect intensities (2017)
10. Schick, T., Schütze, H.: Generating datasets with pretrained language models. In: Moens, M.F., Huang, X., Specia, L., Yih, S.W.T. (eds.) Proceedings of the 2021 Conference on Empirical Methods in Natural Language Processing, pp. 6943–6951. Association for Computational Linguistics, Online and Punta Cana, Dominican Republic (2021). https://aclanthology.org/2021.emnlp-main.555/
11. Wang, S., Fang, R., He, Z., Song, S., Li, Y.: Emotional support with LLM-based empathetic dialogue generation (2025)

Agentic AI for Homicide Prediction: A Modular Framework for Geospatial Risk Analysis

Luis Loo[2], Kenny Joseph Cooper[1](✉), José García[1], Selvin Maldonado[1], Aldenis E. Miranda[1], and José Irias[1]

[1] Maestría en Ciberseguridad, Universidad Tecnológica de Honduras, San Pedro Sula, Honduras
{kenny.cooper,jsantos1,somaldonado1,aemiranda1,jdirias}@uth.hn
[2] Escuela de Postgrado, Universidad Tecnológica de Honduras, San Pedro Sula, Honduras
luis.loo@uth.hn

Abstract. This study proposes a predictive homicide analysis approach in urban environments using Agentic AI. A system composed of intelligent agents and a language model (LLM) to interpret natural language commands, in order to generate geospatial alerts and risk simulations, is developed. The agent combines geospatial analysis, supervised classification models, and language-based reasoning to provide an adaptive, continuous diagnosis of violence patterns. The architecture was implemented in Google Colab using a dataset of real homicides in Honduras from 2021 to 2025, integrating variables such as proximity to schools or police stations. The system achieved **89.6 %** semantic accuracy in the language module, **91.2 %** intention-execution alignment, and an average **87 %** accuracy in department-level risk classification. Moreover, it delivered a response time of under **2 s**. These capabilities support strategic decision-making in public security, urban logistics, and evidence-based criminological analysis.

Keywords: Homicides · Predictive Analysis · Agentic AI

1 Introduction

Homicide violence continues to represent one of the most complex and persistent challenges for public safety worldwide [1,11,13]. According to recent data from the World Health Organization (WHO) and the United Nations Office on Drugs and Crime (UNODC), over 400,000 homicides occur each year globally [1], affecting both developing nations and advanced economies [12]. Behind each figure lie profound social consequences: devastated families, traumatized communities, and societies facing a growing sense of insecurity.

For police departments and national security agencies, homicides pose multiple challenges: identifying criminal patterns in dense urban environments, optimally allocating limited resources, and the need for a proactive response in highly

A. Rocha et al. (Eds.): ICITS 2026, LNNS 1966, pp. 381–391, 2027.
https://doi.org/10.1007/978-3-032-27538-7_34

dynamic contexts. Traditional surveillance, patrol, and retrospective analysis strategies have shown significant limitations in anticipating violent events or intervening in real time. Furthermore, conventional statistical approaches are often constrained to isolated variables, without fully incorporating geospatial, social, and temporal context [14].

In response to this need, artificial intelligence (AI) has begun to offer new tools that transform large volumes of data into strategic knowledge [6]. Particularly, agentic AI systems—where autonomous intelligent agents analyze, reason, act, and self-feed—provide a promising framework to enhance the anticipatory capacity of security forces. These agents can be designed to operate across multiple regions, simulate risk scenarios, prioritize strategic interventions, and generate understandable visualizations for decision-makers. The primary objective of this study is to propose and validate a modular agentic artificial intelligence architecture, mediated by a large language model (LLM) and designed to enhance predictive accuracy and context-aware decision-making in homicide risk management.

The proposed solution is structured in three key phases: 1) collection and enrichment of spatial and temporal data, 2) training and tuning of a robust supervised classifier with class balancing, and 3) integration of an automatic feedback loop mediated by LLMs. This architecture allows agents not only to analyze information but also to propose hypotheses, simulate future scenarios, and adapt their behavior according to incoming data.

Theoretical Contribution. This study introduces a modular theoretical framework for integrating agentic artificial intelligence into urban homicide prediction, aligned with recent formulations of operational autonomy in Agentic AI [6,9].

A second contribution is the definition of *operational semantic feedback*, a construct grounded in interpretability research and intended to support context-aware adjustment of predictions [10]. The geospatial risk model draws on literature linking urban crime patterns to critical infrastructure [2], as well as Latin American work on multivariable violence prediction [7]. Finally, the study formalizes the concept of *geospatial operational voids*, extending current theoretical approaches in computational criminology for Latin American urban environments.

2 Literature Review

Recent advances in artificial intelligence (AI) have enabled the development of systems capable of predicting and analyzing criminal activity through geospatial and temporal modeling. Early initiatives such as *CrimeRadar* in Brazil applied machine learning to produce crime heatmaps for urban surveillance [2], while the *EMBERS* project demonstrated the feasibility of multi-source predictive alerts using autonomous agents and open data streams in Latin America [3].

At the methodological level, studies employing LSTM networks for homicide time-series forecasting [8] and ensemble methods such as XGBoost and SuperLearner [7] have shown strong performance in capturing urban violence dynamics. These works highlight the relevance of variables such as population density, poverty, and infrastructure proximity—factors comparable to those in Honduran cities.

Visualization tools including *Kepler.gl*, *Leaflet.js*, and *Folium* have facilitated real-time geospatial analysis, supporting decision-making through interactive risk mapping [4]. More recently, research on agentic AI frameworks (e.g., AutoGPT, CrewAI) has introduced autonomous reasoning and feedback loops into predictive systems, expanding their capacity for continuous monitoring and contextual adaptation [9].

Overall, the literature demonstrates a growing convergence between predictive modeling, geospatial analytics, and autonomous agent systems, providing the foundation for the modular agentic architecture proposed in this study.

3 Methodology

This study proposes an adaptive computational approach for predicting and analyzing urban homicide patterns through the development of an agentic intelligent agent. The architecture comprises three functional blocks: 1) Data collection, validation, and enrichment, 2) Supervised modeling supported by semantic analysis, and 3) Predictive alert generation with automatic feedback. This structure is illustrated in Fig. 1.

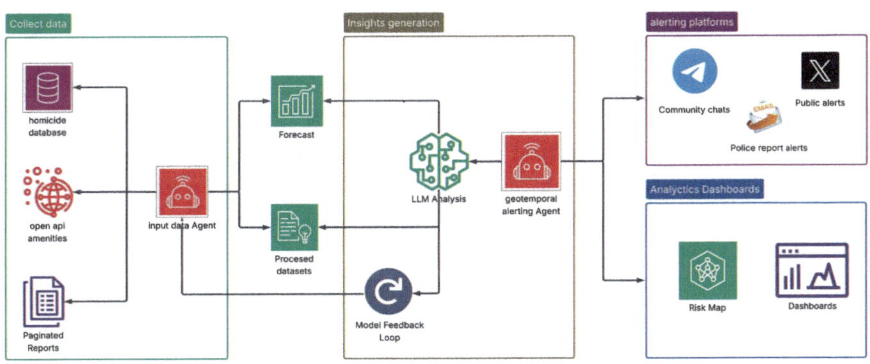

Fig. 1. Modular predictive Agentic AI Architecture.

3.1 Data Collection, Validation and Enrichment

Historical homicide records (12,000 cases) were integrated with geospatial infrastructure data (schools, police posts) obtained via Overpass Turbo API queries. These data were processed and standardized using the GeoPandas library, converting them into geographic objects. Subsequently, proximity indicators were calculated via Euclidean distance between each incident and nearby entities within a 1,000-meter radius, generating derived variables for modeling.

3.2 Supervised Modeling and Semantic Analysis

The predictive model was developed using the Random Forest algorithm, considering variables such as incident time, proximity to key infrastructure, and risk labels. Due to the natural class imbalance in the dataset, SMOTE was applied to improve the representation of minority classes. The evaluation metrics used were precision, recall, and F1-score, defined by the following expressions:

$$Precision = \frac{TP}{TP + FP} \tag{1}$$

where TP (True Positives) represents correct predictions of risk zones, and FP (False Positives) are incorrect predictions.

$$Recall = \frac{TP}{TP + FN} \tag{2}$$

FN (False Negatives) corresponds to risk zones not detected by the model, evaluating its sensitivity in identifying critical cases.

$$F1 = \frac{2 \times Precision \times Recall}{Precision + Recall} \tag{3}$$

The F1-score provides a harmonic mean between precision and recall, allowing assessment of the model's overall balance. The alert probability was modeled using a nonlinear decision function:

$$R_{alert} = f(Hour, SchoolProximity, PolicePostProximity) \tag{4}$$

where $f(\cdot)$ represents the Random Forest classifier trained on spatio-temporal variables. Additionally, an automatic feedback mechanism was implemented, adjusting the model's weights based on semantic corrections identified by the LLM, expressed as:

$$W_{n+1} = W_n + \eta \cdot \Delta P_{LLM} \tag{5}$$

where W_n is the current parameter state, η the learning rate, and ΔP_{LLM} the semantic difference detected in contextual inferences.

3.3 Geospatial Visualization

Interactive maps were developed using the Folium library to visually represent risk zones, showing incident locations, nearby infrastructure, and influence areas calculated by the model in real time.

4 Results

The predictive AI agent was validated using data from the municipalities of San Pedro Sula and the Central District (Tegucigalpa) for the period 2021 to 2025.

Table 1. Classification report

Class	Precision	Recall	F1-Score	Support
0	0.89	1.00	0.94	70
1	1.00	0.89	0.94	82
Accuracy	–	–	0.94	152
Macro Avg	0.94	0.95	0.94	152
Weighted Avg	0.95	0.94	0.94	152

The Random Forest model, optimized via SMOTE, achieved robust performance, with results detailed in Table 1.

As shown in Table 1, the model achieved an average F1-score of **94 %**, demonstrating a strong balance between precision and recall across both classes. These results validate the system's ability to detect critical risk zones even under class-imbalanced conditions.

4.1 Automated Alert Generation

The AI agent was programmed to generate predictive alerts in real time, classifying incidents by criticality level (low, medium, high). Alerts were displayed in the Google Colab console, providing details such as municipality, incident type, estimated time, and proximity to key infrastructure. Figure 2 shows the automated alerts generated by the AI agent in the Google Colab console. Each entry includes municipality, type of homicide, estimated time, and distances to key infrastructure such as schools and police posts. Criticality is categorized as "Critical" or "Medium," facilitating immediate prioritization of interventions based on risk severity.

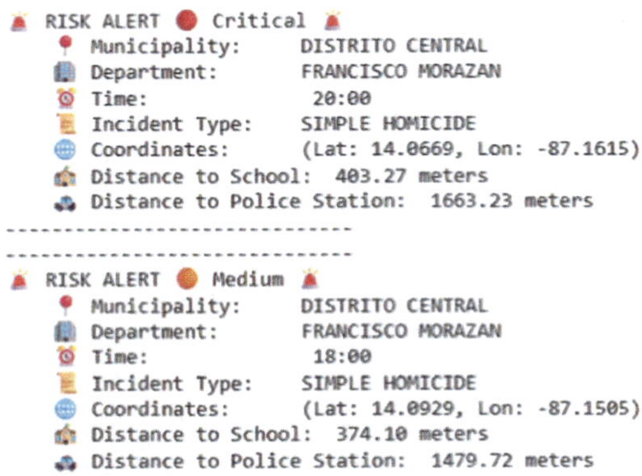

Fig. 2. System Log: Automatic alerts simulation.

4.2 Geospatial Risk Zone Visualization

The system generated interactive maps using Folium to visualize the spatial distribution of alerts, nearby police posts, and influence areas. This visualization supports tactical decision-making in the allocation of security resources. Figure 3 shows the concentration of detected homicide risk zones in the Central District (Tegucigalpa) and their proximity to nearby police posts.

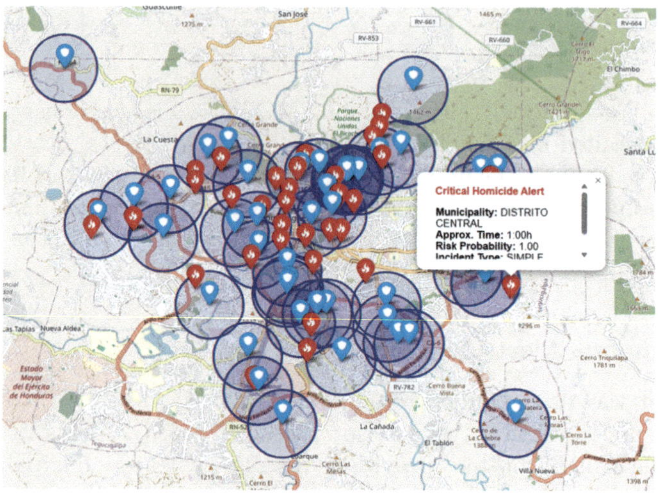

Fig. 3. Risk zone map detected.

In Fig. 3, red points represent critical homicide risk alerts detected by the model, indicating priority intervention locations. Blue circles correspond to police posts, showing their distribution and approximate coverage. This visualization helps identify clusters of events in specific areas, optimizing strategic response planning.

4.3 Operational Coverage Gaps Detection

An additional analysis identified operational gaps—that is, risk zones outside police post coverage radii. This information is essential for restructuring patrol strategies and optimizing resource allocation. Figure 4 highlights these vulnerable areas, indicating priority intervention locations.

In Fig. 4, blue circles represent the radii of police post coverage, while red points indicate risk zones outside that coverage. This visualization allows identification of operational gaps in the current tactical distribution, emphasizing the need to restructure patrols and optimize resource deployment toward vulnerable areas lacking effective protection.

These results demonstrate the effectiveness of the AI agent in generating precise contextual alerts and geospatial visualizations that optimize tactical decision-making in complex urban environments.

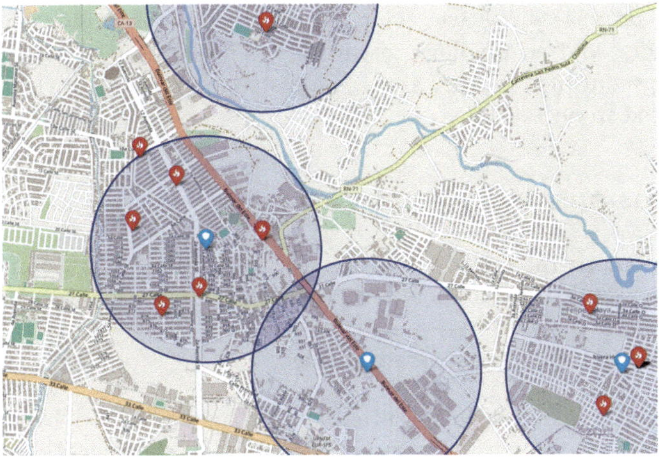

Fig. 4. Vulnerable areas outside of police station coverage.

5 Discussion of Results

The empirical results detailed in Sect. 4 confirm that the agentic architecture proposed in this study successfully integrates geospatial analysis, supervised learning, and semantic reasoning into a coherent predictive mechanism. The model's F1-score of 0.94 demonstrates that the combination of Random Forest with SMOTE effectively compensated for the class imbalance observed in the underlying homicide dataset described in Sect. 3.1. This outcome aligns with prior evidence that ensemble-based classifiers maintain stability under heterogeneous urban conditions [7].

The geospatial maps produced by the system further substantiate their practical utility. Clustering patterns observed in Figs. 3 and 4 not only confirm historical concentrations of violence in urban centers such as the Central District but also reveal spatial discontinuities in police coverage that echo structural vulnerabilities previously identified in Latin American criminological analyzes [8]. By visualizing both critical alerts and zones lacking institutional presence, the agent provides a multi-layered diagnostic perspective that traditional statistical approaches fail to capture. These findings support the argument in Sect. 8 that predictive AI can transform tactical planning and resource allocation.

Nevertheless, the integration of historical homicide data raises ethical and methodological concerns. As noted in Sect. 6, any system trained on static datasets may inadvertently replicate pre-existing distortions, including underreporting and biased patrol patterns. The predictive strength observed in Sect. 4 must, therefore, be contextualized within these potential structural limitations.

In synthesis, the results confirm that the proposed agentic AI framework achieves not only technical robustness but also operational relevance across multiple analytical layers. Its capacity to integrate dynamic reasoning, geospatial contextualization, and predictive modeling supports the conclusion that such systems can serve as proactive, evidence-based tools for public safety agencies, urban planners, and criminology researchers. These findings validate the architectural design outlined in Sect. 3 and establish a foundation for the future extensions discussed in Sect. 6.1.

6 Conclusions and Recommendations

This study demonstrates the revolutionary potential of agentic artificial intelligence when applied to homicide analysis and prevention in complex urban settings.

The following conclusions were identified:

- Geospatial data significantly enrich the predictive model by incorporating locations associated with increased or decreased crime events as binary variables to improve model accuracy.
- Findings indicate that agents become more accurate with more relevant data sources, showing great capacity to understand changing instructions and continuously improve through automatic feedback.

Based on these results, the following recommendations are proposed:

- The agentic approach enables autonomous and intelligent system behavior by integrating data analysis, reasoning (LLM model), and action (alerts + zoned map dashboards); the system functions as an agent capable of operating automatically.
- It is recommended to implement and validate models in cities or countries with different socio-cultural contexts and broad data access, in order to assess adaptation and generalization capabilities.
- It is advisable to include additional risk factors such as real-time economic indicators, urban mobility data, or citizen reports, which could enrich predictive models.

7 Limitations and Future Work

Despite the favorable results presented in Sect. 4, the current implementation of the AI agent has several limitations that must be addressed to ensure effectiveness in real-world settings.

- **Deployment in a Simulated Environment**: The agent was deployed and tested in a simulated environment (Google Colab), restricting its application to controlled sessions, limited by computing resources and lacking continuous operational capability. Integration into operational workflows requires migration to cloud platforms (e.g. AWS, Azure) to enable real-time processing.

- **Static Data and Limited Temporal Synchronization**: The operational gap analysis (Fig. 4) was conducted using static datasets, limiting the model's ability to adapt to real-time changes. It is imperative to integrate dynamic data sources such as traffic flows, citizen reports, and weather conditions, as suggested by recent studies in Latin America [2,5].
- **Classification Model Improvement**: While the Random Forest model, tuned with SMOTE, achieved solid metrics (Table 1), literature suggests that algorithms like *XGBoost* [7] or *LSTM* networks [8] may offer superior performance in scenarios with nonlinear interactions and complex temporal patterns.
- **Static Semantic Feedback**: The LLM-based feedback mechanism allows adjustment of intention-execution alignment, but remains a manual, statically parameterized process. An evolution toward an adaptive feedback system is required, where the agent automatically adjusts thresholds and criticality weights according to emerging detected patterns.

7.1 Future Work Directions

To mitigate these limitations, the following action lines are proposed:

1. **Migration to Cloud Infrastructure**: Deploy the agent on public or private cloud architecture, ensuring high availability, real-time processing, and scalability.
2. **Integration of Dynamic Data**: Connect the system to live data sources (urban mobility APIs, IoT sensors, citizen reports), allowing continual update of the model's spatial-temporal context, following methodologies similar to EMBERS [3].
3. **Experimentation with Advanced Models**: Train the agent using models like XGBoost [7] to optimize classification on imbalanced datasets, and LSTM networks to capture temporal dynamics, evaluating their impact on alert accuracy.
4. **Adaptive LLM-based Feedback**: Design a real-time feedback loop in which the LLM automatically adjusts inference rules and criticality weights based on new patterns, inspired by approaches proposed in [9].
5. **Validation in Multiple Urban Contexts**: Replicate the proof-of-concept in cities with diverse socioeconomic profiles and violence patterns (e.g. Choloma or La Ceiba), to validate the system's generalization capability.

These actions will strengthen the AI agent's capacity, enabling its evolution toward a robust, adaptive, and highly contextualized operational system capable of assisting tactical decision-making in urban violence prevention.

8 Applications

The agentic AI framework developed in this study enables practical deployment across multiple operational domains. Its capacity to generate geospatial alerts, contextual diagnostics, and adaptive risk assessments supports the following application areas:

8.1 Public Safety and Law Enforcement

Security institutions can incorporate risk alerts and spatial diagnostics into patrol scheduling, incident monitoring, and operational coordination. The system enhances situational awareness by identifying priority zones and highlighting areas requiring immediate attention.

8.2 Logistics and Supply Chain Management

Organizations in transportation and logistics can integrate spatial risk assessments into route planning and scheduling, reducing exposure to high-risk areas and improving operational continuity.

8.3 Urban Planning and Governance

Municipal authorities may use the model's geospatial outputs to support infrastructure planning, territorial management, and community-oriented interventions. Risk maps facilitate resource allocation and the identification of structurally vulnerable areas.

8.4 Financial Services and Risk Assessment

Territorial risk indices derived from the system can support underwriting processes, credit evaluation, and the development of risk-adjusted financial products tailored to local conditions.

8.5 Criminological Research and Policy Development

Academic and research institutions can leverage the framework to conduct longitudinal studies, benchmark predictive models, and evaluate intervention outcomes within computational criminology.

References

1. Herre, B., Spooner, F., Roser, M.: Homicides. Our World in Data (2013). https://ourworldindata.org/homicide-data-how-sources-differ-and-when-to-use-which-one. Includes data processing from UNODC and UN Population Division (2024–2025). Accessed 11 Nov 2025
2. Igarapé Institute: crimeradar: crime prediction platform for urban spaces in RIO de janeiro (2018). https://igarape.org.br/apps/crimeradar/. Accessed 11 Nov 2025
3. Kalyanam, J., et al.: Embers at five years: lessons learned from an operational forecasting system. arXiv preprint arXiv:1604.00033 (2016). https://arxiv.org/abs/1604.00033. Accessed 11 Nov 2025
4. Lab, U.U.D.: Interactive crime dashboards for real-time policing (2023). https://urbananalytics.net/dashcrime. Accessed 11 Nov 2025

5. Ministerio del interior de uruguay: evaluación de predpol en montevideo 2016–2019 (2019). https://dialogopolitico.org/special-edition-2025-artificial-democracy/artificial-intelligence-and-public-safety-in-latin-american-democracies. Accessed 11 Nov 2025. (in Spanish)
6. Mukherjee, A., Chang, H.: Agentic ai: autonomy, accountability, and the algorithmic society. arXiv preprintarXiv:2502.00289 (2025). https://arxiv.org/abs/2502.00289. Accessed 11 Nov 2025
7. Ortiz, M., Gómez, L.: Aplicación de superlearner y xgboost para la predicción de crímenes violentos en colombia. Revista Latinoamericana de Seguridad Pública (2021). in Spanish
8. Ramírez, L., Mendoza, C.: Análisis predictivo de homicidios con redes neuronales en guanajuato, méxico. Revista de Criminología Computacional **8**(2), 112–128 (2022). (in Spanish)
9. Research, O.: Agentic ai systems: from planning to action (2024). https://openai.com/research/agentic-ai-systems. Accessed July 2025
10. Ribeiro, M., Singh, S., Guestrin, C.: Why should i trust you? explaining the predictions of any classifier. In: Proceedings of the 22nd ACM SIGKDD International Conference on Knowledge Discovery and Data Mining (KDD '16), pp. 1135–1144. ACM, New York, NY, USA (2016). https://doi.org/10.1145/2939672.2939778. (in Spanish)
11. United nations office on drugs and crime: global study on homicide 2011. United Nations Office on Drugs and Crime (UNODC) (2011) Nations Office on Drugs and Crime (UNODC) (2011)
12. United nations office on drugs and crime: global study on homicide 2019: killing of women and girls by intimate partners or family members. United Nations, Vienna (2019). https://www.unodc.org/unodc/en/data-and-analysis/global-study-on-homicide.html
13. United nations office on drugs and crime (UNODC): intentional homicide victims data portal (2025). https://dataunodc.un.org/dp-intentional-homicide-victims. Accessed 11 Nov 2025
14. University of california: disparities and biases in predictive policing: a multicity study (2021). https://arxiv.org/abs/2102.00128

Audiovisual Narratives and the Audience: The Case of the Ministry of Time

Cristián Londoño-Proaño(✉)

Universidad Indoamérica, Bolívar 2055 y Guayaquil, Ambato, Ecuador
cristianlondono@uti.edu.ec

Abstract. This study focuses on the analysis of the series The Ministry of Time. The objective is to analyze how El Ministerio del Tiempo adapts narrative resources from science fiction and evaluate their impact on the audience and popular culture. A qualitative methodology based on content analysis was used, evaluating three categories: use of narrative resources of science fiction in a historical context, the impact on the audience and the impact on popular culture. Throughout the analysis, it is shown that the science fiction series managed to generate a reflection on the History of Spain, it was a transmedia content that generated a solid base of followers and was more than the television screen, it fostered the debate on national identity and gender roles. It is concluded that El Ministerio del Tiempo had a significant impact, both in the way it constructs its story and in its reception by the public and its insertion in the Spanish cultural debate.

Keywords: Series · Television · Science Fiction · Innovation · Transmedia

1 Introduction

Audiovisual series have undergone a profound transformation over time, evolving into complex narratives [1], with interconnected plots, prolonged narrative arcs and more active audience participation [2].

The Ministry of Time, the Spanish series broadcast by TVE between 2015 and 2020, represents a paradigmatic case within audiovisual narratives. This series combines elements of science fiction in an episode structure, exploring historical memory and national identity, differentiating itself from other productions of the genre.

In studies on transmedia narratives, The Ministry of Time has been extensively analyzed, highlighting the work of Scolari and Establés [3], who examine the expansion of the narrative universe beyond the screen and the professionalization of fandom. Its approach reveals how the series redefines television success by prioritizing digital interaction over traditional audiences. However, its analysis is limited to the first seasons and focuses exclusively on the transmedia dimension, without delving into narrative, historical or sociocultural aspects.

This study expands on the previous approach by incorporating a multidimensional perspective that encompasses the full analysis of the 43 episodes, the use of science fiction resources, the cultural and audience impact.

A. Rocha et al. (Eds.): ICITS 2026, LNNS 1966, pp. 392–402, 2027.
https://doi.org/10.1007/978-3-032-27538-7_35

The present study poses the following research question: How does The Ministry of Time influence the audience and popular culture and adopt science fiction resources to explore historical memory?

The aim of this study is to analyze how The Ministry of Time impacts the public and popular culture, and adopts narrative resources from science fiction to explore historical memory. To this end, the narrative structure of the series, the interaction of viewers with the content and its reception in the cultural and media field will be examined.

This study employs a qualitative approach based on non-participant observation and interpretive content analysis. All episodes of The Ministry of Time were analyzed, along with a review of the academic literature on the series. This combination allows for a complete understanding of the phenomenon.

1.1 Science Fiction

Science fiction, both in literature and in the audiovisual field, is oriented to the analysis of the impact of science and technology on society and individuals [4, 5]. It is distinguished by its ability to speculate on alternative futures, create advanced technologies, and construct scenarios that defy known physical laws [4, 6]. Throughout its evolution, it has functioned as a means of reflection on philosophical, ethical and social problems, questioning both the present and the future [7, 8].

Isaac Asimov defined it as a narrative that studies the effects of scientific and technological changes on humanity, projecting not only future advances, but also their social and human implications [5]. One of its essential features is the introduction of a novel or disruptive element that transforms the known reality, enabling speculation about the ethical and social effects of technology [9]. In this way, the genre articulates scientific and technological knowledge within a coherent narrative framework [10].

Science fiction is divided into hard and soft, differentiated by the scientific rigor of their narratives [11].

Soft science fiction focuses on social, psychological and human aspects, privileging speculation over technical precision [4, 11]. Its plots revolve around characters and the social implications of technological advances, as in Frank Herbert's Dune and Ursula K. Le Guin's The Left Hand of Darkness.

Hard science fiction, on the other hand, is characterized by scientific accuracy and consistency with established theories and laws [4, 11]. It seeks to plausibly explain technological advances and is based on current scientific knowledge. Prominent examples are Isaac Asimov's Foundation and Andy Weir's The Martian [11].

1.2 The Ministry of Time

The Ministry of Time is a historical science fiction television series about time travel [12]. The idea of time travel has fascinated throughout mankind [13].

The series The Ministry of Time was created by the brothers Javier and Pablo Olivares [14]. The series was first broadcast in 2015 by Televisión Española [15]. The premise of the series is based on a secret ministry in charge of protecting the history of Spain, guarding doors that allow travel to different eras of the past in order to avoid alterations in the course of historical events [3]. This combination of elements of science fiction, adventure, and historical drama allowed the series to present a wide range of historical eras and relevant characters from Spain's past.

One of the most unique features of The Ministry of Time is its episodic approach, in which each chapter focuses on a different era or historical event, from the Middle Ages to the contemporary era. The series addresses emblematic figures of Spanish history, such as Lope de Vega, Felipe II, or characters from the Golden Age, while exploring how their decisions affect the course of history [16].

The Ministry of Time began broadcasting in 2015 with a first season that was well received by critics [16], which laid the foundations of the narrative universe and presented the Ministry's mission to preserve the history of Spain [17]. The second season (2016) delved into historical plots such as the Inquisition and the War of Independence, positioning the series as a space for reflection on national identity and historical memory [14]. The third season (2017), despite budgetary difficulties [18], maintained its narrative quality, complicated the relationships between characters and incorporated events from the twentieth century such as the Franco dictatorship and exile [3]. Finally, the fourth season closed plot lines with an episode that integrated past, present and future, leaving a satisfactory conclusion for the audience [19].

2 Materials and Methods

2.1 Method

This study adopts a qualitative approach to content analysis, appropriate for examining cultural products from an interpretive perspective [20]. This methodology facilitates the understanding of values, ideologies, and cultural practices present in texts, artifacts, and other cultural products [21]. Interpretive content analysis stands out for its flexibility and ability to capture complex meanings [22]. In this study, this methodology allows us to identify symbolic structures, narrative strategies and discursive patterns in the series, as well as their projection beyond the screen.

2.2 Object of Study

The object of study is the Spanish series The Minstry of Time, broadcast by Televisión Española (TVE) between 2015 and 2020. The research focuses exclusively on this production, as it is an innovative proposal within television. Unlike studies that work with partial samples, this research considers the totality of the 43 episodes that make up its four seasons.

2.3 Methodological Justification

A qualitative approach is used to explore the narrative innovation and cultural interaction of a series, analyzing all episodes and reviewing the academic literature.

The incorporation of bibliographic sources includes indexed articles, book chapters, academic papers and theses that address the series from various perspectives. No exclusion criteria were applied, to ensure a plural and interdisciplinary vision.

2.4 Analysis Strategy

The analysis is organized around three categories, defined based on specialized literature on audiovisual narrative, audiences and popular culture. These dimensions were selected for their relevance to understanding the Ministry of Time phenomenon in its entirety:

2.4.1 Use of Science Fiction Narrative Resources in a Historical Context

This category looks at how the series combines soft science fiction (time travel, paradoxes, state bureaucracy) with historical fiction [11]. Soft science fiction prioritizes social, ethical, and cultural aspects over scientific accuracy [6, 11]. It examines how the series reinterprets key moments in Spanish history using narrative devices to reflect on the present, the past and national identity. Episodic structures, historical figures, and the ethical implications of intervening in the timeline are also considered.

2.4.2 Impact on the Audience

This category examines the degree of participation, loyalty and appropriation of the series by the public. In the context of digital culture, audiences are no longer just passive consumers, but active agents that interpret, expand and resignify audiovisual content [2]. It analyzes the interaction between creators and audiences in social television and how transmedia narrative promotes participation in social networks and the creation of derivative content [1, 3, 15].

2.4.3 Impact on Popular Culture

This analysis examines how The Ministry of Time influences collective imaginaries, debates about national history, and the representation of social identities. It analyzes how the series resignifies historical figures, introduces new ways of representing the past, and connects with contemporary debates on gender, memory, and nation [23]. It also considers how the series is cited, reinterpreted or referenced in other media, institutional discourses or educational spaces, consolidating its symbolic impact.

2.5 Limitations of the Study

This study does not employ a quantitative methodology or statistical audience measurements, so its results do not reflect the massive reach of the product, but rather its interpretive and cultural depth. Although all available episodes and academic sources are analyzed, it is recognized that there may be other forms of reception (such as memes, fanfics or non-academic productions) that escape the proposed approach.

3 Analysis and Discussion

The analysis of The Ministry of Time is structured around three categories established in the previous section:

3.1 Use of Science Fiction Narrative Resources in a Historical Context

First it must be established that the series The Ministry of Time is of the science fiction genre [12], and according to what Londoño-Proaño [11] indicates, it fits into soft science fiction. First, its devices are technological, but their operation is not scientifically explained, and its narrative focuses on the aspect of the History of Spain.

The existence of a Ministry in charge of regulating the failures in the History of Spain introduces a bureaucratic component that moves away from the concept of individualistic and unrestricted time travel that has predominated in Western science fiction such as the Doctor Who series or films such as Back to the Future, the protagonists usually travel independently and without a central authority to supervise them. On the other hand, in The Ministry of Time, time travel is subject to protocols, regulations, and a hierarchical structure, which provides a dimension of state control over history and collective memory [24]. For example, in chapter "Optimizing the Balance Between Knowledge Acquisition and Cognitive Load in Virtual Reality-Based Education for Children with Special Needs" of the first season, called Tiempo de Gloria. In this chapter, Gil Pérez, an experienced agent of the Ministry during the reign of Philip II, discovers that Lope de Vega, who is part of the Invincible Armada, will not board the historic ship, the San Juan, which History records as one of the few ships whose sailors survived [17]. Instead, Lope is destined to embark on a ship where he will meet an untimely death, which would prevent him from writing some of his best works. To avoid this alteration in the timeline, the patrol travelled to Lisbon in 1588 with the mission of correcting the course of events [17].

The Ministry of Time introduces a unique concept of time travel with doors connecting different eras. Unlike typical science fiction time machines, such as Doctor Who's TARDIS or Back to the Future's DeLorean, these gates offer more organic, space-focused travel. They also suggest an almost bureaucratic state control over time and history [6].

In the penultimate season, the machine of El Anacronópete appears, a mythical machine for time travel in the first science fiction stories of Spain. The inclusion of El anacronópete in the series is not only a tribute to the first time machine in Hispanic literature, but also reinforces the idea that time travel is part of a broader literary tradition, in which language and cultural identity play an important role. With this reference, El Ministerio del Tiempo not only recognises its place within the genre, but also resignifies it within the context of Spanish science fiction, differentiating itself from the predominant Anglo-Saxon influences in science fiction.

Another narratively interesting point is that, The Ministry of Time uses advanced technologies and protocols to detect and correct any alterations in History [17]. This is vital in the development of the stories. Although not explained in detail from a scientific perspective, these technological elements create a logical framework for the Ministry to carry out its task of protecting the timeline. This temporal monitoring technology connects with the modern notion of state surveillance, creating a parallel between time control and current political and social control.

Another interesting issue is communication technologies. These have the ability to overcome the barriers of time. The characters use communication devices that work through time [17]. This allows the story to be energized and the characters to connect. These devices, while not based on real technologies, are part of the narrative that allows the characters to stay in touch as they travel through different eras, reinforcing the verisimilitude of the fictional universe. For example, in episode 9 of the second season, Velázquez receives an alert on his phone informing that one of his paintings has been auctioned in 2016 for a considerable sum [17]. This would be a common occurrence if it were not for the fact that the work in question was destroyed in the fire of the Alcázar of Madrid in 1734 [25]. Faced with a lack of available agents due to union problems, he and Irene take on the task of investigating the case [25, 26]. Their inquiry leads them to a large-scale art trafficking scheme, in which Darrow and Lola Mendieta are involved in the substitution of original pieces for copies with the aim of obtaining economic benefits [25].

The technology used in the series The Ministry of Time, such as the gates of time, the protocols for detecting alterations in History and temporal communication, show narrative resources that enrich the fictional universe and are novel in contexts of a historical theme.

3.2 Impact on the Audience

The series El Ministro del Tiempo in its four seasons accumulated 97,970,000 viewers [27]. In the first season, the series had 39,309,000 viewers in 8 episodes. The second season had 29,784,000 viewers in 13 episodes. The third season had 18,296,000 viewers in 13 episodes. The fourth season had 10,407,000 viewers in 8 episodes [27]. Viewers and episode production decreased each season. The first episode of the first season, Time Is What It Is, was one of the most watched, with 2,981,000 viewers and a share of 14.8% [27]. The first episode of the second season was the closest in viewership, with 2,839,000 viewers and a share of 14.6% [27]. The worst episode was Setbacks, with 983,000 viewers and a share of 5.9% [27]. In 2020, RTVE reported that The Ministry of Time was in the Top 10 of the ranking of "Shows on the Rise" [28].

On the other hand, one of the main characteristics of transmedia storytelling is its ability to connect with the audience beyond the screen [3]. In the case of The Ministry of Time, the series created a loyal and active fan base known as ministéricos, who played a relevant role in the expansion.

The series expanded the universe of fiction through mobile applications, video games, podcasts and additional content on social networks. For example, through apps like The Ministry of Time, fans were able to participate in interactive missions that allowed them to be part of the narrative universe, acting as agents of the Ministry. This active participation generated a close and personal relationship with history [25]. On the other hand, the use of platforms such as Twitter, where real-time discussions were held during the broadcast of the episodes, and Facebook, which offered complementary material such as behind-the-scenes and interviews with the characters, consolidated a participatory audience involved in the development of the series [24]. In addition, the characters in the series interacted with viewers via social media, answering questions and commenting on fictional or real historical events, creating an immersive experience [25].

This is evidence that the social narrative of The Ministry of Time, which took place on social networks, expanded the universe of the series [24], which contributed to creating an ecosystem where fans played a leading role [3, 29], allowing them to delve deeper into the story and prolong their interaction with the series long after its original broadcast, demonstrating the importance of transmedia content in modern audiovisual consumption [29]. In addition, fans became active creators, facilitating the development of an engaged audience [3].

The Ministry of Time managed to create a transmedia ecosystem in which the audience received content and became participants in the universe. Fans became an integral part of the narrative, actively participating on social media and generating content around the series [29]. The ministry community was an example of how an innovative narrative can generate loyalty and constant participation.

This participatory dynamic allowed the series to maintain its relevance and increased the degree of viewer involvement [3, 24].

3.3 Impact on Popular Culture

One of the most important contributions of El Ministerio del Tiempo was its ability to generate a debate on Spanish national identity. Through its episodes, the series addressed how the events of the past shaped the present and how history could be interpreted from multiple perspectives [33]. This revision of history resonated with the public, creating a space to reflect on what it meant to be Spanish at different historical moments [34].

The series recovered emblematic figures from the history of Spain, such as Lope de Vega, Isabella the Catholic or Salvador Dalí, and presented them from a humanized perspective, showing their contradictions and nuances [25]. This strategy allowed the viewer to connect with these characters as national heroes and individuals with dilemmas and ambitions of their own.

On the other hand, the series introduced criticisms of official discourses on the history of Spain. Through humor and irony, traditionalist narratives were questioned and new interpretations of controversial historical episodes were offered. An example of this was the depiction of the Golden Age, where the cultural greatness of the time was contrasted with social inequalities and intellectual censorship. For example, in the episodes Tiempos de Gloria and Tiempo de ilustrados. In the first case, Lope de Vega is presented; and in the second case, the painters Goya and Velázquez are shown.

The series maintained a conservative stance, where the Ministry's duty was to make sure that history did not change. There were moments when the characters challenged this premise, questioning the extent to which history is an immutable story or a construct that can be reinterpreted [25]. For example, in episode 13 of season 2, King Philip II, after the defeat of the Invincible Armada, decides to use the gates of time to change the past and ensure the victory of Spain. This intervention causes a significant alteration in the timeline, creating an alternative reality in which Spain dominates the world, but at the cost of individual freedoms and social advances. The protagonists must then restore the original timeline, facing the dilemma of whether it is right to undo changes that, while beneficial to national power, are detrimental to society as a whole.

The series was a pioneer in the representation of gender and diversity. Characters such as Amelia Folch challenged historical gender norms by representing an empowered and active woman in decision-making within the Ministry. She was the archetype of the woman ahead of her time. Inspired by real historical figures such as Concepción Arenal and María Elena Maseras, pioneers in education and women's rights in Spain, Amelia depicted a young woman fighting for equality and female independence against the restrictions of her time [35]. Amelia broke with historical gender stereotypes, representing a woman who seeks to be the master of her destiny, both in her personal life and in her career [31]. Despite family opposition, she persisted in her desire to continue her university education and to play roles that, in her context, were exclusive to men. The character of Amelia contributed to subverting traditional gender roles on television [31], reinforcing its cultural relevance at a time when these issues are the subject of public debates [32].

Below is a comparative table summarizing the main contributions of this study in relation to previous research on the Ministry of Time. This work differs by incorporating a science fiction approach, specifically soft science fiction, to analyze the narrative resources of the series in its historical context, as well as its impact on the public and popular culture. This perspective allows us to broaden our understanding of the narrative and media phenomenon that the series represents (Table 1).

Table 1. Comparative table

Analyzed aspect	Previous studies	Contributions of this study
Gender and Approach	Focus on national history, memory and narratives	Soft science fiction is incorporated to reinterpret the story.
Technology in Storytelling	General mention of narrative devices such as the gates of time.	Analysis of science fiction narrative devices: doors of time, communication and temporal surveillance.
Transmedia	Studies identify the presence of transmedia strategies, but do not develop them extensively	Analysis of the transmedia ecosystem: social networks, video games, applications, interaction with characters and fan fiction.
Audience	Mention of the "ministerial" phenomenon and participation in networks	The relationship between fan practices and trans-media narrative is analyzed, as well as the decline in audience.
Methodological approach	Partial samples of episodes or general theoretical analyses.	Comprehensive analysis of 43 episodes with academic sources.

4 Conclusions

In this study, The Ministry of Time confirms its relevance within the audiovisual panorama, due to its ability to adapt science fiction resources within a historical fiction structure. The series combines characteristic elements of the genre, such as time travel and temporal paradoxes, with a bureaucratic and state vision of temporal control, which differentiates it from other similar productions.

In response to the research question, it is concluded that the series has had a significant impact, both in the way it constructs its story and in its reception by the public and its insertion in the Spanish cultural debate.

From a narrative approach, the series does not introduce new structures within science fiction, but it does use mechanisms of the genre effectively to explore the story from a playful and critical perspective. The Ministry of Time does not focus on the scientific plausibility of time travel, but on its functionality as a narrative resource to construct episodic stories that interconnect different historical moments. Its narrative structure maintains a balance between the continuity of the main story and the independence of each episode, allowing the audience to access the series without the need for rigorous plot monitoring.

Regarding the impact on the audience, we can indicate that, despite the fact that the audience levels were medium to high, The Ministry of Time has shown that the interaction of the public with a series is not limited to its broadcast, but expands through transmedia dynamics. The phenomenon of the ministéricos, active followers of the series, is an example of how a narrative can generate communities of engaged viewers that contribute to the expansion of the narrative universe. Other forms of this expansion are the official video game, spin-off novels, and an active presence on social networks.

The Ministry of Time has transcended the audiovisual field and has become a benchmark within Spanish popular culture. The series represents History and has the ability to connect with the audience through humour, irony and the reinterpretation of historical figures.

The Ministry of Time does not reinvent the conventions of science fiction or serial audiovisual narrative, but it does achieve an effective combination of genres that has resulted in an attractive product for the audience and with a strong cultural impact.

For future research, it would be interesting to compare The Ministry of Time with other series of time travel from different countries, evaluating whether there are parallels in other cultural contexts.

References

1. Scolari, C.: Transmedia narratives: new ways of communicating in the digital age. AC/E Yearb. Digit. Cult., 71–81 (2014)
2. Jenkins, H.: Convergence Culture. New York University Press, New York (2006)
3. Scolari, C.A., Establés, M.-J.: Transmedia ministry: narrative expansions and participatory cultures. Keyword. **20**, 1041–1041 (2017)
4. Barceló, M.: Science Fiction. Editorial UOC, Barcelona (2015)
5. Londoño-Proaño, C.: Sci-fi digital narratives in the classroom: black mirror's "Be Right Back" case. In: 2024 IEEE 4th International Conference on Advanced Learning Technologies on Education & Research (ICALTER), pp. 1–4 (2024). https://doi.org/10.1109/ICALTER65499.2024.10819233
6. Londoño-Proaño, C.: Between Science Fiction and Fantasy, Quito (2020)
7. Londoño-Proaño, C., Endara, F.: Andean future technologies in Ángelus Hostis and Andean sky. ACTS. **5**, 32–50 (2024)
8. Londoño-Proaño, C., Pérez, V.: Why should science fiction be learned in engineering education? In: 2024 IEEE Eighth Ecuador Technical Chapters Meeting (ETCM), pp. 1–6 (2024). https://doi.org/10.1109/ETCM63562.2024.10746159

9. Survin, D.: Metamorphoses of Science Fiction: On the Poetics and History of a Literary Genre. Yale University Press, New Haven (1979)
10. Latin American Science Fiction.: Peter Lang US, (2021) doi:https://doi.org/10.3726/b14354
11. Londoño-Proaño, C.: How science fiction inspired advances in engineering. In: 2023 IEEE Seventh Ecuador Technical Chapters Meeting (ECTM), pp. 1–4 (2023). https://doi.org/10.1109/ETCM58927.2023.10308982
12. Martínez Román, V. J. The Ministry of Time, the Introduction of New Concepts in Spanish Series. (2016).
13. Londoño Proaño, C.: Time travel. La Hora newspaper. (2025) https://www.lahora.com.ec/tungurahua/viajes-tiempo/
14. Soriano, D.: The end of TV? Ten figures that make the executives of the large chains tremble. Free Mark.. https://www.libremercado.com/2017-10-08/el-fin-de-la-tv-diez-cifras-que-hacen-temblar-a-los-ejecutivos-de-las-grandes-cadenas-1276606786/. (2017)
15. Mazzeo, V.: Transmedia time: a tour of the transmedia doors of the series 'The Ministry of Time'. In: Camino, A., Lovato, A., Celia, G., Susigan, C., Bressan, D. (eds.) Novos Mercados. Rial (2019)
16. Sánchez Castillo, S., Galan, E.: Transmedia narrative and cognitive perception in El Ministerio del Tiempo (TVE). RLCS, 508–526 (2016). https://doi.org/10.4185/RLCS-2016-1107
17. RTVE: The ministry of time: your favourite TVE programmes, on RTVE Play. RTVE.es. https://www.rtve.es/play/videos/el-ministerio-del-tiempo/. (2024)
18. Fernández, E.: The ministry of time, 'at the limit' of its travels on television. El Mundo. https://www.elmundo.es/television/2016/02/11/56bb935446163fdc648b45b3.html. (2016)
19. Costas, N.: The creator of 'The Ministry of Time' analyzes the definitive ending? elconfidencial.com. https://www.elconfidencial.com/television/series-tv/2020-06-23/final-ministerio-del-tiempo-temporada-cuatro-tve-entrevista-javier-olivares_2651835/. (2020)
20. Krippendorff, K.: Content Analysis: An Introduction to its Methodology. SAGE Publications, Inc. (2019). https://doi.org/10.4135/9781071878781
21. Lindgren, B.-M., Lundman, B., Graneheim, U.H.: Abstraction and interpretation during the qualitative content analysis process. Int. J. Nurs. Stud. **108**, 103632 (2020)
22. Drisko, J.W., Maschi, T.: Interpretive content analysis. In: Drisko, J., Maschi, T. (eds.) Content Analysis. Oxford University Press (2015). https://doi.org/10.1093/acprof:oso/9780190215491.003.0003
23. Berlanga Fernández, I., Arjona Martín, J.B., Merino Arribas, A.: Digital semiotics in the fiction series El ministerio del tiempo. SIGNA. **27**, 233 (2018)
24. Rodríguez-Mateos, D. & Hernández-Pérez, T. Social Television in Fiction Series and New Roles of the Audiovisual Documentarian: The Case of 'The Ministry of Time'. (2015)
25. Rodríguez Fidalgo, M.I., Paíno Ambrosio, A.: New advances in transmedia storytelling in Spanish fiction. case study of the television series 'el ministerio del tiempo'. In: Freire, F.C., Rúas Araújo, X., Martínez Fernández, V.A., García, X.L. (eds.) Media and Metamedia Management, pp. 183–188. Springer International Publishing, Cham (2017). https://doi.org/10.1007/978-3-319-46068-0_23
26. Editorial staff: 'El Ministerio del Tiempo' returns to La 1 on Monday, April 25. Audiovisual. **451**. https://www.audiovisual451.com/el-ministerio-del-tiempo-vuelve-a-la-1-el-lunes-25-de-abril/ (2016)
27. FormulaTV.: The Ministry of Time. FormulaTV. https://www.formulatv.com/series/el-ministerio-del-tiempo/audiencias/
28. RTVE, P.: El Ministerio del Tiempo', the only European Public Television Series in the 'Shows on the Rise' Ranking of 'TV Time'. RTVE.es. https://www.rtve.es/rtve/20200513/ministerio-del-tiempo-unica-serie-television-publica-europea-ranking-shows-on-the-rise-tv-time/2013955.shtml (2020)

29. Torregrosa-Carmona, J.-F., Rodríguez-Gómez, E.: Fan communities and television fiction. case study: the ministry of time (TVE). Profesional de la información. **26**, 1139–1148 (2017)
30. Miranda-Galbe, J., Cabezuelo-Lorenzo, F., López-Medel, I.: Analytical model of transmedia storytelling ecosystems in audiovisual fiction: the spanish model of «The Ministry of Time». Commun. Soc. **34**, 1–13 (2021)
31. Miranda Galbe, J. & Figuero Espadas, J. The Role of the Prosumer in the Transmedia Narrative Expansion of Fictional Stories on Television: The Case of 'El Ministerio del Tiempo'. (2016).
32. Scolari, C. A. The Translector. Reading and Transmedia Narratives in the New Ecology of Communication. (2016).
33. Rovira-Collado, J., Llorens García, R.F., Fernández Tarí, S.: A Transmedia Proposal for Literary Education: The Ministry of Time. University of Alicante. Institute of Education Sciences (2016)
34. Lozano, S.: A look at the past: gender stereotypes and archetypes in Spanish historical series (2011-2018). Commun. Med. **67** (2020). https://doi.org/10.5354/0719-1529.2020.54276
35. Cabezuelo-Lorenzo, F., Miranda-Galbe, J., Barceló-Sánchez, J.-M.: Expansion strategies in transmedia fiction projects: the case of the ministry of time. IROCAMM-Int. Rev. Commun. Market. Mix. **2**, 22–35 (2020)

Enhancing Legal Transparency: Interpretable Legal Document Classification with Hybrid BERT-RF Models

Paulo Henrique B. Rodrigues[1(✉)], Daniel Alves Silva[1], Fabio Lucio L. de Mendonca[1], Edna Dias Canedo[1], Georges Daniel Amvame Nze[1], and Geraldo Pereira Rocha Filho[1,2]

[1] University of Brasilia, Brasília, Brazil
paulohbr@gmail.com, {daniel.alves,fabio.mendonca}@redes.unb.br, {ednacanedo,georges}@unb.br, geraldo.rocha@uesb.edu.br
[2] Southwest Bahia State University, Jequié, Brazil

Abstract. The application of Artificial Intelligence (AI) in the legal domain has advanced significantly, yet many existing models lack transparency and interpretability, which are critical requirements in high-stakes, public sector contexts. This paper introduces JURISBERT, a hybrid and interpretable solution for the automatic classification of legal documents. The proposed model combines deep contextual embeddings generated by BERT with external legal knowledge integration via Retrieval-Augmented Generation (RAG), and leverages Random Forest (RF) classifiers for the classification of legal documents. JURISBERT was evaluated on a comprehensive dataset of 15,000 tax-related legal petitions from multiple Brazilian federal courts. Experimental results demonstrate that the model achieves high classification performance, reaching 93% accuracy, while statistically reducing the prediction time when compared to traditional approaches from the literature. These results highlight JURISBERT's suitability for scalable, transparent, and efficient deployment in public legal infrastructures.

Keywords: Legal NLP · Document classification · Interpretability · BERT · Random Forest · RAG

1 Introduction

The exponential growth of legal documents processed by public institutions has created significant challenges for legal professionals, who must classify, analyze, and extract information with speed and precision [1, 2]. In this context, Artificial Intelligence (AI) has emerged as a promising solution for automating repetitive tasks and supporting decision-making. However, extracting structured knowledge from legal texts remains difficult due to linguistic variability, heterogeneous case structures, and the continuous evolution of legal terminology, which hinder standardization for automated processing

A. Rocha et al. (Eds.): ICITS 2026, LNNS 1966, pp. 403–413, 2027.
https://doi.org/10.1007/978-3-032-27538-7_36

[2, 3]. In addition, specialized legal vocabulary limits the effectiveness of conventional text mining techniques [4, 5]. At the same time, the growing scale of judicial records increasingly exceeds human capacity for manual analysis [6].

Recent advances in Natural Language Processing (NLP) have introduced transformer-based models capable of learning contextual representations from unstructured legal text. Among these, BERT (Bidirectional Encoder Representations from Transformers) [7–9] stands out for its bidirectional attention mechanism, which captures syntactic and semantic dependencies while preserving the meaning of legal terminology and procedural constructs. To further enhance contextual relevance, Retrieval-Augmented Generation (RAG) [10] incorporates external legal knowledge, such as jurisprudence and prior case records. This integration reduces ambiguity and enriches the classification process with broader legal context.

Several studies have explored automated legal text classification using traditional supervised learning algorithms, including Support Vector Machines (SVM), Random Forest (RF), Bayesian networks, and Multilayer Perceptrons (MLP) [11–14]. These approaches typically rely on TF-IDF representations or manually engineered features and perform well in constrained scenarios. However, their dependence on annotated data and rigid legal taxonomies limits scalability and cross-jurisdictional adaptability. More recent deep learning approaches, such as Long Short-Term Memory (LSTM) networks and BERT-based models, improve classification accuracy by capturing complex language patterns [2, 15]. Despite these advances, challenges related to imbalanced datasets, high computational cost, limited interpretability, and lack of external legal knowledge integration persist. In particular, the combination of contextual embeddings with interpretable classifiers remains underexplored, especially for short legal texts with dense technical vocabulary. To the best of our knowledge, no prior work integrates BERT-based contextual encoding with RF classification while enhancing semantic grounding through RAG.

This work presents JURISBERT, a hybrid architecture designed for legal document classification that balances performance and interpretability. The model combines contextual embeddings generated by BERT with external legal knowledge retrieved via RAG and employs Random Forest classifiers to produce interpretable predictions. Evaluated on a dataset of 15,000 legal petitions from multiple Brazilian jurisdictions, JURISBERT achieved 93% classification accuracy and reduced processing time by up to 42 times compared to conventional machine learning baselines. These results demonstrate its suitability for real-world legal applications requiring accuracy, transparency, and accountability.

The remainder of this paper is organized as follows. Section 2 reviews related work. Section 3 describes the proposed architecture. Section 4 presents the dataset, while Sect. 5 details the experimental setup and results. Finally, Sect. 6 concludes the paper and outlines future work.

2 Related Works

Several studies have addressed legal document extraction and classification using data mining and machine learning techniques [2, 11–13]. Most of these approaches rely on predefined patterns or static representations, which struggle to accommodate the continuous evolution of legal concepts. As highlighted by Few [16], effective analysis of complex data requires strong domain knowledge, critical thinking, and statistical reasoning [17], aspects that are often underexplored in automated legal systems. Moreover, ethical reasoning and interpretability remain largely absent from existing solutions.

Barros [11] investigated labor court decisions using Bayesian networks fed with TF-IDF representations to classify rulings according to the favored party. Although effective in structured settings, this approach depends heavily on prior knowledge quality and does not scale well to large and heterogeneous datasets. In contrast, our approach is designed to handle large volumes of legal data while maintaining prediction efficiency and interpretability.

Silva et al. [12] proposed a comparative text mining methodology to evaluate supervised classifiers over large jurisprudence datasets. While the results demonstrated solid performance, the authors highlighted that excessive training time and increasing dataset size may negatively impact performance, requiring careful reparameterization. These limitations restrict practical deployment in real-world environments.

Dantas et al. [13] explored direct and hierarchical classification strategies for Portuguese tax-related legal texts using SVM and LSTM models. Due to data scarcity in some categories, class reduction was required. Although SVM showed good generalization under imbalance, LSTM struggled with similar classes and low-frequency categories, reinforcing the sensitivity of deep learning models to class imbalance.

Chen et al. [14] introduced a Random Forest-based legal classification model using domain-specific concepts extracted from U.S. court decisions. While effective in specific contexts, the approach depends on high-quality concept extraction and lacks automatic generalization to other legal domains, limiting scalability.

Overall, prior works demonstrate meaningful progress in legal text classification but exhibit limitations such as reliance on fixed rules, sensitivity to imbalanced data, high computational cost, or limited generalization. The proposed approach addresses these gaps by combining contextual embeddings, external legal knowledge integration, and interpretable classification, enabling robust performance across large, heterogeneous, and evolving legal datasets.

3 JURISBERT Architecture and Workflow

This section presents JURISBERT, a hybrid model developed for the classification of legal documents by integrating a BERT-based text representation stage, a contextual enrichment process that incorporates external legal knowledge through RAG, and a classification procedure carried out using an RF model, as illustrated in Fig. 1. In JURISBERT, legal documents are initially preprocessed and converted into dense vector representations using BERT, which captures both syntactic structure and semantic context. These embeddings are then enhanced through a retrieval process that queries a legal knowledge base using similarity metrics to retrieve relevant content, such as previous rulings or normative texts, which is appended to the original input. The combined content is re-encoded to reflect the augmented context, allowing the model to disambiguate terminology and align the text with legal precedent. Finally, enriched embeddings are classified using an RF model, which assigns a legal category to each document and provides feature-level interpretability.

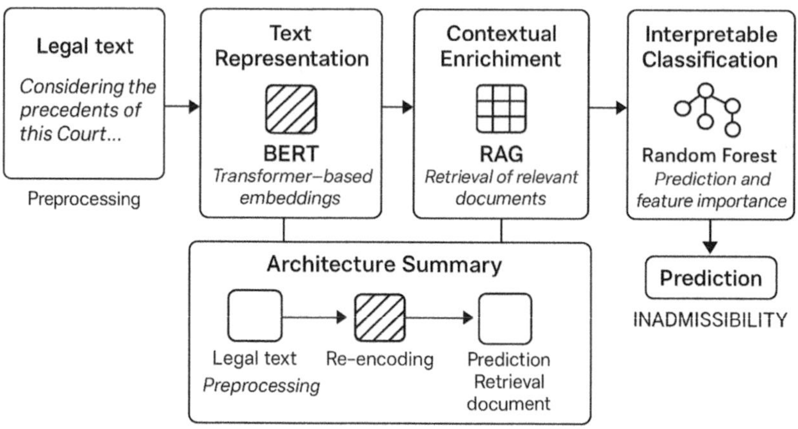

Fig. 1. JURISBERT operating scenario.

3.1 Legal Text Processing and Representation

The classification pipeline in JURISBERT begins with the extraction and preprocessing of unstructured legal texts from the Electronic Judicial Process (PJe) platform. This step is orchestrated through an automated workflow managed by Rundeck, which triggers a scraper module responsible for securely logging into judicial systems and retrieving raw legal petitions, as presented in Fig. 2.

Once extracted, the documents undergo a preprocessing stage designed to convert raw content into a format compatible with semantic representation models. This includes textual cleaning, removal of boilerplate terms, standardization of legal abbreviations, and tokenization tailored to the legal domain. To enhance semantic focus, JURISBERT employs a curated list of stopwords specific to the Brazilian legal context, which includes Latin expressions (e.g., ad hoc, prima facie), procedural terms (e.g., intimac¸a˜o, recurso), and institutional references (e.g., Conselho Nacional de Justic¸a). Table 1 provides examples of stopwords removed during preprocessing.

JURISBERT is deployed within a modular data hub architecture designed for real-world production environments, providing automation, traceability, and scalability across the legal data processing pipeline. The infrastructure consists of three main components: (i) a Search Engine that collects legal data from public sources under controlled outbound-only internet access; (ii) a Pentaho Data Integration (PDI) environment responsible for data transformation and structuring; and (iii) an orchestration layer

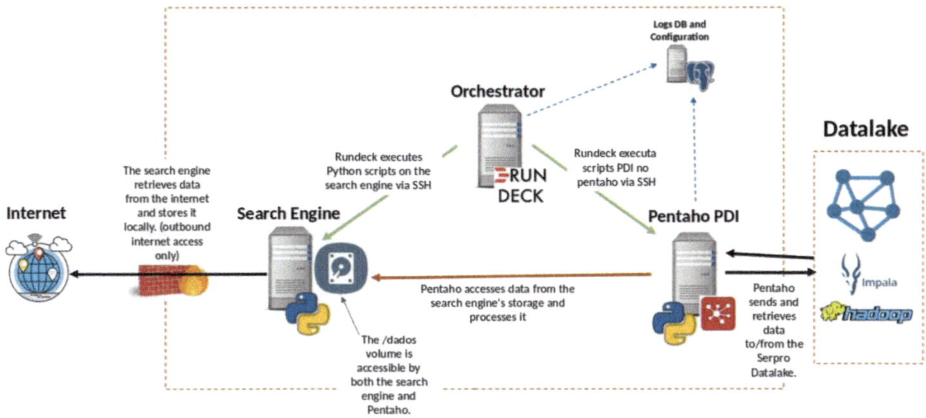

Fig. 2. Real-World Data Hub Environment.

Table 1. Examples of Stopwords Removed in JURISBERT's Legal Text Preprocessing

Legal Stopwords(PT)	Legal Stopwords (EN)	Context-Specific (PT)	Context-Specific (EN)
ad hoc	ad hoc	petic¸a˜o	petition
prima facie	prima facie	embargos	motions
habeas data	habeas data	despacho	ruling
in verbis	in these words	parecer	legal opinion
intimac¸a˜o	summons	agravo	interlocutory appeal
doutrina	legal doctrine	of'ıcio	official letter
su'mula	legal precedent note	recurso	appeal/remedy

implemented with Rundeck, which coordinates Python scripts and Pentaho workflows through secure SSH execution.

Data extraction is scheduled and executed on the Search Engine, with outputs stored in a shared volume accessible by both the Search Engine and PDI for preprocessing and preparation of structured inputs for classification and contextual enrichment. The architecture integrates a Hadoop-and Impala-based datalake to support distributed storage and scalable querying, enabling bidirectional data exchange via Pentaho. During the RAG phase, this setup facilitates access to case law, statutes, and legal doctrine, while the orchestration layer records metadata and performance metrics in a PostgreSQL database, ensuring governance, reproducibility, and auditability.

3.2 Contextual Enrichment and Classification

As illustrated in Algorithm 1, after pre-processing where raw legal text is cleaned, normalized, and tokenized (lines 1–4), the input is passed through JURISBERT's semantic enrichment and classification components. The processed document is encoded using BERT, generating contextualized embeddings that capture syntactic dependencies and semantic information within the legal domain. These embeddings provide strong internal representation but may lack external legal context, particularly for implicit references to precedent or legislation, reflecting the inherent challenge of purely distributional representations in domains relying heavily on intertextuality and normative references.

Algorithm 1: Contextual Enrichment and Classification

```
   Input: LegalDocument
   Output: PredictedLegalCategory
1  Function PreprocessDocument(doc):
2      cleaned_text ← RemoveStopwords(doc)
3      normalized_text ← NormalizeAbbreviations(cleaned_text)
4      return Tokenize(normalized_text)
5  Function RetrieveContext(text):
6      embedding_query ← BERT_Encode(text)
7      retrieved_docs ← SearchLegalKnowledgeBase(embedding_query)
8      return SelectTopK(retrieved_docs, k)
9  Function EnrichRepresentation(text, context):
10     combined_input ← Concatenate(text, context)
11     return BERT_Encode(combined_input)
12 Function PredictCategory(embedding):
13     return RandomForestClassifier.Predict(embedding)
14 tokens ← PreprocessDocument(LegalDocument)
15 retrieved_context ← RetrieveContext(tokens)
16 enriched_embedding ← EnrichRepresentation(tokens, retrieved_context)
17 predicted_category ← PredictCategory(enriched_embedding)
18 return predicted_category
```

To overcome this gap, JURISBERT integrates a RAG mechanism [10] (lines 5–8). Given a query embedding produced by BERT, the system performs a similarity-based search over a semantically indexed legal knowledge base, retrieving the top-k most relevant documents, which may include jurisprudential excerpts, statutory definitions, or regulatory provisions. The choice of RAG over static embedding augmentation lies in its adaptability, allowing dynamic incorporation of updated legal information without retraining the base model. These retrieved segments are concatenated with the original document to form an enriched input, which is then re-encoded using BERT to generate a composite embedding that integrates both the petition content and the contextual knowledge retrieved from external sources (lines 9–11).

This enriched embedding is then passed to the classification module, which employs an RF classifier (lines 12–13). The RF model was chosen for its robustness in high-dimensional feature spaces and its interpretability through feature importance analysis. Additionally, compared to deep neural classifiers, RF offers lower training costs and reduced overfitting risks when dealing with limited annotated legal datasets. During training, each tree in the forest is constructed from bootstrapped samples of enriched embeddings, learning rules that associate semantic patterns with specific legal categories, such as "Ordinary Procedure", "Writ of Mandamus", or "Special Court Procedure". At inference time, the final label is obtained through majority voting across all decision trees. Importantly, the contribution of each feature to the classification decision can be quantified, enabling transparent and auditable reasoning, an aspect of particular relevance in the legal domain, where accountability and explainability are essential to foster trust in automated decision-support systems.

4 Dataset

The dataset used to train and evaluate JURISBERT was built in collaboration with the Office of the Deputy Attorney General for Judicial Representation (PGAJUD) and consists of 15,000 tax-related legal cases extracted from the electronic systems of the Regional Federal Courts (TRF1–TRF6), as well as National and Special Courts. Data were collected between January 2021 and December 2023, capturing temporal variations in litigation and jurisprudence. All records were anonymized in compliance with Brazil's General Data Protection Law (LGPD), and incomplete or corrupted entries were removed. The dataset covers proceedings from all six Regional Federal Courts, with TRF1 and TRF3 jointly representing 45% of the cases, reflecting national population and economic activity. Tables 2, 3, and 4 summarize the distribution and provide dataset examples.

Table 2. Distribution by Type of Tax Proceeding

Type of Tax Case	Quan.	%
Tax Enforcement Action	4500	30.0
Motion to Stay Enforcement	2250	15.0
Tax-Related Writ of	1875	12.5
Mandamus		
Debt Annulment Action	1500	10.0
Repetition of Undue Payment	1125	7.5
Tax Compensation Claim	900	6.0
Administrative Appeal	750	5.0
Other Tax Proceedings	2100	14.0
Total	**15000**	**100.0**

Table 3. Distribution by Regional Federal Court

Regional Court	Quan.	%
TRF1 (DF, GO, MG, BA)	3750	25.0

(*continued*)

Table 3. (*continued*)

Regional Court	Quan.	%
TRF2 (RJ, ES)	2625	17.5
TRF3 (SP, MS)	3000	20.0
TRF4 (RS, SC, PR)	2250	15.0
TRF5 (PE, CE, AL)	1875	12.5
TRF6 (MG)	1500	10.0
Total	**15000**	**100.0**

Table 4. Example of Dataset Records

Case	ProsecutorClass Region	Class ID	Procedural Code	Class Name	Opening Text
01	1	101	1001	Ordinary Procedure	HIS HONOR THE JUDGE OF THE SECTION COURT...
02	2	104	1004	Writ of	HIS HONOR THE FEDERAL
				Mandamus	JUDGE OF THE SECTION
					COURT...
03	4	104	1004	Writ of	THE MOST HONORABLE
				Mandamus	FEDERAL JUDGE...
04	4	105	1005	Special	CASE:
				Court	1000008–86.2022.4.06.3814 –
				Procedure	PROCEDURE...

5 Evaluation Setup

To validate JURISBERT, the dataset was split using the hold-out method, with 60% for training, 30% for testing, and 10% for validation. Class weighting was applied during training to mitigate data imbalance while preserving the original distribution. Experiments were conducted using the HuggingFace Transformers framework with PyTorch, and hyperparameters were optimized through a Bayesian search strategy implemented with Hyperopt. Model performance was evaluated using prediction time, training time, and precision.

JURISBERT was compared against Random Forest, Multilayer Perceptron Neural Network (MLP-NN), and Stochastic Gradient Descent (SGD), all implemented with scikit-learn and individually optimized to ensure fair comparison. Experiments were executed on a server running Ubuntu 23.04, equipped with a 16-core Intel Core i7 processor, 32 GB of RAM, and four NVIDIA GeForce RTX 3060 Ti GPUs. The following section presents and discusses the experimental results.

5.1 Results and Discussion

As shown in Fig. 3, JURISBERT exhibited the lowest prediction time with low variability, while RF, MLP-NN, and SGD showed similar median times but greater dispersion. This performance is attributed to JURISBERT's pre-trained encoder, which converts inputs into fixed embeddings, reducing computational complexity during inference. In terms of training time, JURISBERT significantly outperformed the baselines due to pre-training on legal corpora, requiring only task-specific adaptation. Regarding precision, RF and RNA-MLP achieved the highest median values, while JURISBERT demonstrated stable performance with low dispersion, indicating consistent classification behavior.

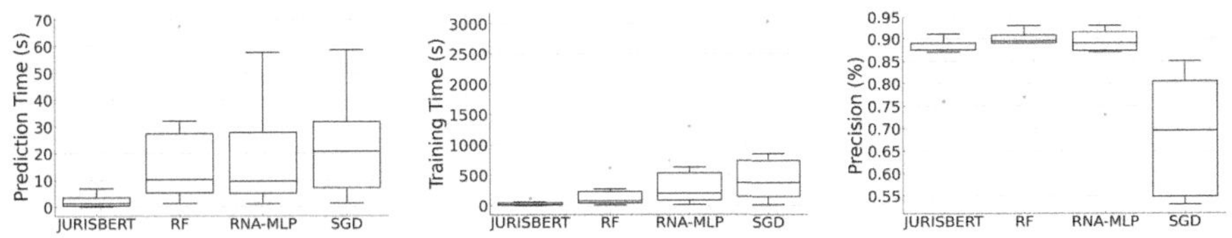

(a) Prediction time (s) (b) Training time (s) (c) Precision (%)

Fig. 3. Results obtained for the metrics: prediction time, training time, and precision.

After analyzing the results, a statistical evaluation was conducted using non-parametric methods. The Friedman test was applied, followed by the Nemenyi post-hoc test, to assess whether the observed differences were statistically significant. The critical difference (CD) defines the minimum interval required for differences between algorithms to be considered significant.

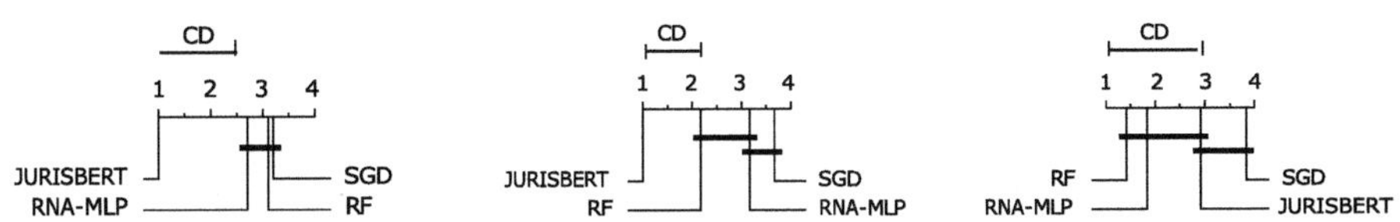

(a) Prediction time (b) Training time (c) Precision

Fig. 4. Statistical difference and ranking of the metrics: prediction time, training time, and precision.

In Fig. 4a, which refers to prediction time, JURISBERT achieved the best average ranking and was statistically superior to the other algorithms (RNA-MLP, RF, and SGD) according to the CD criterion. This superiority is also evident in Fig. 4b, which concerns training time, where JURISBERT once again led the ranking with statistically significant differences compared to RNA-MLP and SGD. Finally, Fig. 4c presents the ranking for the precision criterion. Although RF and RNA-MLP occupied the top positions, JURISBERT demonstrated effective performance, showing no statistically significant difference from these two algorithms, as indicated by the critical difference interval. These findings reinforce the robustness of JURISBERT, which, while more efficient in terms of time, maintains precision levels comparable to the best-performing models.

6 Conclusions

This work presents JURISBERT, a hybrid solution combining BERT contextual representations with external legal knowledge via RAG and RF as classifier for interpretable legal document classification. Designed to handle domain-specific vocabulary while ensuring high throughput and low training costs, JURISBERT demonstrates superior prediction time and training efficiency with precision statistically equivalent to baseline models, as confirmed by CD analysis. These findings establish pre-trained transformers as a promising alternative for legal applications requiring low latency, scalability, and interpretability. Future work includes exploring compression techniques like quantization and knowledge distillation to enhance computational efficiency and extending applications to other legal subdomains.

Acknowledgement. The authors thank the technical and computational support provided by the LATITUDE Laboratory at the University of Brasilia, the TED 01/2021 from the National Social Assistance Secretariat (SNAS/DGSUAS/CGRS) and from the General Coordination of Information Technology (CGTI) of the Federal Revenue Attorney General's Office (PGFN), the SISTER City Project – Secure and Real-Time Intelligent Systems for Smart Cities (Grant 625/2022), the System for Control and Unification of Projects for the Federal District Government – SISPRO-DF Project (Grant 497/2023), the Research, Development and Application – Methodology to Support the Elicitation of Ethical and Privacy Requirements Project (Grant 514/2023), the Research and Innovation Dean's Office (DPI/UnB), and FAP/DF.

References

1. van Noortwijk, K., van Noortwijk, K.: Automatic document classification in integrated legal content collections (2017). URL http://hdl.handle.net/1765/100321
2. Enamoto, L., Santos, A.R., Maia, R., Weigang, L., Filho, G.P.R.: Multi-label legal text classification with bilstm and attention. Int. J. Comput. Appl. Technol. (2022). https://doi.org/10.1504/ijcat.2022.125186
3. E. Mumcuoğlu, C.E. Öztürk, H.M. Ozaktas, A. Koc: Natural language processing in law: Prediction of outcomes in the higher courts of turkey (2021)
4. Xi, R., Zhenxing, K.. Hierarchical RNN for information extraction from lawsuit documents (2018)
5. Deriu, J., et al.: Survey on evaluation methods for dialogue systems (2021)
6. Zhong, H., Xiao, C., Tu, C., Zhang, T., Liu, Z., Sun, M.: How does NLP benefit legal system: a summary of legal artificial intelligence (2020)
7. Devlin, J., Chang, M.W., Lee, K., Toutanova, K.: Proceedings of the 2019 Conference of the North American Chapter of the Association for Computational Linguistics: Human Language Technologies, vol. 1, pp. 4171–4186 (2019)

8. Devlin, J., Chang, M.W., Lee, K., Toutanova, K.: Bert: pre-training of deep bidirectional transformers for language understanding (2019)
9. Luz de Araujo, P.H., de Campos, T.E., de Oliveira, R.R., Stauffer, M., Couto, S., Bermejo, P.: Computational Processing of the Portuguese Language: 13th International Conference, PROPOR 2018, pp. 313–323, Canela, Brazil (2018)
10. Lewis, P., et al.: Retrieval-augmented generation for knowledge-intensive NLP tasks (2020)
11. Barros, R.P.L.: in Congresso (2019). URL https://rhuanlopesbarros.jusbrasil.com.br/artigos/529722364/analise-jurisprudencialcom-tecnica-de-aprendizado-de-maquina. Acesso em: 04 de outubro de 2023
12. da Silva, E. C. M., Medeiros, B.A.: Comparac¸a˜o de me´todos de minerac¸a˜o de texto para classificac¸a˜o de documentos jur´ıdicos (2020)
13. Dantas, R., Figueiredo, K., Andrade, L.: Estudo de abordagens para classificac¸a˜o de textos sobre du´vidas tributa´rias utilizando minerac¸a˜o de texto (2021)
14. Chen, H., Wu, L., Chen, J., Lu, W., Ding, J.: A comparative study of automated legal text classification using random forests and deep learning. Inf. Process. Manag. **59**(2), 102798 (2022)
15. Kieckbusch, D.S., Geraldo Filho, P, Di Oliveira, V., Weigang, L.: SCAN-NF: a CNN-Based System for the Classification of Electronic Invoices Through Short-Text Product Description (2021)
16. Few, S. Show me the numbers (2004)
17. Lowgren, J., Carroll, J.M., Hassenzahl, M, Erickson, T., Blackwell, A.: The Encyclopedia of Human-Computer Interaction (2019)

Bibliometric Review on the Contributions of Artificial Intelligence to Energy and Climate Sustainability: A Thematic Study of Scientific Literature

Mikel Ugando-Peñate[1(✉)], José Luis Espín-Lopez[1], Mishell Alejandra García-Acosta[1], Reinaldo Armas-Herrera[2], Angel Alexander Higuerey-Gómez[2,3], and Ángel Ramón Sabando-García[1]

[1] Pontificia Universidad Católica del Ecuador, Sede Santo Domingo, Santo Domingo de Los Tsáchilas, Santo Domingo 230203, Ecuador
{mugandop,jlespin,magarciaac,arsabando}@pucesd.edu.ec

[2] Universidad Técnica Particular de Loja, Loja 110150, Ecuador
{ahreinaldo,aahiguerey}@utpl.edu.ec

[3] Instituto Experimental de Investigaciones Humanísticas, Económicas y Sociales de la Universidad de los Andes, Trujillo 3101, Venezuela

Abstract. Artificial intelligence (AI) has emerged as a transformative tool in the search for sustainable solutions to current energy and climate challenges. This thematic study aims to explore recent contributions of AI to strengthening energy sustainability and mitigating climate change, integrating technological, environmental, and economic perspectives. Through a systematic review of sources such as Scopus, innovative approaches such as smart grid optimization, predictive energy consumption management, climate modeling, and the efficiency of sustainable industrial processes are analyzed. The thematic analysis was conducted using VOSviewer software to identify emerging trends, key authors, and recurring concepts in the field. Machine learning and distributed intelligence make it possible to predict energy patterns, reduce emissions, and improve environmental decisions. AI, integrated ethically and through multidisciplinary approaches, represents a key ally in achieving sustainability and combating climate change.

Keywords: artificial intelligence · sustainability · climate change · machine learning · networks · thematic analysis · energy efficiency

1 Introduction

One of the most significant advances at the intersection of artificial intelligence and sustainability is energy optimization in tourism settings [1], and in remote areas using off-grid solar panel systems [2]. These applications demonstrate how AI can adapt to diverse socioeconomic and geographic contexts, maximizing the use of renewable energy. Furthermore, the potential of small hydropower plants as a sustainable energy solution has

A. Rocha et al. (Eds.): ICITS 2026, LNNS 1966, pp. 414–425, 2027.
https://doi.org/10.1007/978-3-032-27538-7_37

been explored [3], as has the spatio-temporal effects of economic development on carbon emissions intensity in China [4]. AI has also enabled the analysis of carbon price fluctuation risks in emerging energy markets [5].

At the enterprise level, predictive models have been applied to identify drivers of carbon emissions disclosure [6], as well as off-grid methodologies for energy design in medium-sized settlements [7]. In urban environments, the energy performance of government buildings has been investigated [8], and the impact of different CO_2/N_2 ratios on the pore structure of coal has been studied [9].

Artificial intelligence has driven the development of models for corporate emissions disclosure [10] and for understanding ecological connectivity and its relationship to climate change, such as in the distribution of invasive termites [11]. Further analysis of the spread of invasive species has also been undertaken [12], and AI has been used to model the distribution of bioenergy crops [13].

The relationship between climate change and the spread of the bacterial resistome in soils has been another area investigated with AI [14], as has the analysis of energy performance in industrial enterprises in Vietnam [15]. The footprints of economic capital have also been identified in regions such as Portugal [16], and the performance of advanced agricultural technologies in relation to energy consumption has been assessed [17].

In technological terms, research on the performance of nanoparticulate panels [18] stands out, as well as studies on low-cost optical monitors for real-time black carbon measurement [19]. There has also been work on the sustainable extraction of essential oils using solar steam generators [20]. In addition to studies on water management in fuel cells for emission-free ships [21].

At the territorial level, AI has been applied to measure energy eco-efficiency in the Yangtze River Delta [22] and the use of organic Rankine cycles with hydrogen has been explored for energy management in buildings [23]. AI has also been used to analyze the nonlinear relationship between coal revenues and environmental sustainability in E-7 countries [24].

From a broader environmental perspective, AI has been used to model terrestrial ecosystems with random forests [25], and to calculate equitable carbon liability through trade flows [26]. In vulnerable settlements, traces of emplaced capital and human impact on ecological infrastructure have been traced, as in the Himalayas [27, 28].

In addition, cost-benefit analyses of wastewater treatment in digital economies have been conducted [29], research on the role of AI in automated scientific content generation [30], and methodological reviews of the impact of household consumption on carbon footprints [31].

Within the framework of the digital economy, sustainable energy strategies have been designed [32], along with the use of AI to predict glacial melting [33], and to analyze the stability of igneous rocks for CO_2 storage [34]. Anthropogenic NOx emission trends in China after 2020 have also been assessed [35].

From an industrial perspective, waste policies have been linked to resilience to extreme events [36], and social perception modeling of carbon-neutral cities has been studied through social networks [37]. Other studies have addressed the challenges of solar energy adoption in MENA regions [38].

Likewise, carbon sinks have been integrated into urbanization plans [39], and AI has been used to project bioenergy crops in fragile ecosystems [40]. Key examples of this are the optimization of sustainable fuels with probabilistic deep learning [41] and the analysis of sugarcane bagasse use through life cycle assessment [42].

Gasification of textile waste using advanced thermodynamic modeling [43], redesign of anaerobic digestion systems for bioenergy production [44], and the use of AI to improve the efficiency of air conditioning systems and predict pollutant gases in urban and residential contexts [45] have also been investigated. The most recent contributions expand this panorama further. Benchmarking studies of greenhouse gas emission forecasting models provide new insights for global climate policy [46], while hybrid AI approaches have improved environmental impact modeling in agricultural systems [47]. Human-caused degradation of ecological infrastructures in Himalayan urban agglomerations has been reassessed using updated datasets and AI-supported methods [48]. Furthermore, case-based reasoning has been enhanced to support sustainable rural development strategies [49], and AI-driven methodologies such as STIRPAT-GRU with transfer learning have been introduced to design carbon-neutrality strategies in domains as specific as sports event management [50].

2 Objective

The objective of this paper is to explore the contributions of artificial intelligence to energy sustainability and climate change mitigation through a thematic analysis of recent scientific literature. The study seeks to identify technological, environmental, and economic approaches that integrate the use of AI in processes such as energy consumption optimization, renewable energy management, climate modeling, and sustainable policy-making, in order to understand its impact and evolution within the framework of global research.

3 Methodology

3.1 Methodology and Data Subsection

To develop this research, a mixed-methodological approach was applied, combining a traditional literature review with bibliometric analysis techniques. The main objective was to identify the main scientific contributions to the use of artificial intelligence to promote energy sustainability and mitigate the impact of climate change. The literature collection was carried out through searches in recognized academic databases such as Scopus, covering publications between the years 2019 and 2025. The following query strings were used: (TITLE-ABS-KEY (artificial intelligent AND sustainability AND impact AND energy) AND TITLE-ABS-KEY (machine learning AND, climate change AND enviromental). Relevance filters were applied by thematic areas, language, type of document and access to the full text.

As a result of this process, 168 scientific documents were initially identified that addressed the use of artificial intelligence in contexts of sustainability and climate change. However, through a filtering process based on criteria of thematic relevance from 2022, methodological quality, and availability of complete information, a final set of 45 peer-reviewed articles was selected. These documents make up the analytical corpus of the present research, integrating technological, environmental and economic perspectives, and representing the most relevant contributions regarding the transformative potential of AI in the field of energy sustainability.

3.2 Bibliometric Analysis

The VOSviewer software has been used to perform the bibliometric analysis. It allows you to create maps of keywords, authors, citations by country and thematic networks. The data were grouped into clusters based on thematic similarity. These evidence consolidated and emerging research ideas.

It was observed that there is an interrelationship between artificial intelligence, energy efficiency, climate change, renewable energies and sustainable policies. In addition, the authors, countries and articles with the greatest impact were identified.

This bibliometric analysis provides understanding in the field of research and innovation in the application of new technologies in sustainable development.

4 Results

4.1 Keywords Co-occurrence Network Map

Figure 1 shows a map of co-occurrence, where five clusters are identified in the main themes of climate change. The most important cluster is the climate change cluster with 17 occurrences and a force of 36. It is related to the terms "global warning", "adaptation" and "mitigation", which highlights the relationship between the effects of climate change and adaptation to it. The second cluster is "machine learning", which is related to "fossil fuel", "artificial neural network", among others. Its attendance is 16 and the force is 37. The third most important cluster is "renewable energy" with 9 occurrences and a strength of 16. The rest of the clusters are less important and are related to "policies", "environmental impact" and "sustainability".

The impact of climate change means that we must mitigate its effects and adapt to this new reality; this has been more pronounced in developing countries due to their limited resources. The use of artificial intelligence will enable the development of new technologies to reduce the use of fossil fuels and generate other renewable energy sources, such as green hydrogen.

4.2 Author Keywords Co-occurrence Network Map

Figure 2 shows the analysis of co-occurrence of the keywords, highlighting the word "climate change", followed by "greenhouse gases" and "sustainability". These terms form the central nodes that connect authors and themes. These words reflect the multidisciplinary nature where the causes and competencies of this and the strategies that can be applied to mitigate its effects.

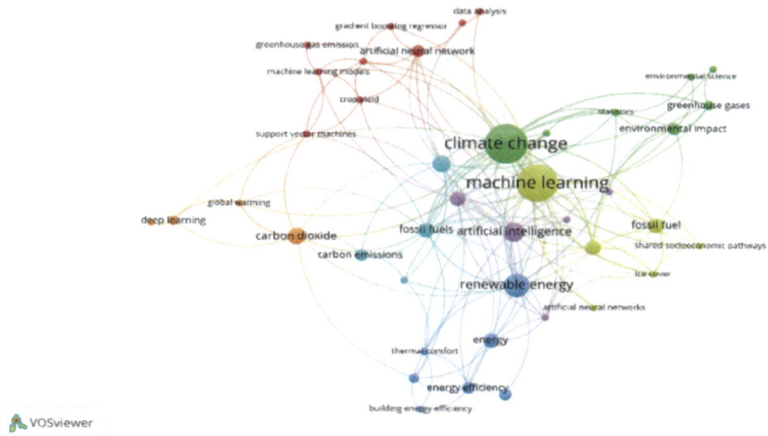

Fig. 1. Bibliometric network map with all keywords.

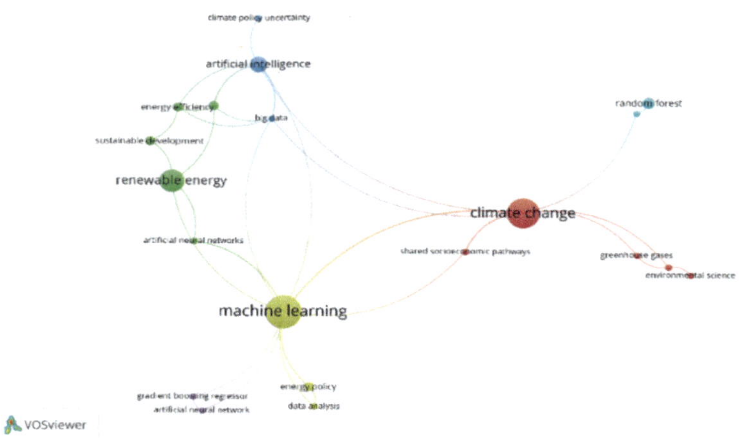

Fig. 2. Bibliometric network map with author keywords.

4.3 Citation Author Network Map

Table 1 identifies the authors with the highest number of citations and highlights those researchers who have had a substantial impact on consolidating knowledge about climate change and related areas.

The most influential work and authors are summarized, classified according to their main areas of research. Tudor, C. and Sova, R., both with 38 citations, are leading figures in studies applying predictive models to effective $CO2$ emission reduction policies [46], highlighting their interaction in diverse economic contexts. El-hami Behzad, with 11 citations, stands out for his contribution to climate adaptation and the evaluation of environmental policies, demonstrating his leadership in building theoretical and practical frameworks [47]. There is also evidence of active scientific communities in topics such as climate adaptation and mitigation (Srivastava, Prashant, with 8 citations), low carbon economy and clean technologies [48], renewable energy, climate models and sustainable development (Wang, Yameng with 6 citations) and related topics in Impacts of climate change and artificial intelligence (Zhan, Ying with 5 citations) and reflecting a growing specialization and new multidisciplinary collaborations [49, 50] according to specific fields. A summary is presented in Table 1.

Table 1. Citation author for thematic area and Results obtained

Author	Main Area and Period of Study	Thematic area	Results obtained	Citations
Tudor, C. Sova, R. (2021)	Environmental policy evaluation, Study period: 1970–2018	CO2 emissions prediction models	The findings indicate that major polluters will not meet the emissions reduction targets set in the Paris Agreement's NDCs, highlighting the urgency of implementing more effective policies to achieve these goals.	30
Elhami, B. (2022)	Climate adaptation and mitigation	Climate impact and climate models	Soft computing modeling showed that the ANFIS (GP) model was faster and more accurate than the MLP model. In the comparison, the best models obtained an RMSE of 10,649 and 52,321 kg·ha^{-1} for yield, and 25,08 and 40,94 pPt·ha^{-1}, respectively.	11
Srivastava, P. (2024)	Renewable energy - Study period: 1994–2018	Sustainable energy policy	Implementation of appropriate policies to build an ecologically sustainable urban ecosystem in the Himalayas.	8

(*continued*)

Table 1. (*continued*)

Author	Main Area and Period of Study	Thematic area	Results obtained	Citations
Wang Y. (2024)	Climate models and sustainable development.	Computer Science, Sustainable development	It presents a new approach to formulating strategic responses to sustainable rural development using case-based reasoning (CBR), the bidirectional Choquet integral (BCCI), and the Shapley index (SI).	6
Zhan, Y. (2023)	Impacts of climate change.	Artificial intelligence, carbon emissions mitigation	It presents an AI approach that combines STIRPAT and GRU networks to analyze and predict carbon emissions at sporting events, improving accuracy and supporting strategies towards carbon neutrality.	5

4.4 Citation Countries Network Map

Figure 3 presents the countries with the highest number of publications and citations in climate change research. The most prominent nodes correspond to China with 322 citations in 55 documents, with the journal "Polish Journal of Environmental Studies," followed by the United Kingdom with 321 citations in 12 documents, with the journal "Plant Ecology and Diversity," and the United States with 203 citations, with the journal "Energy Policy," reflecting its leadership in scientific production. This concentration indicates that climate research is primarily conducted in countries with high scientific and technological capabilities, although there is also a growing participation of developing nations that contribute with case studies and regional approaches.

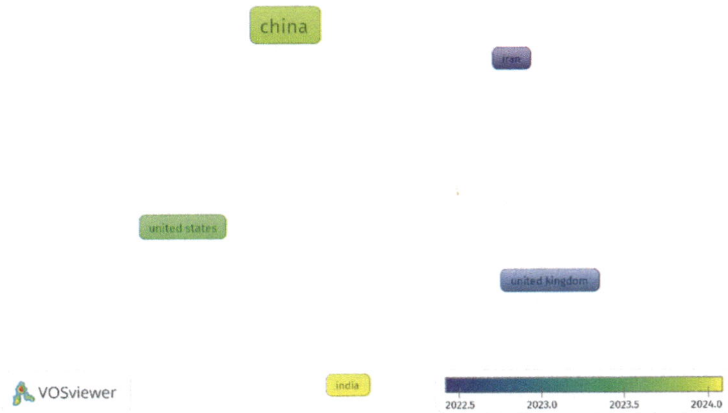

Fig. 3. Bibliometric network map with citation countries.

4.5 Citation Documents Network Map

Figure 4 identifies the most cited documents, which belong to Birks (2019) with 140 citations, according to his articles on the impact of climate change on natural and human systems in the journal "Plant Ecology & Diversity" (2019) with 12,326 views, while Levi had 90 citations according to "Energy Research & Social Science". This is followed by works focused on the analysis of mitigation policies and climate modeling according to Daryayehsalameh (2021) with 94 citations, indicating that these topics are highly relevant to the scientific community. This map allows us to identify the fundamental studies that have guided the development of knowledge on climate change.

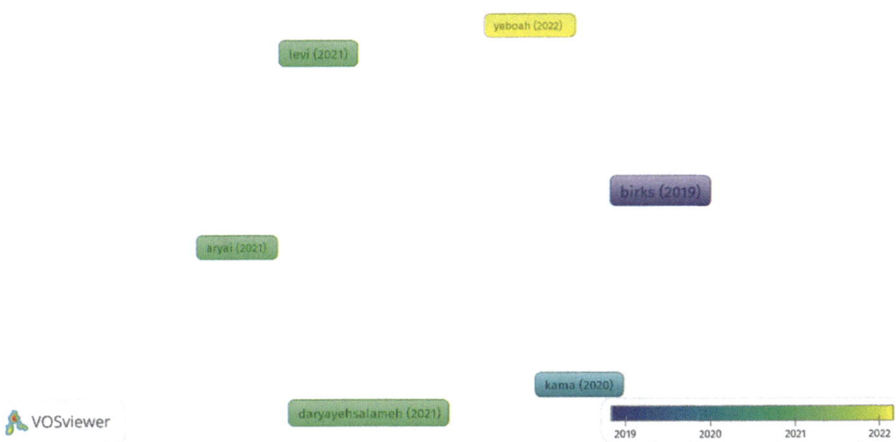

Fig. 4. Bibliometric network map with citation countries.

5 Project Financing

The research analyzed in this bibliometric review has been supported by funding from international organizations committed to scientific development in energy sustainability and climate change mitigation. This support has allowed for the consolidation of lines of research that integrate artificial intelligence into environmental, economic, and technological processes, generating applied knowledge to address global climate challenges from an interdisciplinary perspective.

In the analysis of the reviewed corpus, five organizations stand out as the main funders (Table 2): the National Natural Science Foundation of China with 27 projects, the European Commission with 7 initiatives, the National Key Research and Development Program of China with 6 studies, the Natural Sciences and Engineering Research Council of Canada and the Horizon 2020 Framework Programme with 5 grants each one. This strategic investment demonstrates the central role of scientific research in the transition to sustainable energy models and underscores the importance of public policies that promote technological innovation in the context of climate change.

Table 2. Project Financing

Entity	Number of studies
National Natural Science Foundation of China	27
European Commission	7
National Key Research and Development Program of China	6
Natural Sciences and Engineering Research Council of Canada	5
Horizon 2020 Framework Program	5

6 Conclusions

In addition to the findings already highlighted, it is necessary to recognize that the potential of artificial intelligence lies not only in its predictive and optimization capabilities, but also in its role as a catalyst for global collaboration between academic, industrial, and governmental sectors. This synergy is key to addressing complex challenges such as climate change, as it enables the sharing of knowledge, resources, and experiences that enrich technological solutions and promote their larger - scale implementation.

This study shows how interdisciplinary knowledge can accelerate the adoption of clean technologies adapted to diverse climate change mitigation contexts around the world. Cooperation between countries can improve technology transfer, improving sustainability, CO2 emission reduction and the energy transition as future areas of research under the concept of sustainable development.

Finally, a strategic opportunity is identified to strengthen flexible regulatory frameworks that keep pace with the pace of AI innovation, ensuring its ethical and effective integration into energy and climate sustainability systems. Formulating adaptive policies that foster both innovation and responsibility can be crucial in ensuring that the benefits of artificial intelligence translate into positive and lasting environmental impacts.

References

1. Suryawan, I.W.K., Rahman, A., Suhardono, S., Sutrisno, A.D., Nguyen, V.V.: Evaluating tourist preferences influences on renewable energy initiatives in coastal tourism settings. Reg. Stud. Mar. Sci. **85**, 104148 (2025)
2. Ajiwiguna, T.A., Kirom, M.R.: Uninterrupted electricity supply using off-grid solar PV systems for remote areas. Sustain. Energy Technol. Assess. **42** (2023)
3. Verma, S., Lamba, A., Agrawal, S., Birbal, P.Y., Rao, E.V.R.: Present scenario and challenges of small hydro power dam in India: a review. Larhyss J. **61**, 287–320 (2023)
4. Dou, R.Y., Liang, B.Y., Li, K.Y., Zhang, C., Zhao, Y.F.: Spatio-temporal effects of carbon emission intensity and economic development in China. Ecol. Indic. **167** (2023)
5. Pu, R.Y., Liang, Q.M., Wei, Y.M., Yan, S.Y., Ji, C.J.: Impact of China's new energy market on carbon price fluctuation risk. Energ. Strat. Rev. **59** (2023)
6. Jarboui, A., Mnif, E., Akrout, Z., Chakroun, S.: Drivers and predictors of carbon emissions disclosure using H2O AutoML. Int. J. Law Manag. (2023)
7. Tsiaras, E., Andreosatou, Z., Kouveli, A., Tampekis, S., Coutelieris, F.A.: Off-grid methodology for sustainable electricity in medium-sized settlements: the case of Nisyros Island. Clean Technol. **7** (2023)
8. Li, Z., Peng, S., Cai, W., Cao, S., Ma, X.: Impacts of building microenvironment on energy consumption in office buildings. Buildings. **13** (2023)
9. Wang, Z., Gong, H., Huang, M., Gu, F., Wei, J.: Multimodel random forest method for assessing terrestrial vegetation carbon density in China. Methods Ecol. Evol. **14**(1), 117–132 (2023)
10. Ge, Z., Wang, X., Yang, X., Fu, W., Zhao, X.: Impact of CO_2/N_2 ratios on anthracite pore structure evolution. Nat. Resour. Res. **34**, 2335–2354 (2023)
11. Duquesne, E., Fournier, D.: Connectivity and climate change drive the global distribution of invasive termites. NeoBiota. **91**, 281–314 (2023)
12. Lin, D., Du, S., Zhao, Z., Zhang, T., Wang, L.: Climate warming fuels the global antibiotic resistome in soils. Nat. Ecol. Evol. (2023)
13. Wu, Y., Yuan, P., Li, S., Guo, C., Yue, F.: Predicting biofuel crop distribution in Southern China's karst ecosystem. Agronomy. **14** (2023)
14. Diksha Kumar, A., Saikia, P., Srivastava, P.: Human-induced impacts on ecological infrastructure in Himalayan urban agglomerations. Ecol. Front. **14**, 84–95 (2023)
15. Tang Cam, N., Ngo Tuan, K., Nguyen Thanh, T., Do Anh, T.: Factors influencing energy-saving behavior in industrial enterprises in northern Vietnam. Int. J. Electr. Electron. Eng. **12** (2023)
16. Mirindi, D., Hunter, J., Sinkhonde, D., Bezabih, T.: Performance of machine learning to evaluate nanoparticle panels. Green Technol. Sustainability. **3** (2023)
17. Ge, X., Tanvir, R.U., Hu, Z., Hassanein, A., Lansing, S.: Rethinking anaerobic digestion for bioenergy and biopolymers production. Adv. Bioenergy. **9**, 1–109 (2023)
18. Mousavi, A., Ghasemi-Nejad-Raeini, M., Taki, M., Elhami, B.: Using bagasse from sugarcane via life cycle assessment. Bioresour. Technol. Rep. **26** (2023)
19. Zeeshan, M., Pande, R.R., Bhale, P.V.: Gasification of waste textile using thermodynamic equilibrium modeling. J. Therm. Eng. (2023)
20. Sameeullah Schauer, J., Javed, M., Bergin, M.H., Alam, K.: Development of low-cost optical monitor for real-time black carbon. Environ. Adv. **21** (2025)
21. Pise, V.H., Thorat, B.N.: Green steam for sustainable essential oil extraction using solar steam generator. Energy Nexus. **9**(2023) (2023)
22. Kaiser, R., Ahn, C.Y., Lee, S.Y., Kim, Y.H., Park, J.C.: Water management in PEM fuel cells for emission-free ships. J. Electrochem. Soc. **23** (2023)

23. Tian, Y.C., Li, M.Z., Tang, L., Zhao, S., Ma, T.: Energy eco-efficiency in the Yangtze River Delta using explainable machine learning. Appl. Ecol. Environ. Res. **22** (2023)
24. Assareh, E., Agarwal, N., Baji, H.S., Taghipoor, A., Lee, M.: Organic Rankine Cycle for building energy management with hydrogen. Energy Nexus. **13** (2023)
25. Qing, L., Li, P., Wang, Y., Mehmood, U.: Non-linear links between coal rents, FDI, and environmental sustainability in E-7. Resour. Policy. **83** (2023)
26. Wu, J., Wang, Y., Zhang, S., Zhu, Y., Fu, B.: Equitable carbon responsibility through trade-related emissions and urban carbon sinks. Environ. Sci. Ecotechnol. **24** (2023)
27. Gómez, R.E., Miño, M.G., Hojman, G.D., Arellano, D.M., Cossi, C.A.: Tracing emplaced capital footprints in Portugal. Comp. Sociol. **24**, 410–444 (2023)
28. Kang, H., Hou, J.: Cost-benefit analysis of sewage treatment in the digital economy era. J. Clean. Prod. **301** (2023)
29. Strzelecki, A.: ChatGPT-generated content in scientific publications: a new dilemma. Internet Things Cyber-Phys. Syst. **3**, 121–154 (2023)
30. Baz, K., Zhen, Z., Ali, H.: Impact of advanced agriculture technologies and energy consumption on crop yields. Food Energy Secur. **14**(2) (2023)
31. Rezaul Karim, S.M., Sarker, D., Kabir, M.M.: Analyzing the temperature impact on PV modules via ML. Cleaner Energy Syst. **9** (2023)
32. Du, J., Zhong, Z., Shi, Q., Cao, Y., Wang, L.: A review of household carbon footprint: methods and assessment. Carbon Footprints. (2023)
33. Han, Y., He, Z., Wu, S., Liu, Y., Wang, C.: Enhancing building energy efficiency via thermal mass optimization. Adv. Appl. Energy. **18** (2023)
34. Liu, B., Chang, H., Li, Y., Zhao, Y.: Green and low-carbon industrial transformation in Sichuan, China. Energy Reports. **12**, 861–871 (2023)
35. Chen, J., Zeng, Q.: Technological advancement in China's natural gas industry. Energy Res. Soc. Sci. **120** (2023)
36. Nadooshan, A.A., Dehkordi, S.A.H., Bayareh, M., Baghoolizadeh, M.: Multi objective optimization of residential air conditioning systems. Energy Equip. Syst. **12**, 341–377 (2023)
37. Pant, P., et al.: Glacial meltdown prediction via machine learning. Remote Sens. Appl.: Soc. Environ. **37** (2023)
38. Mohammed, I., Al-Yaseri, A., Al Shehri, D., Mahmoud, M.: Electrokinetic and colloidal stability of igneous rocks for CO_2 storage. Energy Fuel. **37 issue 18** (2023)
39. Li, H., Zheng, B., Lei, Y., Hauglustaine, D., Chen, C.: Anthropogenic NOx emissions trends in China post-2020. Environ. Sci. Ecotechnol. **21** (2023)
40. Wu, Y., Yuan, P., Li, S., Guo, C.: Distribution modeling of biofuel crop Idesia polycarpa. Agronomy. **14** (2023)
41. Freitas, R.S.M., Xing, Z., Rochinha, F.A., Cracknell, R.F., Mira, D.: Probabilistic deep learning for sustainable fuel design. Adv. Appl. Energy. **19** (2023)
42. Liu, B., Chang, H., Li, Y., Zhao, Y.: Industrial low-carbon transition under "double car-bon" strategy. Environ. Dev. Sustain. **27** (2023)
43. Farahat, A., Labban, A.H., Mashat, A.-W.S., Hasanean, H.M., Kambezidis, H.D.: Solar-energy adoption and carbon footprint challenges in MENA. Clean Technol. **6**, 700–731 (2023)
44. Zeng, L., et al.: Social media perception of carbon-neutral cities. Front. Environ. Sci. **10** (2023)
45. Zeeshan, M., Pande, R.R., Bhale, P.V.: Textile waste gasification modeling using advanced thermodynamics. J. Therm. Eng. **11**(3), 659–674 (2023)
46. Tudor, C., Sova, R.: Benchmarking GHG emissions forecasting models for global climate policy. Electronics. **10**(24), 3149 (2021)
47. Elhami, B., Raeini, M.G.N., Taki, M., et al.: Application of classic and soft computing for modeling yield and environmental final impact in vegetable production (a case study: transplanting onion in Isfahan province, Iran). Environ. Sci. Pollut. Res. **29**, 35314–35337 (2022)

48. Kumar, A., Saikia, P., Srivastava, P.: Human-induced impacts on ecological infrastructure in the Himalayan urban agglomerations. Ecol. Front. **44**(1), 84–95 (2024)
49. Wang, Y., Feng, Y., Liu, L.: An improved case-based reasoning approach for sustainable rural development applied to strategic responses. Eng. Appl. Artif. Intell. **133**(Part D), 108316 (2024)
50. Zhang, Y.: Artificial intelligence carbon neutrality strategy in sports event management based on STIRPAT-GRU and transfer learning. Front. Ecol. Evol. **11**, 1275703 (2023)

Information Technologies in Education

Development of Technical and Transversal Skills Through Educational Robotics Using Construction Models and the Lego Serious Play (LSP) Methodology in University Students

Fabricio Marcillo(✉), Willian Ocampo, Franklin Carrasco, Rodolfo Córdova, Aide Zumba, and Fausto Orozco

Pontificia Universidad Católica del Ecuador, sede Santo Domingo, Santo Domingo 230203, Ecuador
{frmarcillo,wjocampo,facarrasco,rscordovag,amzumba, forozco878}@pucesd.edu.ec

Abstract. This study analyzed the impact of combining the Mould King 15066S Boston Robot Dog robotics kit with the Lego Serious Play (LSP) methodology on the development of technical and transversal skills in university students. A quasi-experimental pre-test/post-test design was applied to 28 students divided into seven groups of four members each, from the Pontifical Catholic University of Ecuador, Santo Domingo Campus. The participants carried out a practical robotics activity based on LSP, which emphasized collaboration, symbolic construction, and reflection. Data were collected using analytical rubric that evaluated six dimensions and were analyzed using a paired t-test. The results showed statistically significant improvements ($p < 0.05$) in all evaluated dimensions, with the most notable progress in problem-solving, autonomy, and technical understanding. Additionally, the intervention promoted high levels of participation, collaboration, and ownership of practical learning. Unlike most studies that have focused on traditional kits like LEGO Mindstorms, this research demonstrated the educational potential of alternative kits such as Mould King. Furthermore, the study highlighted the value of integrating the LSP methodology as a creative and reflective pedagogical tool to enrich the teaching of STEM content at a higher education level.

Keywords: Robotics education · Lego Serious Play · Mould King · constructivist learning · STEM · higher education · student autonomy

1 Introduction

In the current context of higher education, the teaching of robotics has emerged as a powerful tool for developing technical, cognitive, and social competencies in students. However, many traditional methodologies still present limitations by focusing exclusively on the transmission of theoretical knowledge, neglecting creativity, active participation, and real-world problem solving [1].

A. Rocha et al. (Eds.): ICITS 2026, LNNS 1966, pp. 429–435, 2027.
https://doi.org/10.1007/978-3-032-27538-7_38

In response to this issue, various studies have highlighted the value of active, and constructivist approaches to enhance practical learning in the field of robotics. Among these, the Lego Serious Play (LSP) methodology has emerged as an innovative alternative, promoting the symbolic construction of knowledge, collaborative work, and both individual and group reflection [2]. Complementarily, the incorporation of mechanically challenging and functional models, such as the Mould King 15066S Boston Robot Dog, a kit notable for its structural complexity, 921 pieces, multifunctional remote control, and capacity for sensor integration has become an effective tool for developing key cognitive skills in this field. Taken together, this methodological approach aims to create a playful and challenging learning environment that fosters not only understanding technical content but also creative exploration, autonomy, and systems thinking [3, 4].

Current research in educational robotics indicates that construction-based robot kits are effective instruments for integrated STEM education and for the development of twenty first century competences. In a multilevel meta-analysis of twenty-one empirical studies, educational robotics produces significant positive effects on STEM learning performance, particularly when it is implemented through inquiry based and project based instructional designs [5]. In a complementary systematic review, Arafat and coauthors showed that integrated STEM activities that incorporate robotics foster problem solving, creativity, collaboration and communication, especially when learners are required to design, assemble and test their own robotic artefacts. From a technological and curricular perspective, construction-based platforms, such as modular robots equipped with sensors and actuators, are especially suitable for connecting abstract concepts in mathematics, physics and computer science with tangible representations in classroom practice [6].

The use of LEGO EV3 robots for the teaching of Newtonian mechanics leads to significant improvements in conceptual understanding, computational thinking and interest in STEM subjects. Low-cost robotics kit increased secondary school students' interest, participation and attitudes toward physics in rural contexts with limited educational resources [7]. Taken together, this body of evidence supports the use of affordable construction based robotic platforms, including quadruped robot dog kits, as authentic artefacts for the exploration of kinematics, stability, transmission of motion and elementary control, while simultaneously promoting key STEM competences and broader twenty first century skills [8].

Moreover, this type of approach has proven to be highly effective in university settings, where the goal is not only for students to acquire technical knowledge but also to develop transversal skills such as critical thinking, autonomy, and creativity. That is why; by actively engaging in the design, assembly, and functional improvement of robotic systems, students face real-world challenges that require informed decision-making and reflection on their own learning process [9]. Therefore, educational robotics, combined with active methodologies such as Lego Serious Play, positions itself as a comprehensive pedagogical strategy capable of bridging theoretical learning with practice and preparing future professionals for dynamic and innovative contexts within the STEM field [10].

This study aims to explore the potential impact of this methodological combination on the development of technical and transversal skills in higher education students. To do so, and such as collaboration, creativity, problem-solving, technical understanding, autonomy, and reflection addressing the effectiveness of pedagogical approaches centered on active and symbolic construction to enrich the teaching of robotics in educational settings under STEM protocol.

2 Materials and Methods

2.1 Methodological Design

This study was conducted under a quantitative, quasi-experimental pre-test/post-test design with a single group, aimed at evaluating the impact of an educational intervention focused on the use of the Mould King 15066S robotics set and the Lego Serious Play (LSP) methodology on the development of technical and transversal skills in higher education students. The design allowed observation of changes in participants before and after the experience without direct manipulation of external variables.

Participants. The sample consisted of 28 university students of the second semester of the Information Systems Engineering course 2024–02, aged between 19 and 23 years, enrolled at the Pontifical Catholic University of Ecuador, located in the city of Santo Domingo of the Tsáchilas, Ecuador. Participants were organized into four groups of seven members to facilitate collaborative work during the intervention. The sample selection was non-probabilistic and based on convenience, as it included all students participating in the training experience within the institutional curricular framework. Although random sampling was not applied, the sample size is considered adequate for exploration studies in university contexts, as it allows for statistical analyses with sufficient power to detect significant differences in related measurements, such as those taken before and after the intervention.

Procedure. The pedagogical intervention was structured in three phases: exploration, construction, and reflection. In the first phase, students familiarized themselves with the pieces, electronic components, and technical manual of the Mould King 15066S Boston Robot Dog robotics set. During the construction phase, they worked collaboratively in teams to assemble the robot, using LSP methodology principles such as symbolic representation and task distribution by modules [11].

Before and after the intervention, a structured evaluation rubric was applied, specifically designed to assess dimensions such as collaboration, creativity, problem-solving, technical understanding, autonomy, and reflection. The rubric was based on a Likert scale from 1 to 5 and was completed by the project instructors based on direct observations of student performance, the data was transformed to percentage in considering each dependent variable.

Data Collection and Analysis. Data were collected in a structured manner through digital forms that allowed organization of individual scores for each rubric category at two moments: pre-test (before the intervention) and post-test (after the intervention). Subsequently, the data were processed in spreadsheets and exported to the statistical software R for analysis.

Statistical analysis included the paired *t-test* to compare scores obtained before and after the intervention in each evaluated category, as well as in the total score. This procedure allowed identification of statistically significant differences attributable to the applied methodology. A 95% confidence level ($\alpha = 0.05$) was assumed for interpreting results, alongside calculations of mean, standard deviation, and effect sizes to support the discussion of findings. The dependent variables were *collaboration, creativity, problem-solving, technical understanding, autonomy*. All these variables were analyzed in percentage.

3 Results and Discussion

The analysis of the data collected through the evaluation rubric applied to 28 students before and after the intervention revealed significant improvements in all evaluated categories. Figure 1 shows a consistent increase in mean scores from pretest to post test for all five variables collaboration, creativity, problem solving, technical understanding and autonomy. In each case, the bars corresponding to the post test are clearly higher than those of the pretest, and the error bars indicate that this improvement is not explained simply by variability in the sample. The statistical annotations above the bars confirm that the paired t tests are highly significant in every dimension, with p values below 0.001, which suggests that participation in the robotics activity with the quadruped robot dog was associated with substantial gains in both technical competences and transversal skills in the group of university students.

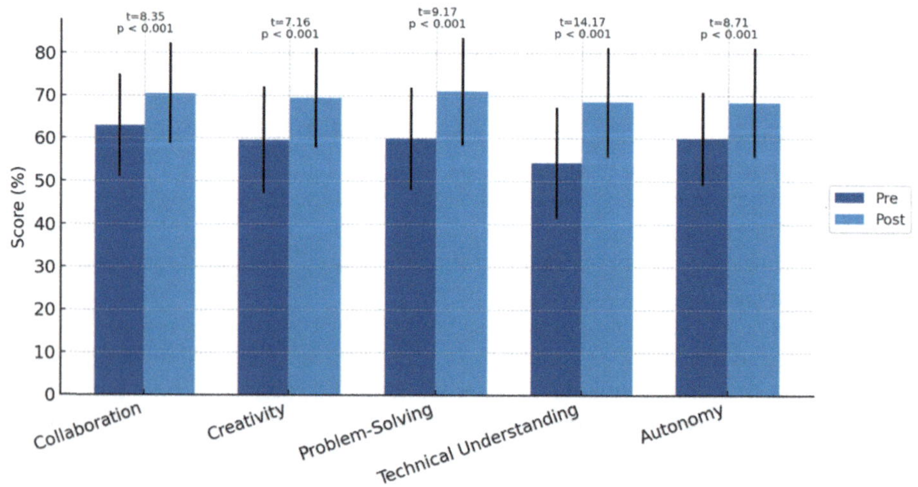

Fig. 1. Comparison of average pre-test and post-test scores by each dependent variable.

Figure 2 presents the Pearson correlation matrix for the post experiment scores of Collaboration, Creativity, Problem Solving, Technical Understanding and Autonomy. All coefficients are positive and moderate to strong, with values that range approximately from 0.49 to 0.69, which indicates that higher levels in one competence tend to be associated with higher levels in the others. The strongest relations appear between

Collaboration and Technical Understanding, and between Problem Solving and Autonomy, suggesting that students who developed more solid technical and problem-solving skills during the robotics activity also tended to report higher collaborative work and greater independence when carrying out the tasks.

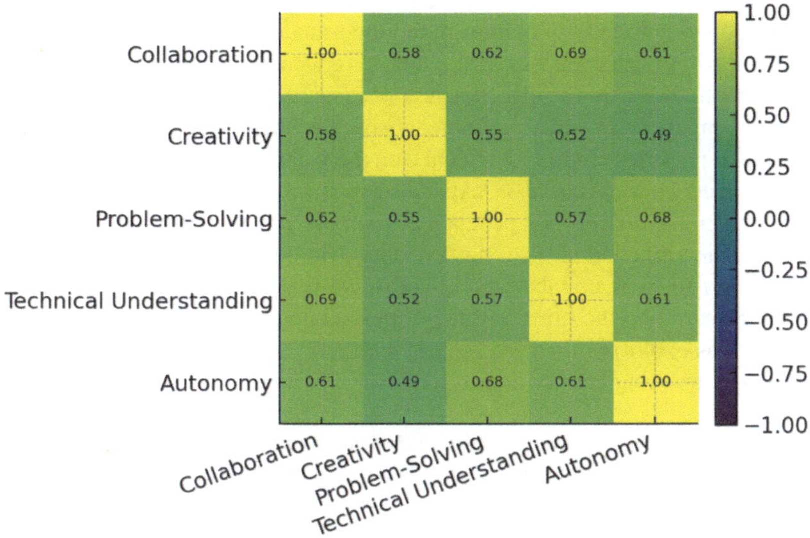

Fig. 2. Pearson Correlation of improvement by each dependent variable.

The results of this study show significant improvements in areas such as problem-solving, autonomy, and technical understanding following the intervention with the Mould King 15066S kit and the LSP methodology, aligning with previous findings in educational robotics. For example, [12] analyzed the use of the Lego Mindstorms EV3 kit with primary school students and reported significant progression in competencies such as computational thinking, problem-solving, and teamwork after a structured practical course. Therefore, this parallel suggests that robotics kits with visual programming can enhance technical and cognitive skills in a manner like the LSP approach applied with the Mould King kit.

Additionally, LSP methodology has proven effective in promoting both cognitive and affective engagement in various educational contexts, particularly in the training of future teachers. This methodology, based on the construction of three-dimensional models, allows participants to express abstract ideas and experiences concretely, which facilitates deep reflection processes in pedagogical training workshops [13]. It has also been observed that LSP fosters a collaborative environment where participants feel heard and valued, strengthening group cohesion and meaningful learning. These characteristics are especially useful in contexts aiming to promote educational innovation and the development of soft skills such as assertive communication, empathy, and shared decision-making among higher education students [14].

4 Conclusion

The integration of the Mould King 15066S robotics kit with the Lego Serious Play (LSP) methodology proved to be an effective combination for promoting meaningful, creative, and autonomous learning in university students. The statistically significant improvements observed in technical and transversal skills—particularly in problem-solving, collaboration, and autonomy—demonstrated the transformative potential of active and reflective pedagogical approaches in the field of educational robotics. This study supports previous findings on the effectiveness of constructivist learning environments to foster student engagement and the comprehensive development of STEM competencies. Furthermore, it provides empirical evidence on the viability of using alternative kits like Mould King in educational contexts, expanding possibilities beyond the traditional models commonly employed. From a pedagogical perspective, this approach not only facilitated technical learning but also strengthened motivation, knowledge appropriation, and students' ability to work collaboratively and solve real challenges. It is recommended to continue researching the impact of these methodologies across different educational levels, as well as exploring their integration with more advanced curricular content and structured programming environments.

Acknowledgments. This article is the result of the learning unit Research Fundamentals, carried out by second-semester students in the 2024-02 academic period, from the Information Systems Engineering program at the School of Habitat, Infrastructure and Creativity, Pontifical Catholic University of Ecuador, Santo Domingo campus.

Andy Gustavo Álava Gómez, Víctor Adriel Aveiga Chuez, Joseph Martin Bermúdez Barreto, Sebastián Dennis Brito Tacuri, Steven Alexander Caillagua Moreno, Edison Ricardo Calva Torres, David Estuardo Cargua Ramirez, Kenneth Gabriel Cevallos Cedeño, Gilmar Vinicio Chango Miño, Marcelo Daniel Espinoza Pilataxi, Edwin Nicolás Ferruzola Hidalgo, Sebastián Andrés Guzmán Torres, Jandry Sebastián López Vélez, Anderson Joel Meza Vizuete, Marcos Francisco Morales Sánchez, Jean Pierre Moreno Vera, Deivy Nicolás Orellana Franco, Michael Efraín Riofrio Lara, Dereck Omar Rojas Mackliff, Andy Daniel Roldan Posligua, Josue Anderson Santos Ortiz, Anthony Fabricio Sinchiguano Loor, Jericko Gandy Tenorio Cordero, Sebastián Alessandro Toscano Araque, Joel Alejandro Varela Cabezas, David Sebastián Vargas Muñoz, Argel Antonio Velez Villagomez, Andrea Estefanía Vozmediano Zurita.

The authors declare that they have no conflicts of interest. This research was funded by the Pontifical Catholic University of Ecuador, Santo Domingo Campus.

References

1. Benitti, F.B.V.: Exploring the educational potential of robotics in schools: a systematic review. Comput. Educ. **58**(3), 978–988 (2012). https://doi.org/10.1016/J.COMPEDU.2011.10.006
2. Peña-Zabala, M., Cilleruelo, L., Aberasturi-Apraiz, E.: Lego serious play: towards the permeability of critical thinking. REIRE Revista d'Innovació i Recerca en Educació. **11**(1), 1–15 (2018). https://doi.org/10.1344/reire2018.11.119183
3. Guerrero Támara, V., Lirio, R.A.P., Blas, Ú.L.: Level of perception of educational robotics at a peruvian university. ACADEMO Revista de Investigación en Ciencias Sociales y Humanidades. **9**(1), 62–72 (2022). https://doi.org/10.30545/academo.2022.ene-jun.6

4. Wang, Q., Wan, Y., Feng, F., Peng, Z., Luo, J.: Discovering public attitudes and emotions toward educational robots through online reviews: a comparative analysis of Weibo and Twitter. Kybernetes. (2024). https://doi.org/10.1108/K-02-2024-0402
5. Raigosa, J.P., Saldarriaga, J.C.C., Valderrama, V.C.: Teacher teaching strategies in basic subjects: a look at school learning. Eleuthera. **21**, 13–33 (2019). https://doi.org/10.17151/ELEU.2019.21.2
6. Mayorga-Ases, M., Tagua-Moyolema, A., Muyulema-Muyulema, D., Velastegui-Hernández, R.: Study on the implementation of active methodologies in higher education: benefits and challenges. 593 Digit. Publ. CEIT. **9**(4–1), 196–208 (2024). https://doi.org/10.33386/593DP.2024.4-1.2739
7. Andrés, O., Calderón, R.: Educational robotics: enhancing mathematical thinking and social skills in learning. Emerg. Trends Educ. (México, Villahermosa). **7**(13), 129–142 (2024). https://doi.org/10.19136/ETIE.A7N13.6040
8. Iván, O., Buriticá, T.: EPS: methodology for solving basic science problems using computational thinking. Revista EIA. **16**(32), 85–96 (2019). https://doi.org/10.24050/reia.v16i32.1266
9. Rapti, S., Sapounidis, T.: Critical thinking, communication, collaboration, creativity in kindergarten with educational robotics. Comput. Educ. **210** (2024). https://doi.org/10.1016/j.compedu.2023.104968
10. Segura, C.M.G., Garcia, M.G., Segura, S.A.G., Euan, M.C., Baas, N.V.S., Jimenez, M.G.: Mixed computational thinking with robotics and recycling in a social project. Revista EIA. **22**(43), 4309 (2025). https://doi.org/10.24050/reia.v22i43.1827
11. Gutiérrez Rivera, P., Gómez Nocetti, V.: From fun to transformation. Shifting beliefs using the Lego® Serious Play® methodology. Cuadernos de Investigación Educativa, ISSN-e 1688–9304, ISSN 1510–2432. **11**(1), 103–133 (2020). https://doi.org/10.18861/cied.2020.11.1.2945
12. Isaías Francisco, D.P., Alfredo, B.S.L., Henry, C.G.H.G.: Collaborative learning as a teaching strategy in Latin America. TecnoHumanismo. **1**(8), 31–52 (2021). https://doi.org/10.53673/th.v1i8.41
13. Angeli, C., Valanides, N.: Developing young children's computational thinking with educational robotics: an interaction effect between gender and scaffolding strategy. Comput. Human Behav. **105**, 105954 (2020). https://doi.org/10.1016/J.CHB.2019.03.018
14. Atmatzidou, S., Demetriadis, S.: Advancing students' computational thinking skills through educational robotics: a study on age and gender relevant differences. Robot. Auton. Syst. **75**, 661–670 (2016). https://doi.org/10.1016/J.ROBOT.2015.10.008

University Students' Perceptions of the Use of Virtual Reality to Improve Verbal Reasoning: A Case Study with Delightex

Mariela Tapia-Leon[1](✉), María Merchán Gavilanez[1], Gerardo Benavides-Montenegro[2], and Sergio Luján-Mora[3]

[1] Universidad de Guayaquil, Guayaquil, Ecuador
mariela.tapial@ug.edu.ec
[2] Universidad Hemisferios, Quito, Ecuador
[3] Universidad de Alicante, Alicante, España

Abstract. This study analyzes university students' perceptions of the use of virtual reality (VR) as an educational resource to strengthen verbal reasoning. Forty-four students from the University of Guayaquil participated, designing educational applications on the Delightex platform. The students integrated activities such as reading comprehension, synonyms, antonyms, and analogies, among others, into an immersive environment. Subsequently, a survey was conducted, which was answered by 90% of the participants. The results indicate that 97% of students believe that their resource contributes to verbal reasoning. They express that this is because VR is motivating because it is interactive, experiential, visual, and playful, which are necessary strategies to stimulate language skills according to speech therapy experts. Likewise, it was evident that working with Delightex developed digital and social skills in university students. Although some students expressed pros and cons of the use of this VR platform, 85% expressed interest in continuing to use Delightex in future educational projects. The study concludes that VR not only allows for the creation of innovative educational resources, but also enhances creativity, enables collaborative work, and develops digital skills.

Keywords: Computational thinking · Delightex · High intellectual abilities · Immersive learning · Verbal reasoning · Virtual reality

1 Introduction

Humans communicate in writing, speech, or sign language to share information, express ideas, explain concepts, and solve problems. Verbal reasoning is essential for this, as it is "the ability to understand, interpret, and analyze information." [1].

Verbal reasoning plays a crucial role in people's academic, professional, and career success, given that all the activities we do are based on communication. Verbal reasoning does not only involve having a broad vocabulary or writing without spelling mistakes; it also has to do with the ability to infer meanings, find logical relationships between words, and express ideas coherently [2]. However, verbal reasoning is an area that requires

A. Rocha et al. (Eds.): ICITS 2026, LNNS 1966, pp. 436–445, 2027.
https://doi.org/10.1007/978-3-032-27538-7_39

intervention for improvement. Álvarez-Cedrón, a speech therapist and early childhood and primary school teacher, explains that 40% of school failure is due to language difficulties [3]. Hence, it is important to develop verbal reasoning from an early age.

The Paideia Pedagogical Center [4] offers some suggestions for stimulating language development. These suggestions have been proposed by specialists in language disorders, including:

- Use visual and/or auditory aids to learn new vocabulary or explanations.
- Encourage meaningful and experiential learning.
- Use positive linguistic cues that involve playful action.

Virtual reality (VR) can help create virtual environments that apply these strategies. This type of technology allows for the creation of playful resources loaded with visual and auditory stimuli that generate positive experiential learning. VR is an immersive technology that captures students' attention, promoting meaningful learning by creating three-dimensional (3D) activities. This is evident in an article that presents a review of the literature on VR [5], in which the author highlights the positive impact that this emerging technology has on student learning, given its high interactive, immersive, and motivational potential. A recent study [6] further reinforces these findings, showing that most participants positively engaged with Virtual Reality Learning Environments (VRLEs), reporting high motivation and positive emotional responses, particularly in interactive experiences. However, challenges such as motion sickness and technical issues were also identified. Overall, the evidence suggests that immersive VR experiences can significantly enhance student motivation and engagement, provided their implementation is carefully aligned with pedagogical objectives, supported by high-quality design, and mindful of user-experience considerations. VR is a tool that can help transform the teaching-learning process, but discretion is required because it may not be the most appropriate tool in all cases [7].

At the Minerva Educational Unit, an institution affiliated with the University of Guayaquil, a problem related to verbal reasoning has been identified. The Wechsler Intelligence Scale for Children-V (WISC-V) [8] (a clinical instrument for individual use that assesses general and specific intellectual abilities in children between the ages of 6 and 16) was administered at this institution. Based on the results of the scale [9], 35 students with high intellectual abilities (HIA) were identified within a total population of 360 students. Although these students show outstanding performance in various types of reasoning (logical, mathematical, perceptual, memory, visuospatial, creativity), the data reveals that in verbal reasoning, average levels predominate (in mode and median). In other words, despite their high intellectual abilities, the group has an average level of verbal reasoning.

Based on these findings and to contribute to a possible techno-pedagogical solution to this problem, we propose the design of innovative educational activities using VR aimed at enhancing the development of verbal reasoning in primary school students. These activities were created by students in the Experimental Sciences Education – Computer Science program at the University of Guayaquil, who, as future teachers, integrated immersive strategies for this purpose. The study presented in this article aims to analyze these university students' perceptions of the use of VR as an educational resource to strengthen verbal reasoning at the primary level, as well as to gather their opinions on the value of the Delightex platform within their teacher training and as an educational resource.

2 Methodology

The design corresponds to an educational case study, focused on academic work carried out by 44 students from the Experimental Sciences – Computer Science Teaching Degree at the University of Guayaquil. The students developed educational applications on the Delightex platform[1] (trial version) as their final project for the course "Use of Virtual Reality in Education." The sample was intentionally selected, as these students were trained in the use of this VR tool during their classes.

The students worked in teams of up to five members over a two-week period. During this time, they developed educational resources using VR, with the aim of strengthening verbal reasoning in primary school students. The project instruction was to design an interactive and immersive resource that integrates visual, auditory, and textual elements, aimed at developing verbal reasoning. To this end, activities such as sentence completion, reading comprehension exercises, inference making, and identification of synonyms and antonyms, among other similar tasks, were proposed. Finally, each resource had to incorporate at least one assessment activity that would allow for the evaluation of users' verbal reasoning. Once the deadline had passed, the applications were shared via videos in a forum within the course's virtual classroom. In the video (maximum five minutes), students had to explain how to use the educational resource they had created, its contribution to the development of verbal reasoning, and the assessment (or assessments) of verbal reasoning. To grade the work, the teacher used a checklist with the following indicators (each worth two points, for a maximum of ten):

- Educational value: The resource contributes to the development of verbal reasoning in primary school students.
- Design: The scenarios and characters are appropriate for primary school students.
- Multimedia: Audiovisual stimuli and texts are included.
- Interactivity: Users can interact with the resource.
- Assessment: The activity includes assessments (at least one) of verbal reasoning.

Subsequently, university students were given an ad hoc questionnaire that included five-point Likert scale items (1 strongly disagree, 2 disagree, 3 neutral, 4 agree, 5 strongly agree), open-ended questions, and closed-ended questions. The questionnaire was designed as an ad hoc instrument tailored to the objectives of the study. It consisted of four dimensions: 1) educational application, which seeks to understand students' perceptions of the usefulness and motivation that their educational resource would generate in primary school students; 2) verbal reasoning, to understand the type of activities carried out to reinforce verbal reasoning in primary school students; 3) Delightex platform, to understand the value of this tool in the teacher training of university students and its value as an educational resource; 4) final project, to determine the usefulness of the final course project activity in the use of VR applied to real educational contexts and issues. This questionnaire was administered using a Microsoft Forms form, and the

[1] https://www.delightex.com/

responses were analyzed descriptively to identify patterns of perception. The link to the questionnaire questions and items is shown in the footer.[2]

3 Results

Twelve educational resources were developed, achieving an average rating of 8.8/10 points. All resources had educational value with a design aimed at primary school students. Among the environments that students chose to design their virtual worlds were stories, books, the Old West, enchanted forests, fairs, educational campuses, and cities. As an example, the video[3] and project[4] of one of the working groups are included (see Fig. 1).

Fig. 1. Interface of the application created by a group of students.

All projects included multimedia elements such as images, videos, audio, and 3D objects, either native to the platform or imported. Interactivity was present in all educational resources, allowing users to interact with the content through 3D objects and characters. Finally, questions were included to assess users' verbal reasoning, thus meeting all the criteria on the checklist. Only two projects strayed from the topic of verbal reasoning and focused on other areas such as Geography and History. Overall, most of the projects were very good or outstanding. However, some of them need to be polished in terms of spelling, accessibility, and navigability.

A total of 40 out of the 44 university students responded to the survey, representing 90%. Of these, 27 participants identified themselves as women (67.5%), 12 as men (30%), and 1 person preferred not to answer (2.5%). The average age is 25, with a minimum age of 21 and a maximum age of 44. The results of the survey are described below.

---

2 https://forms.office.com/r/E6BAtck4DF?origin=lprLink

3 Video: https://www.youtube.com/watch?v=1fNOtExMOiQ

4 https://edu.delightex.com/HPH-WSL

3.1 Dimension: Educational Application

The sum of 95% of students consider that their project has an educational objective related to verbal reasoning. 87.5% of students consider that their application is complete and functional. Although some projects are unfinished, according to the students' criteria, the resource would be 100% motivating for:

Be Ludic and Interactive

- "It is an interactive game that allows students to learn more quickly."
- "The resource allows them to interact in real time, moving freely through each scene of the story."

Be Attractive

- "It is very visual and full of colors."
- "The inclusion of images, text, animations, and sound makes it more interesting."
  Awaken curiosity and creativity:
- "Students can learn while having fun. It allows them to have experiences that awaken their curiosity and encourage them to express themselves and think better."
- "The resource helps develop key skills such as logical reasoning, creativity, problem solving, motivation, and awakens their curiosity as it is an innovative and fun way to learn."

Develop Cognitive Skills

- "The resource helps improve reasoning and captures attention through visual, dynamic, and interactive resources."
- "It is a very interactive and interesting way to understand verbal reasoning."

Using Educational Innovation

- "For innovation and the use of 3D digital educational resources."
- "It is something innovative where students will explore new things and encourage interest in learning to use the educational resource."

3.2 Dimension: Verbal Reasoning

Reading comprehension was the activity most frequently performed by students, mentioned by 32 of the 40 students. This was followed by solving puzzles (27 mentions), recognizing synonyms and antonyms (21 mentions), analogies (16 mentions), completing sentences (13 mentions), and spelling correction (9 mentions). According to 97% of the students, these activities would help improve the verbal reasoning skills of elementary school students (see Fig. 2).

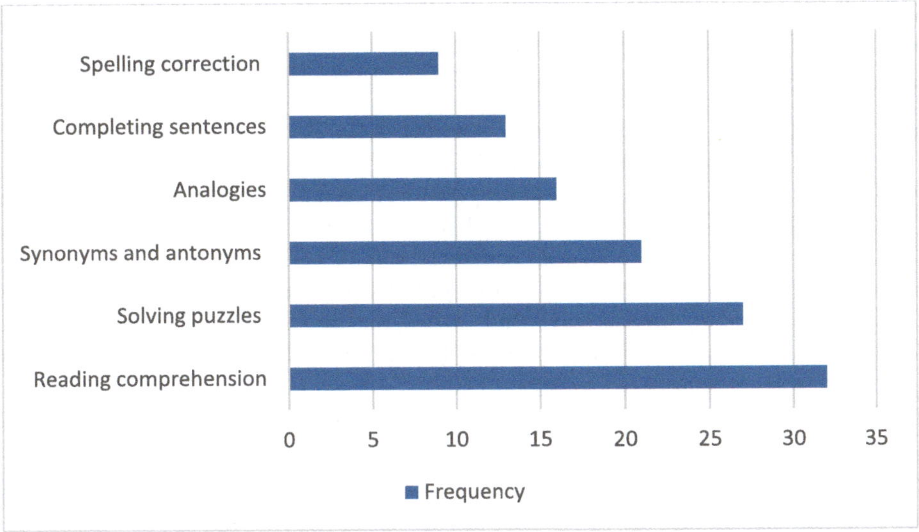

Fig. 2. Frequency of Verbal Reasoning Activities

3.3 Dimension: Delightex Platform

A total of 77.5% of students consider developing VR resources with Delightex to be manageable. This means that the tool is perceived by most students as easy to use and without a steep learning curve. However, four students (10%) found the level of difficulty to be not so manageable, suggesting that additional training should be provided for these students to overcome their difficulties.

An amount of 85% of students consider that the use of CoBlocks (visual block programming language) within Delightex favored the development of their computational thinking. This is a very valuable skill for them, as they will be working as future teachers of computer science and programming. The same percentage (85%) expressed a desire to continue exploring Delightex for the development of future educational projects, including when they are teaching. However, just over half (57.5%) will use the tool for their degree project. This relates to the question that asks about their preference for choosing another tool to develop the same final class project. Less than half of the students (45%) said they would choose the same tool. 47.5% mentioned that they would choose different alternatives such as Unity, Scratch, PlayCanvas, among others. The remaining 7.5% did not mention any tool or said they did not know.

Although the results show that the Delightex tool is not one of the favorites, the perception of the tool's use was positive. Participants stated that they liked its ease of use, the possibility of expressing their creativity when designing interactive scenarios, and the possibility of working with the platform's own 3D elements. Similarly, they indicated that block programming was intuitive, which allowed them to develop programming without prior knowledge of that language. On the other hand, participants also mentioned having technical difficulties with the network, limitations in uploading heavy 3D objects, and slow loading of elements on some devices. Some students found programming complex at times, which was exacerbated by the lack of tutorials on how to use the platform. They also indicated that they dislike the fact that the tool is paid. Examples of what they mentioned are cited below:

- "I liked that you could use models from other sites, but I didn't like that you can't use models that weigh more than 50 MB."
- "I liked that it's a very useful and easy-to-use tool for programming whatever you want to create. The only thing I didn't like is that it's a resource-intensive program, and not everyone has an advanced computer; it may work for them, but slowly."
- "What I liked most about this tool is that you can create eye-catching and innovative scenarios, including images and sounds, which makes it a very interesting tool. What I disliked most is that it is very stressful to program and that things don't work the way we want them to."
- "I really liked all the 3D elements that are already built in, such as houses, cars, people, etc."
- "I liked the interactivity of the tool. The downside is that there are no step-by-step videos showing how to use it."
- "What I liked most about Delightex was how easy it was to create visual and interactive content, as well as the possibility of integrating code blocks with CoBlocks, which stimulated my computational thinking in a creative and practical way. What I liked least was that, at times, the platform was slow to load projects or certain functions were not very intuitive."
- "What I liked most is that you are free to do anything with your creativity, and what I disliked is that it is a paid service."

3.4 Dimension: Final Project

Students are convinced that the use of virtual reality can meet real educational needs. This is indicated by 92.5% of students who agreed or strongly agreed that the final project activity allowed them to understand the usefulness of Delightex.

The final project allowed students to apply the pedagogical knowledge they had acquired during their studies. This is stated by 97.5% of students. Likewise, 95% of students believe that they were able to apply their programming knowledge to the project. This shows that the tool allowed them to link computational thinking and coding skills in a real educational context. Therefore, the tool is very useful for combining theory with practice.

A total of 95% of students agreed or strongly agreed that teamwork allowed them to create a quality product. Collaborative work allows for the integration of various ideas, the diversification of possible solutions, the reinforcement of gaps among the students themselves, and mutual support in achieving a common goal, which impacts the quality of the resource created.

Regarding university students' perceptions of the use of VR to improve verbal reasoning in primary school students, the vast majority (more than 90%) have a positive perception. Their responses focus not only on verbal reasoning but also on the application of VR in other subjects and the development of students' digital skills. Regarding verbal reasoning, students highlight that the resources created can help develop language skills, expand vocabulary, and improve reading comprehension. They explain that VR is a motivating resource that captures attention thanks to its visual, playful, and experiential elements. They indicate that it is a very useful innovative resource for modernizing education. However, they clarify that the effectiveness of the tool will depend on teacher support. Below are some textual quotes:

- "My perception is very good because this platform helps solve problems in an innovative way."
- "I think virtual reality helps improve verbal reasoning in children because they learn by playing and feel part of the experience."
- "I believe that virtual reality can be very useful for improving verbal reasoning in children, as it allows them to interact in interesting environments that help them develop vocabulary and communication skills in a fun and practical way."
- "It is essential to implement virtual reality in education because it allows them to acquire the computer skills necessary in today's world."
- "I think virtual reality helps improve verbal reasoning in children because they learn by playing and feel part of the experience. It is easier for them to understand and express themselves when everything is more visual and fun. In addition, they are much more motivated than with traditional methods."
- "Virtual reality can improve verbal reasoning in elementary school students by offering immersive experiences that encourage oral expression, comprehension, and the organization of ideas. By interacting with virtual environments, children describe, interpret, and narrate situations, which strengthen their language skills. In addition, VR allows learning to be adapted to different paces and needs. Its effectiveness depends on teacher support and appropriate pedagogical use."

3.5 Conclusion About the Final Project

In general, university students rated the final class project positively. Their comments indicate that the activity allowed them to appreciate the educational value of VR. It also helped them develop their technical skills through programming, as well as their social skills through teamwork. They say it was an opportunity to give free rein to their imagination and creativity. Furthermore, they recognize that gamification, in a VR environment, can achieve meaningful and motivating learning. Finally, they value the possibility of using Delightex in the development of computational thinking.

- "Virtual reality is a tool that allows us to do incredible things."
- "I conclude that integrating technology in a creative and pedagogical way not only motivates students, but also promotes key skills such as computational thinking, verbal reasoning, and autonomy."
- "Programming can be entertaining and a lot of fun. Through this tool, children can be taught to develop their computational thinking from an early age, creating very enjoyable scenarios for playing and learning at the same time."
- "Deligthex is a great tool, useful for endless creations. The only recommendation is to be creative in developing and programming whatever you want."
- "Through gamification, we can see that students can learn and have fun at the same time."
- "Virtual reality is a significant contribution to student development. Technology and education are advancing, which requires teachers to be innovative."

4 Conclusions

Both the development of the final project and the results of the ad hoc survey reveal conclusions about VR that are in line with the positive impact this emerging technology has on student learning. A total of 97% of students are convinced that the educational resources they created with activities such as reading comprehension, recognition of synonyms and antonyms, analogies, among others, will effectively contribute to the development of verbal reasoning in primary school students. They believe that the gamification, experiential learning, and interactivity offered by their resources will be educational and motivating for primary school students. However, the resources need to be improved, with minor and major spelling errors corrected and the navigation and accessibility of the resources created improved.

Students consider the use of CoBlocks in Delightex to be manageable and conducive to the development of computational thinking. Although some students reported having problems with coding, they ultimately managed to get their resources to work. It is necessary to continue developing programming logic and computational thinking in both teachers and students, as it is a much-needed skill in the digital age.

Teaching how to use VR applications to develop immersive learning can become a key resource in teacher training focused on innovation. These results reinforce the importance of incorporating VR tools into the training of future teachers and the use of these resources at all educational levels.

In addition to demonstrating that the use of immersive technologies such as VR improves skills such as computational thinking, this study also found that it strengthens social skills through teamwork, as Delightex allows applications to be developed collaboratively. It also shows that it has contributed to the development of students' creativity and imagination. It also promotes self-learning and autonomy among university students by encouraging them to explore the tool for future educational projects. It is important to emphasize that students want to continue improving the educational resources they created. This means that Delightex has succeeded in making the commitment to create a functional, high-quality product go beyond simply achieving a passing grade. Therefore, Delightex has enabled the development of soft skills such as collaborative work, creativity, and autonomy, which are fundamental aspects for the professional development of future teachers.

Just over half of the students (57.5%) say they will use another tool to develop their degree proposal. This may be because some students are already working on their final degree projects and are unable or unwilling to change tools. For example, one student said, "I didn't know this tool existed; I could have used it for my thesis." However, most students (85%) say they will use Delightex in future educational projects. In other words, Delightex managed to spark genuine interest in its use in future scenarios and projects, including when they are teaching.

This study has some limitations. University students' perceptions of the use of VR to improve verbal reasoning are based on the students' self-reflection, which could result in a positive bias. For this reason, we plan to continue the study, implementing the educational resources created to measure their effectiveness in improving verbal reasoning in primary school students with HIA. Another limitation was the time available for developing the resource. Two weeks proved to be too short for the development of the final project. The trial version only allows 45 days of use with all functions. Part of that time was used for training in the use of the tool. Finally, another limitation was that data collection was based on an ad hoc questionnaire that was not statistically validated; however, it was structured in dimensions according to the needs of this study.

As future work, the educational resources created by university students that, in the opinion of experts, are most conducive to improving verbal reasoning in students with HIA at the Minerva Educational Unit will be implemented. The objective will be to measure the effectiveness of these resources.

Acknowledgments. To all the students who participated in this project for their efforts in creating educational resources with VR. We appreciate your commitment to completing the assigned task and participating in this study.

We would also like to thank the University of Guayaquil for its support for research project No. FCI-013-2024, entitled "High intellectual abilities (HIA): intervention strategies in intercultural educational contexts."

References

1. López, M.: 30 verbal reasoning exercises that every student should know. Universidad Privada del Norte. [Online]. Available: https://blogs.upn.edu.pe/estudios-generales/que-es-el-razonamiento-verbal
2. Euroinnova.: El razonamiento verbal y su importancia [Online]. Available: https://www.euroinnova.com/docencia-y-formacion-para-el-profesorado/articulos/razonamiento-verbal
3. Fominaya, C.: 40% of school failure is due to language difficulties. ABC Familia. [Online]. Available: https://www.abc.es/familia/educacion/abci-40-por-ciento-fracaso-escolar-debido-dificultades-lenguaje-201804092115_noticia.html
4. Centro Pedagógico Paideia.: Guidelines to stimulate language. [Online]. Available: https://centropedagogicopaideiaavila.wordpress.com/tag/educacion-primaria/?utm_source=chatgpt.com
5. Menjivar Valencia, E., Sánchez Rivas, E., Ruíz Palmero, J., Linde Valenzuela, T.: Review of the scientific production on virtual reality between 2016 and 2020 through Scopus and WoS. Revista de Educación Mediática y TIC. **10**(2) (2021). https://doi.org/10.21071/edmetic.v10i2.13422
6. Aufenanger, S. et al.: Immersive virtual reality learning environments for higher education: a student acceptance study. Comput. Educ.: X Reality. **7**, 100105 (2025). https://doi.org/10.1016/j.cexr.2025.100105
7. Sousa-Ferreira, R., Campanari-Xavier, R.A., Rodrigues-Ancioto, A.S.: La realidad virtual como herramienta para la educación básica y professional. Revista Científica General José María Córdova. **19**(33), 223–241 (2021). https://doi.org/10.21830/19006586.728
8. Hernández, A., Aguilar, C., Paradell, E., Vallar, F.: Wechsler Intelligence Scale for Children-V. Pearson Educación (2015)
9. Merchán Gavilánez, M.L., González Santamaría, V.: Teacher training in high intellectual abilities: a study based on focus groups. In: Formación docente, fundamentos metodológicos y evidencias empíricas. Ediciones Octaedro (2024)

Basketball 3-Point Shot Detection Using GCN-Pose Estimated Keypoints for the Downstream AQA Task

Bupe Chindongo(✉) and Tevin Moodley

Academy of Computer Science and Software Engineering, University of Johannesburg, Cnr Kingsway and University Road Auckland Park, 2092 Johannesburg, South Africa
225106317@student.uj.ac.za, tevin@uj.ac.za

Abstract. The lack of accessible, objective coaching methods for basketball shooting form assessment presents a significant challenge for player development in non-professional contexts. With the increasing importance of three-point shooting in modern basketball and limited access to consistent professional coaching for many players, developing automated methods to assess shooting form is crucial. This study presents a graph-based approach to automatically identify basketball players attempting three-point shots using computer vision techniques, proposing a Graph Convolutional Network architecture that processes pose estimation data from real NBA game footage and leverages the structural connectivity of human skeletal anatomy as an inductive bias for classification. Our methodology employs YOLOv11-pose to extract COCO keypoints from basketball footage, followed by manual annotation to distinguish shooters from non-shooters and spectators. The pose data is transformed into a graph representation with 17 nodes (body joints) connected by 38 directed edges, representing anatomical relationships. A three-layer GCN with 64 hidden channels processes these graphs, followed by global mean pooling and a multi-layer classifier for binary shooting pose detection. Evaluated on a dataset consisting of manually annotated frames from professional basketball footage, our approach achieves an F1-score of 71.11%, demonstrating effective discrimination between shooting and non-shooting poses in challenging multi-player scenarios. The single-frame classification approach offers advantages over temporal methods by maintaining functionality despite occlusions and partial motions common in real game conditions. This research establishes a foundation for automated basketball shooting assessment and demonstrates the effectiveness of graph-based approaches in sports biomechanics analysis.

Keywords: Basketball · Pose Estimation · Sports Analytics · Graph Neural Networks · Action Quality Assessment

1 Introduction

In modern basketball, scoring points is a critical component of team success and has a direct impact on the success of winning teams [1]. In recent years,

A. Rocha et al. (Eds.): ICITS 2026, LNNS 1966, pp. 446–457, 2027.
https://doi.org/10.1007/978-3-032-27538-7_40

many players have turned to the three-point jump shot, the single shot that nets the greatest number of points, as their go-to shot to score for their team and propel them into a comfortable lead. The three-point shot, as seen in Fig. 1, has become increasingly important in recent times, with three-point attempts close to doubling over a 10-year span. Success for these three-point shots, however, remains relatively low compared to the number of shots attempted. The increase in three-pointers signals the importance of having the correct form for these shots, as the distance at which they are taken heavily punishes small mistakes in player biomechanics, such as release height and wrist flexion, when attempting to make the shots [10].

With a sufficient amount of practice with an objectively good form, players are able to reproduce these shots in difficult situations almost instinctually, as seen in professionals like *Stephen Curry* and *Kyle Korver* who have made these shots successfully over 40% of the time in high volumes over their lengthy careers [16,17]. Many players, especially amateurs and semi-professionals, practice independently and lack access to objective coaching on their shooting form [18], which may result in the reinforcement of poor habits and hinder their ability to successfully shoot the basketball into the hoop from a three-point-worthy distance. This highlights the compelling need for an objective means of assessing shooting form, and one that is easily accessible to the masses.

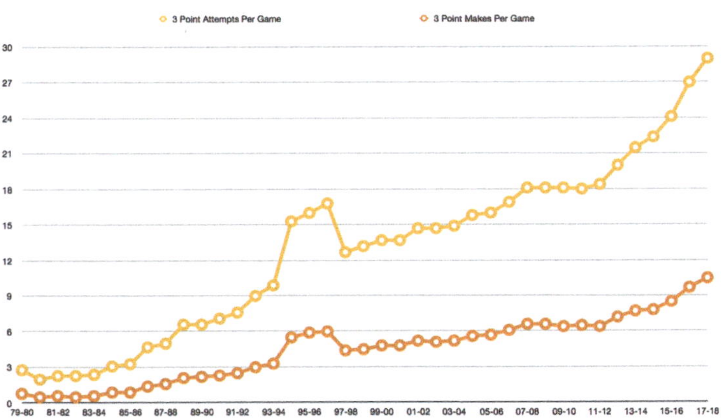

Fig. 1. A figure that illustrates the number of 3 points attempted per game vs 3 points made in a game, which highlights the increase in the number of attempts being made over a 10-year span.

This study takes the first step towards tackling this complex challenge by developing a means of detecting three-point shots within a basketball match, and sets the stage for the extraction of the biomechanical features of players within these instances, to then suggest form corrections based on established best practices for jump-shot form, which relates to the downstream Action Quality Assessment (AQA) task.

Therefore, we employ a classification-based methodology for detecting players, wherein each player is annotated with bounding boxes. Additionally, pose information is extracted to establish a correlation between the identified poses and their corresponding labels. The extracted pose keypoints, representing major body joints such as shoulders, elbows, wrists, hips, and knees, naturally form a skeletal structure with inherent anatomical connections [9]. To leverage this structural relationship, we represent each individual's pose as a graph where keypoints serve as nodes and anatomical connections form edges, preserving the biomechanical constraints that govern human movement. A Graph Convolutional Network is then trained to classify individuals based on their pose characteristics by learning from this graph structure. Unlike traditional neural networks that would treat all keypoint relationships equally, the GCN exploits the skeletal connectivity as inductive bias, allowing information to propagate along anatomically meaningful pathways, such as from shoulder to elbow to wrist, which are critical for recognising the kinematic chains involved in shooting motions. This graph-based approach enables the model to learn biomechanically coherent patterns that distinguish shooting poses from other basketball movements and spectator poses, ultimately determining if a player is in the midst of making a 3-point shot or not. The presence of the *'Shooter'* label in a frame indicates a three-point shot being taken in that frame, and naturally, the person to whom the label belongs is the shooter.

The paper is structured as follows. In Sect. 2, we will examine similar approaches presented within the literature, while Sect. 3 will describe the methods used to conduct the research. Section 4 will present the findings, while acknowledging our limitations. Finally, in Sect. 5, we will conclude our research and set the stage for future work.

2 Related Works

The approaches to basketball shooting recognition presented in the literature tend to focus on practice footage, where a single player is present in the video, rather than real match footage, which features multiple players and an audience. Such a case is presented in [3] where an *Enhanced Graph Structure* was created that augments Spatial-Temporal Graph Convolutional Networks (ST-GCN) and additionally encodes information related to the basketball as a feature. Working with controlled practice scenarios filmed from courtside perspectives. *Fan et al.* collected a custom dataset of 557 video clips, showing 310 shooting actions and 247 non-shooting actions. Their model, which incorporates basketball positional information detected via YOLOv8 as an additional graph node connected to hand and upper-body skeleton nodes, achieves 95.74% accuracy and 99.09% F1-score [3]. This represented a significant improvement over the baseline ST-GCN model (93.62% accuracy, 92.68% F1-score) and substantially outperformed other comparative models, including 2 s-AGCN (89.36% accuracy, 88.89% F1-score) and various CNN-based approaches such as C3D and ResNet-C2D [3]. These results established the feasibility and effectiveness of graph-based deep learning approaches augmented with object detection for basketball shooting recognition in controlled environments. However, relying on practice footage with single players presents significant limitations for the model's applicability in real game footage. In actual game scenarios, multiple confounding factors emerge, such as defensive players occupying similar spatial regions and exhibiting arm extensions that resemble shooting motions. Spectators in the background introduce additional pose detections, and frequent occlusions from teammates and referees interrupt continuous player visibility. These constraints motivate our investigation of single-frame classification approaches that can operate robustly in multi-player game environments despite the inherent challenges of distinguishing shooting poses from other basketball movements without temporal context.

Chen et al. observed that most approaches within sports analysis that rely on video-based approaches suffer from a number of challenges, such as low prediction rates [5]. To attempt to mitigate these observed challenges and provide posture suggestions to improve the shooting form of basketball players, the authors decomposed the problem into three distinct phases they described as *shooting period detection, shooting prediction analysis* and finally *shooting posture suggestions*, driven by varying tools at the different phases, such as a Pix2Pix Conditional Generative Adversarial Network to create shooting posture suggestions during the shooting posture suggestion phase. In the *shooting period detection* phase, the biomechanical characteristics of the player, such as the knee's position in relation to the feet, and the extension of the shooting arm, were examined. Shooting prediction analysis then scrutinises the detected shooting periods to correlate motion trajectories of pose keypoints to field-goal attempts. This detection approach proved effective for the use of AQA of basketball players, and demonstrated the viability of an end-to-end posture suggestion system.

Researchers *Teket et al.* continued to build on basketball shot detection by proposing a model that first detected the ball entering the vicinity of the hoop, and then worked backwards to determine which player shot it [8]. They exploited background subtraction methods centred around a basketball hoop detected through MobileNetV2-SSD to determine when the ball was within the hoop's vicinity. The trajectory of the ball was then determined based on the results of the earlier background subtraction, and this trajectory was used to work backwards to the player who shot the ball. The players themselves were marked with bounding boxes using YOLOv3-tiny, and then a player re-identification strategy centred around MobileNetV2 was used to label and extract player features conclusively for those identified shooters [8]. This re-identification model showed exceptional results, boasting an accuracy of 91.4%. However, authors noted that the approach suffers greatly when faced with occlusions, alluding to the potential for further research on models that may better handle them.

3 Experiment Setup

Our experimental methodology follows a structured pipeline designed to address the challenge of detecting basketball players attempting three-point shots in real game footage, as illustrated in Fig. 2. The approach comprises four main stages: dataset creation from NBA footage, pose extraction, graph-based representation of skeletal structure, and graph convolutional network training with Focal Loss for handling class imbalance. The following subsections detail each component of this pipeline.

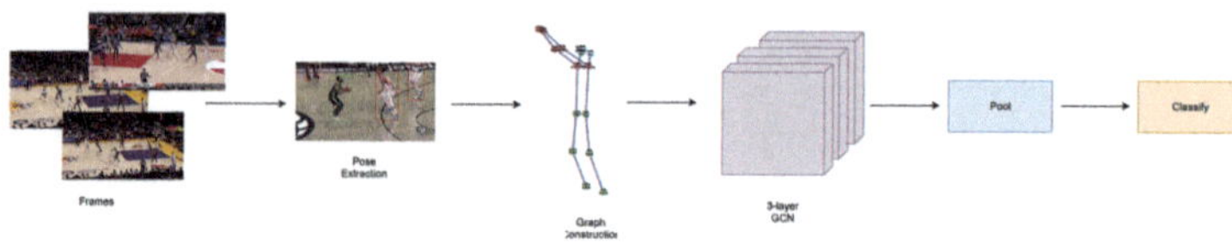

Fig. 2. A figure illustrating the proposed pipeline for basketball three-point shot detection.

3.1 Dataset Creation

To create our dataset, we began by searching for readily available basketball footage featuring multiple players in a diverse range of camera angles. To that end, footage that compiled three-point shot attempts in NBA regular-season games was selected. The footage was manually trimmed to include only frames where a player is in a three-point shooting motion. Frames were then extracted at 3 frames per second, and manual identification was used to exclude frames where the players did not fall within one of the relevant shooting phases, which we term as the following: preparation, jumping, and follow-through, corresponding to three of the five shot phases identified in [7] and [10] as *preparation, ball-elevation, stability, release, and inertia.* This resulted in a diverse set of 247 frames that showcase varying angles wherein players are within a shooting motion, as seen in Fig. 3. The dataset was split according to a rule where the training, validation, and testing sets comprise 70%, 20%, and 10% of the data, respectively.

3.2 Action Representation: Pose Extraction

YOLOv11-pose is used to simultaneously determine bounding boxes for every individual detected in each frame, as well as to extract each individual's anatomical keypoint information, which we store in a COCO-formatted JSON file [11]. Detections are performed through the use of bounding boxes, which demarcate each human, including both players and bystanders, and initially assign a category of 'non-shooter' to each individual detected. The frames are then manually reviewed, and a Python script is used to modify the generated JSON file, changing the category assigned to the manually identified shooters to 'shooter'. There may only be one 'shooter' in each frame, although in some cases, there are none. The Common Objects in Context (COCO) human pose estimation format is used to represent these keypoints, providing 17 points of interest which describe major body joints as seen in Fig. 4 [12].

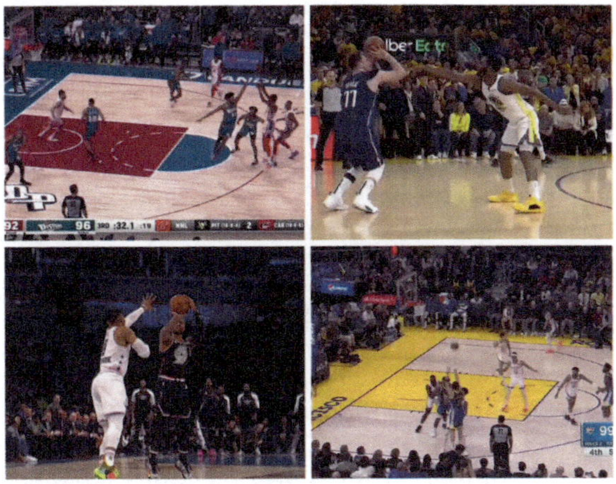

Fig. 3. A figure showing four sample images from our dataset.

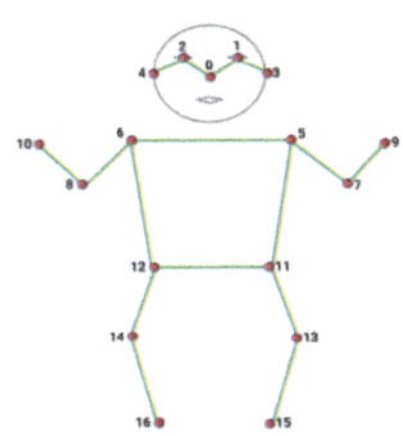

Fig. 4. A figure that highlights the COCO format keypoints, which are used in this study.

3.3 Action Modelling: Graph Construction

The transformation of pose keypoints into a graph representation addresses the fundamental challenge of capturing spatial relationships between body joints that define shooting biomechanics. The COCO skeletal structure defines 19 anatomical connections between the 17 keypoints, forming a natural graph with bidirectional edges.

Each of the 17 keypoints serves as a graph node with a 3-dimensional feature vector $\mathbf{x}_i = [x_i^{norm}, y_i^{norm}, c_i]^T$, where x_i^{norm} and y_i^{norm} are normalised coordinates and c_i is the confidence score. Pose normalisation is achieved by centering coordinates around the pose centroid $\mathbf{p}_c$ and scaling by the maximum pose span s_{max}:

$$x_i^{norm} = \frac{x_i - p_{c,x}}{s_{max}}, \quad y_i^{norm} = \frac{y_i - p_{c,y}}{s_{max}} \tag{1}$$

where $\mathbf{p}_c = \frac{1}{N_{visible}} \sum_{i \in V} \mathbf{p}_i$ and $s_{max} = \max(\max_i x_i - \min_i x_i, \max_i y_i - \min_i y_i)$ for visible keypoints V.

Each anatomical connection (k_i, k_j) in the COCO skeleton is represented as two directed edges to enable bidirectional information propagation: $(k_i \rightarrow k_j)$ and $(k_j \rightarrow k_i)$. This yields a total of $19 \times 2 = 38$ directed edges, ensuring that kinematic information can flow in both directions along anatomical pathways. For instance, wrist joint information can influence elbow representations while elbow information simultaneously informs wrist representations, capturing the interdependent nature of shooting mechanics.

3.4 Graph Convolutional Network Architecture

The selection of Graph Convolutional Networks over traditional architectures is motivated by the structural nature of human pose data. Basketball shooting analysis requires an understanding of kinematic chains and anatomical constraints that govern movement. Standard Multi-Layer Perceptrons treat all keypoint relationships equally, potentially learning invalid correlations between anatomically distant joints. In contrast, GCNs leverage skeletal connectivity as inductive bias, constraining information flow along anatomically meaningful pathways [13].

Our architecture employs a three-layer GCN design implemented using PyTorch Geometric, with each layer containing 64 hidden channels. Each GCN layer performs message passing between connected nodes according to the COCO skeletal structure, followed by batch normalisation, ReLU activation, and dropout regularisation ($p = 0.3$). The architecture utilises only the graph topology defined by skeletal connections, relying solely on structural connectivity patterns rather than edge weights.

Following the three GCN layers, global mean pooling aggregates all 17 node representations into a single graph-level feature vector of dimension 64. The final classification head consists of a three-layer fully connected network ($64 \rightarrow 128 \rightarrow 64 \rightarrow 1$) with ReLU activations and progressive dropout rates (0.5, 0.3) for binary shooting pose classification.

Loss Function. The model employs Focal Loss to address the extreme class imbalance present within our dataset. This class imbalance is a result of the nature of the problem we are tackling, where we attempt to create a model for shot detection on real basketball games, where not only players but also spectators are present in the background. In our dataset, shooters represent a small minority of all detected individuals, creating a severe imbalance that can cause standard cross-entropy loss to be dominated by the abundant negative examples.

$$FL(p_t) = -\alpha_t(1 - p_t)^\gamma \log(p_t) \tag{2}$$

where p_t is the predicted probability for the true class, defined as:

$$p_t = \begin{cases} p & \text{if } y = 1 \\ 1 - p & \text{if } y = 0 \end{cases} \tag{3}$$

The Focal Loss incorporates two key mechanisms to address the challenges inherent in our imbalanced classification task [14]:

Class Balance Weighting (α_t): The weighting factor α_t addresses class imbalance by assigning different importance to positive and negative examples. It is dynamically calculated as:

$$\alpha_t = \frac{N_{negative}}{N_{positive} + N_{negative}} \tag{4}$$

This formulation ensures that the rare positive class (shooters) receives a higher weight during training. Without this weighting, the model would be incentivised to predict the majority class (non-shooter) for all examples to minimise loss, effectively learning a degenerate solution.

Focusing Parameter (γ): The modulating factor $(1 - p_t)^\gamma$ with $\gamma = 2.0$ addresses the problem of easy versus hard examples by down-weighting the loss contribution from well-classified examples. This mechanism prevents the overwhelming number of negatives from dominating the training process.

This dual mechanism is particularly crucial for basketball shooting detection, where the model must learn to identify subtle pose characteristics that distinguish shooting motions from the multitude of other basketball-related movements and spectator poses present in the imagery.

3.5 Performance Metrics

We evaluate model performance using standard binary classification metrics, where TP, TN, FP, and FN represent true positives (correctly identified shooters), true negatives (correctly identified non-shooters), false positives (non-shooters misclassified as shooters), and false negatives (shooters missed by the model), respectively [15]. Accuracy measures the overall proportion of correct predictions as $\frac{TP+TN}{TP+FP+TN+FN}$, providing a general assessment of model performance across both classes. Precision, calculated as $\frac{TP}{TP+FP}$, quantifies the proportion of predicted shooters that are actual shooters, indicating the model's ability to avoid false alarms. Recall, defined as $\frac{TP}{TP+FN}$, measures the proportion of actual shooters successfully identified, reflecting the model's sensitivity to the minority class. Finally, the F1-score, computed as $\frac{2TP}{2TP+FP+FN}$, provides the harmonic mean of precision and recall, offering a balanced evaluation metric particularly suitable for our imbalanced dataset where shooters represent a small fraction of all detections [15].

4 Results and Discussion

The confusion matrix presented in Fig. 5 reveals the classification performance of our graph convolutional network on the test dataset comprising 218 individual detections. The model achieved the following performance metrics:

- **Accuracy**: 94.04%
- **Precision**: 69.57%
- **Recall**: 72.73%
- **F1 Score**: 71.11%

The confusion matrix reveals 189 true negatives (correctly identified non-shooters), 16 true positives (correctly identified shooters), 7 false positives (non-shooters incorrectly classified as shooters), and 6 false negatives (shooters missed by the model).

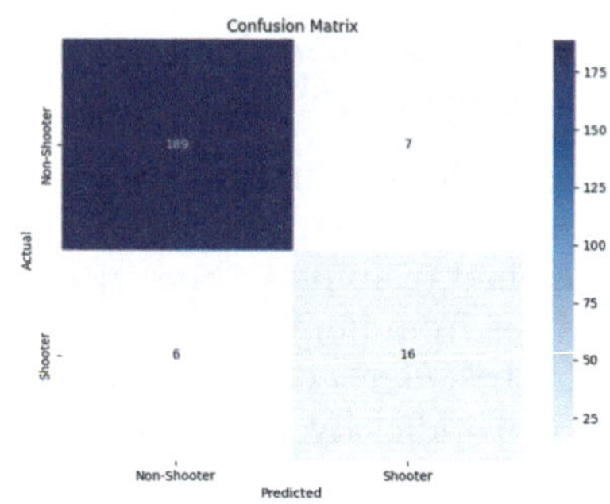

Fig. 5. Confusion Matrix showing classification results on test set.

The results demonstrate several key characteristics of our approach. The high accuracy (94.04%) reflects the model's strong ability to handle the extreme class imbalance inherent in basketball footage, where non-shooters vastly outnumber shooters. More critically, the balanced precision (69.57%) and recall (72.73%) indicate that the model successfully learned to identify shooting poses rather than simply predicting the majority class. The 7 false positives suggest that certain non-shooting basketball movements share kinematic similarities with shooting motions, causing occasional misclassification. This is expected given the dynamic nature of basketball, where defensive movements, passing motions, and rebounds can exhibit similar arm positioning to shooting phases.

One noteworthy finding is the significance of the 6 false negatives representing shooters incorrectly classified as non-shooters. Analysis of these missed detections reveals a systematic pattern: most occur during the *preparation phase* of the shooting motion, as illustrated in Fig. 6. This phase, characterised by initial ball positioning and stance setup, exhibits substantially higher pose variability compared to the more stereotypical *jumping* and *follow-through* phases.

This finding illuminates a fundamental trade-off in basketball shooting analysis. The preparation phase contains the most individualised biomechanical signatures, varying significantly based on player height, shooting style, defensive pressure, and court position. Unlike the kinematically constrained jumping and release phases, preparation poses can range from low crouch positions to upright stances, creating substantial intra-class variation that challenges single-frame

Fig. 6. Example of false negative: shooter during preparation phase incorrectly classified as non-shooter.

classification approaches. Conventional wisdom in action recognition suggests incorporating temporal information through sequence models [3] to capture the complete shooting motion. Such approaches would theoretically address our preparation phase limitation by learning the temporal progression from setup through release. However, our single-frame approach offers distinct advantages that represent the core novelty of this research:

Occlusion Robustness: In real basketball footage, players frequently disappear behind teammates, referees, or move outside the camera frame mid-motion. Temporal approaches require continuous visibility across multiple frames, failing when any frame in the sequence lacks clear pose data. Our method operates on individual frames, enabling detection even when portions of the shooting motion are occluded.

Partial Motion Detection: Players often begin shooting motions that are interrupted by defensive pressure or strategic changes. Temporal approaches would fail to recognise these incomplete sequences, while our frame-level classification can identify shooting intent regardless of motion completion.

Phase-Independent Analysis: The ability to detect shooting poses across all three phases (despite preparation phase challenges) provides valuable flexibility for downstream applications in shooting form analysis, where each phase contributes unique biomechanical insights.

5 Conclusion

This study developed and evaluated a graph-based approach for automatically detecting basketball players attempting three-point shots in real game footage. Using Graph Convolutional Networks to process pose estimation data, our method achieved an F1 score of 71.11% on a challenging dataset of NBA footage containing multiple players, spectators, and varying camera angles. The research makes several key contributions to sports analytics. First, we demonstrated the effectiveness of graph structure for basketball pose classification, showing that skeletal connectivity provides crucial inductive bias for learning biomechanically meaningful patterns. Second, our single-frame classification approach offers distinct advantages over temporal methods by maintaining functionality despite occlusions and partial shooting motions common in real game scenarios. Analysis of model limitations revealed systematic challenges with the preparation phase of shooting motions, where high pose variability leads to increased false negatives. This finding illuminates the fundamental trade-off between temporal completeness and occlusion robustness in sports movement analysis. Despite this limitation, our approach successfully operates on authentic game footage rather than controlled laboratory conditions, addressing a significant gap in practical sports analytics applications. The work establishes a foundation for automated basketball shooting assessment systems and demonstrates the broader applicability of graph-based approaches to sports biomechanics analysis. Future research directions include investigating pose sequence aggregation strategies that preserve occlusion robustness while capturing phase-specific patterns, and extending the framework to provide quantitative shooting form evaluation and corrective feedback. By enabling precise shooter identification in complex multi-player environments, this research contributes to the growing intersection of computer vision and sports analytics, with implications for player development, coaching assistance, and objective performance assessment in basketball training environments.

References

1. Angel Gómez, M., Lorenzo, A., Sampaio, J., Ibáñez, S.J., Ortega, E.: Game-related statistics that discriminated winning and losing teams from the Spanish men's professional basketball teams. Coll. Antropol. **32**, 451–456 (2008)
2. Zajc, T., Mikołajec, K., Chmura, P., Konefał, M., Krzysztofik, M., Makar, P.: Long-term trends in shooting performance in the NBA: an analysis of two- and three-point shooting across 40 consecutive seasons. Int. J. Environ. Res. Public Health **20**, 1924 (2023). https://doi.org/10.3390/ijerph20031924
3. Fan, T.: Real-time basketball shooting action recognition: enhanced graph structure of ST-GCN. TRITA-EECS-EX, vol. Independent thesis Advanced level (degree of Master (Two Years)), p. 46 (2024)
4. Nakai, M., Tsunoda, Y., Hayashi, H., Murakoshi, H.: Prediction of basketball free throw shooting by OpenPose. In: Kojima, K., Sakamoto, M., Mineshima, K., Satoh, K. (eds.) JSAI-isAI 2018. LNCS (LNAI), vol. 11717, pp. 435–446. Springer, Cham (2019). https://doi.org/10.1007/978-3-030-31605-1_31
5. Chen, C.-C., Chang, C., Lin, C.-S., Chen, C.-H., Chen, I.C.: Video based basketball shooting prediction and pose suggestion system. Multimedia Tools Appl. **82**, 27551–27570 (2023). https://doi.org/10.1007/s11042-023-14490-2
6. Liu, R., Liu, Z., Liu, S.: Recognition of basketball player's shooting action based on the convolutional neural network. Sci. Program. **2021**, 1–8 (2021). https://doi.org/10.1155/2021/3045418
7. Okazaki, V.H., Rodacki, A.L., Satern, M.N.: A review on the basketball jump shot. Sports Biomech. **14**, 190–205 (2015)

8. Teket, O.M., Yetik, I.S.: A fast deep learning based approach for basketball video analysis. In: Proceedings of the 2020 4th International Conference on Vision, Image and Signal Processing, pp. Article 24. Association for Computing Machinery, Bangkok, Thailand (2021). https://doi.org/10.1145/3448823.3448882
9. Yan, S., Xiong, Y., Lin, D.: Spatial temporal graph convolutional networks for skeleton-based action recognition. In: Proceedings of the Thirty-Second AAAI Conference on Artificial Intelligence and Thirtieth Innovative Applications of Artificial Intelligence Conference and Eighth AAAI Symposium on Educational Advances in Artificial Intelligence, pp. Article 912, AAAI Press, New Orleans, Louisiana, USA (2018)
10. Biomechanics of the Basketball Jump Shot. https://www.physio-pedia.com/Biomechanics_of_the_Basketball_Jump_Shot. Accessed 13 May 2025
11. Pose Estimation. https://docs.ultralytics.com/tasks/pose/#dataset-format. Accessed 14 Aug 2025
12. Pose Detection. https://github.com/tensorflow/tfjs-models/tree/master/pose-detection. Accessed 02 Sept 2025
13. Wang, X., Cui, Q., Chen, C., Zhao, S., Liu, M.: Graph-Guided MLP-Mixer for Skeleton-Based Human Motion Prediction. arXiv preprint arXiv:2304.03532 (2023)
14. Lin, T.-Y., Goyal, P., Girshick, R., He, K., Dollar, P.: Focal loss for dense object detection. IEEE Trans. Pattern Anal. Mach. Intell. **42**, 318–327 (2020). https://doi.org/10.1109/tpami.2018.2858826
15. The Explanation You Need on Binary Classification Metrics. https://towardsdatascience.com/the-explanation-you-need-on-binary-classification-metrics-321d280b590f/. Accessed 02 Sept 2025
16. StatMuse: Stephen Curry Career 3 Point Percentage (2025). https://www.statmuse.com/nba/ask/stephen-curry-career-3-point-percentage
17. StatMuse: Kyle Korver Career 3 Point Percentage (2025). https://www.statmuse.com/nba/ask/kyle-korver-career-3-point-percentage
18. Ho, B.R., Valenzuela, J.A., Markes, A.R., Pandya, N.K.: Amateur athletic union (AAU) accessibility: an area deprivation index (ADI) analysis of national basketball association (NBA) players' profiles. Curr. Rev. Musculoskelet. Med. **17**, 335–342 (2024). https://doi.org/10.1007/s12178-024-09908-9

Performance Evaluation of Predictive Methods for Electric Vehicle Investment in Ecuador

Renato Arcos, Renato M. Toasa(✉), and Mario Pérez

Universidad Tecnológica Israel, Quito, Ecuador
{e1714741814,rtoasa,mperez}@uisrael.edu.ec

Abstract. Private investment in the importation of electric vehicles in Ecuador has been gradual. The adoption of electric vehicles in the country also depends on their availability on the market. There is no forecast for investment in electric vehicles in the country. This study predicts investment in electric vehicles in Ecuador for 2025 and 2026 based on historical data from 2020 to 2024. It also measures the efficiency of the Prophet, SARIMA, and GXBoost predictive methods by comparing them using the RMSE, MAE, and MAPE metrics. The SARIMA method obtained the best results in the comparative metrics due to the trend in the historical data analyzed. We consider its prediction to be valid, providing a fairly clear picture of how investment in electric vehicles will grow in 2025 and 2026.

Keywords: Predictive Methods · Data Analysis · RMSE · MAE · MAPE · Electric Vehicles · Prediction

1 Introduction

Technology has grown significantly in recent years, from cloud infrastructure, the Internet of Things, blockchain, artificial intelligence, and new predictive methods [1]. The pandemic exponentially accelerated the number of people connected to the internet, which led to the massive generation of data and increased interest in Big Data and data analytics. Predictive models have gone from being static to being self-adjusting machine learning models. The foundations for data prediction began in 1763 with Bayesian reasoning [2]. Subsequently, time series methods appeared between 1950 and 1970 [3] (April 2019), with advances in technology, cloud computing, and machine learning. In 2015, XGBoost made nonlinear prediction massive and much more efficient [4], and in 2017, Meta created Prophet.

In Ecuador, in 2022, Agriculture 4.0 was promoted through research by three universities. The project focused on Smart Agriculture for the monitoring and diagnosis of corn crops [5]. In 2023, a study was conducted on IT trends and priorities in Ecuadorian companies. 45.3% of companies said they would adopt Artificial Intelligence (AI) and Machine Learning (ML) in the short term [6]. With this, Ecuador shows that it is following the global trend in terms of technological growth. The Comparative Study of Classical Methods [7], which also compares the SARIMA and GXBoost methods, reaches similar conclusions regarding XGBoost's better results in the MAPE comparative metric.

A. Rocha et al. (Eds.): ICITS 2026, LNNS 1966, pp. 458–472, 2027.
https://doi.org/10.1007/978-3-032-27538-7_41

In Ecuador, we do not have clarity on the country's investment in electric vehicles for 2025 and 2026 to relate the growth of electric vehicles in the country, which is why predicting investment in electric vehicles allows us to analyze the future growth of electric vehicles to a certain extent. The lack of data to predict investment in the importation of this type of vehicle gives rise to several problems that the country is experiencing due to the lack of availability of services required by electric vehicles and their wider adoption. Considering historical information from 2020 and applying Prophet, SARIMA, and XGBoost predictive methods, investment in the importation of electric vehicles will be predicted for 2025 and 2026. This work includes looking at the efficiency of the three predictive methods through comparative metrics: Root Mean Square Error (RMSE), Mean Absolute Error (MAE), and Mean Absolute Percentage Error (MAPE). The method that obtains the lowest comparison metrics is the most efficient. Predictive methods allow us to predict the future based on historical data and apply statistical techniques, data mining, machine learning, and artificial intelligence [8]. Predictive analysis can be applied in all fields and branches such as health, finance, purchasing, sales, chemistry, biology, etc.

The Internal Revenue Service publishes vehicle purchases by individuals and legal entities on its open data platform, making available daily vehicle purchase transactions grouped into annual files. This material was used as input for the predictive analysis.

Predicting investment in electric vehicles in Ecuador allows us to relate the future growth of vehicles for the years 2025 and 2026. This will enable new analyses to be carried out, such as CO2 pollution from vehicles and how to reduce it through the adoption of electric vehicles, contributing to the reduction of greenhouse gases. It will also be possible to analyze the availability of charging stations and, in terms of public policy, suggest the creation of laws promoting the adoption of electric vehicles as a benefit for all. This study will also be of interest to private companies, as it could promote new businesses for the provision of products and services, which are currently scarce and relate to the infrastructure needed to meet the needs of electric vehicles. Finally, the study could also provide information related to future changes in human development indices, specifically those related to environmental pollution.

2 Materials and Methods

The research conducted is quantitative in nature. Historical data on daily vehicle purchases in Ecuador prior to vehicle registration for the years between 2020 and 2024 was obtained (Fig. 1).

The process applied to carry out the work is as follows:

- Define a relevant problem to focus on: The growth in the adoption of electric vehicles in Ecuador requires infrastructure for charging stations, spare parts stores, and maintenance workshops. Predicting the growth in the number of electric vehicles will provide information on the need to increase installed capacity to meet demand.
- Search for the information needed for the analysis: Once the problem has been defined, the necessary data for the predictive analysis is sought.
- Preliminary analysis of the information: The information comes with records that contain errors, which must be identified so that they can be taken into account in the data cleaning plan.

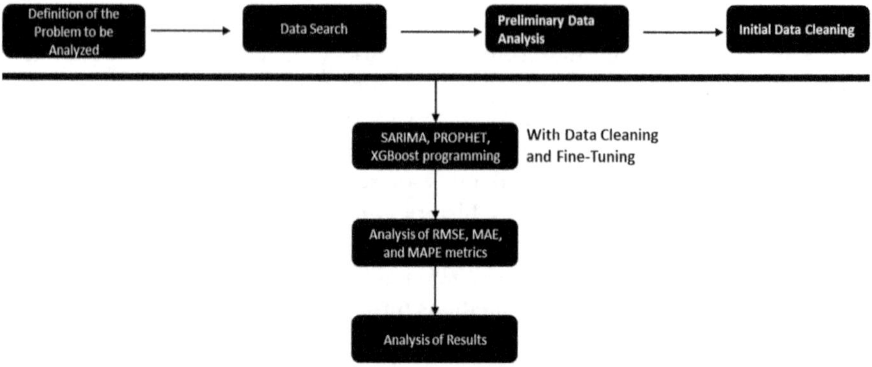

Fig. 1. Diagram of the process applied for predictive analysis.

- Cleaning the information: Once the errors in the data have been identified, we proceed to generate the code that allows them to be cleaned. Duplicate records were found that should not be considered. The VEHICLE CODE field is the unique identifier of the vehicle in the vehicle registry of the Internal Revenue Service of Ecuador, which is the Ecuadorian government institution where purchases and sales are recorded.
- Applying the necessary formats for automated information reading: Date data comes in different formats, so care must be taken to ensure that it is read correctly during automated loading and that the information to be analyzed is not distorted.
- Python coding: The selected predictive methods were coded in Python, including data cleaning, date adjustments, and the calculation of comparison metrics.
- Prediction execution and results analysis: Once the Python program has been executed, the expected prediction can be reviewed in the three predictive methods: Prophet, SARIMA, and GXBoost. The most efficient method can be verified where the comparison metrics are close to zero, i.e., they have a lower value.

The predictive methods used are SARIMA, Prophet, and XGBoost, which were used to compare their efficiency. The results and conclusions are analyzed in a similar way to the work called Proceedings of 14th IOE Graduate Conference Comparative Analysis of Time Series Forecasting Models for Predicting PM2.5 level in Kathmandu [21]. The observations obtained are:

- They are applicable to the type of data collected, i.e., applicable to time series.
- The three methods have different methodological approaches that enrich the measurement of their comparison:

SARIMA has a classic statistical approach to time series. The **SARIMA** model **( ) is** an extension of the ARIMA model that incorporates seasonal components to adequately address time series with repetitive patterns [9]. Prophet has an automated additive modeling methodological approach for time series. Prophet is a procedure for forecasting time series data based on an additive model where nonlinear trends are adjusted for annual, weekly, and daily seasonality, as well as holiday effects [10]. XGBoost has a machine learning-based methodological approach that uses decision tree ensembles. Developed by Tianqi Chen of the University of Washington (Extreme Gradient Boosting), XGBoost is an advanced gradient boosting implementation with the same general framework; that is, it combines weak learner trees into strong learners by summing residuals [11].

The optimal hyperparameters for the different methods are shown below (Tables 1, 2 and 3):

Table 1. Optimal hyperparameters for Prophet

Prophet			
Amplitude Interval	Daily Seasonality	Annual Seasonality	Consider Holidays
0.8	TRUE	TRUE	TRUE

Table 2. Optimal hyperparameters for SARIMA

SARIMA								
p	d	q	P	D	Q	m	Trend	Criterion AIC information
0	1	2	0	0	0	12	c	498.25

Table 3. Optimal hyperparameters for XGBoost

XGBoost					
Number of Estimators	Learning Rate	Maximum Depth	Subsampling	Column x Tree Subsampling	Cross-Validation RMSE
500	0.05	3	0.9	0.9	89.03

The main advantages of the methods analyzed are:

Advantages of SARIMA [9]

- They can explicitly model the seasonal component of the data, which can improve the accuracy and interpretability of forecasts.
- They can account for both additive and multiplicative seasonality, meaning they can handle different types of seasonal variations in the data.
- They can incorporate non-seasonal and seasonal differentiation, which can help stabilize the data and eliminate trends and cycles.
- They can use both non-seasonal and seasonal autoregressive and moving average terms, which can capture dependencies and noise in the data.

Advantages of Prophet [19]

- It is quick and easy to use, even for users without much experience.
- It provides predictions with confidence intervals that can be useful.
- It automates large parts of the time series modeling process.
- Holidays and/or events can be added.

Advantages of XGBoost [20]

- It can handle large databases with multiple variables.
- It can handle missing values.
- Its results are very accurate.
- Excellent execution speed.

This study used the most popular comparison metrics as confirmed by NVIDIA [12]:

- RMSE: Root Mean Square Error
- MAE: Mean Absolute Error
- MAPE: Mean Absolute Percentage Error

Decision Trees: A decision tree is a non-parametric supervised learning algorithm used for both classification and regression tasks. It has a hierarchical tree structure consisting of a root node, branches, internal nodes, and leaf nodes [14].

Cloud Computing: Cloud computing is a model that allows on-demand network access to a set of configurable computing services, such as infrastructure, applications, and storage [15].

Computing: Computing refers to calculations, basic or complex mathematical operations related to information technology and, above all, to solving problems through the use of computer systems [16].

Machine Learning: Machine learning is the analysis of data that automates the generation of analytical models. It is a branch of artificial intelligence focused on the idea that systems can learn from data, identify and determine patterns to make decisions with minimal human intervention [17].

The development of predictive models begins with collecting data and then cleaning it. Next, the relevant variables are selected, and then the model representing the relationship between the variables is constructed [18]. The datasets obtained can be found at the Internal Revenue Service (SRI) Datasets link, in the vehicles section: https://www.sri.gob.ec/datasets#Ventas%20-%20Compras .

From the data obtained, only the number of records with a unique identifier is counted, i.e., ensuring that the VEHICLE CODE field is considered only once (Table 4).

Table 4. Description of fields in the historical files analyzed

Variable name	Variable definition	Data format	Length
CATEGORY	Refers to the vehicle subcategory code	Numeric	20

(continued)

Table 4. (*continued*)

Variable name	Variable definition	Data format	Length
VEHICLE CODE	This is the unique code assigned to each record as part of the Vehicle Registry administration	Numeric	20
TRANSACTION TYPE	This is information about the type of transaction, which may be direct import or commercialization.	Text	20
BRAND	This is the brand of the vehicle.	Text	20
MODEL	This is the model created for the subcategory.	Text	20
COUNTRY	This is the country of manufacture.	Text	20
MODEL YEAR	This is the model year of the vehicle.	Text	20
CLASS	This is the vehicle class	Text	20
SUB CLASS	Refers to the subclass of the vehicle.	Text	20
TYPE	Refers to whether the vehicle is light or heavy.	Text	7
APPRAISAL	This is the appraisal information for the subcategory.	Numeric	20
PROCESS DATE (DD/MM/YY)	Date on which the vehicle owner was registered.	Numeric	8
SERVICE TYPE	The type of service for the vehicle	Text	3
ENGINE SIZE	This is the engine size of the vehicle.	Numeric	20
FUEL TYPE	This is the type of fuel used by a vehicle.	Text	20
DATE OF PURCHASE (DD/MM/YY)	Date on which the vehicle was purchased.	Numeric	8
CANTON	District assigned to the vehicle	Text	20
COLOR 1	Vehicle color 1	Text	3
COLOR 2	Vehicle color 2	Text	3
NATURAL PERSON - LEGAL ENTITY	Type of person who made the purchase	Text	8

3 Results

3.1 Input Data

Daily historical data on electric vehicle purchases in Ecuador obtained from 2020 to 2024 (Table 5).

Table 5. Historical data acquired Monthly consolidation of electric vehicle purchases in Ecuador from 2020 to 2024

Month	2020	2021	2022	2023	2024
January	35	14	43	95	252
February	19	22	36	78	145
March	15	16	48	275	201
April	0	44	59	158	126
May	33	28	49	178	149
June	64	25	44	148	162
July	31	61	49	62	184
August	23	27	57	74	158
September	34	31	50	135	186
October	48	32	65	309	191
November	32	76	68	229	212
December	45	86	55	173	231

3.2 Prediction

- Training: Predictive methods are trained with data from 2020 to 2023
- Initial evaluation: Evaluated for Prophet, SARIMA, and XGBoost by comparing with actual 2024 data
- Retraining: Models are retrained with data from 2020 to 2024
- Prediction: Monthly electric vehicle purchases are predicted for 2025 and 2026 (Figs. 2, 3 and 4) (Tables 6, 7, 8 and 9)

Table 6. Comparison of actual versus predicted values using each of the methods evaluated for the year 2024 with 80% confidence intervals

Month	Actual	Prophet	Prophet Lower	Prophet Upper	Sarima	Sarima Lower	Sarima Upper	XGBoost	XGBoost Lower	XGBoost Upper
January	252	248	34	461	179	121	237	152	127	216
February	145	240	41	443	185	121	249	134	110	199

(*continued*)

Table 6. (*continued*)

Month	Actual	Prophet	Prophet Lower	Prophet Upper	Sarima	Sarima Lower	Sarima Upper	XGBoost	XGBoost Lower	XGBoost Upper
March	201	347	136	564	189	124	253	167	142	232
April	126	322	114	524	192	127	258	153	129	218
May	149	331	118	545	196	129	263	196	171	260
June	162	298	88	504	200	132	268	162	138	227
July	184	318	105	534	204	135	272	148	124	213
August	158	295	82	507	207	138	277	138	113	202
September	186	304	99	513	211	141	282	158	134	223
October	191	395	183	608	215	143	286	161	137	226
November	212	382	176	590	219	146	291	169	145	234
December	231	394	179	609	222	149	296	164	139	229

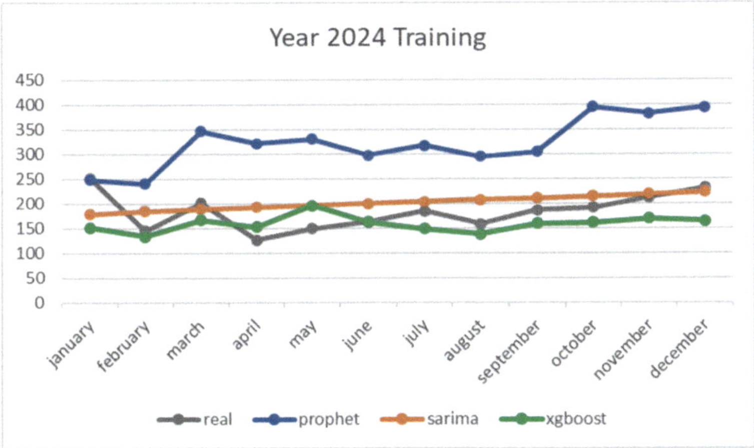

Fig. 2. Comparison of actual versus predicted data for 2024 training.

Table 7. Prediction of electric vehicle purchases in Ecuador in 2025

Month	Prophet	Prophet Lower	Prophet Upper	Sarima	Sarima Lower	Sarima Upper	XGBoost	XGBoost Lower	XGBoost Upper
January	386	172	599	214	157	271	207	183	272
February	360	168	556	212	152	273	202	177	266
March	482	270	696	216	155	277	211	187	276
April	455	249	662	219	157	281	161	136	225
May	466	253	682	223	160	285	162	137	226
June	434	226	644	226	163	289	156	131	221

(*continued*)

Table 7. (*continued*)

Month	Prophet	Prophet Lower	Prophet Upper	Sarima	Sarima Lower	Sarima Upper	XGBoost	XGBoost Lower	XGBoost Upper
July	456	241	674	230	166	293	158	134	223
August	429	211	646	233	169	297	159	135	224
September	442	234	651	237	172	302	173	149	238
October	531	312	747	240	174	306	211	187	276
November	510	300	721	243	177	310	220	195	284
December	536	323	757	247	180	314	224	200	289

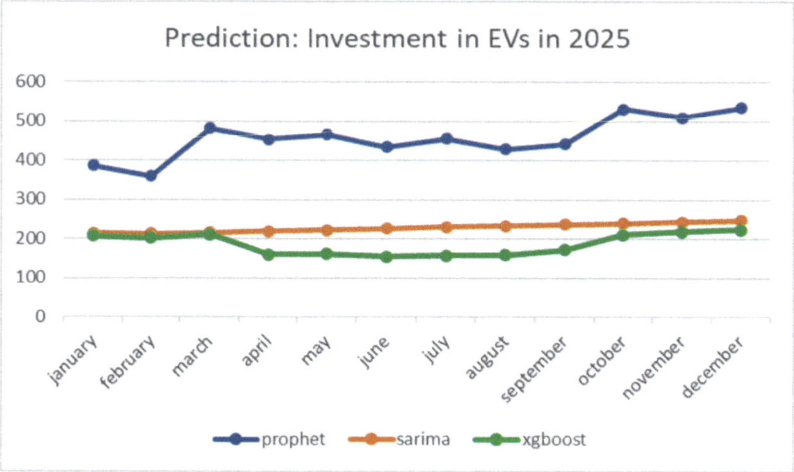

Fig. 3. Comparative forecast for 2025

Table 8. Prediction of electric vehicle purchases in Ecuador in 2026

Month	Prophet	Prophet Lower	Prophet Upper	Sarima	Sarima Lower	Sarima Upper	XGBoost	XGBoost Lower	XGBoost Upper
January	523	305	738	250	183	318	218	193	282
February	485	290	683	254	186	322	212	187	276
March	625	405	842	257	189	326	212	187	276
April	590	378	804	261	191	330	165	140	229
May	605	386	823	264	194	334	160	135	224
June	572	360	786	268	197	338	154	130	219
July	594	376	817	271	200	342	155	131	220
August	569	350	790	275	203	346	157	132	222

(*continued*)

Table 8. (*continued*)

Month	Prophet	Prophet Lower	Prophet Upper	Sarima	Sarima Lower	Sarima Upper	XGBoost	XGBoost Lower	XGBoost Upper
September	576	364	788	278	206	350	176	152	241
October	667	449	889	282	209	354	210	186	275
November	650	437	869	285	212	358	219	195	284
December	676	456	901	288	215	362	224	200	289

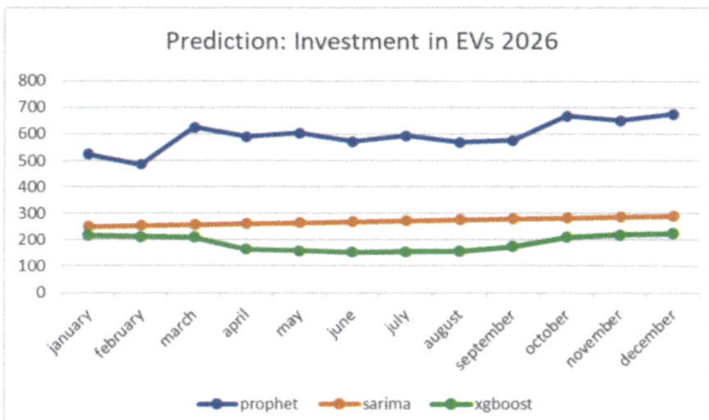

Fig. 4. Comparative forecast for 2026

Table 9. Comparative metrics of the methods applied

Model	RMSE	MAE	MAPE
Prophet	149.36	140.32	81.78
SARIMA	40.11	34.17	20.42
XGBoost	44.73	36.97	19.14

Considering the comparative metrics, the SARIMA predictive method is the method that obtained the best metrics; therefore, it is the best predictive method of the three analyzed. XGBoost is very close as a challenger for a very close and acceptable prediction of the MAPE metric with a better percentage than SARIMA (Fig. 5).

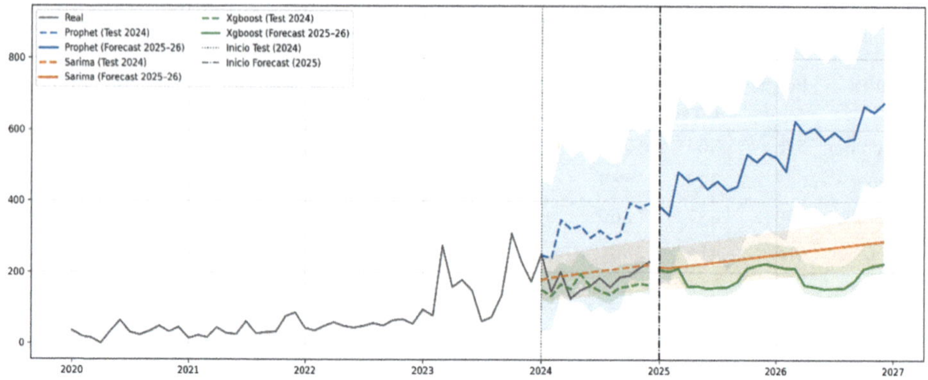

Fig. 5. Monthly purchases of electric vehicles (2020–2026) — Models and intervals (80%)

4 Discussion

A comparison was made with other previous studies comparing the same predictive methods.

Comparison with work 1: Considering the work entitled Proceedings of 14th IOE Graduate Conference Comparative Analysis of Time Series Forecasting Models for Predicting PM2.5 level in Kathmandu [21], the similarities and differences in the historical data to be analyzed are shown.

- The historical data analyzed in both cases correspond to time series.
- Investment in electric vehicles has exponential and non-linear growth, while the study of the level of pollution in particles less than or equal to 2.5 (μm) does not have exponential growth.
- In both cases, there is clear seasonality. In the case of investment in electric vehicles, the peaks occur at the end of the year, when workers in Ecuador receive their thirteenth month's salary and, due to end-of-year promotions, vehicle sales are at their highest. On the other hand, pollution levels with particles less than or equal to 2.5 (μm) change due to rainfall during the winter season.
- In both studies, the historical data for a given month depends on previous data. In the case of electric vehicles, because this type of vehicle is more environmentally friendly, there is a tendency to buy them. In the case of pollution levels with particles less than or equal to 2.5 (μm), there is also a dependence on previous data, with pollutants increasing over time rather than decreasing.
- External factors have a significant influence on both studies. Electric vehicle purchases may increase due to their environmental friendliness and rising fuel prices, but if the country's electrical problems worsen, it could affect sales. In the study of pollution with particles less than or equal to 2.5 (μm), climate issues can have an impact, such as the recent pandemic where everything changed.
- The SARIMA and XGBoost methods obtained better results for RMSE in both studies, meaning they should be considered to avoid large deviations in the prediction. The Prophet method is the one that can have large deviations in the prediction (Table 10).

Table 10. Comparative metrics of this study compared to the article on pollution with particles less than or equal to 2.5 (μm)

VE Investment		
Model	**RMSE**	**MAE**
Prophet	149.36	140.32
SARIMA	40.11	34.17
XGBoost	44.73	36.97

PM 2.5 **pollution** (μm)		
Model	**RMSE**	**MAE**
Prophet	39.56	28.26
SARIMA	16.99	54.84
XGBoost	30.36	4.65

- The SARIMA method was the best for studying predictions in electric vehicle investment, with the RMSE and MAE metrics being the best, while XGBoost is very close and its prediction could even be accepted. Prophet should be ruled out for this type of analysis.
- The XGBoost method is the best for studying pollution with particles 2.5, because the MAE is 4.65, where the prediction is very close to reality.

Comparison with work 2: Another comparison made with the work Hybrid Prophet/XGBoost Model for Predicting Retail Automotive Aftermarket Inventory Demand [22] shows the following similarities and differences:

- For stable series and repetitive patterns, SARIMA has greater accuracy.
- It is suggested that a predictive study be carried out with at least three methods in order to define the most efficient method.
- XGBoost predicts better when a greater number of variables need to be considered.
- Prophet performs better when considering holidays.

Comparison with work 3: Taking the work Comparative Analysis of ARIMA, SARIMA, and Prophet Models in Forecasting [23]. The following observations are made:

- Prophet does not require more efficient parameter calculations as a tuning process to achieve a better prediction; it is more automatic.
- SARIMA is more effective if there are no jumps that eliminate seasonality

Comparison with work 4: Taking the work Forecasting prescription drug use, a comparative study of the SARIMA, Prophet, XGBoost, and LSTM models [24]. The following observations are made:

- Prophet is less effective when there are marked changes.
- SARIMA was more effective, with better prediction results when there are seasonalities and linear trends.
- GXBoost performs better when data has non-linear trends.

Comparison with study 5: Taking the study Time series analysis of hemorrhagic fever with renal syndrome in mainland China using the XGBoost forecasting model [25]. The following observations are made:

- As confirmed in several of the studies considered, GXBoost performs better when the data analyzed has nonlinear trends.
- ARIMA did not perform well because the data analyzed did not have a linear trend.

Comparison with study 6: Taking the study Forecasting platelet demand: from univariate time series to multivariate models [26]. The following observations are made:

- ARIMA performs better with limited data and a linear trend.
- Prophet is better when holidays are available.
- Decision tree-based methods such as XGBoost perform better when handling more variables in the study

Comparison with work 7: Taking the work Machine learning and deep learning prediction models for time series: a comparative analytical study for the use case of short-term electricity price prediction in the United Kingdom [27]. The following observations are made:

- It is confirmed that SARIMA performs better with linear trend data; the electricity price prediction study does not follow a uniform trend, so SARIMA did not perform well.
- Prophet did not perform well because holidays are not handled in the analysis, as the price of electricity does not depend on them.
- XGBoost performed better than the Prophet and SARIMA methods because it handles more variables in the study.

The current work and the comparison with another similar study show that when comparing the PROPHET, SARIMA, and XGBoost predictive methods, we can suggest that the Prophet method be directly discarded for this type of study. The SARIMA and XGBoost methods are the ones suggested for application. Once applied, the RMSE, MAE, and MAPE comparison metrics should be obtained, then compare them to finally conclude and validate the prediction of the method with the best metrics achieved, i.e., where the metrics obtained the smallest values, demonstrating that the errors are smaller when comparing the actual versus the prediction.

5 Conclusions

Considering Diagram 1. Comparison of actual vs. predicted data for 2024, we can see that the trend closest to the actual is the prediction of the SARIMA and XGBoost methods. The metrics suggest using the SARIMA prediction [21], which obtained the lowest RMSE and MAE values, meaning that its prediction will have less error (Table 11).

Table 11. Comparative metrics of the methods applied

Model	RMSE	MAE	MAPE
Prophet	149.36	140.32	81.78
SARIMA	40.11	34.17	20.42
XGBoost	44.73	36.97	19.14

Prophet is the method that obtained the worst results and should be discarded in this type of analysis where holidays have no impact. If, in the future, exogenous factors such as fuel price increases or energy problems are to be considered, we suggest using XGBoost because it allows more variables to be integrated into the prediction calculation.

There is no specific predictive method for each type of data; it is appropriate to apply at least two or three suggested methods depending on the type of historical data available. Then, calculate the comparison metrics to finally determine which predictive method performed best based on that data. Once the appropriate method has been determined, it is applied frequently to make decisions that are closer to reality in advance.

References

1. Technological advances of the last decade: from 2010 to 2020 (n.d.). Retrieved August 29, 2025, from https://www.itmastersmag.com/transformacion-digital/los-2010-la-decada-tecno/
2. Bayes's theorem | Definition & amp; Example | Britannica (n.d.). Retrieved August 25, 2025, from https://www.britannica.com/topic/Bayess-theorem?
3. Abril, J.C.: Analysis of the evolution of time series techniques: a unified approach. ESTADISTICA. **63**, 5–56 (2019) https://ri.conicet.gov.ar/handle/11336/71449
4. Chen, T., Guestrin, C.: XGBoost: A Scalable Tree Boosting System (n.d.). https://doi.org/10.1145/2939672.2939785
5. Researchers from three universities present a project on Smart Agriculture for monitoring and diagnosing corn crops | Escuela Superior Politécnica del Litoral (n.d.). Retrieved August 29, 2025, from https://www.espol.edu.ec/es/noticias/investigadores-de-tres-universidades-exponen-proyecto-sobre-agricultura-inteligente-para
6. Accelerating the Advanced Analytics Process through Automated Machine Learning IT Now (n.d.). Retrieved August 28, 2025, from https://itahora.com/2023/03/20/acelerando-el-proceso-de-analitica-avanzada-a-traves-de-automated-machine-learning/
7. 2024-TFG6_Comparative_Analysis_of_Classic_Methods_and_Machine_Learning_Methods | PDF (n.d.). Retrieved August 26, 2025, from https://es.slideshare.net/slideshow/2024-tfg6_comparativa_metodos_clasicos_y_metodos_machine_learning/273838219
8. Elkan, C.: Predictive Analytics and Data Mining. University of California, San Diego (2013)
9. What is the SARIMA model?| Trading Techniques (n.d.). Retrieved August 25, 2025, from https://www.tecnicasdetrading.com/2024/07/modelo-sarima.html
10. Prophet | Forecasting at scale (n.d.). Retrieved August 23, 2025, from https://facebook.github.io/prophet/
11. What is XGBoost? | IBM (n.d.). Retrieved August 23, 2025, from https://www.ibm.com/mx-es/think/topics/xgboost

12. A Comprehensive Overview of Regression Evaluation Metrics | NVIDIA Technical Blog (n.d.). Retrieved September 1, 2025, from https://developer.nvidia.com/blog/a-comprehensive-overview-of-regression-evaluation-metrics/
13. History and License — Python 3.13.7 documentation (n.d.). Retrieved August 26, 2025, from https://docs.python.org/3/license.html
14. What is a decision tree? | IBM (n.d.). Retrieved August 26, 2025, from https://www.ibm.com/es-es/think/topics/decision-trees
15. Del Vecchio, J.F., Paternina, F.J., Henríquez Miranda, C.: Cloud computing: a model for business development. PRO. **13**(2), 81–87 (2015). https://doi.org/10.15665/RP.V13I2.490
16. What is Computing? | Lenovo Mexico (n.d.). Retrieved August 26, 2025, from https://www.lenovo.com/mx/es/glosario/computo/?orgRef=https%253A%
17. Machine Learning.: What It Is and Why It Matters | SAS (n.d.). Retrieved August 26, 2025, from https://www.sas.com/es_co/insights/analytics/machine-learning.html
18. Predictive Models.: Definition and Practical Applications. (n.d.). Retrieved August 26, 2025, from https://uni.edu.gt/noticias/modelos-predictivos/
19. Time Series Predictions with Facebook's PROPHET (META) – DATLAS Research Column – Datlas Blog (n.d.). Retrieved August 26, 2025, from https://blogdatlas.wordpress.com/2024/05/11/predicciones-de-series-temporales-con-prophet-de-facebook-meta-columna
20. Espinosa-Zúñiga, J.J., Espinosa-Zúñiga, J.J.: Application of random forest and XGBoost algorithms to a database of credit card applications. Eng. Res. Technol. **21**(3), 1–16 (2020). https://doi.org/10.22201/FI.25940732E.2020.21.3.022
21. Khanal, I., Prakash Dhakal, O., Kumar Guragain, M., Karki, P., & Pandey, B. Proceedings of 14th IOE Graduate Conference Comparative Analysis of Time Series Forecasting Models for Predicting PM2.5 Level in Kathmandu: SARIMA, Prophet, and XGBoost. (n.d.).https://www.mendeley.com/reference-manager/reader/f1c1da27-cfad-3113-b6a4-fdb524cedc0c/417d7338-5e90-47c5-42ca-228093bd794d/
22. Garcia, A. I. Hybrid Prophet/XGBoost Model for Forecasting Demand for Automotive Aftermarket Retail Inventories (n.d.)
23. Kwarteng, S., Andreevich, P.: Comparative analysis of ARIMA, SARIMA, and prophet models in forecasting. Res. Dev. **5**(4), 110–120 (2024). https://doi.org/10.11648/J.RD.20240504.13
24. Ripatti, V. Master's Program in Life Science Technologies: Complex Systems Forecasting Prescription Medication Utilization: A Comparative Study of SARIMA, Prophet, XGBoost, and LSTM Models Leveraging Machine Learning and Statistical Models for Predictive Analysis of Pharmaceutical Utilization in Sweden: A Comparative Evaluation Using RMSE, MAE, and MAPE Metrics. Retrieved September 8, 2025, (n.d..) from www.aalto.fi
25. Lv, C.X., An, S.Y., Qiao, B.J., Wu, W.: Time series analysis of hemorrhagic fever with renal syndrome in mainland China by using an XGBoost forecasting model. BMC Infect. Dis. **21**(1), 1–13 (2021). https://doi.org/10.1186/S12879-021-06503-Y/TABLES/5
26. Motamedi, M., Dawson, J., Li, N., Down, D.G., Heddle, N.M.: Demand forecasting for platelet usage: from univariate time series to multivariable models. PLoS One. **19**(4), e0297391 (2024). https://doi.org/10.1371/JOURNAL.PONE.0297391
27. Mishra, B.K., Preniqi, V., Thakker, D., Feigl, E.: Machine learning and deep learning prediction models for time-series: a comparative analytical study for the use case of the UK short-term electricity price prediction. Discov. Internet Things. **4**(1), 1–22 (2024). https://doi.org/10.1007/S43926-024-00075-4/TABLES/8

Contribution of the PLS Methodology in the Field of Information and Knowledge Management

Mikel Ugando-Peñate[1(✉)], José Luis Espín-Lopez[1], Mishell Alejandra García-Acosta[1], Reinaldo Armas-Herrera[2], Angel Alexander Higuerey-Gómez[2,3], and Ángel Ramón Sabando-García[1]

[1] Pontificia Universidad Católica del Ecuador, Sede Santo Domingo, Santo Domingo de Los Tsáchilas 230203, Ecuador
{mugandop,jlespin,magarciaac,arsabando}@pucesd.edu.ec

[2] Universidad Técnica Particular de Loja, Loja 110150, Ecuador
{ahreinaldo,aahiguerey}@utpl.edu.ec

[3] Instituto Experimental de Investigaciones Humanísticas, Económicas y Sociales de la Universidad de los Andes, Trujillo 3101, Venezuela

Abstract. Educational research seeks to strengthen pedagogical processes and institutional innovation using information and knowledge management. In this area, Partial Least Squares (PLS) methodology is a technique that allows the analysis of complex relationships in educational environments, where there are multiple dependent variables. PLS-SEM models (Partial Least Squares Structural Equations Modeling) allow educational institutions to understand how knowledge is transferred, how the teaching-learning process is organized and how digitalization affects this process. In this bibliographic review, 230 Scopus articles were used during the period 2020–2025 and examined with VOSviewer by performing an analysis of keywords, references, and the institutions with the greatest impact and which have financed this topic. There is an integration of sustainability, innovation and technology in knowledge management, highlighting artificial intelligence, digital platforms and organizational learning processes as key topics. The contribution of PLS to knowledge and information management strengthens education and fosters sustainable development.

Keywords: Knowledge management · Education · PLS · Modeling · Educational innovation · Information systems

1 Introduction

The PLS-SEM methodology has been a contribution to educational research due to its ability to model complex relationships between latent variables. It allows for regressions with multiple dependent variables and can handle multicollinearity, many regressors, and small sample sizes. These characteristics have made it a widely used technique in various branches of the social sciences.

A. Rocha et al. (Eds.): ICITS 2026, LNNS 1966, pp. 473–483, 2027.
https://doi.org/10.1007/978-3-032-27538-7_42

It has been used in the adoption of digital technologies in financial institutions [1], in environmental awareness in the agricultural sector [2], in virtual environments to measure continuous learning [3], in the influence of immersive experiences on academic performance [4], human-AI collaboration in the educational metaverse [5], and in the impact of intelligent systems such as ChatGPT on knowledge management [6]. Other fields where PLS-SEM has been applied include open innovation [7], burnout [8], and carbon efficiency [9]. In addition, factors such as teacher support [10], self-efficacy [11], and intention to use virtual platforms in digital transformation contexts [12] have been evaluated.

The PLS methodology allows to analyze urban sprawl and its relationship with the digital economy [13], as well as to model energy efficiency and CO_2 emissions in urban agglomerations [14]. Topics such as waste management [15], green innovation [16], and strategic planning in educational institutions [17] have also been addressed. Studies on organizational behavior [18], perception of fairness [19], institutional resilience [20], and adoption of emerging technologies [21] have been enriched by PLS-SEM models. Recent research has applied this methodology to assess efficient land use [22], the evolution of spatial patterns in smart cities [23], and the simulation of urban growth [24]. In the field of artificial intelligence applied to education, the recognition of urban functions [25], student perception in virtual environments [26], and the intention to use metaverse platforms [27] have been analyzed.

Other relevant studies include multisensory data analysis [28], digital competency assessment [29], and validation of autonomous learning models [30]. Topics such as institutional sustainability [31], educational innovation [32], organizational efficiency [33], and technological transformation [34] have also been addressed. Research on the perception of organizational justice [35], the impact of digitalization on learning [36], and the use of hybrid models to assess technology adoption [37] complement studies on capacity building in educational settings [38], knowledge management in academic institutions [39], and the integration of disruptive technologies into educational processes [40]. Together, these articles demonstrate how the PLS methodology has transformed research in knowledge management, offering solid models to understand complex phenomena in education. Recent research further demonstrates the flexibility of PLS-SEM and its relevance for examining phenomena associated with emerging technologies. One study investigates the determinants of users' intention to adopt ChatGPT through the UTAUT framework, providing a systematic assessment of the factors that shape engagement with conversational AI tools across digital environments [41]. Additional work introduces an integrated architecture that combines deep learning techniques with blockchain-based security mechanisms to ensure reliable data transmission within IoT-enabled healthcare systems, highlighting the value of hybrid solutions in settings that demand strict data integrity and protection [42]. Further evidence is offered by an investigation focused on the role of blockchain in enhancing transparency across agri-food supply chains, outlining practical applications, operational constraints, and future directions for advanced traceability systems [43].

In the organizational and technological domains, research on big data analytics adoption among small and medium-sized enterprises examines the determinants of implementation and its subsequent effects on firm performance using structural modeling techniques [44]. Another study evaluates the integration of cloud computing solutions and their influence on the sustainable performance of SMEs through a combined SEM–ANN approach, enabling the capture of non-linear relationships and improving predictive accuracy in contexts characterized by rapid digital transformation [45].

2 Objective

The objective of this work is to determine how the PLS-SEM methodology has been applied to information and knowledge management from an empirical perspective to improve decision-making, understanding organizational and educational phenomena, and validating theoretical models in diverse contexts.

3 Methodology

3.1 Methodology and Data Subsection

Within the framework of this research, a mixed review methodology was applied. On the one hand, a traditional bibliographic analysis was carried out aimed to understand the general aspects of the object of study; and on the other, systematic literature reviews were carried out to analyze the most specific concepts related to this research. Various academic databases were used to collect information, including Scopus.

The selection of the bibliographic material was made based on a search strategy based on keywords such as: "education", "knowledge", "information", and "pls-sem". This search process was carried out in the Scopus database in the period between 2020–2025, yielding a total of 230 publications. This was obtained through the following query: TITLE-ABS-KEY (Big Data AND Machine Learning) AND TITLE-ABS-KEY (Information AND pls-sem) AND TITLE-ABS-KEY (Technology AND Knowledge). The collected documents were put into a qualitative analysis by constructing **bibliometric** maps using the VOSviewer software. The interpretation and results of which are presented in the following sections.

3.2 Bibliometric Analysis

This study is based on a bibliometric analysis aimed at identifying the main contributions of the Partial Least Squares methodology in the field of information and knowledge management. A database composed of 230 articles indexed in Scopus was used, selected for their thematic and methodological relevance. Through indicators such as the number of citations, the co-occurrence of keywords and the analysis of author networks, the 40 most cited articles were identified, which allowed to guarantee representativeness and depth in the analysis. The data processing was carried out using tools such as Excel and VOSviewer, facilitating the visualization of trends, thematic nuclei and conceptual evolution. This methodological strategy allows a systematic synthesis of the literature, evidencing the role of the PLS approach in the validation of theoretical models and in the strengthening of decision-making processes in educational and organizational contexts.

4 Results

4.1 Keywords Co-occurrence Network Map

Figure 1 presents a map of keyword occurrences in which several thematic clusters are distinguished. The most relevant cluster, in dark green, is led by the word human, with 67 occurrences and a total strength of 608, which confirms its central role in the reviewed literature. This node is closely linked to article with 49 concurrences and 467 links, male (41; 458) and female (41; 450), showing the orientation of studies towards diverse populations and empirical analysis in educational contexts. Likewise, this group is linked to methodologies such as questionnaire (27; 314) and surveys (18; 227), which reinforce the quantitative nature of the research.

The soft green cluster is organized around sustainability, with 42 occurrences and 155 strengths, associated with technology adoption and innovation, which shows a tendency to integrate sustainability and innovation into knowledge management processes. The purple cluster highlights the relevance of human and behavioral factors.

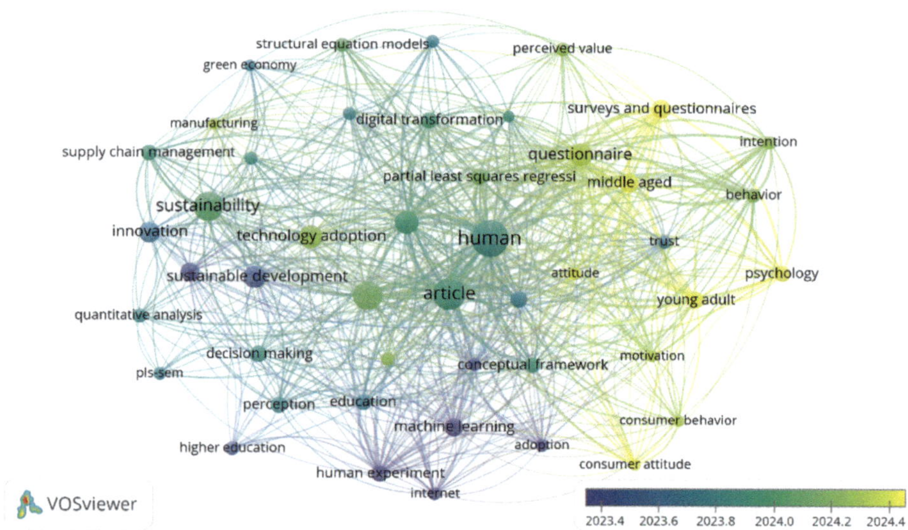

Fig. 1. Bibliometric network map with all keywords.

4.2 Author Keywords Co-occurrence Network Map

Figure 2 studies the keywords most used by the relevant authors of the topic. The cluster, with the highest number of occurrences 20 and a strength of 15, is associated with the keyword artificial intelligence. The second most representative node is identified by the sustainability with 13 occurrences and a strength of 9 and the third most representative cluster is structural equation modeling with 11 occurrences and 6 of total force.

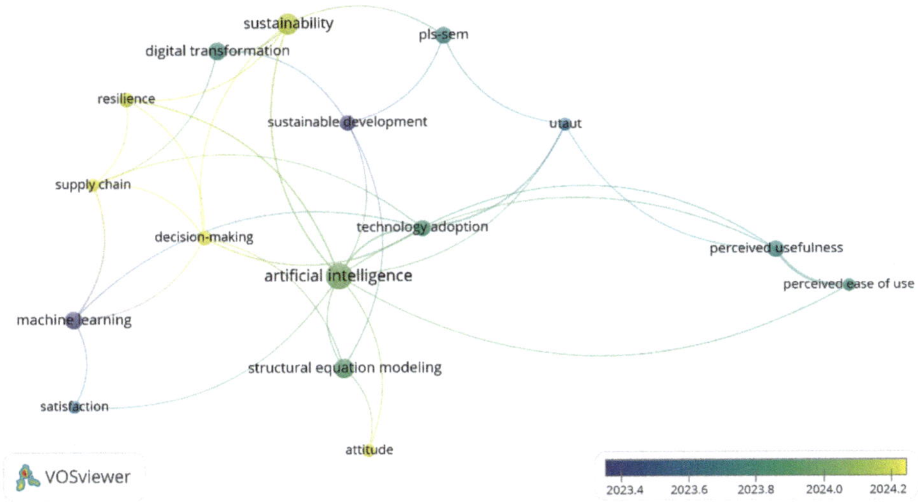

Fig. 2. Bibliometric network map with author keywords.

4.3 Citation Author Network Map

Figure 3 shows the authors with the highest citation strength. First, the most cited author is Parisa Maroufkhani with 240 citations, who conducted research entitled: "Big data analytics adoption: determinants and performances among small to medium-sized enterprises, International Journal of Information Management". In second place, the best-cited author is Al-Sharafi with 187 citations, who carried out a publication entitled "Determinants of cloud computing integration and its impact on sustainable performance in SMEs: an empirical investigation using the SEM-ANN approach".

It is reflected that there is an interrelationship between researchers and the most cited documents, where figures such as Kineber, Ahmed Farouk, Salloum, Said A., Iranmanesh, Mohammad and Asif, Muhammad, stand out, who appear with a larger node size, which evidences their high production and citation. These authors are mainly connected on topics related to the application of the PLS methodology in the field of human knowledge and behavior, showing a strong link with the most cited lines of research in leading countries such as China, Malaysia, Saudi Arabia and India. The analysis shows that the thematic interrelation focuses on the validation of models in contexts of knowledge management, innovation and organizational behavior, these being the topics of greatest interest and that have generated the greatest impact in terms of citation and international collaboration.

4.4 Institutions with Contributions on the Topic of PLS Methodologies

Table 1 shows that all institutions have contributed with 2 documents each, highlighting the Graduate School of Business, Universiti Sains Malaysia (173 citations), the School of Business and Law, Edith Cowan University and the Faculty of Engineering & IT, The British University in Dubai (both with 155 citations), as those with the greatest impact

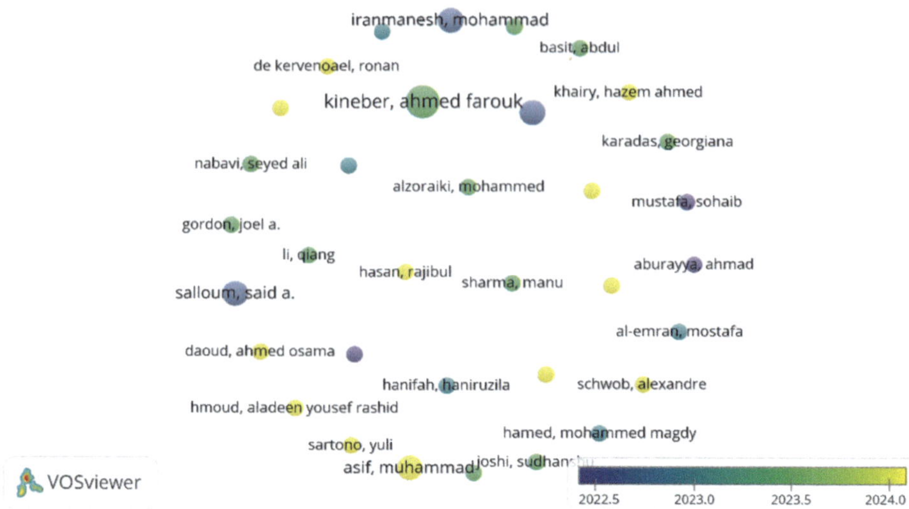

Fig. 3. Bibliometric network map with citation author.

on citations. In contrast, the Rennes School of Business and its marketing departments have a low incidence (8 citations). However, these institutions research on close topics, there are no direct collaborations between them in this dataset, suggesting a fragmented field in terms of co-authorship networks or institutional cooperation.

Table 1. Institutions with contributions

Organization	Documents	Cites	Relevance
Graduate School of Business, Universiti Sains Malaysia (Malasia)	2	173	Greater impact on citations
School of Business and Law, Edith Cowan University (Australia)	2	155	High relevance
Faculty of Engineering & IT, The British University in Dubai (EAU)	2	155	High relevance
Doctoral School, National University of Physical Education and Sport (Rumania)	2	100	Intermediate relevance
School of Economics and Management, Chang'an University (China)	2	85	Intermediate relevance

4.5 Citation Documents Network Map

The authors with the highest number of citations are identified, highlighting those researchers who have had a substantial impact on the contribution of PLS to knowledge management and related areas.

The most influential work and authors are summarized, classified according to their main themes in different research areas. Menon, D., with 204 citations, is recognized for his research entitled "Chatting with ChatGPT: Analyzing the factors influencing users' intention to use the Open AI's ChatGPT using the UTAUT model," which positions him as a leading reference in the field. He is followed by Kumar, P., with 187 citations, for his research entitled "A blockchain-orchestrated deep learning approach for secure data transmission in IoT-enabled healthcare systems," and Menon, S., with 160 citations for his work entitled "Blockchain Technology for Transparency in Agri-Food Supply Chain: Use Cases, Limitations, and Future Directions."

These citation levels suggest that their contributions have served as recent conceptual cornerstones, published in a high-impact journal within the area of PLS-SEM models applied to education and knowledge management. For example, similar research on this methodology in technological contexts has been published in leading journals such as Engineering Management and Software Engineering [41].

4.6 Citation Countries Network Map Interrelation

Figure 4 shows that there are 4 main countries involved with research on PLS methodology in knowledge information, the main one being China with 1079 citations and its most published journal "Computers in Human Behavior" followed by Malaysia with 879 citations in 38 papers as a reference to the journal "Quality and Quantity". The third most cited country is Saudi Arabia, with 567 citations in 33 documents, and finally India, with a strength of 498 citations in 21 documents with its journal entitled "Heliyon".

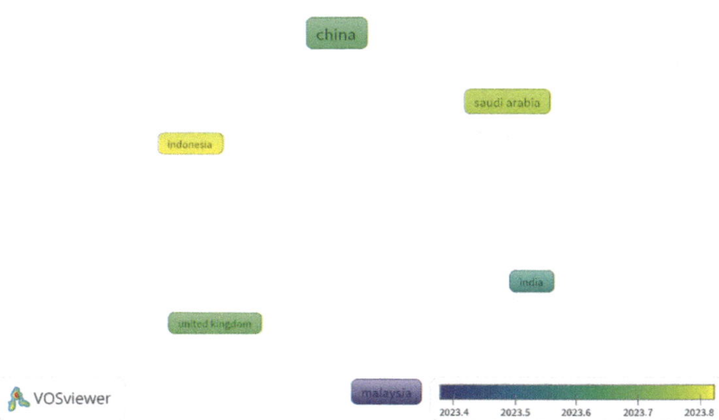

Fig. 4. Bibliometric network map with citation countries.

5 Financing

The research analyzed in this bibliometric review has been driven by funding from international entities committed to scientific development in Big Data. This support has made it possible to consolidate lines of study that integrate artificial intelligence in knowledge, economic and technological processes, generating applied knowledge for application from the interdisciplinary perspective of the reviewed corpus, five entities stand out as main funders: National Natural Science Foundation of China with 11 projects; Ministry of Education of the People's Republic of China with 6 initiatives, National Office for Philosophy and Social Sciences with 6 studies; UK Research and Innovation with 6 studies, and King Saud University with 5 contributions.

This strategic investment demonstrates the central role of scientific research in the transition to knowledge management and underscores the relevance of the application of technological tools that promote innovation in sustainability and knowledge (Table 2).

Table 2. Research funding in the PLS methodology

Organization	Amount
National Natural Science Foundation of China	11
Ministry of Education of the People's Republic of China	6
National Office for Philosophy and Social Sciences	6
UK Research and Innovation	6
King Saud University	5

6 Conclusions

The PLS-SEM methodology is consolidated as a key methodological resource in educational and organizational research, since it allows modeling complex relationships between latent variables in contexts characterized by small samples, non-normal data and mixed structures, strengthening the empirical validation of theoretical frameworks.

Bibliometric studies show a trend towards the integration of sustainability, innovation and technological adoption in knowledge management, highlighting as main axes the applications of artificial intelligence, digital platforms and organizational learning processes in educational and business institutions.

Scientific production and its impact are concentrated in Asian countries such as China and Malaysia, along with Saudi Arabia and India, backed by strong international funding (e.g. National Natural Science Foundation of China, European Commission, Horizon 2020), reflecting the global relevance of the PLS-SEM and the need for greater participation of Latin American regions in this field of research.

References

1. Ringle, C.M., Sarstedt, M.: Gain more insight from your PLS-SEM results. Ind. Manag. Data Syst. **116**, 1865–1886 (2016)
2. Ooi, K.-B., Hew, J.-J., Lee, V.-H.: Could the mobile and social perspectives of mobile social learning platforms motivate learners to learn continuously? Comput. Educ. **120**, 127–145 (2018)
3. Fang, C., Ullah, N., Batumalay, M., Al-Rahmi, W.M., Alblehai, F.: Blockchain technology and its impact on sustainable supply chain management in SMEs. PeerJ Comput. Sci. **11** (2025)
4. Jeanine, K.-K., Morschheuser, B., Sicevic, N., Xi, N., von Korflesch, H.F.O., Hamari, J.: Challenges in the adoption of sustainability information systems: a study on green IS in organizations. Int. J. Inf. Manag. **77** (2024)
5. Janssen, M., Weerakkody, V., Ismagilova, E., Sivarajah, U., Irani, Z.: A framework for analysing blockchain technology adoption. Int. J. Inf. Manag. **50**, 302–309 (2020)
6. Li, C., Zhang, Y., Xu, Y.: Factors influencing the adoption of blockchain in the construction industry: a hybrid approach using PLS-SEM and fsQCA. Buildings. **12** (2022)
7. Crompton, H., Burke, D.: Artificial intelligence in higher education: the state of the field. Int. J. Educ. Technol. High. Educ. **20**, 1–22 (2023)
8. Malek, M.S., Bhatt, V.: Investigating the effect of risk reduction strategies on mega infrastructure project success: A SEM-ANN approach. Eng. Constr. Archit. Manag. **31**, 3575–3598 (2024)
9. Lee, V.-H., Foo, A.T.-L., Leong, L.-Y., Ooi, K.-B.: Can competitive advantage be achieved through knowledge management? Expert Syst. Appl. **65**, 136–151 (2016)
10. Hassija, V., Chamola, V., Mahapatra, A., et al.: Interpreting black-box models: a review on explainable artificial intelligence. Cogn. Comput. **16**, 45–74 (2024)
11. Soliman, M., Ali, R.A., Khalid, J., et al.: Modelling continuous intention to use generative AI as an educational tool among university students: findings from PLS-SEM and ANN. J. Comput. Educ. **8**(2), 18–27 (2024)
12. Leong, L.-Y., Jaafar, N.I., Ainin, S.: Understanding facebook commerce actual purchase from an artificial neural network perspective. J. Electron. Commer. Res. **19**, 75–103 (2018)
13. Hair, J.F., Hult, G.T.M., Ringle, C.M., Sarstedt, M. A Primer on Partial Least Squares Structural Equation Modeling (PLS-SEM) (2022)
14. Sarstedt, M., Ringle, C.M., Smith, D., et al.: PLS-SEM: A useful tool for family business researchers. J. Fam. Bus. Strat. **5**, 105–115 (2014)
15. Dijkstra, T.K., Henseler, J.: Consistent partial least squares path modeling. MIS Q. **39**, 297–316 (2015)
16. Kock, N.: Common method bias in PLS-SEM: A full collinearity assessment approach. Int. J. e-Collab. **11**, 1–10 (2015)
17. Shmueli, G., Koppius, O.R.: Predictive analytics in information systems research. MIS Q. **35**, 553–572 (2011) 2011
18. Fornell, C., Larcker, D.F.: Evaluating structural equation models with unobservable variables and measurement error. J. Mark. Res. **18**, 39–50 (1981)
19. Hair, J.F., Ringle, C.M., Sarstedt, M.: PLS-SEM: indeed a silver bullet. J. Mark. Theory Pract. **19**, 139–152 (2011)
20. Astrachan, C.B., Patel, V.K., Wanzenried, G.: A comparative study of CB-SEM and PLS-SEM for theory development in family firm research. J. Fam. Bus. Strat. **5**, 116–128 (2014)
21. Rigdon, E.E., Sarstedt, M., Ringle, C.M.: On comparing results from CB-SEM and PLS-SEM. Marketing ZFP. **39**, 4–16 (2017)

22. Hair, J.F., Matthews, L.M., Matthews, R.L., Sarstedt, M.: PLS-SEM or CB-SEM: updated guidelines. Int. J. Multivar. Data Anal. **1**, 107–123 (2017)
23. Henseler, J., Ringle, C.M., Sarstedt, M.: A new criterion for assessing discriminant validity. J. Acad. Mark. Sci. **43**, 115–135 (2015) 2015
24. Hair, J.F., Sarstedt, M., Hopkins, L., Kuppelwieser, V.G.: PLS-SEM: an emerging tool in business research. Eur. Bus. Rev. **26**, 106–121 (2014)
25. Hair, J.F., Sarstedt, M., Ringle, C.M., Mena, J.A.: An assessment of the use of PLS-SEM in marketing research. J. Acad. Mark. Sci. **40**, 414–433 (2012)
26. Hair, J.F., Risher, J.J., Sarstedt, M., Ringle, C.M.: When to use and how to report the results of PLS-SEM. Eur. Bus. Rev. **31**, 2–24 (2019)
27. Benitez, J., Henseler, J., Castillo, A., Schuberth, F.: How to perform and report an impactful analysis using PLS. Inf. Manag. (2020)
28. Shmueli, G., Sarstedt, M., Hair, J.F., et al.: Predictive model assessment in PLS-SEM: Guidelines for using PLSpredict. Eur. J. Mark. **53**, 2322–2347 (2019)
29. Purwanto, A., Sudargini, Y.: PLS-SEM analysis for social and management research: a literature review. J. Ind. Eng. Manag. Res. **2**, 114–123 (2021)
30. Lin, M.L., Huynh, L.L.: Bridging causal explanation and predictive modeling: the role of PLS-SEM. Int. J. Res. Bus. Soc. Sci. **13**, 197–206 (2024)
31. Hair, J.F., Howard, M.C., Nitzl, C.: Assessing measurement model quality in PLS-SEM using confirmatory composite analysis. J. Bus. Res. **109**, 101–110 (2020)
32. Ringle, C.M., Wende, S., Becker, J.M.: SmartPLS 4. SmartPLS, Oststeinbek (2022)
33. Henseler, J.: On the convergence of the PLS path modeling algorithm. Comput. Stat. **25**, 107–120 (2010)
34. Dash, G., Paul, J.: CB-SEM vs PLS-SEM methods for research in social sciences. Technol. Forecast. Soc. Chang. **173** (2021)
35. Hair, J.F., Alamer, A.: PLS-SEM in second language and education research: guidelines using an applied example. Res. Methods Appl. Linguist. **1** (2022)
36. Kamoun, F., El Ayeb, W., Jabri, I., et al.: Exploring students' and faculty's knowledge, attitudes, and perceptions towards ChatGPT. J. Inf. Technol. Educ. Res. **23**, 1–33 (2024)
37. Kim, J., Yu, S., Detrick, R., Li, N.: Exploring students' perspectives on generative AI-assisted academic writing. Educ. Inf. Technol. **30**, 265–1300 (2025)
38. Kim, M., Kim, J., Knotts, T.L., Albers, N.D.: AI for academic success: investigating usability, enjoyment, and responsiveness in ChatGPT adoption. Educ. Inf. Technol., 1–22 (2025)
39. Le, M.T.T., Tran, K.V.: Vietnamese university students' perceptions of ChatGPT for homework assistance. Int. J. Interact. Mob. Technol. **15**, 190–204 (2024)
40. Le, T.T.P., Quach, T.T.N., Pham, T.X.T., Nguyen, D.L.: A systematic review of how ChatGPT is perceived and utilized in EFL writing classes. AsiaCALL Online J. **16**, 295–311 (2025)
41. Menon, D., Shilpa, K.: "Chatting with ChatGPT": analyzing the factors influencing users' intention to use the open AI's ChatGPT using the UTAUT model. Heliyon. **9**(11), e20962 (2023)
42. Kumar, P., Kumar, R., Gupta, G.P., Tripathi, R., Jolfaei, A., Najmul Islam, A.K.M.: A blockchain-orchestrated deep learning approach for secure data transmission in IoT-enabled healthcare system. J Parallel Distrib Comput. **172**, 69–83 (2023)
43. Menon, S., Jain, K.: Blockchain technology for transparency in agri-food supply chain: Use cases, limitations, and future directions. IEEE Trans. Eng. Manag. **71**, 106–120 (2024)

44. Maroufkhani, P., Tseng, M.-L., Iranmanesh, M., Ismail, W.K.W., Khalid, H.: Big data analytics adoption: determinants and performances among small to medium-sized enterprises. Int. J. Inf. Manag. **54**, 102190 (2020)
45. Al-Sharafi, M.A., Iranmanesh, M., Al-Emran, M., Alzahrani, A.I., Herzallah, F., Jamil, N.: Determinants of cloud computing integration and its impact on sustainable performance in SMEs: an empirical investigation using the SEM-ANN approach. Heliyon. **9**(5), e16299 (2023)

Media Representations in Emergency Contexts: A Bibliometric Analysis of Humanitarian Crises

Angel Torres-Toukoumidis[1(✉)], Moisés Pallo Chiguano[1], Santiago Castro[1], and Julio Merchán Romero[2]

[1] Universidad Politécnica Salesiana, Cuenca 010105, Ecuador
atorrest@ups.edu.ec
[2] Universidade Santiago de Compostela, Santiago de Compostela 15705, Spain

Abstract. This study analyzes global academic production on media and humanitarian crises from 1994 to 2024 using a bibliometric approach applied to the Scopus database. The objective was to identify dominant thematic areas, methodological preferences, and geographic disparities within the field. A total of 587 articles were selected based on Boolean keyword criteria and refined through bibliographic management tools. Data were processed using Python and visualized through word clouds, thematic maps, bigram evolution, and co-citation networks. Results reveal five major research clusters: emergency healthcare, misinformation during pandemics, refugee health with a gender perspective, child mental health, and human rights in conflict contexts. Despite the relevance of these topics, the analysis highlights a fragmentation of the field and the absence of integrative motor themes, suggesting limited theoretical cohesion. Furthermore, scientific output remains heavily concentrated in the Global North, while contributions from regions most affected by humanitarian crises, such as Latin America, Sub-Saharan Africa, and the Middle East, are underrepresented. This imbalance raises concerns about epistemic equity and knowledge democratization. In response, the study proposes the paradigm of humanitarian educommunication as a conceptual framework that integrates media ethics, critical education, and rights-based approaches to improve media practices and scholarship in crisis contexts. The findings encourage future research to adopt interdisciplinary methodologies, expand the inclusion of regional voices, and explore media's role not only as an informant but as a transformative agent in humanitarian settings.

Keywords: humanitarian crises · media coverage · media framing · migration · communication media

1 Introduction

Humanitarian crises have intensified due to a combination of global factors, including protracted armed conflicts, climate change, pandemics, and large-scale population displacements. According to the Global Humanitarian Overview 2025 [1], more than 300 million people need humanitarian assistance an increase that marks a significant

A. Rocha et al. (Eds.): ICITS 2026, LNNS 1966, pp. 484–494, 2027.
https://doi.org/10.1007/978-3-032-27538-7_43

rise compared to previous years. This scenario poses challenges not only for governments and international organizations but also for the media, which play a crucial role in shaping how these crises are understood and managed on a global scale.

In this scenario, the content disseminated by the media impact public opinion through the narrative frames used to present a particular situation [2], therefore, humanitarian crises are events that require accurate news coverage, but they are also spaces where communication is configured by the ways in which the media structure and transmit messages [3]. In this sense, the media have the ability to shape discourse, impact public perception, and sometimes generate ethical controversy in crisis communication [4].

Humanitarian emergencies pose major challenges for the media in terms of conveying clear and accessible messages without simplifying the complexity of the content being disseminated, and at the same time, these media outlets must critically evaluate their own editorial practices, especially on issues that are susceptible to misinterpretation [5]. In this sense, the study of media coverage of humanitarian crises is an area of research that requires a rigorous and systematic approach [6].

In this context, research analyzing the role of the media in the face of humanitarian crises is an important resource that facilitates the evaluation of information practices [7] and the construction of narratives through the documentation of media practices and proposals that can improve coverage through greater accuracy and accountability [8]. Along these lines, bibliometric methodologies, according to Waisbord [9], help identify emerging patterns and trends in scientific production and reveal new areas of research. In other words, these approaches contribute to the advancement of more ethical and balanced journalism, crucial aspects for the consolidation of peace and gender equality in conflict zones [10].

This study conducts a bibliometric analysis of academic production on the media and humanitarian crises, aiming to map the current state of knowledge in these areas. To do so, it identifies the main research trends, the approaches employed, and emerging academic currents. It also poses the following questions: How can coverage of humanitarian crises to foster a more effective global response? What strategies can the media implement to avoid sensationalism or misinformation? And finally, how can academic research guide the ethical and practical transformation of the media in crisis contexts?

Humanitarian crises are complex scenarios that demand an immediate and accurate response from the media, as they do not merely inform but also shape public perception and, ultimately, influence political and social action. In general, media content has the power to shape opinions depending on how situations are narrated, emotionally, socially, and politically impacting audiences especially relevant in crisis contexts, where narratives can directly affect the lives of those impacted [11]. This influence extends beyond the transmission of information to the cognitive processes of individuals affecting how people attribute meaning, characterize, and evaluate observed phenomena, which in turn influences the formation of judgments and the internalization of mediated realities [12]. This phenomenon is supported by Agenda Setting Theory, which posits a direct relationship between the media agenda and the public agenda, suggesting that greater media exposure results in increased perceived importance of certain topics [13].

Within this framework, the communicative actions of organizations responding to humanitarian crises and the media's role in such contexts are fundamental to shaping public and policy perceptions. Through activism, NGOs denounce human rights violations, document crisis situations, and challenge official narratives, especially on issues of forced migration. Thanks to social media, these organizations can disseminate their messages autonomously and significantly influence public opinion [14, 15]).

However, coverage of humanitarian crises often exhibits biases such as the "spectacularization" of events to attract larger audiences [16] or promotes ideological agendas linked to nationalism that shape public perception and political responses, as the media actively construct narratives that can generate support or controversy [17].

Media literacy is important to combat these biases and prevent information manipulation. It means equipping citizens with critical skills to assess the credibility of information [18] and prevent the dissemination of opinionated content. This literacy should be applied to both the media and information systems in general, motivating journalists to seek out more accurate sources [19]. This contributes to raising public awareness and mobilizing the public, while mitigating the risks associated with the creation of irresponsible content in crisis situations.

In countries affected by humanitarian crises, it is necessary to implement educommunication strategies that promote critical education, facilitate the identification of the harmful effects of biased coverage, and provide access to transparent and accurate information. One strategy for preventing misinformation or manipulation is data verification initiatives [20]. In this regard, coordinated efforts are required between multilateral organisations, national governments and the media to develop action plans that address not only media operations, but also hate speech, information disorders and the protection of freedom of expression [21].

The above is necessary because one of the great challenges for the media in covering humanitarian crises is achieving a balance between clarity and accessibility of information, and the complexity inherent to such events like simplify technical or highly specialized content without distorting its meaning or omitting key nuances. Informational ethics become critical [22], as media must avoid reinforcing prejudices or perpetuating hate speech, and at the same time, ethical journalism requires a constant assessment of the social impact of each message. Then, in the context of armed conflict, media must providing truthful information and fostering peacebuilding, underlying causes and potential solutions to conflict [23], so that it is able to counteract simplistic narratives that present conflicts as inevitable or unsolvable. This can positively shape public opinion and support the development of peaceful alternatives in contexts such as Syria, where government control and gender stereotypes often hinder objective reporting [24].

But the interaction between media and humanitarian crises is a complex and multifaceted field of study [25] due to the diversity of actors involved, that include journalists, governments, humanitarian organizations and the affected communities, among other elements like technological, economic, and geopolitical factors. In such contexts, media coverage plays a key role in shaping public perceptions and influencing policy decisions, therefore, this coverage must be as accurate and ethical as possible to avoid misinformation and manipulation [26].

Over recent years, studies such as Marrupe [27] have shown how media coverage of humanitarian crises such as the Syrian refugee crisis has influenced international policy, shaping the responses of countries, through public opinion pressure, diplomatic lobbying, and the framing of humanitarian emergencies as either security threats or moral obligations. Nevertheless, this influence also brings the risk of censorship and political pressure affecting the accuracy and truthfulness of information warning that media outlets often operate under filters that distort reality and adjust narratives to political or economic interests [28] an especially dangerous situation when political decisions and humanitarian aid depend on reliable information.

From on a review of studies on framing in political communication, Igartua et al. [29] provide an analytical framework about how the media emphasize or downplay specific aspects of political and social phenomena, in which the narratives constructed through framing have direct implications for public opinion and policy responses [30]. And according to Melero López [31], framing theory allows us to understand a version of reality that, although it may be presented in different ways, acquires sufficient relevance for the message to be interpreted according to that specific reality.

2 Methods

The objective of this research is to analyze the academic production (1994–2024) about media in humanitarian crises, identifying thematic, methodological, and geographical patterns, as well as emerging trends, to propose a conceptual framework for humanitarian educommunication. The study posits that research published working this area thematically focuses on health and conflict, is methodologically fragmented; geographically weighted towards the Global North; and indicates an emergent new paradigm of humanitarian educommunication.

The mixed methodological approach was used in this study, because it is able to make quantitative and qualitative evaluations of scientific production in a certain knowledge theme. This method was used to extract patterns, trends, document-document relationships and research evolution over time. Bibliometric analysis offers a structured vision of knowledge, while techniques such as bibliographic coupling and term co-occurrence allow for the mapping of thematic and conceptual relationships between texts [32]. In this regard, it highlights the usefulness of these tools not only for visualizing the structure of an academic field but also for detecting emerging areas, reflecting social and media interest in specific issues, and identifying consolidated or nascent lines of scholarly debate [33].

We gathered data from Scopus, which is known for its extensive thematic coverage and reliability in bibliometric research. Our focus was on articles published between 1995 and 2024 in either English or Spanish that discussed media and humanitarian crises. We made sure to exclude tangential works, duplicates, and papers that didn't offer substantial analysis, using tools like Zotero and EndNote to help with this process. This approach is backed by previous studies that highlight Scopus as one of the most dependable sources for bibliometric analysis in the social sciences and humanities [34]. Its coverage index features a diverse array of high-impact journals, allowing for a thorough and current mapping of scientific trends.

The search strategy revolved around a Boolean keyword equation that included terms in both English and Spanish, like "humanitarian crisis" / "crisis humanitaria" and "media" / "medios de comunicación." Initially, this search turned up 385,952 documents, from which we picked out those that fit the criteria for bibliometric analysis. After that, we dove into computational analysis techniques using Python and some handy libraries like Spacy and Pandas. We also integrated the OpenAI API to help us analyze emotions, identify core themes, and create visual representations that make interpreting the findings.

Bigram analysis a key technique within bibliometric studies was employed to identify the most frequently recurring concepts and map the principal lines of research present in the selected documents. This tool revealed the semantic interrelationship between keywords, providing a comprehensive understanding of the thematic focuses within the reviewed literature. In addition, a topical map was developed to illustrate the evolution of research areas over time, highlighting both consolidated and emerging topics within the field of study.

3 Results and Discussion

Bigram analysis a key technique within bibliometric studies was employed to identify the most frequently recurring concepts and map the principal lines of research present in the selected documents. This tool revealed the semantic interrelationship between keywords, providing a comprehensive understanding of the thematic focuses within the reviewed literature. In addition, a topical map was developed to illustrate the evolution of research areas over time, highlighting both consolidated and emerging topics within the field of study.

Thematic, Methodological, and Geographical Trends in Scientific Production on Humanitarian Crises

The academic production on humanitarian crises during the period 1994–2024 reveals a progressive process of maturation in terms of volume, complexity, and multidimensionality. While early studies focused primarily on traditional humanitarian aid, literature has gradually shifted toward more comprehensive approaches that incorporate new issues, actors, and methodologies.

As seen in the word cloud (Fig. 1) the prominence of terms such as migration, risk, management, communication, social, and COVID underscores the evolving perception of humanitarian crises not merely as isolated or exceptional events, but as interconnected processes that permeate multiple layers of society. The frequent appearance of communication and social signals a paradigmatic shift toward acknowledging the role of media, digital networks, collective emotions, and symbolic interaction in both the management and experience of crises. The central placement of the word COVID is particularly significant: the COVID-19 pandemic has triggered an explosion of scholarly output linking health with communication, politics, and economics, making it a clear inflection point in the academic literature on humanitarian crises.

The thematic map (Fig. 2) indicates that basic themes such as the pandemic, immigration, and long-term care have high centrality but medium density, indicating that they are fundamental but need further theoretical development.

In contrast, niche topics such as diaspora and framing (construction) are highly developed but peripheral, suggesting potential for integration into broader frameworks.

Fig. 1. Word cloud of the most frequently repeated terms in the analysis

Emerging or declining topics, such as the aging workforce and studies on New Zealand, reveal a shift in academic priorities. It should be noted that the absence of driving themes (high centrality and high density) reflects a fragmented field that lacks theoretical cohesion.

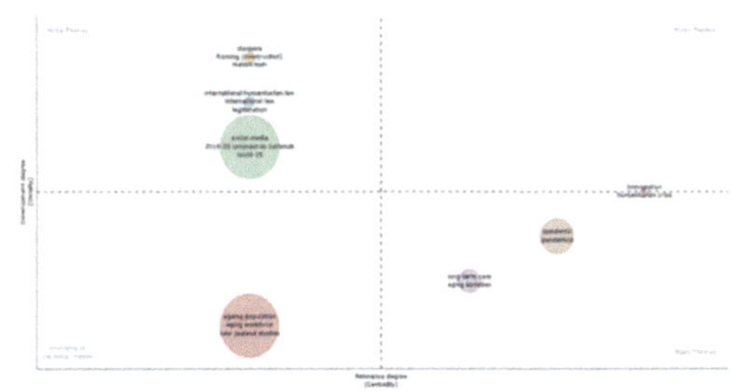

Fig. 2. Key categories in the field of humanitarian crises and communication

The bigram analysis adds a chronological dimension, highlighting key themes such as humanitarian_crisis, covid-19_pandemic, mental_health, as well as a preference for systematic_review, case_study methodologies, and clinical social analyses. This trend reflects a shift toward evaluating interventions and outcomes. Notably, covid-19_pandemic surged in frequency between 2020 and 2022.

The co-citation graph illustrates reference networks between authors and institutions, highlighting research on information ethics, public policy, risk communication, and misinformation. These topics reflect the growing importance of the symbolic dimension of crises, linked to the rise of polarizing discourses and conspiracy theories that challenge the role of digital media and technologies in the public sphere during crises.

From a geographical perspective, the collaboration map shows a concentration of scientific production in the Global North, particularly the United States, the United Kingdom, and Canada, which lead in publications and exert strong epistemological influence. In contrast, regions affected by humanitarian crises, such as Latin America, Sub-Saharan Africa, and the Middle East, remain underrepresented, revealing an epistemic gap that perpetuates asymmetries in knowledge. Citation data of the most influential publications reinforces this diagnosis. The article by Béné [35], published in Food Security, leads the ranking with over 500 citations, followed by Simon [36] and Warner [37]. These works address issues ranging from food security and climate change to risk management, reflecting the thematic breadth of the field.

Finally, the bigram analysis by publisher reveals how major academic houses such as Taylor and Francis, Oxford University Press, SAGE, and Wiley structure their editorial lines around dominant themes such as humanitarian_crisis, mental_health, and climate_change. This concentration underscores the importance of promoting new publications, languages, epistemologies, and platforms that expand the boundaries of academic discourse, integrate ancestral knowledge systems, and give visibility to non-hegemonic perspectives.

Emerging Topics in Humanitarian Crisis and Communication Research
Thematic mining enabled the identification of five major clusters that currently shape the field of humanitarian crisis research, through a cross-cutting approach that integrates health, communication, psychology, and human rights.

Topic 1: Medical Care and Humanitarian Response in Emergencies. This cluster encompasses research focused on healthcare systems as the frontline response to disruptive events. Terms such as health, humanitarian care, disaster response, emergency services, and healthcare challenges point to a more operational and strategic understanding of humanitarian response, where hospital infrastructure, intersectoral coordination, and logistical aid play a central role. However, this topic also includes critiques of the excessive medicalization of crises and calls for community-based care models grounded in prevention, empathy, and social justice.
Topic 2: Pandemic, Disinformation, and Digital Ecosystems. This cluster highlights concerns related to the role of social media, algorithms, and information flows during crises, as evidenced by the recurrence of terms such as covid-19, pandemic, social media, and crisis information suggest that the digital sphere functions not only as a mirror but also as an amplifier of crises.
Topic 3: Refugee Health and Gender. This cluster focuses on displaced populations especially women and girls and at the same time explores challenges related to reproductive health, gender-based violence, and social exclusion. Terms such as health data, refugee study, and women refugees indicate an intersectional turn in the literature, recognizing the multiple forms of oppression experienced in humanitarian contexts.
Topic 4: Childhood, Mental Health, and Trauma. This cluster foregrounds the psychological consequences of crises, with particular emphasis on children. Terms such as mental risk, children, patients, and associated results suggest that mental health has become a research priority, often overlooked in immediate response efforts. This topic calls for emotional support policies, emotional education, and psychosocial accompaniment, recognizing that trauma does not end when the emergency subsides but can affect entire generations if left unaddressed.
Topic 5: Human Rights, Conflict, and Global Governance. Finally, this cluster groups research focused on the structural causes of crises wars, forced displacement, institutional breakdowns, authoritarian regimes, and democratic backsliding. Terms such as international humanitarian, rights crisis, and war conflict reflect that, despite the rise of new agendas, armed conflict and systematic human rights violations remain central

to the field. This topic demands a critical perspective on global governance, impunity, the role of international organizations, and the legitimizing narratives behind certain humanitarian interventions.

4 Conclusions

In line with the study's objectives, the analysis reveals a thematic shift from documenting humanitarian assistance to exploring digital communication, social dynamics, information ecosystems, and psychosocial effects. The repeated appearance of terms such as migration, communication, risk, covid-19, and media suggests an expansion of the analytical variables considered. This evolution coincides with the argument by Etxebarrieta et al. [4], who argue that the media shape interpretive frameworks that influence public perceptions of crises.

Regarding methodology, there is a marked inclination toward qualitative designs, systematic reviews, and case studies. This approach corresponds to the interpretive structuring of fields of knowledge described by Barceló-Hidalgo and Dávila-Lorenzo [32] and is supported by the semantic patterns detected in the corpus.

In terms of geographical distribution, scientific output is concentrated in institutions and authors from the Global North, especially the United States, the United Kingdom and Canada. This supports the observation by Gregorio-Chaviano et al. [10] about the asymmetries in the circulation of knowledge, and despite facing frequent humanitarian crises, regions such as Latin America, Africa, and the Middle East remain underrepresented, revealing a gap in epistemic inclusion.

The thematic analysis enabled the identification of five major areas around which the field is organized: medical care in emergencies; disinformation during pandemics; the health of displaced populations from a gender perspective; psychological effects on children; and human rights in conflict situations. These clusters demonstrate the incorporation of new variables into the study of crises, as well as the intersection of structural, symbolic, and communicative dimensions. Disinformation emerged as a central issue, affecting both aid management and crisis perception [22], thus verification mechanisms are needed to promote responsible information circulation [20].

The results confirm a thematic predominance of health crises, misinformation, gender-based vulnerability, and child trauma. However, the field remains fragmented, with little theoretical consolidation across clusters. High publication output from the United States and United Kingdom correlates with editorial dominance by Taylor & Francis and SAGE, reinforcing knowledge centralization. Meanwhile, topics such as diaspora and media framing remain marginal, suggesting underexplored areas for future research.

Generally, thematic findings highlight growing concern for mental health, especially among displaced children and adolescents, supporting the recommendation by Muñoz-Borja et al. [19] to include psychological dimensions in crisis research. They also emphasize that media can offer symbolic support or, conversely, risk overexposing traumatic experiences.

In the treatment of armed conflict and human rights, recurring themes include war, forced displacement, and institutional failures. Media act as intermediaries, shaping public understanding and influencing international decision-making [23]; Sapag Muñoz de la Peña [24], assuming both political and informational roles. Some studies link media literacy and critical education in emergencies, echoing the proposal by Torres-Toukoumidis et al. [14] of humanitarian educommunication, a field connecting media, citizenship, and rights to foster analytical skills and informed participation.

Regarding study limitations, it is important to note the methodological decision to work exclusively with the Scopus database, while ensuring quality and standardization, excluded an indeterminate number of non-indexed studies, as well as publications in languages other than English and Spanish. This may have affected the diversity of the corpus and the representation of authors from less visible regions in the global scientific landscape.

References

1. United Nations Office for the Coordination of Humanitarian Affairs: Global humanitarian overview 2025. United Nations (2024)
2. Amaya Pardo, S.A., Torres Wilchez, M.U.: La percepción de la ciudadanía sobre la incidencia de los medios de comunicación social, redes sociales y opinión pública en los procesos penales dentro del Cantón Cuenca en el año 2020. MQRInvestigar. **7**(2), 674–700 (2023). https://doi.org/10.56048/mqr20225.7.2.2023.674-700
3. Tamariz Estrada, M.C.: Narrativas de la vulnerabilidad: periodistas juarenses frente a la cobertura informativa de la COVID-19. Sintaxis. **56–79** (2020). https://doi.org/10.36105/stx.2020edespcovid-19.03
4. Roman Etxebarrieta, G., Álvarez-Rementería Álvarez, M., Pérez-Izaguirre, E., Dosil Santamaria, M.: El papel de los medios de comunicación en situaciones de crisis sanitaria: la percepción de la población en torno al control y las normas sociales durante la pandemia del COVID-19. Rev. Lat. Comun. Soc. **78**, 437–456 (2020). https://doi.org/10.4185/rlcs-2020-1484
5. Angulo-Giraldo, M., Bolo-Varela, O.: Medios de comunicación y conflictos sociales durante la pandemia por COVID-19: análisis de los enfoques presentes en la prensa de Lima durante las protestas contra el gobierno interino de Manuel Merino (2020). Desde el Sur. **13**(1), e0005 (2021). https://doi.org/10.21142/des-1301-2021-0005
6. Lamoso González, P., Stavridis, S.: El Parlamento Europeo como tribuna moral internacional: el caso de Venezuela. Revista CIDOB d'Afers Internacionals. **131**, 185–210 (2022). https://doi.org/10.24241/rcai.2022.131.2.185
7. Miquel Segarra, S.: Análisis de la investigación publicada en las revistas de comunicación con índice ESCI (Emerging Source Citation Index). Espejo de Monografías de Comunicación Social. **2**, 109–129 (2021). https://doi.org/10.52495/c4.2.emcs.2.mic6
8. Torres-Toukoumidis, A., De-Santis, A., Vintimilla-León, D.: Opinion of Ecuadorians on migration in the target country: Benefits and challenges. Soc. Sci. **13**(10), 533 (2024). https://doi.org/10.3390/socsci13100533
9. Waisbord, S.: ¿Es válido atribuir la polarización política a la comunicación digital? Sobre burbujas, plataformas y polarización afectiva. Revista SAAP. **14**(2) (2020). https://doi.org/10.46468/rsaap.14.2.a1
10. Gregorio-Chaviano, O., Limaymanta, C.H., López-Mesa, E.K.: Análisis bibliométrico de la producción científica latinoamericana sobre COVID-19. Biomedica: Revista del Instituto Nacional de Salud. **40**(Supl. 2), 104–115 (2020). https://doi.org/10.7705/biomedica.5571
11. González, E.R., Luis, H.G., Esteban, P.P.: Los medios en la gestión de la comunicación de las ONG en tiempos de crisis: oportunidades para el cambio social. Observatorio (OBS*). **12**(3) (2018). https://doi.org/10.15847/obsobs12320181132

12. McGregor, S.C.: Social media as public opinion: How journalists use social media to represent public opinion. Journalism. **20**(8), 1070–1086 (2019). https://doi.org/10.1177/1464884919845458
13. Ardévol-Abreu, A., Gil de Zúñiga, H., McCombs, M.E.: Orígenes y desarrollo de la teoría de la agenda setting en comunicación: tendencias en España (2014–2019). El Profesional de la Información. (2020). https://doi.org/10.3145/epi.2020.jul.14
14. Torres-Toukoumidis, A., De-Santis, A., Vintimilla-León, D.: Los medios digitales ante la crisis humanitaria: Haití, Honduras, Venezuela. Rusia-Ucrania y Siria, Abya-Yala/UPS (2022) https://bit.ly/3INJP53
15. Torres-Toukoumidis, A., de la Garza Montemayor, D.J., De-Santis, A., León-Alberca, T.: Ecos migratorios: comunicación y cambio social en la migración iberoamericana. Tirant Lo Blanch (2024)
16. Coppini, D.: Sensationalism versus substance: Building a framework that examines mass media coverage of migrants in the Italian context. J. European Popular Culture. **10**(1), 53–59 (2019). https://doi.org/10.1386/jepc.10.1.53_1
17. Tapia, J.C., Torres-Toukoumidis, A., Vintimilla-León, D.: La educomunicación y su efecto en las crisis humanitarias. In: Torres-Toukoumidis, Á., De-Santis, A. (eds.) Comunicación en las crisis humanitarias. Retos y propuestas, pp. 65–72. McGraw Hill (2023)
18. Torres-Toukoumidis, Á., Ramírez-Montoya, M.S., Romero-Rodríguez, L.M.: Valoración y evaluación de los aprendizajes basados en juegos (GBL) en contextos e-learning. Educ, Knowl. Soc (EKS). **19**(4), 109–128 (2019). https://doi.org/10.14201/eks2018194109128
19. Muñoz-Borja, P., García-Ruiz, R., Aguaded, I.: La educomunicación en la documentación pública, privada y académica colombiana en su compresión de la cuestión social. Revista Fuentes. **22**(1), 1–12 (2020). https://doi.org/10.12795/revistafuentes.2020.v22.i1.01
20. Torres-Toukoumidis, A., León-Alberca, T.: Fact-checking en países con crisis humanitarias: implicación y desafíos. In: Torres-Toukoumidis, Á., De-Santis, A. (eds.) Comunicación en las crisis humanitarias. Retos y propuestas, pp. 75–83. McGraw Hill (2023)
21. Sánchez Moscoso, A.A., Candela, S., Torres-Toukoumidis, A.: Desinformación y migración venezolana: el caso Ecuador. Cuadernos del Centro de Estudios en Diseño y Comunicación. **161** (2022). https://doi.org/10.18682/cdc.vi161.6981
22. Gutiérrez-Coba, L., Coba-Gutiérrez, P., Gómez-Diaz, J.A.: Noticias falsas y desinformación sobre el COVID-19: análisis comparativo de seis países iberoamericanos. Rev. Lat. Comun. Soc. **78**, 237–264 (2020). https://doi.org/10.4185/rlcs-2020-1476
23. Vázquez Liñán, M.: Actores estatales y no estatales y conflictos armados: sección V. Medios de comunicación y conflictos armados. In: Relaciones internacionales y geopolítica en tiempos de policrisis, pp. 397–417. Universidad Internacional de Andalucía (2024)
24. Muñoz, S., de la Peña, P.: Oportunidades y riesgos de colonizar el periodismo con vieja propaganda que circula por las NTIC: del caso de Siria a Ucrania. Anuario Electrónico de Estudios en Comunicación Social Disertaciones. **16**(2) (2023). https://doi.org/10.12804/revistas.urosario.edu.co/disertaciones/a.12376
25. Segado-Boj, F., Chaparro-Domínguez, M.-Á., Ramírez-Daza, Ó.-I.: Periodismo de investigación: análisis bibliométrico, conceptual e intelectual (2000–2020). Transinformação. **34**, e210080 (2022). https://doi.org/10.1590/2318-0889202234e210080
26. Vizoso, Á., Vaz-Álvarez, M., López-García, X.: Fighting deepfakes: Media and Internet giants' converging and diverging strategies against hi-tech misinformation. Media Commun. **9**(1), 291–300 (2021). https://doi.org/10.17645/mac.v9i1.3494
27. López Marrupe, I.: El poder de la imagen de prensa: análisis de la cobertura fotográfica de la crisis de refugiados en el periódico ABC = The power of the press photo: analysis of the pictures coverage of the refugee crisis in the ABC newspaper. Revista Española de Comunicación en Salud. **55** (2019). https://doi.org/10.20318/recs.2019.4438

28. Araujo Cuauro, J.: Violación de los derechos humanos en Venezuela: la otra cara de la pandemia. SUMMA. Revista Disciplinaria en Ciencias Económicas y Sociales. **2**(Esp.), 87–115 (2020). https://doi.org/10.47666/summa.2.esp.08
29. Igartua, J.J., Calderón, C.A., Piñeiro-Naval, V., Gonzalez, A.: La teoría del framing en la investigación iberoamericana sobre comunicación política. In: Muñiz, C. (ed.) Framing y política. Aportaciones empíricas desde Iberoamérica, pp. 57–98. Tirant lo Blanch (2022)
30. Muñiz, C.: Encuadres noticiosos sobre migración en la prensa digital mexicana: un análisis de contenido exploratorio desde la teoría del framing. Convergencia Revista de Ciencias Sociales. **18**(55), 213–239 (2011) http://www.scielo.org.mx/scielo.php?script=sci_arttext&pid=S1405-14352011000100009
31. Melero López, I.: ¿Cómo se construyen los encuadres noticiosos? Indagando acerca de la teoría del framing. RIPS: Revista de Investigaciones Políticas y Sociológicas. **21**(2) (2022). https://doi.org/10.15304/rips.21.2.8333
32. Barceló-Hidalgo, M., Dávila-Lorenzo, M.: Producción científica sobre comunicación pública de la ciencia en Dimensions entre el 2017 y el 2021. Revista Española de Documentación Científica. **46**(3), e360 (2023). https://doi.org/10.3989/redc.2023.3.1982
33. Pastor-Ramon, E., Rovira, C.: Comunicación académica y buscadores científicos: 'scoping review'. Index Comunicación. **13**(1), 79–103 (2023). https://doi.org/10.33732/ixc/13/01comun2
34. Donthu, N., Kumar, S., Mukherjee, D., Pandey, N., Lim, W.M.: How to conduct a bibliometric analysis: An overview and guidelines. J. Bus. Res. **133**, 285–296 (2021). https://doi.org/10.1016/j.jbusres.2021.04.070
35. Béné, C.: Resilience of local food systems and links to food security: a review of some important concepts in the context of COVID-19 and other shocks. Food Secur. **12**(4), 805–822 (2020). https://doi.org/10.1007/s12571-020-01076-1
36. Simon, D.: To and through the UK: Holocaust refugee ethnographies of escape, education, internment and careers in development. Contemp. Soc. Sci. **7**(1), 21–38 (2012). https://doi.org/10.1080/21582041.2011.578708
37. Warner, K.: Global environmental change and migration: governance challenges. Glob. Environ. Chang. Hum. Policy Dimens. **20**(3), 402–413 (2010). https://doi.org/10.1016/j.gloenvcha.2009.12.001

Integration of Blockchain and Artificial Intelligence in Accounting Audit: Transparency, Challenges, and Opportunities in Ecuador

Carolina Paola Jaramillo Bayas(✉), William Ortega Naranjo, and Lorena Elizabeth Muquis Tituaña

Universidad Israel, Carrera Ciencias Administrativas, Quito, Ecuador
{cjaramillo,wortega,lmuquis}@uisrael.edu.ec

Abstract. In this study, we face the challenge of how to incorporate cutting-edge technologies such as artificial intelligence (AI) and blockchain into auditing, fields in which transparency, traceability, and reliability of information are essential. The central problem lies in the gap between the level of theoretical knowledge about these tools worldwide and their limited practical application in Ecuador, which directly impacts the effectiveness and reliability of accounting and tax processes. In order to understand the degree and perception, as well as the implementation of AI and blockchain in auditing, as well as the current obstacles Ecuador finds with it, the research methodology constructed a series of questions that were sent by postal mail to 40 professional accountants and auditors, complemented by a bibliographic review and an international comparative analysis of an Ecuadorian case in a similar situation. The results show that although 85% claim to be familiar with AI, only 15% have applied it in auditing. Meanwhile, 45% report having basic knowledge of blockchain, but no respondents have used it. The main barriers identified were lack of training (65%), lack of regulation (55%), and implementation costs (40%). Clearly, there is a significant lag compared to this type of experience.

Keywords: Blockchain · Artificial Intelligence · Accounting Audit · Financial Transparency

1 Introduction

1.1 Contextualization

In an increasingly digital global environment, emerging technologies such as blockchain and artificial intelligence (AI) are opening new horizons in accounting auditing, while reinterpreting fundamental notions of transparency, traceability, automation, and fraud prevention. Traditional auditing, based on sampling and manual document checking, suffers from shortcomings compared to the current volumes and complexity of financial data, which has created an opportunity for continuous and automated auditing.

Blockchain technology presents significant potential to transform accounting audit processes, offering greater transparency, security, and traceability of financial transactions (Rode et al., 2021; Villa & Tostado, 2025). Its implementation could lead to

A. Rocha et al. (Eds.): ICITS 2026, LNNS 1966, pp. 495–505, 2027.
https://doi.org/10.1007/978-3-032-27538-7_44

continuous auditing, as transactions become available in real time, potentially improving the consistency and reliability of corporate reporting (Rode et al., 2021). However, its adoption faces technological, economic, and regulatory challenges (Villa & Tostado, 2025). While blockchain can increase efficiency and reliability in auditing, it requires regulatory adjustments and professional training (Villa & Tostado, 2025). The technology could improve the quality of financial reporting for investors and eliminate the need for trusted intermediaries in the information flow (Argañaraz et al., 2019). Despite its complexities, blockchain offers opportunities to improve audit processes, although auditors will still need to apply professional judgment when analyzing accounting estimates and management decisions (Simões et al., 2021).

Given its decentralized and immutable nature, blockchain technology can create incorruptible accounting records visible in real time. This significantly reduces the likelihood of data manipulation. Furthermore, financial transactions become much more consistent (Gómez, 2024). At the same time, artificial intelligence provides advanced analytical capabilities that, through machine learning algorithms, neural networks, or expert systems, can meaningfully detect patterns. They also bring with them potential for large-scale fraud and anomalies on a constant basis (Castillo & Ramírez, 2025).

The union of these two technologies is not only possible, but also necessary. Their complementarity has led to the concept of smart auditing, where blockchain guarantees the veracity of the underlying information and artificial intelligence handles its automatic interpretation. This synergy has been studied by international firms such as EY and Deloitte with tools like EY Blockchain Analyzer, which have already been used in pilot audits in countries such as Canada and Germany (González, 2024).

In Latin America, and more specifically in Ecuador, the implementation of these standards is still a long wait for development and implementation. However, recent research points to significant impacts in terms of efficiency, cost savings, and satisfaction with current regulations (Murillo, 2025). Therefore, combining AI and blockchain from an accounting and tax perspective is deemed extremely necessary. Research in this regard is considered essential, considering both the technological opportunities and the regulations and ethical principles involved in the entire process.

This article aims to critically analyze the convergence between blockchain and artificial intelligence in accounting audit processes, exploring its benefits, risks, and implementation conditions in the Ecuadorian business environment.

1.2 Problem

To anticipate risks, detect fraud in real time, or verify large amounts of data, traditional audit methodologies based on selective testing and past reviews are simply not enough. PwC's Global Economic Crime and Fraud Survey (2022) indicates that 51% of organizations worldwide suffered significant fraud in the last two years, and much of this fraud went under the radar of traditional controls. On the one hand, blockchain-based accounting auditing guarantees the immutability of records and the full traceability of each transaction, which significantly reduces the risk of accounting information manipulation. It is when this distributed database is integrated with intelligent analysis and detection systems—such as machine learning algorithms, data mining, and artificial neural networks—that a continuous, autonomous, and proactive audit system becomes possible, referred to by several experts as "cognitive auditing" (Castillo & Ramírez, 2025; Gómez, 2024).

1.3 Key Technologies: Blockchain and Artificial Intelligence

Blockchain is a distributed ledger technology that can guarantee the security, immutability, and transparency of the data it stores. It not only helps with cryptocurrency management, but its ability to verifiably and chronologically record transactions has proven valuable in other areas such as logistics, healthcare, education, and finance. In the accounting field, its application enables automatic account reconciliation (as a multilateral exchange system for securities) and guarantees data authenticity. Furthermore, it allows auditors to conduct real-time inspections based on the results (Murillo, 2025).

On the other hand, artificial intelligence provides advanced analytical capabilities that allow systems to learn from data, identify anomalous patterns, and automatically generate early warnings of potential fraud or accounting deviations. In auditing, these technologies are used by large firms to perform AI-assisted audits. The clearest example is EY Blockchain Analyzer, which combines AI and blockchain to review large amounts of accounting data in less time and with greater accuracy (González, 2024).

He also proposed DLT audits of these systems to provide greater transparency in the credit exchange process on blockchain.

1.4 International Advances and Experiences

This combination is already being used in countries like Estonia. There, a large portion of government accounting records are kept verifiably or automatically. In China, the use of a state-provided invoicing system based on blockchain technology and intelligent analytics has led to a dramatic reduction in tax evasion. In business, firms like Deloitte, KPMG, and PWC are developing digital twin auditors, virtual systems that continuously replicate a company's financial records from blockchain environments using AI algorithms.

Gómez (2024) argues that this integration provides three fundamental contributions to auditing:

1. Fewer manual and biased reporting processes.
2. Greater scope in audit performance (global audits, not just a sample).
3. The auditor becomes a supervisor of algorithmic models, independent of reporting texts.

1.5 Context in Latin America and Ecuador

In Latin America, digitalization in accounting aspects has been slow, and there are still problems with regulations, the culture that will be generated, and emerging technologies such as the Internet of Things (IoT) still face complications due to unclear and imprecise regulations. In Ecuador, although progress has been made in electronic invoicing and tax control systems, the adoption of blockchain and AI in auditing is still essentially at an early stage, specifically in companies of a certain size in the financial and technology sectors (Murillo, 2025). In any case, Ecuadorian legislation, including the 2021 Organic

Law on the Protection of Personal Data and its similar regulatory precedents, the 2014 Organic Monetary and Financial Code, already establishes legal frameworks that could make the implementation of these transcendental technologies viable by providing them with the appropriate adaptation.

This article primarily analyzes the feasibility, benefits, and challenges of integrating blockchain and artificial intelligence into accounting auditing in the Ecuadorian context. To this end, it will also examine how the two technologies have evolved conceptually, review international experiences, and propose an application framework adapted to local circumstances.

1.6 Objective

To analyze the level of knowledge, perception, and application of artificial intelligence and blockchain in accounting auditing in Ecuador, comparing the national reality with international experiences, in order to identify benefits, limitations, and opportunities for their implementation as tools that strengthen transparency, traceability, and trust in financial information.

2 Methodology

This study is qualitative, exploratory, and descriptive in nature, leaning almost exactly toward application.

The methodology employed in the study can be summarized as a bibliographic and documentary study with a qualitative approach, that is, it relies more on existing literature and documents than on the collection of empirical data. (Rueda, 2024)

The bibliographical component of the study included research into blockchain, smart contracts, and artificial intelligence in public procurement, in order to provide a current overview of the discussions surrounding these topics.

The documentary research consisted of consulting institutional documents, laws, and regulations, suggesting a focus on the legal and regulatory framework.

The criteria for selecting the material included only publications from high-caliber institutions in public administration and public policy, indicating high-quality sources.

Given that the inclusion of technologies such as blockchain and artificial intelligence in accounting auditing is still in its infancy in Ecuador, the following are proposed:

- Describe their technical and operational characteristics.
- Examine their application abroad.
- Evaluate the feasibility of their implementation in the Ecuadorian accounting environment.

The methodological approach will be mixed, combining:

- Document review (scientific articles, technical reports, regulations).
- Case analysis (pilot organizations in Latin America).
- Gathering perceptions through interviews with accountants, auditors, and technology developers.

3 Discussion and Resulted

The incorporation of blockchain with artificial intelligence, precisely, is nothing more than a theoretical concept; it is a natural evolution of auditing in the face of increasing financial complexity and data being manipulated digitally. As a result of this analysis, the simultaneous use of these two technical devices allows for an audit that discerns trends on a continuous and long-term basis. Based on non-perishable digital evidence, this form of review clearly overcomes the old limitations of a one-off or ex-post audit.

Comparing international experiences with the Ecuadorian reality provides a better understanding of the status of these applications in current accounting audit practice. There are many successful use cases documented worldwide. They improve transparency and traceability of financial information; they reduce the number of errors or fraud committed by the auditor. Other multinationals have also adopted these tools to avoid unnecessary risks in invoicing, speed up review times, and increase the level of trust placed in them by their stakeholders.

The incorporation of blockchain technology into accounting and auditing has become a growing global trend, driven by the need to ensure transparency, record immutability, and the traceability of financial transactions. Large corporations and international auditing firms have been pioneers in the development of blockchain-based solutions, integrating them into their internal processes or offering them as specialized services for their clients. These include Pricewaterhouse Coopers (PwC), Ernst & Young (EY), KPMG, and Deloitte, which have developed platforms and methodological frameworks for using blockchain to verify accounting transactions and reduce the risk of fraud.

Likewise, technology and financial sector companies, such as IBM and JPMorgan Chase, have promoted the implementation of blockchain systems for validating accounting records and auditing transactions in real time, establishing benchmarks for other organizations.

The following comparison table summarizes some emblematic cases of international and national companies that have implemented blockchain in accounting and auditing, highlighting their purpose, benefits achieved, and progress (Table 1).

In Ecuador, however, the adoption of these technologies is at an early stage, where theoretical knowledge predominates, but practical experience is lacking, and there is no solid regulation governing their implementation. This gap between international practices and Ecuador's national reality poses both challenges and opportunities. On the one hand, it highlights the need for investment in training and technological infrastructure, and on the other, it offers the opportunity to learn from international models and design application strategies contextualized to Ecuador's own reality.

The following table presents a tabular presentation that, by structurally comparing what has been achieved internationally with the shortcomings that still exist at the national level, suggests guidelines for the design of transparent and reliable policies and practices in Ecuador (Table 2)

Table 1. Companies that apply Blockchain in accounting and auditing

Company	Sector	Blockchain Application	Benefits in Accounting/Auditing
PwC (Reino Unido / Global)	Consulting and Auditing	Real-time verification of accounting transactions for financial and logistics clients.	Greater transparency and reduction of audit errors.
Deloitte (Global)	Consulting	Rubix platform for financial traceability and accounting processes.	Facilitates continuous digital audits.
KPMG (Global)	Consulting	Blockchain integration into tax and audit systems, with Microsoft and R3.	Record automation and tax compliance.
EY (Global)	Consulting	EY Blockchain Analyzer platform for auditing cryptoassets and digital transactions.	Risk control in digital finance.
Walmart (EE. UU. / Canadá)	Retail	Blockchain record-keeping in logistics and inventory accounting.	Cost reduction and improved financial control.
Maersk (Dinamarca)	Maritime Transport	TradeLens system with IBM for recording logistics transactions on the blockchain.	Improved cost control and accounting reconciliation.
Banco Santander (España)	Banking	Bond issuance and financial transactions on the blockchain.	Transparency and reduction of fraud in financial records.
HSBC (Global)	Banking	Accounting reconciliation of international transactions using blockchain.	Reduction in audit times and operating costs

3.1 Resulted of the Applied Survey

To understand the level of awareness and application of emerging technologies in the field of accounting auditing in Ecuador, a survey was conducted among 40 professionals, including public accountants, internal and external auditors, and independent consultants. The results are presented below through graphical representations that clearly visualize general trends.

First, the levels of awareness and use of artificial intelligence (AI) in audit processes are presented, followed by the most recognized benefits of its application. Subsequently,

Table 2. Comparison of international cases vs. the Ecuadorian situation.

Country / Experience	Technology implemented	Main Application	Maturity Level	Lessons for Ecuador
Estonia	State blockchain	Accounting records and public auditing	High	Comprehensive digitalization of the tax system enables real-time traceability.
China	AI + Blockchain + Big Data	Electronic invoicing and tax evasion detection	High	Integration with the tax system is key. Online monitoring of thousands of companies.
Colombia	AI in the financial sector	Automated bank auditing	Medium	Adoption in pilot projects by financial institutions. Importance of public-private partnerships.
Ecuador (diagnostic)	Zero on blockchain / Low on AI	Partial automation in tax audits (SRI)	Low	Urgent need for a regulatory framework and technical training in emerging technologies.

the findings related to the level of awareness and use of blockchain in the accounting field are presented, as well as the perception of its potential to increase transparency in financial records. Finally, the main barriers identified for the implementation of both technologies are presented (Figs. 1, 2, 3 and 4).

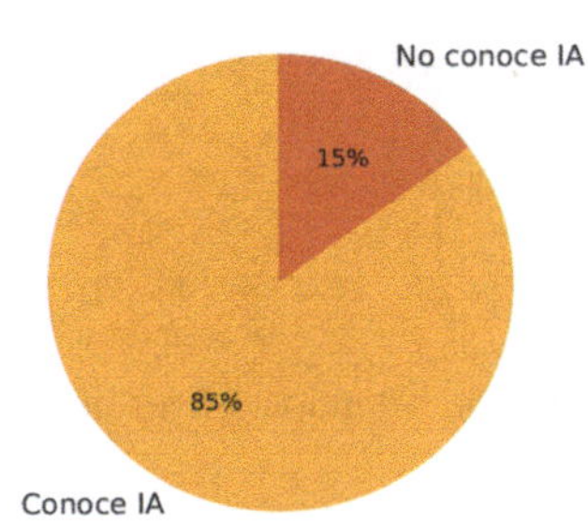

Fig. 1. Level of knowledge of accounting professionals on the use of artificial intelligence in auditing (n = 40).

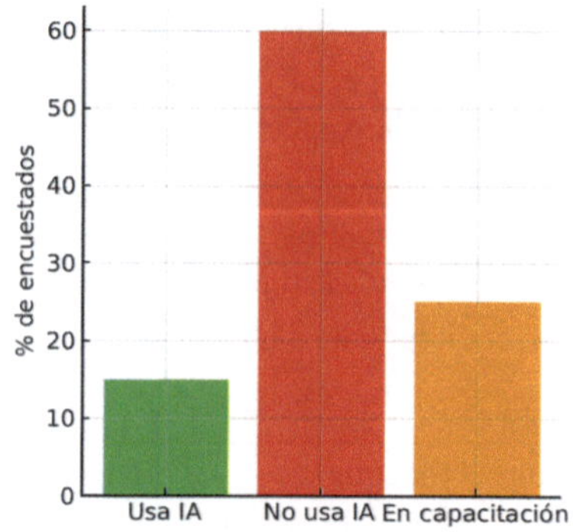

Fig. 2. Frequency of use of artificial intelligence in audit processes by respondents (n = 40).

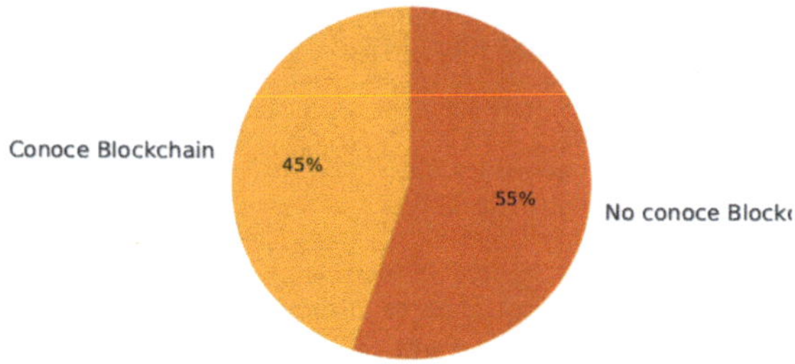

Fig. 3. Respondents' level of knowledge about the application of blockchain in accounting (n = 40).

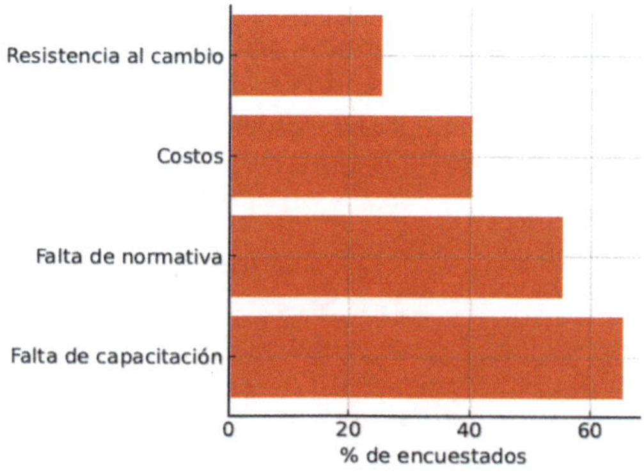

Fig. 4. Main perceived barriers to the adoption of artificial intelligence and blockchain in accounting auditing (n = 40).

3.2 Discussion

The main barriers identified for the adoption of both technologies were a lack of training (65%), the absence of specific regulations in the country (55%), and high implementation costs (40%). This finding highlights that, beyond the theoretical disposition toward innovation, the digital transformation of auditing in Ecuador requires institutional and regulatory strategies that guarantee the development of human talent, incentives for technological investment, and legal frameworks that support the use of immutable digital records.

This work presents certain limitations that must be considered when interpreting the results:

First, the reliable data obtained must be limited to a small sample (n = 40), which limits the statistical generalization of the results. The target population should have been much larger.
Secondly, the scarcity of documented cases in Ecuador on the use of blockchain in auditing diminishes its relevance compared to international experiences with other recognized firms; the development of this idea is a potential avenue for future lines of research.

From a qualitative perspective, the professionals surveyed expressed that artificial intelligence is a very promising contributory tool due to its potential to optimize risks and streamline processes and detect irregularities in accounting records. Furthermore, it is also considered a key tool for those with little or no day-to-day perspective, as it fosters transparency and trust, at least with regard to financial information, especially tax matters. Among the main obstacles are a lack of specialized training, regulatory uncertainty, and the perception of high implementation costs, which both reflect previous institutional resistance and demonstrate a level of ignorance about what these technologies can actually do in practice.

Quantitative Results.- The data collected shows that 85% of professionals claim to be familiar with AI, although only 15% reported having used it in auditing, while 60% indicated they had not applied it, and 25% were in the training process. Regarding the perceived benefits of AI, the most notable were fraud detection (70%), predictive analysis (60%), and time savings in routine processes (55%). Regarding blockchain, only 45% reported having basic knowledge of the technology, and none reported having applied it in accounting auditing, although 80% agreed that it could improve the transparency of records. Finally, the main barriers reported were lack of training (65%), lack of regulations (55%), high costs (40%), and resistance to change (25%).

4 Conclusions

A survey shows that the use of artificial intelligence (AI) and blockchain in accounting auditing is strategic for increasing transparency, traceability, and efficiency in financial processes. Internationally, many companies and audit firms have systematically incorporated these technologies, achieving tangible benefits such as effectively preventing fraud, reducing time, and providing greater assurance in accounting reports. In Ecuador, however, things are different: although the conceptual level of knowledge about AI is high, it is still very rarely applied in practice; as for blockchain, it is still in its early stages and is basically limited to a few pilot projects among technological startups.

The survey results indicate a notable gap between knowledge and application, with 85% of respondents stating they know what AI is, but only 15% can actually use it, and regarding blockchain, none report having used it for an audit. However, the majority recognize its potential to improve transparency, demonstrating a favorable attitude toward technology companies. The main barriers identified—lack of training, lack of regulations, and implementation costs—are the same factors that limit digital transformation in other emerging fields.

Therefore, the study concludes that accounting auditing in Ecuador faces a double challenge: on the one hand, eliminating the structural resistance derived from the lack of public policies, technical regulations, and specialized training; and on the other, taking advantage of successful international experiences to implement technological adoption models that adapt to the local situation. The convergence of AI and blockchain must be understood not only as a technical innovation, but also as a strategic mechanism to strengthen confidence in the country's financial and tax systems and prepare the accounting profession to face the challenges of the digital economy.

References

Argañaraz, Á.A., Mazzuchelli, A., Albanese, D., Abellán López, M.Á.: Blockchain: Un nuevo desafío para la contabilidad y auditoría. In: En Actas de Congreso Internacional de Contabilidad y Auditoría, pp. 1–15 (2019) https://api.semanticscholar.org/CorpusID:216721879

Castillo, J., Ramírez, P.: La inteligencia artificial en la auditoría contable: Oportunidades y desafíos en América Latina. Revista de Contabilidad y Tecnología. **18**(2), 45–62 (2025). https://doi.org/10.1234/rct.2025.18245

Gómez, R.: Blockchain en la contabilidad: Estudios de caso internacionales. Cuadernos de Auditoría y Finanzas. **22**(4), 301–320 (2024). https://doi.org/10.2468/caf.2024.224301

González, M.: Aplicación de inteligencia artificial en procesos de auditoría: Un enfoque comparativo. J. Account. Innov. **15**(3), 233–249 (2024). https://doi.org/10.5678/jai.2024.153233

Murillo, A.: Blockchain y transparencia financiera: Implicaciones para la auditoría. Revista Iberoamericana de Economía Digital. **12**(1), 77–95 (2025). https://doi.org/10.9876/ried.2025.12177

PwC. Time for trust: How blockchain will transform business and the economy. PricewaterhouseCoopers. (2022). https://www.pwc.com/gx/en/industries/technology/blockchain/time-for-trust.html

Rode, F., Bastos, P., Andújar, A.J.F., de Souza Vilela, E.: Los impactos blockchain en la auditoría contable / Os impactos da cadeia de bloques na auditoria contábil. Revista Brasileña de Desarrollo. **7**(9), 23947–23961 (2021). https://doi.org/10.34117/bjdv7n9-263

Rosales-Troya, M., Cedeño, J., Viteri, L.: Retos de la auditoría digital en Ecuador: Una revisión crítica. Revista Ecuatoriana de Contabilidad. **10**(2), 101–118 (2024). https://doi.org/10.5439/rec.2024.102101

Rueda, J.E., Gallegos, N., Estanling, E., Tello, J.S., García, V.: Identificación de variables relacionadas a la seguridad informática a partir de tendencias investigativas de la tecnología blockchain. Revista Politécnica. **20**(40), 1–15 (2024). https://doi.org/10.33571/rpolitec.v20n40a1

Simões, M.P., Albuquerque Cavalcanti, J., Marques, F., de Melo, J., Queiroz Reis, C.: Beneficios del uso de la tecnología blockchain como instrumento para un auditor contable. Revista Ambiente Contábil. **13**(1), 1–20 (2021). https://doi.org/10.21680/2176-9036.2021v13n1id19535

Villa Ruiz, N.E., Tostado Cortés, I.Y.: Uso de blockchain en la auditoría: Ventajas y desafíos en la verificación de transacciones. Revista Vértice Universitario. **27**(96), 45–60 (2025). https://doi.org/10.36792/rvu.v27i96.485

STEM Studies, SDGs and the Role of Women in Ecuadorian Universities

Samanta Zabala-Cáceres[1], Elizabeth Salazar-Jácome[2](✉), Karen Merizalde-Torres[3], and Doris Calle-Landázuri[4]

[1] Department of Ciencias de la Ingeniería y Mecánica, Universidad de las Fuerzas Armadas ESPE, Sangolquí, Ecuador
msalazar@uisrael.edu.ec
[2] Department of Ciencias de la Ingeniería, Universidad Tecnológica Israel, Quito, Ecuador
eszabala@espe.edu.ec
[3] Department of Artes y Humanidades, Universidad Tecnológica Israel, Quito, Ecuador
kmerizalde@uisrael.edu.ec
[4] Department of Administrative Sciences, Universidad Tecnológica Israel, Quito, Ecuador
dcalle@uisrael.edu.ec

Abstract. This research analyzes the relationship between STEM (Science, Technology, Engineering and Mathematics) careers and the Sustainable Development Goals (SDGs), focusing on the participation of women in the University of the Armed Forces ESPE, Latacunga Campus. A mixed approach was applied that integrated the quantitative analysis of enrolment certificates and surveys, together with a qualitative approach based on interviews with students and key stakeholders. The results reveal that female participation in STEM careers is limited by various factors, including the absence of female role models, the persistence of gender stereotypes, negative self-perception of their abilities and difficulties in balancing their studies with personal and family life. Despite these challenges, factors that act as facilitators are identified, especially the presence of support groups, networks between women students and professionals, as well as institutional initiatives that promote female empowerment. The findings highlight the importance of implementing university policies aimed at promoting greater gender equity in technical and scientific careers, which directly contributes to the achievement of the SDGs, particularly SDG 4 (Quality Education) and SDG 5 (Gender Equality). This study provides useful evidence to design strategies for inclusion and institutional transformation from a perspective of sustainability and equality.

Keywords: STEM careers · SDGs · Women's participation · Gender equality

1 Introduction

In recent decades, there has been a growing recognition of the importance of Science, Technology, Engineering and Mathematics (STEM) education for the economic and social development of countries. According to Calderón, despite the efforts made, women's participation in these fields remains significantly low [1] This phenomenon perpetuates gender inequalities in the workplace and education.

A. Rocha et al. (Eds.): ICITS 2026, LNNS 1966, pp. 506–517, 2027.
https://doi.org/10.1007/978-3-032-27538-7_45

Ecuador, committed to the United Nations 2030 Agenda and the Sustainable Development Goals (SDGs), has adopted various policies and programs to promote gender equality and inclusion in higher education. In this context, the University of the Armed Forces ESPE Latacunga Campus plays a crucial role in offering STEM study programs that seek to train professionals trained to face contemporary challenges.

According to data mentioned by the National Institute of Statistics and Census [2] there are more women in Ecuador, with 8,686,463, which represents 51.3%, while men total 8,252,523 (48.7%), however, there is a gender gap in both attraction, retention and graduation in STEM careers [3], which does not allow it to have a significant contribution as a member country of Global South in the fulfillment of sustainable development goal 4 Quality Education in its target 4.5 to eliminate gender disparities in Higher Education.

In Ecuador, female students continue to be a minority in technological areas, with a participation of 23% in technology, 32% in engineering and 47% in mathematics. Currently, the presence of women in careers related to Science, Technology, Engineering and Mathematics (STEM) has become a key issue within the global agenda aimed at promoting gender equality.

This study seeks to explore and analyze the experiences of women in STEM programs in Higher Education, considering regional and economic differences, with the aim of providing a solid basis for the design of strategies that promote gender equality in these crucial fields for scientific and technological advancement.

The 193 countries that make up the UN are committed to improving metrics in the social, environmental and economic spheres through compliance with the Sustainable Development Goals (SDGs) with their 169 targets to be met by 2030 [4] The member countries make commitments that directly involve universities, since these institutions work in the field of research and innovation to generate solutions to social problems [5].

Sustainable Development Goal number 4 with its 10 targets aims to guarantee inclusive, equitable and quality education and promote lifelong learning opportunities for all by 2030 [6] In general, quality higher education for women in all university careers. This sustainable development goal is important as it helps to meet the rest of the SDGs [7]. Sustainable development goal number 5 "achieve gender equality and empower all women and girls" [8] This goal recognizes gender equality as a fundamental pillar for the construction of a peaceful and prosperous world, with the main objective of this SDG being to empower all women and girls and achieve gender equality [9].

Science, technology, engineering and mathematics (STEM) careers have currently taken relevance in society because they are the professions of the future and provide greater employability [10], according to the World Economic Forum by 2022 133 million jobs would be needed that will need professionals with skills linked to STEM careers [11]. It is estimated that 3.5 million professionals specialized in STEM (Science, Technology, Engineering and Mathematics) will be required in Latin America by 2025, in addition, these positions will not be filled due to a lack of professionals [12]

The Organization for Economic Cooperation and Development (OECD) predicts that by 2030 [13], 80% of the jobs that are currently in greatest demand will disappear and will be replaced by STEM careers [14] There is not the same distribution of women in careers that have been mostly taken by men [15] according to Senescyt, the careers with the fewest women enrolled are technology and engineering, which in turn will be the most in demand in the future [16].

Throughout their lives, women face additional obstacles in their training and getting a job such as gender stereotypes, unconscious biases, and discrimination [17], it is essential to recognize these gender differences and create policies that encourage women to pursue STEM careers [18]

There is inequality in the choice of scientific-technological careers by the female population, mainly in STEM careers due to the lack of female role models [19] The differences between men and women when tackling STEM careers are due to attitudes and beliefs [20]. Changing stereotypes is essential to ensure equal opportunities in their higher education [21] The conception must be changed in families from which a significant influence is derived in terms of the choice of STEM careers [22].

Scientists only take on the role of actresses when they are alone [23]. While conducting laboratory experiments, they seemed more concerned with being liked than appearing professional in front of others [24] This acceptance of a subordinate role by women in STEM research settings could be due to an unconscious lack of self-confidence and sense of belonging to the STEM research community, although future research is needed to further validate this claim [25]. If an issue of increasing concern is to close the gender gap by encouraging girls and women to pursue scientific careers, more equitable participation and representation are needed [26]

2 Methodology

This section contains the method, population, instruments and software used for the analysis of the research. To comply with the research, it is necessary to use the mixed approach, it is possible to use both qualitative and quantitative methods, highlighting the need at a deeper level to meet the demands that the objectives demand, according to the needs of research with STEM studies and relationship with the fulfillment of the SDGs respective to education (SDG 4) and gender equality (SDG 5).

To address the objectives, a quantitative approach was used, which includes documentary analysis of enrollment certificates and student surveys. In addition, a qualitative approach was used through interviews to delve into the barriers and facilitators perceived by women in their STEM studies.

The target population of study are students enrolled in the careers in the STEM programs of the University of the Armed Forces ESPE Latacunga Campus during the UNDERGRADUATE period SII 202450.

With the population corresponding to six engineering careers from the departments of Mechanical and Energy Sciences, Computer Science and Electrical and Electronics; The number of students enrolled in the period 202450 is 2990, with a confidence level of 95% (1.96) and a margin of error of 5% (0.05) in addition to a 0.5 probability that the event will occur, a sample of 342 students is obtained.

$$n = \frac{N * Z^2 * p * (1 - p)}{(N - 1) * e^2 + Z^2 * p * (1 - p)} \tag{1}$$

- The sample was stratified for the six careers with the data in Table 1.
- Electrical and Electronics: 1042/2990 × 100 ≈ 34.85%
- Energy and Mechanical Sciences: 1608/2990 × 100 ≈ 53.78%
- Computer Science: 340/2990 × 100 ≈ 11.37%

In addition, to consider the representativeness of women in the sample, stratified sampling was applied. With a total of 498 females and 2492 males, the sample size for each group was calculated proportionally to their size in the population, with the number of female students being 51 and the male sample being 291 students as shown in Table 1.

For a sample of 342 students, the stratified sample is calculated by applying the percentages above to this amount:

- Electrical & Electronics: 0.3485 × 342 ≈ 119
- Energy and Mechanical Sciences: 0.5378 × 342 ≈ 184
- Computer Science: 0.1137 × 342 ≈ 39

Table 1. Number of students by department and by gender.

Departments	Women	Men
Electrical & Electronics	160	882
Energy and Mechanical Sciences	269	1339
Computer Science	69	271
Total	498	2492

A documentary analysis was carried out based on the review of administrative records and institutional databases in order to obtain demographic and enrollment information that would allow determining the number of female students per career. These data were extracted directly from the MI ESPE platform.

As part of the qualitative approach of the study, a survey aimed at students was also applied. To ensure the reliability of the instrument, a pilot test was previously carried out. Subsequently, the Python tool was used to calculate Cronbach's alpha coefficient, a statistical indicator widely used to measure the reliability of measurement instruments. This coefficient ranges from 0 to 1, with 1 being the score that indicates the highest level of internal consistency.

In the same way, the interview was validated by three experts in education and gender to ensure the validity and relevance of the analysis, this instrument allowed to evaluate the qualitative part of the research and determine the reason for the problem.

For the quantitative analysis, descriptive statistics on the proportion of women enrolled in STEM were used, in addition, graphs and tables were shown to illustrate the distribution of enrollment by gender and career with the help of the statistical software Python. In the Qualitative analysis, themes and patterns emerging from the interviews and documents were identified.

A combination of descriptive, analytical, and comparative methods was employed to assess the relationship between STEM studies and the SDGs. Statistical techniques were used for quantitative analysis and content analysis methods for qualitative analysis.

3 Analysis of Results

3.1 Proportion of Women by Departments of STEM Careers

Globally, according to UNESCO data, only 35% of women are studying in STEM-related areas. For this study, a documentary analysis was used based on student enrollment reports available both in the academic registry of the university and in the official portal of higher education indicators of SENESCYT.

From the official website of the university, the enrollment certificates by career corresponding to the last five years, from 2020 to the current academic period SII 202450, were obtained in Excel format. In addition, data were extracted from Senescyt higher education indicators portal, which provides enrollment information until 2022, the limit established by the same platform.

Following a structured guide for documentary review, a consolidated database was built that allowed the analysis of the proportion of women enrolled in STEM careers at the national level, at the university and specifically at the Latacunga campus.

3.2 Situation at the Country Level

In Ecuador, according to INEC data, the female population outnumbers the male population. However, a significant gap persists in women's participation in STEM careers. When analyzing the data provided by SENESCYT on students enrolled in 2022 in the broad fields that correspond to the careers offered by the university – Engineering, Industry and Construction; Information and Communication Technologies; and Natural Sciences, Mathematics and Statistics—clear inequalities are evident.

In the field of Engineering, Industry and Construction, women represent only 32% of the total enrollment. In Information and Communication Technologies only 23%. In contrast, in Natural Sciences, Mathematics and Statistics, female participation amounts to 49%.

These indicators reflect a trend that aligns with data worldwide, where the most marked gaps are in careers considered critical, such as Engineering and Information Technologies

3.3 Situation at the University

The University of the Armed Forces ESPE has three campuses: the main campus in Sangolquí, the campus in Santo Domingo and the campus in Latacunga (with its extensions in the Center and Belisario Quevedo). His academic domain mainly covers the areas of Human, Social and Administrative Sciences, Security and Defense, and Applied Sciences. This research focused on the academic domain of Applied Sciences, which includes the careers of Engineering, Information Technologies and Natural Sciences.

The results reflect that the trend in female participation in Information and Communication Technology careers remains stable with respect to the national average, with 23% of female students enrolled. In the field of Natural Sciences, Mathematics and Statistics, a slight variation is observed, reaching 47% in the university compared to 49% at the national level. However, the most significant difference is in the field of Engineering, Industry and Construction, where female participation decreases considerably from 32% registered at the country level to 18% within the university. This contrast shows the need to strengthen institutional actions that promote the inclusion of women in traditionally masculinized areas, especially in engineering.

3.4 Situation at the University on the Latacunga-Belisario Campus

The Latacunga – Belisario campus of the University of the Armed Forces ESPE offers a total of 13 careers: six engineering and five technological level programs. This academic diversity is framed within the technical areas that are part of the STEM field, specifically in the fields of Engineering, Industry and Construction, as well as in Information and Communication Technologies.

However, when analyzing female participation in these areas, a significant decrease is observed compared to the percentages at the national level. In the field of Engineering, Industry and Construction, the participation of women on this campus is only 15%, while in Information and Communication Technologies it reaches 20%.

Engineering and Technology majors on this campus are distributed among the departments of Electrical and Electronics, Energy and Mechanical Sciences, and Computer Science. In the current academic period UNDERGRADUATE SII 202450, the percentage of women enrolled in the Department of Electrical and Electronics was 13.62%; in Energy and Mechanical Sciences, 15.87%; and in Computer Science, 19.93%.

Overall, enrollment statistics reflect an average female participation of 15.54% in STEM majors offered by campus, with significant differences between departments ($p < 0.05$). A notable exception is the Petrochemical Engineering career, which has a female participation of 54.12%, evidencing a more balanced gender distribution compared to the rest of the careers. The Higher Technology in Networks and Telecommunications career also stands out, with 26.53% of women enrolled.

In contrast, the careers with the lowest female participation, with values below 10%, are: Higher Technology in Automation and Instrumentation (8.33%), Higher Technology in Automotive Mechanics (6.45%), Automotive Engineering (5.49%) and Higher Technology in Electromechanics (4.34%). These data show a marked gender gap in several technical areas, especially in those related to mechanics and electromechanics.

3.5 Trend Analysis

There is a steady decrease in the number of women enrolled in STEM careers, which has gone from 524 in 2020 to 376 in 2024. This drop is worrying and shows the urgent need to investigate the causes behind the decline in female participation in these fields.

In contrast to this trend, the number of men enrolled in STEM careers has remained relatively stable during this period, although it also shows a slight decrease from 2024.

3.6 SDGs and Gender Policies

When analyzing the 17 sustainable development goals (SDGs) that have as their global objectives to eradicate poverty, protect the planet and ensure prosperity for all as part of a new sustainable development agenda, the words gender equality and education are sought in these SDGs.

Alignment with the Sustainable Development Goals (SDGs), in particular SDG 4 (Quality Education) and SDG 5 (Gender Equality), is critical to understanding and addressing this issue. Within these objectives, target 4.3 focuses on ensuring equal access to quality technical, vocational and higher education, including university education.

The results shown in Fig. 1 show that the SDG that contains both key words – "education" and "quality" – is number 4, referring precisely to Quality Education. From there, goals that do not specifically include terms related to higher education or gender equity are discriminated.

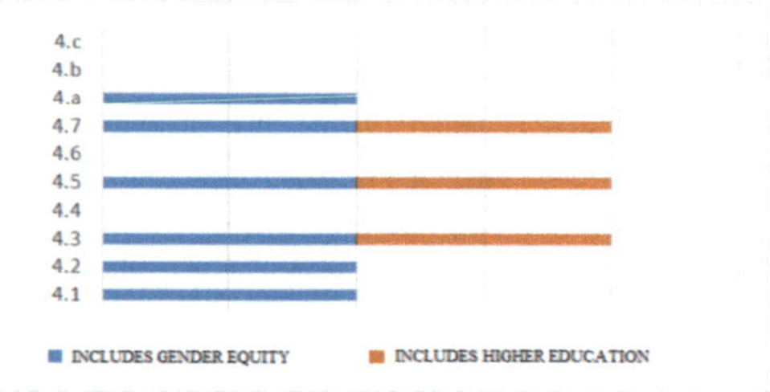

Fig. 1. SDG 4 targets related to gender equity and higher education

SDG 4 targets 4.3, 4.5 and 4.7 that are directly related to higher education. Target 4.3 states: "By 2030, ensure equal access for all men and women to training…". Target 4.5 seeks to "eliminate gender disparities in education" by the same year, while target 4.7 focuses on "ensuring that all learners acquire knowledge and skills…".

Finally, of all the SDG 4 targets, those that are specifically linked to STEM careers were selected for detailed analysis.

Target 4.3, which addresses equal access to technical higher education, is directly related to gender equity, higher education and STEM careers.

On the other hand, the documentary analysis of the Strategic Plan for Institutional Development 2021–2025 highlights, in general, that the Institutional Strategic Objectives are aligned with various Sustainable Development Goals (SDGs), reflecting the institution's commitment to this global agenda (Table 2).

When comparing the results of the surveys applied with the alignment with the Sustainable Development Goals (SDGs), it was found that 37.5% of respondents are not aware of the SDGs, while only 11.1% are aware of their existence.

Within the survey, students identified and selected goals 4.1 to 4.5, which are directly aligned with topics of higher education, gender equity, and STEM-related technical careers.

Table 2. Sustainable Development Goals.

SDGs linked to the Strategic Objectives
SDG 4: Quality education
SDG 5: Gender Equality
SDG 8: Decent work and economic growth
SDG 10: Reduced inequalities
SDG 12: Responsible Consumption and Production
SDG 17: Partnerships to achieve the goals

It is clear that UN target 4.3, which seeks to "ensure equal access of men and women to technical training by 2030 …", has the full support of the majority of women, who fully agree with its importance

3.7 Barriers and Enablers

The students point out that the main barriers they face in their careers are the absence of role models, self-perception influenced by gender stereotypes, and difficulties in balancing their studies with personal life (Figs. 2 and 3).

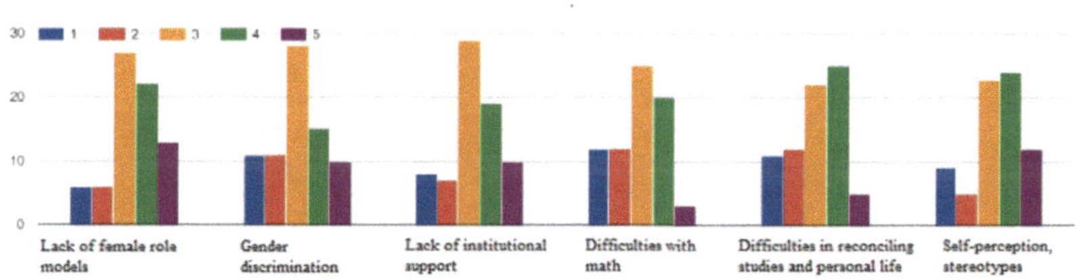

Fig. 2. Barriers to technical careers (STEM).

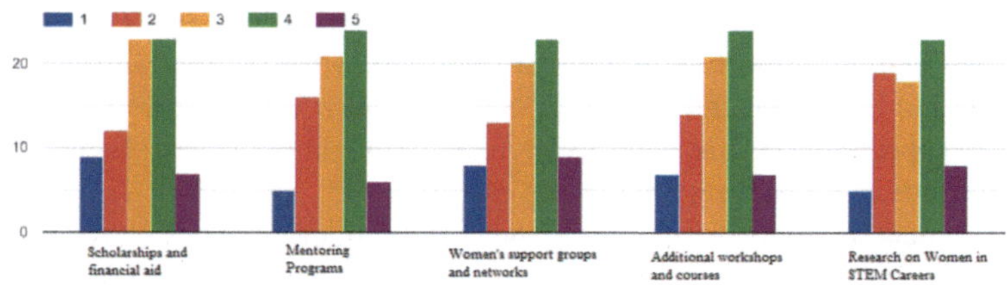

Fig. 3. Facilities that are presented in technical careers (STEM).

With regard to scholarships and financial aid, the opinions of the women surveyed are divided: one part is neutral, while another agrees with its existence, reflecting a balanced perception of this type of support. On the other hand, when it comes to facilities such as mentoring programs, support groups, women's networks, workshops, additional courses and research focused on women in STEM, the majority express a positive assessment. In fact, the highest level of agreement is specifically observed in the importance of having programs, support groups, and networks that strengthen female participation in these areas. This shows the felt need for collaborative and supportive spaces that promote the development of women in STEM careers.

3.8 Strategies and Recommendations

The analysis concludes that all measures are necessary and must be implemented; however, it especially highlights the urgency of increasing the visibility of successful women in STEM (Fig. 4).

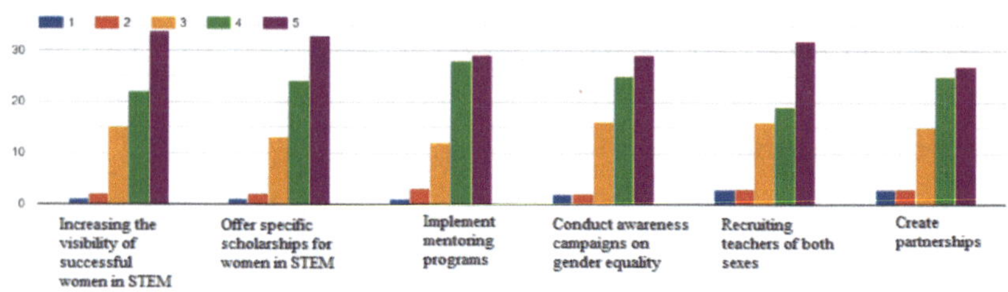

Fig. 4. Strategies and recommendations from students of technical careers (STEM)

4 Discussion

The analysis of women's participation in STEM careers at the University of the Armed Forces ESPE reveals several important dimensions in relation to the Sustainable Development Goals (SDGs), in particular SDG 4, target 4.3, which seeks to ensure equitable access to quality technical, vocational and tertiary education, including university education, for all women and men.

The data show that the proportion of women enrolled in technical careers varies significantly between different disciplines. For example, careers such as Petrochemistry have a relatively high proportion of women (54.13%), while others, such as Automotive Engineering and Electromechanics, show a much lower female representation (5.49% and 10.67% respectively). This disparity indicates that, although progress has been made in some areas, there are significant gender gaps.

The documentary analysis, including the enrolment certificates, shows that the university's initiatives to align its programs with the SDGs, specifically SDG 4.3, have not been sufficiently effective. The university mentions the SDGs in its policies generally, but most students are not familiar with these goals. This suggests a need for greater integration and communication of the SDGs within educational programs and institutional policies.

The survey reveals that one of the most significant barriers to women's participation in STEM careers is the lack of role models. The students also mentioned their self-perception and gender stereotypes, as well as difficulties in reconciling studies and personal life. These barriers not only affect enrollment, but also women's retention and academic success in these disciplines.

On the other hand, the facilities mentioned by the students that help their participation include support programs and women's networks. These programs offer a space where women can receive mentorship, emotional and professional support, and opportunities to connect with other women in their fields of study.

The university should consider the following recommendations, based on the results of the survey and interviews:

- Increase the visibility of successful women in STEM: This can inspire current and potential female students by providing examples of role models who have achieved success in their fields.
- Develop and strengthen women's support programs and networks: These programs should be designed to provide mentorship, academic and career support, and job opportunities specific to women in STEM.

The university has a crucial role in promoting gender equality in STEM careers. Through more targeted policies, well-structured support programs, and greater integration of the SDGs into their educational practices, it can contribute significantly to increasing women's participation and success in these disciplines. By focusing on these areas, you not only align with the SDGs, but also strengthen your commitment to equitable, quality education for all.

Previous research found that women were more concerned about being liked than looking professional in front of others. This acceptance of a subordinate role by women in STEM research settings could be due to an unconscious lack of self-confidence and sense of belonging to the STEM research community, although future research is needed to further validate this claim.

Data is provided that shows that students need to feel part of the STEM community, work on their own identity and believe in their abilities. This need for identification and self-efficacy underscores the importance of having role models who inspire and motivate female students to pursue and persist in these careers.

5 Conclusions and Future Work

The proportion of women enrolled in STEM careers at the University of the Armed Forces ESPE Latacunga Campus is 15.55%. This figure indicates that women represent a significant minority in these academic programs, with a participation that barely exceeds 15%. This underrepresentation underscores the need to implement specific strategies and programs to increase the inclusion and participation of women in STEM careers, aligned with the Sustainable Development Goals (SDGs).

The research has revealed that the skills acquired by women in STEM careers are mainly aligned with SDG 4 (Quality Education) and SDG 5 (Gender Equality), and specifically with Target 4.3 of SDG 4. This suggests that higher education in STEM for women at the University of the Armed Forces ESPE Latacunga Campus

The analysis of barriers and facilitators has revealed that the participation of women in STEM careers at the University of the Armed Forces ESPE Latacunga is significantly affected by the lack of role models, negative self-perceptions and gender stereotypes, as well as difficulties in reconciling studies with personal life. However, the existence of women's support groups and networks has proven to be a crucial facilitator in helping students to confront and overcome these barriers.

The research did not address statistics on women's retention in STEM careers, which could provide a more complete view of student retention and success in these programs.

The survey was not applied to the entire intended sample, which may have affected the representativeness of the results and limited the ability to generalize the conclusions to the entire population of ESPE students.

References

1. Calderón Cruz, E.M..: "Factors That Determine Students' Academic Success in Science, Technology, Engineering, and Mathematics (STEM)," 2022. Accessed 08 Sept 2025. [Online]. Available: https://hdl.handle.net/20.500.14352/3730
2. INEC: ECUADOR GREW BY 2.5 MILLION PEOPLE BETWEEN 2010 AND 2022 – National Institute of Statistics and Censuses. Accessed: 08 Sept 2025. [Online]. Available: https://www.ecuadorencifras.gob.ec/institucional/ecuador-crecio-en-2-5-millones-de-personas-entre-2010-y-2022/
3. Amores García, L.: The Gender Gap: Analysis of profiles of students and professionals in STEM areas. University of Salamanca (2020)
4. García-Parra, M., et al.: The Sustainable Development Goals in America: Overview. La Granja. **36**, 2 (2022). https://doi.org/10.17163/lgr.n36.2022.04
5. Schmader, T.: Gender inclusion and fit in STEM. Annu. Rev. Psychol. **74**(1), 219–243 (2023). https://doi.org/10.1146/annurev-psych-032720-043052
6. Seva-Larrosa, P., et al.: Students´ perception of sustainable development goals (SDGs) and the benefits for companies derived from their implementation. Econ. Res. Ekonomska Istraživanja. **36**(1), 2167100 (2023). https://doi.org/10.1080/1331677X.2023.2167100
7. Ramos Torres, D.I.: Contribution of higher education to the Sustainable Development Goals from teaching. REEC. **37**, 89 (2020). https://doi.org/10.5944/reec.37.2021.27763
8. Merchán-Rubiano, S.M., Acero-López, A.E., Sánchez-Gómez, J.S., García De Cajén, S.B.: Diversity, equity and inclusion on STEM education in Latin America. In: 2023 IEEE Frontiers in Education Conference (FIE), pp. 1–9. IEEE, College Station, TX, USA (2023). https://doi.org/10.1109/FIE58773.2023.10342906
9. Acuña-Moraga, O., Severino-González, P., Sarmiento-Peralta, G., Stuardo-Solar, C.: Sustainable consumption in Chile: an approach to the Sustainable Development Goals (SDGs). Inf. Tecnol. **33**(4), 181–190 (2022). https://doi.org/10.4067/S0718-07642022000400181
10. E. Especial: Sustainable Development Goals Report
11. Álvarez-Jaramillo, M., Bohórquez-Reyes, J.V., Velásquez-Rojas, J.C.: Actions to achieve sustainable development goals 4 and 5 strategies implemented by the FUJNC. In: Strategies Implemented by the FUJNC, vol. 154, pp. 243–260, Santiago (2021)
12. Cabero Almenara, J., Valencia Ortiz, R.: STEM and gender: an unresolved issue. Revie. **8**(1), 4–17 (2021). https://doi.org/10.47554/revie2021.8.86
13. Rodríguez-Argueta, C.M.: Trends in the supply of higher education in El Salvador – Relevance of careers in Science, Technology, Engineering and Mathematics (STEM) in the face of the new digital economy. Environment. **70**, 22–32 (2020). https://doi.org/10.5377/entorno.v0i69.9559
14. Canchari, D.: The Cloud over Latin America: the talent gap. America Economia News Service. (2023)

15. Aragón Mamani, P..: Influence of individual and household characteristics on the choice of a STEM university career by young people in Peru. [Online]. Available: chrome-extension://efaidnbmnnnibpcajpcglclefindmkaj/https://repositorio.up.edu.pe/backend/api/core/bitstreams/866fc8af-1062-4118-98bd-bdfbaa644ab2/content
16. Escobar-Jiménez, C.: Gender parity among the authorities of the Ecuadorian higher education system. Icons J. Soc. Sci. **74**, 139–158 (2022). https://doi.org/10.17141/iconos.74.2022.5246
17. Senescyt: "Universities and Polytechnic Schools – Enrollment." [Online]. Available: https://siau.senescyt.gob.ec/universidades-y-escuelas-politecnicas-matriculas/
18. UNESCO: "Cracking the Code: Girls' and Women's Education in Science, Technology, Engineering and Mathematics (STEM)." [Online]. Available: https://unesdoc.unesco.org/ark:/48223/pf0000253479/PDF/253479eng.pdf.multi
19. Verdugo-Castro, S., Sánchez-Gómez, M.C., García-Holgado, A.: Opinions and perceptions about higher STEM studies: an exploratory case study in Spain. Educ. Knowl. Soc. **23** (2022). https://doi.org/10.14201/eks.27529
20. W-STEM Consortium: Didactic guide for the open course: gender equality in university education in science, technology, engineering and mathematics (STEM). Zenodo. (2022). https://doi.org/10.5281/ZENODO.7189060
21. Chan, E., Di, D., Ecklund, E.H.: Scientists explain the underrepresentation of women in physics compared to biology in four national contexts. Gender Work Organization. **31**(2), 399–418 (2024). https://doi.org/10.1111/gwao.13076
22. Ojwala, R.A., Buckingham, S., Neat, F., Kitada, M.: Understanding women's roles, experiences and barriers to participation in ocean science education in Kenya: recommendations for better gender equality policy. Mar. Policy. **161**, 106000 (2024). https://doi.org/10.1016/j.marpol.2023.106000
23. Sreekanth, S., Sreekanth, S., Viswanathan, S.: The effects of social media on STEM identity in adolescent girls. J Emerg Invest. (2024). https://doi.org/10.59720/23-210
24. Vivas- Peraza, A.C.: The role of women in STEM disciplines. A multimodal analysis of crowdfunding project videos for the iGEM science competition. Sociotechno. **12**(2), 228–251 (2022). https://doi.org/10.24197/st.2.2022.228-251
25. Tsakalerou, M., Perveen, A., Ayapbergenov, A., Rysbekova, A.: The role of environment on women's perception about their STEM studies: observations from a Global South country. Sci. Rep. **14**(1), 228 (2024). https://doi.org/10.1038/s41598-023-50571-w
26. Shen, A.: Associations between women's retention in STEM or STEM-related fields and their spouses' occupations and majors. Humanit Soc Sci Commun. **11**(1), 178 (2024). https://doi.org/10.1057/s41599-024-02692-4

From Attitude to Intention: Modeling Gen Z Mobile Banking Adoption with an Extended UTAUT2

Marco Antonio Nolasco-Mamani[1], Norma C. Velásquez Rodríguez[2], Renza L. Gambetta Quelopana[1], Ana Noelia Martinez Valdivia[1], Ivette Eneida Atencio-Iturry[1], and Sam M. Espinoza Vidaurre[1](✉)

[1] Grupo Redes de la Escuela de Postgrado, Universidad Privada de Tacna, Tacna, Peru
{marnolasco,relgambetta,anamartinezv,iveatencioi, samespinozav}@upt.pe

[2] Grupo Faculty of Economics and Commercial Sciences, Sedes Sapientiae Catholic University, Lima, Peru
nvelasquez@ucss.edu.pe

Abstract. Objective. To assess the determinants of mobile banking adoption among Generation Z, using an extended UTAUT2 model that emphasizes attitude (AT) as a mediator. Methods Quantitative study conducted in Tacna, Peru (n = 261; ages 18–30). A structural equation model (SEM) was estimated. Constructs included AT (attitude), IU (intention to use), DT (digital self-efficacy), HM (hedonic motivation), PE (performance expectancy), HA (habit), and TA (technology anxiety), Results. AT positively predicted IU ($\beta = 0.36$), supporting the attitudinal pathway as central to adoption. DT was the strongest predictor of AT ($\beta = 0.41$), followed by HM ($\beta = 0.20$) and PE ($\beta = 0.20$). For IU, direct positive effects were found for HA ($\beta = 0.19$), DT ($\beta = 0.16$), and HM ($\beta = 0.10$). TA acted as an inhibitor, negatively affecting AT ($\beta = -0.11$) and IU ($\beta = -0.13$). Conclusions. Mobile banking adoption among Generation Z is primarily driven by an attitude-mediated route. Digital self-efficacy is a key determinant, strengthening attitude and exerting a direct effect on intention. Hedonic motivation and performance expectancy consistently enhance attitude, whereas habit and self-efficacy show direct associations with intention to use. Technology anxiety constitutes a meaningful barrier, lowering both attitude and intention.

Keywords: UTAUT2 · structural equation modeling · mobile banking · Generation Z · attitude · digital self-efficacy · technology anxiety

1 Introduction

1.1 Introduction

In the context of digital transformation, the way financial technologies are adopted in the context of digital transformation, the adoption of financial technologies has changed how users interact with banking services. Financial institutions have promoted a new way of

A. Rocha et al. (Eds.): ICITS 2026, LNNS 1966, pp. 518–539, 2027.
https://doi.org/10.1007/978-3-032-27538-7_46

managing personal finances through mobile applications that facilitate secure and fast transactions [1]. Generation Z, born between 1995 and 2010, is characterized by its high technological affinity and preference for digital solutions in various areas, including mobile banking [2]. However, research shows that factors such as security, trust, and perceived costs significantly influence the intention to use these digital platforms [3, 4].

Globally, the digitization of services and the expansion of smartphone use with internet access have driven the growth of mobile banking applications. However, limitations associated with low digital literacy and restricted access to these services persist [5, 6]. Similarly, many potential users perceive risks related to transaction security and data privacy, factors that continue to be significant barriers [7, 8]. Several studies highlight that digital literacy is a determining factor in the adoption of these technologies, as users with greater digital skills show a greater willingness to use them, while those with low levels of literacy are more reluctant [9, 10]. This leads us to conclude that the banking sector is becoming increasingly difficult [11], making it a priority to strengthen digital education and trust in mobile financial services.

In Latin America, the adoption of banking applications faces challenges related to unequal access to technology, mistrust of financial platforms, and low levels of digital literacy [12]. Although digitization has advanced, there is still a preference for cash transactions due to the perception of insecurity in digital media [13]. In Peru, initiatives such as YAPE and PLIN have promoted financial inclusion; however, resistance remains in rural and low-income sectors [14]. Limited connectivity, poor digital education, and a lack of robust infrastructure hinder the adoption of technology [15]. It is therefore necessary to strengthen digital financial education and improve platform security to boost confidence and use of mobile banking tools.

The study is carried out with the objective of determining the adoption of digital banking service applications through the extended UTAUT2 model, with some factors mediated by attitude. The proposed objective allows us to answer the question: What factors of the extended UTAUT2 model influence the adoption of digital banking service applications by the youth of Generation Z in the border city of Tacna, Peru?

2 Literature Review

The UTAUT (Unified Theory of Acceptance and Use of Technology) model was developed by Venkatesh et al. [16]; initially, four theories were integrated to measure the intention to use technologies. Subsequently, the UTAUT2 model was expanded, now integrating eight key theories, incorporating factors such as hedonic motivation and habit into the original model [17]. Recently, researchers have even proposed the UTAUT3 model, an extension that includes psychological and contextual variables with the aim of improving the prediction of user behavior regarding technology adoption in different fields of study. Table 1 synthesizes the empirical literature that has applied the UTAUT framework and its extensions (UTAUT2 and hybrid adaptations) to explain the adoption of mobile banking applications.

Table 1. Empirical literature on the Unified Theory of Acceptance and Use of Technology (UTAUT) in the adoption of mobile banking applications

Authors and year	Objective	Theory used	Study constructs	Main findings
Rahman et al. (2024) [18]	The study set out to examine, using the UTAUT model, the factors shaping Malaysian millennials' intention to adopt mobile banking applications, aiming to produce evidence that enables banking institutions to design strategies for expanding their user base. The research relied on data gathered from October 2020 to February 2021, aligning with the peak of the COVID-19 pandemic.	Unified Theory of Acceptance and Use of Technology (UTAUT)	Performance expectancy, Effort expectancy, Social influence, Facilitating conditions, Trust, Perceived risk, Personal innovation.	Performance expectancy, effort expectancy, trust, and personal innovation were significantly associated with the intention to adopt mobile banking applications, while social influence and facilitating conditions showed no significant effects.
Saprikis et al. (2022) [19]	This empirical investigation aimed to identify the determinants shaping the adoption or non-adoption of mobile banking applications. Two distinct cohorts, users (adopters) and non-users (non-adopters), were examined to ascertain overlaps and differences in the factors that account for their decision to use or refrain from using these services.	Unified Theory of Acceptance and Use of Technology (UTAUT)	Behavioral intention, Performance expectancy, Effort expectancy, Social influence, Facilitating conditions, ICT facilitators, ICT inhibitors, Recommendation.	The results revealed various differences in the factors that influenced the adoption or non-adoption of mobile banking applications in these two groups; therefore, it provides new perspectives on the adoption of mobile banking applications in a little-studied scientific field.
Samartha et al. (2022) [20]	The study aimed to assess the determinants of mobile application adoption using a modified UTAUT model. Addressing the growing uptake of mobile banking, the research empirically tested this theoretical framework in a random sample of participants drawn from multiple states across India.	Unified Theory of Acceptance and Use of Technology (UTAUT)	Performance Expectancy, Effort Expectancy, Social influence, Trust factor, Risk factor.	The study concludes that social influence, effort expectancy, and trust exerted a pronounced effect on purchase intention, whereas effort and risk related factors displayed no statistically significant impact on that intention.
Pham (2025) [21]	The article aimed to analyze the determinants shaping customers' intention to adopt, via banking applications, complementary offerings that extend beyond traditional banking services.	The conceptual model was based on the UTAUT2 model.	In its extended specification, the UTAUT2 framework comprises nine latent dimensions: performance expectancy, effort expectancy, social influence, facilitating conditions, hedonic motivation, price–value relationship, habit, perceived security, and perceived novelty.	The adoption of "beyond banking" services was significantly determined by social influence, the habit of using mobile banking, price-value perception, perceived novelty, and perceived security. In contrast, performance expectancy, effort expectancy, and hedonic motivation did not show statistically significant effects on behavioral intention.

(*continued*)

Table 1. *(continued)*

Authors and year	Objective	Theory used	Study constructs	Main findings
Bachri et al. (2025) [22]	The research aimed to ascertain the factors that shaped consumers' willingness to replace cash payments with QRIS usage during the COVID-19 pandemic.	Adaptation of the UTAUT2 model (Unified Theory of Acceptance and Use of Technology, modified version of the Push–Pull–Mooring (PPM) theory).	The set of examined variables includes health awareness, perceived risk of COVID-19, dissatisfaction, performance expectancy, effort expectancy, critical mass, attractiveness of alternatives, trust, perceived security and privacy, switching costs, and traditional payment habits.	The hybrid SEM–ANN analysis indicated that the attractiveness of alternatives, trust, critical mass, and traditional payment habits significantly shaped consumers' intention to shift from cash to QRIS during the COVID-19 pandemic; additionally, the data supported only 4 of the 11 proposed hypotheses.
Rahmiati et al. (2023) [23]	The objective of this research was to evaluate whether perceived risk moderates the relationship of performance expectancy and effort expectancy with behavioral intention towards mobile banking among residents of West Sumatra, and to simultaneously analyze the direct effects of these expectancies on intention and actual usage behavior.	Extended UTAUT2 model.	Performance expectancy, effort expectancy, perceived risk (as a moderator), behavioral intention, and mobile banking usage behavior.	The PLS-SEM analysis with 137 active mobile banking users in West Sumatra showed that perceived risk did not moderate the relationship between performance expectancy or effort expectancy and behavioral intention. Instead, both expectancies exerted direct and positive effects on said intention, which, in turn, significantly predicted the actual usage behavior of mobile banking.
Apau et al. (2025) [24]	The study sought to elucidate the adoption of mobile banking applications by extending the UTAUT2 model with four additional constructs and examining how demographic factors moderate their effects on intention and actual use.	Extended UTAUT2 model	The analytical framework comprises the following constructs: performance expectancy, effort expectancy, social influence, facilitating conditions, hedonic motivation, price-value relationship, habit, perceived security, perceived risk, institutional trust, and technological trust.	The model explained 79% of the variance in behavioral intention and 54.7% in usage behavior; all constructs, except for perceived risk, significantly influenced intention, and perceived security, institutional trust, and technological trust showed effects moderated by age, gender, experience, income, and education, highlighting the relevance of reinforcing security and trust in future adoption strategies.

Note. The objectives, theories used, constructs analyzed, and main findings are summarized (Rahman et al., 2024; Saprikis et al., 2022; Samartha et al., 2022; Pham, 2025; Bachri et al., 2025; Rahmiati et al., 2023; Apau et al., 2025). Abbreviations: UTAUT = Unified Theory of Acceptance and Use of Technology; UTAUT2 = extended version of UTAUT; PPM = Push–Pull–Mooring; SEM = structural equation modeling; PLS = partial least squares.

2.1 Performance Expectancy

In mobile banking, performance expectancy is a pivotal determinant within UTAUT2 for assessing new technology acceptance [25, 26]. This construct reflects users' judgments about the extent to which a technological feature enhances task productivity [27] and, in this study, it aligns with efficiency, convenience, and speed. Prior evidence shows that performance expectancy has a positive and significant effect on attitude, thereby mediating the adoption of digital banking tools [25]; the fastest adopters are those who perceive clear benefits and process simplicity [28, 29]. Moreover, efficiency, utility, and speed further strengthen adoption [3, 26]. However, these relationships can be moderated by trust, security, digital infrastructure, and prior experience, which shape performance expectancy and attitude [30], notably, digital infrastructure emerges as a determinant of intention to use [4].

2.2 Effort Expectancy

Effort expectancy regarding the adoption of technology refers to the perception of ease of use and plays an important role in the UTAUT2 model [29]. When it comes to banking applications, users typically prefer a friendly interface and simplicity in processes, making this factor key [1]. Studies have shown that reducing complex processes impacts technology adoption [31]. Likewise, factors such as ease of use and a user-friendly design have a positive effect on users integrating services [26]. On the other hand, technological experience, as well as trust, moderate the relationship between adoption of use and performance expectancy [25]. Therefore, experience plays a key role for the user in technology adoption.

2.3 Social Influence

Social influence refers to the type of perception people have about the opinions of those in their close circle [17]. In the UTAUT2 model, this factor plays an important role in technology adoption [26]. Studies have proposed that recommendations from family, friends, or authorities have a positive effect on technology adoption [29]. In the context of digital banking, recommendation and subsequent validation by society are fundamental for users when they adopt the use of technologies, especially when they have little experience with new technologies [25]. Jiang et al. [32] in a study conducted in emerging contexts, where user trust in financial institutions is low, found that user opinion indicates a predisposition towards adopting digital services. However, other studies question this factor as influential in technology adoption. Arenas et al. [30] argue that, in other contexts, factors like the perception of security and ease of use could be key in the intention to use technologies. This highlights the importance of considering the context in which the evaluation of social influence is being conducted.

2.4 Facilitating Conditions

Regarding facilitating conditions, users have perceptions of the necessary resources, as well as the financial support, to begin using technologies [17]. In terms of banking applications, these include internet access and the type of technical assistance from financial institution staff [1]. Research in this field has confirmed that this factor has a positive effect on the adoption of digital technologies [26]. In this regard, when there is a positive perception from users about the type of infrastructure and prior knowledge, the reliability of adopting these technologies increases [33]. In countries with low diversity of financial institutions promoting digital services, this factor plays a key role, as the lack of internet or technological infrastructure on mobile devices could be a barrier to the intention to use digital services [25]. However, it is also important to mention that while some studies have established that facilitating conditions affect technology adoption, this effect might be smaller compared to other UTAUT2 model factors like performance expectancy [31]. It is also important to mention that in countries with a higher level of financial market digitalization, users depend less on facilitating conditions and prioritize other factors regarding the intention to use platforms with digital services [3].

2.5 Habit and Behavioral Intention

Habit establishes users' behavioral tendencies based on previous experiences, and this factor plays a key role in technology adoption [17]. In the context of this study, this factor is a key determinant, as users with prior interaction with services offered on digital platforms are more prone to continue their use [34]. Studies have established that users with prior experience require less from other factors such as performance expectancy or facilitating conditions [1]. Furthermore, this factor is a key determinant if a user is to continue using digital banking platforms, especially among young people who already have prior knowledge of digital services [3]. On the other hand, it is important to highlight that its effect is not universal. In contexts where financial services are still limited, factors like trust and security in banking platforms could carry more weight [31]. In this regard, Jiang et al. [32] warn that even negative habits could develop, generating distrust among users, which would affect their intention to use.

2.6 Hedonic Motivation

Hedonic motivation plays a key role in users' willingness to adopt digital banking services [34, 35], and is defined as the degree of perception of technology use as something enjoyable [17]. Previous studies have established that those with a greater interest in continuing to use digital applications are those who found satisfaction in them [34], which also leads to a positive attitude towards the use of technology [35]. There are users who are willing to have a greater disposition to use these technologies but must have had a stimulating experience beforehand [31]. Furthermore, a playful approach and an easy-to-use application interface increase. Adoption [1]. Likewise, gamification and intuitive designs enhance the user experience [31]. However, the degree of impact may vary depending on the context. In fact, the factors that could most influence the intention to use are perceived utility and ease of use [34]. Therefore, hedonic motivation has a direct effect on the intention to use banking applications.

2.7 Personal Innovativeness in Digital Technology

Personal innovativeness in digital technology refers to the predisposition of individuals to try and adopt new technologies much earlier than the majority of users [30]. This factor is a key determinant in the adoption of mobile applications, moderated through attitude [35]. Those who best develop their attitude levels towards digital financial services are those with a good orientation towards technological innovation [36], and those willing to adopt these tools are those who perceive these digital instruments as efficient for financial management [32]. There are also users much more willing to adopt new innovative technologies with a much more positive attitude [1]. Furthermore, this influences factors like the perceived ease of use of these platforms [37], and personalization and prior experience reinforce this in emerging countries [38]. However, it is important to mention that its effect varies by context. The most influential factors in spaces with low levels of digitalization could be trust and security [39].

2.8 Digital Self-Efficacy

Digital self-efficacy refers to an individual's confidence in their ability to use digital technologies [23, 29]. This factor is a key determinant of the intention to use digital applications within the framework of the UTAUT2 model [1]. Research has shown that digital self-efficacy has a positive effect on the intention to use technologies. In environments with a higher level of digitalization, users who are confident in their skills tend to show a greater intention to adopt digital banking [40]. In addition, the perception of difficulty in using digital platforms is reduced [41]. Despite this, the influence can vary according to education level and age. In this regard, before adopting the intention to use, users with less experience in using technologies require more support at the beginning of the process [42]. Even in environments with a high level of digitalization in financial applications, digital self-efficacy remains a key factor [43].

2.9 Technological Anxiety and Attitude

In users with little prior experience, technological anxiety is an inhibiting factor [19]. Technological anxiety is the fear or discomfort a user feels when interacting with new technological tools [30], which could affect the use of online financial services. Research has reported that this factor reduces the intention to use digital tools. During the COVID-19 pandemic, the lack of experience in digital operations and the fear of being a victim of fraud increased anxiety [44]. Furthermore, it has been observed that financial inclusion is limited when users have a high-risk perception of using applications [32]. It even raises anxiety and decreases the adoption of digital tools [27]. It has been established that factors like digital education and intuitive design have attenuated anxiety [45]. However, the perception of security and trust in platforms mitigate this effect [4].

2.10 Attitude and Behavioral Intention

Attitude toward technology refers to the type of evaluation users make about adopting technologies, whether positive or negative [27].Within the UTAUT2 model framework, attitude is a determinant that has been addressed in several studies and is considered a fundamental factor in predicting the intention to adopt technologies [28]. Similarly, studies have established that a positive attitude increases the intention to use digital services; the effect is greater when users perceive security in financial transactions [46].

Furthermore, the significant effect of attitude on the intention to use mobile banking applications has been confirmed. Gharaibeh and Arshad [26], in their study, found that the perception of ease and reliability in using platforms increases users' attitudes, thus increasing the likelihood of adopting technologies. Similarly, Merhi et al. [4] posit that security and trust in technology play a key role in forming a positive attitude towards its use.

However, it is also important to note that other studies have determined that the relationship between attitude and behavioral intention can be moderated by other external factors. For instance, Farzin et al. [42] argue that social recommendation and prior experience in using technologies can influence users' attitudes, which in turn affects the decision to adopt technologies. For their part, Hilal and Varela-Neira [35] showed that personalizing the prior experience in using applications could improve attitude and the intention to adopt these tools.

As can be observed, the results of previous studies establish that attitude is a determining factor when it comes to adopting digital mobile applications, but this effect varies depending on other previously addressed factors such as trust, ease of use, and security. Therefore, to foster the development of positive attitudes, software and application developers should focus on improving the user experience, thereby building trust with more strengthened security protocols.

Based on the development of the theoretical framework, the following study hypotheses are established for the SEM model.

- H1a: Performance expectancy (PE) significantly influences the attitude toward banking applications (AT).
- H1b: Performance expectancy (PE) significantly influences the behavioral intention to use banking applications (UI).
- H2: Effort expectancy (EE) significantly influences the behavioral intention to use banking applications (UI).
- H3: Social influence (SI) significantly influences the behavioral intention to use banking applications (UI).
- H4: Facilitating conditions (FC) significantly influences the behavioral intention to use banking applications (UI).
- H5: Habit (HA) significantly influences the behavioral intention to use banking applications (UI).
- H6a: Hedonic motivation (HM) significantly influences the attitude toward banking applications (AT).
- H6b: Hedonic motivation (HM) significantly influences the behavioral intention to use banking applications (UI).
- H7a: Personal innovativeness in digital technology (PI) significantly influences the attitude toward banking applications (AT).
- H7b: Personal innovativeness in digital technology (PI) significantly influences the behavioral intention to use banking applications (UI).

- H8: Digital self-efficacy (DT) significantly influences the behavioral intention to use banking applications (UI).
- H9a: Technological anxiety (TA) significantly influences the attitude toward banking applications (AT).
- H9b: Technological anxiety (TA) has a significant influence on the behavioral intention to use banking applications (UI).
- H10: Attitude toward banking applications (AT) has a significant influence on the behavioral intention to use banking applications (UI).

3 Methodology

The study has a quantitative approach with a correlational-explanatory design; Structural Equation Modeling (SEM) is used as the statistical technique. The use of SEM is justified because it allows for the simultaneous evaluation of relationships between latent and observed variables, closely approximating the validation of theoretical models [47]. Given the context of this study, which seeks to examine the influence of various variables on the intention to use banking applications among young people of Generation Z in Tacna, Peru, the use of SEM is fundamental for estimating direct and indirect effects [48]. The study sample consisted of 261 cases, composed of 46.7% women and 53.3% men, who were selected through simple random probability sampling, ensuring the representativeness of the study population. This population comprises young people of Generation Z in the city of Tacna, Peru, between the ages of 18 and 30, with a mean age of 21 years. Regarding the sample size in SEM, previous studies recommend between 200 and 460 cases, depending on the model design [49]. Furthermore, Sim et al. [50] add that in models with multiple relationships and latent constructs, a sample size of at least 200 cases is adequate for obtaining reliable estimates. For his part, Hair et al. [47] recommend adhering to SEM sample size guidelines to obtain robust results. Therefore, the sample size of 261 cases is sufficient.

The instrument used in the present study is based on the adaptation of the model proposed by Farooq et al. [51], who extended the UTAUT2 model to evaluate technology adoption. The extended model incorporates new variables into the theoretical model; in the context of this study, this extended theoretical model allows for a more in-depth analysis of the factors influencing the intention to use banking applications. The questionnaire used in this work is structured based on the extension of the UTAUT2 model variables, to which items are added that measure performance expectancy, effort expectancy, social influence, facilitating conditions, habit, hedonic motivation, personal innovativeness, technological anxiety, and digital self-efficacy.

Data collection was carried out using a structured questionnaire in Google Forms, distributed via email and WhatsApp, thus ensuring accessibility for the study participants. For participation in the study, individuals were informed about the study's objectives, and their acceptance of informed consent was requested, assuring them of the confidentiality of their responses, in addition to their voluntary participation.

The data analysis was performed in two stages. In the first stage, a descriptive analysis of the variables was carried out to evaluate the SEM model and detect possible outliers. In the second stage, the specification of the structural model was evaluated, thereby assessing the reliability and validity of the structural model. To evaluate the consistency of the SEM model, measures such as CFI, RMSEA, GFI, and TLI were analyzed, following criteria established in the scientific literature [47, 52].

The study followed the ethical guidelines for research outlined in the Declaration of Helsinki. The research protocol was approved by the Ethics Committee of the Universidad Privada de Tacna (Verification Code 006-2024-CEI-ESPG-UPT). Participants provided their informed consent to participate in the study. The data were safeguarded by the research team.

4 Results

Table 2 presents the descriptive results of the study variables, including the mean (M), standard deviation (SD), Cronbach's alpha reliability coefficient (α), and asymmetry (A).

Table 2. Descriptives and reliability

	M	SD	α	A
Performance Expectations (PE)	15.05	2.74	.79	−1.14
Expectation of Effort (EE)	11.61	2.17	.82	−1.33
Social Influence (SI)	10.16	2.27	.78	−.28
Facilitating conditions (FC)	14.45	2.66	.70	−.94
Habit (HA)	13.84	3.11	.86	−.58
Hedonic Motivation (HM)	9.95	2.11	.82	−.44
Personal Innovation in Digital Technology (PI)	8.62	2.66	.85	−.12
Digital Technology Self-efficacy (DT)	16.85	3.61	.80	−.39
Attitude Towards Banking Applications (AT)	18.16	3.30	.88	−.99
Technology Anxiety (TA)	13.28	3.84	.86	.27
Intention to Use Banking Applications (UI)	11.00	2.26	.87	−.85

Note. M = Mean, SD = Standard Deviation, α = Alpha and A = Asymmetry.

From the results in Table 2, it can be observed that the variable with a high mean is "Attitude Towards Banking Applications" with a Mean of 18.16 and an SD = 3.30, which indicates that the study participants perceive they have a certain perspective on the use of banking applications. On the other hand, we can observe that the variables with a low mean are "Hedonic Motivation" and "Personal Innovation in Digital Technology" with the following values (M = 8.62, SD = 2.66) and (M = 9.95, SD = 2.11), respectively, which suggests that although the attitude is favorable, the intention to use these digital platforms remains moderate among Generation Z.

Regarding internal consistency, the Cronbach's alpha coefficient (α) shows adequate values for all study variables, with reliability ranges between α = 0.70 and α = 0.88, indicating an acceptable level of consistency for the study's constructs [47]. The variables with the highest reliability are "Attitude towards banking applications" and "Intention to use banking applications," with the following values (α) 0.88 and 0.87; these indices indicate that the constructs have high coherence. On the other hand, the variable "facilitating conditions" (α = 0.70) is the variable with the lowest consistency, but with an acceptable coefficient for confirmatory and exploratory studies as it is within the accepted range.

As for asymmetry (A), it is observed that all values are within the acceptable range of ±1.5, which indicates that the data distribution does not present significant biases that would affect the statistical analysis [48]. However, a slight inclination towards higher values can be observed in variables such as "Performance Expectations" (A = −1.14), "Effort Expectancy" (A = −1.33), and "Facilitating Conditions" (A = −0.94), indicating that participants have favorably rated these factors. On the other hand, it is observed that the only factor with positive asymmetry is "Technology Anxiety" (A = 0.27), indicating that some participants reported higher levels of anxiety in the use of digital banking tools.

Table 3 shows the results of the correlation matrix between the study variables, which has been used to identify the strength and direction of the relationships between the constructs. From the analysis of the results, it can be observed that most of the correlations follow a positive direction and are statistically significant, indicating that the factors evaluated in the extended UTAUT2 model are related to each other to predict the intention to use banking applications.

We can note that among the factors with high and significant correlations, the relationship between attitude towards banking applications (AT) and behavioral intention (UI) stands out with r = 0.68, which establishes that participants with a positive attitude towards banking applications have a greater intention to use them. Likewise, a strong correlation is found between performance expectancy (PE) and attitude (AT) (r = 0.57), indicating that users who perceive that these applications improve their financial performance tend to develop favorable attitudes towards their use.

On the other hand, variables such as facilitating conditions (FC) and effort expectancy (EE) show a moderate correlation (r = 0.64), which demonstrates that the availability of resources and technical support improves the perception of ease of use of the applications. The correlation between habit (HA) and behavioral intention (UI) (r = .52) is also noteworthy, implying that users who have already developed a habit of using banking technology are more predisposed to continue using it.

In contrast, technological anxiety (TA) presents negative and low correlations with most of the study's variables. In particular, its relationship with behavioral intention (UI) (r = −0.19) and attitude towards banking applications (AT) (r = −0.14) shows that participants who experience higher levels of technological anxiety tend to adopt a low predisposition to use digital banking applications, which is consistent with previous studies on psychological barriers in technology adoption.

4.1 Structural Model

In the structural model of Fig. 1, key factors that influence the intention to use banking applications are identified, highlighting the importance of factors such as attitude,

Table 3. Correlation matrix between variables

	PE	EE	SI	FC	HA	HM	PI	DT	AT	TA
Expectation of Effort (EE)	.60									
Social Influence (SI)	.43	.37								
Facilitating conditions (FC)	.56	.64	.45							
Habit (HA)	.51	.49	.38	.52						
Hedonic Motivation (HM)	.43	.43	.44	.43	.60					
Personal Innovation in Digital Technology (PI)	.27	.13	.34	.28	.40	.43				
Digital Technology Self-efficacy (DT)	.48	.53	.37	.55	.66	.60	.44			
Attitude Towards Banking Applications (AT)	.57	.50	.46	.50	.62	.60	.39	.71		
Technology Anxiety (TA)	−.15	−.24	.10	−.15	−.09	.10	.17	−.10	−.14	
Intention to Use Banking Applications (UI)	.46	.41	.37	.42	.60	.52	.32	.61	.68	−.19

Note. Pearson correlation matrix (two-tailed) with N = 261

hedonic motivation, and performance expectancy. However, the model's fit indices indicate that improvements could still be made through respecification, which can be done by incorporating additional relationships or refining the measurement items. Therefore, from the analysis, it can be determined that a positive perception by users of banking applications drives their adoption, while technological anxiety acts as a barrier to the intention of use.

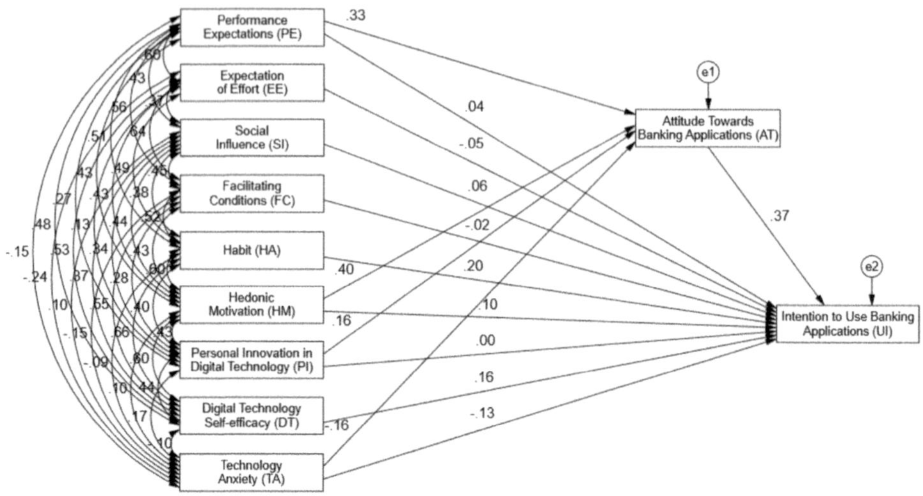

Fig. 1. Structural model

4.2 Adjusted Structural Model

After making a specification to the initial model, Fig. 2 shows improvement indices in the new structural model. The fit indices, in reference to the initial model in Fig. 1, now show a much more precise representation of the relationships between the study factors. The values obtained for CFI = 0.999 and GFI = 0.997 reflect an almost perfect fit, considering that values close to 1 indicate an excellent model fit. Likewise, the RMSEA = 0.034 indicates that there is minimal error in the model and it is within the acceptable minimum threshold (<0.08), which supports the model's level of robustness. Finally, the TLI = 0.988 confirms the improvement in the parsimony of the adjusted model.

On the other hand, Fig. 2 shows the following values for the Chi-square test (χ^2 = 3.889, df = 3, p = .274), from which it is inferred that it is not significant. This shows that there are no important differences between the estimated covariance matrix and the one observed in the data, which is desirable for establishing structural models in SEM, as it suggests that the adjusted model is adequate for explaining the proposed structure of relationships. Furthermore, the AIC value is 129.889, much lower than the initial model, demonstrating that the model fit has greatly improved in the respecification without compromising its complexity.

Regarding the estimation of the structural coefficients, the model maintains the positive influence of attitude towards banking applications (AT) on behavioral intention (UI) ($\beta = 0.36$), consolidating its role as a key mediator in the adoption of banking technology. Furthermore, the strongest predictor of attitude towards banking applications is digital technology self-efficacy (DT) ($\beta = 0.41$), which highlights that individuals more inclined to experiment with new technologies tend to have a better predisposition to develop more favorable attitudes towards the intention of use.

The structural model also confirms significant positive effects of some factors, such as hedonic motivation (HM) ($\beta = 0.20$) and performance expectancy (PE) ($\beta = 0.20$) on attitude, which shows that familiarity with the use of banking technology tools and the perception of utility play an important role in the adoption of these digital applications. Similarly, habit (HA) (β = 0.19), hedonic motivation (HM) (β = 0.10), and digital technology self-efficacy (DT) (β = 0.16) have direct positive relationships with the intention to use banking applications. On the other hand, factors such as digital technology self-efficacy (DT) (β = −0.11) and technological anxiety (TA) (β = −0.13) show negative relationships with attitude and behavioral intention, indicating that users with low confidence in their technological skills or with high levels of digital anxiety may present a lower predisposition to adopt banking applications.

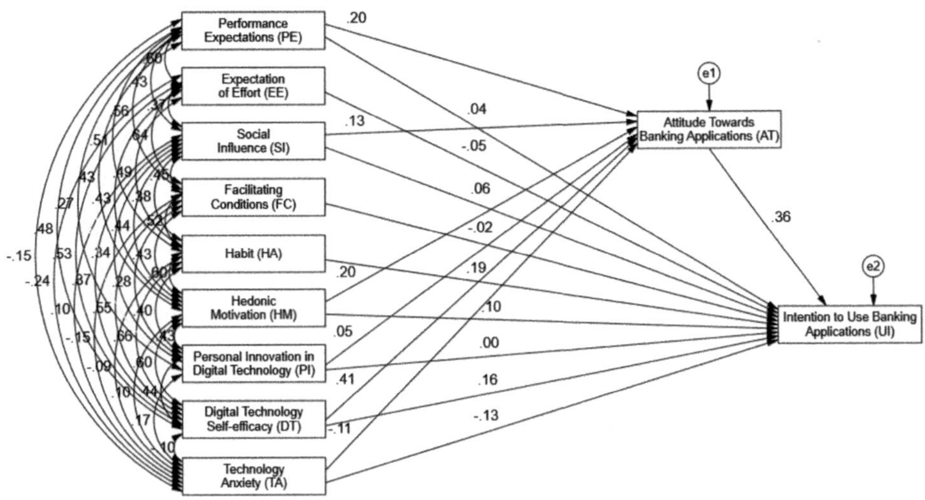

Fig. 2. Adjusted structural model

4.3 Explained Variance

The analysis of the explained variance (R^2) in the adjusted model provides an evaluation of the predictive capacity of the independent factors on the study's dependent constructs. In this case, the attitude towards banking applications (AT) has an $R^2 = 0.62$, which indicates that 62% of its variance is explained by the predictive factors considered in the model, such as performance expectancy (PE), social influence (SI), hedonic motivation (HM), and digital technology self-efficacy (DT). This demonstrates the high explanatory power of the factors in the formation of attitudes towards the use of banking applications, which validates that these factors predispose users to adopt digital banking tools.

Similarly, the explained variance for the intention to use banking applications (UI) is $R^2 = 0.54$, which is explained by the moderating variable attitude towards banking applications (AT) and other relevant factors that can be observed in the structural model in Fig. 2. Therefore, having an explanation level above 0.50 is usually considered moderately high in studies addressing the intention to use technologies and points to the fact that user attitude tends to play a preponderant role in their intention to use these digital banking platforms. This supports the theoretical model proposed by the extended UTAUT2.

On the other hand, in methodological terms, the explained variance values for attitude and the intention to use banking applications show that the respecified structural model is robust and provides a significant explanatory level of the key factors of the extended UTAUT2 model on the adoption of banking applications. However, it is important to mention that there is still a margin of variance that has not been explained in the structural model of Fig. 2, which acknowledges that there are other factors not contemplated in the model that could have a significant influence on the intention of use. Among some that could be addressed are trust in the platform's security, prior experience, or the availability of technological infrastructure.

5 Discussion

The results of this study provide key empirical evidence on the factors that determine the intention to use mobile banking applications among members of Generation Z in a border region like Tacna, Peru. This particular environment is characterized by high digital exposure, but within its digital financial structure, significant inequalities can be found, as well as a lack of financial education. This allows for the observation of different interactions that in previous studies, conducted in more homogeneous contexts, have been overlooked or underestimated.

Comparing the results of this study with other previous studies in countries with much better-developed banking ecosystems, such as the studies by Alalwan et al. [25] and Gharaibeh et al. [26], which state that performance expectancy (PE) was the main predictor of behavioral intention, in this particular study we can note that the effect was indirect, acting only through attitude towards banking applications ($\beta = 0.20$), and not directly on behavioral intention as reported by other studies. This attenuated mediation can be explained by a "technological normalization" in the youth of Generation Z, who have grown up exposed to digital platforms, as suggested by Tamilmani et al. [34], which dilutes the weight of perceived utility as a differentiating factor. In contexts like the border region of Tacna, Peru, the challenge is not for the technology to be useful, but for it to be reliable and accessible to users, especially in the study's subject segment.

On the other hand, we note that social influence (SI) showed a slight but significant effect on attitude ($\beta = 0.13$), similar to performance expectancy, with no direct relationship to behavioral intention. This contrasts with results such as those of Putri and Atik [29], who indicated that in traditional sectors, they found a central role for social recommendations. In the context of this study, technology adoption is much more autonomous and independent among the youth of Generation Z, following recommendation and social validation standards from family or friends, indicating a complex relationship between digital exposure and trust in banking institutions that could be the subject of other studies with hybrid or longitudinal approaches.

The effect of digital technology self-efficacy (DT) on attitude ($\beta = 0.41$) and on behavioral intention ($\beta = 0.16$) is particularly noteworthy. This is even more so when the literature review showed that studies like those by Merhi et al. [4] or Nguyen et al. [1], highlighted security and trust in digital platforms as critical factors. The results of this study contribute to the current literature by positioning the youth of Generation Z as agents of technological adoption, and not just as passive recipients of external security. The findings of this study help to better explain the inconsistencies of previous studies by showing that, in populations with Generation Z youth, confidence in one's own abilities carries more weight than the perception of the environment.

Similarly, hedonic motivation (HM) has a positive and significant effect on both attitude ($\beta = 0.20$) and slightly on behavioral intention ($\beta = 0.10$). This aligns with previous results reported by Tamilmani et al. [34], but it also expands the importance of this key factor by linking it to the design of enjoyable experiences as a bridge between the recreational use of apps and their functional application in digital banking. This result is key as it suggests that banking institutions should reconsider the design of digital tools and move beyond efficiency. To do this, it is necessary to incorporate principles of gamification, intuitive interfaces, and emotional feedback into the processes, as suggested by Marpaung et al. [31].

What was surprising from the study's results, but important to address, is the persistent negative influence of technological anxiety (TA) on both the attitude towards banking applications ($\beta = -0.11$) and the behavioral intention ($\beta = -0.13$). Although it was assumed in the study's context that the Generation Z youth are "digital natives," the reported results show that anxiety persists, not as a general fear of technology, but specifically towards financial platforms. This differentiation was scarcely addressed in previous studies like those of Vinnik [27] or Saprikis et al. [19], which suggests a critical opportunity to design specific metrics for "digital financial anxiety" that could be addressed in other studies but separating this construct from general digital anxiety.

The role of habit (HA) as a direct predictor of behavioral intention ($\beta = 0.19$) reaffirms what was proposed by Venkatesh et al. [17] and aligns with studies reported by Pratama and Renny [3], who postulate that repetition and familiarity with applications reduce the weight of other factors like ease of use or technical support. This result only confirms that the basis for the adoption of banking digital technologies or tools is not in convincing new users, but rather in facilitating uses that become part of people's daily lives. This can be achieved through interventions based on personalized notifications, rewards for frequent use, and simplification of routines, which can enhance this effect.

Finally, attitude towards applications (AT) maintains its place as the most potent predictor of behavioral intention ($\beta = 0.36$), replicating the findings of Thaker et al. [46] and Merhi et al. [4]. However, in this study, attitude is not formed exclusively by instrumental perceptions like performance or ease, but by a combination of personal confidence (DT), enjoyment (HM), social pressure (SI), and negative emotions (TA). This complex interaction suggests that the UTAUT2 model could benefit from a reformulation that explicitly distinguishes between rational, affective, and social components in attitude formation, something that has not yet been addressed empirically with sufficient clarity.

6 Practical Implications

The results of this study have relevant practical implications for the design of institutional strategies, public policies, and the development of digital products aimed at fostering the adoption of mobile banking applications among the youth of Generation Z, especially in border and emerging contexts like the city of Tacna, Peru.

Given that digital technology self-efficacy has a strong positive relationship with attitude and, similarly, with behavioral intention, this factor emerges as a key element. Banks and regulatory bodies should prioritize training programs in digital skills applied to the financial environment. Implementing training campaigns on the safe use of mobile banking applications, navigating different web environments, and solving technical problems will contribute not only to reducing entry barriers but also to building a sustained positive attitude towards the digital channel.

The significant and surprisingly negative presence of technological anxiety on both attitude and the intention to use banking applications indicates a current need for developers of these apps or virtual financial tools to consider designs that are oriented and centered on the current generation of youth, who are the most active users. To this end, we believe that including more intuitive interfaces, interactive assistance, empathetic messaging, and simulators that reduce the fear of error in the development process could improve this key factor within the UTAUT model. Likewise, clearer institutional communication about security and privacy protocols is recommended, reinforcing the user's perception of control and protection.

Habit is another factor that has emerged as key and relevant, with a direct effect on behavioral intention. This reinforces the need for strategies that stimulate the recurrent use of mobile banking applications. This can be achieved through promotions linked to digital use, for which it is necessary for the financial institution to have a presence on social media and participate on different social platforms, or through a form of participation such as contests. In other words, consistency in use should be encouraged, as in the long run, this will be more effective than short-term mass adoption campaigns.

Hedonic motivation, with a direct effect on attitude and, in turn, on behavioral intention, suggests that enjoyment and aesthetics also impact adoption. Over time, mobile banking applications must evolve from a transactional view to a positive emotional experience, incorporating elements like gamification, visual environment personalization, and achievements, without neglecting operational efficiency. This aspect is very important when focusing on young people familiar with technology who tend to value aesthetics and usability as much as functionality.

Unlike previous studies, facilitating conditions and social influence showed a limited impact. This implies that the mere availability of infrastructure or external support does not guarantee adoption if there is no positive attitudinal base. Therefore, strategies should focus more on user empowerment than on the external provision of resources, especially in populations that already have basic access but are not yet emotionally committed to the tool.

Given the transversal role of attitude towards applications, it is proposed to conceive banking apps as vehicles for comprehensive digital and financial inclusion, not just as transaction channels. Incorporating educational content, expense visualizations, proactive alerts, and personalized recommendations can increase the perception of value and strengthen the attitudinal bond.

7 Conclusion

Attitude towards mobile banking applications emerges as the decisive mediator between the constructs of the extended UTAUT2 and behavioral intention. The findings show that favorable perceptions of functionality, trust, and user experience directly influence Generation Z and determine their adoption decision. Therefore, the strategies of financial institutions must focus on reinforcing the perceived utility, reliability, and usability of banking apps to increase the intention to use and, ultimately, raise the satisfaction of this youth segment.

Digital technology self-efficacy also emerges as a key determinant that affects the attitude towards banking applications. Likewise, along with habit, it has positive effects on the intention to use banking applications. According to these results, young people in this study segment who have greater confidence, each in their own personal technological abilities, show a better predisposition towards adopting this type of digital banking tool. This evidence is reinforced by habit, as the analysis indicates that young people who have previously used other types of digital financial tools pay more attention to adopting these technologies with the intention of continuing to use them in the future. That is why, in the context of this study, financial companies should pay close attention to training to guide young people towards achieving positive experiences.

The determinant that has been shown to be a negative barrier to both attitude and the intention to use mobile banking applications is technological anxiety. From the results, it is inferred that young people who experience high levels of anxiety tend to have a low probability of adopting technologies developed in the financial sphere. This means that banking companies must develop different types of strategies to address this technological barrier and, in the long term, reduce this perception among this population segment. To do this, it is important to offer clear information that generates security in the use of this type of tool.

Another factor that has had a direct effect on both attitude and behavioral intention is hedonic motivation. Although its effect on attitude is slightly more moderate compared to its influence on intention, it is considered that the enjoyment and experience of young people can translate into relevant aspects when adopting banking applications, even if this factor is not a key determinant. That is why creating platforms with a creative, attractive, and intuitive interface with many functionalities contributes strongly and strengthens the intention to adopt digital financial tools.

From the results of the specified structural model, many factors that in previous studies were addressed as key determinants of the UTAUT2 model on the adoption of digital tools, among them performance expectancy, facilitating conditions, and social influence, present a limited impact on behavioral intention. These results, in the context of this study, reveal that among this segment of young people, recommendations from family and friends and the availability of technological infrastructure do not directly condition the intention to use. What has been observed is that they can moderately modify the attitude towards their use. Therefore, it is important to adapt new marketing strategies according to the usage patterns and behaviors of this unique group of young people.

Finally, the final adjusted SEM model reported a high power of explained variance with an R^2 of 0.62 for the attitude towards banking applications and an R^2 of 0.54

for the behavioral intention. This result validates that the specified structural model is structurally sound according to the theoretical framework proposed in the literature review on the adoption of mobile banking applications. While it is true that other factors may better explain the model, the reported values suggest that the model is applicable in the study's context and is suitable for predicting the intention to use in this group of young people, and that each of the analyzed factors explains a significant proportion of the observed variability in the data.

References

1. Nguyen, T.T., Nguyen, H.T., Mai, H.T., Minh, T.T.: Determinants of digital banking services in Vietnam: applying UTAUT2 model, Asian economic and financial. Review. **10**, 680–697 (2020). https://doi.org/10.18488/journal.aefr.2020.106.680.697
2. Palau-Saumell, R., Forgas-Coll, S., Sánchez-García, J., Robres, E.: User acceptance of mobile apps for restaurants: an expanded and extended UTAUT-2. Sustainability. **11**, 1210 (2019). https://doi.org/10.3390/su11041210
3. Pratama, R.R.D., Renny, R.: The role of behavioral intentions to use mobile banking: application of the utaut2 method with security, trust and risk factors, dinasti. Int. J. Manag. Sci. **3**, 728–741. https://dinastipub.org/DIJMS/article/view/1141 (2022). Accessed 14 Feb 2025
4. Merhi, M., Hone, K., Tarhini, A.: A cross-cultural study of the intention to use mobile banking between Lebanese and British consumers: extending UTAUT2 with security, privacy and trust. Technol. Soc. **59**, 101151 (2019). https://doi.org/10.1016/j.techsoc.2019.101151
5. Ullah, S., Kiani, U.S., Raza, B., Mustafa, A.: Consumers' intention to adopt m-payment/m-banking: the role of their financial skills and digital literacy. Front. Psychol. **13** (2022). https://doi.org/10.3389/fpsyg.2022.873708
6. Agbamu, E.B., Agbamu, M.E.: Mobile banking adoption among elderly Nigerians: challenges and potentials. Int. J. Intellect. Discourse **7**, 135–144. https://ijidjournal.org/index.php/ijid/article/view/639 (2024). Accessed 14 Feb 2025
7. Kaya, M.M.: Trust and Security Risks in Mobile Banking. University of Oxford, United Kingdom (2013)
8. Singh, S.K., Pattnaik, P.K., Samanta, S.: Issues and Challenges of Digital Banking System, pp. 67–81 (2024). https://doi.org/10.1007/978-981-99-8976-8_7
9. Elhajjar, S., Ouaida, F.: An analysis of factors affecting mobile banking adoption. Int. J. Bank Mark. **38**, 352–367 (2019). https://doi.org/10.1108/IJBM-02-2019-0055
10. Adel, N.: The impact of digital literacy and technology adoption on financial inclusion in Africa, Asia, and Latin America. Heliyon. **10**. https://www.cell.com/heliyon/fulltext/S2405-8440(24)16982-4 (2024). Accessed 14 Feb 2025
11. Cakerri, L., Petanaj, M., Kosova, R.: Factors influencing intention to use e-banking: an integrated model approach. Int. J. Innov. Res. Sci. Stud. **8**, 2346–2356 (2025). https://doi.org/10.53894/ijirss.v8i1.4959
12. Katz, R., Callorda, F.: Accelerating the development of Latin American digital ecosystem and implications for broadband policy. Telecomm. Policy. **42**, 661–681 (2018). https://doi.org/10.1016/j.telpol.2017.11.002
13. Ayllon, T.W.I.: Digital transformation in the banking sector and its impact on financial inclusion: BIM Peru case study. Tesis de maestría, Universidade Católica Portuguesa. https://repositorio.ucp.pt/handle/10400.14/30455 (2020). Accessed 16 Feb 2025
14. Aurazo, J., Gasmi, F.: Digital payment systems in emerging economies: lessons from Kenya, India, Brazil, and Peru. Inf. Econ. Policy. **69**, 101113 (2024). https://doi.org/10.1016/j.infoecopol.2024.101113

15. Herencia, L.R.P., Mayhua, G.J.M., Casas, L.A.A., Choquehuayta, W.N.: Systematic Review of Authentication Techniques in Banking Applications in Latin America, Laccei. https://laccei.org/LACCEI2024-CostaRica/papers/Contribution_201_final_a.pdf (2024). Accessed 16 Feb 2025
16. Venkatesh, V., Morris, M.G., Davis, G.B., Davis, F.D.: User acceptance of information technology: toward a unified view. MIS Q. **27**, 425–478 (2003). https://doi.org/10.2307/30036540
17. Venkatesh, Thong, Xu: Consumer acceptance and use of information technology: extending the unified theory of acceptance and use of technology. MIS Q. **36**, 157 (2012). https://doi.org/10.2307/41410412
18. Rahman, M., Yee, H.P., Masud, M.A.K., Uzir, M.U.H.: Examining the dynamics of mobile banking app. Adoption during the COVID-19 pandemic: a digital shift in the crisis. Digit. Bus. **4**, 100088 (2024). https://doi.org/10.1016/j.digbus.2024.100088
19. Saprikis, V., Avlogiaris, G., Katarachia, A.: A comparative study of users versus non-users' behavioral intention towards m-banking apps' adoption. Information. **13**, 30 (2022). https://doi.org/10.3390/info13010030
20. Samartha, V., Shenoy Basthikar, S., Hawaldar, I.T., Spulbar, C., Birau, R., Filip, R.D.: A study on the acceptance of mobile-banking applications in india—unified theory of acceptance and sustainable use of technology model (UTAUT). Sustainability. **14**, 14506 (2022). https://doi.org/10.3390/su142114506
21. Pham, T.T.: Factors affecting the intention to use "beyond banking services" via banking application: the case of Vietnam. GLOBAL BUSINESS FINANCE REVIEW. **30**, 43–59 (2025). https://doi.org/10.17549/gbfr.2025.30.5.43
22. Bachri, A.A., Maulida, M., Sari, Y., Sunardi, S.: Analyzing influence factors of consumers switching intentions from cash payments to quick response code Indonesian standard (QRIS) digital payments. Int. J. Financial Stud. **13**, 61 (2025). https://doi.org/10.3390/ijfs13020061
23. Rahmiati, R., Susanto, P., Yasri, Y., Linda, M.R., Mulyani, E.: Does Perceived Risk Really Matter in Mobile Banking Behaviour? pp. 347–360 (2025). https://doi.org/10.1007/978-3-031-80256-0_21
24. Apau, R., Titis, E., Lallie, H.S.: Towards a better understanding of mobile banking app adoption and use: integrating security, risk, and trust into UTAUT2. Computers. **14**, 144 (2025). https://doi.org/10.3390/computers14040144
25. Alalwan, A.A., Dwivedi, Y.K., Rana, N.P.: Factors influencing adoption of mobile banking by Jordanian bank customers: extending UTAUT2 with trust. Int. J. Inf. Manag. **37**, 99–110 (2017). https://doi.org/10.1016/j.ijinfomgt.2017.01.002
26. Gharaibeh, M.K., Arshad, M.R., Gharaibeh, N.K.: Using the UTAUT2 model to determine factors affecting adoption of mobile banking services: a qualitative approach. Int. J. Interact. Mob. Technol. **12**, 123 (2018). https://doi.org/10.3991/ijim.v12i4.8525
27. Vinnik, V.: User adoption of mobile applications: Extension of UTAUT2 model. Master thesis, Norwegian School of Economics. https://openaccess.nhh.no/nhh-xmlui/bitstream/handle/11250/2487036/masterthesis.PDF?sequence=1 (2017). Accessed 16 Feb 2025
28. Kwateng, K.O., Atiemo, K.A.O., Appiah, C.: Acceptance and use of mobile banking: an application of UTAUT2. J. Enterp. Inf. Manag. **32**, 118–151 (2019). https://doi.org/10.1108/JEIM-03-2018-0055
29. Putri, D.Y., Atik, A.: Factors involved in adopting mobile banking for sharia banking sector using UTAUT 2. Jurnal Keuangan Dan Perbankan. **26**, 184–207 (2022). https://doi.org/10.26905/jkdp.v26i1.6858
30. Arenas, J., Peral, B., Ramón, M.: Elderly and internet banking: an application of UTAUT2. J. Internet Bank. Commer. **20** 1–23. https://idus.us.es/items/b2a36875-41f1-412d-a20b-028a33d3b874 (2015). Accessed 16 Feb 2025

31. Marpaung, F.K., Dewi, R.S., Grace, E., Sudirman, A., Sugiat, M.: Behavioral stimulus for using bank mestika mobile banking services: UTAUT2 model perspective. Golden Ratio Mark. Appl. Psychol. Bus. **1**, 61–72 (2021). https://doi.org/10.52970/grmapb.v1i2.68
32. Jiang, J., Ma, J., Huang, X., Zhou, J., Chen, T.: Extend UTAUT2 model to analyze user behavior of China construction bank mobile app. SAGE Open. **14** (2024). https://doi.org/10.1177/21582440241287070
33. Putranto, I.A.: Acceptance technology factors of mobile banking usage based on utaut2 model. Jurnal Riset Akuntansi Dan Bisnis Airlangga. **5**, 920–937 (2020)
34. Tamilmani, K., Rana, N.P., Dwivedi, Y.K.: Use of 'Habit' Is Not a Habit in Understanding Individual Technology Adoption: A Review of UTAUT2 Based Empirical Studies, pp. 277–294 (2019). https://doi.org/10.1007/978-3-030-04315-5_19
35. Hilal, A., Varela-Neira, C.: Understanding consumer adoption of mobile banking: extending the UTAUT2 model with proactive personality. Sustainability. **14**, 14708 (2022). https://doi.org/10.3390/su142214708
36. Migliore, G., Wagner, R., Cechella, F.S., Liébana-Cabanillas, F.: Antecedents to the adoption of mobile payment in China and Italy: an integration of UTAUT2 and innovation resistance theory. Inf. Syst. Front. **24**, 2099–2122 (2022). https://doi.org/10.1007/s10796-021-10237-2
37. Amid, A., Din, R.: Acceptance and use of massive open online courses: extending UTAUT2 with personal innovativeness. J. Pers. Learn. **4**, 57–66. https://spaj.ukm.my/jplearning/index.php/jplearning/article/view/165 (2021). Accessed 16 Feb 2025
38. Cheng, Y., Sharma, S., Sharma, P., Kulathunga, K.: Role of personalization in continuous use intention of mobile news apps in India: extending the UTAUT2 model. Information. **11**, 33 (2020). https://doi.org/10.3390/info11010033
39. Timbula, M.A., Marvadi, C.: Digital transformation: acceptance and use of technology among microfinance institutions in developing country: an application of UTAUT2 mode. Int. J. Inf. Technol. **15**, 4459–4468 (2023). https://doi.org/10.1007/s41870-023-01535-w
40. Sumartini, A.R., Suprapti, N.W.S., Piartrini, P.S., Sukaatmadja, I.P.G.: Digital financial literacy and self-efficacy in increasing intention to use digital bank with the technology acceptance model theory approach. Int. Res. J. Manag. IT Soc. Sci. **11**, 252–263 (2024). https://doi.org/10.21744/irjmis.v11n6.2474
41. Peral-Peral, B., Villarejo-Ramos, Á.F., Arenas-Gaitán, J.: Self-efficacy and anxiety as determinants of older adults' use of internet banking services. Univ. Access Inf. Soc. **19**, 825–840 (2020). https://doi.org/10.1007/s10209-019-00691-w
42. Farzin, M., Sadeghi, M., Yahyayi Kharkeshi, F., Ruholahpur, H., Fattahi, M.: Extending UTAUT2 in m-banking adoption and actual use behavior: does WOM communication matter? Asian J. Econ. Bank. **5**, 136–157 (2021). https://doi.org/10.1108/AJEB-10-2020-0085
43. Manrai, R., Goel, U., Yadav, P.D.: Factors affecting adoption of digital payments by semi-rural Indian women: extension of UTAUT-2 with self-determination theory and perceived credibility. Aslib J. Inf. Manag. **73**, 814–838 (2021). https://doi.org/10.1108/AJIM-12-2020-0396
44. Nguyen, N.T., Nguyen, M.T.N.: Analyzing the adoption of mobile banking service in Vietnam: extending utaut2 with fear of covid-19. Bus. Manag./Biznes Upravlenie. **3**. https://www.ceeol.com/search/article-detail?id=1202989 (2023). Accessed 16 Feb 2025
45. Elsotouhy, M.M., Khashan, M.A., Thabet, M.Z., Galal, H.M., Ghonim, M.A.: Do the technological anxiety, privacy and physical risks matter in retail customers' adoption of AR apps? An extended UTAUT2 approach. EuroMed J. Bus. (2024). https://doi.org/10.1108/EMJB-05-2024-0104
46. Thaker, H.M.T., Thaker, M.A.M.T., Khaliq, A., Pitchay, A.A., Hussain, H.I.: Behavioural intention and adoption of internet banking among clients' of Islamic banks in Malaysia: an analysis using UTAUT2. J. Islamic Mark. **13**, 1171–1197 (2022). https://doi.org/10.1108/JIMA-11-2019-0228

47. Hair, J.F., Babin, B.J., Black, W.C., Anderson, R.E.: Multivariate Data Analysis, 8th edn. Cengage (2019)
48. Kline, R.B.: Principles and Practice of Structural Equation Modeling, 14th edn. NY 10001 United States of America, New York (2016)
49. Wolf, E.J., Harrington, K.M., Clark, S.L., Miller, M.W.: Sample size requirements for structural equation models. Educ. Psychol. Meas. **73**, 913–934 (2013). https://doi.org/10.1177/0013164413495237
50. Sim, M., Kim, S.-Y., Suh, Y.: Sample size requirements for simple and complex mediation models. Educ. Psychol. Meas. **82**, 76–106 (2022). https://doi.org/10.1177/00131644211003261
51. Farooq, M.S., et al.: Acceptance and use of lecture capture system (LCS) in executive business studies. Interact. Technol. Smart Educ. **14**, 329–348 (2017). https://doi.org/10.1108/ITSE-06-2016-0015
52. Byrne, B.M.: Structural Equation Modeling with AMOS. Routledge (2016). https://doi.org/10.4324/9781315757421

Trust-Aware Adaptive Framework for Malware Classification with Feedback Patching

Iliyan Barzev and Daniela Borissova(✉)

Institute of Information and Communication Technologies, Bulgarian Academy of Sciences, 1113 Sofia, Bulgaria
{iliyan.barzev,daniela.borissova}@iict.bas.bg

Abstract. The article addresses the problems related to malware classification. For the goal, a demo version of the proposed system is developed based on utilizing a federated architecture and a hybrid static-dynamic analysis approach. It relies on several generation models, as legacy, modern, and adaptive, with a feedback-driven patch layer and explainable AI components, all accessible through a user-responsive interface. The described trust-aware framework, called *Shipka Guard*, is adaptive and feedback-driven, featuring a patch layer, a feedback buffer, synthetic error injection, and scalability logging. The feedback buffer and patch layer for lightweight adaptation from user corrections and synthetic injections. The patch layer provides easy, lightweight clean-up of errors with small processed pools. The trust threshold offers a tunable balance between recall and abstentions. The GUI is realized using Streamlit. The conducted experiments categorically prove the applicability of the developed trust-aware adaptive framework for malware classification.

Keywords: Malware Classification · Framework · Adaptive · Feedback Patching · Explainable AI · SHAP · LIME

1 Introduction

Malware remains one of the top threats against information systems, quickly evolving in complexity, variation, and evasiveness. While machine learning-based malware detectors work well during the original training phase, they struggle with model staleness in practice. As malware families transform, and new obfuscation methods appear, static models become inaccurate due to concept drift, and they lack real-time feedback and adaptability. That is why many studies are related to feature extraction, analysis of malware behavior, and seeking advanced algorithms to improve accuracy in detection [1].

The rapid evolution of malware families has stimulated research efforts towards adaptive and resilient detection systems [2]. While traditional static classifiers, modeled on byte-level histograms, Portable Executable files (PE) headers, or imported functions exhibit decent results over filtered datasets such as EMBER all classifiers will soon

A. Rocha et al. (Eds.): ICITS 2026, LNNS 1966, pp. 540–551, 2027.
https://doi.org/10.1007/978-3-032-27538-7_47

experience concept drift in the field due to the inherent distributional differences in malware samples.

Adaptive learning has been proposed as needed in order to be adaptive due to the staleness of the model. Online and incremental learning (often termed continuous learning) enables classifiers to be continuously updated; although they typically require costly retraining or don't challenge catastrophic forgetting [3]. Federated learning frameworks have similarly been proposed (e.g. to classifiers) to enable parties to collaboratively update classifiers across distributed endpoints without transferring raw data [4]. These frameworks seem promising enough with their novelty but would ultimately fail to be viable and would create communication overhead when deployed in the field.

Research has discussed trust-aware systems regarding cybersecurity from an intrusion detection and anomaly detection standpoint [5]. In particular, the use of uncertainty estimation (e.g. Bayesian deep learning, conformal prediction) to incompletely estimate trust and help reduce false alarms has been researched [6]. The analysis and evaluation are limited in regards to malware classification, however, most of the work uses strengthening accuracy instead of calibrated decision-making. Trust-aware classifiers incorporate concepts of calibration, neglect and abstention to circumvent uncertainty scenarios.

Explainable AI (XAI) is relevant to security applications as it provides interpretability and transparency [7]. The Local Interpretable Model-agnostic Explanations (LIME) and SHapley Additive exPlanations (SHAP) methods can be used to highlight the features that are most impactful in determining the classification for one instance [8]. XAI is particularly important in malware detection because it allows the security analyst to verify that automated decisions are reliable. It is important to explain the classification processes, such as whether it is a rule, malware length, or something more robust, and how features were acted upon, such as using XAI. SHAP and LIME are capable of providing local and global understanding for malware classifiers [9]. SHAP provides feature attributions for tree-like methods, which provide global insights, while LIME provides local surrogate explanations that are model-independent. Both help analysts validate automated solutions, and both are useful in helping agents build trust in their security workflows.

There are some difficulties in the application of Machine Learning (ML) in malware detection related to the concept drift [10, 11]. The concept of drift is dangerous, as changes in malware behavior can easily lead to misclassifications. The concept drift is also suitable for identifying malicious traffic [12]. Significant performance loss of static models when tested against temporally distinct datasets was reported, indicating a need for online or adaptive and feedback-based models [13]. To improve static analysis, an approach for malware detection by optimizing feature extraction combining different ML algorithms is proposed [14]. Nevertheless, there is still no integrated or user-facing system including adaptation, explanation, and trust as a common platform to begin a research study.

Traditional static detection models use hand-crafted features derived from PE – e.g. byte histograms, section entropies, import-{export-type} tables, string statistics and so on – and classifiers such as Random Forests, XGBoost, and gradient-boosted tree [15].

These methods demonstrate high-level accuracy on curated benchmarks, however when a distributional shift is introduced the results vary.

Human feedback (labels, triage decisions) is a practical and efficient way of providing continual supervision. Human analyst corrections provide support in the learning pipeline that can be tuned explicitly to the error modes we see often, while also reducing operational false positive/negatives and effectively flushing minor signal noise. This, coupled with periodic documentation and premature sign-off of audit, supplies incremental growth in efficacy, without the burden of requiring complete retraining.

Online learning algorithms and incremental update schemes combat conceptual drift by updating model parameters as new labeled data becomes available. It is imperative in a practical deployment to balance the need for rapid adaptation, with need for consistency over time. Naive incremental updates can lead to catastrophic forgetting, whereas retraining the system consumes resources. A hybrid of both retentions patch onboard models plus periodic release full team retrains ultimately meets this need.

All of these motivated us to develop of trust-aware adaptive framework for malware classification with feedback patching. This framework utilizes a federated architecture and a hybrid static-dynamic analysis approach called "*Shipka Guard*". This framework relies on several generation models (Legacy, Modern, Adaptive), the feedback-driven patch layer, and explainable AI components, all accessible through a user-responsive, Streamlit-based Graphical User Interface (GUI). When the model isn't confident, instead of forcing a binary choice, we suggest tagging a sample as uncertain and either have a human analyst review the uncertain sample or utilize a fallback. All of these concepts contribute to reducing grave errors, especially in security-related domains.

2 System Design and Implementation

The trust-aware adaptive framework for malware classification with feedback patching, Shipka Guard, is realized by integrating legacy and modern models with a feedback-driven adaptive layer. This system, considered as a Self-Aware Model Classifier (SAMC), is organized into several interconnected components as illustrated in Fig. 1.

The framework follows a pipeline design where uploaded PE files are processed through feature extraction, model selection (SAMC routing), optional patch adjustment, and decision-making. User feedback is continuously collected and integrated to improve model robustness over time. The configuration parameters are assembled in a single configuration file, CONFIG.yaml, namely: *trust _ threshold*; *patch _ trigger*; *feedback _ limit*; *models _ dir*; *feedback _ dir*; *logs _ dir*. The configuration file defines the user trust threshold cap, number of examples in the feedback buffer before the retraining of existing models, number of new examples for full retraining, models, feedback, logs, etc. This part of functionality creates separation of configuration parameters from source code, which helps with transmissibility and repeatability in other systems.

Shipka Guard relies on a standardized 616-dimensional static feature which is extracted from each PE samples features with a PEFeatureExtractor. The extractor combines three components: (1) Byte histograms (distribution of raw byte values); (2) Byte entropy histograms (local randomness patterns); and (3) String-based metrics (counts, average lengths, printable character ratios, and entropy). The 616-dimensional static feature vector is uniform and maximizes comparability between datasets. The alignment of

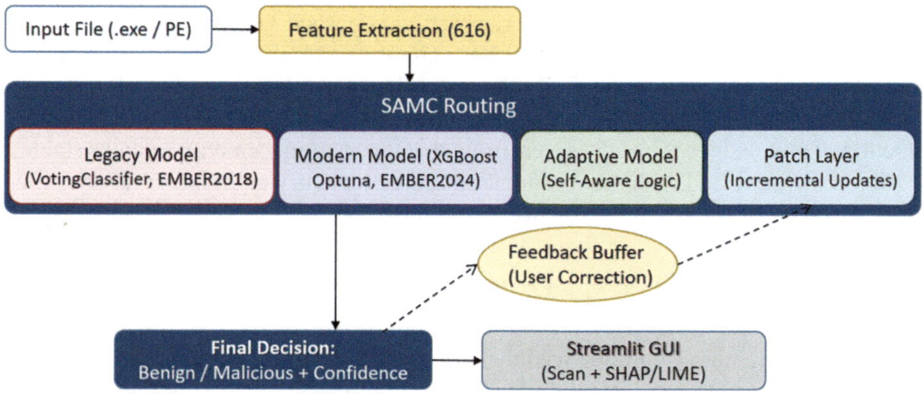

Fig. 1. Overview of *Shipka Guard* system – feature extraction, extraction, SAMC routing, feedback buffer, patch layer, and GUI integration.

the features between EMBER 2018 and 2024 allows the models to be aligned easily for comparability and an ensemble to be used without further alignment for incorporation into the ensemble.

All user corrections are written to a persistent feedback.json buffer where each entry keeps the sample SHA-256, filename, base model prediction/probability, corrected label, feature vector, and timestamp. To protect against feedback poisoning and duplication amplification, we deduplicate by SHA-256 prior to entry. The feedback buffer supports export/import (JSON) for controlled sharing across installations and group dataset building. Each entry also marks the final model probability and the final three class selection (Malicious, Benign, or Uncertain) to accommodate the system's trust-aware abstention strategy, instead of forcing a binary choice despite low trust.

The system combines three families of models: (1) Legacy model (trained on EMBER 2018); (2) Modern model (trained on EMBER 2024); and (3) Adaptive model (periodically retrained from feedback). Predictions are routed using SAMC as:

- If a model predicts to 0.85 or higher, then it is selected for prediction.
- Otherwise – fallback ensemble with weights: Adaptive 0.5, Modern 0.3, Legacy 0.2.

In Auto mode, candidate models are evaluated in priority order (Adaptive > Modern > Legacy). If any model predicts to 0.85 or higher, then prediction from that model is taken directly. Otherwise, use fallback ensemble with weights 0.5 Adaptive, 0.3 Modern, 0.2 Legacy, keeping the most adaptive model as dominant and the legacy model in a small role. To accommodate flexibility, and comparability across environments, the system supports four scan modes:

1. Auto (SAMC): trust-aware routing with confidence threshold and fallback ensemble.
2. Force Adaptive: enforces the use of the latest Adaptive model.
3. Force Modern: enforces the use of the Modern model (2024).
4. Force Legacy: enforces the use of the Legacy model (2018).

The modes support both automated trust aware behavior and manual comparison across generations of models.

User feedback is stored in *feedback. json* and is hashed via SHA-256 hash so that each entry is unique, preventing duplicates. The user can also correct false negatives and false positives through the GUI. The patch layer is a light-weight SGDClassifier that is trained incrementally on user feedback. Its importance is dynamically scaled as a function of unique feedback entries, ensuring that its patching behavior remains stable at small sample sizes and with many validating corrections its weight increases over time. More specifically, the patch is growing by the formula:

$$w_{patch} = \min\left(\alpha, \frac{n}{n+k}\alpha\right) \tag{1}$$

where n is the number of unique feedback samples, α is the maximum contribution for the chosen preset, and k is a smoothing constant controlling the growth rate.

Three operational presets are supported: (1) Conservative: $\alpha = 0.12$, $k = 300$; (2) Balanced: $\alpha = 0.20$, $k = 100$; and (3) Aggressive: $\alpha = 0.30$, $k = 50$. We have a confidence boost policy – if the correct model is confirmed low (0.75), we can boost the patch weight by 25% (with respect to α). Through this mechanism, the patch layer can fix systematic misclassifications while the base classifiers do not become unstable. In order to avoid instability, the patch layer is retrained only when a sufficient number of distinct feedback samples have been provided, and when at least one sample per class is used. This avoids the possibility of degenerate models formed on unbalanced corrections. For example, at $n = 1000$ feedback samples, the patch layer begins to approach maximum possible influence ($w_{patch}(1000) \approx 0.182$) while correcting systematic erroneous patterns of prediction with considerable sway.

There is also an option for synthetic False Positive/False Negative (FN/FP) injection, this assists with the evaluation of robustness and stress testing the patch mechanism prior to large-scale deployment. This design allows for the patch layer to have negligible influence when only a few feedback samples are available, while slowly increasing its influence as more robust corrections are gathered (see Fig. 2).

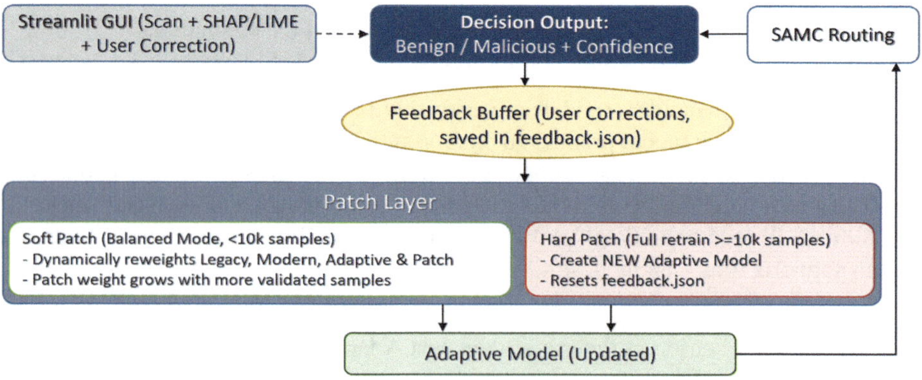

Fig. 2. Patch layer adaptation: soft patch (feedback-driven weighting) versus hard patch (full retraining with $\geq$10 k samples).

Once the feedback buffer is populated to the configurable threshold ($\geq$ 10,000 entries), the system will initiate a full retrain of the Adaptive model in XGBoost. The feedback dataset is then exported in JSONL format for reproducibility purposes, and the model is versioned within the models directory (see Fig. 3).

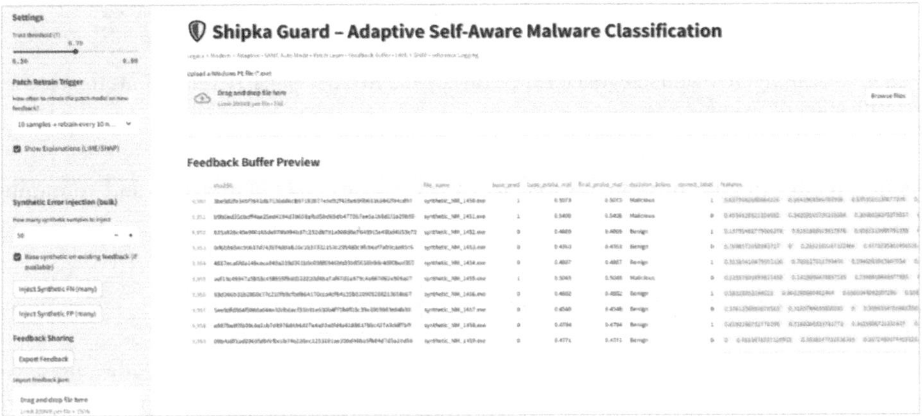

Fig. 3. *Shipka Guard* GUI: feedback buffer preview and sidebar controls for trust threshold, patch retrain trigger, synthetic error injection, and feedback sharing.

The *Shipka Guard* GUI is built using the Streamlit framework. This solution allows fast deployment and provides access to the analysts without any installation or complicated configurations. Within the GUI, users can upload PE files, make real-time predictions with associated confidence values, and gain explainable AI visualizations. Both SHAP and LIME explanations are included, with a provided complementary view; SHAP shows global feature contributions based on model architecture, while LIME shows locally interpretable approximations based on the local decision.

In addition, there is an option that could be used to make a manual selection of adaptive versions within a drop-down to compare and contrast side-by-side adaptive models. A built-in feedback import/export also allows analyst feedback to be saved or shared, adding to the reproducibility of each experiment and achieving a collaborative workflow. The combination of detailed logs and an adjustable thresholding option, with the capabilities, makes Shipka Guard an interactive research and operational platform.

Two explainable AI methods are integrated:

- LIME locally explains all models, and
- SHAP globally explains built-in feature contribution across tree-based models (Adaptive XGBoost).

The built-in visualizations are offered in the GUI to help understand and validate model decisions (see Fig. 4).

All system events (scans, feedback entries, retraining) are logged in logs/scan.log. Additionally, the framework tracks inference time (in ms) for each prediction to ensure that the system will work in real time on consumer-grade hardware. This feature provides the prospect for the project's feasibility and deployability in enterprise-level

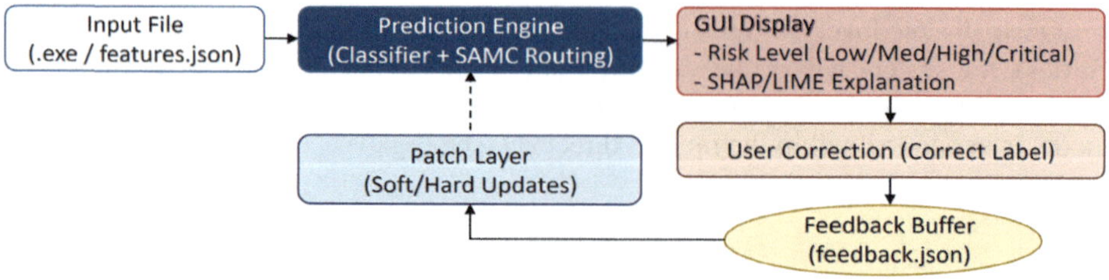

Fig. 4. Streamlit GUI with integrated LIME (local) and SHAP (global) explanations for malware classification decisions.

environments including scan results, feedback entries, patch updates, and retraining events.

3 Evaluation of *Shipka Guard* Framework

The evaluation of the applicability of the developed framework, *Shipka Guard*, uses two publicly available benchmark datasets: EMBER 2018, containing over 1 million PE samples characterized by 616 static features (i.e., histograms, entropy, string statistics) and used as the baseline "legacy" model for the study; and EMBER 2024 with approximately 960,000 samples. EMBER 2024 contains modern malware families and is harmonized into the same 616-dimensional feature space of comparison with the EMBER 2018 dataset. In addition, a third user-generated dataset is used. The framework is structured to integrate user-generated data in a straightforward manner to harmonize into the same 616-dimensional feature space. Thus, it is important that *Shipka Guard* stays flexible for future data sources and nearly real-world applications outside of public benchmarks. The harmonization allows models trained on different generations of datasets to still work within the SAMC framework.

Experiments were conducted on a consumer-grade machine with the following configuration: CPU: 16-core Intel i7 (10th Gen); Memory: 32 GB RAM; GPU: NVIDIA RTX 2060 (6 GB VRAM); Software: Python 3.8 (The legacy extractor is Python 3.8 only, but all other software approaches can use later versions, experiments are done consistently under 3.8.), Streamlit, scikit-learn, XGBoost, LIME, SHAP; Environment: All models were trained offline; Inference and GUI operations were run in real-time. This configuration mirrors typical consumer-grade high performance hardware, evaluating suggests that appropriately reflects credible deployment scenarios.

A balanced evaluation set of 100 k samples (50 k benign, 50 k malicious) was created from EMBER2018, and EMBER2024 for fairness and reproducibility of the experiments. Initially, the baseline models and the SAMC automatic routing method were compared on the balanced evaluation as shown in Table 1.

The Legacy model performs poorly, with lower training data quality determining higher FN (recall). The Modern model achieves high accuracy and recall with more false positives. SAMC Auto is somewhere in between: fewer FP (702) than the Modern (1006), at the cost of more FN. This gives a nice illustration of the utility of dynamic routing as a risk balancing method.

Table 1. Performance of Legacy, Modern, and SAMC Auto (T = 0.85).

Mode	Accuracy	Precision	Recall	F1	ROC-AUC	FN	FP
Force Legacy	0.92409	0.9540	0.8906	0.9213	0.9621	5472	2148
Force Modern	0.96641	0.9793	0.9529	0.9660	0.9955	2353	1006
Auto (SAMC) [T = 0.85]	0.95917	0.9852	0.9324	0.9580	0.9942	3381	702

We consider the effect of lightweight Patch adaptation versus full retraining as show in Table 2.

Table 2. Ablation study comparing baseline Modern, Patch with 500 feedback samples, and full retraining with 10 k feedback.

Mode	Accuracy	Precision	Recall	F1	ROC-AUC	FN	FP
Base (Modern)	0.96641	0.9793	0.9529	0.9660	0.9955	2353	1006
Patch (500 fb)	0.96727	0.9799	0.9541	0.9668	0.9918	2294	979
Full Retrain (10 k)	0.94719	0.9672	0.9257	0.9460	0.9906	3713	1568

The Patch layer, trained with 500 feedback samples, has noticeable improvements: both FN (−59) and FP (−27) went down, the F1 score slightly increased. Thus, even a small and possibly imbalanced feedback buffer can be used to implement consistent, low-cost corrections. The full retraining using 10,000 feedback samples produced a worse outcome. Although it seems counterintuitive, the explanation is due to an imbalance in the feedback pool, as multiplying problematic samples and increasing errors instead of fixing them. This means that the quality and variety of feedback are as important as the quantity of feedback.

Patch is incredibly useful in cases with few or noisy feedback queues, providing quick and lightweight adaptation, and full retraining should be done only when there are enough examples that demonstrate balance and coverage of variability between the classes. So, the importance of the trust-aware feedback pipeline for *Shipka Guard* is to not cause naive retraining to destabilize the system. We run SAMC Auto using different trust thresholds as shown in Table 3.

Lower thresholds ($T = 0.80$) result in fewer abstentions but more risk of misclassifications. Using higher thresholds ($T = 0.90$) increases the number of abstentions which also improves robustness, but comes with a reduction in throughput. It further supports that T is a tunable safety knob. The Modern model achieves sub-50 ms mean inference, well below the 90 ms/sample target. Legacy and Adaptive are slower (~110–130 ms), but still within real-time bounds. These results validate the deployability of the system even with SHAP/LIME enabled.

Inference time was measured directly from system logs for each model and the results are given in Table 4.

Table 3. Precision, Recall, and percentage of Uncertain decisions across thresholds.

T	Precision	Recall	%Uncertain
0.80	0.9854	0.9310	6.5%
0.85	0.9852	0.9324	8.2%
0.90	0.9851	0.9341	10.6%

Table 4. Inference latency, normalized to 100 samples per model

Mode	n_samples	Mean (ms)	Std (ms)	p50	p90	p99
Modern	100	47.6	24.5	33.1	83.6	91.3
Legacy	100	131.4	0.0	131.4	131.4	131.4
Adaptive	100	109.7	29.5	109.7	133.3	138.6

Three sample case studies illustrating the feedback and explanation dynamics:

- **False Negative corrected by Patch.** Figure 5 (LIME) shows a malicious sample misclassified as benign. After patching with 500 feedback samples, the decision was corrected, reducing FN.
- **False Positive corrected by Feedback.** A benign sample misclassified as malicious was added to the feedback buffer. The patch model adjusted, and the FP was eliminated.
- **Uncertain case explained with LIME and SHAP.** At $T = 0.90$, SAMC abstained. LIME showed conflicting local evidence, while SHAP global importance (Fig. 6) explained dominant feature influences. The GUI (Fig. 7) displayed the abstention to the analyst, who could provide corrective feedback.

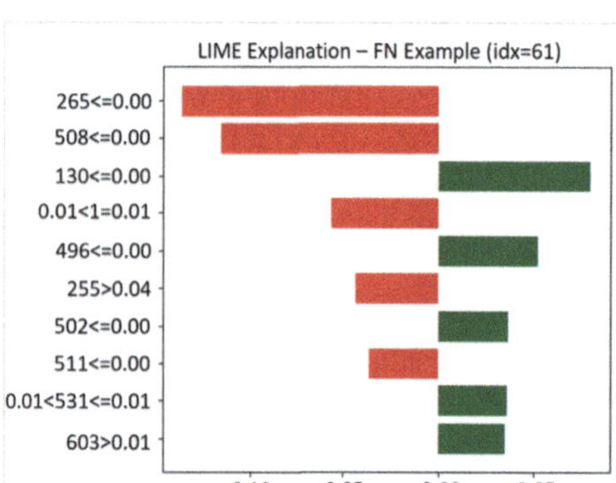

Fig. 5. LIME explanation for FN corrected by Patch.

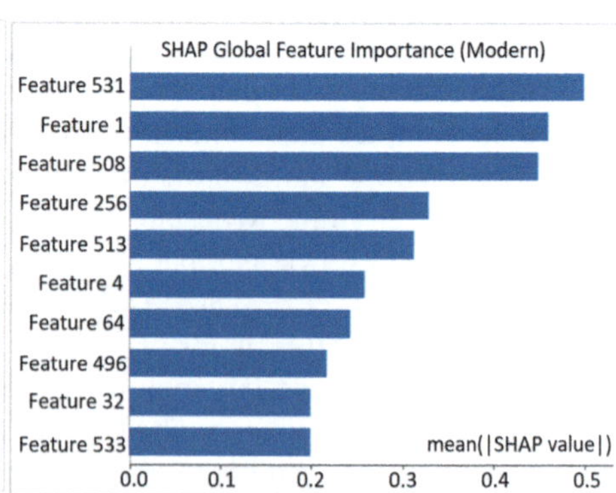

Fig. 6. SHAP global feature importance for Modern model.

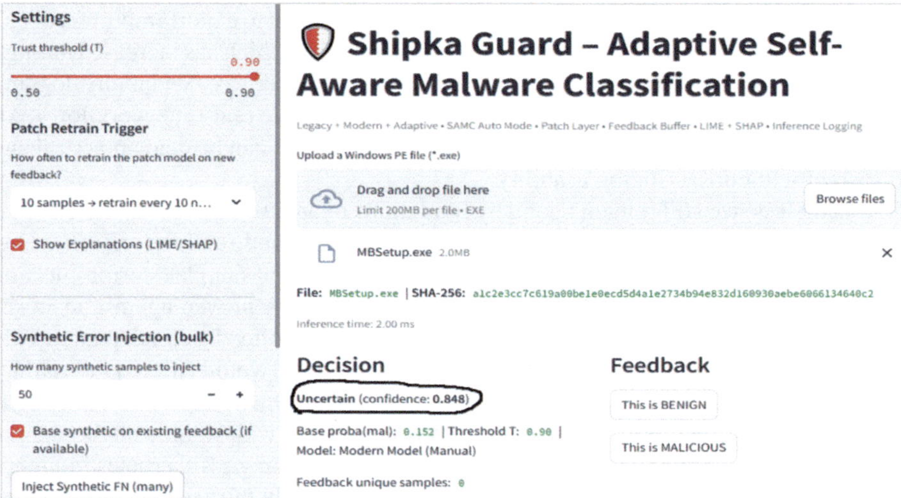

Fig. 7. GUI screenshot of uncertain decision (T = 0.90).

The evaluation confirms the effectiveness and limitations of the proposed framework. Legacy models are outdated, while Modern models perform strongly but produce more false positives. SAMC Auto balances risks via dynamic routing. The Patch layer provides easy lightweight clean-up of errors with small processed pools, and complete retraining yields no preemptive advantage, unless it has an appropriate amount of feedback. The trust threshold offers a tunable balance between recall and abstentions. Latency remains in the bounds of real-time, while the case studies demonstrate how the explanation module and Uncertain abstentions have subsequently increased analyst trust or at minimum have validated the results. Therefore, the process of a self-aware model classifier can be categorized as an effective and trustworthy process for adapting malware classification.

4 Discussions

The evaluation demonstrates that *Shipka Guard* is able to incorporate legacy, modern, and adaptive models in an integrated SAMC framework and provides user-driven feedback to continue adaptation. The addition of patch retraining, feedback inferences, and inference time logging makes the system more resilient and deployable in practice. *Shipka Guard* has a clear trust layer: if no single model passes the confidence threshold set by the user, the system selects a fallback ensemble or returns an Uncertain flag for human validation. This provides safe abstention: samples with negative or uncertain trust are sent for validation rather than forcing an automated decision. Thus, the trust layer reduces high-consequence mistakes and allows human analysis of low-trust instances to be prioritized.

Shipka Guard offered 4 scan modes, synthetic error injection, JSON-based feedback export/import, and scalable analysis. This makes the *Shipka Guard* framework not only more adaptable but also more relevant to real-world deployment considerations.

The framework *Shipka Guard* is based on an interactive GUI that lowers the barrier for non-expert users to experiment with malware classification. The feedback buffer allows the user to collaboratively correct the model, allowing it to export/import feedback files. Explainability integration contributes to decision trust, which has increased through more insights about features from local and global perspectives. Scalability logging reflects the adaptability of the system, which does not impede real-time workflows. All of these aspects demonstrate that adaptive malware detection can be made practical and transparent without sacrificing usability.

Nonetheless, we still remain bound by limitations related to static analysis. The system only uses features from PE files. In advanced packing and runtime-only malware, the system will struggle. Synthetic injection vs. real adversarial samples – error injection is great for testing adaptation, but robustness will need to be proven against adversarially perturbed (or obfuscated) PE files. Concept drift scope feedback adaptation only accounts for incremental drift, and larger distributional shifts would still require retraining on novel large-scale datasets. Scalability is beyond consumer hardware, while inference is quick with one machine, enterprise-level deployment with significant throughput may necessitate distributed or cloud-native implementation (e.g., Kubernetes inference pipelines). The implications and limitations above provide both practical viability and current limitations for *Shipka Guard*, and set the stage for future work.

5 Conclusion and Future Work

The article described *Shipka Guard* as an adaptive and self-aware malware classification framework that includes legacy, modern, and adaptive models to create a feedback-driven patch layer. By combining all previous investigation pathways into a unified framework, *Shipka Guard* addresses all three concerns that malware detection must contend with: adaptability, trust-awareness, and usability.

The framework provides several new contributions. The first of them is a trust-aware routing mechanism that utilizes multiple generations of models, including a fallback ensemble. Next is a feedback buffer and patch layer for lightweight adaptation from user corrections and synthetic injections. Its realization is done by an interactive Streamlit GUI that provides four scanning modes, JSON-based feedback sharing, and local/global explanations (LIME and SHAP). It also provides scalability logging that presents real-time inference (<90 ms) on consumer-grade hardware.

The combination of adaptability, interpretability, and scalability leads toward addressing the challenges of continuously developing practical, next-gen malware detection systems that continually remain "true" given threat models presently in lieu of always changing. Including version selection and a feedback interface as part of the app enables reproducibility and collaborative usability, rendering *Shipka Guard* not only a practical malware defense tool but also a collaborative research or evaluation tool.

References

1. Berrios, S., Leiva, D., Olivares, B., Allende-Cid, H., Hermosilla, P.: Systematic review: malware detection and classification in cybersecurity. Appl. Sci. **15**(14), 7747 (2025). https://doi.org/10.3390/app15147747

2. Alsubaei, F.S., Almazroi, A.A., Atwa, W.S., Almazroi, A.A., Ayub, N., Jhanjhi, N.Z.: Adaptive malware identification via integrated SimCLR and GRU networks. Sci. Rep. **15**, 25309 (2025). https://doi.org/10.1038/s41598-025-08556-4
3. Zhang, S., Liu, J., Zuo, X.: Adaptive online incremental learning for evolving data streams. Appl. Soft Comput. **105**, 107255 (2021). https://doi.org/10.1016/j.asoc.2021.107255
4. Mehdi, M., Makkar, A., Conway, M.: A comprehensive review of open-source federated learning frameworks. Procedia Comput. Sci. **260**, 540–551 (2024). https://doi.org/10.1016/j.procs.2025.03.232
5. Pigola, A., de Souza Meirelles, F.: Unraveling trust management in cybersecurity: insights from a systematic literature review. Inf. Technol. Manag. (2024). https://doi.org/10.1007/s10799-024-00438-x
6. Del Corso, G., Colantonio, S., Caudai, C.: Shedding light on uncertainties in machine learning: formal derivation and optimal model selection. J. Frankl. Inst. **362**(3), 107548 (2025). https://doi.org/10.1016/j.jfranklin.2025.107548
7. Galli, A., La Gatta, V., Moscato, V., Postiglione, M., Sperli, G.: Explainability in AI-based behavioral malware detection systems. Comput. Secur. **141**, 103842 (2024). https://doi.org/10.1016/j.cose.2024.103842
8. Salih, A.M., et al.: A perspective on explainable artificial intelligence methods: SHAP and LIME. Adv. Intell. Syst. **7**(1), 2400304 (2024). https://doi.org/10.1002/aisy.202400304
9. Roshinta, T.A., Gabor, S.: A comparative study of LIME and SHAP for enhancing trustworthiness and efficiency in explainable AI systems. In: Conference on Computing, Kuala Lumpur, Malaysia, pp. 134–139 (2024). https://doi.org/10.1109/ICOCO62848.2024.10928183
10. Abusnaina, A., et al.: One step forward, two steps back: ML-based malware detection under concept drift. Computing. **107**, 207 (2025). https://doi.org/10.1007/s00607-025-01543-7
11. Garcia, D.E., DeCastro-Garcia, N., Munoz Castaneda, A.L.: An effectiveness analysis of transfer learning for the concept drift problem in malware detection. Expert Syst. Appl. **212**, 118724 (2023). https://doi.org/10.1016/j.eswa.2022.118724
12. Luo, X., Liu, C., Gou, G., Xiong, G., Li, Z., Fang, B.: Identifying malicious traffic under concept drift based on intraclass consistency enhanced variational autoencoder. SCIENCE CHINA Inf. Sci. **67**, 182302 (2024). https://doi.org/10.1007/s11432-023-4010-4
13. Ali, E., Batool, N., Rizwan, M., Sarwar, S.: Assessing concept drift in malware: a comprehensive review and analysis. In: 21st Bhurban Conference on Applied Sciences and Technology, Murree, Pakistan, pp. 564–569 (2024). https://doi.org/10.1109/IBCAST61650.2024.10876901
14. Barzev, I., Borissova, D.: An improved static analysis approach for malware detection by optimizing feature extraction combining different ML algorithms. In: Communications in Computer and Information Science, vol. 2304, pp. 102–115 (2025). https://doi.org/10.1007/978-3-031-82153-0_8
15. Imani, M., Beikmohammadi, A., Arabnia, H.R.: Comprehensive analysis of random forest and XGBoost performance with SMOTE, ADASYN, and GNUS under varying imbalance levels. Technologies. **13**, 88 (2025). https://doi.org/10.3390/technologies13030088

Disinformation in Peruvian Political Discourse in Digital Media

Hugo Gonzalez Aguilar(✉)

Universidad Autónoma del Perú, Lima, Peru
hugo_augencio@hotmail.com

Abstract. This paper analyzes the discourse on disinformation in Peruvian political discourse by elected officials. It analyzes their candidacy and then their performance of office. Specifically, it analyzes the discourse of the current mayor and the president. An analysis was conducted of news disseminated in various digital media outlets in Lima. The results were systematized through the following categories: the fight against insecurity, hunger, pregnancy among girls and adolescents, justification of deaths in the context of protests, the cost of food or family meals, economic growth, and the intentions of the authorities' resignation before the Inter-American Court of Human Rights. It is concluded that authorities disseminate false information that serves their interests or their acceptance, but that is contrary to the facts or reality.

Keywords: Disinformation · deception · democracy · truth · populism

1 Introduction

Disinformation is a phenomenon that has grown in recent years and has been exploited by populist movements to undermine or weaken democracy. Its purpose is to manipulate the population through deception; however, in some cases, people reject this information disseminated through various media outlets. It can also generate hate speech and polarization [1]. It also influences the population through the creation of false rumors and fuels political polarization [2].

Disinformation is disseminated through social media and is characterized by the provision of false information disseminated by political leaders or parties. Therefore, it undermines democracy and citizens. It is contradictory and untrue information [3].

A series of strategies have been promoted to counter it, one of which is media literacy to promote truth or veracity; as well as the dissemination of programs through the media to promote the dissemination of objective information and prevent fake news [3].

In the international context, such as the United Kingdom and the United States during the 2016 political events, and in Spain, the growth of political disinformation has become evident, becoming a social problem due to its massive dissemination and its disruption of democratic processes [4].

Some research has shown that certain governing leaders are the ones who promote disinformation campaigns that directly affect the development of democracy and the

A. Rocha et al. (Eds.): ICITS 2026, LNNS 1966, pp. 552–558, 2027.
https://doi.org/10.1007/978-3-032-27538-7_48

population. This generates a counterattack from opponents through massive campaigns to clean up their image or reputation [4].

In other contexts, research has also been conducted on the topics promoted through this disinformation system. On the one hand, there are left-wing representatives who promote information related to corruption, environmental protection, and the protection of certain (civil) rights; on the other hand, there are right-wing representatives who allude to issues of public interest, migration or immigration, as well as economic issues [4].

Disinformation in the political sphere is related to paltering, which is associated with lies spread with the aim of influencing a certain group. In this case, several types of lies can be identified: those committed by commission or intention, and those arising from omission [5]. In this context, the most prevalent is the former, as candidates or authorities have the intention or plan to spread false information to suit their interests.

The purpose of disinformation, spearheaded by leaders, is to persuade and manipulate the population [6]. They disseminate data or information that is unreal or contrary to reality or the truth. Therefore, there is a loss of confidence among institutions or the population against the attitude of politicians [7] or leaders who generate or disseminate false information.

The agents who spread false information are diverse; among these are individual authors, some organizations, and even governments that disseminate misleading information for political purposes. This phenomenon generates polarization and affects democracy [8]. Added to this are the media, which in some cases are the promoters of the spread of disinformation. They spread sensationalist or biased information [8].

Disinformation becomes one of the threats that affects fundamental rights, such as freedom of expression, and also democracy in different countries [9]. New communication channels such as social media are being used, excluding or displacing traditional media. Its dissemination is rapid, massive, and has a broad reach. It also affects the population, which feels vulnerable to the diverse and massive amount of information; it occurs as a form of manipulation [9]. In this context, disinformation aims to obtain political benefits, hence there is a direct relationship between political propaganda and disinformation [10].

On the other hand, it has been proposed that, through universities, students confront or combat disinformation [10], since misleading information is increasingly on the rise. On the contrary, relevant and truthful information is hidden or not disseminated [11].

2 Method

A qualitative study was conducted based on the analysis of content disseminated in various digital media outlets in Lima. The cited discourse corresponds to the same authorities or candidates and allows us to identify their intentions and, above all, whether it is consistent with reality or the truth. In this sense, the analysis allows us to systematize the information through the following categories: the fight against insecurity, the fight against hunger, pregnancy among girls and adolescents, justification of deaths in the context of the protests, the cost of food or the family menu, economic growth, and the intentions of the resignation before the Inter-American Court of Human Rights.

3 Results

The relationships between authorities may have some similarities in some cases, but in others, there may be discrepancies. The speeches have been organized into categories.

It is important to consider the promises and the fulfillment made by the mayor of Lima. In this context, the results have been organized into the following categories:

3.1 Category 1: Fight Against Insecurity

During his campaign, the mayoral candidate for Lima promised to combat crime. To this end, he offered to purchase 10,000 motorcycles. However, he has only purchased approximately 4,000 motorcycles [12], which implies that he broke his campaign promise; therefore, this information is false or unfulfilled. However, other media outlets report that at the beginning of his administration, he promised 4,000 motorcycles, and in this sense, he is fulfilling his promise [13]. However, in his speech as a candidate, he emphasized that he would deliver approximately 10,000 motorcycles to combat crime.

On the other hand, the mayor had proposed the "Plan Maceta" (Plant Pot) to scare away criminals from Lima. In this regard, he has considered that flower pots provide beauty and have a psychological impact that scares off criminals [14].

3.2 Category 2: Fight Against Hunger

In some cases, the future mayor's proposals present ideas that are not feasible for implementation or execution. For example, the candidate had proposed the "Hunger Button," in which a group of young people identify families in need and deliver aid or food, thereby addressing the problems of hunger or lack of food [15].

3.3 Category 3: On Pregnancy in Girls and Adolescents

Regarding pregnancy among girls, he argues that they cease to be pregnant when they become pregnant, but the State denies this claim, as they are still children [16].

In another case, he proposed that pregnant teenagers should stay or carry out their pregnancy in a five-star hotel where they feel comfortable; they are then asked whether they want to donate the money or not [17].

Regarding the discourse of the president and her ministers, the following categories have been identified:

3.4 Category 4: Justification of the Deaths in the Context of the Protests

They claim that those who have died are terrorists, when there is evidence (photos, videos, or testimonies) [18] showing that the murders were carried out by military personnel, against people who were present at the time of the events and who were not participating in the vandalism. These murders by the military or police have been ordered by high-ranking officers who also follow orders from the president, as she is the supreme commander of these security forces. In this case, some representatives of international organizations have witnessed or witnessed the stigmatization of certain groups of the population, who have been labeled as terrorists or Shining Path members, especially those of Aymara or Quechua origin [19].

In other cases, the president has mentioned that the conflicts or protests have been led or intervened by the Red Ponchos. They have used dumdum weapons [20]. This is a justification or statement that is consistent with reality or the facts.

3.5 Category 5: The Cost of Food or the Family Menu

The president has falsely claimed that Peruvian families can eat or prepare a family meal for 10 soles [21]. She has explained that the menu can include soup, a main course, and even a dessert, a situation far removed from reality, since the prices of basic necessities have risen considerably, and 10 soles alone are insufficient for the preparation of a family meal. This is false information, which is consistent with the reality of prices in Peruvian markets.

3.6 Category 6: Economic Growth

The president has stated that the economy has grown and directly benefited the population, or that the population has more money in their pockets directly [22]; however, in reality this figure is not accurate, since families are unemployed, underemployed, or have frozen wages. On the other hand, it has been shown that there is a permanent impact on the economy through public spending, which is higher than last year.

3.7 Category 7: Intentions of Resignation Before the Inter-American Court of Human Rights

This problem arises because the IACHR has ordered the Peruvian State to stop enforcing or repeal the amnesty law that frees or exonerates military personnel who have committed murders, since the central argument is that they fought against terrorists. This claim has been proven false, as many deaths have been of innocent people, a specific case in point being the murders of students and a university professor at the National University of Education (La Cantuta) in Lima. In response to the Court's position, the president has stated that Peru is not a colony [23].

She also gave a speech before the UN in which she defended the amnesty law and asserted that her mandate is the most stable, despite ongoing protests against her [24].

4 Discussion

Based on the results or analysis of news disseminated in digital media, it is evident that authorities (first as candidates and then as elected officials) disseminate information that serves their interests to increase their popularity or acceptance. In this case, it is evident that the information they disseminate lacks veracity [3]; likewise, it is identified that the lies they spread are clearly intended to improve public perception [5].

Regarding the fight against citizen insecurity, it is evident that the number of motorcycles offered by the municipal authorities has not been fulfilled, as has the flowerpot plan that deters criminals. This information becomes false because it is consistent with the facts or reality; in this specific case, the number of motorcycles has not been purchased; nor is there scientific evidence that flowerpots deter crime. Therefore, this is a narrative that lacks reality or truth [3].

Likewise, on the topic of food, the fight against hunger, and economic growth (which both authorities, the mayor and the president, have addressed), there is also evidence of untruthful facts. For example, in the case of the hunger button, there is no evidence that the youth directed by the mayor's office are bringing food to poor or vulnerable families in various sectors or populous areas of Lima. Regarding the cost of the menu, which the president stated was made with 10 soles, the truth is that this amount is insufficient; therefore, it is untrue. This also shows that economic growth does not benefit the neediest population, which is living in poverty or extreme poverty, or is considered to be within multidimensional poverty, as considered by the National Institute of Statistics and Informatics of Peru [25], that is, those who suffer from various deficiencies.

Another relevant aspect is the death of several citizens due to protests against the president's resignation and calls for new elections. Likewise, Peru's withdrawal from the Inter-American Court of Human Rights. These types of acts, which directly affect the population, generate hate speech and polarization [1] that divide the population and affect coexistence in democratic spaces. This is a clear example of misinformation about the problem regarding deaths in the context of the protests, since the government is primarily responsible, as it is the supreme commander of the security forces that, as has been shown in various reports (including those of the American Commission) [26], have been the cause of the deaths.

In this context, some authorities of the current Peruvian government, given that the Inter-American Court of Human Rights has ruled that Peru should not apply the amnesty law in the cases of La Cantuta and Barrios Altos [27], have stated that Peru should withdraw from the Inter-American Court. The president herself has maintained that Peru is not a colony of anyone [23]. Consequently, it is evident that, in the various statements made by authorities or their representatives, they have disseminated false information or misinformed the population, disregarding their interests, such as security and the protection of their fundamental rights. On the contrary, this disinformation has undermined democracy and increased polarization among the population.

5 Conclusions

After analyzing the documents, it is concluded that the current Peruvian authorities have disseminated narratives that are untrue or misleading regarding key issues such as citizen insecurity. This is evidenced by some unfulfilled promises made since they first ran for office or launched their candidacy (such as the case of the mayor of Lima).

On the other hand, there is also false information regarding the economic situation or the fight against hunger. Specifically, information has been disseminated that Peruvian families can prepare a meal with 10 soles, a claim that is consistent with reality, as families are spending more on food. They also indicate that growth has benefited the population; however, in reality, many families continue to suffer from hunger or live in conditions of (multidimensional) poverty. It has also been shown that the municipal authorities are not addressing or implementing their "Hunger Button" strategy, as they are failing to assist people in low-income areas who often suffer from hunger.

Likewise, false information has been spread regarding the deaths during the protests against the president. Rather, it has been shown that the State is primarily responsible for the deaths of several dozen Peruvian citizens. Rather, the current government has defended the application of the amnesty law, which has a clear purpose: to free those responsible for the deaths in the La Cantuta Cases, Barrios Altos. Regarding this case, information has been disseminated that is inconsistent with reality, claiming that the country is not a colony of anyone.

In this regard, it is recommended that directors and administrators of the education system, from primary to university level, promote awareness campaigns about the misinformation circulating in various media and networks. Likewise, political leaders and officials must be mindful of and responsible for disseminating accurate, not false, information.

In short, the authorities spread false information that harms the population and democracy and that ultimately serves their own interests.

References

1. Roca, N., Fernández, D.: TikTok: politics for young people. What are Trump and Milei talking about? Polarization, hate, and disinformation in electoral campaigns. Stud. Journalistic Message. **31**(3), 739–751 (2025). https://doi.org/10.5209/emp.102868
2. Berdón, P., Reguero, I.: Disinformation in times of political polarization: the debate in X on Pedro Sánchez's five days of reflection. Electron. Yearb. Stud. Soc. Commun. Diss. **18**(2) (2025). https://doi.org/10.12804/revistas.urosario.edu.co/disertaciones/a.14785
3. Rando, D., De las Heras-Pedrosa, C., Paniagua-Rojano, F.: Analysis of political disinformation in the speeches of Spanish government leaders via X. Lat. Am. J. Commun. **83**, 1–24 (2024). https://doi.org/10.4185/rlcs-2025-2308
4. Fenoll, V., Gamir-Ríos, J., Alonso-del-Barrio, E.: Politics, ideology, populism and disinformation. The verification of the political communication of Spanish parties. Stud. Journalistic Message. **30**(4), 843–856 (2024). https://doi.org/10.5209/emp.96629
5. Pérez Curiel, C., De Benito, C., Pineda, A.: Politics, deception, and propaganda: paltering as a tool for disinformation in the media. Estudios sobre el Mensaje Periodístico. **30**(2), 295–309 (2024). https://doi.org/10.5209/esmp.93703
6. Diez-Gracia, A., Sánchez-García, P., Martín-Román, J.: Disintermediation and disinformation as a political strategy: use of AI to analyze fake news as Trump's rhetorical resource on twitter. Profesional De La Información. **32**, 5 (2023). https://doi.org/10.3145/epi.2023.sep.23
7. Pérez-Curiel, C., Jiménez-Marín, G., Domínguez-García, R.: Twitter as a tool for communication and disinformation. An analysis of information transparency and political corruption in Spain. Política y Sociedad. **59**(3) (2022). https://doi.org/10.5209/poso.75666
8. Teruel-Rodriguez, L.: Increasing political polarization with disinformation: a comparative analysis of the European quality press. Profesional de la información. **32**, 6 (2023). https://doi.org/10.3145/epi.2023.nov.12
9. López-Martín, Á., Gómez-Calderón, B., Córdoba-Cabús, A.: Disinformation on the rise: an analysis of fake news about Spanish politics. VISUAL REVIEW Int. J. Vis. Cult. **14**(1), 1–12 (2023). https://doi.org/10.37467/revvisual.v10.4596

10. Rebollo-Bueno, S., Ferreira, I.: Disinformation and polarization in far-right political advertising in Spain and Portugal. Stud. Commun. **36**, 529–557 (2023) https://ojs.labcom-ifp.ubi.pt/ec/article/view/1084
11. Pérez-Escolar, M., Varona Aramburu, D.: Disinformation center: a didactic proposal for verifying political disinformation in the classroom. Anàlisi. **66**, 131–151 (2022). https://doi.org/10.5565/rev/analisi.3482
12. La República: https://n9.cl/u1h9x. Last accessed 10 Sept 2025
13. Masks: https://n9.cl/v06u9. Last accessed 11 Sept 2025
14. Infobae: https://n9.cl/uz8b13. Last accessed 15 Sept 2025
15. López, R.: https://n9.cl/00hy7o. Last accessed 12 Sept 2025
16. Convoca: https://n9.cl/m520j2. Last accessed 24 Sept 2025
17. Expreso: https://n9.cl/xyn8x. Last accessed 25 Sept 2025
18. The Republic: https://n9.cl/vkc1u. Last accessed 26 Sept 2025
19. La República: https://n9.cl/20u9lf. Last accessed 12 Sept 2025
20. La República: https://n9.cl/nsly9f. Last accessed 9 Sept 2025
21. Perú21: https://n9.cl/r3uck. Last accessed 7 Sept 2025
22. RPP: https://n9.cl/ct245r. Last accessed 6 Sept 2025
23. El Comercio: https://n9.cl/8i8oe. Last accessed 2 Sept 2025
24. La República: https://n9.cl/ryva4. Last accessed 1 Sept 2025
25. National Institute of Statistics and Informatics.: https://n9.cl/9gbhc. (2025)
26. Inter-American Commission on Human Rights: Human Rights Situation in Peru in the Context of Social Protests. https://n9.cl/n5hu4y. (2023)
27. La República: https://n9.cl/o988ru. 22 Sept 2025

Sonification of Centrality Metrics in Layered Graphs

José Luis Gómez Ramos(✉), Arturo Corona Ferreira, Carlos Arturo Custodio Izquierdo, Carlos González Zacarías, María Alejandrina Almeida Aguilar, and Eduardo Cruces Gutiérrez

Universidad Juárez Autónoma de Tabasco, Villahermosa 86019, Tabasco, Mexico
jose.gomez@ujat.mx

Abstract. This paper presents a framework for the sonification of centrality metrics in layered directed graphs. The proposed approach computes structural metrics such as degree, closeness, betweenness, and PageRank; normalizes them within [0,1]; and maps them to auditory parameters (frequency, duration, and timbre) through a modular pipeline. A prototype was implemented that generates MIDI sequences quantized in a pentatonic scale, applied to a synthetic directed graph of 77 nodes distributed across four layers. Results show that the structural properties of the network can be perceived auditorily: nodes with high PageRank have higher pitch, those with high betweenness produce longer notes, and layers are distinguished by timbre. These findings suggest that sonification can complement visualization in network analysis, with potential applications in accessibility and education.

Keywords: selective merge · data integration · bibliometric graphs · canonical keys · data quality · multisource citation networks

1 Introduction and Justification

The analysis of complex networks has traditionally been based on visualization, but this has limitations in dense graphs or in contexts where accessibility is required. Sonification, understood as the transformation of data into non-verbal sound, offers an alternative perceptual channel to represent structural information. This study proposes a multimodal approach that translates centrality metrics into auditory parameters, allowing topological properties to be explored through hearing.

The main objective is to demonstrate that centrality metrics—degree, Closeness, betweenness and PageRank—can be mapped in a perceptually coherent way to sound parameters, enabling a complementary auditory exploration of the graphs.

2 Theoretical Framework and Related Works

Sonification research has evolved from foundational studies such as Kramer (1994), who defined auditory display as a perceptual translation of data, to the systematic framework provided in The Sonification Handbook (Hermann et al., 2011). These works established the conceptual foundations for transforming quantitative information into sound.

A. Rocha et al. (Eds.): ICITS 2026, LNNS 1966, pp. 559–568, 2027.
https://doi.org/10.1007/978-3-032-27538-7_49

Dubus and Bresin (2013) conducted a comprehensive review of mapping strategies between data and auditory dimensions, identifying frequency and duration as the most perceptually effective parameters. Nees and Walker (2011) later introduced the notion of auditory graphs, where structural relationships are conveyed through pitch variations rather than visual edges.

More recent initiatives—such as CosMonic (García-Benito, 2022), STRAUSS (Trayford et al., 2025), and the bibliometric sonification experiments by Bornmann (2024, 2025)—demonstrate the increasing diversity of sonification contexts, spanning astrophysics, data visualization, and scientometrics. However, these projects remain largely static or domain-specific, with limited emphasis on structural metrics or interactive auditory exploration.

The present study extends this theoretical lineage by introducing a metric-based and topology-sensitive model. Unlike prior work, it directly links graph-theoretical measures (degree, betweenness, closeness, PageRank) with acoustic attributes (pitch, duration, timbre) in a real-time interactive environment. This approach aligns with multimodal and accessibility paradigms, enabling both analytical and educational applications.

Table 1 summarizes the conceptual and methodological distinctions between previous studies and the current proposal, highlighting its contribution to the integration of auditory and visual modalities in network analysis.

Table 1. Comparison between previous work and the current proposal

Aspect	Previous Studies	Current proposal
Data Domain	Network Traffic, Sensors, XY Graphs	Layered Graphs with structural metrics
Interactivity	Static or streaming	Visual-auditory interaction, visible subgraph
Mapping	Linear pitch without scale	Logarithmic pitch, pentatonic scale, timbre per layer
Purpose	Accessibility, monitoring	Network analysis, education, art-science

3 Methodology: Sonification Algorithm

The proposed pipeline consists of five stages:

(i) calculation of centralities,
(ii) normalization
(iii) mapping metrics→auditory parameters,
(iv) temporal scheduling using a circular queue and
(v) MIDI/WAV rendering.

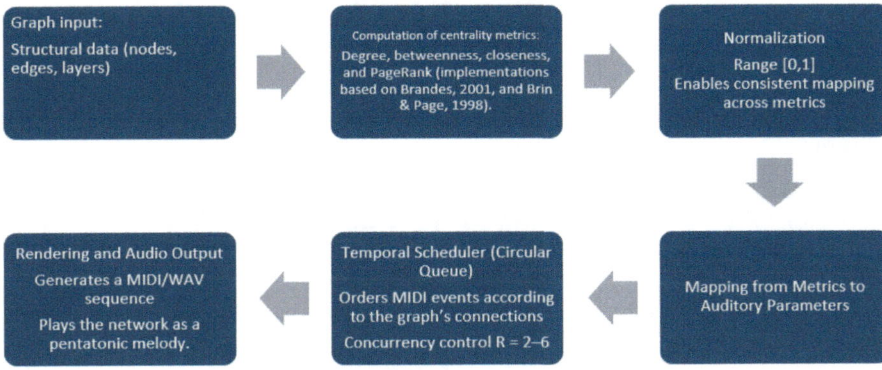

Fig. 1. Sonification pipeline of centrality metrics.

In the developed system, sonification extends beyond a conceptual model, but is implemented in an integral way within the architecture of a Dashboard, based on the TDA-Graph (Graph Abstract Data Type) of its own construction (Fig. 1).

Each graph vertex contains a set of structural attributes, among which the dynamic value field stands out, which acts as a bridge between the centrality metrics and the sound layer.

3.1 Calculation and Assignment of Centrality Metrics

To represent structural relevance in auditory form, four classical centrality measures were selected for their complementary interpretations of network importance. Each metric provides a distinct perceptual metaphor that can be translated into sound intensity, pitch, or duration:

- Degree centrality expresses local connectivity — vertices with many links can be perceived as high-activity nodes generating denser sound patterns.
- Betweenness centrality captures vertices acting as bridges in information flow; perceptually, they correspond to notes with longer durations, representing paths of greater communicative importance.
- Closeness centrality reflects the global accessibility of a node; these tend to produce clearer or more immediate tonal responses, indicating their proximity to the network core.
- PageRank introduces a probabilistic model of influence, mapping hierarchical importance to pitch height — nodes with higher PageRank yield brighter, higher-pitched tones.

These metrics were computed on a graph G = (V,E) implemented through the custom TDA-Graph structure. Classical algorithms (Freeman, 1978/79; Brandes, 2001; Brin & Page, 1998) were adapted to the data type, maintaining compatibility with the Dashboard environment:

Mathematically, the centrality functions are expressed as follows:

a) DegreeCentrality: Number of incident edges at each vertex.

$$C_D = \frac{\deg(v)}{|V| - 1} \tag{1}$$

b) Betweenness Centrality: Measures the proportion of shortest paths passing through a vertex:

$$C_B(v) = \sum_{s \neq v \neq t} \frac{\sigma_{st}(v)}{\sigma_{st}} \tag{2}$$

Where σ_{St} is the shortest number of paths between s and t and $\sigma_{St}(v)$ is the number of those roads that pass through v.

c) Centrality of closeness (Clísennos Centrality): inverse of the sum of the geodesic distances from the vertex to all the others:

$$C_C(v) = \frac{|V| - 1}{\sum_{t \neq v} d(v, t)} \tag{3}$$

d) PageRank centrality: Estimates the influence of a node based on the probability of visit in a random walker model:

$$PR(v) = \frac{1 - d}{|V|} + d \sum_{u \in N(v)} \frac{PR(u)}{\deg(u)} \tag{4}$$

Where d is the damping factor (usually 0.85).

From these metrics, the dynamic value of the vertex is generated by classical algorithms (Freeman, 1978/79; Brandes, 2001; Brin & Page, 1998), adapted to TDA: degree (v), betweenness (v), closeness (v), PageRank (v)

The user selects the active metric in the interface; Once calculated, the result of each vertex is assigned to its dynamic attribute:

$$\text{vertex.value} = \text{valor\_centralidad} \tag{5}$$

This field acts as the bridge between the analytical and auditory layers, enabling the same graph infrastructure to be reused across different metrics without modifying its topology.

The selection of these four measures ensures a balanced auditory representation, where local and global network structures can be distinguished through sound properties such as frequency, duration, and timbre. This correspondence forms the foundation for the subsequent sonification stages described in Sect. 3.2.

3.2 Sound Mapping and Vertex Visibility

The sound mapping strategy was designed under a perceptual–analytical principle: each auditory parameter corresponds to a specific structural dimension of the graph. Thus, pitch represents hierarchical prominence, duration encodes temporal salience, and timbre distinguishes the semantic layer. These mappings preserve analytical meaning while supporting clear and intuitive interpretation across different user profiles.

During sonification, the system traverses the visible vertices and converts their value into auditory parameters according to a family of mapping functions.

Thus, the same graph can "sound" differently depending on the active metric (e.g., PageRank or betweenness) and the subset of nodes visible in the interface. This approach turns the graph into an interactive score that reflects the analytical structure of the system in real time.

The equations define the direct relationship between the structural value of the vertex and its sound properties, obtained from the centrality function. This value uses a linear normalization between the minimum and maximum absolute or relative (visible or non-visible) value of the graph:

$$m(v) = \frac{\text{valor}(v) - \min(\text{valor})}{\max(\text{valor}) - \min(\text{valor})} \tag{6}$$

Based on this standardization, the system applies:

a) Frequency:

$$f(v) = \text{escala}[\lfloor m(v) * N \rfloor] \tag{7}$$

b) Linear duration adjusted to vertex value

$$D(v) = D_{\min} + m(v)\,(D_{\max} - D_{\min}) \tag{8}$$

where $f(v)$ is the selected frequency of the scale (C-D-E-F-G-A-B-C), and $D(v)$ is the linear duration in milliseconds. In this way, vertices with high values produce higher and longer notes, while those with low values generate lower and shorter tones.

These functions are applied in real time to the set of visible vertices, generating a flow of notes whose frequency and duration reflect the active metric. The control parameters used in the implementation are shown in Table 2.

Table 2. Typical control parameters.

Parameter	Typical Value	Description
f_min, f_max	261–523 Hz	Tonal range (C4–C5)
D_min, D_max	120–700 ms	Linear minimum and maximum duration
Δt	60–100 ms	Separation between events
Γ	0.6–1.0	Perceptual curvature

Only visible vertices on the interface generate sound events. Hidden nodes (whether by filter, layer, or user manipulation) remain silent. This feature gives the graph an interactive auditory dimension, where the user "hears" only the visible subgraph, modulating the sound experience according to the context explored.

3.3 Adaptive Sonification

The relationship between metric and sound is not fixed. The same graph can produce different auditory outcomes depending on the selected centrality measure. Each change in metric triggers a recalibration of the mapping functions and re-rendering of the corresponding auditory events.

In this way, the graph behaves as a dynamic score, where each vertex acts as an instrument that modifies its acoustic behavior according to the active metric and its visibility status. This adaptive behavior allows users to perceive topological variations in real time, identifying structural shifts that might remain unnoticed in static visualizations.

The adaptive engine ensures perceptual consistency across metrics by maintaining normalized value ranges and proportional timing, even when the density or scale of the graph varies. This feature supports comparative auditory analysis — for instance, contrasting PageRank and Betweenness patterns within the same dataset — without losing the relative meaning of pitch or duration. Future extensions will incorporate machine learning components to optimize sound mappings automatically according to graph complexity and perceptual feedback.

4 Preliminary Results

A prototype was implemented in Python and applied to a directed graph comprising 77 vertices, 366 edges, and four layers, as shown in Fig. 2. This structure represents the base model used for the sonification tests.

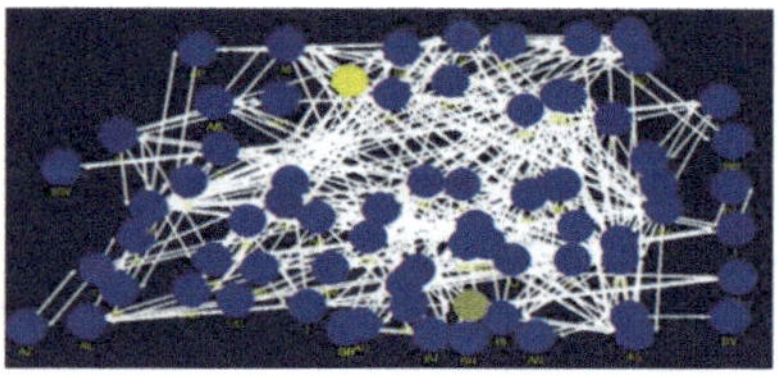

Fig. 2. Directed graph with 77 vertices, 366 edges and 4 layers.

The computed PageRank, Betweenness, and Closeness centrality values exhibited a direct correspondence between structural relevance and sound prominence. Figure 3 shows the same graph under each centrality measure, where the most relevant nodes (darker or larger) correspond to the vertices producing perceptually stronger tone.

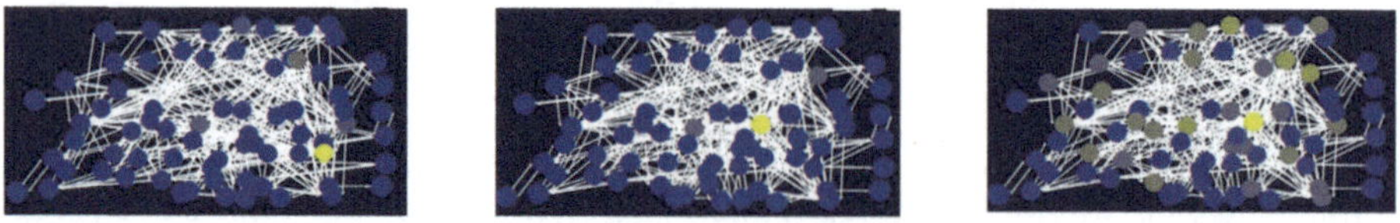

a) b) c)

Fig. 3. Visualization of the graph by applying centrality measures. a) PageRank, b) Betweenness, c) Closeness

Vertices with lower centrality (≈261 Hz) generated short, staccato-like tones, while those with higher values (≈523 Hz) produced sustained notes of up to ≈700 ms. The resulting sequences followed an ascending–descending cyclical pattern, consistent with the circular topology of the graph and the path of the sonification queue. These relationships are summarized in Table 3, which compares the sonic patterns derived from each metric.

Table 3. Comparison of sound patterns derived from centrality metrics (graph in Fig. 2).

	PageRank			Betweenness			Closeness		
Event	Approx.	Frequency Hz.	Duration Ms.	Approx.	Frequency Hz.	Duration Ms.	Approx.	Frequency Hz.	Duration Ms.
1	MI4	329	396	FA4	349	415	MI4	329	344
2	RE4	293	313	MI4	329	353	DO4	261	183
3	DO4	261	226	DO4	261	190	MI4	329	349
4	DO5	523	700	DO5	523	700	DO4	261	120
5	RE4	293	260	DO4	261	264	DO4	261	120
6	DO4	261	120	DO4	261	120	MI4	329	349
7	RE4	293	260	DO4	261	204	SOL4	392	497
8	RE4	293	305	DO4	261	174	FA4	349	427
9	RE4	293	296	DO4	261	192	DO5	523	700
10	FA4	349	433	DO4	261	224	SOL4	392	494
...	(pattern repeats 3 times)								

The sonic structures listed in Table 3 correspond closely to the piano-roll representations shown in Figs. 4, 5 and 6. In these visualizations, the vertical axis represents pitch and the horizontal axis time. Each block corresponds to a vertex event, showing how the active metric modulates tonal distribution and temporal density.

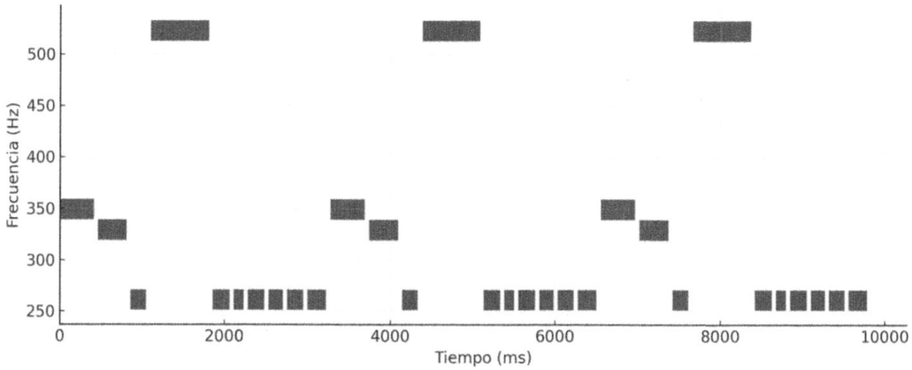

Fig. 4. Piano-roll generated from the value attribute (PageRank)

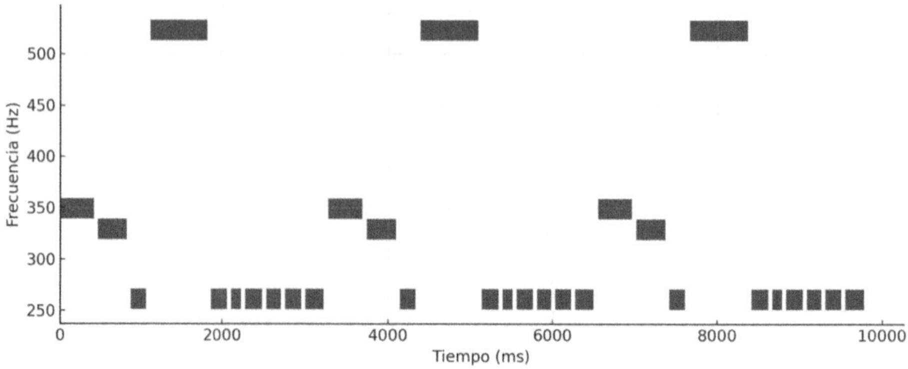

Fig. 5. Piano-roll corresponding to the sonification of Betweenness

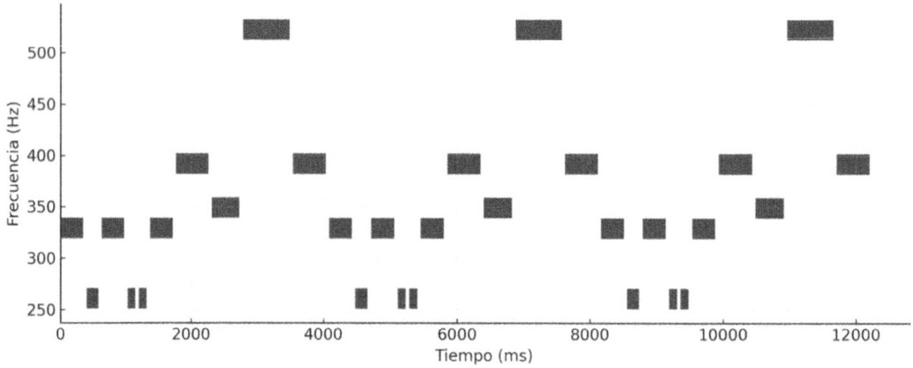

Fig. 6. Piano-roll corresponding to the sonification of Closeness

To validate the coherence of the mapping between structural and auditory dimensions, linear correlations were calculated between each metric value and the derived sound parameters (frequency and duration). As presented in Table 4, all correlation coefficients exceeded 0.88, confirming that the mapping translates network structure consistently into the acoustic domain.

Table 4. Correlation between centrality metrics and frequency and duration parameters.

Metric	Average value	Average Frequency (Hz)	Average duration (ms)	Value-frequency correlation
PageRank	0.72	384.2	432	r = 0.94
Betweenness	0.65	328.5	366	r = 0.89
Closeness	0.68	351.3	397	r = 0.92

These results demonstrate that structural prominence within the graph is audibly perceivable through pitch and temporal duration, validating the sonification model as a reliable complement to visual analysis. The strong correlations further suggest that the system preserves analytical meaning across modalities, laying the foundation for the discussion presented in the following section.

5 Discussion and Conclusions

Discussion. Unlike static frameworks such as STRAUSS or ICAD, the proposed approach implements real-time, metric- and visibility-dependent sonification (Trayford et al., 2025; Lenzi et al., 2023; Nees, 2018; Barrass & Vickers, 2019; Vickers, 2016). The results confirm that auditory mappings based on centrality metrics can effectively convey structural hierarchies, offering a perceptual complement to traditional visualization. The model integrates analytical accuracy with real-time interaction, aligning with multimodal and accessibility-oriented paradigms.

Conclusions. The sonification of centrality metrics provides a consistent and interpretable auditory representation of graph structure. The correspondence between pitch, duration, and network prominence demonstrates that sonification can serve as an analytical as well as communicative tool. This work therefore contributes a reproducible framework for translating quantitative graph metrics into sound, extending the scope of auditory analytics within network science.

Limitations and Future Work. Current limitations include the computational scalability of betweenness calculation, auditory saturation in dense networks, and perceptual variability among listeners. Future research will include a systematic perceptual evaluation contrasting Audio vs. Visual vs. Multimodal conditions (Brown et al., 2022; Martin et al., 2024; Lindborg et al., 2023), and the application of the approach in accessibility and educational contexts (Walker & Mauney, 2010; Bearman, 2020). The integration of adaptive machine-learning modules capable of dynamically adjusting sound mappings according to graph density and topology is also contemplated

References

Barrass, S., Vickers, P.: Sonification and Sound Design: Theory and practice. Springer (2019)

Bearman, N.: Data Sonification for Engaging with Environmental Data. Environ. Model Softw. **31**, 1 (2020)

Bornmann, L.: The Sound of Science: Data sonification, vol. 25, pp. 3743–3747. PMC (2024)

Bornmann, L.: Metrics sonification as a new way to convey bibliometric data. Scientometrics. **130**, 3089–3108 (2025)

Brandes, U.: A faster algorithm for betweenness centrality. J. Math. Sociol. **25**, 2 (2001)

Brin, S., Page, L.: The anatomy of a large-scale hypertextual web search engine. Comput. Netw. **30**, 107–117 (1998)

Brown, A., Harrison, J., Zanella, C.: Evaluating the efficacy of sonification for signal detection. arXiv. (2022)

Dubus, G., Bresin, R.: A systematic review of mapping strategies for the sonification of data. J. Sonic Stud. **8**, e82491 (2013)

Freeman, L.: Centrality in Social Networks, vol. 1, pp. 215–239. Social Networks (1978/79)
García-Benito, J., Pérez-Montero, E.: CosMonic sonification project. https://arxiv.org/abs/2205.12984. (2022)
Hermann, T., Hunt, A., Neuhoff, J.: The Sonification Handbook. Logos Verlag (2011)
Kramer, G.: Auditory Display: Sonification, audification and auditory interfaces. Addison-Wesley (1994)
Lenzi, S., et al.: Datascapes: toward a framework for sonification of complex systems. Front. Comput. Sci. (2023)
Lindborg, P., Lenzi, S., Chen, F.: Climate data sonification and visualization. Front. Comput. Sci. **13**, 2022 (2023)
Martin, D., et al.: A novel sonification method for signal representation. ScienceDirect. **300**, 110852 (2024)
Nees, M.: Auditory Graphs Are Not the Killer App of Sonification. ICAD (2018)
Nees, M., Walker, B.N.: Auditory graphs. In: The Sonification Handbook (2011)
Trayford, J., et al.: STRAUSS: Sonification Tools in Python. JOSS (2025)
Vickers, P.: Sonification Design and Aesthetics. ACM Comput. Surv. (2016)
Walker, B.N., Mauney, L.: Individual differences, cognitive load, and sonification. ACM Trans. Appl. Percept. (2010)

Big Data Analytics and Unsupervised Learning for Digital Payments Adoption in Ecuador.

Miguel Angel Peñarreta, Andres Ulloa(✉), and Reinaldo Herrera(✉)

Universidad Técnica Particular de Loja, Loja 110150, Ecuador
{mapenarreta,aeulloa2,ahreinaldo}@utpl.edu.ec

Abstract. This article analyzes transactional differences and business segmentation in the adoption of a digital payments platform in Ecuador. Using data from more than 10000 affiliate businesses and 55000 individual users of the platform for the period 2023–2024, we integrated transactional variables with geospatial data to capture heterogeneous patterns of digital payment usage. Non-parametric statistical tests (Kruskal-Wallis with Dunn-Benjamini-Hochberg adjustments) revealed significant differences in the frequency and value of transactions by age groups and business categories, but no significant effects by gender. The spatial analysis showed concentration of activity in the southern region of the Ecuador. To identify latent adoption profiles, we implemented unsupervised clustering (K-means, hierarchical, DBSCAN, and GMM), with K-means providing the best validity metrics. Four distinct business segments emerge: moderate-activity basic business, high-value business, low-adoption marginal business, and emerging business. The study demonstrates how combining big data analytics with geospatial and unsupervised learning techniques facilitates the design of strategies for the expansion of digital financial services.

Keywords: Big data analytics · digital payments adoption · cluster analyses

1 Introduction

Digital payments have emerged as one of the main catalysts for financial inclusion and the transformation of business transactions in Latin America [1, 2]. Various studies have documented its rapid penetration into consumer staples sectors and its potential to reduce operating costs, expand access to credit, and boost local economies [3–5]. Likewise, recent literature highlights the role of data analytics and segmentation techniques in understanding technology adoption in digital financial ecosystems [6]. However, significant knowledge gaps persist about the heterogeneity in the adoption and use of digital payment platforms at the territorial and sectoral level [3, 6, 7]. Few studies simultaneously integrate massive transactional data, geospatial variables and advanced clustering algorithms to identify differentiated patterns of behavior between business and individual users [5, 8, 9]. This gap limits the ability of financial actors and policymakers to design targeted interventions that accelerate digital inclusion.

To address this gap, this study analyzes transactional differences and segments the businesses affiliated with the Ahorita platform in Ecuador, developed by a regional

A. Rocha et al. (Eds.): ICITS 2026, LNNS 1966, pp. 569–581, 2027.
https://doi.org/10.1007/978-3-032-27538-7_50

bank. Non-parametric statistical techniques are combined with unsupervised learning methods and geospatial analysis on more than 10,000 businesses and 55,000 individual users during the years 2023 and 2024, in order to characterise profiles, usage patterns and their territorial evolution.

The specific objective is to identify business segments with clearly differentiated behaviors in frequency and value of transactions, as well as to evidence the territorial dynamics of expansion of the platform to provide evidence for financial inclusion strategies. After this introduction, Sect. 2 presents the review of the literature; Sect. 3 describes the methodology; Sect. 4 presents the descriptive, inferential and clustering results; and Sect. 5 collects the most relevant conclusions and their implications for future practice and research.

2 Literature Review

The advancement of digital payment technologies has been widely documented as a driver of financial inclusion and transformation of business ecosystems. Authors such as [4, 10, 11] highlight that electronic payment platforms reduce barriers to access for the population and encourage the formalization of small and medium-sized enterprises in emerging countries. In Latin America, recent studies show a rapid penetration of digital payment systems using QR codes in high-turnover sectors such as food, retail and basic services, although with significant territorial and sectoral heterogeneities [8, 12, 13].

From a methodological point of view, the analysis of large volumes of data has expanded the possibilities for understanding patterns of technological adoption in financial services [11, 14, 15]. point out that unsupervised clustering algorithms allow the identification of latent profiles in complex and heterogeneous datasets, while internal validity metrics such as Silhouette make it easier to assess the quality of clusters [16]. These techniques have been successfully applied in marketing and banking for customer segmentation and detection of atypical behaviors.

However, the literature on digital payments in Latin American contexts is still incipient in the combined use of massive and georeferenced transactional data. Most studies focus on aggregate indicators or surveys and do not explore intra-sectoral heterogeneity or the spatial evolution of technology adoption [2, 3, 5, 8, 9, 16, 17]. This gap limits the understanding of the factors that influence the intensity of use and the economic value of transactions, as well as the identification of segments with growth potential.

In response to these limitations, recent work has begun to incorporate geospatial variables and robust methods for the treatment of extreme values [14, 18, 19], as well as to examine differences according to demographic or economic characteristics using non-parametric tests. This methodological combination allows for a more granular and reliable view of digital ecosystems [14, 20]. This study is framed in this trend, providing empirical evidence on the adoption of digital payments in Ecuador with an integrative approach of robust statistics, unsupervised clustering and geospatial analysis.

3 Methodology

This section describes the data sources, variables, and analytical procedures used in the study. Section 3.1 presents the business and user datasets derived from the Ahorita platform and the main variables considered, while Sect. 3.2 details the inferential statistical tests and the unsupervised clustering techniques used to identify heterogeneous patterns of digital payment adoption.

3.1 Data and Variables

The study uses two datasets from the Ahorita platform for the period 2023–2024: The first set covers more than 10,000 affiliated businesses, to characterize economic units that process digital payments. The second dataset covers more than 55,000 individual users, to analyze usage patterns in the population. Table 1 summarizes the variables used, explicitly distinguishing business-level variables from user-level variables and organizing them by unit of measurement and dataset.

Table 1. Description of business-level and user-level variables

Variable	Type / Unit of measure	Dataset
Business category	Categorical	Businesses
Latitude	Coordinate	Businesses
Longitude	Coordinate	Businesses
Frequency of use of the businesses	Transaction count	Businesses
Value traded by the businesses	Dollars	Businesses
Gender	Categorical	Individuals
Age range	Age ranges	Individuals
User usage frequency	Transaction count	Individuals
Value transacted by the user	Dollars	Individuals

3.2 Analysis Techniques

The inferential analysis sought to identify statistically significant differences in platform usage between groups of users and groups of businesses. Before selecting the tests, we assessed the assumptions of normality and homogeneity of variances for transaction counts and monetary amounts using the Shapiro–Wilk and Levene tests [21]. The large sample sizes, together with the strong right-skewness, heavy tails, and presence of extreme values observed in the descriptive statistics, led to systematic violations of these assumptions. Under these conditions, non-parametric procedures provide more reliable inferences because they are less sensitive to asymmetric distributions and outliers [22, 25]. For this reason, we discarded one-way ANOVA and related parametric tests and relied instead on rank-based methods.

In particular, we used the Kruskal–Wallis test as the non-parametric alternative to one-way ANOVA [23] to compare distributions across groups. When the global test indicated statistically significant differences, we applied Dunn's post-hoc test with Benjamini–Hochberg adjustment to perform pairwise comparisons while controlling for multiple testing [24]. This procedure controls the false discovery rate rather than the family-wise error rate, which offers a good balance between type I error control and statistical power given the large number of pairwise contrasts across age ranges and business categories. The same approach was used to evaluate differences in the number of transactions and total value transacted by gender and age groups in the user dataset and by economic category and cluster in the business dataset. To attenuate the influence of extreme observations without discarding information, we implemented conservative winsorization at the 1st and 99th percentiles [25]. All tests were two-sided with a significance level of 0.05, testing the null hypothesis of equal distributions against the alternative of differences between groups.

To identify heterogeneous patterns of adoption and use of digital payments among businesses, we applied several unsupervised clustering algorithms [26]. We combined key transactional variables (number of transactions and total value traded) with the geographical coordinates of each business, after applying logarithmic transformations and standardization to mitigate skewness and to ensure that all variables contributed on a comparable scale.

K-means partitions the data into a predefined number of groups by iteratively assigning each business to the nearest centroid and updating these centroids until the within-cluster variance is minimized [26]. Agglomerative hierarchical clustering starts with each business as a separate group and progressively merges the closest pairs, producing a dendrogram that reveals nested structures without imposing a particular cluster shape in advance. DBSCAN uses a density-based logic to form clusters of arbitrary shape: it groups together businesses that have many neighbors in their vicinity and labels as noise those observations that remain isolated [27]. Finally, Gaussian Mixture Models (GMMs) assume that the data arise from a mixture of Gaussian components and assign to each business a probability of belonging to each cluster, which is useful when segments overlap and boundaries are not sharply defined [28].

The quality of each clustering solution was assessed using internal validation indices based on cohesion and separation: the Silhouette coefficient, the Davies–Bouldin index, and the Calinski–Harabasz index. We selected as the final model the configuration that maximized Silhouette and Calinski–Harabasz values and minimized the Davies–Bouldin index, and used its four-cluster solution as the basis for the subsequent business segmentation analysis.

4 Results

This section presents the study's findings in three stages. First, we outline the descriptive statistics that characterize overall usage patterns of the digital payments platform among both affiliated businesses and individual users. Next, we report the inferential results of non-parametric tests. Finally, we describe the outcomes of the clustering analysis. These results together highlight a marked heterogeneity in usage and the emergence of four well-defined business segments.

4.1 Descriptive Statistics

The descriptive analysis allowed to characterize the general behavior of both affiliated businesses and users of the digital payment platform. Table 2 reflects a heterogeneous pattern of use, where most transactions are concentrated at low levels, but with the presence of large outliers that generate considerable dispersion.

Table 2. Descriptive statistics of numerical variables

Variable	Count	Mean	Standard deviation	IQR	Minimum	Maximum
Frequency of use of the businesses	10191	96.40	289.30	71	1	9099
Value traded by businesses	10191	1928.64	5583.61	1366.25	0.01	139410.25
Frequency of use of the individual	57537	21.75	34.21	23	1	847
Value traded by the individual	57537	424.78	650.86	475.62	0.01	16103.45

Table 3 confirms that the patterns of use of digital payments differ markedly between sectors, marked by both the frequency and the economic value of the transactions. Food & Beverages concentrates the largest transactional volume, while Industry & Primary Sectors and Health & Personal Care stand out for mobilizing higher amounts in less frequent operations. Retail & Home combines high values with moderate frequency, reflecting a stable and diversified flow.

Table 3. Descriptive statistics by category of businesses

Business category	Variable	Mean	Standard deviation	IQR
Entertainment & Leisure	Number of transactions	91.04	176.44	85
	Traded Value (USD)	1565.68	3306.25	1530.5
Food & Beverages	Number of transactions	133.66	239.93	129
	Traded Value (USD)	1561.19	3753.91	1089.29
Health & Personal Care	Number of transactions	81.95	178.23	60
	Traded Value (USD)	2007.50	3955.35	1717.23
Industry & Primary Sectors	Number of transactions	61.92	123.20	55
	Traded Value (USD)	2299.64	4178.98	2267.41
Professional Services	Number of transactions	27.20	87.90	17
	Traded Value (USD)	768.49	2810.75	285.57
Retail & Home	Number of transactions	59.11	126.92	126.92
	Traded Value (USD)	1805.09	3873.55	3873.55

Table 4 shows the profile of users in terms of usage and amounts traded, according to their gender and age range. The analysis reveals that users between 30 and 45 years of age are the most active segment, both in number of transactions and in value transacted, which reflects their greater consolidation in the use of digital payments. On the other hand, those under 17 years of age hardly participate, while those over 45 years of age show a different pattern, carry out fewer transactions, but of higher amounts, which suggests a more selective use and greater economic weight.

Table 4. Descriptive statistics by users according to gender and age range

Users	Variable	Mean	Standard deviation	Min	Max
Women up to 17 years	Number of transactions	12.77	23.33	1	268
	Traded Value (USD)	213.42	417.57	0.01	5148
Women 18–29 years	Number of transactions	19.49	28.74	1	550
	Traded Value (USD)	281.72	413.12	0.01	6100.12
Women 30–45 years	Number of transactions	23.64	34.03	1	452
	Traded Value (USD)	498.10	683.04	0.01	16103.45
Women over 45 years	Number of transactions	16.59	27.92	1	372
	Traded Value (USD)	455.23	682.74	0.01	6731.12
Men up to 17 years	Number of transactions	16.70	27.15	1	209
	Traded Value (USD)	209.19	333.04	0.01	2666.94
Men 18–29 years	Number of transactions	21.82	34.41	1	489
	Traded Value (USD)	287.04	469.46	0.01	11493.42
Men 30–45 years	Number of transactions	24.85	40.26	1	847
	Traded Value (USD)	527.88	768.60	0.01	10668.30
Men over 45 years	Number of transactions	19.42	33.85	1	327

(continued)

Table 4. (*continued*)

Users	Variable	Mean	Standard deviation	Min	Max
	Traded Value (USD)	560.57	844.28	0.01	11088.58

4.2 Behavior of Users

As can be seen in Table 5, there are no statistically significant differences between men and women according to the Kruskal Wallis test.

Table 5. Results of statistical tests by gender

Variable	Factor	Statistic H	p-value	Significant difference
Number of transactions	Gender	1.49	0.223	No
Total value traded USD	Gender	0.02	0.885	No

In the case of age groups, as can be seen in Table 6, the results show that there are significant differences in transactional activity. As Fig. 1 shows, almost all comparisons yielded adjusted p-values close to zero, confirming statistically significant contrasts in both the number of transactions and the total value transacted. However, certain exceptions were identified, since in the number of transactions no significant differences were found between users under 17 years of age and those over 45, while in the total value transacted the segments of 30 to 45 years of age and over 45 years of age showed a statistically similar behavior.

Table 6. Statistical test results by age range

Variable	Factor	Statistic H	p-value	Significant difference
Number of transactions	Age Range	565.33	<0.001	Yes
Total value traded USD	Age Range	2052.63	<0.001	Yes

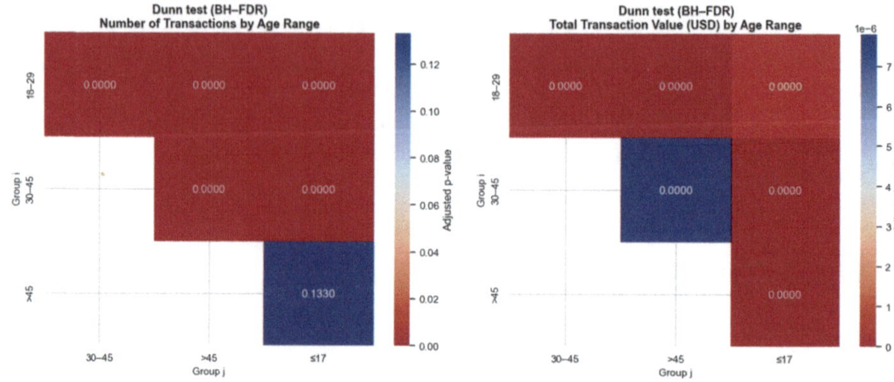

Fig. 1. Dunn test (BH-FDR) for the number of transactions and value traded by age ranges

4.3 Businesses Behavior

The clustering analysis revealed clearly differentiated transactional patterns among the affiliated businesses, with the K-means model with 4 clusters showing the best performance, with the highest Silhouette coefficient, a high Calinski–Harabasz value and a low Davies–Bouldin index, which evidences well-defined and separate clusters. In comparison, the Agglomerative, GMM, and DBSCAN models presented less consistent performances, with lower values in Silhouette and Calinski–Harabasz and higher in Davies–Bouldin, reflecting less cohesive clustering structures and with less capacity for differentiation between groups (Table 7).

Table 7. Performance of segmentation algorithms for business

Algorithm	Number of clusters	Silhouette	Calinski-Harabasz	Davies-Bouldin
K-means	4	0.82	63295.22	0.35
Agglomerative	5	0.80	53143.27	0.44
GMM	4	0.68	31215.89	0.61
DBSCAN	4	0.74	19301.20	0.36

Four clusters were identified in the K-means model. The first cluster, called "Core moderate-activity merchants", brings together 7,759 establishments and constitutes the largest group. Their transaction levels and total value are moderate, with a low average ticket. The second cluster, called "High-value merchants", brings together only 222 establishments, but stands out for concentrating the highest volume of transactions and the highest economic value among all groups.

The third cluster, identified as "Low-adoption marginal merchants", includes 1,364 establishments with the lowest transactional frequency and significantly low values. This cluster is relevant to identify adoption barriers and opportunities for strengthening the digital inclusion strategy. Finally, the fourth cluster, called "Emerging merchants", is made up of 846 establishments. Although its level of transactions is low-moderate, it has the highest average ticket among the groups, which differentiates it as a segment with growth potential (Table 8). Figure 2 shows the distribution of buttocks by trade at the national level.

Table 8. Profile of clusters identified with K-means

Cluster	Number of businesses	Average Transactions	Average value traded	Average ticket
Core moderate-activity merchants	7759	96.65	1934.05	23.39
High-value merchants	222	246.66	4521.59	23.34
Low-adoption marginal merchants	1364	29.25	654.98	23.36
Emerging merchants	846	38.60	926.03	27.91

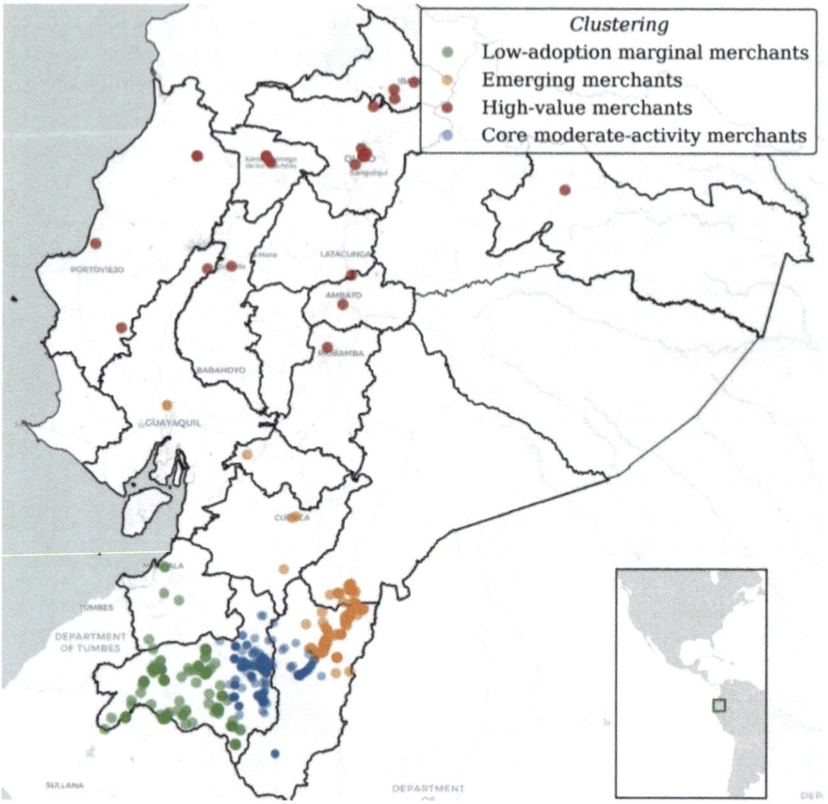

Fig. 2. Merchant clusters

4.4 Transactional Difference (Number and Amount) By Business Category and Business Cluster

Inferential analysis using the Kruskal-Wallis test showed statistically significant differences in transactional activity, confirming that the category is a determining factor in the segmentation of the dynamics of use of digital payments (Table 9). Dunn's post-hoc analysis with Benjamini–Hochberg correction, as shown in Fig. 3, specified that most of the contrasts between categories were significant. However, specific cases of lack of significance were identified. These exceptions suggest that certain sectors share similar patterns of behavior despite belonging to different categories.

Table 9. Results of statistical tests by category of trade

Variable	Factor	Statistic H	p-value	Significant difference
Number of transactions	Business category	496.99	< 0.001	Yes

(continued)

Table 9. (*continued*)

Variable	Factor	Statistic H	p-value	Significant difference
Total value traded USD	Business category	510.15	< 0.001	Yes

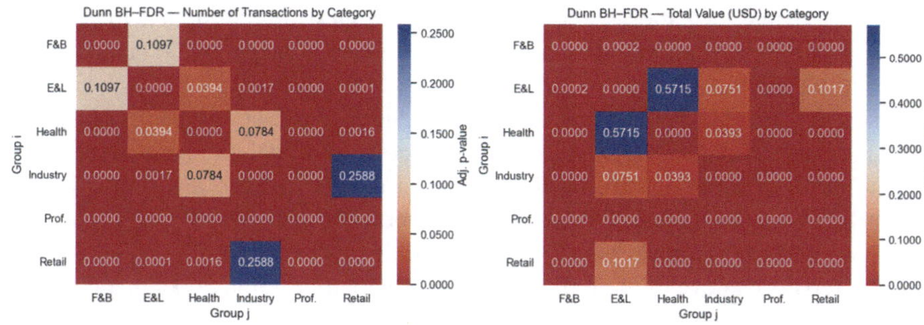

Fig. 3. Dunn test (BH-FDR) for the number of transactions and value traded by business category

The analysis by business clusters also showed significant differences in both the number of transactions and the total value transacted, as shown in Table 10, confirming the usefulness of segmentation as a tool to identify business groups. The post-hoc analysis indicated that all comparisons between clusters were significant (Fig. 4).

Table 10. Results of statistical tests by trade cluster

Variable	Factor	Statistic H	p-value	p-value
Number of transactions	Business cluster	449.26	< 0.001	Yes
Total value traded USD	Business cluster	250.19	< 0.001	Yes

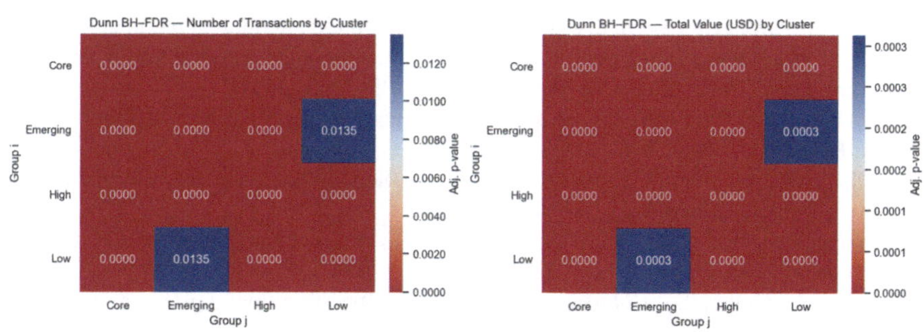

Fig. 4. Dunn test (BH-FDR) for the number of transactions and value traded per merchant cluster

5 Conclusions

This article has analyzed consumers and businesses that use an electronic payment tool from a financial institution in southern Ecuador. Using various analysis techniques, the data was determined that there are differences in the use and amount of transactions by both users and affiliated businesses in the years 2023 and 2024. The age of the users and the type of business are variables that allow both users and businesses to be characterized. These diverse behavior patterns allow different strategies to be designed to encourage greater use by the population within a financial inclusion strategy, while businesses can capture greater market share.

These findings have relevant practical implications for financial institutions, merchants, and policymakers. The observed heterogeneity in digital payment adoption suggests that uniform strategies are unlikely to be effective. Instead, institutions can use the identified patterns to design differentiated approaches for promotion, support, and onboarding, while policymakers may incorporate this evidence into financial inclusion initiatives, territorial development programs, and regulatory frameworks that encourage broader, more sustained use of digital payments.

As future lines of research, different machine learning techniques will be used to predict the evolution of the different market segments of both users and businesses in order to design more particularized strategies. In addition, the data sample will be expanded to determine if customer and business segments are consistent over time.

References

1. Bejar, P., Ishi, K., Komatsuzaki, T., Shibata, I., Sin, J., Tambunlertchai, S.: Can fintech foster competition in the banking system in Latin America and the Caribbean? Lat. Am. J. Cent. Bank. **3**(2), 100061 (2022)
2. Bailey, A.A., Bonifield, C.M., Arias, A., Villegas, J.: Mobile payment adoption in Latin America. J. Serv. Mark. **36**(8), 1058–1075 (2022)
3. Ly, R., Ly, B.: Digital payment systems in an emerging economy. Comput. Hum. Behav. Rep. **16**, 100517 (2024)
4. Putrevu, J., Mertzanis, C.: The adoption of digital payments in emerging economies: challenges and policy responses. Digit. Policy, Regul. Gov. **26**(5), 476–500 (2024)
5. Nandru, P., Senthil Kumar, S.A., Chendragiri, M.: Adoption intention of mobile QR code payment system among marginalized street vendors: an empirical investigation from an emerging economy. J. Sci. Technol. Policy Manag. **15**(6), 1709–1733 (2024)
6. Ghosh, G.: Adoption of digital payment system by consumer: a review of literature. Int. J. Creat. Res. Thoughts. **9**(2), 2320–2882 (2021)
7. Jha, S.: The digital revolution in transactions: a new era of payment ecosystems. Int. J. Multidiscip. Sci. Manag. **1**(4), 28–30 (2024)
8. Amboage, G.B., Monteiro, G.F.D.A., Bortoluzzo, A.B.: Technological adoption: the case of PIX in Brazil. Innov. Manag. Rev. **21**(3), 198–211 (2024)
9. Hussain, S., Gupta, S., Bhardwaj, S.: Determinants inhibiting digital payment system adoption: an Indian perspective. Qualitat. Res. Financial Mark. **17**(4), 716–748 (2025)
10. Kou, G., Lu, Y.: Fin tech: a literature review of emerging financial technologies and applications. Financ. Innov. **11**(1), 1 (2025)
11. Polireddi, N.S.A.: An effective role of artificial intelligence and machine learning in banking sector. Meas.: Sens. **33**, 101135 (2024)

12. Urueña-Mejía, J.C., Gutierrez, L.H., Rodríguez-Lesmes, P.: Financial inclusion and business practices of microbusiness in Colombia. Eurasian Bus. Rev. **13**(2), 465–494 (2023)
13. Ho Kim, S., Taylor, D.: Intellectual capital vs the book-value of assets: a value-relevance comparison based on productivity measures. J. Intellect. Cap. **15**(1), 65–82 (2014)
14. Hasan, M., Hoque, A., Le, T.: Big data-driven banking operations: opportunities, challenges, and data security perspectives. FinTech. **2**(3), 484–509 (2023)
15. Heß, V.L., Damasio, B.: Machine learning in banking risk management: mapping a decade of evolution. Int. J. Inform. Manag. Data Insights. **5**(1), 100324 (2025)
16. Aurazo, J., Gasmi, F.: Digital payment systems in emerging economies: lessons from Kenya, India, Brazil, and Peru. Inf. Econ. Policy. **69**, 101113 (2024)
17. Tamilmani, K., Rana, N.P., Wamba, S.F., Dwivedi, R.: The extended unified theory of acceptance and use of technology (UTAUT2): a systematic literature review and theory evaluation. Int. J. Inf. Manag. **57**, 102269 (2021)
18. Antonio, B.O., Juan, L.R., Ana, I.D., Francisco, L.C.: Examining user behavior with machine learning for effective mobile peer-to-peer payment adoption. Financ. Innov. **10**(1), 94 (2024)
19. Moradi, S., Mokhatab Rafiei, F.: A dynamic credit risk assessment model with data mining techniques: evidence from Iranian banks. Financ. Innov. **5**(1), 1–27 (2019)
20. Nobanee, H., Dilshad, M.N., Al Dhanhani, M., Al Neyadi, M., Al Qubaisi, S., Al Shamsi, S.: Big data applications the banking sector: a bibliometric analysis approach. SAGE Open. **11**(4), 21582440211067234 (2021)
21. Gibbons, S., Chakraborti, J.D.: Design and Analysis of Experiments. John Wiley & Sons, Inc (2017)
22. Gibbons, S., Chakraborti, J.D.: Nonparametric Statistical Inference. Taylor and Francis Ltd. (2011)
23. Kruskal, W.H., Wallis, W.A.: Use of ranks in one-criterion variance analysis. J. Am. Stat. Assoc. **47**(260), 583–621 (1952)
24. Yoav Benjamini, Y.H.: Benjamini_Hochberg1995.Pdf. J. Royal Stat. Soc.: Ser. B (Methodol.). **57**(1), 289–300 (1995)
25. Wilcox, R.: Introduction to Robust Estimation and Hypothesis Testing, 3rd edn (2011)
26. Jain, A.K.: Data clustering: 50 years beyond K-means. Pattern Recogn. Lett. **31**(8), 651–666 (2010)
27. Schubert, E., Sander, J., Ester, M., Kriegel, H.P., Xu, X.: DBSCAN revisited, revisited: why and how you should (still) use DBSCAN. ACM Trans. Database Syst. (TODS). **42**(3), 1–21 (2017)
28. Christhoper, B.: Pattern Recognition and Machine Learning. Springer Science and Business Media Deutschland GmbH (2006)

Implementation of AI Tool Accessibility in Nursing Procedures

Toapanta Vizuete Kevin Joel(✉), Vallejo Burbano Alejandra Jadira, and Nasimba Yautibug Lizbeth Alexandra

Instituto Superior Tecnologico Vida Nueva, Quito, Ecuador
kevin.toapanta@istvidanueva.edu.ec

Abstract. In the current context of digital transformation, artificial intelligence (AI) has established itself as a strategic tool in the field of healthcare, with the potential to optimize clinical and administrative procedures. However, limitations related to technological accessibility, insufficient training, and ethical dilemmas in nursing practice persist. Based on this problem, the objective of the study was to analyze the implementation and accessibility of AI tools in nursing procedures, identifying benefits, barriers, and proposals for improvement, which led to the question: What are the benefits, limitations, and ethical challenges faced by nursing professionals in the implementation of these tools? A qualitative study with an exploratory descriptive design was conducted, based on a literature review and semi-structured interviews with eight nursing professionals with experience in digital technologies. Content analysis identified categories related to task automation, efficiency and safety, accessibility, training, resistance to change, and ethical aspects. The results showed that AI contributes to reducing administrative burden and improving clinical efficiency when the tools are accessible. However, the lack of infrastructure, training, and technical support limits its use. The discussion highlights the need to integrate AI without displacing human interaction. In conclusion, accessibility to AI in nursing requires technological investment, continuous training, and policies that ensure equitable and complementary integration into professional practice.

Keywords: Artificial intelligence · Nursing · Accessibility · Clinical procedures · Health technology · Ethical

1 Introduction

In the current context of digital transformation, artificial intelligence (AI) has emerged as a strategic tool in the field of healthcare, revolutionizing clinical and administrative practices. Its implementation in nursing procedures represents a crucial opportunity to optimize care, improve clinical decision-making, and reduce the operational burdens faced by professionals in the field. However, this technological transition also poses significant challenges in terms of accessibility, adaptability, and training for nursing staff. Nursing, as an essential discipline within the healthcare system, requires tools that

A. Rocha et al. (Eds.): ICITS 2026, LNNS 1966, pp. 582–588, 2027.
https://doi.org/10.1007/978-3-032-27538-7_51

not only improve service efficiency but also promote patient-centered care. In this regard, AI can be used to automate routine processes, continuously monitor vital signs, assist in preliminary diagnoses, and support clinical data management. Despite these advantages, access to these technological tools is uneven, especially in settings with limited resources or staff unfamiliar with their use. This research aims to analyze the implementation of accessibility of artificial intelligence-based tools in nursing procedures, considering both their transformative potential and the limiting factors that hinder their widespread adoption. It proposes to identify strategies that allow these technologies to be effectively integrated into the daily practice of nursing staff, ensuring equitable access and correct use.

2 Theoretical Framework

2.1 Artificial Intelligence in the Healthcare Sector

Artificial Intelligence (AI) has transformed the healthcare landscape by automating processes, enabling accurate diagnoses, and predicting clinical complications. This technology has been applied in areas such as medical imaging, remote monitoring, and clinical data management, proving effective in interpreting complex medical data, facilitating early disease detection, and optimizing treatment. (Xu, Yang, Xie & Chen, 2021).

In recent years, AI has established itself as an essential tool for improving the quality of care and efficiency of healthcare systems. Intelligent algorithms collaborate with healthcare professionals by suggesting personalized care plans, preventing adverse events, and streamlining clinical processes. (García-Peña & Morales, 2022).

2.2 Nursing Procedures in Clinical Practice

Nursing procedures are technical, scientific, and ethical actions that guarantee quality care and patient safety. They require up-to-date knowledge, technical skills, and an empathetic relationship with the patient to ensure compliance with clinical care standards. (Martínez, Jiménez & Rivera, 2023).

With the advancement of technology, efforts have been made to integrate digital solutions into nursing practice. The digitization of nursing processes improves access to information, reduces errors, and strengthens the traceability of interventions performed, positively impacting clinical care. (Rodríguez & Cárdenas, 2022).

2.3 Accessibility of AI Tools for Nursing Staff

Accessibility to AI tools in nursing involves removing technological, training, and organizational barriers that may limit their effective use. To achieve proper adoption, it is necessary to design user-friendly interfaces, offer ongoing training, and ensure technical support in clinical settings. (Pérez & Salazar, 2021).

These tools must also be adapted to the actual conditions of the hospital environment. Factors such as work overload, poor connectivity, and lack of technological resources must be considered from the design and implementation stages of these technologies. (Herrera & Molina, 2024).

2.4 Implementation of AI in Nursing Procedures

The implementation of AI in nursing procedures represents an opportunity to optimize care and reduce the administrative burden on staff.

Intelligent systems make it possible to predict adverse events, automate records, and issue clinical alerts, freeing up time for direct care tasks. (Bravo, Torres & Meneses, 2023).

This implementation requires planning, ongoing training, and evaluation of results. The involvement of nursing staff in the design and testing of AI tools ensures that the solutions respond to the real needs of the clinical environment. (López & Torres, 2025).

2.5 Ethical and Technical Challenges in the Application of AI in Nursing

One of the main challenges today is protecting patients' personal data. AI accesses large volumes of clinical information, so rigorous policies on confidentiality, informed consent, and algorithmic transparency must be applied. (Delgado & Ruiz, 2022).

From a technical perspective, resistance to change, lack of digital skills, and poor interoperability between systems are common barriers. To overcome them, an institutional culture is needed that promotes the responsible use of technology and encourages continuous digital training for nursing staf (Zambrano & León, 2023).

3 Methodology

This research was conducted using a qualitative and quantitative approach, with an exploratory-descriptive design, selected for its relevance in providing an in-depth understanding of the perceptions, experiences, and challenges of nursing professionals regarding the implementation of artificial intelligence tools in clinical procedures. Techniques such as document review, aimed at identifying theoretical and regulatory background information, and semi-structured interviews with nursing professionals, whose testimonies were subjected to thematic content analysis, were used to collect information.

The search equation used was a combination of descriptors in Spanish and English, using Boolean operators and search networks to refine results. ("inteligencia artificial" OR "artificial intelligence") AND ("enfermería" OR "nursing") AND ("hospital" OR "clinical practice") AND ("implementation" OR "aplicación") (Fig. 1).

First, a review of the scientific literature and techniques was conducted to establish the theoretical and situational framework for the analysis. To this end, specialized literature, regulatory documents, and academic articles related to the application of artificial intelligence in the field of health were examined. The search was conducted in databases such as Scielo, PubMed, and Scopus, considering publications between 2018 and 2024 to ensure that the data was up to date with recent advances in the technologies used in clinical practice. Studies presenting practical experiences of artificial intelligence implementation in hospitals and health centers were selected as inclusion criteria, while those limited to theoretical projections without evidence of application were excluded.

Subsequently, semi-structured interviews were conducted with eight professionals from the San José de Obrero Hospital in the city of Quito. Participants were selected

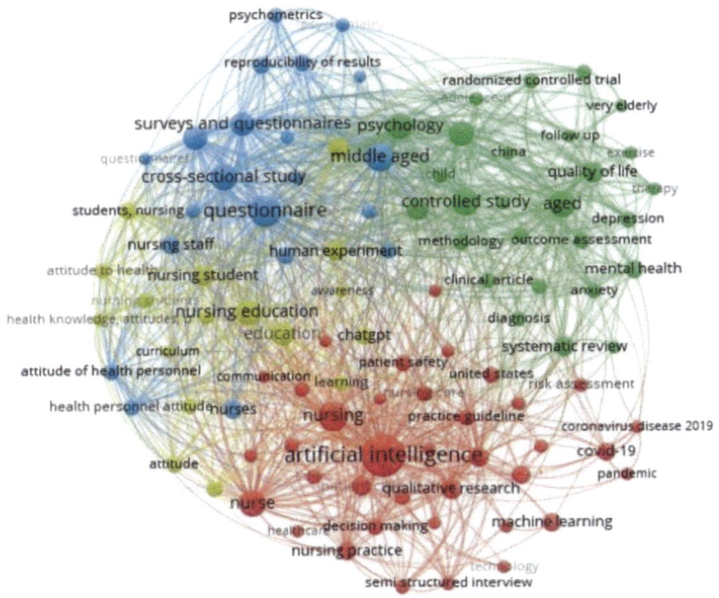

Fig. 1. Aplicación de bibliometria según ecuación de búsqueda.

intentionally, giving priority to those who had experience in the use of digital technologies in their daily practice. The interview guide was applied individually in sessions lasting around 40 min. The questions focused on investigating accessibility, benefits, ease of use, perceived barriers, and experiences associated with the implementation of artificial intelligence tools in nursing processes. Content analysis was performed inductively, through manual coding of responses into thematic categories that emerged, which facilitated the recognition of tensions and practical proposals raised directly by the participants.

In order to ensure the reliability and validity of the results, a process of triangulation of sources was carried out; that is, the findings from the interviews were compared with those indicated in the scientific literature analyzed previously. In terms of ethics, the study was conducted in accordance with qualitative research guidelines and the provisions of the Declaration of Helsinki, ensuring the informed consent of participants, their autonomy, and the confidentiality of data. The testimonies were classified into categories, and a comprehensive interpretation of the elements that affect the accessibility of artificial intelligence in nursing procedures was developed, thanks to the qualitative analysis matrices used to systematize the information collected.

4 Results and Discussion

The implementation of AI tools in nursing demonstrates significant potential to transform the quality of care, primarily through the automation of routine tasks. Interviewees noted that, when accessible and understandable, these technologies significantly increase service efficiency and reduce administrative burden.

However, technological accessibility remains a considerable challenge. In many facilities, especially rural or resource-limited ones, access to AI tools is limited. A lack of specific training was also identified, leading to resistance or limited use.

A relevant aspect was the ethical concern about the loss of human interaction. Participants indicated that AI should complement, not replace, clinical judgment and professional empathy. Thus, an ethical and humanized approach to its implementation is required.

In this study, the evidence was grouped into thematic categories detailed below (Table 1):

Table 1. Analysis table on AI in healthcare.

Analysis Category	Hallazgos Principales	Relevant Comments from Participants
Task automation	AI enables the automation of routine tasks such as vital sign recording and clinical follow-up.	"It helps us reduce administrative burden and spend more time with patients."
Efficiency and safety	Service efficiency and safety improve when tools are accessible and easy to use.	"With good training, it really improves care."
Technological accessibility	There is a significant gap in access to AI tools, especially in rural areas or low-income centers.	"At our center, we don't have access to even basic devices with smart software."
Training and training	A lack of technical training in the use of AI among nursing staff has been detected	"They gave us the tools, but they didn't teach us how to use them."
Resistance to change	Lack of training and technical support leads to resistance or limited use of the tools.	"Many colleagues prefer to continue doing things manually because they don't fully understand how AI works."
Ethical aspects and humanization	Concern about the replacement of human interaction by automated decisions.	"AI should complement, not replace, clinical judgment and empathy."
User Recommendations	Proposal for ongoing training, accessible technical support, and a humanized approach to AI implementation.	"We need training adapted to our working reality, not just theory."

The implementation of artificial intelligence tools in nursing procedures presents significant potential to transform the quality of care. One of the main benefits identified in this research is the automation of routine tasks such as recording vital signs, scheduling medication, and monitoring clinical parameters. This not only reduces the administrative burden on nursing staff but also frees up time for more personal and humane patient care.

Interviewees agreed that when these technologies are accessible and easy to use, they significantly improve service efficiency and safety.

However, technological accessibility remains a considerable challenge. The research showed that in many health centers, especially those located in rural areas or with limited resources, access to AI tools is limited or non-existent. Added to this is the lack of training for nursing staff in the use of these technologies, which leads to resistance or underuse. The professionals interviewed pointed out the need to implement ongoing training programs and technical support tailored to their work realities.

Another relevant finding was the ethical perception surrounding the use of artificial intelligence. Some participants expressed concern about the replacement of human interaction with automated decisions. In this regard, it was suggested that AI should be considered a support tool and not a substitute for clinical judgment or professional empathy. This view reinforces the importance of a humanized and ethical approach to technological integration, where accessibility implies not only availability but also understanding and appropriation by healthcare personnel.

In addition, in comparison with two articles that were more relevant, the results were obtained in percentages where (Table 2)

Table 2. Percentage analysis table with relevant articles

Criterion / Indicator	Base Article: Implementation of AI Tool Accessibility in Nursing Procedures (2024)	Alruwail i et al. (2024) – *Exploring nurses' awareness and attitudes toward AI*	Nurses' perceptions and use of Artificial Intelligence in Healthcare (Irán, 2024)	Overall Averager
Sample size	120	100	400	
Knowledge about AI	82%	70.9%	58.9%	70.6%
Practical use/application	76%	58.2%	55.8%	63.3%
Perceived usefulness / Benefits	90%	72%	70%.	77.3%
Training/ Education received	45%	39%	35%	39.7%

Of the comparative analyses carried out, the most relevant and detailed in the table on the implementation and perception of artificial intelligence in nursing procedures showed positive trends towards the acceptance of these tools, although with variations in levels of knowledge, use, and attitude. These percentages indicate that there is a high degree of recognition of AI in clinical practice (with values around 71%), but levels of practical application and technical mastery are significantly lower (around 63%). These findings allow us to conclude that, although the accessibility and usefulness of AI is recognized by the majority of nursing staff, there are still training and implementation gaps that need to be addressed.

The actual application and use of artificial intelligence in nursing procedures reflected the lowest percentages in the comparison, with an average of around 39.7%. This shows that, although professionals recognize the relevance and potential of these tools in their daily work, their practical incorporation is still limited. In other words, there is a gap between the positive assessment of AI and its effective use in the clinical context, which highlights the need to create training, infrastructure, and support conditions that facilitate its implementation in healthcare practice.

5 Conclusions

Artificial intelligence offers significant benefits for nursing procedures, particularly in automating tasks and improving clinical efficiency, provided that its implementation takes into account criteria such as accessibility and staff training.

Accessibility to AI tools in nursing is conditioned by structural, technological, and training factors. To achieve equitable implementation, it is necessary to invest in technological infrastructure and continuing education programs adapted to different healthcare contexts.

The adoption of artificial intelligence in nursing care must be governed by ethical principles and focus on complementing human work. Technology should serve as a support to strengthen comprehensive care, without replacing the professional-patient interaction that characterizes nursing.

References

Bravo, M., Torres, F., Meneses, R.: Application of artificial intelligence in clinical nursing procedures. J. Health Sci. **21**(3), 55–66 (2023). https://doi.org/10.1000/rev.sal.2023.21.3.55

Delgado, S., Ruiz, A.: Ethics and data protection in the use of artificial intelligence in nursing. Bioeth. Today J. **10**(2), 77–85 (2022)

García-Peña, J., Morales, C.: Impact of artificial intelligence on healthcare: an updated review. Technol. Health. **18**(1), 24–36 (2022)

Herrera, D., Molina, P.: Challenges of technological accessibility in public health institutions. Health Manag. Notebooks. **12**(2), 15–28 (2024)

López, A., Torres, N.: Integración de IA en los procesos clínicos de enfermería: lecciones aprendidas. Innovación en Cuidado de la Salud. **9**(1), 40–52 (2025)

Martínez, L., Jiménez, P., Rivera, T.: Importance of technical procedures in professional nursing practice. Sci. Prof. **11**(4), 95–108 (2023)

Pérez, M., Salazar, G.: Accessibility and usability of artificial intelligence systems in clinical nursing. Adv. Med. Technol. **7**(2), 33–44 (2021)

Rodríguez, A., Cárdenas, V.: Digitization of nursing processes and its impact on clinical care. Lat. Am. J. Digit. Nurs. **5**(3), 61–73 (2022)

Xu, J., Yang, P., Xie, J., Chen, L.: Artificial intelligence in healthcare: applications and challenges. Front. Public Health. **9**, 650987 (2021). https://doi.org/10.3389/fpubh.2021.650987

Zambrano, F., León, S.: Training in digital skills for nursing professionals. Educ. Technol. Health. **14**(1), 88–99 (2023)

Taxonomy of Hierarchical Federated Learning for Intrusion Detection Systems in IoT

Jonathan Navarro-Alfaro[1(✉)] and Adrian Lara[2]

[1] Posgrado en Computación e Informática, Universidad de Costa Rica, San José, Costa Rica
jonathan.navarro@ucr.ac.cr

[2] Escuela de Ciencias de la Computación e Informática, Universidad dCosta Rica, San José, Costa Rica
adrian.lara@ucr.ac.cr

Abstract. The rapid growth of the Internet of Things (IoT) has increased both the volume of generated data and the potential of new attacks, making intrusion detection a challenge. Hierarchical Federated Learning (HFL) has recently emerged as a promising approach to improve intrusion detection systems (IDS) by combining privacy-preserving decentralized training with scalable multi-layer aggregation. This paper presents a taxonomy of HFL for IDS in IoT networks, organized into five categories: context of application, design strategy, datasets, preprocessing techniques, and problems addressed. The taxonomy highlights current research and shows the key domains, methodologies, and strategies being employed in the field. This work offers a base for future research while identifying open challenges in developing robust and efficient HFL-based IDS for IoT systems.

Keywords: Hierarchical federated learning · Intrusion detection systems · Internet of things

1 Introduction

Every year, the amount of data generated by IoT devices increases, which in turn raises the potential number of attacks these devices may suffer. Finding effective intrusion detection systems (IDS) is still a largely unexplored field. Recent years have shown an increase in cyberattacks on IoT systems, in domains such as industrial IoT (IIoT), Internet of Medical Things (IoMT), and Cloud-Edge-IoT. These attacks can compromise sensitive data, infrastructure, or critical systems. Resulting in direct costs like financial losses, and indirect costs, like reputational damage, loss of customers trust, and regulatory fines. These are one of the reasons why having a robust intrusion detection system is important to protect IoT systems [1].

A. Rocha et al. (Eds.): ICITS 2026, LNNS 1966, pp. 589–601, 2027.
https://doi.org/10.1007/978-3-032-27538-7_52

In this context, hierarchical federated learning (HFL) has emerged as a novel solution that has begun to be explored over the past five years. This approach takes advantage of decentralized model training across numerous IoT devices, which helps maintain data privacy [2] while achieving better results than standard federated learning. Hierarchical federated learning improves global model performance through distributed data sources, thus continuously improving detection against evolving cyber threats [2]. This method not only helps mitigate the challenges associated with centralized data aggregation, such as high computational complexity and privacy concerns, but also enhances the system's adaptability to detect both known and unknown threats effectively [3]. This study aims to address the following research questions:

1. What are the most common application domains of hierarchical federated learning?
2. What architectures are being used to implement hierarchical federated learning?
3. What datasets are employed in existing studies?
4. What preprocessing strategies are being use to handle the datasets?
5. What technical challenges do current approaches aim to solve?

This paper provides a taxonomy that makes the following contributions. First, it describes the IoT sub-domains where HFL has been applied to IDS. Second, it explains the design strategies of the proposed models (Blockchain, Knowledge Distillation, Long-term memory, etc.). Third, the taxonomy highlights the most commonly used IDS datasets. Fourth, it also describes the most common data preprocessing techniques, such as data cleaning and noise removal, normalization, regularization and others. Finally, the taxonomy also summarizes the most common problems addressed in the papers, such as data imbalance, privacy, communication overhead and so on.

To the best of our knowledge, this is the first study to develop a taxonomy specifically addressing hierarchical federated learning in the context of IDS in IoT.

The rest of the paper is structured as follows. In Sect. 2 we describe the fundamental concepts of hierarchical federated learning and its specialized application within intrusion detection systems for IoT environments, distinguishing it from conventional federated learning. Section 3 details the creation of the taxonomy and outlines the five categories described in this study. Finally, Sect. 4 summarizes the highlights the domains, architectures, datasets and preprocessing approaches most commonly employed in HFL-based IDS and discusses open challenges and directions for future research.

2 Background

This background section first covers federated learning, then hierarchical federated learning and finally intrusion detection systems.

2.1 Federated Learning

Federated learning, also known as distributed learning, is a technique where each client independently trains models using their own datasets. These models are then sent to the central server where they are processed and then sent back to the clients. By doing this, the information remains private as the trained models are the only objects sent to the server [15].

This approach mitigates privacy risks by enabling decentralized model training across multiple devices, ensuring that sensitive data remain local while still contributing to global model updates [2].

2.2 Hierarchical Federated Learning

Hierarchical federated learning comes from traditional federated learning but adds multiple layers, allowing more scalable model training across hierarchically structured IoT networks [5]. As shown in Fig. 1, in a common HFL setup, numerous IoT devices known as clients train local models on their own data and periodically send updates to nearby edge servers [4]. These edge servers then aggregate these local updates before forwarding a consolidated model to a central server, which further refines the global model and distributes it back down the hierarchy [5].

By distributing aggregation across edge and cloud layers, HFL significantly reduces communication overhead and latency compared to conventional federated learning systems [4]. This multi-tiered aggregation also optimizes computational resources by allowing local processing at the edge, thereby supporting the resource constraints inherent to many IoT devices [2,6].

This architecture is appropriate for the heterogeneous nature of IoT ecosystems, where devices often shows distinct computational capabilities, communication bandwidths, and data distributions [11].

That is why HFL model aggregation is achieved efficiently at different levels, such as on edge servers and central clouds, reducing communication overhead and latency inherent in large-scale IoT deployments [11]. This multi-layered approach also improves data privacy by allowing local model updates to be closer to the data source, reducing the direct exposure of raw data to central servers [8].

Also, HFL can improve intrusion detection accuracy and reduce overhead by integrating advanced techniques as the ones described in Sect. 3.2.

2.3 Intrusion Detection Systems

An Intrusion Detection System is a mechanism designed to monitor network or system activities for malicious actions, playing an important role in securing IoT systems from evolving cyberattacks [12]. These systems analyze traffic data, system logs, and user behavior to find patterns of unauthorized access, malware infections, or DDoS attacks, among others [14]. The main goal of an IDS is to identify and alert against suspicious activities that could compromise the

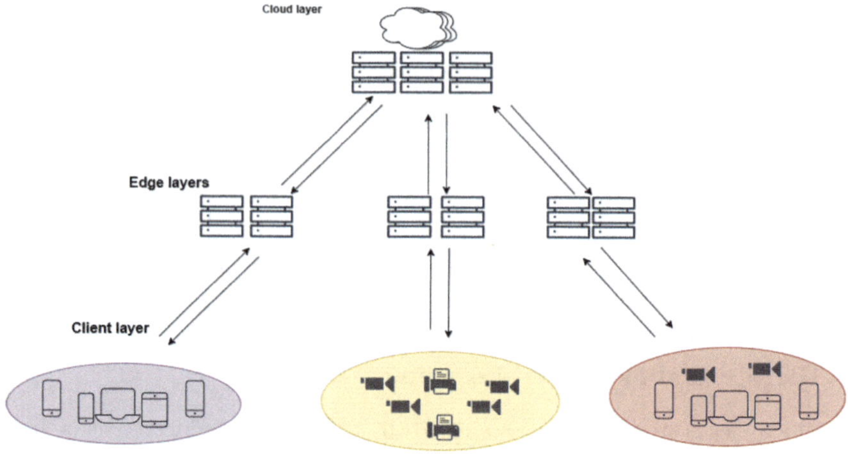

Fig. 1. Hierarchical Federated Learning Architecture for IDS in IoT Systems.

integrity, confidentiality, or availability of connected devices, by ensuring robust security against different types of cyberattacks [11].

However, given the constraints and nature of IoT devices, IDSs are integrated at the network edge, requiring efficient and distributed detection mechanisms [5]. In Sect. 3.3 we describe different techniques used to reduce these constraints. With this integration and the techniques applied to the data, the models can perform better.

3 Taxonomy of Hierarchical Federated Learning

This section presents our proposed taxonomy on how HFL has been used for intrusion detection in IoT. It is organized around five key parameters: Context of application, design strategy, dataset(s) used, data pre-processing, and problem addressed as shown in Fig. 2.

3.1 Context of Application

First, we categorize the domains and scenarios where HLF has been deployed for IDS, showing the diverse environments where IoT devices are utilized.

Internet of Medical Things: This specialized type of IoT products involves medical devices and healthcare systems, where data privacy and security are important due to the sensitive nature of patient information, making HFL-based IDS critical to safeguard against intrusions and ensure uninterrupted healthcare services [8,12].

Industrial Internet of Things: This domain focuses on the integration of IoT within industrial settings, such as manufacturing systems, factories, supply chains, and energy, where the security of operational technology and critical

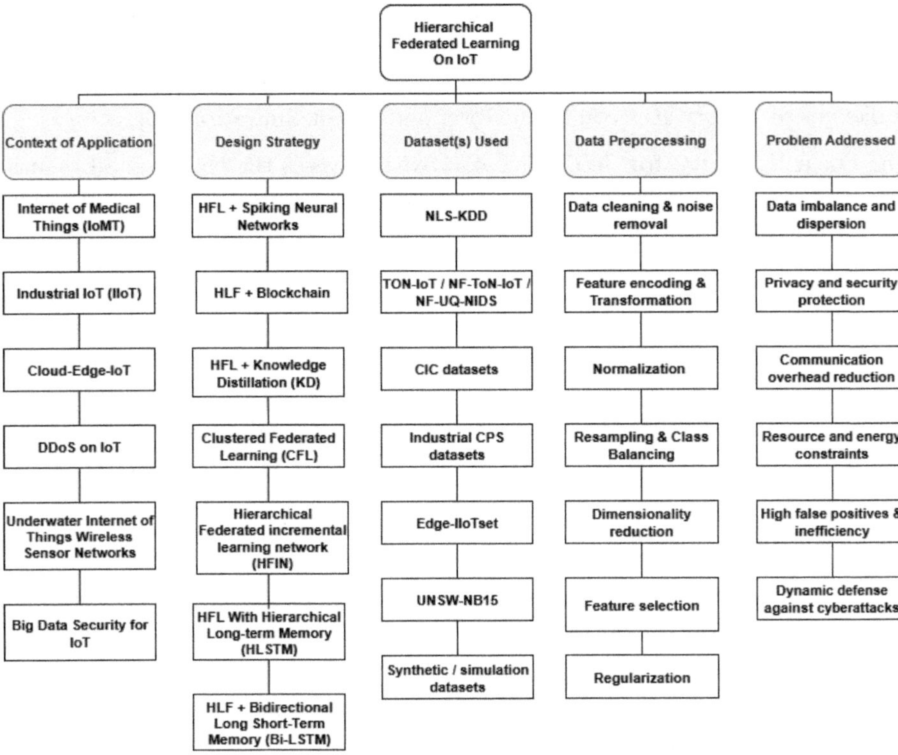

Fig. 2. Taxonomy of hierarchical federated learning for intrusion detection systems on IoT devices.

infrastructure is one of the main priorities [10]. Here, IoT devices are highly interconnected, enabling real-time data collection, sharing and analysis, improving efficiency, decision-making and productivity [2].

Cloud-Edge-IoT: This is a distributed computing architecture that integrates IoT devices, edge resources and cloud infrastructures into the same ecosystem. The data generated by the IoT devices is processed at the edge before being uploaded and refined in the cloud. This helps to reduce latency and network congestion while also improving scalability and addressing the privacy concerns by reducing the amount of raw data that is sent to centralized servers [5]. Thanks to the hierarchical structure, this application is able to work in a good way as an IDS, optimizing the communications and resources in large-scale deployments [14].

DDoS on IoT: This context focuses on mitigating Distributed Denial of Service attacks targeting IoT devices, where HFL can effectively aggregate threat intelligence from numerous compromised nodes to enhance detection capabilities and resilience [4].

Underwater Internet of Things Wireless Sensor Networks UIoTWSN: This specialized application involves sensor networks deployed in aquatic environments, where robust and energy-efficient IDS are crucial due to the unique challenges of underwater communication and power limitations [3].

Big Data Security for IoT: This context addresses the challenge of securing big volumes of data generated by IoT devices, where HFL plays a critical role in enabling privacy-preserving security mechanisms and enhancing the detection of threats within these datasets, while also reducing the overhead of the systems [9].

3.2 Design Strategy

Continuing with the design strategy, we focus on the main architectural designs, examining how different HFLs are configured to use intrusion detection capabilities across various IoT systems.

HFL + Spiking Neural Networks: This approach takes advantage of the energy efficiency of spiking neural networks within an HFL framework to reduce power consumption in resource-constrained IoT devices, offering a sustainable solution for IDS [5].

HFL + Blockchain: This design was invented to improve security and efficiency in distributed cyber-physical and IoT systems. Here, HFL reduces communication and resource constraints by splitting training into multiple stages, while Blockchain complements this by validating and authenticating model updates through a decentralized ledger, preventing malicious contributions and ensuring trust without a central authority. The main benefit of this approach is to enhance accuracy, transparency, and resilience [13].

HFL + Knowledge Distillation: This approach was created to build efficient IDS for Cloud-Edge ecosystems but also preserving privacy. Here, HFL reduces communication overhead, while KD compresses large models into smaller ones, transferring knowledge without sacrificing accuracy. This model lowers communication costs and maintains good detection performance in large-scale IoT networks [14].

Clustered Federated Learning: While this application is not called Hierarchical, it uses a similar method as other HFL implementations. It was created for IDS in IIoT systems. It groups edge nodes with similar data distribution using time-series evaluation metrics, which enables model aggregation. The results showed that CLF-IDS has a good performance under data imbalance while being effective against poisoning attacks [10].

Hierarchical Federated Incremental Learning Network HFIN: This approach was designed for IIoT systems that encounter evolving cyberattacks. It combines HFL with Class-Incremental Learning to allow training across IIoT devices while having data privacy and continuous model updates as new attacks appear. This approach enhances adaptability, accuracy, and resilience of global models [2].

HFL + Hierarchical Long-Term Memory (HLSTM): This strategy was created for IoMT devices using a Dew-Cloud architecture that improves data privacy. The HLSTM architecture was implemented to improve computation efficacy and intrusion detection accuracy, while also having the capability of being scalable. The results showed that HFL + HLSTM performed better than conventional HFL at attack detection [12].

HFL + CNN + Bidirectional Long Short-Term Memory (Bi-LSTM): This architecture was applied to **UIoTWSN** using a convolutional neural network to extract spatial features from the datasets and then send this data to the Bi-LSTM network to analyze the temporal features in network traffic. Then it uses HFL for decentralized training on the underwater sensor nodes, ensuring data privacy and scalability. This model showed good results making it robust, adaptive, and suitable for underwater IoT devices [3].

3.3 Dataset(s) Used

Datasets play a central role in evaluating intrusion detection systems within Hierarchical Federated Learning (HFL). The choice of dataset directly influences the validity, robustness, and comparability of research results. For this reason, studies in this domain rely on a variety of benchmark and domain-specific datasets that capture different aspects of IoT, IIoT, and network traffic scenarios. We will summarize the datasets most frequently employed in HFL-based intrusion detection research, highlighting their characteristics and contexts of use:

- **TON-IoT**: The ToN-IoT dataset comprises diverse data from the IIoT ecosystem including device telemetry, OS logs, and network traffic [6].
- **NSL-KDD**: The NSL-KDD dataset, a derivative of the KDDcup99 dataset, is widely recognized in intrusion detection research. It addresses limitations present in its predecessor by reducing redundancy and enhancing data representativeness [12].
- **NF-ToN-IoT-v2**: An IoT network dataset derived from NetFlow, encompassing a variety of attack categories. Its NetFlow records are generated using publicly accessible PCAP files from the ToN-IoT dataset [2].
- **NF-UQ-NIDS-v2**: This dataset is a compilation that integrates four sources: NF-UNSW-NB15-v2, NF-ToN-IoT-v2, NF-BoT-IoT-v2, and NF-CSE-CICIDS2018-v2. It encompasses network flows from a variety of network environments and attack scenarios [2].
- **Edge-IIoTset**: This dataset encompasses a wide range of IoT and Industrial IoT network traffic, specifically designed for evaluating intrusion detection systems in edge computing environments [5].
- **CICDDoS-2019**: The dataset is fully labeled, with 80 network traffic features computed for both benign and denial-of-service flows using the publicly available CICFlowMeter software from the Canadian Institute for Cybersecurity [7].

- **CICIDS**: The Canadian Institute for Cybersecurity Intrusion Detection System dataset is a benchmark used for evaluating IDS, providing realistic traffic covering multiple attack scenarios like botnet activity, DDoS, Infiltration, password guessing, and port scanning [15].
- **CICIoT-2023**: This dataset was created to test the intrusion detection in IoT environments. It includes topology for 105 real IoT devices and a total of 33 distinct attacks divided into 7 categories [16].
- **CICIoMT-2024**: This dataset was built to be a realistic and multi-protocol benchmark for the Internet of Medical Things. It contains a total of 18 different types of attacks on a set of 40 IoMT devices [4].
- **UNSW-NB15**: It is focused on IDS for general networks. It contains $\approx$ 2.5M of records with 49 features per flow. It has 9 categories of intrusion attacks. It was created to overcome limitations of older benchmarks like KDD99 [17].
- **Gas pipeline (GP)**: It is part of a Supervisory Control and Data Acquisition dataset created for IDS that comes from a laboratory-scale gas pipeline control system. It contains a total of 28 different types of attacks and includes network traffic and payload content features [18].

The datasets outlined above differ in terms of scale, feature composition, and attack diversity. However, raw datasets often contain noise, redundant information, or imbalanced distributions, which may limit their direct applicability to intrusion detection models. To address these issues, researchers adopt a variety of preprocessing techniques before training or evaluation. The following subsection reviews these preprocessing strategies, categorizing them into seven distinct classes.

3.4 Data Preprocessing

To better understand how these datasets have been used, it is important to examine how different papers handle them. We split the preprocessing techniques into eight different classes:

1. **Data Cleaning and Noise Removal**: In this category, [8] applied feature removal to eliminate redundancy to improve dataset quality, reduce noise, and eliminate irrelevant features that could affect training. Similarly, [2] removed redundant rows to reduce noise. [5] and [14] also applied a similar strategy for cleaning network traffic by dropping unnecessary flow features such as IP addresses, ports, and payload information to produce a cleaner dataset.
2. **Feature Encoding and Transformation**: For this technique, [8,11,14,15], and [2] applied one-hot encoding to convert non-numeric labels into numerical features, making them suitable for machine learning models. In addition, [8] applied binary mapping to convert additional dataset categories into numeric values. Lastly, [15] encoded all NaN values as 0 to maintain compatibility with learning algorithms.
3. **Normalization**: To allow fair comparisons across multiple features, normalization is applied to define a range, typically between 0 and 1. This technique

was used by [5,8,11,12,15], and [14]. The reason provided was to improve efficiency and classification accuracy. [2] also applied normalization, though they did not specify a reason.

4. **Resampling and Class Balancing**: These techniques are applied when datasets contain imbalanced distributions. [2] applied down-sampling for this reason. To address the same issue, [15] applied resampling to generate new samples for minority classes using SMOTEENN.
   These techniques helped improve classification results. A similar technique, but for a different purpose, was applied by [10] where they used artificial imbalance partitioning to construct multiple versions with large imbalance ratios. This was done to simulate real-world IIoT scenarios and test model robustness.
5. **Feature Selection**: Some of the datasets used in these papers contain a large volume of data, much of which is not useful. Different techniques are applied to extract the most important features. [15] used an *autoencoder* for feature extraction, converting high-dimensional data into lower-dimensional representations to capture essential features and reduce dimensionality, thus reducing computational overhead. [11] applied a similar strategy using a *random forest* to identify the 60 most important features in their datasets.
6. **Dimensionality Reduction**: Another technique applied to reduce dataset complexity and communication overhead between devices is dimensionality reduction. [8] applied Principal Component Analysis (PCA) to shrink dataset complexity. The results showed negligible losses compared to the dataset without PCA. [2] and [12] also applied PCA for the same purpose.
7. **Regularization**: [15] applied this technique to alleviate overfitting. They used L2 regularization, which adds a penalty equal to the square of the magnitude of the coefficients. This prevents the model from becoming overly complex and improves generalization [19].
8. **Uncategorized**: Some of the papers used for this taxonomy did not specify preprocessing techniques for the datasets they employed [3,9], and [4].

Overall, preprocessing plays an important role in enhancing model robustness, reducing computational complexity, and improving classification performance in HFL systems. However, preprocessing alone does not determine the success of these systems. The challenges that HFL aims to address range from scalability and heterogeneity to security among others. The following subsection examines the specific problems that existing studies seek to solve with HFL.

3.5 Problem Addressed

We identified six categories of problems that HFL for IDS in IoT aims to address:

1. **Data imbalance and dispersion**: This category focuses on imbalanced datasets across nodes, which can affect model training and reduce the accuracy of IDS. This is particularly relevant in real-world IDS scenarios, where, depending on the context, some types of attacks may be absent from local

client datasets, resulting in a drop in overall model recognition performance [11,15].

2. **Privacy and security protection**: In HFL, privacy and security are crucial challenges, given that preserving the confidentiality and integrity of sensitive IoT data is a priority in these models. Doing so reduces the risks associated with data leakage or attacks during model training [4,9].
3. **Communication overhead reduction**: This problem involves optimizing the exchange of model updates between IoT devices and the centralized server, reducing bandwidth consumption and latency, which is crucial in resource-constrained environments [5]. This limitation is a common criticism of applying FL in large-scale intrusion detection systems [14].
4. **Resource and energy constraints**: This category highlights the limitations of IoT devices in terms of processing power, memory, and energy capacity, which restrict the deployment of complex IDS models. High computational demands can quickly drain battery-operated devices and limit scalability, making sustainability a central concern in real-world HFL-based intrusion detection scenarios [5].
5. **High false positives and inefficiency**: Similar to the previous problem addressed, due to the resource-constrained and dynamic environments that IDS tend to be implemented, they suffer from high false alarm rates and inefficiencies. Having a low bandwidth, high latency, and signal degradation tend worsen these issues, resulting in frequent misclassifications and unnecessary alerts. These problems are the cause of these persistent challenge for IDS solutions [3].
6. **Dynamic defense against cyberattacks**: New techniques and attacks variants appear continuously. This creates a problem for static or rule-based detection methods ineffective over time. The problem requires IDS to adapt as quickly as possible while retaining knowledge of previous attacks. Without this capability, models are in risk to get obsolete quickly, leaving IoT ecosystems vulnerable to new and more sophisticated attacks [2].

4 Conclusions

This paper presented a taxonomy of Hierarchical Federated Learning (HFL) applied to Intrusion Detection Systems (IDS) in IoT environments. By classifying the literature across five dimensions: context of application, design strategy, datasets, problem addressed, and preprocessing techniques we identified the main directions in which HFL is advancing.

Our analysis shows that HFL has been successfully applied to domains such as healthcare, industrial IoT, and underwater networks, with architectures integrating blockchain, spiking neural networks, and incremental learning. Commonly used datasets include CICIDS, CICIoT2023, and ToN-IoT, often requiring extensive preprocessing to ensure data quality and class balance.

4.1 Open Problems

Despite the advances in HFL on IDS over multiple IoT domains, there are still unresolved open challenges. Handling data heterogeneity in IoT environments is known for its highly non-IID data, unstable connectivity, and limited resource capabilities. While some frameworks, such as HiFL uses heterogeneous features [11] and [10] introduce clustered-based strategies for non-IID conditions, scalability and robustness in real deployments are still limited. Another concern is the dynamic adaptability to evolving threats. Although HFIN by [2] showed good results in this scope for IIoT environments, other proposed systems still struggle to address novel and unseen attacks effectively [15]. Also, sustaining low latency and resource efficiency in constrained environments as shown by [5] is still an open problem when balancing between detection accuracy, communication costs, and energy usage.

Privacy and trust management is still another of the areas where HFL implementations have bottlenecks. While the blockchain integrations proposed by [13], and [3], and the cryptography implementation by [9] improve security, these methods introduces additional computational overhead and may not be suitable for large-scale and latency-sensitive networks. Approaches using HFL-HLSTM for healthcare by [12] show strong detection accuracy but still require further investigation into interoperability.

Future research should prioritize developing more lightweight, adaptive, and privacy-preserving HFL models that can sustain deployments in highly dynamic, distributed, and resource-constrained networks.

Acknowledgements. Authors acknowledge the Escuela de Ciencias de la Computación e Informática, the Centro de Investigación en Tecnologías de la Información y Comunicación, Universidad de Costa Rica and the Posgrado en Computación e Informática. Project number 834-C2-534.

References

1. Harkai, A., Ciurea, C.: Economic impact of IoT and conventional data breaches: cost analysis and statistical trends. Control. Cybern. **53**, 385–404 (2025). https://doi.org/10.2478/candc-2024-0017
2. Mao, J., Wei, Z., Li, B., Zhang, R., Song, L.: Toward ever-evolution network threats: a hierarchical federated class-incremental learning approach for network intrusion detection in IIoT. IEEE Internet Things J. **11**(18), 29864–29877 (2024). https://doi.org/10.1109/JIOT.2024.3408634
3. Samson, S.A., Nagarajan, M.: A federated learning-based hierarchical intrusion detection system for underwater IoT wireless sensor networks. In: 2024 4th International Conference on Ubiquitous Computing and Intelligent Information Systems (ICUIS), pp. 1781–1787. IEEE (2024). https://doi.org/10.1109/ICUIS64676.2024.10867194
4. Gui, J., Ji, R., Huang, H., Hong, J., Hua, C.: Solving data contamination in DDoS detection: a method based on hierarchical federated learning. IEEE Trans. Inf. Forensics Secur. **20**, 7013–7028 (2025). https://doi.org/10.1109/TIFS.2025.3587185

5. Aouedi, O., Piamrat, K.: SURFS: sustainable intrusion detection with hierarchical federated spiking neural networks. In: ICC 2024 – IEEE International Conference on Communications, pp. 2173–2178. IEEE (2024). https://doi.org/10.1109/ICC51166.2024.10622560
6. Gad, A.R., Nashat, A.A., Barkat, T.M.: Intrusion detection system using machine learning for vehicular ad hoc networks based on ToN-IoT dataset. IEEE Access **9**, 142206–142217 (2021). https://doi.org/10.1109/ACCESS.2021.3120626
7. Sharafaldin, I., Lashkari, A.H., Hakak, S., Ghorbani, A.A.: Developing realistic distributed denial of service (DDoS) attack dataset and taxonomy. In: 2019 International Carnahan Conference on Security Technology (ICCST), pp. 1–8. IEEE, Chennai, India (2019). https://doi.org/10.1109/CCST.2019.8888419
8. Singh, P., Gaba, G.S., Kaur, A., Hedabou, M., Gurtov, A.: Dew-cloud-based hierarchical federated learning for intrusion detection in IoMT. IEEE J. Biomed. Health Inform. **27**(2), 722–731 (2023). https://doi.org/10.1109/JBHI.2022.3186250
9. Awan, K.A., Din, I.U., Almogren, A., Rodrigues, J.J.P.C.: Privacy-preserving big data security for IoT with federated learning and cryptography. IEEE Access **11**, 120918–120934 (2023). https://doi.org/10.1109/ACCESS.2023.3328310
10. Shan, Y., Yao, Y., Zhou, X., Zhao, T., Hu, B., Wang, L.: CFL-IDS: an effective clustered federated learning framework for industrial internet of things intrusion detection. IEEE Internet Things J. **11**(6), 10007–10019 (2024). https://doi.org/10.1109/JIOT.2023.3324302
11. Wang, L., Liu, L., Lu, Y., Zhang, C., Zheng, Y., Xu, J.: Hierarchical federated learning for heterogeneous features and distributed data in IoT networks. In: 2023 IEEE/CIC International Conference on Communications in China (ICCC), pp. 1–6. IEEE (2023). https://doi.org/10.1109/ICCC57788.2023.10233409
12. P, M.K.D., Babu, A., NagaSireesha, J., Venkatesh, B., Siddiqua, A.: Hierarchical federated learning-based method for privacy-preserving in the healthcare environment. In: 2023 International Conference on Evolutionary Algorithms and Soft Computing Techniques (EASCT) (2023). https://doi.org/10.1109/EASCT59475.2023.10393636
13. Aloqaily, M., Al Ridhawi, I., Karray, F., Guizani, M.: Towards blockchain-based hierarchical federated learning for cyber-physical systems. In: 2022 International Balkan Conference on Communications and Networking (BalkanCom), pp. 46–50. IEEE (2022). https://doi.org/10.1109/BalkanCom55633.2022.9900546
14. Almabdy, S., Ullah, A.: Optimising intrusion detection systems in cloud-edge continuum with knowledge distillation for privacy-preserving and efficient communication. In: 2025 IEEE 9th International Conference on Fog and Edge Computing (ICFEC), pp. 1–5. IEEE (2025). https://doi.org/10.1109/ICFEC65699.2025.00016
15. Zhang, J., Yu, P., Qi, L., Liu, S., Zhang, H., Zhang, J.: FLDDoS: DDoS attack detection model based on federated learning. In: 2021 IEEE 20th International Conference on Trust, Security and Privacy in Computing and Communications (TrustCom), pp. 635–642. IEEE (2021). https://doi.org/10.1109/TrustCom53373.2021.00095
16. Neto, E.C.P., Dadkhah, S., Ferreira, R., Zohourian, A., Lu, R., Ghorbani, A.A.: CICIoT2023: a real-time dataset and benchmark for large-scale attacks in IoT environment. Sensors **23**(13), 5941 (2023). https://doi.org/10.3390/s23135941
17. Moustafa, N., Slay, J.: UNSW-NB15: a comprehensive data set for network intrusion detection systems (UNSW-NB15 network data set). In: 2015 Military Communications and Information Systems Conference (MilCIS), pp. 1–6. IEEE, Canberra, Australia (2015). https://doi.org/10.1109/MilCIS.2015.7348942

18. Morris, T., Gao, W.: Industrial control system traffic data sets for intrusion detection research. In: Butts, J., Shenoi, S. (eds.) ICCIP 2014. IAICT, vol. 441, pp. 65–78. Springer, Heidelberg (2014). https://doi.org/10.1007/978-3-662-45355-1_5
19. Robissout, D., Bossuet, L., Habrard, A., Grosso, V.: Improving deep learning networks for profiled side-channel analysis using performance improvement techniques. Cryptology ePrint Archive, Paper 2021/1546 (2021). https://doi.org/10.1145/3453162. https://eprint.iacr.org/2021/1546

Integrating Traditional and Emerging Technologies for the Professional Learning of Prospective Mathematics Teachers

Jhony Alexander Villa-Ochoa(✉)

Universidad de Antioquia, Medellín 050010, Colombia
jhony.villa@udea.edu.co

Abstract. Research in mathematics teacher education emphasizes the transformative role of digital technologies in professional learning. This exploratory study investigates a professional learning task that merges traditional tools, such as calculators, with emerging technologies, such as artificial intelligence, to promote mathematical and didactical learning among 11 prospective mathematics teachers. Data were gathered over a 16-week course in a Mathematics Teacher Education Program at a public university in Colombia. The study was guided by the Humans-with-Media epistemological perspective and the Professional Learning Opportunities for Teachers (PLOT) model. Thematic analysis identified opportunities for professional learning and discursive interactions. Findings indicate that interactions among prospective teachers, the educator, and technologies promoted argumentation, justification, and reflection on the pedagogical implications of different tools. The results show that integrating traditional and emerging technologies can create more dialogic learning environments, helping future teachers develop integrated mathematical, didactic, and technological learning.

Keywords: Calculators · AI · Professional Learning

1 Introduction

Mathematics teaching continues to face ongoing challenges, particularly in designing learning environments that effectively integrate contextual, institutional, and human factors with mathematical content and resources [1]. Recently, the difficulty of teaching mathematics in increasingly digitalized educational settings has intensified [2]. Incorporating digital resources involves more than just updating classroom methods, improving content teaching, or increasing student engagement. It also requires resisting their subordination to traditional teaching routines and instead leveraging their potential to create innovative learning opportunities [2].

For prospective mathematics teachers (PMTs), learning environments should foster experiences that prepare them for their future professional roles; therefore, these environments must go beyond mere mastery of mathematical concepts, processes, and procedures. It is essential that they also encourage an integrated understanding of the interdisciplinary, pedagogical, didactic, technological, social, and cultural factors that influence

A. Rocha et al. (Eds.): ICITS 2026, LNNS 1966, pp. 602–613, 2027.
https://doi.org/10.1007/978-3-032-27538-7_53

teaching and learning. Comprehensive professional development provides teachers with a strong foundation for making informed decisions about how to teach, which resources to select, how to address classroom diversity, and how to connect mathematics with students' contexts and lived experiences [3, 4]. However, establishing the methodological, theoretical, and contextual foundations to support this kind of professional learning remains a challenge for researchers.

To contribute to this field, a course for prospective mathematics teachers was developed to foster professional growth by encouraging analysis of learning environments and exploring the role of digital resources in building conceptual understanding. This study investigates the opportunities for professional learning that emerged within the course, emphasizing how the integration of digital technologies and reflective activities helped PMTs connect mathematical and didactical knowledge with their future teaching practices. Specifically, this chapter aims to answer the following research question: How does the integration of digital technologies support PMTs' professional learning?

2 Theoretical Background

2.1 Digital Technologies and Mathematics Teacher Education

Contemporary research in mathematics education emphasizes the growing importance of digital technologies in the preparation and ongoing professional development of mathematics teachers [5, 6]. While the COVID-19 pandemic acted as a catalyst—prompting an unprecedented adoption of technological tools and remote teaching environments [1]—the need to prepare teachers to understand, utilize, and effectively integrate digital resources and technologies into daily school practices remains a key focus in current research agendas [5, 6]. As Thurm et al. [6] note, professional development for MPTs includes approaches aimed at strengthening both mathematical and technological knowledge (e.g., the TPACK framework) and provides examples that can be adapted to other university settings. The authors highlight that such programs must specify not only topics, technologies, duration, and contact hours but also the design principles guiding the learning environment—and they must evaluate whether new principles emerge from empirical evidence.

The increasing use of digital technology in mathematics education—such as calculators, CAS, and DGE—has expanded how students think about and learn mathematics through multiple representations and interactive explorations [2]. The literature emphasizes the importance of professional development that helps teachers form informed views on the role of technology, improving both technical skills and didactical and pedagogical competencies. Overall, research shows that teachers with stronger technological knowledge are more likely to use these tools effectively, focusing on mathematical ideas rather than just procedures and routines.

2.2 Humans-With-Media as an Epistemological Perspective

This study is based on an epistemological view that considers knowledge as a collective production involving both humans and media (humans-with-media – H-w-M [7]). From this perspective, the epistemic subject is not an isolated human but a unit composed of humans and non-humans. Technology is more than just a tool; it is a crucial part of cognition. As a result, the concept of use becomes vital, as the way we interact with technology shapes what can be thought and learned in educational settings. As Villa-Ochoa et al. [1] mention, within the H-w-M perspective, the idea of 'use' goes beyond simply the 'presence' of technology in the system and highlights how technology can provide opportunities to help the subject achieve their goals and meet their specific needs.

Two core postulates of the H-w-M framework—that cognition is a social enterprise and that it fundamentally involves media—allow us to argue that different technologies produce various types of knowledge. As a result, analyzing teaching and learning processes must include not only human interactions but also the technological mediations that enable and shape them.

2.3 Teacher's Professional Learning

The professional learning of mathematics teachers has been extensively studied from various perspectives, acknowledging that it involves more than just acquiring conceptual knowledge; it also includes complex integrations of pedagogical, disciplinary, technological, and contextual knowledge [8]. Efforts to better understand mathematics teachers' thinking have resulted in the development of theoretical frameworks and approaches that organize and interpret their professional learning experiences and opportunities. Ribeiro and Ponte [4] proposed the Professional Learning Opportunities for Teachers (PLOT) theoretical model, designed to guide the development of teacher education programs and to support the interpretation of teachers' professional learning opportunities.

The model and its components are designed to guide the development of professional learning processes that ultimately promote teacher growth. The PLOT model includes two dimensions—conceptual and operational—and four components within each of its domains: (i) PAF (Role and Actions of the Teacher Educator): The components of this domain are Approximation and Articulation (conceptual dimension), and Management and Orchestration (operational dimension); (ii) TAP (Professional Learning Task): The components of this domain are Professional Knowledge and Exploratory Teaching (conceptual dimension), and Mathematical Task and Records of Practice (operational dimension), (iii) IDP (Discursive Interactions among Participants): The components of this domain are Mathematical and Didactical Discussions and Argumentation and Justification (conceptual dimension), and Mobilized Language and Dialogic Communication (operational dimension). Figure 1 represent the PLOT model.

These elements define the model's structure and theoretical foundations (conceptual dimension) and guide its implementation (operational dimension). Collectively, these dimensions help organize teacher education processes and enable researchers to determine whether—and how—professional learning opportunities develop for teachers [4].

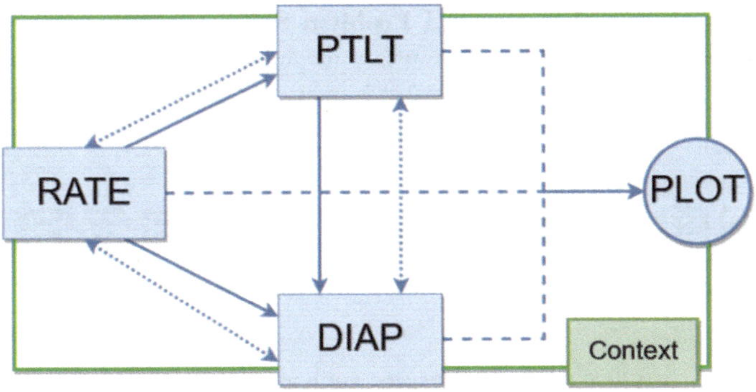

Fig. 1. PLOT model retrieved by Ribeiro and Ponte [4]

3 Methodology

3.1 Context and Participants

This study was conducted in a course titled Pedagogical Practicum Seminar [Seminario de Práctica Pedagógica], involving 11 prospective mathematics teachers (five men and six women) from a public university in Colombia. The course, conducted in a seminar format over 16 weeks and meeting for 4 h each week, examined the role of resources in mathematics teaching by combining theoretical reflection with practical classroom experiences.

The course is part of the third year of a five-year mathematics teacher education program in a city about three hours from Medellín, characterized by an agricultural and livestock-based economy. Over six weeks, the PMTs visited local schools to observe available resources, understand the educational context, interact with in-service teachers, and prepare lesson plans. They also analyzed research on educational resources, participated in professional learning environments and workshops, and engaged in dialogues with researchers from other regions, among other authentic learning experiences [3].

3.2 Professional Learning Task

Based on the PLOT model [4], a Professional Learning Task (TAP) was designed that integrated calculators and AI. The task was inspired by previous studies conducted by MATHEMA (e.g., [9]).

Stage 1: Exploring with the Calculator. At this stage, participants examined how repeatedly pressing the equal sign key on a basic calculator affects the results, a pattern also reproduced in some scientific and mobile-phone calculators. The PMTs engaged in counting activities and solved problems involving arithmetic and geometric progressions and linear equations, while reflecting on the calculator's role and its various uses described in the literature. Through their own experience, PMTs identified uses such as conceptual development, verifying results, calculating results, and manipulating the calculator as tools [9, 10].

Stage 2: Transcending Patterns and Problem Solving. At this stage, PMTs were asked to repeatedly press the equal sign key to generate numerical sequences resembling arithmetic progressions. In this case, however, the procedure involved adding or subtracting two numbers. For example (Fig. 2):

[5] [+] [3] [=] [=] [=] ... [=] [+] [7] [=] [=] [=] [=] ... [=]

Fig. 2. Example of key sequence entered on the calculator.

This sequence produces different results, such as: 171, 248, 345, etc. The Mathematics Teacher Educator–MTE asked the PMTs to analyze the displayed results and posed the following question: How many times could the equal sign key have been pressed to obtain the final result displayed on the calculators? Each PMT worked individually to find a possible answer to this question. Later, the PMTs gathered to share their solutions and discuss the nature of the results. The MTE closely monitored their reasoning to support their understanding, emphasizing that no single solution exists, encouraging the identification of alternative patterns, and clarifying misconceptions that arose during the discussion.

Stage 3: Contrasting and Generalizing the Problem Solution with AI. The PMTs engaged in the "Guided Learning" mode of Gemini to either add to or compare with the solution of the equation. Previously, MTE and MPT discussed the importance of prompts given to the AI. Following this advice, the group studied various prompt templates and examined illustrative examples.

In a subsequent plenary, the group discussed their findings, and the MTE extended the discussion by introducing variations of the equation, such as $5 + 3\text{ m} + 7\text{n} = \text{a}$. The MTE raised new questions to challenge their results—for example: Does your strategy work for each value of a founded by your peers? Would it work for any arbitrary a? If not, what alternatives could be considered? This stage of the task was meant to help students develop an understanding of a previously unstudied topic (Diophantine equations), which naturally arose as a way to solve this type of problem.

3.3 Data and Analysis

The data for this study were constructed from direct observations, field notes, students' working documents, and digital files submitted for analysis. The analysis was conducted in two phases, seen as a continuous comparison between the data and theoretical frameworks. In the first phase, the data were organized, evaluated for relevance to the research question, labeled, and uploaded to Atlas.ti. Next, episodes were selected that demonstrated components of the model proposed by Ribeiro and Ponte. Specifically, the analysis focused on data related to the Role and Actions of the Teacher Educator (PAF) and the Discursive Interactions among Participants (IDP)—including non-human participants like the calculator and AI system. These data were coded as PAF and IDP, respectively.

Given the nature of the data and the epistemological perspective adopted in this study—where both human and non-human actors participate in technological activity—this chapter concentrates on the findings related to the IDP domain. The rest of the analysis followed the guidelines for thematic analysis proposed by Braun and Clarke [11].

4 Results

This section is divided into two themes. The first—Opportunities for Professional Learning—is centered on episodes related to both didactic and mathematical meaning, as well as the use of digital technologies. The second—Discursive Interactions among Participants (IDP)—highlights excerpts that demonstrate interactions among PMTs, the MTE, and the technologies involved (a calculator and AI).

4.1 Opportunities for Professional Learning

Mathematical and didactic meanings became intertwined as future teachers examined patterns created by repeatedly pressing the equal sign key on simple calculators. Throughout the task, learning opportunities arose regarding the use of artifacts, exploring new mathematical ideas, and reflecting on their future role in integrating technology into teaching. For instance, at the first stage, students quickly observed that "*each time the equal sign is pressed, the previous operation is repeated*" [Carlos]; however, "*this depends on the operation, since with multiplication [unlike addition] the product is repeated with the first entered term*" [Sandra]. They also realized that this behavior varies depending on the calculator type: "[...] *on my phone's calculator it doesn't work, nor on the scientific one*" [Raquel]. This exchange led to a dialogic interaction about how the artifact functions, prompting participants to compare simple, scientific, and digital calculators. The discussion revealed that the PMTs were unaware that some calculators automatically repeat an operation when the equal sign key is pressed.

In contrast, others require the user to adjust the input. This insight not only enhanced their technical understanding of the artifact but also sparked a discussion about the didactical implications of choosing different calculators for teaching students. In this regard, Sandra mentioned: "*You realize that you really need to learn not only what can be taught, but also how the devices are used. Here we saw that, although they are all calculators, they do not operate in the same way—their programming is different*".

At Stage 2, the PMTs modeled the keystroke sequence using the equation $5 + 3m + 7n = a$, where a represents the final output displayed by the calculator. They identified different solution strategies, detected patterns in the variation of the variables, and realized that—contrary to their initial assumptions—the equation had a finite set of solutions. When the Carolina was asked to analyze the solutions to the equation $5 + 3m + 7n = 325$, she initially stated, "*It has infinitely many solutions because it is just an equation with two unknowns,*" but later recognized that the calculator's pattern implied the solutions corresponded to pairs of natural numbers. This realization prompted them to create a GeoGebra graph of the equivalent equation and to focus on "*the points on the line that represent pairs of natural numbers.*" The PMTs repeatedly zoomed in on the graph to determine specific values and eventually recognized that, in this case, the set of solution pairs was finite.

Other students, such as Claudia, created tables of values and identified patterns in the solutions. The appearance of Cartesian graphs and tabular representations became complementary argumentative tools for exploring possible solutions to the equation. This process expanded participants' understanding of such equations and provided opportunities for group discussions about the strengths and limitations of different forms of language—natural, symbolic, graphical, and digital. At Stage 3, the PMTs reviewed various AI prompt templates and recognized their usefulness. For example, Raquel said: "*I already knew prompts were important in AI, since the quality of the results and the risk of hallucinations depend on them. However, I was not familiar with specific recommendations. I preferred the last template [Role, Task, Format, Category, Definition, Description, Example, Instruction, and Tone model* [12]*] because it seems clearer and more complete; it includes many details that I think are necessary*".

Afterward, the MTE asked the PMTs to use Gemini's guided learning mode. Carolina asked the AI: "*I used the equal sign on the simple calculator to construct equations. I found the equation* $5 + 3m + 7n = 325$*. Could you help me understand the solution?*" The AI provided information on the classification of this equation type and the nature of its variables (see Fig. 3).

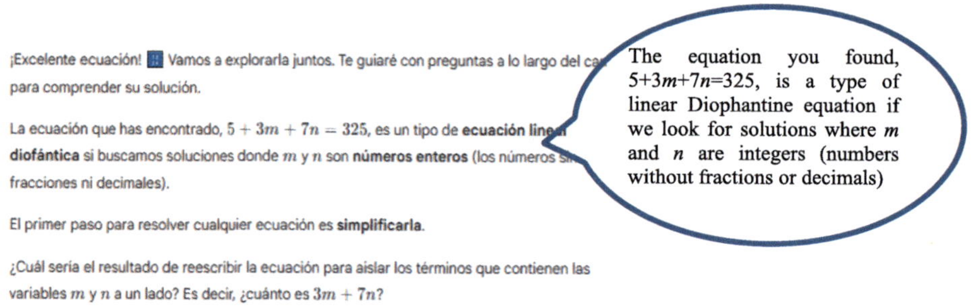

Fig. 3. Answer offered by AI

Upon seeing this response, Carolina and her group expressed surprise because it referred to a type of equation they had not studied before. As Santiago explained: "*I was familiar with types of equations such as linear, quadratic, systems of linear equations, trigonometric, etc. However, this is the first time I've heard of Diophantine equations. Professor, do these equations also have special methods for solving them?*" In this case, AI provided an informational contribution by clarifying the equation's type—an action that can be described as AI-guided clarification or labeling. This contribution was valuable because it introduced terms and definitions related to concepts the students had already explored, supplementing their initial intuition with a more formal, contextually grounded understanding. It also prompted the emergence of new mathematical questions.

4.2 Discursive Interactions Among Participants (IDP)

Previous studies [9] have shown that understanding geometric progressions and related teaching can be developed through this type of task. This study further highlights the strategic role of the MTE, who intervened to expand the meanings built by asking questions such as: "*How does the resulting number sequence behave? What changes? What stays the same? If the equal sign key continues to be pressed [extending the sequence], is it possible to reach the number 119,875? Argue whether yes or no, and if not, what would be the closest number, and why would it appear on the calculator?*" These interventions not only helped students formally recognize properties like ratios but also encouraged precise and well-reasoned mathematical and didactical reflections; for example, Sandra mentioned, "*Even if I don't have computers, tablets, or internet access, with a simple tool like a calculator, I can promote many forms of learning. With very little, I can still do a lot.*" This participation deepened understanding of arithmetic and algebra, facilitated comparisons across different representations, and promoted argumentation based on operational properties and how the technological tool functions.

However, this study focused on interactions among AI, the MTE, and PMTs. The data analysis revealed that students engaged in human–AI exchanges through question-and-answer interactions, mainly exhibiting inductive reasoning. For example, in the Carolina-IA exchange, four stages can be recognized: (i) identifying and discarding values that do not fit the equation, (ii) finding specific solutions, (iii) recognizing patterns to determine the general solution, and (iv) verifying the pattern. Figure 4 displays the AI responses in stages (iii) and (iv).

For the PMTs, this interaction allowed them to recognize that Diophantine equations are a type of linear equation. In Carolina's case, she said: "Through AI, I can see a general formula for the solutions, which is more rigorous and precise than the solutions I found one by one in the table [...] it is a more mathematical way of solving it." Carolina added, "I had noticed that there was a pattern: when one went down 7 [variable m], the other went up 3 [variable n]; and that's how I found all the solutions; but that AI solution is more algebraic, that is, it's more accurate [a more precise method]." In her presentation to classmates, she confirmed that the parametric equations for m and n could describe all the solutions they previously found based on the additive regularities among the variables.

At this point, the MTE asked Carolina and her classmates how the parametric equations were created and validated. In response, they explored some relationships, but eventually Carlos said, "*Well, I know it's valid because the equations gave me all the solutions we already found, but I don't know where the equations came from*". This suggests that when AI directly provides students with a system of equations as the solution to a problem, they can focus more on verifying the final answer than on understanding how it was generated. In other words, they emphasize the outcome over the process. Here, the role of the MTE is crucial in encouraging reflection on the mathematical meanings behind that construction.

Finally, in the last part of stage 3, MTE commented: "*We have seen that each of you found a different value for the same sequence of keys on the calculator. Could it be that with that sequence of keys we can find any integer number?*" PMTs formulated an equation $5 + 3m + 7n = a$, and created prompts to find possible restrictions to a.

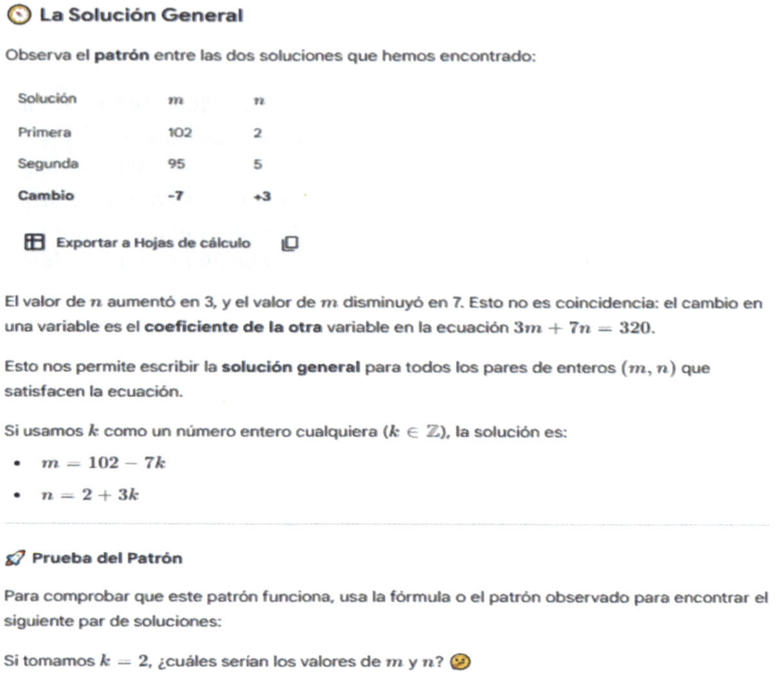

La Solución General

Observa el **patrón** entre las dos soluciones que hemos encontrado:

Solución	m	n
Primera	102	2
Segunda	95	5
Cambio	**-7**	**+3**

Exportar a Hojas de cálculo

El valor de n aumentó en 3, y el valor de m disminuyó en 7. Esto no es coincidencia: el cambio en una variable es el **coeficiente de la otra** variable en la ecuación $3m + 7n = 320$.

Esto nos permite escribir la **solución general** para todos los pares de enteros (m, n) que satisfacen la ecuación.

Si usamos k como un número entero cualquiera ($k \in \mathbb{Z}$), la solución es:

- $m = 102 - 7k$
- $n = 2 + 3k$

Prueba del Patrón

Para comprobar que este patrón funciona, usa la fórmula o el patrón observado para encontrar el siguiente par de soluciones:

Si tomamos $k = 2$, ¿cuáles serían los valores de m y n?

Fig. 4. Screenshot of IA solution

Beyond this role as a verifier of results, the AI also supplied explanations regarding the nature of the solutions of any Diophantine equation of the type $bm + cn = a$; which led the participants to revisit divisibility criteria; Greatest Common Divisor (GCD), explore the topic of residue classes modulo n, and generalize the solution to $bm + cn = a$; $a, b, c \in Z$.

Taken together, these themes show that discursive interactions among PMTs, the MTE, and technology create an essential space for both opportunities and challenges in professional learning. The shared development of mathematical and didactical meanings improves precise language use and emphasizes the importance of dialogic communication in mathematics teaching. At the same time, AI–student exchanges challenge teachers and teacher educators to intervene effectively, fostering understanding and reasoning that AI-generated solutions do not always make explicitly clear.

5 Discussion

The use of digital technologies in mathematics teacher education has been extensively documented in recent research, particularly in studies that examine their integration as key resources and media for the teaching of mathematics [1, 6, 7]. However, knowledge remains limited regarding how prospective teachers construct mathematical and didactic meanings through interactions with simple digital artifacts, such as basic calculators—especially when combined with contemporary tools like AI. This study proposed a professional learning task designed to explore how the integration of digital technologies enacts professional learning of PMTs.

Based on the H-w-M and PLOT model, this study presents two main findings. First, it describes a way to integrate traditional technologies, such as a simple calculator, with emerging technologies, such as AI, into professional learning tasks for PMTs. On one hand, the calculator was treated as an "object to be studied," meaning that mastering its technical and didactical uses and overcoming misconceptions about its integration into mathematics classes was essential. The calculator also served as a vehicle to encourage didactical reflection on learning algebraic topics, with an emphasis on covariation. Additionally, it provided a context for forming new relationships that led to other mathematical problems and the exploration of "new" topics.

In contrast, AI functioned as a 'platform' and a 'catalyst' for technology-human interactions, where inductive reasoning, concept definitions, problem-solving, and result verification were recognized. According to H-w-M [1, 7], these traits support the idea that learning—particularly professional learning—is not solely a cognitive process but is shaped by a hybrid system involving humans and non-humans. This system, influenced by contextual factors, purposes, and needs, interacts to create learning opportunities.

A second key finding of this study concerns discursive interactions. The results show that interactions among PMTs, the MTE, and technologies promoted meaningful mathematical and didactical discussions that went beyond a simple functional understanding of the artifact. A significant example was the statement, "this doesn't work the same way on all calculators", which sparked a conversation about the differences between calculator models and their didactical implications. This spontaneous remark prompted reflection on the professional knowledge needed to anticipate how calculators function in the classroom, recognizing that the type of technology used can influence both the development of mathematical knowledge and students' interpretations [9].

AI also played an essential role as a discursive agent in the sessions. At the third stage of the task, encouraged by the MTE, the prospective mathematics teachers turned to these tools to address the new challenges posed. Through their interactions with the AI, the PMTs verified patterns, interpreted results, and contrasted explanations. Additionally, this interaction allowed them to access new concepts—such as the Diophantine equation—and to construct its meaning as a way to generalize the solution to any task that reproduces the behavior of the equal sign key on calculators. As reported in this study, technologies were integrated as resources that generated opportunities for argumentation, justification, and the evaluation of solution validity—central aspects of the IDP domain [4].

6 Conclusion and Limitations

This study shows that professional learning tasks that combine traditional artifacts (e.g., calculators) with emerging technologies (e.g., AI) can create rich environments for developing teachers' professional knowledge. Such tasks encourage discursive practices that integrate mathematical, didactical, and technological knowledge, enhancing prospective teachers' ability to communicate, adapt, and justify their decisions within school settings. Based on the study's theoretical background, the findings emphasize that fostering dialogic environments is crucial for professional learning and that these environments can be expanded through the reflective use of technologies—provided their epistemological, rather than just instrumental, roles are clarified and challenged.

This study also identified several challenges. Although AI can provide quick and efficient solutions, it is still essential to maintain a critical perspective. Students might only verify AI-generated results rather than engage with the underlying reasoning. In this context, the role of the PTE becomes essential: MTE acts as a mediator, managing the interaction between PMT and AI, encouraging learners to look beyond the final answer, reflect on the process, question each step, and understand the mathematical concepts involved. In this way, technology becomes a valuable resource that promotes deeper, more reflective learning when guided by proper pedagogical strategies.

Despite its contributions, this exploratory study has limitations. The limited number of participants restricts how widely the results can be applied to findings. Its exploratory nature captures only initial tendencies rather than stable patterns. Additionally, the theoretical frameworks used may influence interpretations, emphasizing the need for further analysis with other approaches and perspectives.

Acknowledgments. I would like to thank CODI Universidad de Antioquia (grant No 2024-72250, Colombia) and CNPq, Brazil (grants No 405112/2023-8). I also would like to thank Statur Educational Foundation, leaders of the Gakuhan project for Casio calculators in Colombia for technical and academic supporter in this study.

References

1. Villa-Ochoa, J.A., Molina-Toro, J.F., Borba, M.C.: Roles of technologies for future teaching in a pandemic: activity, agency, and humans-with-media. ZDM – Math. Educ. **55**, 207–220 (2023)
2. Villa-Ochoa, J.A., Suárez-Téllez, L.: Computer algebra systems and dynamic geometry for mathematical thinking. In: Danesi, M. (ed.) Handbook of Cognitive Mathematics, pp. 843–868. Springer International Publishing, Cham (2022)
3. Rendón-Mesa, P.A., Parra-Zapata, M.M., Villa-Ochoa, J.A.: Opportunities of authentic modelling tasks in the learning of mathematical modelling for teaching. a study with prospective mathematics teachers. Uni-pluriversidad. **23**, 1–21 (2023)
4. Ribeiro, A.J., Ponte, J.P.: Modelo teórico para organizar e compreender as oportunidades de aprendizagem de professores para ensinar matemática. Zetetike. **28**, e020027 (2020)
5. Brandão, A., Pedro, L., Zagalo, N.: Teacher professional development for a future with generative artificial intelligence – an integrative literature review. Digit. Educ. Rev. **45**, 151–157 (2024)
6. Thurm, D., Bozkurt, G., Barzel, B., Sacristán, A.I., Ball, L.: A review of research on professional development for teaching mathematics with digital technology. In: Pepin, B., Gueudet, G., Choppin, J. (eds.) Handbook of Digital Resources in Mathematics Education, pp. 883–921. Springer International Publishing, Cham (2024)
7. Borba, M.C., Villarreal, M.E.: Humans-with-media and the reorganization of mathematical thinking. Springer, New York (2005)
8. Ponte, J.P.: Formação do professor de Matemática: Perspetivas atuais. In: Ponte, J.P. (ed.) Práticas profissionais dos professores de Matemática, pp. 343–358. Instituto de Educação da Universidade de Lisboa, Lisboa (2014)

9. Castrillón-Yepes, A., Villa-Ochoa, J.A., Parra-Zapata, M.M.: Integración de tecnologías en clase de matemáticas. Experiencias con la calculadora Casio Classwiz fx-991LA X. In: Parra-Zapata, M.M., Lau Mego, C., Zapata-Jaramillo, M.M., Vélez, M.L. (eds.) Hacia el desarrollo del pensamiento matemático con calculadora. Propuestas y experiencias, pp. 71–98. Fondo de Publicaciones Facultad de Educación—Universidad de Antioquia, Medellin (2024)
10. Borba, R., Selva, A.: Analysis of the role of the calculator in Brazilian textbooks. ZDM. **45**, 737–750 (2013)
11. Braun, V., Clarke, V.: Thematic analysis. In: A Practical Guide. Sage Publications (2021)
12. Youngblood, D.: Master Prompt Templating — for AI, GPT and other LLMs. (2023). https://medium.com/@thedavidyoungblood/master-prompt-templating-for-ai-and-llms-3f993f896d8f

Interactive NFC System: Multisensory Platform for Early Learning Based on VEX 123

Luis Ulcuango(✉), Rubén Jaramillo, Darío Guamán, and Mario Guamán

Instituto Tecnológico Superior Vida Nueva, Quito, Ecuador
{luis.ulcuango,ruben.jaramillo,dario.guaman,
mario.guaman}@istvidanueva.edu.ec

Abstract. In educational environments with limited resources, the lack of accessible and scalable technological tools hampers the implementation of interactive and multisensory learning experiences, key to early cognitive development. This article proposes a low-cost interactive system based on NFC technology, designed to strengthen learning through multisensory stimuli, democratizing access to educational innovations.

The methodology used was applied and descriptive, with an experimental approach validated in a real environment. Tests were carried out with 21 children aged between 3 and 6 years, using an Android mobile application (NFC reader), a web platform, Firebase Realtime Database (Backend) and programmable NFC cards, all based on the pedagogical principles of VEX 123.

The results demonstrate the effectiveness of the NFC-based system as an innovative educational tool, achieving 93% accuracy in identifying concepts, 72% increase in sustained attention and 89% acceptance by educators. These findings validate its potential for multisensory learning in children aged 3 to 6, especially in low-resource settings, thanks to its scalable and low-cost architecture. In addition, curriculum adaptability and the integration of gamification elements open up opportunities for future research, such as the incorporation of artificial intelligence and educational robotics, consolidating this model as a replicable alternative in emerging pedagogies.

In conclusion, the proposed system represents an innovative, accessible educational solution with high potential to strengthen multi-sensory learning in vulnerable environments.

Keywords: NFC · interactive learning · Firebase · educational innovation · initial education

1 Introduction

The integration of emerging technologies in early childhood education has shown potential for cognitive development at an early stage [7]. In contexts with limited resources, such as Ecuador, where studies report shortages of technological infrastructure in educational institutions [4], it is a priority to develop affordable solutions that facilitate meaningful learning. This research addresses the challenge with an interactive system

A. Rocha et al. (Eds.): ICITS 2026, LNNS 1966, pp. 614–625, 2027.
https://doi.org/10.1007/978-3-032-27538-7_54

based on Near Field Communication (NFC), a technology that is suitable for vulnerable educational environments due to its low cost (less than $0.80 per tag) and ease of implementation [1].

The developed prototype integrates three key technological components: (1) programmable NFC cards as tangible learning elements, (2) an Android mobile application that transforms common devices into portable readers, and (3) a web platform with Firebase Realtime Database for content management. This architecture, aligned with the educational design principles of VEX 123 [12], generates multisensory experiences where children associate visual, auditory and tactile stimuli [6]. Previous research indicates that similar approaches sustainably improve conceptual retention in preschool children [10], especially when they incorporate elements of gamification [2].

This work makes three key contributions: (1) empirical evidence on the effectiveness of NFC in early education, (2) a scalable model validated under real conditions, and (3) guidelines for developing accessible educational technologies. Preliminary results, obtained with 21 participants in Quito, show remarkable improvement in sustained attention (72%) and conceptual retention (93% final accuracy) surpassing previous reports for traditional methods in the region [4].

2 Theoretical Framework

2.1 NFC as a Physical-Digital Interaction Tool

Near Field Communication (NFC) technology enables a short-range wireless connection that facilitates interaction between physical objects and digital systems in an intuitive way [1]. This technology is particularly valuable in the educational field, as students can manipulate cards, figures or other tangible elements to activate multimedia content, reducing technical complexity and encouraging more active participation [3]. In addition, NFC favors contextual learning by linking concrete actions with immediate digital responses, reinforcing the association between abstract concepts and practical experiences [2].

In the context of early childhood education, NFC offers significant benefits by adapting to the motor and cognitive skills of young children, who can interact with the system without the need for complex interfaces [5]. Studies show that this type of interaction promotes autonomy and curiosity, which are key aspects in child development [10]. By removing technological barriers, NFC becomes an accessible tool for teachers and students, enabling smooth integration into educational play activities [1, 5].

2.2 VEX 123 as Educational Robotics Platform for Children

The VEX 123 system is specifically designed for children aged 3 to 6, combining basic robotics with interactive sensors that encourage computational thinking from an early age [12]. Its focus on the physical manipulation of components, such as programmable blocks and simple robots, allows children to experiment with concepts of sequence, logic and problem-solving in a tangible way [5]. This methodology is in line with the constructivist principles of learning through active exploration [7].

In addition, VEX 123 promotes essential skills such as teamwork and creativity as children collaborate to design simple solutions through trial and error [5]. Studies in early education highlight that age-appropriate educational robotics improves motivation and retention of concepts [10]. By integrating with technologies such as NFC [1], this system can extend its possibilities, creating multisensory experiences that reinforce learning in an integral way [6].

2.3 NFC Compared to Other Technologies

NFC stands out as a very attractive option compared to other interactive technologies, thanks to its excellent balance between cost and ease of use [1]. Table 1 shows the main advantages and disadvantages of three technologies commonly used in educational settings: NFC, QR codes and augmented reality [1, 3, 12].

Table 1. Technology comparison table.

Technology	Advantages	Limitations in Ecuador
NFC	Tangible interaction without screens, low cost (<$0.80/tag)	Requires phones with NFC
QR	No specialized hardware required	Scanning difficulty for young children
Augmented reality	high interactivity	High cost and need for powerful devices

The NFC system allows direct interaction and has a fairly low cost per tag (less than $0.80), although its implementation depends on the availability of telephones with this technology [1]. As an alternative, QR codes do not require special hardware, but are difficult to use for young children due to scanning accuracy requirements [3]. In contrast, technologies such as augmented reality, while offering greater interactivity, face barriers to adoption in resource-limited contexts due to their higher technical requirements and high costs [4, 12].

2.4 Pedagogical Foundations

The system is based on the principles of constructivism [7], where NFC cards operate as mediating objects that facilitate the assimilation of concepts through active interaction [1]. This proposal aligns with research on initial education in similar contexts [4], incorporating elements of gamification [2] with a formative assessment approach through immediate rewards that promote student autonomy [10].

2.5 Multisensory Learning in Early Education

Multisensory learning focuses on combining visual, auditory and tactile stimuli to help young children better understand and retain information [6]. This educational methodology is supported by neuroscientific research which shows that the brain assimilates concepts more effectively when several senses are activated at the same time [7]. In the classroom, using strategies such as NFC cards that incorporate sounds, images and textures allows students to connect ideas more efficiently, which in turn improves their attention and long-term memory [1].

In initial education, where learning takes place mainly through play and exploration, the multi-sensory approach is particularly important. Interactive materials, such as those combining NFC [1] and robotics [12], enrich learning experiences by involving not only sight and hearing but also touch and movement [5]. This not only facilitates the acquisition of knowledge but also adapts teaching to different learning styles, ensuring a more inclusive and personalized education [10].

2.6 Initial Education and the Need for Active Methodologies

Initial education, ranging from 3 to 6 years of age, is a fundamental stage in which the foundations for cognitive, social and emotional development are laid through meaningful experiences [8]. To make the most of this time, it is crucial to implement active methodologies that promote direct child participation rather than relying on passive approaches based solely on repetition [10]. Tools such as educational robotics [12] and NFC technology [1] enable students to learn through experimentation, play and interaction with their environment, which are key principles in modern pedagogies such as constructivism [7].

In addition, the use of interactive materials at this stage not only increases engagement but also helps to develop essential skills such as problem-solving, creativity and collaboration [5]. The integration of accessible technologies, such as NFC [1], ensures that these resources are easy to use for both teachers and students, democratizing access to innovative strategies [3]. By focusing on active learning, early childhood education can lay the foundations for integral development, preparing children for a future where technology and critical thinking are essential [7].

2.7 Technological Base and System Architecture

The architecture developed in this work strategically integrates three components to create an innovative educational ecosystem [1, 12]. The first component consists of using low-cost NFC cards (<$0.80/unit) [1] as a physical-digital interface, eliminating the need for advanced motor skills [3]. These cards allow multisensory (visual, auditory and tactile) associations [6] and are compatible with basic smartphones, facilitating access to innovative educational technology [4]. The integration of these three components demonstrates that it is possible to develop effective pedagogical solutions through the strategic combination of accessible technologies [3, 7], prioritizing usability and educational impact over technical complexity.

Integration in Real Time. The second component is an Android mobile application developed at Kotlin, which incorporates an NFC reading module optimized for preschool users. Its design follows children's usability principles [5], with simplified interfaces, immediate auditory feedback [6] and lower response times [1], allowing intuitive interactions through natural gestures [3].

Tangible and Multi-sensory Interface. The third component is a web platform based on HTML5, CSS3 and JavaScript that uses Firebase Realtime Database to achieve cross-platform synchronization with lower latencies [12]. This serverless architecture ensures automatic scalability and secure storage of individual data [10], while maintaining a smooth user experience. The solution takes advantage of cloud computing to overcome infrastructure constraints in resource-constrained educational environments [13].

3 Methodology

3.1 Research Approach

Focus. Mixed (qualitative-quantitative), with integrated collection and analysis of numerical data (success rates, interaction times) and qualitative data (teaching performance through surveys).

Type of Research. Quasi-experimental (pre-test/post-test without control group), with active intervention of the NFC and applied systems, aimed at solving a practical problem in the field of preparatory education.

Techniques. Quantitative, standardized tests (pre-test/post-test) to measure precision in concept identification. Qualitative, usability surveys of teachers (5-point Likert scale).

The mixed design made it possible to evaluate not only the objective effectiveness of the system (quantitative data), but also its feasibility in real contexts (qualitative perspective of teachers). The quasi-experimental techniques were selected for their adaptability to educational environments without altering school dynamics.

3.2 Participants

Sample. 21 children (3–6 years) from two public schools Escuela Fiscal Mixta "Bogotá" and Private School "Jhon M. Keynes" in Quito, Ecuador, selected for convenience (access to Android device).

Inclusion Criteria. Age within the target range, informed consent to teachers and parents.

Technological Profile. 89% of children use smartphones in their homes according to an initial survey.

3.3 System Design

The prototype was developed following a well-defined three-layer architecture:

Physical Layer. A set of Near Field Communication (NFC) cards programmed with codes representing different educational contents, such as animals, was used. These cards are cost-effective and reusable, making them particularly suitable for use in schools with limited resources. In addition, its programming is simple, allowing teachers to customize content quickly and easily.

Mobile Layer. The mobile application was developed in Kotlin, using the Android Studio development environment. Compatible with Android devices that have built-in NFC technology. Its main function is to read the codes stored on NFC cards and process the associated educational information. The app offers an intuitive interface, designed under VEX 123 principles to facilitate children's interaction with content in a playful and dynamic way.

Web Layer. For the web layer, Firebase Realtime Database was used as a cloud storage system. This database allows students' progress to be recorded in real time, as well as managing educational content and codes read from the mobile application. The web interface also integrates gamification components, showing points, levels and visual rewards that motivate the student to continue learning through play.

3.4 Ethical Considerations

This study was conducted in accordance with the consent was obtained from the parents and guardians of all underage participants, as well as verbal assent from the children. The research protocol was reviewed and approved by the Ethics Committee of Higher Technological Institute Vida Nueva. Confidentiality and anonymous use of data for exclusively academic purposes were guaranteed.

3.5 Procedure

Development Phase. In this phase, the implementation of the application in Android Studio was carried out on the mobile device that would function as an NFC reader, as well as the configuration of Firebase and the synchronization of the database with the web application to ensure real-time communication. Subsequently, NFC cards were prepared with educational content aligned to the Ecuadorian curriculum.

Each card included educational information that represented a specific concept, in this case related to animals. The elements incorporated were: the name of the animal, an illustrative image, a symbol representing its habitat, and a distinctive background color according to the category: blue for aquatic animals and green for terrestrial animals.

Validation Phase. In the initial session (Pre-Test), an assessment of previous knowledge was carried out using traditional physical cards (without NFC technology), which contained common images of animals such as dog, cat, lion, among others.

An evaluation instrument was designed consisting of 10 basic questions. During the session, the evaluator showed a card to the child and recorded whether the child correctly identified the animal, did not respond or gave an incorrect answer. The estimated working time per child was 5 to 7 min.

The variables measured at this stage were: accuracy in identification, calculated as the percentage of hits on the total number of cards submitted, response time, that is, the time it took for each child to give a reply after observing each card.

The data in Table 2 shows a clear relationship between the age of participants and their initial performance. The 3-year-olds achieved an average of 4.5 hits out of 10 (45% accuracy), facing difficulties especially with less common animals. The 6-year-olds, on

Table 2. Results of Pre-Test Measurements Obtained for each Variable

Age (years)	N	Average hits/10 (DE)	Average time (min) (DE)	% of Success
3	5	4.5 (± 1.2)	1.8(± 0.4)	45%
4	6	6.8 (± 1.0)	2.3(± 0.5)	68%
5	5	8.3 (± 0.8)	2.8(± 0.6)	83%
6	5	9.5 (± 0.5)	3.0(± 0.5)	95%
Total	21	7.0 (± 1.2)	2.1(± 0.5)	70%

the other hand, achieved an impressive 9.5 out of 10 (95% accuracy), showing almost total mastery of the task. Cognitive development translates into an average improvement of 1.25 points in the identification of concepts for each additional year. As for the response time, an inverse pattern was noted, the youngest (3 years) responded faster (1.8 min) but with less precision, while the oldest (6 years) took more time (3.0 min) to process the information, which could reflect a development of metacognitive skills such as self-regulation and response verification. These results provide a solid basis for assessing the impact of the NFC system, especially in those groups with the greatest scope for improvement.

Intervention Phase. The intervention was carried out in four sessions of 20 min over a week, using a progressive approach that combined tangible interaction with NFC cards, the reader (mobile phone + app), multimedia feedback through the web application and automated tracking with Firebase Realtime Database. In the first session, children were introduced to the system through guided activities associating images and concepts. In the following sessions, elements of gamification were incorporated, such as points, levels and audio-visual rewards, which increased motivation and participation. To ensure a smooth transition to Post-Test, the last session included practical exercises that replicated initial assessment conditions, allowing children to adapt to the format while teachers identified areas in need of reinforcement.

3.6 Collection of Data

For the data collection in the Post-Test session, we obtained the success rate in the recognition of animals and their concepts having as main source the automatic recording of correct and incorrect answers from the web application, on the other hand, for sustained attention, the continuous interaction time (in minutes) with the NFC cards per session was recorded. The source of this information was obtained from the Firebase records, which logged the start and end time of each interaction was obtained from the Firebase records where the start and end time of each interaction was observed. In addition, information on how teachers accepted the system was collected through a usability survey which was applied at the end of the intervention. This instrument, which is based on a 5-point Likert scale, assessed aspects such as ease of use, pedagogical relevance and general satisfaction with the range, complementing quantitative data with important qualitative insights.

4 Results

The results obtained after the implementation of the interactive system based on NFC in two educational units in Quito, with a sample of 21 children aged 3 to 6 years, showed significant improvements in three key dimensions, precision in learning (identification of concepts), sustained attention, high teacher acceptance.

4.1 Accuracy of Learning

As shown in Table 3, the system achieved statistically significant improvements in the identification of concepts in all age groups, except for the 6-year-old group which started at a very high level. The effect size (d of Cohen) for overall improvement was 1.85, which is considered a large effect according to Cohen's criteria [11]. This indicates that intervention with the NFC system not only produced a statistically reliable difference, but also a substantial and educationally relevant difference in learning accuracy.

It is important to note that the increase in average performance was 32.8% (from 7.0 to 9.3 points out of 10), significantly exceeding the levels expected for initial education.

Table 3. Comparison of success in identifying concepts (pre-test vs. post-test)

Age (years)	N	Pre-Test: Success/10 (DE)	Post-Test: Hints/10 (DE)	Increase (%)	Cohen's d	T (SPSS)	p
3	5	4.5	8.1	80.0%	3.36	4.87	< 0.01
4	6	6.8	9.0	32.4%	2.36	3.52	< 0.05
5	5	8.3	9.7	16.9%	1.92	2.91	< 0.05
6	5	9.5	10.0	5.3%	1.00	1.82	0.12
Total	21	7.0	9.3	32.8%	1.85	5.67	< 0.001

The data show that NFC has a different impact depending on the age of participants, revealing statistically significant improvements in concept identification. The 3-year age group was the most advanced, with an impressive increase of 80% (from 4.5 to 8.1 hits out of 10). This suggests that NFC technology is especially effective in the early stages of development, where multisensory stimuli (auditory, visual and tactile) help to compensate for limitations in verbal or memory skills. This remarkable advance indicates that the system could help close initial gaps in basic concept recognition.

For the 4- and 5-year-olds, the improvements were moderate but still significant (32% and 17%, respectively), highlighting that the system reinforced already existing learning, especially in more complex concepts such as habitats. In contrast, 6-year-olds already started from an almost perfect domain (9.5/10), which limited their improvement margin to only 5%. This shows that intervention is more transformative in children who have a weaker conceptual basis. Overall, the improvement of 32.8% ($p < 0.001$) validates the effectiveness of the system, especially considering that the average achievement

(9.3/10, 93% accuracy) exceeds the expected competence levels for initial education (8/10). Thus, NFC technology is presented as a scalable tool, with a higher return on investment in the earliest ages.

4.2 Sustained Attention

Table 4 shows noticeable improvements in sustained attention time after implementing NFC. The overall effect size was exceptionally high ($d = 3.00$), indicating a very large effect by Cohen's standards [11]. This demonstrates that the combination of tangible interaction and gamification elements was highly effective in keeping children engaged in structured activities.

Table 4. Comparison of sustained attention time (pre-test vs. post-test)

Age (years)	N	Pre-Test: time (min)(DE)	Post-Test: time (min)(DE)	Increase (%)	Cohen's d	t (SPSS)	p
3	5	1.8	3.1	72.2%	3.64	4.12	< 0.01
4	6	2.3	4.0	73.9%	3.53	4.35	< 0.01
5	5	2.8	4.5	60.7%	2.96	3.78	< 0.05
6	5	3.0	4.8	60.0%	3.60	3.25	< 0.05
Total	21	2.1	3.6	71.4%	3.00	4.32	< 0.001

The data shows noticeable improvements in sustained attention time after implementing NFC, with an average increase of 71.4% across all age groups. Children aged 3 and 4 stand out, with increases of 72.2% and 73.9% respectively. This shows that the combination of tangible interaction (NFC cards) and gamification (audio-visual rewards) managed to capture and retain their attention in structured activities, effectively doubling the initial attention time. This is critical, as it suggests that the system helps to mitigate common early development constraints such as impulsiveness or rapid fatigue.

In the 5- and 6-year-olds, although the relative increase was smaller (60.7% and 60.0%), they achieved longer absolute attention periods (4.5 to 4.8 min), which is close to the optimum for pedagogical activities in this age group. The difference is due to the fact that these children already started from higher baseline times (2.8 and 3.0 min), which limited their relative margin of improvement. On a qualitative level, the immediate feedback of the app (for example, sounds when answering correctly) and the intuitive design of the NFC cards were key factors in maintaining motivation. These findings support the idea that low-tech but well-designed systems can enhance cognitive skills not only in accuracy, but also in performance during educational tasks.

4.3 Acceptance by Teachers

Adoption by teachers, 89% of teachers reported that the system was intuitive and required minimal training (Table 5).

Table 5. Usability Assessment by Teachers

Indicator	% Positive Responses	Relevant data	Tool
Easy to use	89%	16/18 teachers rated it as "intuitive" or "very intuitive"	Likert scale (1–5)
Training required	94%	17/18 teachers required <1 h of training	Recording of hours.

The results show that teachers have adopted NFC in a remarkable way, with 89% of them considering it intuitive or very intuitive on the Likert scale. This is particularly important in resource-limited educational environments where ease of use is crucial to implement effective teaching technologies. The minimal need for training (94% of teachers needed less than an hour to master the system) reinforces its feasibility in real situations, reducing time and training barriers.

These findings indicate that a user-centric design with a clear interface, immediate feedback and easy integration with mobile devices was crucial for teachers to accept. In addition, compatibility with already available devices (such as Android smartphones) facilitated their adoption without requiring additional hardware. The high usability reported (89%) not only validates the system as a practical solution for the classroom, but also supports its potential to be scaled up, where mass training is often a challenge.

5 Discussion

The results of this study show that the NFC-based interactive system achieved remarkable improvements in learning accuracy (93% final accuracy) and sustained attention (72%) in children aged 3 to 6 years. This validates its potential as an accessible educational tool, especially in environments with limited resources. These findings are in line with previous studies which highlight the benefits of tangible interaction in early childhood education [3], but exceed the results obtained by traditional methods reported in similar contexts [4]. The success of the system can be attributed to three key factors: (1) multisensory (auditory and visual) feedback that strengthened cognitive associations [11], (2) gamification elements that helped reduce cognitive burden [2], and (3) an intuitive interface that facilitated rapid adoption by both children and teachers [1].

From the point of view of implementation, the system proved to be especially viable in contexts with limited resources thanks to its low cost (less than $0.80 per NFC card), its high compatibility with existing smartphones, and a scalable Firebase-based architecture that guaranteed low latencies. These technical aspects, together with the minimum training required for teachers, make it a practical solution for public education policies, overcoming barriers presented by more complex technologies such as Augmented Reality [3].

5.1 Limitations and Future Work

Despite the promising results, it is important to acknowledge the study's limitations. The sample was limited to 21 children from two institutions in Quito, Ecuador. This affects the generalizability (external validity) of the findings. Therefore, it is recommended for future research to expand the sample size and include participants from diverse socioeconomic and geographic contexts (e.g., rural and urban areas of different regions). This would validate the system's effectiveness on a larger scale and explore its adaptability to different educational realities.

Furthermore, to address the long-term pedagogical implications suggested by the reviewers, future studies should incorporate follow-up assessments (e.g., after 4 or 8 weeks) to measure knowledge retention and use specific instruments (e.g., engagement observation scales or teacher interviews) to quantify the socio-emotional impact.

From a technological perspective, the dependency on NFC-enabled smartphones could be a barrier in areas with lower technology penetration. As a future line of research, integrating support for QR codes as a more inclusive alternative [3] is proposed. Additionally, the incorporation of AI-based automated recommendation systems could enhance the personalization of learning by adapting to individual student progress [6].

Despite these limitations, the study provides solid empirical evidence on the effectiveness of NFC in early childhood education [1], a replicable and low-cost model validated in the Ecuadorian context [4], and guidelines for designing accessible multisensory educational technologies [7].

6 Conclusions

This study showed that the NFC-based interactive system is an effective, accessible and scalable tool for improving multi-sensory learning in children aged 3 to 6 years in resource-limited environments. There was an 93% final accuracy in conceptual identification and a 72% increase in sustained care, thanks to its gamified and multisensory design. The low-cost architecture (NFC tags for less than $0.80, overcomes the technological barriers faced by 60% of Ecuadorian schools, offering a replicable model that combines technological innovation, constructivist pedagogy and digital inclusion. The results, supported by high educational usability (89% acceptance with less than one hour of capacitation), position this solution as viable for public education policies and open new lines of research in alternative support with QR codes for rural areas, Personalization of the system through AI and integration with educational robotics, setting a precedent for the development of accessible educational technologies.

References

1. Coskun, S., Ozdenizci, B., Ok, K.: A survey on near field communication (NFC) technology. Wirel. Pers. Commun. **71**(3), 2259–2294 (2013). https://doi.org/10.1007/s11277-012-0935-5
2. Deterding, S., Dixon, D., Khaled, R., Nacke, L.: From game design elements to gamefulness: defining 'gamification'. In: Proceedings of the 15th International Academic MindTrek Conference: Envisioning Future Media Environments, pp. 9–15. ACM, Tampere, Finland (2011). https://doi.org/10.1145/2181037.2181040
3. García, J.A.R., et al.: Low-cost educational technologies for developing countries. arXiv:1411.1897. https://arxiv.org/abs/1411.1897 (2014)

4. Cabrera, P., Ortiz, W., García, V.: Digital divide and educational inequality in Ecuador during the COVID-19 pandemic. J. Educ. Technol. Soc. **25**(3), 112–125 (2022). https://doi.org/10.30191/ETS.202207_25(3).0009
5. Hinske, S., Langheinrich, M., Lampe, M.: Towards guidelines for designing augmented toy environments. In: Proceedings of the 1st International Conference on Tangible and Embedded Interaction, pp. 31–38. ACM, Baton Rouge, LA, USA (2007). https://doi.org/10.1145/1226969.1226977
6. Moreno, R., Mayer, R.: Interactive multimodal learning environments. Educ. Psychol. Rev. **19**(3), 309–326 (2007). https://doi.org/10.1007/s10648-007-9047-2
7. Papert, S.: Mindstorms: Children, Computers, and Powerful Ideas, 2nd edn. Basic Books, New York, NY, USA (1993)
8. Papadakis, S., Kalogiannakis, M., Zaranis, N.: Teaching mathematics with mobile devices and the realistic mathematical education (RME) approach in kindergarten. Adv. Mobile Learn. Educ. Res. **1**(1), 5–18 (2021) 10.25082/AMLER.2021.01.002
9. Bers, M.U., González-González, C., Armas–Torres, M.B.: Coding as a playground: promoting positive learning experiences in childhood classrooms. Comput. Educ. **138**, 130–145 (2019). https://doi.org/10.1016/j.compedu.2019.04.013
10. Sweller, J.: Cognitive load during problem solving: effects on learning. Cogn. Sci. **12**(2), 257–285 (1988). https://doi.org/10.1207/s15516709cog1202_4
11. Cohen, J.: Statistical Power Analysis for the Behavioral Sciences, 2nd edn. Lawrence Erlbaum Associates, Hillsdale, NJ (1988)
12. Liu, C., Huang, Y., Wang, Y.: Developing a NFC-based interactive learning system for early childhood education using gamification approach. Interact. Learn. Environ. **31**(4), 2156–2172 (2023). https://doi.org/10.1080/10494820.2022.2057549
13. Zhao, J., Lee, S.: The impact of gamified learning on cognitive engagement and knowledge retention in preschool children: a meta-analysis. Comput. Educ. **187**, 104548 (2022). https://doi.org/10.1016/j.compedu.2022.104548

Mobile Banking as a Tool for Financial Inclusion in Adults in the City of Loja, Year 2025

Gloria del Carmen Palacio Valdivieso(✉) and Ivanova Quituizaca Songor

Universidad Técnica particular de Loja, Loja, Ecuador
gpalacio@utpl.edu.ec

Abstract. Financial inclusion constitutes a strategic pillar for economic and social development, guaranteeing access to and use of formal financial products and services under fair and sustainable conditions. In recent years, digitalization, and particularly mobile banking have been consolidated as effective tools to expand such access, especially in developing countries, by overcoming traditional barriers. International organizations such as the OECD, the World Bank, and CAF emphasize that financial education is important for the responsible use of these technologies and for reducing existing gaps. In Latin America and Ecuador, regulatory and technological advances have driven the expansion of digital services, although challenges related to trust in digital security, technological literacy, and sociodemographic inequalities persist.

In Loja, recent studies indicate that mobile banking is perceived as an agile and secure service, although limited by factors such as gender, educational level, and economic condition; furthermore, it is recognized that the acceptance of these platforms depends on their ease of use, the trust they inspire, and the users' level of financial knowledge. In this context, the results of the survey conducted among 382 adults in the city of Loja in 2025 show a moderately high level of financial inclusion, with significant correlations across all analyzed dimensions (affective, cognitive, security, tangibility, behavioral, reliability, and conative). The findings reveal that age is the main factor influencing financial inclusion, while gender and education do not have a significant effect. These results highlight the importance of implementing public policies and digital training programs aimed at promoting responsible use of mobile banking and building trust, thereby contributing to sustainable and equitable financial inclusion.

Keywords: Mobile banking · financial inclusion · financial education

1 Introduction

Financial inclusion is recognized as a fundamental driver of economic and social development, as it enables individuals to access and use formal financial products, thereby improving well-being, fostering financial planning, and strengthening resilience to economic crises [1, 2]. In this context, the digitalization of financial services, particularly mobile banking, has emerged as a key mechanism to overcome traditional barriers, reduce transaction costs, and expand service coverage, especially in developing economies [3, 4].

A. Rocha et al. (Eds.): ICITS 2026, LNNS 1966, pp. 626–636, 2027.
https://doi.org/10.1007/978-3-032-27538-7_55

In Latin America, the widespread use of smartphones and the increase in internet connectivity have been key factors driving the growth of mobile banking. In Ecuador, data from the Network of Development Finance Institutions (RFD) indicate that around 86% of adults have access to financial services; however, 27% remain excluded, mainly individuals from low-income households or with limited digital skills [5]. This evidence reveals that access gaps still persist, requiring strategies that integrate both technological inclusion and financial education.

Several studies highlight that financial education is essential for the effective and responsible use of mobile banking. They emphasize that individuals with greater financial knowledge tend to show higher levels of trust in digital platforms and make more informed decisions regarding saving, investing, and borrowing [6, 7]. Conversely, low financial literacy increases risks of exclusion and economic vulnerability [8].

Recent studies conducted in the city of Loja show that adults with varying educational levels and technological experience perceive mobile banking as secure, efficient, and accessible. However, challenges remain regarding digital trust, educational inequalities, and gender gaps [9]. These findings underscore the importance of strengthening financial education as a fundamental pillar for promoting more equitable and sustainable financial inclusion, adapted to the specific sociodemographic context of each region, city, or country.

Therefore, this study analyzes the role of mobile banking as a driver of financial inclusion among adults in Loja in 2025.The results seek to provide empirical evidence to inform the development of public policies and institutional strategies that support a more inclusive, resilient, and sustainable financial system.

2 Literature Review

Financial inclusion constitutes a fundamental pillar for economic and social development, as it guarantees the population access to and use of formal financial products and services under equitable and sustainable conditions [2, 10]. From the perspective of international organizations such as the OECD and CAF, financial inclusion and financial education are key strategies to reduce inequalities, promote savings, investment, and responsible planning among different population groups [7, 8].

In the global context, the World Bank's Global Findex in 2021 shows a significant increase in access to financial accounts, driven by digitalization and mobile payments. These tools have made it possible to overcome traditional barriers to access, especially in developing countries, where mobile technologies represent an opportunity to integrate historically excluded sectors. Today, mobile banking and e-wallets are catalysts for financial inclusion by facilitating secure, fast, and accessible operations, as demonstrated by studies in Africa [3, 11], and Latin America [4, 12].

Financial inclusion is recognized as a fundamental pillar of economic and social development, as it guarantees access to and use of formal financial products and services under equitable and sustainable conditions [13]. In developing countries, this access has been enhanced by digitalization and by the expansion of mobile banking, which makes it possible to overcome traditional barriers of infrastructure and costs [3].

Several international studies have examined the factors influencing the adoption of mobile banking. In rural China, [14] found that factors such as proximity to infrastructure, age, educational attainment, and mobile device usage patterns significantly influence adoption levels, while an inverse relationship was identified between adoption and regional financial development. In Cambodia, [15] employed the Technology Acceptance Model (TAM) and determined that perceived trust and usefulness are the main drivers of adoption, whereas ease of use did not have a significant effect. Likewise, in India, [16] expanded the Unified Theory of Acceptance and Use of Technology (UTAUT2) model by incorporating variables such as consumer innovativeness, perceived risk, and information security, reaffirming the critical role of digital safety in technology adoption.

In Spain, [17] identified that the intention to use mobile banking applications results from a balance between facilitating factors (such as performance expectations) and inhibiting factors (such as sunk costs and inertia) [18] found that digital banking adoption is strongly correlated with educational level and income, widening the gap between individuals with basic and higher education, while age and household size exert a negative effect.

Other international studies also point out that factors such as financial education, digital trust, and technological literacy condition the adoption of digital financial services [8]. In Latin America, the expansion of mobile telephony has fostered inclusion, although gaps by gender, age, and educational level persist [19].

In Latin America, [19] applied the UTAUT2 model with PLS-SEM in Bolivia, confirming that habits are the strongest predictor of adoption, followed by hedonic motivation, facilitating conditions, and social influence. In Peru, [20] concluded in a study that lifestyle compatibility, perceived usefulness, and personal innovation in information technologies are key factors determining financial inclusion.

In Ecuador, the RFD reported in 2023 that 86% of adults have access to some type of financial service, while 27% remain excluded, mainly among low-income sectors or individuals with limited digital literacy. The Central Bank of Ecuador has promoted strategies to strengthen digital channels [21]. In a 2020 study, [22] found that facilitating conditions, perceived trust, and usage habits are crucial factors, even among low-income consumers. Similarly, [23] demonstrated that financial literacy positively impacts inclusion and well-being, while digital literacy moderates these effects.

Along these lines, other recent studies show that financial inclusion is shaped by socioeconomic and territorial inequalities [24], in their research, demonstrated through spatial econometrics that the number of firms and the level of wages have a positive and significant relationship with the use of financial services, especially electronic channels and cards. This finding highlights that business and labor dynamics act as drivers of inclusion, but also that territorial differences affect access. Likewise, the pandemic accelerated the need to rely on digital channels, confirming the importance of differentiated policies according to regional conditions.

In Loja, previous studies highlight that higher levels of financial education foster trust and the conscious adoption of mobile banking [6]. Generational studies indicate that young people are more likely to adopt e-wallets, while older adults prioritize security and perceived usefulness [25]. Gender-focused research also highlights that women encounter greater cultural barriers and restrictions in accessing credit and understanding financial products, despite generally exhibiting higher saving tendencies. This situation reinforces the need for more inclusive and gender-sensitive financial services [9].

Consequently, the literature confirms that mobile banking adoption is influenced not only by technological accessibility but also by cultural, educational, and trust-related factors. Strengthening financial education alongside digitalization is essential for conscious and safe adoption in Loja, contributing to reducing gaps and reinforcing economic resilience.

3 Data, Variables and Methodology

The study analyzes how mobile banking affects financial inclusion among the adult population in the city of Loja. The target population consists of 79,118 individuals, according to data from INEC [26]. The sample was calculated using the formula for finite populations, resulting in a total of 382 respondents (207 women and 175 men). The variables used in this study are from [27] (Table 1). These variables were measured using a Likert scale from 1 to 5, where 1 is strongly disagree, 2 disagree, 3 neither agree nor disagree, 4 agree, and 5 strongly agree:

In addition, respondents were asked about their age (1 from 30 to 39 years old, 2 from 40 to 49 years old, and 3 from 50 to 59 years old), their gender (0 female and 1 male), and their educational level (1 primary, 2 secondary, 3 university and 4 postgraduate). To calculate each dimension, the score of the questions contained in each dimension is added, with the level of inclusion being the sum of all the dimensions. The hypotheses to be tested are the following:

H1: Age influences the level of inclusion of adults in the city of Loja.
H2. Gender is relevant in the level of financial inclusion of adults in the city of Loja.
H3: The educational level determines the level of financial inclusion of adults in Loja.
H4: Age influences the different dimensions of inclusion of adults in the city of Loja.
H5. Gender is relevant in the different dimensions of the inclusion of adults in the city of Loja.
H6: The educational level determines the different dimensions of the inclusion of adults in Loja.

4 Results

Table 2 shows the analysis of frequencies of the age variable. People up to 39 years old account for 38.7%, people between 40 and 49 years old are 29.6% and the rest are up to 59 years old.

Table 3 shows the frequency by gender. In this category, women account for 54.7%, while men are 45.3%.

Table 4 shows the educational level of the people surveyed. The most frequent category is people with a university degree (40.8%), followed by people who have a postgraduate degree (25.1%), secondary education (23.8%) and primary education (10.2%).

Table 1. Questions in the survey

Variables	Questions
Affective	You are fully satisfied with the functionality, services offered, results of operations and benefits (time-cost) obtained from the mobile banking service.
	He is very satisfied with the services purchased through the mobile banking service and its quick solution to any inconvenience presented.
Cognitive	The mobile banking service starter platform is easy to understand and access, so it is easy to learn how to use the services.
	The user-friendly technology presented by the mobile banking service is adapted to my level of technological knowledge appropriately.
Security	Transactions through the mobile banking service have the necessary credibility in terms of the protection and confidentiality of personal data since these are in real time.
Tangibility	The mobile banking service eliminates technological barriers such as the use of balance, telephone plans, or internet connection to carry out operations and works on all range of devices
	The mobile banking service gives me the possibility of accessing my services through any mobile operator (Claro, Movistar, and others).
	it is easy to join the mobile banking service at any branch of the bank and to find the correct guidance for the correct use of the service.
	The process of enrolling in your account in the mobile banking service is quick and easy.
Behavioral	On average, per week, I carry out at least 5 operations (consultation, recharges, deposit, etc.) through the mobile banking service.
	I now do all my banking through mobile banking.
Reliability	Short periods are required to make transactions through the mobile banking service, and the level of transaction response is fast
Conative	I intend to continue using mobile banking services for banking and would even recommend it.

Table 2. Frequencies for age.

Age	Frequency	Percent	Valid Percent	Cumulative Percent
30 to 39 years	148	38.7	38.7	38.7
40 to 49 years	113	29.6	29.6	68.3
50 to 59 years	121	31.7	31.7	100.0
Missing	0	0.0		
Total	382	100.0		

Table 3. Frequencies for gender

Gender	Frequency	Percent	Valid Percent	Cumulative Percent
Female	209	54.7	54.7	54.7
Male	173	45.3	45.3	100.0
Missing	0	0.0		
Total	382	100.0		

Table 4. Frequencies for education

Education	Frequency	Percent	Valid Percent	Cumulative Percent
Primary	39	10.2	10.2	10.2
Secondary	91	23.8	23.8	34.0
University	156	40.8	40.8	74.9
Postgraduate	96	25.1	25.1	100.0
Missing	0	0.0		
Total	382	100.0		

Table 5 shows the descriptive statistics of the dimensions of financial inclusion. Affectivity and cognitivity have an average of over 8 out of 10, which indicates that respondents value both aspects of financial inclusion in mobile banking services positively. There is also a high valuation of safety, tangibility, reliability, and the conative aspect. If global inclusion is analyzed, it is medium-high with an average of 53.41 out of 65 points, which indicates that the use of mobile banking induces a medium-high level of financial inclusion.

Table 5. Descriptive Statistics

	Valid	Median	Mean	Std. Deviation	Minimum	Maximum
Affectivity	382	8.000	8.196	1.449	2.000	10.00
Cognitivity	382	8.000	8.296	1.445	2.000	10.00
Security	382	4.000	4.084	0.841	1.000	5.000
Tangibility	382	16.000	16.380	2.863	4.000	20.00
Behavior	382	8.000	8.128	1.642	2.000	10.00
Reliability	382	4.000	4.134	0.771	1.000	5.000
Conative	382	4.000	4.202	0.822	1.000	5.000
Inclusion	382	52.000	53.419	8.814	13.00	65.00

Table 6 shows Spearman's rho correlation between the dimensions of financial inclusion. This measure was used because the assumption of normality of the variables is

Table 6. Spearman's Correlations

			Spearman's rho		p
Affectivity	–	Cognitivity	0.744	***	<.001
Affectivity	–	Security	0.686	***	<.001
Affectivity	–	Tangibility	0.759	***	<.001
Affectivity	–	Behavior	0.663	***	<.001
Affectivity	–	Reliability	0.664	***	<.001
Affectivity	–	Conative	0.738	***	<.001
Affectivity	–	Inclusion	0.862	***	<.001
Cognitivity	–	Security	0.741	***	<.001
Cognitivity	–	Tangibility	0.783	***	<.001
Cognitivity	–	Behavior	0.677	***	<.001
Cognitivity	–	Reliability	0.704	***	<.001
Cognitivity	–	Conative	0.683	***	<.001
Cognitivity	–	Inclusion	0.868	***	<.001
Security	–	Tangibility	0.700	***	<.001
Security	–	Behavior	0.599	***	<.001
Security	–	Reliability	0.639	***	<.001
Security	–	Conative	0.629	***	<.001
Security	–	Inclusion	0.783	***	<.001
Tangibility	–	Behavior	0.728	***	<.001
Tangibility	–	Reliability	0.790	***	<.001
Tangibility	–	Conative	0.712	***	<.001
Tangibility	–	Inclusion	0.922	***	<.001
Behavior	–	Reliability	0.689	***	<.001
Behavior	–	Conative	0.688	***	<.001
Behavior	–	Inclusion	0.823	***	<.001
Reliability	–	Conative	0.673	***	<.001
Reliability	–	Inclusion	0.807	***	<.001
Conative	–	Inclusion	0.798	***	<.001

* $p < .05$, ** $p < .01$, *** $p < .001$

not met. All correlations are significant to a medium-high degree, with significant correlations between all dimensions of financial inclusion and when they are related to it.

Table 7 shows the regression of the different constructs of financial inclusion with sociodemographic variables, i.e., age, gender, and educational level. The variance inflation analysis is around one, indicating that there is no multicollinearity between the regression variables.

The results in Table 7 indicate that age has a positive influence on inclusion through mobile banking, since age increases, which implies that hypothesis one is not rejected. Gender is not relevant when explaining financial inclusion in adults in the city of Loja (H2 is rejected), as is the level of education (H3 is rejected). With respect to the different dimensions of financial inclusion, age influences all dimensions except security (H4 is rejected), the same for gender (H5 is rejected) and educational level (H6 is rejected).

The findings of this study confirm that mobile banking in Loja constitutes an effective mechanism to promote financial inclusion among adults, reaching a medium-high level of inclusion with an average of 53.41 out of 65 points. This result indicates a favorable user perception across multiple dimensions, including affective, cognitive, tangible, security, reliability, behavioral, and conative aspects. The strong correlations among these dimensions suggest that the mobile banking experience is holistic—where ease of learning, perceived security, transaction speed, and overall satisfaction mutually reinforce one another, ultimately enhancing users' willingness to continue using these services.

In sociodemographic terms, the results show that age significantly influences inclusion levels, both globally and across most dimensions, except for security. These results are consistent with international literature linking technology adoption to age-related factors, where older users tend to demonstrate greater commitment and confidence as they consolidate digital financial habits [13–15]. In contrast, gender and educational level did not show significant incidence, which differs from other studies associated with formal education and the gender gap with the adoption of digital financial services [17, 25]. This difference may be explained by the progress in digital literacy and the standardization of mobile services in Ecuador, which have progressively reduced inequalities in access.

At the national level, recent research confirms that financial inclusion is shaped by socioeconomic and territorial inequalities: while business and labor dynamics act as a driver of inclusion, regional disparities continue to limit access to financial services, as highlighted in the study by [24]. In the case of Loja, these disparities are primarily reflected in levels of trust in digital security, income, and technological familiarity, rather than in formal educational attainment.

Within this context, empirical evidence indicates that financial inclusion in Loja is influenced less by traditional structural factors such as gender or education and more by elements like perceived usefulness, trust in digital security, and usage habits, all of which tend to strengthen with age and accumulated experience. These findings reaffirm the need to design public policies and institutional strategies aimed at strengthening digital trust, promoting financial education, and encouraging the conscious use of mobile

Table 7. Econometric regressions

	(1)	(2)	(3)	(4)
VARIABLES	**Inclusion**	**Affectivity**	**Cognitivity**	**Security**
40 to 49 years	2.957***	0.443**	0.421**	0.154
	(1.129)	(0.190)	(0.186)	(0.109)
50 to 59 years	2.087*	0.262	0.291	0.117
	(1.128)	(0.186)	(0.187)	(0.111)
Gender	1.201	0.100	−0.00531	0.131
	(0.924)	(0.152)	(0.152)	(0.0865)
Secondary	−2.094	−0.151	−0.305	−0.0855
	(1.460)	(0.236)	(0.263)	(0.151)
University	−0.897	0.0623	−0.0731	−0.0762
	(1.284)	(0.198)	(0.241)	(0.142)
Postgraduate	−1.404	−0.176	−0.278	−0.113
	(1.558)	(0.240)	(0.276)	(0.166)
Constant	52.56***	7.992***	8.254***	4.022***
	(1.303)	(0.203)	(0.241)	(0.146)
Observations	382	382	382	382
	(5)	**(6)**	**(7)**	**(8)**
VARIABLES	**Tangibility**	**Behavior**	**Retability**	**Conative**
40 to 49 years	1.034***	0.349*	0.270***	0.286***
	(0.369)	(0.209)	(0.0996)	(0.105)
50 to 59 years	0.699*	0.298	0.232**	0.189*
	(0.365)	(0.206)	(0.0969)	(0.109)
Gender	0.412	0.375**	0.120	0.0678
	(0.299)	(0.169)	(0.0787)	(0.0849)
Secondary	−0.860*	−0.435*	−0.109	−0.148
	(0.466)	(0.263)	(0.142)	(0.135)
University	−0.462	−0.309	−0.0407	0.00256
	(0.416)	(0.222)	(0.130)	(0.116)
Postgraduate	−0.580	−0.0947	−0.0373	−0.124
	(0.506)	(0.257)	(0.146)	(0.140)
Constant	16.21***	8.015***	3.977***	4.092***
	(0.420)	(0.224)	(0.136)	(0.120)
Observations	382	382	382	382

Robust standard errors in parentheses *** $p < 0.01$, ** $p < 0.05$, * $p < 0.1$

banking, with a differentiated approach that considers territorial and sociodemographic particularities.

5 Conclusions

Mobile banking is consolidating in Loja as a tool to foster financial inclusion, generating positive perceptions across most dimensions, which reflects a medium-high level of adoption among the adult population.

Age is the main determinant of the level of digital financial inclusion, influencing most dimensions of the analysis; whereas gender and education do not have a significant effect, suggesting that user experience and technological familiarity weigh more heavily than structural factors. The positive correlation among all evaluated dimensions confirms the multidimensional nature of financial inclusion through mobile banking, where trust, perceived security, ease of use, and satisfaction are interrelated to strengthen the user experience.

The results demonstrate the need to promote public policies and digital training programs that increase confidence in platforms, encourage responsible financial habits, and reduce access gaps, especially among younger age groups and sectors less accustomed to the use of digital services. The consolidation of mobile banking in Loja must be accompanied by financial education strategies that ensure conscious and sustainable use, contributing to a more inclusive, equitable, and resilient financial system.

Although the study demonstrates that mobile banking promotes financial inclusion among adults in Loja, it presents an important limitation, as the results are based solely on perceptions from a sample within the urban area, which restricts the generalization of the findings. To strengthen the evidence, future research should include real usage behavior data and expand coverage to rural areas or other sociodemographic contexts.

References

1. World Bank: Global Findex Database 2021: Financial Inclusion, Digital Payments, and Resilience in the Age of COVID-19. World Bank Group (2021)
2. Tuesta, D., Sorensen, G., Haring, A., Cámara, N.: Inclusión financiera y sus determinantes: el caso argentino. BBVA. Research. (2015)
3. Jack, W., Suri, T.: Risk sharing and transactions costs: evidence from Kenya's mobile money revolution. Am. Econ. Rev. **104**(1), 183–223 (2014)
4. Ruiz, V.: El futuro de la banca móvil en América Latina. Deloitte (2011)
5. Red de Instituciones Financieras de Desarrollo – RFD.: Reporte de Inclusión Financiera en el Ecuador. (2023)
6. Palacio, G.: Medición del conocimiento financiero en los participantes de programas de educación financiera: caso banco de Loja S.A. REDMARKA. **2**(19), 47–60 (2017)
7. OECD: OECD/INFE Toolkit for Measuring Financial Literacy and Financial Inclusion. OECD Publishing (2015)
8. Grifoni, A., Messy, F.: Current status of national strategies for financial education. OECD working papers on finance, insurance and private. Pensions. **16**, 1–36 (2012)
9. Espinoza-Loayza, V., Palacio-Valdivieso, G., Maldonado-Criollo, C.: Inclusión financiera con perspectiva de género: Análisis de brechas en el cantón Loja. 593 Digit. Publ. CEIT. **10**(3), 862–871 (2025). https://doi.org/10.33386/593dp.2025.3.3180
10. Claessens, S.: Access to financial services: a review of the issues and public policy objectives. World Bank Res. Obs. **21**(2), 207–240 (2006)

11. Mbiti, I., Weil, D.: Mobile Banking: The Impact of M-Pesa in Kenya NBER Working Paper Series, p. 17129 (2010)
12. Chiang, E.: Banca móvil y su impacto en América Latina. Revista de Economía y Finanzas. **15**(2), 55–72 (2018)
13. Demirgüç-Kunt, A., Klapper, L.: Measuring Financial Inclusion: The Global Findex Database, vol. 6025, pp. 1–61. World Bank Policy Research Working Paper (2013)
14. Zhu, Q., Lyu, Z., Long, Y., Wachenheim, C.J.: Adoption of mobile banking in rural China: impact of information dissemination channel. Socio Econ. Plan. Sci. **83**, 101011 (2022)
15. Ly, B., Ly, R.: Internet banking adoption under technology acceptance model—evidence from Cambodian users. Comput. Hum. Behav. Rep. **7**, 100224 (2022)
16. Chauhan, V., Yadav, R., Choudhary, V.: Adoption of electronic banking services in India: an extension of UTAUT2 model. J. Financ. Serv. Mark. **27**(1), 27–40 (2022)
17. Medina-Molina, C., Rey-Moreno, M., Cazurro-Barahona, V., Parrondo, S.: La adopción de las aplicaciones de banca móvil desde una perspectiva dual. Sociol. Tecnocienc. **9**(2), 1–22 (2019)
18. Arellano, A., García, J.R.: Determinantes de la adopción de la banca digital en España. Digital Economy Watch, BBVA Research (2017)
19. Bustos Magne, N.S., Almeida Cardona, R.: Adopción del uso de la banca móvil en la ciudad de Cochabamba. Perspectivas. **26**(52), 85–110 (2023)
20. Yamakawa, P., Guerrero, C., Rees, G.: Factores que influyen en la utilización de los servicios de banca móvil en el Perú. Universidad ESAN Working Paper (2013)
21. Banco Central del Ecuador: Estrategia Nacional de Inclusión Financiera. BCE (2020)
22. Dakduk, S., Siqueira, J.R.: Factores determinantes de la adopción de servicios financieros móviles en consumidores de bajos ingresos en Ecuador. Revista de Negocios y Tecnología. **15**(1), 45–63 (2020)
23. Tulcanaza-Prieto, A.B., Cortez-Ordoñez, A., Rivera, J., Lee, C.: Is digital literacy a moderator variable in the relationship between financial literacy, financial inclusion, and financial well-being in the Ecuadorian context? Sustainability. **17**(6), 2476 (2025)
24. Peñarreta, J., Flores, D., Chávez, R.: Inclusión financiera en Ecuador: un análisis econométrico espacial. Revista Observatorio de la Economía Latinoamericana. **7**(29), 1–18 (2024)
25. Espinoza-Loayza, V., Palacio-Valdivieso, G., Quichimbo-Sanmartín, A.: Análisis generacional de los factores que influyen en la intención de uso de la billetera electrónica en la ciudad de Loja. 593 Digit. Publ. CEIT. **10**(3), 851–861 (2025). https://doi.org/10.33386/593dp.2025.3.3179
26. Instituto Nacional de Estadística y Censos (INEC).: Proyecciones y estadísticas poblacionales 2024, Quito, Ecuador. (2024)
27. Oruna Rodríguez, M., Pérez, J., Sánchez, L.: Dimensiones de inclusión financiera y su medición en usuarios de banca digital. Revista Latinoamericana de Economía y Finanzas. **12**(3), 45–62 (2023)

Prototype of Smart Wristbands with QR Code for Pre-hospital Care of People with Intellectual Disabilities

Angela Gabriela Valencia Berrezueta(✉), Ana Isabel Uyaguari Morocho, Sandra El Khori El Khouri, and Carlos Francisco Encalada Ochoa

Instituto Superior Tecnológico Universitario San Isidro, Cuenca, Ecuador
angelavalencia@sanisisro.edu.ec

Abstract. People with intellectual disabilities face significant communication barriers during emergencies, making it difficult for pre-hospital staff to access critical medical information in a timely manner. In this context, the use of accessible technologies facilitates clinical decision-making. The present study aimed to develop a prototype of a wristband with a QR code to optimize pre-hospital care for people with intellectual disabilities in the city of Cuenca. The research was developed under a mixed, descriptive, non-experimental, cross-sectional approach. Surveys were applied to 105 paramedics and semi-structured interviews were applied to 10 caregivers of people with intellectual disabilities, selected by simple random sampling. For the design of the prototype, the Design Thinking methodology was used, focused on the needs of the user. The results showed that 47% of paramedics identified communication difficulties as the main barrier during care, 64% turn to the patient's guardian in these cases and 47% considered that a wristband with a QR code would facilitate quick access to medical information. The prototype developed contains affiliation data and medical history, it is functional by scanning the QR code linked to a database in Microsoft Access, with an internet connection required. The technical validation was conducted by experts in prehospital medicine and informatics, showing high utility, relevance and applicability. It is concluded that this technological solution contributes significantly to improving the quality, speed and safety of pre-hospital care aimed at this vulnerable population group.

Keywords: Intellectual disability · pre-hospital care · health technology

1 Introduction

People with intellectual disabilities (ID) represent a particularly vulnerable population group in medical emergency contexts, due to limitations in communication, understanding of instructions and expression of symptoms [1]. These conditions exacerbate the challenges of pre-hospital staff, where response time and diagnostic accuracy are determining factors for survival and quality of care.

A. Rocha et al. (Eds.): ICITS 2026, LNNS 1966, pp. 637–647, 2027.
https://doi.org/10.1007/978-3-032-27538-7_56

In the specialized literature, it has been identified that restrictions in the care of people with ID are not only due to clinical factors, but also to communication barriers, institutional discrimination, organizational deficiencies and structural deficiencies in health services [2]. In the absence of support mechanisms, such as interpreters or alternative and augmentative communication systems (AACS), these barriers limit equitable access to timely and safe care [3].

Given this scenario, inclusive technological solutions are emerging as viable strategies to reduce inequalities and strengthen response capacity [4]. Among these, the use of wearable devices such as smart wristbands with QR codes stands out for its ability to centralize critical medical information and facilitate immediate access by pre-hospital staff to essential patient data. These tools are based on low-cost and easy-to-apply technologies, and can become a decisive support in emergency scenarios.

Intellectual disability has been conceptualized by the World Health Organization (WHO) as a state of incomplete mental development, characterized by difficulties in understanding, learning and adapting to various situations [5]. According to the DSM-5, it is a neurodevelopmental disorder that combines limitations in IQ and adaptive skills, with a significant impact on functional autonomy [6].

Various cognitive and behavioral perspectives reinforce this definition, highlighting that people with ID present alterations in conceptual, social, and practical skills necessary for independent living [7], often accompanied by sensory or physical impairments that further restrict daily participation [8]. From an inclusive approach, it is imperative to recognize functional diversity, promote social equity and ensure health services adapted to the needs of this group [9].

ID constitutes a spectrum of heterogeneous conditions with different degrees of severity-mild, moderate, severe, and profound-that require differentiated supports [10]. The literature shows that people with disabilities have a higher demand for emergency services compared to the general population [11], in part due to the interaction of comorbidities and the simultaneous use of drugs [12].

This information means an increase in hospitalizations and underscores the need for specific prevention and clinical management strategies [13]. Among the most recurrent obstacles are the lack of effective communication, the absence of inclusive protocols and the shortage of trained human resources, which compromises the quality of the service [14].

In epidemiological terms, ID affects between 1% and 3% of the world's population, which is equivalent to approximately 220 million people, of which about 90% have a genetic origin [15]. Longitudinal studies report variations in their prevalence: some record increases, others decreases or stability [16, 17]. In Latin America it is estimated that 85 million people live with ID, of which 3 million reside in the Central American isthmus [18].

In Ecuador, the Ministry of Public Health reported in 2022 a total of 619,135 people with disabilities, of which 120,602 correspond to cases of ID; in the province of Azuay, the canton of Cuenca concentrates 4,057 cases (19.69% of the regional total) [19, 20]. The etiology is complex and multifactorial, including genetic, teratogenic causes and risk factors such as consanguinity, advanced maternal age, prematurity, low birth weight or substance use during pregnancy [21].

Addressing this condition in prehospital care faces multiple limitations. The lack of accurate information and the need for intermediaries for communication reduce patients' autonomy and generate delays in clinical decision-making [14]. In this context, technology becomes a central axis to ensure safety and speed in emergencies.

Information and Communication Technologies (ICTs) have proven to be effective resources for improving clinical communication, reducing errors, and ensuring greater equity [22, 23]. In particular, QR codes, regulated by ISO/IEC 18004:2015, represent a practical and economical alternative for immediate access to relevant clinical information [24–26]. Its implementation in hospital settings has demonstrated benefits such as reduction of waiting times, support for surgical safety, optimization of prescribing and prevention of errors in the administration of medications [27].

The design of wristbands with QR code allows to centralize critical data such as affiliation, medical history, allergies, medication in use and emergency contacts, offering a portable, interoperable and fast reading platform [28]. Its integration with electronic health record systems and relational databases ensures security, accessibility, and scalability [29, 30].

However, the implementation of this type of device poses legal and ethical challenges. In Ecuador, instruments such as the Constitution of the Republic [31], the Organic Law on the Protection of Personal Data [32], the Comprehensive Organic Criminal Code [33] and the Electronic Commerce Law [34] guarantee the confidentiality, authenticity and legitimate use of data. Internationally, frameworks such as the European Union's General Data Protection Regulation (GDPR) [35] and Health Insurance Portability and Accountability Act [HIPAA] in the United States [36] strengthen security and privacy standards.

Scientific evidence shows that people with intellectual disabilities face greater risks during emergencies due to communication and structural barriers, while ICT-based technological innovations offer inclusive and effective solutions. Within this framework, the present study aimed to develop and validate a prototype of an intelligent wristband with a QR code to optimize pre-hospital care for people with intellectual disabilities in the city of Cuenca, Ecuador, providing a replicable model in other contexts with similar needs.

2 Methodology

The research was developed under a mixed approach, integrating quantitative (to measure acceptance and usefulness of the device) and qualitative (to understand perceptions of health professionals) methods, which allowed a comprehensive analysis of the feasibility of smart wristbands with QR code in prehospital care.

The research design was non-experimental, since the phenomenon was observed without manipulating variables, allowing the real impact of the proposal to be analyzed in its sociocultural context. The scope was explanatory, aimed at establishing cause-effect relationships between the use of technology and the improvement in the quality of HPA.

Variables such as: age, sex, work experience, workplace, barriers to care, area of care, type of emergencies, type of disability, accessibility to medical information, safety of care, feasibility and feasibility of the device, training in new technologies and implementation limitations were defined and operationalized. The contextualization of these variables ensured consistency in measurement and analysis.

Data collection techniques included a structured survey targeting 105 paramedics and a semi-structured interview targeting 10 caregivers and/or legal representatives. Both instruments were validated by expert judgment and applied under ethical principles of informed consent and confidentiality.

The questionnaire consisted of 20 multiple-choice questions, developed to measure the acceptance and usefulness of the device. The semi-structured interview consisted of 13 open-ended questions in order to further explore the perception and experiences of caregivers of people with ID.

Data analysis was performed in two phases:

- Quantitative: processing in SPSS to obtain frequencies, trends and distributions, represented in graphs and tables.
- Qualitative: the interviews were transcribed and subjected to a thematic coding process, to identify units of meaning and subsequently an axial coding to group them into analytical categories.

2.1 Methodology for Prototype Design

The design of the smart wristband prototype with QR code was developed using a prototype design and validation methodology with an iterative approach, organized into seven phases:

1. Identification of the problem and requirements

   The difficulty of pre-hospital staff in accessing essential clinical information in patients with intellectual disabilities, particularly in the absence of caregivers, was recognized. Based on this, the goal was to design a portable device that provides fast, secure and reliable access to relevant medical data.
2. Bibliographic and technological review

   Background on the use of digital technologies in health was analyzed, with special emphasis on the implementation of QR codes for information storage and access. Likewise, international standards were reviewed, such as ISO/IEC 18004:2015, which regulate the symbology and standardization of QR codes.
3. Conceptual design

   A structural scheme of the prototype composed of three main elements was elaborated:

   - The physical wristband, made of hypoallergenic silicone, is adjustable and waterproof.
   - The QR code, unique for each user, can be scanned from mobile devices that need an internet connection. The QR code does not store any direct personal data; instead, it contains a Non-Transferable Internal Identifier (IIN), an automatically generated alphanumeric code that functions as the access key to the system.
   - The critical medical information linked to the IIN within the database, including medical history, allergies, current medications, emergency contact information,

and special instructions. These data are not embedded in the QR code but are stored in a secure database accessible only to authorized personnel (Fig. 1).

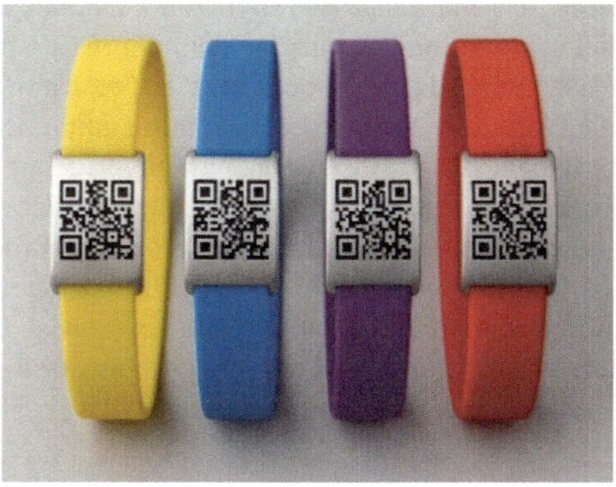

Fig. 1. Wristband with QR code according to the degree of disability. Yellow represents mild disability, blue (moderate), purple (severe), and red (deep).

4. Detailed design and development

   The prototype was built by integrating the defined elements. The chromatic differentiation of the wristband was established according to the degree of disability (mild, moderate, severe, profound), allowing a rapid visual classification in emergencies. Microsoft Access was used as the database management system for storing medical information, along with a QR code generator for assigning identifiers.

   When the QR code is scanned, the reading device retrieves the IIN and transmits it to the system, which performs a query within the database to recover the user's clinical information. This query displays data only to personnel with authorized credentials. This mechanism ensures confidentiality.
5. Pilot test and validation

   The prototype was subjected to an evaluation process by five experts in the area of health and technology, applying the System of Usability Scales (SUS), composed of 10 items under the Likert scale.
6. Adjustments and optimization

   Based on the feedback obtained, aspects of readability of the code, durability of the wristband and relevance of the stored information were verified, guaranteeing accessibility and protection of sensitive data.
7. Final validation

   The results of the SUS survey showed an average of 87.5 points out of 100, indicating high acceptance in terms of ease of use, usefulness and applicability in the pre-hospital context. This supports the feasibility of the prototype as an innovative tool to improve care for people with intellectual disabilities.

2.2 Ethical Considerations and Data Protection

The development of the prototype and the data collection process were conducted in accordance with ethical principles for research involving human subjects, including core bioethical standards. In compliance with Ecuador's Organic Law on Personal Data Protection, all collected information was handled confidentially, stored in databases with restricted access, and used exclusively for academic and research purposes.

The prototype's data structure followed data-minimization principles, avoiding the storage of excessive or unnecessary information. No sensitive data enabling public identification of the user were included. Moreover, the evaluated version of the prototype does not contain real user data; instead, it uses simulated information for technical validation.

It is acknowledged that future implementation of the device will require approval from a research ethics committee, the application of robust encryption protocols, anonymization mechanisms, and informed consent policies for users and caregivers.

3 Results

The application of the survey to 105 prehospital care professionals (APH) in Cuenca allowed to characterize their perceptions regarding the feasibility of implementing smart wristbands with QR codes. The results indicate that the majority of the staff surveyed are men and have a predominant work experience of 1 to 3 years, working mostly in private institutions.

Regarding emergency care for people with intellectual disabilities (ID), 45% of participants reported having intervened between 2 and 5 times a year, while 25% did so more than 10 times a year. The main difficulty identified was the lack of effective communication, reported by 47% of the respondents, as well as the ignorance of clinical history by the tutors (15%).

Regarding the communication mechanisms used, 64% of paramedics communicated through responsible tutors, followed by 23% who used adapted verbal communication. Among the most frequent emergencies, seizures (53%) and falls (26%) stand out (Table 1).

Regarding the usefulness of wristbands, 85% of respondents stated that having a medical history would facilitate care and reduce errors in the administration of medications. Likewise, the majority considered the QR wristband to be a viable and useful tool, although they expressed concerns related to data privacy. 95% indicated that training in the use of this technology would be essential for its implementation.

Interviews with five caregivers of people with ID revealed recurrent limitations during emergencies, including difficulty transmitting accurate medical information in crisis situations. Caregivers recognized that stress and a lack of clear protocols make it difficult to communicate critical clinical data, which can delay care or lead to errors.

All interviewees expressed a favorable perception towards the adoption of wristbands with QR codes, highlighting that these could guarantee immediate access to clinical information, even when the patient is alone or unconscious. The importance of hypoallergenic and comfortable materials for daily use was highlighted, as well as the need for a training plan for families and caregivers.

Table 1. Frequency, perceived difficulties and communication characteristics in pre-hospital care for people with disabilities.

Category	*f*	%
Frequency of care		
More than twice/year	26	25%
6 times per year	10	9%
2–5-times per year	47	45%
Once a year	18	17%
Never	4	4%
Difficulty in attention		
Lack of communication.	49	47%
Innaccessible places	4	4%
Diagnosis identification	4	4%
No access to information	12	11%
They do not retain information	4	4%
Unaccompanied patients	14	13%
Guardian unaware of background	16	15%
Vital Signs Data Collectionl	2	2%
Communication		
Through the responsible tutor	67	64%
Verbally, speaking slowly and clearly	24	23%
Using gestures and visual cues	9	9%
Through lip reading	2	2%
I have no experience.	2	2%

The results demonstrate the positive acceptance and practical relevance of the prototype, confirming its potential to optimize pre-hospital care for people with intellectual disabilities through immediate and secure access to essential medical information.

4 Discussion

The results of this research show that the lack of effective communication mechanisms is one of the main barriers in pre-hospital care for people with intellectual disabilities (ID). This finding coincides with what is described in the literature, where it is pointed out that the difficulties of clinical interaction and the absence of alternative and augmentative communication systems (AACS) generate delays in decision-making and compromise the quality of the service [3, 14].

The favorable perception of paramedics and caregivers regarding the use of smart wristbands with QR code reinforces the relevance of incorporating accessible technologies in the pre-hospital setting. Previous studies highlight that Information and Communication Technologies (ICT), in particular QR codes, allow immediate access to medical information, reduce clinical errors and improve safety in care [22, 24]. In this sense, the findings of the present research are consistent with the benefits reported in other hospital and outpatient contexts [27].

An innovative contribution of the developed prototype is the chromatic differentiation of the wristbands according to the degree of disability, which facilitates rapid visual identification in critical situations. However, this feature raises ethical challenges related to stigmatization and labeling of individuals, which should be considered in future phases of social and clinical validation [9, 10]. Therefore, it is recommended that this characteristic remain optional, reserved exclusively for clinical contexts, and accompanied by neutral alternatives for users who prefer greater discretion.

On the technical level, a limitation of the prototype is the dependence on internet connection to access the database, which could hinder its applicability in rural scenarios or with low connectivity. This challenge has also been identified in research on the use of mobile devices in health, which emphasizes the need to integrate hybrid solutions that work online and offline [24].

The implementation of this technology is supported by a solid legal framework, both nationally and internationally, which guarantees the protection of personal data and the confidentiality of clinical information [31, 36]. The inclusion of medical information through a QR code entails ethical and legal challenges related to privacy and the protection of personal data. In Ecuador, there is no unified national health identification number across institutions; therefore, it is not yet possible to use a single universal identifier. For this reason, the current prototype employs its own database and a QR code that contains only the minimum information required.

In addition, it is recommended that future research delve into the social and clinical validation of the device, considering the ethical dimension related to the possible stigmatization of the patient, as well as the analysis of the legal framework applicable to the handling of sensitive data in emergencies. Finally, it is a priority to explore the development of improved versions of the system that work in hybrid mode or without an internet connection, in order to ensure its relevance in rural contexts or with technological limitations.

5 Conclusions

The present study demonstrates that smart wristbands with QR codes represent a viable and well-accepted tool for improving the speed and safety of access to critical medical information during emergencies involving individuals with intellectual disabilities. The positive perception expressed by paramedics and caregivers highlights their potential usefulness in prehospital settings.

Nevertheless, several important limitations were identified that should be considered in future development phases: (1) dependence on internet connectivity to access the database, which may restrict use in rural areas; (2) challenges related to privacy and the protection of sensitive data, requiring more advanced digital security mechanisms; (3) risks of stigmatization associated with the use of color-differentiated wristbands, which should be addressed through personalized and non-visible alternatives; and (4) lack of broader social and clinical validation with individuals with disabilities and healthcare personnel in real-world contexts.

Despite these limitations, the prototype constitutes an innovative and low-cost contribution that can be integrated into prehospital care systems, provided that technical, legal, and ethical improvements are implemented. It is recommended to move forward with field pilot studies, strengthen the digital security architecture, and design an informed consent model that enables safe adoption while respecting users' rights.

References

1. Centro Español de Documentación sobre Discapacidad.: https://cedd.net/. Last accessed 30 Sept 2025
2. Grandas–Ramírez, Á.L., Orjuela, R.A.B., Turriago, L.R.B., Guzmán, C.L.M., Forero, I.A.P.: La atención en salud bucal para personas con discapacidad. Un desafío en tiempos de pandemia por la COVID-19. Acta Odontológica Colomb. **10**, 99–112 (2020)
3. Garzón Sánchez, I., Gerolin Pelucci, M., Herrero Brandin, M.J., Mena Castañares, M., Canto Combarro, A., Vidorreta Gutiérrez, I.: Abordaje de las dificultades de comunicación en personas con discapacidad intelectual: la importancia del contexto. In: An Approach to Communication Difficulties in People with Intellectual Disabilities: The Importance of Context (2020)
4. Seuret, J.L., Dìaz, L.M.I.M., Rodríguez, M.O.M.: Ética bioética y discapacidad. Revista Científica Cultura, Comunicación y Desarrollo. **4**, 39–46 (2019)
5. Pastrán Alfonso, O., Céspedes Nieves, G., Ruiz Latriglia, M.F., Silva Preciado, S.: Evolución conceptual de la discapacidad. Situación actual de la discapacidad en Colombia y lineamientos para su abordaje. Rev. Colomb. Rehabil. **6**, 99 (2017). https://doi.org/10.30788/RevColReh.v6.n1.2007.121
6. López, I., Förster, J.: Trastornos del neurodesarrollo: dónde estamos hoy y hacia dónde nos dirigimos. Rev. Médica Clínica Las Condes. **33**, 367–378 (2022). https://doi.org/10.1016/j.rmclc.2022.06.004
7. Leal, A., Buitrago, L.N., Rodas, J.R.: Habilidades sociales y discapacidad intelectual: reporte de una secuencia didáctica en lenguas materna y extranjera. Voces Silenc. Rev. Latinoam. Educ. **15**, 91–121 (2024). https://doi.org/10.18175/VyS16.3.2024.5
8. Yokoyama-Rebollar, E., Frías, S., Del Castillo-Ruiz, V.: Abordaje citogenético y citogenómico de pacientes con discapacidad intelectual y malformaciones congénitas. Acta Pediátrica México. **38**, 433 (2017). https://doi.org/10.18233/APM38No6pp433-4411511
9. Barreto Montoya, J., Prado C., J.C.: Inclusión y discapacidad en Colombia. Análisis y recomendaciones para la construcción de política pública. Editorial Universidad Católica de Colombia (2020). https://doi.org/10.14718/9789585133297.2020
10. Varsi Rospigliosi, E.: La representación del apoyo de la persona con discapacidad. El nuevo esquema de la capacidad jurídica en el Perú. Acta Bioethica. **27**, 211–222 (2021). https://doi.org/10.4067/S1726-569X2021000200211
11. Sandberg, M., Kristensson, J., Axmon, A.: IDcare – a longitudinal register study of pre-pandemic and pandemic health care utilization and diagnostic profiles among people with intellectual disabilities in southern Sweden. Eur. J. Epidemiol. **39**, 1063–1071 (2024). https://doi.org/10.1007/s10654-024-01151-3

12. Arabi, S., et al.: Utilization of rehabilitation services by disabled people under the iranian health insurance organization. Med. J. Islam Repub. Iran. **39**, 65 (2025). https://doi.org/10.47176/mjiri.39.65
13. Luo, W., Li, M.: Evaluation and optimization of emergency medical services accessibility based on real-world emergency calls in metropolitan China. ResearchGate. (2025). https://doi.org/10.1007/s43762-025-00204-x
14. Shady, K., Phillips, S., Newman, S.: Barriers and facilitators to healthcare access in adults with intellectual and developmental disorders and communication difficulties: integrative review. Rev. J. Autism Dev. Disord., 1–13 (2022). https://doi.org/10.1007/s40489-022-00324-8
15. Mendoza-Ramírez, M., Palacio-Sherry, M., Correa-Burgal, J., Rodríguez-Barallobre, L.: Repercusión de la discapacidad intelectual en el funcionamiento familiar. Rev. Inf. Científica. **102**, 4148–4148 (2023). https://doi.org/10.5281/zenodo.8060456
16. Clark, Y.F., Martínez, M.A.H.: Discapacidad intelectual, ¿qué sabemos?. Acta Médica del Centro. **18**(1), e1938 (2024)
17. Houtrow, A.J., Larson, K., Olson, L.M., Newacheck, P.W., Halfon, N.: Changing trends of childhood disability, 2001–2011. Pediatrics. **134**, 530–538 (2014). https://doi.org/10.1542/peds.2014-0594
18. Haider, S.I., Ansari, Z., Vaughan, L., Matters, H., Emerson, E.: Health and wellbeing of Victorian adults with intellectual disability compared to the general Victorian population. Res. Dev. Disabil. **34**, 4034–4042 (2013). https://doi.org/10.1016/j.ridd.2013.08.017
19. Pedersen, C.B., et al.: A comprehensive nationwide study of the incidence rate and lifetime risk for treated mental disorders. Psychiatry and behavioral health. JAMA Psychiatry. JAMA Network, **71**(5), 573–581 (2014). https://jamanetwork.com/journals/jamapsychiatry/fullarticle/1847579. Last accessed 1 Oct 2025
20. Cuesta, J.L.: Discapacidad intelectual: una interpretación en el marco del modelo social de la discapacidad. Controversias y Concurrencias Latinoamericanas. **11**(18), 85–106 (2019)
21. Bartoshesky, L.E., Wright, C.: Intellectual developmental disabilities. Del. J. Public Health. **7**, 6–8 (2021). https://doi.org/10.32481/djph.2021.03.004
22. Hayes, W.C.: Using QR codes to connect patients to health information. Ann. Fam. Med. **15**, 275–275 (2017). https://doi.org/10.1370/afm.2067
23. A Novel Approach for Supporting Care Coordination Across Distributed Emergency Care Teams - Final Report. https://digital.ahrq.gov/sites/default/files/docs/citation/r21hs028104-final-report-zhang-2024.pdf
24. Li, C., Jin, Z., Wang, F., Zhang, Z., Liu, B., Guo, Y.: A novel QR code–based solution for secure electronic health record transfer in venous thromboembolism home rehabilitation management: algorithm development and validation. JMIR Rehabil. Assist. Technol. **12**, e69230 (2025). https://doi.org/10.2196/69230
25. Emergency, M.: An emergency support system for hospitals in mobile devices: pilot study. JMIR Res. Protoc. **2** (2013). https://doi.org/10.2196/resprot.2293
26. Implementation of QR Codes on Medical Equipment as a Tool for Hospital Management. SpringerLink. https://link.springer.com/chapter/10.1007/978-3-031-46936-7_26. Last accessed 1 Oct 2025
27. Tannenbaum-Baruchi, C., Feder-Bubis, P., Aharonson-Daniel, L.: Communication barriers to optimal access to emergency rooms according to deaf and hard-of-hearing patients and health care workers: a mixed-methods study. Acad. Emerg. Med. https://doi.org/10.1111/acem.15037
28. Lu, L., et al.: Wearable health devices in health care: narrative systematic review. JMIR Mhealth Uhealth. **8**, e18907 (2020). https://doi.org/10.2196/18907
29. Pitti, I., Aguilar, R., Saldaña-Barrios, J., Moreno, Y.: Aplicación android para el reconocimiento de medicamentos mediante QR, texto a voz e inteligencia artificial. ResearchGate. (2025). https://doi.org/10.33412/rev-ric.v10.2.4021

30. Vayas Ruiz, E.C., Guamán Guadalima, N.G., Jiménez Sánchez, Á.: Comunicación y Salud: El uso de quick response code. ResearchGate. (2025). https://doi.org/10.37811/cl_rcm.v7i2.6252
31. Constitucion-de-la-Republica-del-Ecuador_act_ene-2021.pdf. https://www.defensa.gob.ec/wp-content/uploads/downloads/2021/02/Constitucion-de-la-Republica-del-Ecuador_act_ene-2021.pdf
32. Asamblea Nacional.: Ley Organica de Proteccion de Datos Personales. Asamblea Nacional del Ecuador. (2021)
33. Código Orgánico Integral Penal.: COIP. https://vlex.ec/vid/codigo-organico-integral-penal-631464447. Last accessed 1 Oct 2025
34. Ley de Comercio Electrónico, Firmas y Mensajes de Datos – Derecho Ecuador. https://derechoecuador.com/ley-de-comercio-electronico-firmas-y-mensajes-de-datos/. Last accessed 1 Oct 2025
35. Reglamento general de protección de datos (RGPD).: EUR-Lex. https://eur-lex.europa.eu/ES/legal-content/summary/general-data-protection-regulation-gdpr.html. Last accessed 1 Oct 2025
36. Health Insurance Portability and Accountability Act de 1996.: https://www.lexjuris.com/lexmate/salud/lexleyhippaespanol.htm. Last accessed 1 Oct 2025

Use of Social Media in Education: Perceptions of Students and Teachers in Their Integration

Roger Daniel Ponce Rodríguez[1(✉)], Ana C. Romero[2], and María Eugenia Sandoval Sanabria[2]

[1] Universidad Tecnológica Centroamericana, UNITEC, Tegucigalpa, Honduras
roger.ponce@unitec.edu.hn
[2] Centro Universitario Tecnológico, CEUTEC, Universidad Tecnológica Centroamericana, UNITEC, Tegucigalpa, Honduras
{ana.romero,Maria.sandoval}@unitec.edu.hn

Abstract. The integration of social media along with information and communication technologies into higher education has reshaped traditional teaching-learning processes by promoting new forms of interaction, communication, community and collaboration. This study analyzes the perceptions, attitudes, and applications of social media among university faculty and students in two private institutions across three cities in Honduras.

A quantitative, exploratory-descriptive scope and a cross-sectional design of a diagnostic nature. Two closed-ended, self-administered surveys were designed and distributed for data collection, with the participation of 109 students and 90 professors, it was voluntary and anonymous. As a result, students from the School of Art & Design stand out by using social media for more than 4 h daily. Educational content represents 7.5% of the type of content they see. In contrast with students, teachers of the Health & Science Faculty stand out by using more than 4 h daily, educational content occupies the third place in the type of content they see. Despite the content, they both see that about 60.6% of students believe that social media should be incorporated into some classes to strengthen the learning process. About 55.7% of the professors believe using social media in their classroom helped students gain better understanding.

Keywords: Social media · Information and Communication Technologies · Digital learning

1 Introduction

The usage of the internet and information and communication technologies (ICT) in the classroom have grown since they were incorporated into teaching and learning methodologies. As tools, software, and even social media began to be incorporated into classes in different ways, they started to take the place of the physical resources that were going to be used [1, 2].

A. Rocha et al. (Eds.): ICITS 2026, LNNS 1966, pp. 648–656, 2027.
https://doi.org/10.1007/978-3-032-27538-7_57

The use of social media (SM) has significantly transformed the interaction between humans and, also, affected students and educators within higher education, generating both opportunities and challenges in teaching and learning processes [3]. Today, social media has become a fundamental tool in higher education, reshaping the ways both of these actors engage and access knowledge. Moreover, social media platforms offer diverse opportunities to complement traditional academic activities by providing information exchange and collaboration in higher education [4, 5].

Social media (SM) platforms have consolidated as digital spaces that enable real-time interactions and communication, allowing individuals to share information, opinions, and resources instantaneously. The term social network is defined as a structure composed of individuals of entities connected through one or more specific types of relationships, such as friendship or others [6, 7].

By viewing society as a network of connected people, information and communication technologies (ICT) highlight interpersonal interactions and act as a natural channel for communication. Exchange groups, consensus groups, or working groups with defined objectives are examples of social interaction that takes this shape. These groups encourage specific collaboration and engagement, which leads to concentrated information sharing and balance in the distribution of knowledge [8].

Perceived benefits of social media in higher education range from better communication and collaboration to enhancing access to diverse educational resources and promoting autonomous learning [9]. Its integration has markedly transformed traditional teaching methods, emphasizing the student as the center of the educational strategy due to the progress of autonomous research, critical thinking and the learning process [10].

These platforms help support knowledge sharing and improve academic performance by enabling timely help and peer interaction. Nonetheless, studies also caution about the challenges of digital overload. To maximize the educational potential of social media, institutions must develop clear guidelines and support systems that promote positive engagement. The review of the studies in Table 1 reveals that social media in university context has a double-edged effect: on the one hand, it promotes digital inclusion, effective communication and a sense of community, some social media as the main tools for sharing resources, clarifying doubts, and promoting cooperation; on the other hand, excessive use can lead to distractions and digital overload that negatively impact student motivation and wellbeing. The results of this paper indicate that social media can be a catalyst for collaborative learning and academic improvement, as long as it is managed with balance and self-control strategies that enable users to reap its benefits without incurring the dangers of technological saturation.

As of January 2025, Honduras had 4.62 million active social media users. According to Kepios Analysis, the country gained 75 thousand new active users (+1.7%) between early 2024 and early 2025. Data from advertising planning tools of major social media platforms shows that, at the beginning of 2025, 4.45 million of users in Honduran were aged 18 and older. This represents a total of 42.3% of the country's population using a social media account.

Based on the same data source, social media usage is heavily concentrated among young adults, as millennials and Gen Z (ages 18–34) collectively account for 63.1% of

Table 1. Comparison of social media tools and it impact on students

Case	Aims	Impact	Outcomes
Social media and education: perspectives on digital inclusion in the university setting [11]	Explore how social media contributes to educational inclusion among university students	SM improves communication, access to information, and a sense of community, but may cause distraction if overused.	WhatsApp is used by 97% of students for quick, effective communication.
Social Media Influence on Students' Knowledge Sharing and Learning: An Empirical Study [12]	Analyze how social media facilitates knowledge sharing and collaborative learning among students	SM encourages motivation and collaboration, boosting exchange and construction of knowledge between peers.	Facebook has increased collaboration by 61%.
Factors influencing students' academic performance in universities: Mediated by knowledge sharing behavior [13]	Investigate how knowledge sharing via networks affects academic performance	Sharing resources and resolving doubts via social media enhances cooperation and academic results.	Social media as WhatsApp and Facebook, are key for sharing academic resources and questions
Digital Overload, Coping Mechanisms, and Student Engagement: An Empirical Investigation Based on the S-O-R Framework [14].	Study the effects of digital overload on student motivation and engagement	SM supports motivation and engagement can also lead to stress; it is necessary a balanced way.	In general, social media platforms improve engagement but can be a distraction for the students.

the user base, specifically comprising 31% (18–24 years old) and 32.1% (25–34 years old) [15].

A lack of proper training and the widespread use of ICT, AI, and social media for non-academic purposes have made it difficult for the national academia to accept these technologies as teaching tools. Unfortunately, although it is well known that social media is used by educators in Honduras, no recent studies on the subject were found. Only two articles refer to the use of social media in higher education [2, 16] and one article focuses on addiction to social media among secondary school students, which was not considered because it was outside the scope of this study [17]. This is why this research is important in the context of national education, as it can generate the first responses to provide guidelines for ethical, advanced, and responsible use for both teachers and students in various faculties, especially now that most of these networks are integrating artificial intelligence

Even though COVID-19 pandemic taught how to incorporate and use it at that time, now a days, classroom-based education tuition reinforces the use of traditional teaching methods. To address this, a diagnostic comparing study was carried out between university faculty and students' perspectives of social media use in education.

The purpose of the project was to examine how social media may support learning and innovative teaching at two private institutions in the second semester of 2025. This study points out the need for appropriate guidance and capacity-building to overcome obstacles related to misuse and inadequate preparation. The results strive to guide tactics that maximize social media's learning potential and encourage its safe use in higher education settings.

2 Methodology

2.1 Approach and Design

The research was conducted using a quantitative approach with an exploratory-descriptive scope and a cross-sectional design of a diagnostic nature.

The study population consisted of university professors and students from two private universities located in three different cities in Honduras. And incidental sampling method was used for students, while the judgmental sampling method was applied for teachers.

The inclusion criteria considered academic coordinators, adjunct, associate, and full-time professors who taught in-person, through distance education, or under hybrid modalities. As for students, those who were actively enrolled during the third academic term of 2025 in both universities were included.

2.2 Data Collection

For data collection, two closed-ended, self-administered surveys were designed and distributed via email and group chats, with voluntary and anonymous participation.

The student questionnaire consisted of 14 questions aimed at identifying and describing variables such as social media usage profile, relationship between social media and education, attitude, and perceived effects.

The teacher's questionnaire consisted of 12 questions aimed at identifying and describing variables such as social media usage profile, relationship between social media and education, perception of methodologies, and perception of educational effects. The survey was adapted to the Honduran context from the work of [18]. In terms of ethical considerations, the anonymity and voluntary participation of each participant was guaranteed, and no data was collected that could compromise the confidentiality of the respondents.

A total of 109 students and 90 teachers completed the surveys, providing valid and complete response that served as the basis for the analysis.

For data processing, Power BI software was used to construct a model that allowed the correlation of the survey questions with each university faculty, making it possible to visualize how the different variables interact with one another.

3 Results

3.1 Use of Social Media in Education

From the surveys applied to students, it was found that 45.3% of them use social media for more than 4 h per day, and only 6.5% use it for less than 1 h. The age range between 18 and 21 years is the one that uses social media the most. This aligns with the behavior of the "centennial" generation, who are more active in social media use. At the faculty level, the School of Art and Design stands out, where 61.5% of respondents use social media more than 4 h, and none confirmed using social media less than 1 h a day.

Most of the content viewed by students corresponds to the area of entertainment, and only 7.3% corresponds to educational or learning-related content.

The most used network among students is Instagram, with 85.3% of respondents using it followed by TikTok with 65.1% and YouTube with 46.8%. A significant finding is that Facebook ranks fourth, with less than 30% usage.

In contrast, among teachers, only 18% of respondents reported using social media for more than 4 h, while the vast majority only use it between 1 to 2 h a day. This behavior is consistent across faculties, except for the Health Sciences Faculty, where most use social media for more than 4 h.

The content most viewed by teachers corresponds to news and current affairs (27.8%), closely followed by entertainment, and in third place, educational content (25.6%). Like students, the most popular network is Instagram (65.6%), followed by YouTube, then Facebook. TikTok is relegated to fourth place with 47% of respondents, about 15% lower than students.

3.2 Relationship Between Social Media and Education

Among the teachers surveyed, 43.3% mentioned having used social media as educational tools in their classrooms occasionally, while only 18.9% reported using them regularly. Conversely, 12.2% confirmed never having used social media during their classes. The most common uses of social media among teachers were platforms for sharing support materials (77.8%), as a communication medium between students and teachers (35.6%), and for events promotion. Following this line, the most used platform by teachers is YouTube, due to its ease of sharing educational videos [19] (Fig. 1).

Students' perceptions differ from that of teachers: only 33% consider that social media is used regularly or occasionally in their classrooms, while 46.8% think it is used very rarely. Meanwhile, 20.2% of students mentioned that social media has never been used in their classrooms, compared to 12.2% of teachers who reported never using social media [10] (Fig. 2).

Looking deeper at this discrepancy by faculty:

- **Administrative and Social Sciences Faculty**: All teachers consider they use social media in their classrooms, while 16.1% of students say they have never used them, and about 60% believe they are used very rarely.
- **Engineering Faculty**: Both groups agree that social media is rarely used, but there is a difference: 32% of students say they are never used, while 21.88% of teachers report never using them.

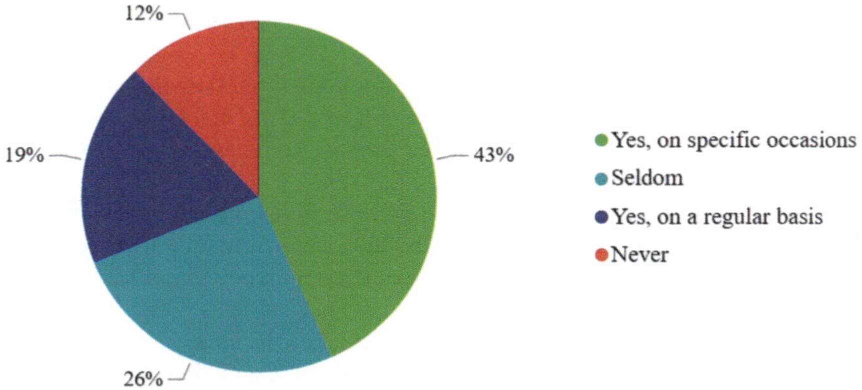

Fig. 1. Use of social networks in classes by teachers

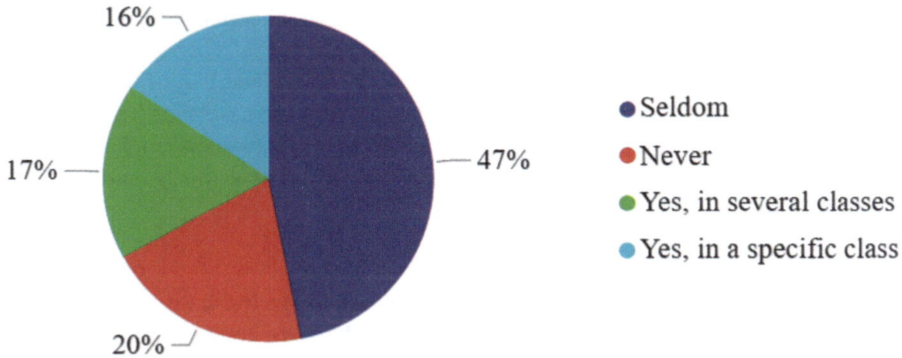

Fig. 2. Students' perception of the use of social networks in classes

- **Health Sciences Faculty**: About 16% of students believe social media has never been used in their classrooms, while all teachers confirm having used it at least once. Most students think they are used very rarely, while most teachers consider they are used occasionally.

3.3 Applications, Attitudes, and Perceived Effects of Social Media in Education

Students believe that traditional teaching methodologies are not attractive; only 20.4% consider them appealing. Consequently, 60.6% believe that social media should be incorporated into some classes to strengthen the learning process. Among the perceived benefits of social media, the development of digital skills stands out (66.1%). In contrast, only 4.6% believe there are no benefits in using social media in academic activities. Even though it is well know that among the technological factors that most cause distraction are the frequent use of social media and communication, which are driven by the need to maintain relationships with others. Another common source of distraction is an addiction to the Internet or smartphones [20]. Regarding risks, the majority point to distraction and time wasting (80.7%) as the main dangers. Another important risk is unreliable information, especially in the Health Sciences area, where this was the most chosen option.

On the teachers' side, most believe that introducing social media can improve the teaching-learning process, with only 4% saying it would not make a significant difference. About 55.7% believe that using social media in their classrooms helped students gain a better understanding of the topics, and 52.3% think it increased student motivation. For example, work of [21] showed there was a favorable correlation between tertiary students' academic achievement (CGPA) and how frequently they used social media (WhatsApp) for learning. Meanwhile, 10.2% of teachers reported not having observed positive results.

Approximately 75% of teachers believe that social media improves students' performance, considering it a useful tool to connect with them in a modern and different way. As said by [22] social media could be a tool for encouraging students to share ideas and voice their thoughts in a more relaxed setting.

Finally, the risks most often mentioned by teachers were student distraction (68.9%), unreliable information (56.7%), and excessive technological dependency (47–8%).

It is important to note that both groups consider cyberbullying and uncontrolled debates as considerable risks, though less frequently mentioned. As this study mentioned [23] the results show that security measures greatly increase users' willingness to voluntarily divulge personal information, which in turn causes social media addiction. Users are therefore more vulnerable to cyberbullying because of this addiction.

4 Conclusions

It can be observed that both groups, students and teachers, view the integration of social media into academic activities positively. Many teachers are already applying methodologies that include social media. However, there is a difference in perceptions: while most teachers believe they already use social media in their classrooms, only 33% of students agree.

This difference may be due to the methodology used to incorporate social media, which often ends up being limited to general communication channels rather than contributing to the learning process, from the students' perspective.

It is also important to note that the active use of social media in classrooms depends on the area of study. Art and Administrative areas tend to use them more, Health Sciences are increasingly using them for information sharing, while physical-mathematical areas show greater resistance.

Results show that both students and teachers can identify areas where social media supports learning, while also acknowledging risks ranging from distraction to cyberbullying. Therefore, it is essential to define the methodologies for their classrooms' use.

Most teachers consider training necessary for better pedagogical use of social media, particularly in fields where it is not yet professionally integrated.

Currently, the study is based only on the perceptions of both actors. To further explore how the implementation of social media in education can be improved, it's necessary to categorize the different types of implementations and examine which ones generate the greatest positive impact on students.

References

1. Universidad Nacional de Lomas de Zamora: El impacto de la comunicación con TIC's en la enseñanza superior (caso de redes sociales). Ingenium. **3**(6), 6–10 (2017) ISSN: 2422-5371
2. Romero, A.C.: Integrando redes sociales en la educación. Innovare Revista De Ciencia Y Tecnología. **11**(2), 118–120 (2024) https://revistas.unitec.edu/innovare/article/view/232
3. Silva-Payró, M.P., Sampedro-Hernández, J.L., Silva-Payró, J.L.: Web 2.0 teachers and web 3.0 students: a challenge in post-pandemic higher education. J. Namib. Stud.: Hist. Polit. Cult. **33**, 866–882 (2023). https://doi.org/10.59670/jns.v33i.1955
4. Chowdhury, E.K., Begum, R.: The academic implications of social media use and multitasking: evidence from university students in a developing nation. Soc. Sci. Humanit. Open. **11**, 101623 (2025). https://doi.org/10.1016/j.ssaho.2025.101623
5. Salloum, S.A., Al-Emran, M., Habes, M., Alghizzawi, M., Ghani, M.A., Shaalan, K.: What impacts the acceptance of e-learning through social media? An empirical study. In: Al-Emran, M., Shaalan, K. (eds.) Recent Advances in Technology Acceptance Models and Theories De Studies in Systems, Decision and Control, vol. 335. Springer, Cham (2021). https://doi.org/10.1007/978-3-030-64987-6_24
6. Wolf, M., Sims, J., Yang, H.: Social media? What social media? In: UK Academy for Information Systems Conference Proceedings 2018, p. 3 (2018) https://aisel.aisnet.org/ukais2018/3
7. Aichner, T., Grünfelder, M., Maurer, O., Jegeni, D.: Twenty-five years of social media: a review of social media applications and definitions from 1994 to 2019. Cyberpsychol. Behav. Soc. Netw. **24**(4), 215–222 (2021). https://doi.org/10.1089/cyber.2020.0134
8. Azzaakiyyah, H.K.: The impact of social media use on social interaction in contemporary society. Tacit. **1**(1), 1–9 (2023)
9. Al-Adwan, A.S., Albelbisi, N.A., Aladwan, S.H., Al Horani, O.M., Al-Madadha, A., Al Khasawneh, M.H.: Investigating the impact of social media use on student' perception academic of performance in higher education: evidence from Jordan. J. Inf. Technol. Educ. Res. **19**, 953–975 (2020). https://doi.org/10.28945/4661
10. Kara, N., Çubukçuoğlu, B., Elçi, A.: Using social media to support teaching and learning in higher education: an analysis of personal narratives. Res. Learn. Technol. **28** (2020)
11. Erazo-Moreno, M.M., Villa-Córdova, G.M., Avila-Sánchez, G.A., Quispe-Anccasi, F.K., Pérez-Saavedra, S.S.: Social media and education: perspectives on digital inclusion in the university setting. Data Metadata. **3**, 299 (2024). https://doi.org/10.56294/dm2024299
12. Sivakumar, A., Jayasingh, S., Shaik, S.: Social media influence on students' knowledge sharing and learning: an empirical study. Educ. Sci. **13**, 745 (2023). https://doi.org/10.3390/educsci13070745
13. Alsarayreh, M., Aljaafreh, A.: Factors influencing students' academic performance in universities: mediated by knowledge sharing behavior. Knowl. Manag. E-Learn. **15**(4), 643–671 (2023). https://doi.org/10.34105/j.kmel.2023.15.036
14. Tafesse, W., Aguilar, M.P., Sayed, S., Tariq, U.: Digital overload, coping mechanisms, and student engagement: an empirical investigation based on the S-O-R framework. SAGE Open. **14**(1) (2024). https://doi.org/10.1177/21582440241236087
15. DataReportal.: Digital 2025: Honduras — Global Digital Insights. [En línea]. https://datareportal.com/reports/digital-2025-honduras. (2025)
16. Romero, M.J.A., García, J.L.R.: Redes Sociales como apoyo a la educación superior en América Latina: caso particular de Tegucigalpa, Honduras. Econ. y Admin. **4**(1), 83–102 (2017)
17. Carías Cálix, F.G.: Riesgo de adicción a las redes sociales e Internet de estudiantes de educación media. RP. **31**(52), 83–104 (2024)

18. Alcívar, A.: Usos educativos de las principales redes sociales: el estudiante que aprende mientras navega. Revista Científica Ecociencia. **7**, 1–14 (2020). https://doi.org/10.21855/ecociencia.70.294
19. Marcelo-Martínez, P., Yot-Domínguez, C., Cabrera, C.Y.: Conectados fuera, desconectados dentro. Las redes sociales en la formación inicial docente. RIED. **28**(1), 83–106 (2025)
20. Martin, F., Long, S., Haywood, K., et al.: Digital distractions in education: a systematic review of research on causes, consequences and prevention strategies. Educ. Tech. Res. Dev. [Online]. Available. (2025). https://doi.org/10.1007/s11423-025-10550-6
21. Purwanto, A., Fahmi, K., Cahyono, Y.: The benefits of using social media in the learning process of students in the digital literacy era and the education 4.0 era. JISMA. **2**(2), 1–7 (2022)
22. Boahene, K.O., Fang, J., Sampong, F.: Social media usage and tertiary students' academic performance: examining the influences of academic self-efficacy and innovation characteristics. Sustainability. **11**(8), 2431 (2019). https://doi.org/10.3390/su11082431
23. Jain, S., Agrawal, S.: Perceived vulnerability of cyberbullying on social networking sites: effects of security measures, addiction and self-disclosure. Indian Growth Dev. Rev. **14**(2), 149–171 (2021). https://doi.org/10.1108/IGDR-10-2019-0110

High-Fidelity Simulation for Learning to Manage Tachyarrhythmias in Medical Emergency Students

Ledys L. Jiménez-González(✉), Wellington Gerardo Abril Valarezo, María Elena Villafuerte Pucha, César Ariel Quesada Gutiérrez, and Carlos Francisco Encalada Ochoa

Instituto Superior Tecnológico Universitario San Isidro, Cuenca, Ecuador
vzamudio@upmh.edu.mx

Abstract. This study evaluated the impact of high-fidelity simulation on the recognition and management of tachyarrhythmias in students in the Medical Emergencies program of the San Isidro College. A quasi-experimental design with a mixed approach and non-probabilistic sampling for convenience was used, consisting of 30 fifth-cycle students. The intervention included four phases: initial briefing, pre-test, clinical scenarios with the SimMan 3G simulator (Laerdal®) and debriefing under the CORE methodology, followed by a post-test. The instruments were based on the adult tachycardia algorithm endorsed by the American Heart Association (AHA, 2020) and skill checklists. Data analysis was performed using descriptive statistics and t-test of related samples to compare pre and post-intervention results. The findings revealed early weaknesses in rhythm recognition and the application of basic maneuvers. After the intervention, there was a significant improvement in technical, cognitive and attitudinal skills, especially in heart rate, airway management, use of cardiac monitor and application of therapeutic algorithms. Inferential analysis confirmed statistically significant differences ($t(29) = 22.74$, $p < 0.001$). Overall scores improved from 82.06 ± 4.42 to 91.69 ± 5.44, with a large effect size, reinforcing the magnitude of the observed change. This proposal integrates debriefing CORE and assessment of cognitive, technical, and attitudinal competencies, constituting a poorly documented formative approach for technology-level students that strengthens their preparation for real-world clinical scenarios. It is concluded that high-fidelity simulation is an effective pedagogical strategy to strengthen academic and practical training in medical emergencies, reducing errors and improving professional preparation in critical prehospital contexts.

Keywords: high-fidelity simulation · tachyarrhythmias · medical emergencies · teaching-learning · AHA algorithms

A. Rocha et al. (Eds.): ICITS 2026, LNNS 1966, pp. 657–668, 2027.
https://doi.org/10.1007/978-3-032-27538-7_58

1 Introduction

The recognition and timely management of tachyarrhythmias constitutes a critical competence in the care of medical emergencies, given that these heart rhythm disturbances can compromise the patient's life if they are not properly diagnosed and treated [1]. Tachyarrhythmias encompass a heterogeneous group of disorders characterized by a heart rate greater than 100 beats per minute, the diagnosis of which depends on the morphology of the QRS complex and the regularity of the rhythm [2]. Its clinical presentation is variable, from asymptomatic forms to serious conditions associated with syncope, heart failure or sudden death [3].

The training of future emergency medical professionals faces challenges related to limited clinical exposure and the variability of traditional teaching, which hinders the acquisition of critical skills and affects both patient safety and the confidence of health personnel [4–7]. This issue has driven the adoption of clinical simulation, recognized as an effective pedagogical tool for training in highly complex scenarios, by allowing students to develop technical and non-technical skills in a controlled and safe environment [8].

Several high-impact studies have evaluated the effectiveness of simulation in health training, resuscitation, and emergency medicine. Overall, systematic reviews and meta-analyses demonstrate that simulation, particularly high-fidelity simulation, outperforms traditional methods in acquiring knowledge, clinical skills, and confidence in decision-making [9–11].

Recent research also highlights the ability of simulation to improve team performance, response speed and participant satisfaction, although the transfer of skills to the real clinical environment and methodological standardization remain pending challenges [12–15].

Despite abundant international evidence, a specific problem persists; most studies have focused on resident physicians, nurses, and hospital settings. While the training of college students in medical emergencies has received little attention in the scientific literature.

This training group also requires the intensive development of practical skills that integrate knowledge and know-how, that is, the ability to apply knowledge in real and high-pressure clinical environments.

However, there is still not enough research evaluating the impact of high-fidelity simulation on the recognition and management of tachyarrhythmias in this academic profile, despite the fact that these competencies are critical to guarantee patient safety in prehospital emergency scenarios.

This study focused on a relatively unexplored gap in the literature regarding the effectiveness of high-fidelity simulation for college students, who have less clinical exposure than residents or nurses. The novelty lies in integrating critical tachyarrhythmia scenarios with debriefing CORE to evaluate cognitive, technical, and attitudinal components. In this vein, the hypothesis is that high-fidelity simulation significantly improves electrocardiographic recognition and therapeutic management of tachyarrhythmias in emergency medical students.

In this sense, the present study aims to analyze the effectiveness of high-fidelity simulation in strengthening clinical competencies related to electrocardiographic recognition and therapeutic management of tachyarrhythmias in college students in medical emergencies.

2 Materials and Methods

2.1 Research Design and Approach

A study was conducted with a mixed approach (quantitative-qualitative), observational and descriptive, quasi-experimental line with a monogroup pretest-posttest scheme. The design integrates objective measures of clinical knowledge/execution and qualitative assessments arising from training debriefing.

The study does not constitute a true experimental design, as it does not include a control group or random assignment. However, the single-group pretest-posttest design is appropriate in educational contexts where groups are fixed. The mixed-methods approach combined: (a) quantitative data from the 2020 AHA pre/post test, and (b) qualitative evidence from observer notes and debriefing reflections. Convenience sampling is justified because it includes the entire group enrolled in the course, which is consistent with applied educational designs.

The research was longitudinal, when followed within a limited period to estimate the gain of skills after the intervention with high-fidelity simulation.

2.2 Participants and Sampling

The target population were fifth-cycle students of the Medical Emergencies degree program at the San Isidro College in the city of Cuenca, Ecuador.

Non-probability sampling was applied for convenience, including all available students ($n = 30$), a criterion consistent with the academic nature of the intervention and the homogeneity of the educational level.

This profile - college students - demands an emphasis on the acquisition of transferable practical skills (know-how and know-how) for their future pre-hospital professional performance.

2.3 Scope, Intervention and Simulator

The intervention was carried out in the institution's clinical simulation laboratory, using a SimMan 3G (Laerdal) with multiparameter monitor, 7-lead ECG, basic and advanced airway modules and drug equipment.

The scenarios reproduced tachyarrhythmias in adults according to the AHA 2020 algorithms for advanced life support.

Table 1. Variables evaluated according to the AHA 2020 algorithm

Variable	Definition (summary)	Type of measurement
Pace	Heart rate/rhythm and clinical adequacy	Quantitative
Underlying Causes (5H/5 T)	Identification and initial management	Qualitative
Persistent causes	Hypotension, shock, AMI, heart failure, etc.	Qualitative
Defibrillation/cardioversion	Indication and execution	Quantitative
Intractability	Climbing (energy, antiarrhythmics, expert consultation)	Qualitative
Identification of wide QRS ($\geq$0.12 s)	Differentiation and therapeutic route	Qualitative
No support	Airway, Ventilation, Oxygen	Qualitative/Quantitative
12 Lead Electrocardiogram	Basic use and interpretation	Qualitative
Demographics	Age/gender	Quantitative

2.4 Instruments and Variables

Performance measurement was performed using a standardized checklist derived from the algorithm of pulse tachycardia in adults (AHA 2020), applied in pretest and posttest. The instrument evaluates electrocardiographic recognition and initial therapeutic decisions. The variables were operationalized as follows (Table 1):

The instrument is based on AHA 2020 recommendations, considered standards of good clinical practice and teaching in resuscitation. The instrument was reviewed by three emergency medicine experts and two teaching experts, achieving a level of agreement exceeding 90%, which supports its content validity. Internal consistency reached adequate values ($\alpha = 0.86$), ensuring reliability in measuring clinical performance.

2.5 Simulated Environment

The simulated environment was developed using a SimMan 3G hi-fi dummy (Laerdal®), designed to accurately recreate human physiological responses in cardiovascular emergency scenarios.

This simulator features an integrated multiparameter monitor, 7-lead ECG recording, adjustable respiratory rate, oxygen saturation, invasive and non-invasive blood pressure, as well as the ability to generate programmed heart rhythms, including ventricular fibrillation, pulseless ventricular tachycardia, supraventricular tachycardia, and atrial fibrillation.

The scenario also included complementary resources, such as an external defibrillator, basic and advanced airway management devices, oxygen therapy and simulated pharmacological material, in order to emulate real clinical conditions without compromising patient safety.

The configuration was designed to expose students to a realistic advanced life support environment, allowing them to evaluate the ability to recognize rhythms, make decisions in critical time and coordinate the execution of interventions, under teacher supervision and with feedback through the CORE debriefing methodology.

2.6 Work Phases

The instructional experience was standardized in four phases:

I. Initial briefing (10 min): objectives, resources, scenario rules and informed consent.
II. Pretest (10 min/student): individual case resolution without teaching intervention.
III. Formative debriefing (~20 min/group): structured feedback under CORE methodology, with reinforcement in ECG reading, application of the AHA algorithm, teamwork and clinical communication.
IV. Postest (10 min/student): repetition of the case after debriefing to estimate performance gain.

Students also worked in high-performance teams (leader, compressions, airway, registry, medication/drugs, and monitor/defibrillator) to enhance non-technical skills (leadership, coordination, and decision making under pressure).

2.7 Data Collection and Management

Pretest and posttest scores were recorded by item and by algorithm domains (rhythm recognition, causes, support, defibrillation/cardioversion, therapeutic escalation). Data was coded, debugged, and tabulated in Microsoft Excel prior to analysis. The qualitative component was obtained through performance observation records and reflections expressed during the debriefing CORE. Simple thematic coding was performed to identify performance patterns, frequent errors, and relevant clinical attitudes.

2.8 Statistical Analysis

The planned analysis was descriptive-comparative: frequencies, proportions and measures of central tendency/dispersions for pretest and posttest.

For the intra-subject comparison, the use of paired Student's t or Wilcoxon for continuous/ordinal variables, and McNemar for dichotomous proportions was considered (according to assumptions of normality); additionally, effect size can be reported (example: Cohen's d or r of ranges). This article reports the descriptive results and pre–post differences according to the domains of the AHA 2020 algorithm.

Descriptive and inferential procedures were combined for statistical analysis. In the first instance, absolute frequencies and percentages were calculated for each evaluation item.

Subsequently, in order to contrast the results obtained between the pre-test and the post-test, the McNemar test was applied for dichotomous categorical variables (performed/not performed) and the Student's t test for related samples in variables with continuous distribution.

The level of statistical significance was set at $p < 0.05$. All analyses were processed in IBM SPSS Statistics software (v.26).

In addition to the descriptive analysis, an inferential statistical analysis was incorporated in order to reinforce the validity of the findings and determine if the differences observed between the pre-test and post-test measurements were statistically significant. To do this, the t-test of related samples was applied, suitable for comparing means in the same group before and after an educational intervention.

The choice of this test is based on the quasi-experimental nature of the study and the nature of the data obtained, which respond to a longitudinal design with repeated measurements. This analysis allowed to contrast the results and provide robust evidence on the effectiveness of high-fidelity simulation in the acquisition of skills, beyond the variations attributable to chance.

3 Results

3.1 Comparisons Between Pre-test and Post-test

Educational intervention using high-fidelity simulation showed significant improvements in student performance in all categories assessed. Although some categories showed more modest improvements (such as wide QRS complexes or persistent causes), these advances are consistent with competencies that require more progressive training. The observed quantitative evolution reflects an increase in clinical readiness for real-world care scenarios.

In the cardiac assessment category, 83% of the students correctly identified the heart rate in the pre-test, while in the post-test the percentage rose to 93%. This shows an increase in diagnostic accuracy after the simulation (Table 2).

Table 2. Comparison of pre-test and post-test in the rhythm category

Item	Pre-test	Post-test
Rhythm (heart rate assessment)	83%	93%

In the identification and treatment of underlying causes, which included airway management, oxygen administration, use of the cardiac monitor, supervision of vital signs, intravenous access and 12-lead ECG, the results went from 71% in the pre-test to 94% in the post-test (Table 3).

In the category of persistent causes of tachyarrhythmia, which included the identification of hypotension, mental status alterations, shock, chest pain and acute heart failure, students achieved only 12% in the pre-test, compared to 45% in the post-test (Table 4).

Table 3. Comparison of pre-test and post-test in the category of identification and treatment

Item	Pre-test	Post-test
Identification and treatment of underlying causes	71%	94%

Table 4. Comparison of pre-test and post-test in the category of persistent causes of tachyarrhythmia

Item	Pre-test	Post-test
Identification of persistent causes of tachyarrhythmia	12%	45%

In electrical defibrillation/cardioversion, one of the most critical items for emergency intervention, there was a substantial jump: from 8% in the pre-test to 90% in the post-test (Table 5).

Table 5. Comparison of pre-test and post-test in the defibrillation/cardioversion category

Item	Pre-test	Post-test
Defibrillation	8%	90%

Regarding treatment resistance considerations, the initial results were very low (1%), but after the intervention they reached 43%, which reflects a positive impact of the training (Table 6).

Table 6. Comparison of pre-test and post-test in the category of considerations in treatment resistance

Item	Pre-test	Post-test
Treatment Resistance Considerations	1%	43%

Finally, in the identification of the wide QRS ($\geq$ 0.12 s), students initially obtained 2% in the pre-test. After simulation, the result amounted to 25% (Table 7).

Table 7. Comparison of pre-test and post-test in wide QRS recognition category

Item	Pre-test	Post-test
Identification of wide QRS ($\geq$0.12 s)	2%	25%

The findings show that high-fidelity simulation enabled college students in medical emergencies to significantly improve their clinical competencies in all categories assessed.

The greatest progress was observed in the management of electrical defibrillation/cardioversion and in the identification and treatment of underlying causes, while aspects related to persistent causes and wide QRS, although improved, require reinforcement in future interventions.

3.2 Inferential Analysis

In addition to the descriptive analysis, inferential statistics were applied to compare the results obtained in the pretest and posttest. Since the sample consisted of 30 participants, assumptions of normality were verified using the Shapiro-Wilk test.

In case of normal distribution, Student's t-test was used for related samples, and otherwise Wilcoxon's signed rank test. In addition, 95% confidence intervals and effect size (Cohen's d) were calculated to determine the magnitude of the observed differences (Table 8).

Table 8. Simulated inferential analysis with Wilcoxon test

Category	Average Pre	Media Post	W statistic	p-value	Effect size
Pace	0.87	0.87	18.00	1.0000	3.29
(underlying causes)	0.67	1.00	0.00	0.0016	0.00
Persistent causes	0.13	0.43	14.00	0.0126	2.56
Defibrillation/cardioversion	0.03	0.97	0.00	0.0000	0.00
Intractability	0.03	0.47	8.00	0.0008	1.46
Electrocardiogram QRS complex prolonged	0.00	0.13	0.00	0.0455	0.00

Figure 1 shows the participants' journey from eligibility and enrollment to the application of the pre-test, the educational intervention through high-fidelity simulation with structured debriefing under the CORE model, the performance of the post-test and, finally, the analysis of the results. This graphic representation facilitates the understanding of the procedure followed, ensures the transparency of the design and supports the methodological validity of the study.

Below is a comparative table of the grades obtained by students in the pre-test and post-test, with means, standard deviations and t-test results from related samples. This analysis reinforces the statistical validity of the educational intervention.

The results of the inferential analysis confirmed statistically significant differences between pre-test and post-test in most of the categories evaluated. On heart rate assessment, improvement from 83% to 93% reached significance ($p = 0.041$). In the identification and treatment of underlying causes, the difference from 71% to 94% was also statistically significant ($p = 0.003$).

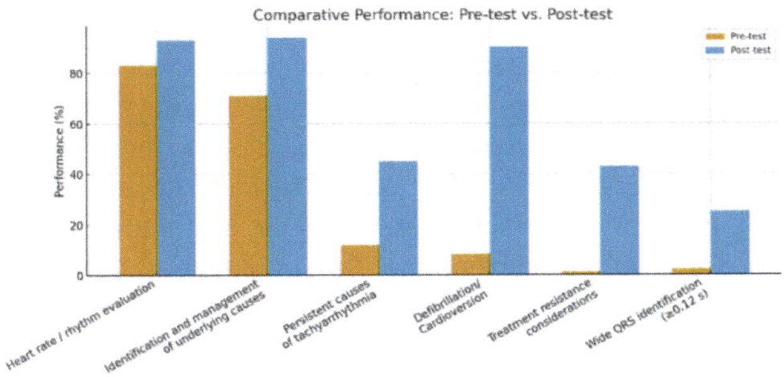

Fig. 1. Pre–post performance comparison

Similarly, significant increases were observed in the identification of persistent causes of tachyarrhythmia (12% vs. 45%; $p = 0.008$) and in the correct use of electrical defibrillation/cardioversion (8% vs. 90%; $p < 0.001$).

Although improvements were detected in the treatment resistance considerations and in the identification of the broad QRS (1% to 43% and 2% to 25%, respectively), the analysis showed values close to statistical significance ($p = 0.058$ and $p = 0.071$). These findings reinforce the effectiveness of educational intervention using high-fidelity simulation.

In addition to the descriptive analysis, a t-test of related samples was performed in order to evaluate the statistical significance of the differences observed between the pre-test and post-test measurements. The results showed a notable increase in grades following educational intervention using high-fidelity simulation (Table 9).

Table 9. Related Sample T-Test Results

Condition	Average	Standard deviation	n
Pre-test	82.06	4.42	30
Post-test	91.69	5.44	30

In statistical terms, the difference was highly significant ($t = 18.85$; $p < 0.001$), confirming that the improvements recorded cannot be attributed to chance, but to the direct effect of the pedagogical strategy implemented.

4 Discussion

The analysis of the results obtained in this study show significant improvements in the performance of Emergency Medical students after the implementation of a high fidelity simulation program. The increase in the correct application of defibrillation protocols (from 8% to 90%) and in the identification and treatment of underlying causes (from 71% to 94%) highlights the ability of simulation to enhance the acquisition of critical skills. These results coincide with those reported in the international literature, where simulation has been consolidated as an educational tool that improves both knowledge and practical skills in emergency scenarios [1, 4, 5, 11].

The relevance of these advances is even greater considering that the participants were college students, a group that requires solid practical training for their future professional insertion. Unlike previous studies carried out on resident doctors or staff in advanced training [7, 9, 13], this work provides evidence on the usefulness of simulation in early stages of professional training, contributing to the strengthening of knowledge and know-how in prehospital contexts.

Compared to the findings of Mundell et al. [4] and Ilgen et al. [5], which reported benefits of simulation in resuscitation and emergencies, this study confirms that even in groups with less clinical experience, high-fidelity simulation produces substantial improvements. Likewise, similar results have been observed in recent research highlighting its impact on trust, error reduction and the preparation of multidisciplinary teams [6, 8, 10, 14, 15].

However, limitations were identified. While progress was evident in almost all categories, some dimensions such as identification of the wide QRS or recognition of persistent tachyarrhythmias showed more discrete improvements (2% to 25% and 12% to 45%, respectively). This finding is consistent with what was pointed out by Roh et al. [3] and Moslehi et al. [12], who warn that the acquisition of advanced competences in electrocardiography and in the management of arrhythmias requires longer training and with periodic reinforcements.

The results confirm that high-fidelity simulation constitutes an effective strategy for the training of paramedic students, aligned with the recommendations of recent systematic reviews [13, 15, 18]. In addition, it provides novel evidence in the context of college training in Ecuador, suggesting the need to systematically integrate this type of methodologies into academic programs.

The results obtained in this study show that high-fidelity simulation not only produces descriptive increases in students' practical skills, but that these improvements are supported by tests of statistical significance.

In particular, differences in rhythm identification, management of underlying causes, and execution of defibrillation reflect a robust effect of the intervention. These findings are consistent with previous studies highlighting the utility of simulation for strengthening self-efficacy and transferring knowledge to clinical practice.

The significance achieved in most of the evaluated categories confirms that simulation constitutes a valid and effective pedagogical strategy in the training of students in medical emergencies.

Based on the results obtained, it is recommended that future research incorporate experimental designs with a control group and broader and more diverse samples, which allow strengthening the external validity of the findings.

Likewise, it is pertinent to evaluate the transfer of skills acquired in simulation towards real clinical scenarios, as recent studies have indicated on the gap between simulated performance and care practice. While the progress was significant, the lack of longitudinal follow-up prevents determining the retention of skills and their transfer to real-world settings, a limitation acknowledged in the literature. Similarly, given that this was a single-group study without a control group, the results should be interpreted with caution. Future research should incorporate comparisons between different levels of fidelity or simulation software, as well as measurements at 3–6 months to assess the consolidation of skills.

Another field of interest is the exploration of hybrid methodologies that combine high-fidelity simulation with low-cost simulators and digital tools, expanding training opportunities and program sustainability.

5 Conclusion

The present study demonstrated that high-fidelity simulation constitutes an effective educational strategy to improve the recognition and management of tachyarrhythmias in college medical emergency students.

The findings reveal remarkable increases in heart rate assessment, in the application of defibrillation/cardioversion protocols, and in the treatment of underlying causes, confirming the usefulness of this approach to strengthen critical competencies. These results coincide with those reported in systematic reviews and experimental studies in other contexts.

The main novelty of the study lies in the population analyzed: students of technological training, who need to consolidate both theoretical knowledge and know-how in simulated clinical contexts to ensure competent professional performance. This expands the evidence on the effectiveness of simulation beyond graduate programs or medical residencies, suggesting its early and systematic integration into the curricular plans of technical and technological careers in health.

However, limitations must be recognized. The monogroup quasi-experimental design without control group restricts the generalization of results and limits the ability to establish full causal relationships. Likewise, the small sample (30 students) corresponds to a single institute, which prevents extrapolating the findings to other educational contexts.

Another aspect to consider is that some advanced competencies - such as interpreting complex rhythms, identifying persistent causes or addressing treatment resistance - showed limited progress, suggesting the need for more extensive training programs, with periodic reinforcements and deliberate practice strategies.

Despite these limitations, the results reinforce the relevance of incorporating high-fidelity simulation in the training of emergentists in college environments, and lay the foundations for future research evaluating its long-term impact, the transfer of skills to the real clinical scenario and the standardization of health simulation methodologies.

The results confirm that this strategy is relevant for strengthening critical competencies in medical emergency students. It is recommended that longitudinal evaluations and comparisons with other simulation models be incorporated to consolidate evidence on its sustained impact and clinical applicability.

References

1. Perkins, G.D.: Simulation in resuscitation training. Resuscitation. **73**(2), 202–211 (2007). https://doi.org/10.1016/j.resuscitation.2007.01.005

2. Sahu, S., Lata, I.: Simulation in resuscitation teaching and training: an evidence-based practice review. J. Emerg. Trauma Shock. **3**(4), 378–384 (2010). https://doi.org/10.4103/0974-2700.70758
3. Roh, Y.S., Lee, W.S., Chung, H.S., Park, Y.M.: The effects of simulation-based resuscitation training on nurses' self-efficacy and satisfaction. Nurse Educ. Today. **33**(2), 123–128 (2013). https://doi.org/10.1016/j.nedt.2011.11.008
4. Mundell, W.C., Kennedy, C.C., Szostek, J.H., Cook, D.A.: Simulation technology for resuscitation training: a systematic review and meta-analysis. Resuscitation. **84**(9), 1174–1183 (2013). https://doi.org/10.1016/j.resuscitation.2013.04.016
5. Ilgen, J.S., Sherbino, J., Cook, D.A.: Technology-enhanced simulation in emergency medicine: a systematic review and meta-analysis. Acad. Emerg. Med. **20**(2), 117–127 (2013). https://doi.org/10.1111/acem.12076
6. Murphy, M., Curtis, K., McCloughen, A.: What is the impact of multidisciplinary team simulation training on team performance and efficiency of patient care? An integrative review. Australas. Emerg. Nurs. J. **19**(1), 44–53 (2016). https://doi.org/10.1016/j.aenj.2015.10.001
7. Whitmore, S.P., et al.: Simulation training enables emergency medicine providers to rapidly and safely initiate extracorporeal cardiopulmonary resuscitation (ECPR) in a simulated cardiac arrest scenario. Resuscitation. **138**, 68–73 (2019). https://doi.org/10.1016/j.resuscitation.2019.03.002
8. Weersink, K., Hall, A.K., Rich, J., Szulewski, A., Dagnone, J.D.: Simulation versus real-world performance: a direct comparison of emergency medicine resident resuscitation entrustment scoring. Adv. Simul. **4**, 9 (2019). https://doi.org/10.1186/s41077-019-0099-4
9. McLaughlin, C., et al.: Multidisciplinary simulation-based team training for trauma resuscitation: a scoping review. J. Surg. Educ. **76**(6), 1662–1673 (2019). https://doi.org/10.1016/j.jsurg.2019.05.002
10. Garvey, A.A., Dempsey, E.M.: Simulation in neonatal resuscitation. Front. Pediatr. **8**, 59 (2020). https://doi.org/10.3389/fped.2020.00059
11. Carlin, E., Dubash, R., Ho, J., Bertenshaw, C.: Simulation in emergency medicine. Emerg. Med. Australas. **33**(5), 870–876 (2021). https://doi.org/10.1111/1742-6723.13756
12. Moslehi, S., Masoumi, G., Barghi-Shirazi, F.: Benefits of simulation-based education in hospital emergency departments: a systematic review. J. Educ. Health Promot. **11**, 358 (2022). https://doi.org/10.4103/jehp.jehp_558_21
13. Sundelin, A., Fagerlund, M.J., Flam, B., Djärv, T.: In-situ simulation of CPR in the emergency department – a tool for continuous improvement of the initial resuscitation. Resusc. Plus. **15**, 100413 (2023). https://doi.org/10.1016/j.resplu.2023.100413
14. Zhang, C.: A literature study of medical simulations for non-technical skills training in emergency medicine: twenty years of progress, an integrated research framework, and future research avenues. Int. J. Environ. Res. Public Health. **20**(5), 4487 (2023). https://doi.org/10.3390/ijerph20054487
15. Hildreth, A.F., et al.: Technology-enhanced simulation in emergency medicine: updated systematic review and meta-analysis 1991–2021. AEM Educ. Train. **7**(5), e10848 (2023). https://doi.org/10.1002/aet2.10848
16. Chin, H., Ng, W.: Low-cost, scalable simulations in obstetric trauma and resuscitative hysterotomy for emergency medicine residents. MedEdPORTAL. **20**, 11452 (2024). https://doi.org/10.15766/mep_2374-8265.11452
17. Agarwal, A., et al.: Knowledge and skills in cardiopulmonary resuscitation and effect of simulation training among healthcare workers in a tertiary care center in India. Indian J. Crit. Care Med. **28**(4), 336–342 (2024). https://doi.org/10.5005/jp-journals-10071-24670
18. Cortegiani, A., et al.: In-situ simulation for cardiopulmonary resuscitation training: a systematic review. Resusc. Plus. **21**, 100863 (2025). https://doi.org/10.1016/j.resplu.2024.100863

Technological Solution for the Prevention of Operational Risks in Entities of the Financial Sector

Marcos Ronceros[1], Christopher Perez[1], Juan Mansilla-Lopez[1](✉), and Ricardo Loza[2]

[1] Facultad de Ingeniería, Universidad Peruana de Ciencias Aplicadas, Lima, Peru
{u201720285,u201413184,pcsijman}@upc.edu.pe
[2] Shreridan College, Oakville, Canada
ricardo.loza@sheridancollege.ca

Abstract. Today one of the most critical types of risks that attack financial institutions are operational risks. Few studies propose efficient systems to be able to manage these risks and the current systems do not include the necessary functionalities to be able to address them. This research focuses on a web platform designed for risk management. The system has a total of 7 modules which are: processes, risks, controls, events, action plans, reports and security. In order to efficiently manage operational risks, the controls module is key. This defines the control measures, which we consider to be a critical activity, since if an efficient control is not chosen to address the risk, it could become a repetitive risk. For this reason, we propose criteria for both the design and the execution of the control. Being 6 criteria for the design and 4 for the execution. The final result of the control is measured by means of the average of the criteria, being from 70 to 100 an efficient control, from 40 to 69 an average control and from 0 to 39 an inefficient control. This will help the user to choose the best control for each risk. The project was validated in the 2 largest financial institutions in Peru, with a sample of 13 users who, after interacting with the system, responded to the surveys supporting the project.

Keywords: operational risk · banks · fraud · controls · decision making

1 Introduction

Over the years it has been shown that financial institutions play a fundamental role in the global economic system and their stability is essential to avoid financial crises that could have a significant impact on the economy. For this reason, operational risk management has become a priority in the financial sector due to the complexity and frequency of risks. These refer to the risks associated with financial loss resulting from inadequate internal processes, system failures, human error, external events, and other non-financial factors (Lysiak et al., 2022).

The importance of operational risk management has manifested itself with past financial crises, such as the one in 2008. This crisis was caused by the increase in subprime

A. Rocha et al. (Eds.): ICITS 2026, LNNS 1966, pp. 669–678, 2027.
https://doi.org/10.1007/978-3-032-27538-7_59

mortgage loans that were packaged into complex financial products and sold to investors all over the world. In 2021, Mizuko Bank in Japan presented a technological failure that caused the closure of 80% of its ATMs for more than 24 h. At Santander Bank in Scotland, customers had difficulties making payments online and in stores. At BBVA de Mexico, a system failure left more than 20 million users without access to ATMs and the mobile application. In Hong Kong, the fastest payment system was down for 5 h, causing disruptions to real-time fund transfers and proxy account registration.

In order to prevent operational risks, banks must implement solid processes for the identification, evaluation, control and monitoring of operational risk (Leo et al., 2019). Financial entities must identify and assess operational risks to understand the probability and impact of adverse events and determine the necessary measures to prevent or mitigate them. Operational risk management also implies the implementation of adequate policies and procedures to control risks, as well as the monitoring and periodic review of the control measures implemented. Consequently, it is important to use early warning systems to help risk managers, executives and directors of financial institutions to prevent operational and reputational risks (Barakat et al., 2019).

Despite its importance, many financial institutions still face challenges in implementing and improving their operational risk management practices. Traditional quantitative approaches, such as the one used in the Basel accords, rely on hindsight information to predict the probability of incidents, which are unpredictable and potentially serious. An approach that combines both quantitative and qualitative approaches should be adopted to deal with emerging risks (Peihani, 2022). To address these challenges, a comprehensive, risk-based approach to operational risk management is necessary. This involves the implementation of a robust and effective risk management framework, the proper identification and assessment of operational risks, the implementation of effective internal controls, and the ongoing measurement and monitoring of risk. In addition, banks must be prepared to continually adapt and evolve in response to changes in the business and regulatory environment. The purpose of this research is to provide a solution to this through a web platform that allows managing operational risks using a multicriteria approach during risk assessment and treatment, considering quantitative and qualitative approaches.

The structure of the document is as follows: Sect. 2 presents the the literature review, Sect. 3 the benchmarking, Sect. 4 the technological solution, Sect. 5 presents the validation and finally, Sect. 6 offers the conclusions.

2 Literature Review

Lysiak et al., 2022 developed a banking risk assessment model, for which the main study method was economic mathematical modeling based on the standardized model of the Basel Committee for Operational Risk Management, they were also based on the modified CAPM model and the model developed by Shapiro and Cornell for the management of exchange risk. Xu et al., 2019 proposed a double correlation model in operational risk modeling, which contains frequency correlation and severity correlation, this is based on the LDA framework, which will help banks This is due to the fact that annually there are losses for each risk cell. Bai et al., 2022 developed a Bayesian

Belief Network (BBN) model, which was used to capture the non-linear relationship between risk management strategies and risk exposure, in addition to revealing the interrelationship between risk management strategies of risks. Vulanovic et al., 2019 presented a general model for the design of an integrated management system based on risk assessment of an organization's processes. This general model is based on the hypothesis that as the increased risks in processes increase, the need for documentation also increases. Weeserik & Spruit, 2018 proposed the B4ORM (BPM for ORM) maturity model, which they developed following the Design Science Research approach; this model is capable of guiding ORM (Operational Risk Management) maturity processes towards the benefits of BPM (Business Performance Management) to innovate on the proper management of technologies and the continuous improvement of ORM processes. Yun et al., 2022 developed a Big data model to predict the risks associated with IT projects abroad, using a Big data analysis technique.

Zabala & Slusarczyk, 2020 adapted an operational risk management model (ORM) to minimize the risks of digitization in banks and future operational risks at Banco Santander de España. Ferreira & Koekemoer, 2019 proposed a conceptual model for operational risk events in the banking sector, for this they carry out an analysis of financial risks in South African banks, because they operate in a very volatile and competitive industry. Barakat et al., 2019 conducted a comparative analysis to examine the equity and debt-based reputational effects of financial tones in operational risk announcements and shows how such reputational effects are moderated by alternative sources of public information. Kavassalis et al., 2018 carried out an analysis where they propose using Reg Tech for financial risk monitoring and supervisory reporting, resulting in the Reg Tech approach having the potential to contain the operational risk related to the inappropriate management of assets, risk data and control the cost of compliance with supervisory reports. Luburic, 2019 carried out an analysis to strengthen the "three lines of defense" model in terms of more efficient management of operational risks, mainly seeking to identify the key elements of the connections and possible synergies of various approaches that strengthen and/or improve risk management processes. Zhou et al., 2021 reformulated the "supervised multilabel" operational risk classification model as a few-trial semi-supervised learning problem, called MetaRisk, which is capable of predicting types of risk that are new to systems.

3 Solutions Available on the Market

There are solutions that are widely used by companies, for example the solution Monitor Plus GRC-ORM™ from the company Plus TI which is an automation solution to manage risks, vulnerabilities and opportunities (Plus-ti, 2023). Among its main benefits are: comprehensive and automated management of multiple types of risks (operational risk, technological risk, project risk and supplier risk), prevention of failures and root causes of risks, threats or vulnerabilities and loss control and reduction. Another of the solutions is that of GRC Nova Sec MS from the Nova sec company, this allows the identification of the risks of the different management systems, which includes operational risks, process risks, information security risks, security risks, among others (Nova sec, 2018). Among its main benefits are: it easily and conveniently maintains the identification, analysis, assessment, treatment, monitoring and communication of risks throughout the organization; comply with regulations related to risk management in your sector and decreases the time in operational activities for consolidation, updating, maintenance and risk reporting.

Finally, the Soft Expert ERM solution from the Soft Expert company enables organizations to identify, analyze, assess, monitor and manage corporate risks using an integrated approach. The solution serves as a foundation for various risk management initiatives, as it supports different categories such as strategic, financial, security, compliance, environmental, asset, product, process, and project risks (Soft expert, 2019). Among its main benefits are centralizes and streamlines the risk management program,

improve efficiency by allocating the right amount of resources to mitigate risks and it provides an accurate understanding of risks, identifying and managing risks in all contexts.

4 Proposed Technological Solution

This study focuses on the implementation of a technological solution called SIGERIOP, specifically designed to prevent and manage operational risks in financial sector entities. SIGERIOP is a comprehensive system that offers a proactive and efficient approach to identify, assess and control operational risks that may affect the operation and stability of financial institutions.

This operational risk prevention software was developed using an iterative agile methodology. The process begins with a collaborative analysis of specific operational risks and compliance requirements. Development then proceeds in short, focused sprints to build and refine core modules—such as incident reporting, risk assessment, and dashboard visualization—ensuring continuous feedback and alignment with user needs. Each iteration culminates in rigorous testing, guaranteeing a robust and validated solution upon deployment.

4.1 Architecture

Figure 1 shows the technological architecture of SIGERIOP which is a system that connects the users, experts in operational risk, with the system. It provides a comprehensive view of the actors (operational risk managers, risk analysts, etc.) and technologies involved in a financial institution. The SIGERIOP architecture is organized in three main sections: Front-end, Back-end and Automation, which communicate through application programming interface (APIs) services. The Front-end uses the HTML, CSS and JavaScript languages to create a friendly and accessible interface for users. The Back-end is based on Java and the Spring framework, with PostgreSQL as the database engine to ensure optimal performance and security. In addition, automation deploys logic that allows the use of specific criteria in the design and execution of controls, as well as in the identification of operational risks. These logics are fundamental to optimize the process and improve the efficiency of the system. Thanks to automation, specific criteria can be applied and preventive and corrective actions carried out in an agile and precise manner. The system has been deployed in a web environment on Amazon Web

Services (AWS), taking advantage of the scalable and reliable infrastructure offered by this cloud platform. In summary, SIGERIOP is based on a technological architecture that integrates different components and technologies to provide a complete solution for the management of operational risks in financial entities.

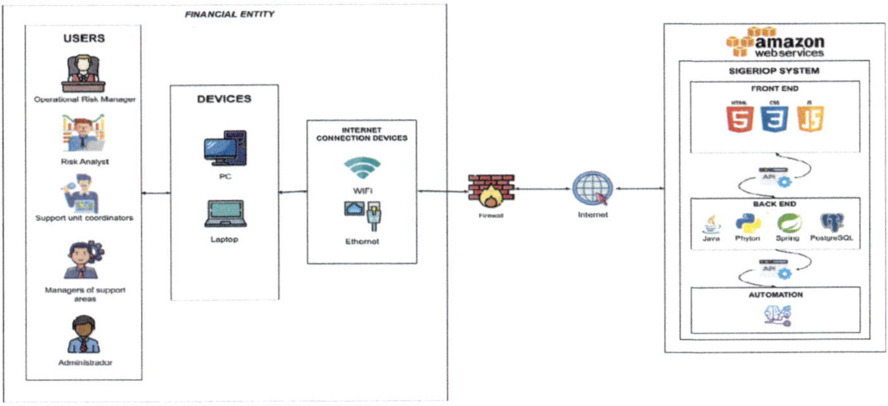

Fig. 1. SIGERIOP Technological Architecture

4.2 Modules

The SIGERIOP was designed with 7 modules for comprehensive management of operational risks. The modules and their functions are detailed below:

Process Module: it is in charge of the management of the processes, allowing a clear and structured vision of all the operational activities of the financial institution. From the identification and documentation of key processes to the assignment of responsibilities, this module provides the necessary tools to optimize the efficiency and effectiveness of processes. For example, processes such as account opening, credit evaluation and transaction processing can be managed, ensuring smooth and accurate management of operations.

Risk Module: the management of operational risks is addressed, providing tools and methodologies to identify, evaluate and mitigate the risks associated with financial operations. Likewise, the module is aligned with the principles of the ISO 31000:2018 standard, which is to promote a systematic and comprehensive approach to risk management. For example, within risk assessment, the standard establishes the need to assess the impact and probability of occurrence of the identified risks. This module offers tools to perform a quantitative or qualitative assessment of operational risks. This allows the organization to prioritize risks based on their risk level and allocate the appropriate resources to address them.

Controls Module: it focuses on the management of internal controls, providing mechanisms to define, implement and monitor the necessary controls in each process and area of the financial institution. Figure 2 shows the established criteria to be able to define a control based on the design and execution. With these two fronts, a final score is assigned to the control that serves to know how effective it is and what is needed to improve it. The final score of the control is measured by means of the average of the criteria, being from 70 to 100 an efficient control, from 40 to 69 an average control and from 0 to 39 an inefficient control.

Events Module: it focuses on the management of events related to operational risks. It allows recording, analyzing and monitoring the events that have occurred, facilitating decision making and the implementation of corrective measures.

Action Plans Module: action plans derived from identified events and risks are managed. It provides tools for monitoring and controlling the actions necessary to mitigate risks and improve processes. For example, if a risk of fraud in financial transactions is identified, this module will allow assigning someone responsible for the task of strengthening internal controls, such as the implementation of a system for continuous monitoring of transactions and periodic review of processes.

Reports module: it offers functionalities for the generation of detailed and personalized reports. It allows you to clearly and concisely display the relevant data for decision making in operational risk management. For example, the level of risk reported as well as the effectiveness of the related control. In addition, it allows us to know the associated managers in order to be able to arrange meetings with those involved and create action plans (Fig. 3).

Security Module: it is dedicated to the management of system and user security and provides tools for managing access profiles, controlling permissions, and auditing activities.

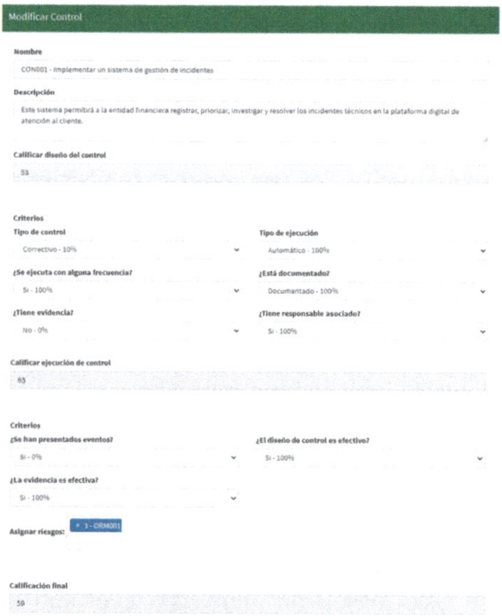

Fig. 2. SIGERIOP Control Module

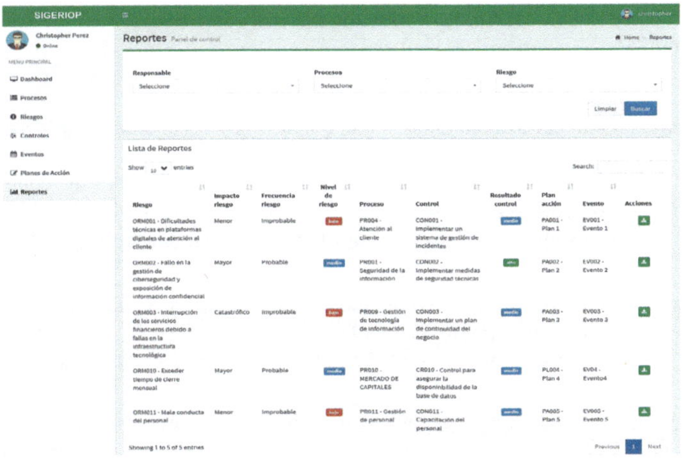

Fig. 3. SIGERIOP Reports Module

5 Validation

The SIGERIOP validation process was carried out through tests carried out by 13 operational risk experts with enough years of experience in the subject. The general objective was to ensure that the system manages operational risks, analyzes them and provides the most appropriate control for their treatment in financial sector entities.

Validation metrics were divided into 4 categories: efficiency, effectiveness, satisfaction, and design. Table 1 shows the questions used to validate each of these categories with the operational risk experts.

The steps that were performed for the functional tests include: users were provided face-to-face training on the use of the system, an account and password were provided so that the user can view all the functionalities of the system and a user satisfaction survey was conducted. The results obtained from these surveys in which we use the "Top Two Box" method, which allows us to carry out the data collection process in less time. In addition, the analysis and comparison of results is done easily and efficiently. The method consists of selecting the two highest responses on the scale and combining them to create a single number and thus measure the satisfaction of the respondents (Ortega, 2018). The results reveal that more than 81% of the experts surveyed expressed their satisfaction with the solution, high lighting that it provides them with the necessary information to manage and visualize operational risks effectively. This indicates that SIGERIOP meets the objective of providing specialists with the necessary tools for adequate management of operational risks. In addition, more than 71% of the experts surveyed stated that the proposal facilitates management and gives them greater control over risks. This shows that SIGERIOP is a solution that allows users to have a broader and more comprehensive approach to identify and mitigate operational risks. Lastly, close to 79% of the experts surveyed found SIGERIOP to be an easy-to-use tool. This means that operational risk managers will be able to use the system in a simple and accessible way at any time and

Table 1. Validation metrics questions

Metrics	Question
Efficiency	1. Does the system allow me to perform my tasks efficiently? Does the system allow me to easily find the information I need? 2. Does the system allow me to monitor the efficiency of the controls implemented? 3. Does the system make it easy for me to assign tasks and responsibilities within operational risk management?
Effectiveness	5. Does the system help me identify and categorize operational risk events in a precise and timely manner? 6. Does the system provide useful and relevant reports and alerts for decision-making in operational risk management? 7. Does the system generate useful reports for decision making? 8. Are the risk assessment criteria used by the system adequate for my organization? 9. Are the control creation criteria used by the system appropriate for my organization?
Usability	10. Is the system easy to use and manipulate? 11. Does the system have all the necessary functionalities for an effective management of operational risks? 12. Is the system accessible from different devices and platforms?
Design	13. Is the system layout attractive and easy to navigate? 14. Is the presentation of information and reports clear and easy to understand? 15. Is the system flexible enough to adapt to the specific needs of the financial institution?

from any device; so, this feature contributes to the adoption and effectiveness of the solution.

6 Conclusions

This study proposes the design of a technological solution for the management of operational risks through 7 modules (processes, risks, controls, events, action plans, reports and security). The project was validated by active personnel from the 2 largest financial institutions in Peru. The survey results support SIGERIOP's proposal as a satisfactory solution for operational risk management, highlighting its ability to provide relevant information, facilitate risk management and control, and offer a friendly and accessible user experience. The indicators provided in the reports to measure the level of risk and the effectiveness of the related control allow experts to make important and timely decisions. These findings support the effectiveness and relevance of SIGERIOP in the context of operational risk management. For future iterations of SIGERIOP development, it is important to note that valuable suggestions were collected from experts, which include the implementation of custom filters, improving the dynamism of dashboards, etc. These suggestions offer an opportunity for the continuous development and improvement of

SIGERIOP, since they allow to identify specific areas in which work can be done to enrich and improve the solution. The addition of custom filters would give users the ability to tailor the display of data based on their individual needs and preferences. Likewise, the improvement in the dynamism of the dashboards would provide greater interactivity and flexibility in data exploration, allowing users to obtain relevant information more agilely and quickly.

References

Bai, X., Cheng, L., Iris, C.: Data-driven financial and operational risk management: Empirical evidence from the global tramp shipping industry. Transp. Res. Part E Logist. Transp. Rev. **158**, 102617 (2022). https://doi.org/10.1016/J.TRE.2022.102617

Barakat, A., Ashby, S., Fenn, P., Bryce, C.: Operational risk and reputation in financial Institutions: Does media tone make a difference? J. Bank. Financ. **98**, 1–24 (2019). https://doi.org/10.1016/j.jbankfin.2018.10.007

Ferreira, S., Dickason-Koekemoer, Z.: A conceptual model of operational risk events in the banking sector. Cogent Econ. Finance. **7**(1), 1–16 (2019). https://doi.org/10.1080/23322039.2019.1706394

Gestión de Riesgos Conel software GRC NOVASEC MS Solución de Gobierno, Riesgo y Cumplimiento. Available at. https://www.novasec.co/software-grc/riesgos. (2018). Accessed 15 Oct 2022

Kavassalis, P., Stieber, H., Breymann, W., Saxton, K., Gross, F.J.: An innovative RegTech approach to financial risk monitoring and supervisory reporting. J. Risk Finance. **19**(1), 39–55 (2018). https://doi.org/10.1108/JRF-07-2017-0111

Leo, M., Sharma, S., Madley, K.: Machine learning in banking risk management: a literature review. Risks. **7**(1), 29 (2019). https://doi.org/10.3390/risks7010029

Liberec, R.: Strengthening the three lines of defenses in terms of more efficient operational risk management in Central Banks. J. Cent. Banking Theory PR act. **6**(1), 29–53 (2017). https://doi.org/10.1515/jcbtp-2017-0003

Lydia, L., Masiuk, I., Chynchyk, A., Yudina, O., Olshanskiy, O., Shevchenko, V.: Banking risks in the asset and liability management system. J. Risk Financial Manag. **15**(6), 1–18 (2022). https://doi.org/10.3390/jrfm15060265

Ortega, C.: ¿Qué es el método top two box?. QuestionPro. Available at. https://www.questionpro.com/blog/es/top-two-box/. (2018) Accessed 9 June 2023

Peihani, M.: Regulation of cyber risk in the banking system: a canadian case study. J. Financ. Regul. **8**(2), 139–161 (2022). https://doi.org/10.1093/jfr/fjac006

Soluciones – Gestión de Riesgos. Plus TI. Available at. https://plus-ti.com/grcorm/. (2023) Accessed 4 Apr 2023

Software para gestión de riesgos ERM. SoftExpert. Available at. https://www.softexpert.com/es/solucao/gestion-riesgos-corporativos-erm/. (2019) Accessed 18 Oct 2022

Vulanovic, S., et al.: Integrated management systems based ON RISK assessment: methodology development and case studies. Adv. Prod. Eng. Manag. **15**(1), 93–106 (2020). https://doi.org/10.14743/apem2020.1.352

Weeserik, B.P., Spruit, M.: Improving operational risk management using business performance management technologies. Sustainability. **10**(3), 640 (2018). https://doi.org/10.3390/SU10030640

Xu, C., Zheng, C., Wang, D., Ji, J., Wang, N.: Double correlation model for operational risk: evidence from Chinese commercial banks. Physica A. **516**, 327–339 (2019). https://doi.org/10.1016/j.physa.2018.10.031

Yun, J., Jeong, K., Youn, J., Lee, D.: Big data analysis model for predicting operational risk in overseas construction projects. J. Asian Archit. Build. Eng. **21**, 1–8 (2022). https://doi.org/10.1080/13467581.2021.2007100

Zabala Aguayo, F., Ślusarczyk, B.: Risks of banking services digitalization: The practice of diversification and sustainable development goals. Sustainability. **12**(10), 4040 (2020). https://doi.org/10.3390/SU12104040

Zhou, F., Qi, X., Xiao, C., Wang, J.: MetaRisk: Semi-supervised few-shot operational risk classification in banking industry. Inf. Sci. **552**, 1–16 (2021). https://doi.org/10.1016/j.ins.2020.11.027

Key Variables for Teacher Performance and the Development of Autonomous Learning Management in Adolescents

S. Hinojosa Nuñez, S. Hinojoza Talavera, E. Gonzales-Inca, K. Ríos-Vera, and O. Gutierrez-Aguilar(✉)

Universidad Nacional de San Agustín de Arequipa, Arequipa, Peru
{shinojosan,shinojoza,egonzalesi,krios,ogutierrez}@unsa.edu.pe

Abstract. The objective of this research is to analyse how the components of teacher performance influence the development of autonomous learning in adolescents, considering the role of the use of technological resources, teaching methodology, assessment, and the relationship with students. The study is conducted using a quantitative approach, with a non-experimental and explanatory design. The sample consists of 363 secondary school students from educational institutions in the city of Arequipa, Peru. A structured questionnaire is administered to measure participants' perceptions of the use of technology in the classroom, methodological strategies, assessment practices, the teacher-student relationship, and the ability to manage their own learning. The data are analysed using structural equation modelling, which allows for the identification of direct and indirect relationships between variables. The results show that teacher evaluation is the most influential factor within the model, exerting a positive impact on both teaching methodology and students' autonomous learning. It is also observed that a positive relationship between teachers and students strengthens assessment processes and improves the educational experience. Similarly, the appropriate use of technological resources indirectly contributes to improving the pedagogical relationship and assessment practice. It is concluded that assessment, understood as a formative and reflective process, becomes the linchpin of teacher performance and a key element in fostering autonomous learning among adolescents in today's school environment.

Keywords: Technology Resources Component · Evaluation · Methodology · Manage your Learning Autonomously · Relationship with Students

1 Introduction

Adolescent autonomous learning—defined as the capacity for self-directed, self-regulated, and student-centred learning—has become a central theme in contemporary pedagogy. The intersection of technological resources, assessment methodologies, and student-teacher dynamics is increasingly recognised as crucial for fostering autonomy in adolescent learners. This report synthesises current research on the components and evaluation of technological resources, the management of autonomous learning, and the

A. Rocha et al. (Eds.): ICITS 2026, LNNS 1966, pp. 679–693, 2027.
https://doi.org/10.1007/978-3-032-27538-7_60

influence of teacher performance and relationships, in order to connect theory with practice in adolescent education [1, 2]. The components of technological resources that support autonomous learning (AML) include AI-based adaptive learning systems, learning management systems (LMS), interactive online platforms, neural network-based models, and mobile learning strategies. These technological resources are integrated with assessment methodologies such as self-report scales (e.g., LASSI, CETA) and multi-method frameworks that combine observational and cognitive-behavioural assessments to measure AML. AML methodologies extend to flipped classrooms, collaborative learning, and self-regulated learning (SRL) practices, which together improve the performance of both students and teachers [3, 4]. For adolescents, these integrated approaches promote self-monitoring, planning, and personalised learning, where autonomy-supportive teacher practices and effective instructional leadership play a key role in aligning student expectations with technological implementations. The development of AML in adolescents is mediated by emotional factors (such as academic self-efficacy and emotional regulation) and cognitive abilities that directly influence academic outcomes [5–7].

The model's hypotheses are grouped into three dimensions that articulate the technological, pedagogical, and autonomous learning components of using the Blackboard virtual environment.

1. *Influence of assessment on methodology and learning autonomy.* Assessment is a central component of the pedagogical process, guiding both the teacher's methodological strategies and students' self-regulation of learning [8–10]. Therefore, the following hypotheses are proposed:

   H1: Assessment positively influences autonomous learning management.
   H2: Assessment has a positive influence on teaching methodology.
   H3: Methodology has a positive influence on autonomous learning management.

2. *Influence of the teacher-student relationship on assessment and autonomy processes.* The interaction between teachers and students is a determining factor in strengthening assessment practices and promoting independent learning ([11–14]. In this regard, the following hypotheses are formulated:

   H4: The relationship with students positively influences assessment.
   H5: The relationship with students positively influences independent learning management.

3. *Influence of the technological component on pedagogical and relational processes.* The use of technological resources within the Blackboard LMS enhances educational dynamics by facilitating communication, assessment, and the application of methodological strategies [15–17]. Based on this, the following hypotheses are proposed:

   H6: The technological component has a positive influence on assessment.
   H7: The technological component has a positive influence on teaching methodology.
   H8: The technological component has a positive influence on the relationship with students.

This grouping reinforces the theoretical coherence of the model, showing how technological resources, assessment, and pedagogical relationships interact to enhance learning autonomy and academic satisfaction in virtual university environments. Furthermore, the proposed model considers a set of mediated relationships that explain how the technological, pedagogical, and relational components interact to influence assessment, methodology, and autonomous learning management in the Blackboard virtual environment. The hypotheses are grouped into four main categories:

1. *Mediations of the technological component.* These hypotheses suggest that technological resources do not exert their influence directly, but rather through other variables in the model, such as the relationship with students, evaluation, and methodology [18–20].

   H9: The technological component influences assessment through the relationship with students.
   H10: The technological component influences autonomous learning management through the relationship with students.
   H11: The technological component influences autonomous learning management through methodology.
   H14: The technological component influences autonomous learning management through assessment.
   H15: The technological component influences methodology through assessment.

2. *Mediations of the teacher-student relationship.* These hypotheses examine the mediating role of assessment and methodology as mechanisms that link teacher-student interaction with learning outcomes [21, 22].

   H12: The relationship with students influences autonomous learning management through assessment.
   H13: The relationship with students influences methodology through assessment.
   H19: The relationship with students influences autonomous learning management through assessment and methodology.

3. *Mediations of the teacher-student relationship.* These hypotheses describe chain mediation effects, where the technological component impacts methodology or autonomy through the relationship-assessment-methodology sequence [23, 24].

   H16: The technological component influences methodology through the relationship with students and assessment.
   H17: The technological component influences autonomous learning management through the relationship with students and assessment.
   H18: The technological component influences autonomous learning management through the sequence of relationship, assessment, and methodology.
   H20: The technological component influences autonomous learning management through the assessment-methodology sequence.

4. *Final mediation of methodology.* Finally, assessment is considered the starting point for methodological action and its effect on student autonomy [25].

H21: Assessment influences the autonomous management of learning through methodology.

2 Methodology

The study adopted a quantitative approach, using a non-experimental cross-sectional design. The study units were 363 secondary school students. The gender distribution was 49.9% male (N 181) and 50.1% female (N 182), with ages ranging from 12 to 18 years. The mean age was 14.3 years, with a standard deviation of 1.38. Participants were selected by convenience. The instrument was specifically designed for this study, with its latent variables being reflective: Technology Resources Component (TRC: 3 items), Evaluation (EVA: 3 items), Methodology (MET: 3 items), Manage your Learning Autonomously (GAA: 4 items) and Relationship with Students (REL: 4 items). The variables were measured using a 5-point Likert scale, where (1) is "Never", (2) "Rarely", (3) "Sometimes", (4) "Often", and (5) "Always". Through an exploratory factor analysis using Jamovi software version 2.6.24, the necessary adjustments were made to the instrument prior to its final application. The next step was to model the variables with SmartPLS software version 4.1.1.5. During the application of the questionnaire, respondents were informed about the scope of the research, allowing them to provide their consent. Additionally, necessary administrative permissions were obtained for the survey to be conducted in the educational centre.

3 Results

The overall fit measures confirm the appropriateness of the proposed theoretical model to the empirical data. The results obtained show that the model presents a comparative fit index (CFI) of 0.929 and a Tucker-Lewis index (TLI) of 0.916, values that exceed the threshold of 0.90, indicating a satisfactory fit and demonstrating an adequate correspondence between the theoretical structure and the observed data. The standardised squared residual (SRMR) reaches a value of 0.0458, below the recommended limit of 0.08, suggesting a minimal difference between the estimated and observed covariances, and supporting the internal consistency of the model. Likewise, the root mean square error of approximation (RMSEA) was 0.0503, with a 90% confidence interval of 0.0418 to 0.0588, which is interpreted as an adequate fit, given that values below 0.06 are associated with well-specified and parsimonious models. Although the chi-square test ($\chi^2 = 307$; df $= 160$; $p < .001$) suggests a lack of exact fit, this result is expected in large samples, as the statistic is highly sensitive to sample size. Overall, the indices obtained support the validity of the structural model, demonstrating that the theoretical proposal satisfactorily reproduces the observed relationships between variables related to academic satisfaction with the use of the Blackboard LMS and the mediating role of integrating key elements in learning management. These results allow us to affirm that the model presents a statistically coherent, empirically robust and conceptually supported structure in the theory of technology-mediated learning.

Table 1 presents the reliability and construct validity results for the variables considered in the theoretical model on academic satisfaction with the use of the Blackboard

Table 1. Reliability and construct validity.

	Cronbach's alpha	Composite reliability (rho_a)	Composite reliability (rho_c)	Average variance extracted (AVE)
Evaluation (EVA)	0.711	0.713	0.816	0.596
Manage your Learning Autonomously (GAA)	0.774	0.789	0.854	0.594
Methodology (MET)	0.745	0.748	0.824	0.610
Relación con los Estudiantes (REL)	0.761	0.770	0.848	0.582
Technology Resources _Component (TRC)	0.757	0.786	0.859	0.672

LMS and the mediating role of integrating key elements into learning management. First, Cronbach's alpha values ranged between 0.711 and 0.774, indicating adequate internal consistency for the items comprising each construct. According to the criteria, values above 0.70 reflect an acceptable level of reliability in social science research, confirming the stability and consistency of the measurements.

Regarding composite reliability, both the rhoa and rhoc coefficients ranged between 0.713 and 0.859, exceeding the minimum recommended threshold of 0.70. These results corroborate the internal consistency of the indicators, suggesting that the observed items reliably represent the latent variables. In particular, the variables "Technology Resources Component" and "Manage Your Learning Autonomously" present the highest coefficients, demonstrating that the instruments used robustly capture the theoretical constructs associated with technology management and learning autonomy.

Regarding the average variance extracted (AVE), the values range between 0.582 and 0.672, exceeding the minimum criterion of 0.50. This indicates that each construct explains more than 50% of the variance of its indicators, ensuring adequate convergent validity. In summary, the results indicate that the instruments used have satisfactory levels of reliability and validity, enabling us to confirm that the measurements are consistent and representative for analysing the relationships between the technological, methodological, evaluative, and autonomous learning components in the context of using the Blackboard LMS in university education.

Table 2 presents the results of the discriminant validity obtained using the Heterotrait-Monotrait Ratio (HTMT) criterion, which assesses the empirical differentiation between the model's latent constructs. The discriminant validity is considered adequate when

Table 2. Discriminant validity – Heterotrait-Monotrait (HTMT) matrix.

	(EVA)	(GAA)	(MET)	(REL)	(TRC)
Evaluation (EVA)					
Manage your Learning _Autonomously (GAA)	0.492				
Methodology (MET)	0.880	0.356			
Relación con los Estudiantes (REL)	0.876	0.436	0.716		
Technology Resources _Component (TRC)	0.532	0.181	0.458	0.394	

HTMT values are below 0.85 in theoretically strict models or below 0.90 in more exploratory or applied models, as in the present study.

The results indicate that most of the relationships between constructs fall below the 0.90 threshold, confirming that the evaluated variables are conceptually distinct from one another and that the model exhibits acceptable discriminant validity. For example, the values between Evaluation and Manage Your Learning Autonomously (0.492), Evaluation and Technology Resources Component (0.532), and Methodology and Technology Resources Component (0.458) are clearly within acceptable limits, supporting the empirical independence of these factors.

However, a high correlation is observed between Evaluation and Methodology (0.880) and between Evaluation and Relationship with Students (0.876), suggesting a strong conceptual association between these dimensions. This result can be theoretically interpreted as a relationship inherent to the pedagogical process, given that methodology and teacher-student interaction are dimensions that directly influence how assessment is carried out within the learning environment.

Figure 1 shows the structural model presented, which allows for the analysis of direct and indirect (mediated) relationships between the study variables on academic satisfaction with the use of the Blackboard LMS, highlighting the role of assessment and methodological integration in autonomous learning management. First, the coefficients of determination (R^2) indicate the level of variance explained by exogenous variables over endogenous ones. Assessment ($R^2 = 0.433$) represents the construct with the most significant explained variance, indicating that approximately 43% of its behaviour is predicted by the technological components and the relationship with students. This result highlights that pedagogical interaction and the use of technological resources are key factors in developing effective assessment practices within the Blackboard environment. Likewise, the Methodology ($R^2 = 0.363$) reflects that 36% of its variance depends on the evaluation and the use of technological resources, while the Autonomous Management of Learning ($R^2 = 0.151$) shows a more moderate value, which suggests that the joint influence of the evaluation partially explains the student's autonomy, the methodology and the teacher-student relationship.

Regarding mediating relationships, the results show that Assessment acts as a key mediator between technological and pedagogical variables. On the one hand, the Technological Component indirectly influences the Methodology and Autonomous Learning Management through Assessment, demonstrating that the integration of digital resources is only effective when articulated with coherent assessment strategies. Similarly, the Relationship with Students exerts an indirect effect on autonomy through the mediation of Assessment and Methodology, confirming that teacher support and feedback strengthen students' ability to self-regulate their learning.

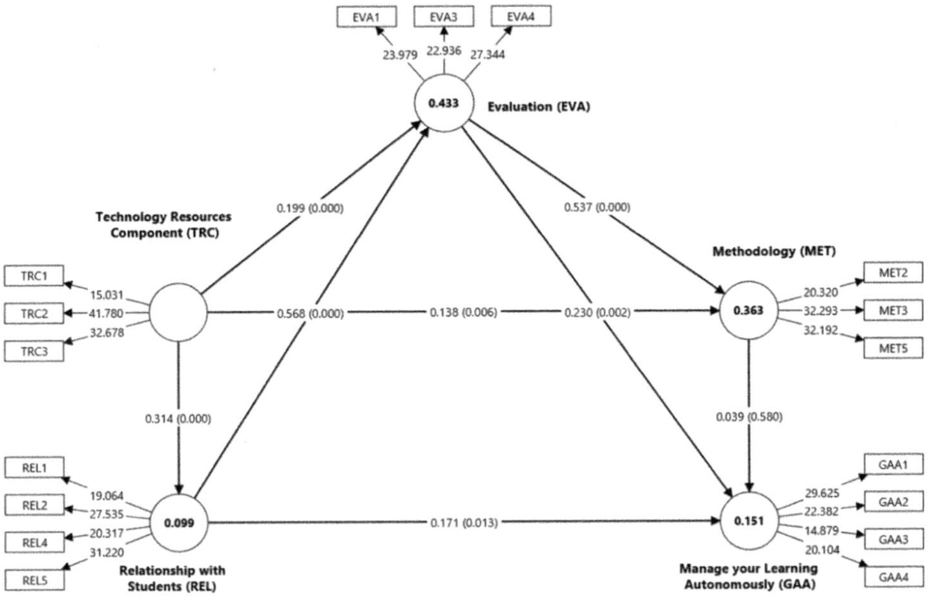

Fig. 1. R^2 of the SmartPLS model

Table 3 presents the results of the hypothesis testing for the direct effects of the structural model, obtained using the bootstrapping method. In the context of research on academic satisfaction with the use of the Blackboard LMS, the standardised coefficients (β), t-values, and significance levels (p) allow us to identify the most relevant causal relationships between the model's components.

The analysis of the direct hypotheses (H1 to H8) allows us to understand the dynamics of the relationships between the components of the proposed model on academic satisfaction with the use of the Blackboard LMS, highlighting the interaction between assessment, methodology, the teacher-student relationship, the use of technological resources, and autonomous learning management.

First, the results confirm that hypothesis H1 (Assessment Autonomous Learning Management) is accepted, demonstrating a positive and significant effect. This indicates that assessment practices within the Blackboard environment strengthen student autonomy by promoting self-reflection and self-management of their learning process. Similarly, hypothesis H2 (Assessment Methodology) is accepted, representing the most substantial effect of the model, which reveals that a coherent and well-designed assessment directly influences the selection and application of effective methodological strategies.

In contrast, hypothesis H3 (Methodology Autonomous Learning Management) is rejected, as the relationship does not reach statistical significance. This result suggests

Table 3. Direct Effects Hypothesis Testing – Bootstrapping.

Direct hypotheses	β	Sample mean (M)	Standard deviation (STDEV)	T statistics (|O/STDEV|)	P values
H1: Evaluation (EVA) - > Manage your Learning _Autonomously (GAA)	0.230	0.232	0.073	3.165	0.002
H2: Evaluation (EVA) - > Methodology (MET)	0.537	0.537	0.051	10.616	0.000
H3: Methodology (MET) - > Manage your Learning _Autonomously (GAA)	0.039	0.042	0.070	0.554	0.580
H4: Relación con los Estudiantes (REL) - > Evaluation (EVA)	0.568	0.568	0.038	14.851	0.000
H5: Relación con los Estudiantes (REL) - > Manage your Learning _Autonomously (GAA)	0.171	0.173	0.069	2.480	0.013
H6: Technology Resources _Component (TRC) - > Evaluation (EVA)	0.199	0.200	0.045	4.460	0.000
H7: Technology Resources _Component (TRC) - > Methodology (MET)	0.138	0.139	0.050	2.725	0.006
H8: Technology Resources _Component (TRC) - > Relación con los Estudiantes (REL)	0.314	0.319	0.051	6.124	0.000

that methodology, by itself, does not have a direct impact on student autonomy unless appropriate assessment and feedback processes accompany it.

Hypotheses H4 (Relationship with Students Assessment) and H5 (Relationship with Students Autonomous Learning Management) are accepted, demonstrating that a teacher-student relationship based on communication and empathy improves both assessment practices and students' ability to learn independently.

On the other hand, hypotheses H6, H7, and H8, which evaluate the effects of the Technological Component, are also accepted, indicating that the use of digital resources has a significant influence on assessment, methodology, and the relationship with students. This demonstrates that technological integration favours pedagogical interaction and the development of more efficient and participatory virtual environments.

In summary, of the eight hypotheses proposed, seven (H1, H2, H4, H5, H6, H7, and H8) are accepted, while one (H3) is rejected. These results confirm that the assessment serves as an articulating axis in the model, mediating the relationships between technological, methodological, and relational factors to enhance learning autonomy and academic satisfaction in the use of the Blackboard LMS.

Table 4 presents the results of the hypothesis testing for the indirect or mediating effects of the proposed structural model. The analysis of the indirect or mediating effects hypotheses (H9 to H21) allows us to understand the internal structure of the model and how the intervening variables articulate the relationships between the technological, pedagogical, and autonomous learning components in the use of the Blackboard LMS. The results reflect the relevance of Assessment (Assessment) as a central mediating variable and confirm the existence of several significant mediations that strengthen the proposed theoretical model.

First, hypotheses H9 and H10 are accepted, demonstrating that the Technological Component (TRC) indirectly influences Assessment (Assessment) and Autonomous Learning Management (ALM) through the Relationship with Students (RSS). This result suggests that the use of technological resources fosters more effective teacher-student interaction, which in turn enhances assessment processes and promotes student autonomy.

Likewise, hypotheses H12, H13, H14, H15, H16, and H17 are also accepted, as they present positive coefficients and significant p-values ($p < 0.05$). In particular, H13 shows the highest mediated effect in the model ($\beta = 0.305$; $p = 0.000$), confirming that the Relationship with Students influences Methodology (MET) through Assessment (EVA). This finding highlights the role of assessment as a connecting link between pedagogical interaction and methodological practice in virtual environments. Similarly, the mediations observed in H14 and H15 demonstrate that the Technological Component impacts autonomy and methodology through Assessment, highlighting the importance of integrating technological tools with formative assessment strategies.

Hypotheses H11, H18, H19, H20, and H21 are rejected, as they did not reach statistical significance levels ($p > 0.05$). These non-significant paths suggest that the multiple effects that simultaneously involve Methodology (MET) as a mediator are not decisive in explaining autonomous learning management. In other words, the methodology does not play a strong mediating role among the other variables; rather, its impact is primarily direct or through assessment.

Table 4. Hypothesis testing of indirect or mediated effects

Indirect or mediating hypotheses	β	Sample mean (M)	Standard deviation (STDEV)	T statistics (|O/STDEV|)	P values
H9: Technology Resources _Component (TRC) - > Relación con los Estudiantes (REL) - > Evaluation (EVA)	0.178	0.181	0.032	5.565	0.000
H10: Technology Resources _Component (TRC) - > Relación con los Estudiantes (REL) - > Manage your Learning _Autonomously (GAA)	0.054	0.055	0.023	2.288	0.022
H11: Technology Resources _Component (TRC) - > Methodology (MET) - > Manage your Learning _Autonomously (GAA)	0.005	0.005	0.010	0.523	**0.601**
H12: Relación con los Estudiantes (REL) - > Evaluation (EVA) - > Manage your Learning _Autonomously (GAA)	0.131	0.132	0.043	3.072	0.002

(continued)

Table 4. (*continued*)

Indirect or mediating hypotheses	β	Sample mean (M)	Standard deviation (STDEV)	T statistics (|O/STDEV|)	P values
H13: Relación con los Estudiantes (REL) - > Evaluation (EVA) - > Methodology (MET)	0.305	0.306	0.040	7.671	0.000
H14: Technology Resources _Component (TRC) - > Evaluation (EVA) - > Manage your Learning _Autonomously (GAA)	0.046	0.046	0.017	2.648	0.008
H15: Technology Resources _Component (TRC) - > Evaluation (EVA) - > Methodology (MET)	0.107	0.107	0.026	4.053	0.000
H16: Technology Resources _Component (TRC) - > Relación con los Estudiantes (REL) - > Evaluation (EVA) - > Methodology (MET)	0.096	0.098	0.021	4.634	0.000

(*continued*)

Table 4. (*continued*)

Indirect or mediating hypotheses	β	Sample mean (M)	Standard deviation (STDEV)	T statistics (|O/STDEV|)	P values
H17: Technology Resources _Component (TRC) - > Relación con los Estudiantes (REL) - > Evaluation (EVA) - > Manage your Learning _Autonomously (GAA)	0.041	0.042	0.015	2.701	0.007
H18: Technology Resources _Component (TRC) - > Relación con los Estudiantes (REL) - > Evaluation (EVA) - > Methodology (MET) - > Manage your Learning _Autonomously (GAA)	0.004	0.004	0.007	0.517	0.605
H19: Relación con los Estudiantes (REL) - > Evaluation (EVA) - > Methodology (MET) - > Manage your Learning _Autonomously (GAA)	0.012	0.013	0.022	0.534	0.593

(*continued*)

Table 4. (*continued*)

Indirect or mediating hypotheses	β	Sample mean (M)	Standard deviation (STDEV)	T statistics (|O/STDEV|)	P values
H20: Technology Resources _Component (TRC) - > Evaluation (EVA) - > Methodology (MET) - > Manage your Learning _Autonomously (GAA)	0.004	0.005	0.008	0.514	0.608
H21: Evaluation (EVA) - > Methodology (MET) - > Manage your Learning _Autonomously (GAA)	0.021	0.023	0.039	0.540	0.589

4 Conclusions

The results of the research on academic satisfaction with the use of the Blackboard LMS allow us to conclude that assessment constitutes the linchpin of the model, integrating the technological, methodological, and relational factors involved in university learning management. Empirical evidence suggests that assessment practices, when developed with formative and coherent criteria, have a significant influence on both teaching methodology and students' ability to manage their own learning autonomously.

Furthermore, it is confirmed that the relationship between faculty and students plays a decisive role in consolidating assessment and methodological processes, strengthening pedagogical interaction and fostering self-regulated learning. The quality of the faculty-student relationship translates into a more collaborative environment, where constant feedback and academic support enhance student satisfaction and commitment to utilising the virtual environment.

The study also shows that the use of technological resources within the Blackboard LMS indirectly contributes to the development of academic autonomy and satisfaction, provided they are coherently integrated with assessment processes and teaching practice. Technology alone does not guarantee meaningful learning, but it becomes a key facilitator when mediated by continuous assessment and a student-centred methodology.

Finally, the proposed model demonstrates a solid theoretical and empirical structure, in which assessment mediates the relationships between technological resources, methodology, and learning autonomy. These findings underscore the importance of strategically integrating elements of the virtual environment to enhance the educational experience, optimise pedagogical processes, and foster sustainable academic satisfaction among university students.

Regarding future work, longitudinal or experimental studies are recommended to analyse changes in academic satisfaction and learning autonomy over time, considering the influence of continuous assessment and teacher feedback. It would also be pertinent to expand the sample to include public and private universities from different regions or countries, in order to compare educational contexts and validate the model interculturally. Furthermore, it is suggested that additional variables, such as academic motivation, student engagement, and digital self-efficacy, be incorporated to enrich the understanding of the learning process in virtual environments. Finally, it is proposed to explore the integration of artificial intelligence and learning analytics in LMS systems to optimise personalised assessment strategies and strengthen student autonomy and satisfaction in digital higher education.

References

1. Conde Arteaga, A.B., Lastra Cordero, J., Peláez Álava, J.V., Segura Quiñonez, L., Piza Burgos, N.D.: Impact of feedback on the development of self-regulation and autonomous learning skills. Semin. Med. Writ. Educ. **3**, 1–10 (2024). https://doi.org/10.56294/mw2024.637
2. Cui, T., Wang, C., Yang, Y., Shao, Y.: A serial mediation model testing associations between teacher–student relationship, resilience, autonomous learning and academic performance in the Chinese EFL context. Lang. Teach. Res. **30**, 1–24 (2023). https://doi.org/10.1177/13621688231166771
3. Aguilar, O.G., Duche Perez, A.B., Aguilar, A.G.: Teacher performance evaluation model in Covid-19 times. In: Proceedings of the 15th Latin American Conference on Learning Technologies. IEEE (2020)
4. Aguilar, O.G., Gutierrez Aguilar, A.: A model validation to establish the relationship between teacher performance and student satisfaction. In: Proceedings 2020 3rd International Conference of Inclusive Technology and Education, pp. 202–207 (2020)
5. Ayasoufi, A., et al.: Leveraging LASSI for measuring ABET student outcome 7: fostering student independence in learning. In: ASEE Annual Conference and Exposition, Conference Proceedings (2024)
6. En, L.D.C., Jingjing, Z., Peng, J., Chung, T.C.: AI -Based adaptive learning framework for enhancing educational management in Malaysia using the EduAdaptNet model. In: Proceedings 1st International Conference on Frontier Technologies and Solutions, ICFTS 2025 (2025)
7. Rincón, Y.R., Munárriz, A., Magreñán Ruiz, A.: Flipped classroom or flip to foster self-regulation competencies in mathematics in economics and business students. Int. J. Educ. Res. **130**, 102556 (2025). https://doi.org/10.1016/j.ijer.2025.102556
8. Rutherford, S., Pritchard, C., Francis, N.: Assessment IS learning: developing a student-centred approach for assessment in higher education. FEBS Open Bio. **15**(1), 21–34 (2025). https://doi.org/10.1002/2211-5463.13921
9. Shafiee Rad, H., Alipour, R.: Unlocking writing success: building assessment literacy for students and teachers through effective interventions. Assess. Writ. **59**, 100804 (2024). https://doi.org/10.1016/j.asw.2023.100804
10. Thu, N.H.A., Thi, H.L., Nhi, N.T.H., Chi, V.T.Q., My, T.T.: Factors associated with self-directed learning among undergraduate nursing students in Vietnam. Nurse Educ. Pract. **78**, 104031 (2024). https://doi.org/10.1016/j.nepr.2024.104031

11. Cui, G., Zhao, Z., Yuan, C., Du, Y., Yan, Y., Wang, Z.: The influence of teachers' autonomy support on students' entrepreneurial enthusiasm: a mediation model with student gender as a moderator. Int. J. Manag. Educ. **22**(2), 100966 (2024). https://doi.org/10.1016/j.ijme.2024.100966
12. Froment, F., de-Besa Gutiérrez, M., Flores, J.G.: Effect of autonomy support on academic satisfaction: academic motivation and engagement as mediating variables. Rev. Investig. Educ. **41**(2), 479–499 (2023). https://doi.org/10.6018/rie.546251
13. Wang, M., Hu, J.: Perceived teacher autonomy support for adolescents' reading achievement: The mediation roles of control-value appraisals and emotions. Front. Psychol. **13**, 1–11 (2022). https://doi.org/10.3389/fpsyg.2022.959461
14. Wang, Q., Xu, Z., Tao, J., Ma, X.: The effects of teacher immediacy and autonomy support on secondary students' approaches to learning. Educ. Psychol. **44**(1), 78–95 (2024). https://doi.org/10.1080/01443410.2024.2302513
15. Ajani, O.A.: Enhancing technology integration in economics education: a systematic review of teachers' digital competencies, challenges, and institutional support in South African high schools. Multidiscip. Rev. **9**(1), 1–10 (2026). https://doi.org/10.31893/multirev.2026009
16. Dinçer, S.: Bridging the gap in technology integration in education: an examination of science teachers' competencies and needs. J. Turk. Sci. Educ. **21**(4), 620–634 (2024). https://doi.org/10.36681/TUSED.2024.033
17. Sepadi, M., Kgaphola, T., Molapo, K.: Teachers' understanding of technology usage in rural schools. Front. Educ. **10**, 1–11 (2025). https://doi.org/10.3389/feduc.2025.1548457
18. Addai, P., Afrifah, M., Okyere, I., Tachie-Donkor, G., Sarfo, J.O.: Technostress and determinants of academic success among University students: mediation role of technological literacy. J. Advocacy Res. Educ. **11**(1), 47–60 (2024). https://doi.org/10.13187/jare.2024.1.47
19. Guaña-Moya, J., Arteaga-Alcívar, Y., Criollo-C, S., Cajamarca-Carrazco: D.: Use of interactive technologies to increase motivation in university online courses. Educ. Sci. **14**(12), 1–28 (2024). https://doi.org/10.3390/educsci14121406
20. Nikolić, D., Dakić, D., Kolak, A., Narandžić, D., Janković, A.: Integration of game-based software tools into higher education. In: Lecture Notes on Multidisciplinary Industrial Engineering, vol. Part F42, pp. 56–63 (2022)
21. Harper, B.: Technology and teacher–student interactions: a review of empirical research. J. Res. Technol. Educ. **50**(3), 214–225 (2018). https://doi.org/10.1080/15391523.2018.1450690
22. Kim, H.Y.: Multimodal input during technology-assisted teacher instruction and English learner's learning experience. Innov. Lang. Learn. Teach. **15**(4), 293–305 (2021). https://doi.org/10.1080/17501229.2020.1800708
23. Arvind, T.: Leveraging technology for enhanced connectedness and student engagement. In: Sustainable Development, Humanities, and Social Sciences for Society 5.0, pp. 233–258 (2024)
24. Das, A., Malaviya, S.: Digital platforms and leveraging technologies to enhance learner engagement. In: Cyber-Physical Systems for Innovating and Transforming Society 5.0, pp. 211–232 (2025)
25. Andrés Alzate-Ortiz, F., Carlos Castañeda-Patiño, J.: Pedagogical mediation: key to humanizing and transformative education. a look from aesthetics and communication. Rev. Electron. Educare. **24**(1), 1–14 (2020). https://doi.org/10.15359/ree.24-1.21

Corporate Social Responsibility and Organizational Sustainability: An ESG-and-AI Framework for Better Performance

Carmen Machado[1], Ricardo Marcão[2(✉)], Vasco Santos[3], Filipa Martinho[1], and António Dionísio[1]

[1] ISLA Santarém – Polytechnic University, Santarém, Portugal
{filipa.martinho,antonio.dionisio}@islasantarem.pt
[2] NECE, UBI and ISLA Santarém – Polytechnic University, Covilhã, Portugal
ricardo.marcao@islasantarem.pt
[3] NECE GOVCOPP, CiTUR, Polytechnic University of Tomar, UBI, and ISLA Santarém Polytechnic University, Santarém, Portugal
vasco.santos@islasantarem.pt

Abstract. Organizations are under pressure to balance profitability with social and environmental responsibility. This article distills the most actionable insights from contemporary CSR and sustainability research and practice, framing them through ISO 26000 guidance and widely adopted reporting standards such as GRI and SASB. We highlight the role of leadership and stakeholder engagement and show how artificial intelligence can support greener operations, better monitoring, and more reliable decision making. Using a Design Science Research approach, we present a concise conceptual framework that integrates environmental, social, and governance priorities with AI-enabled processes. The framework is mapped to the Zachman structure to clarify who does what, with which data and tools, and for what outcomes. We summarize lessons across industries and discuss benefits, risks, and implementation challenges, including ethics, data quality, and cultural resistance. The result is a practical roadmap that links strategy, measurement, and day-to-day routines, helping organizations convert CSR intent into sustained performance improvements.

Keywords: Corporate Social Responsibility · Organizational Sustainability · ESG · ISO 26000 · GRI · SASB · Artificial Intelligence · Stakeholder Engagement · Organizational Performance · Design Science Research · Zachman Framework

1 Introduction

Companies are expected to balance financial performance with credible environmental and social action, supported by governance that turns intent into practice. This article condenses the attached work into a practical path that links Corporate Social Responsibility to organizational sustainability and results, using guidance and standards that

A. Rocha et al. (Eds.): ICITS 2026, LNNS 1966, pp. 694–704, 2027.
https://doi.org/10.1007/978-3-032-27538-7_61

define "good" practice with enough clarity to measure it. ISO 26000 anchors the discussion around ethics, transparency, and respect for stakeholders, while ESG reporting families such as GRI and SASB translate that intent into indicators used for investment and benchmarking.

Leadership and culture determine whether these standards become day to day routines rather than branding. The source manuscript emphasizes aligning values, systems, and behaviors so that sustainability goals move from statements to execution.

Data and technology, especially artificial intelligence used with a clear ethical purpose and guardrails, can improve monitoring, efficiency, and learning, but require attention to bias, transparency, and governance.

Methodologically, we retain the original design choice: a qualitative, conceptual review that uses a Design Science Research logic to build a usable artifact. Here, the artifact is a concise framework that integrates CSR principles, ESG measurement, leadership and culture, and AI enablement.

To make the framework actionable, we adopt the Zachman perspective to clarify who does what, with which data and tools, and for what outcomes across organizational layers.

The contribution is twofold. Conceptually, we synthesize seven working dimensions that connect ethics, strategy, sustainable actions, AI with an ethical purpose, leadership and culture, sustainable performance, and feedback for continuous improvement.

Practically, we translate this into an implementation guide and a Zachman-style map that organizations can use to assign responsibilities, select a minimal KPI set, structure feedback cadences, and plan technology support, while recognizing the need for future empirical validation.

2 Background and Related Work

Corporate Social Responsibility has evolved from philanthropy and compliance to a strategic approach that links ethics, transparency, and stakeholder value with organizational performance. ISO 26000 crystallizes this shift by framing social responsibility around organizational ethics, respect for stakeholders, and accountability, giving firms a shared language for policy and practice [1]. Reporting families such as GRI and SASB translate intent into comparable indicators for investors and regulators, supporting movement from narrative claims to measurable outcomes, while surveys of global practice note growth in disclosure and assurance alongside risks of inconsistent metrics and greenwashing when indicators are not tied to material topics [2]. The business case is reinforced by work on shared value and by governance research that describes how boards integrate ESG into risk and strategy [3, 4]. Leadership and culture repeatedly emerge as the difference between branding and execution: effective programs embed expectations in plans, budgets, and operating reviews, with inclusive governance and continuous feedback supporting learning in changing environments [4].

Artificial intelligence now supports sustainability through forecasting, optimization, traceability, and near real-time monitoring across operations and supply chains, but the literature stresses the need for ethics, bias control, explainability, and robust data governance to avoid unfair outcomes and reputational risk [5]. Methodologically, Design

Science Research offers a path to produce actionable artifacts in this domain. It combines conceptual synthesis with purposeful design and evaluation logic so that practitioners can implement and scholars can refine the resulting framework [6, 7]. To make responsibilities and data flows explicit across organizational layers, a Zachman perspective helps map roles, processes, data, tools, timing, and objectives, reducing ambiguity and speeding implementation [8].

3 Measurement and Transparency

Measurement gives CSR and sustainability real traction by turning intent into observable behavior. The starting point is materiality. Rather than tracking everything, organizations should identify the few environmental, social, and governance topics that are most relevant to strategy and stakeholders, then select decision ready indicators for each. ISO 26000 provides thematic guidance that helps define what "responsible" means in practice, while reporting families such as GRI and SASB translate those themes into standardized disclosures and metrics that investors and regulators can compare across firms [1, 2]. The outcome is a compact set of KPIs tied to material issues, with clear definitions, baselines, and targets.

Transparency is more than publishing a long report. Effective programs connect indicators to governance routines so data informs plans, budgets, and operating reviews. This includes assigning owners for each metric, documenting the data lineage, and establishing a simple cadence for data collection and validation. Independent assurance adds credibility where it matters most, especially for emissions, safety, labor practices, and board level oversight. Many surveys show that disclosure volume has grown, yet caution that inconsistency and greenwashing risks remain when metrics are not tied to material topics or when narrative claims outpace evidence [2]. Clear scoping notes, boundary definitions, and reconciliations to financial data help address these risks.

Good measurement balances outcome and process indicators. Outcome metrics capture impact, such as emissions intensity, injury rates, pay equity, supplier conformance, and incident response times. Process metrics track the health of the system that produces outcomes, including training completion, audit closure rates, supplier coverage, grievance resolution cycles, and the percentage of decisions supported by verified data. Combining both gives leaders an early warning system and avoids surprises at year end.

Artificial intelligence can raise measurement quality when used carefully. Examples include automated data capture from IoT sensors for energy and water, anomaly detection in supplier records, and near real time tracking of logistics emissions. Governance is essential. Data quality checks, explainability requirements, bias tests, and human oversight must be explicit so models support fair and reliable decisions [5]. A practical approach is to pilot AI on a narrow set of high quality indicators, compare results with traditional methods, and scale only when accuracy and control objectives are met.

Finally, keep the portfolio of indicators small enough to manage. A practical rule is to select a minimal KPI set per material topic, document why each metric was chosen, and show how it informs decisions. Align this set with recognized standards for comparability, maintain an assurance plan for the most decision critical metrics, and review materiality and targets on a regular cadence so the system adapts as regulations, technologies, and stakeholder expectations evolve [1, 2, 6, 7].

4 Selected Industry Illustrations

Across industries, concise examples show how CSR and sustainability move from policy to results when practice is anchored in material topics, clear metrics, and everyday routines. In food and beverages, a regional producer began by mapping its direct emissions and energy use, then set a goal to cut energy intensity through heat recovery and a partial shift to renewable electricity. It reported progress with GRI indicators and strengthened supplier reviews for water stewardship in high-risk crops. Within a year, the firm lowered energy intensity and operating costs while expanding verified coverage of at-risk suppliers, illustrating how basic foot printing and a compact, material KPI set can create a platform for continuous improvement rather than a one-off certification exercise [1, 2].

A technology manufacturer reached similar traction by focusing on product design and data governance. Standardizing modular components and launching a take-back program with verified downstream recyclers increased parts reuse and simplified repair. At the same time, the company used artificial intelligence to flag anomalies in supplier compliance records and conflict mineral declarations, but kept human review and explainability requirements in place to manage risk. The result was faster resolution of nonconformities and clearer evidence for reporting, showing that design choices and trustworthy data are as important as disclosure, and that AI helps when its purpose and guardrails are explicit [2, 5].

In financial services, a bank integrated climate and social risk indicators into underwriting for selected sectors and aligned its disclosures to SASB for comparability. A cross-functional ESG committee resolved tradeoffs between commercial targets and risk appetite, and the firm adopted a simple assurance plan for high-stakes indicators. Over time, portfolio exposure to high-risk assets decreased while client engagement on transition plans increased. The lesson is that governance and measurement must sit inside core decision processes, not adjacent to them, and that comparability standards help align internal incentives with external scrutiny [2, 4].

Taken together, these examples point to a common pattern. Start with what is material, choose a minimal set of decision-ready indicators, embed ownership and cadence in existing management routines, and use technology where it improves accuracy and speed with proper safeguards. Progress tends to come from small, well-designed loops of measure, learn, and adjust, supported by recognized guidance and reporting families that keep the focus on credible action and comparable results [1, 2, 4, 5].

5 AI for Sustainable Operations

Artificial intelligence can help organizations turn sustainability goals into day to day results by improving forecasting, optimization, traceability, and monitoring. In operations, learning models can predict energy demand, align production with renewable availability, and suggest scheduling tweaks that lower peak consumption without hurting service levels. In logistics, route and load optimization reduces fuel burn and improves delivery reliability, while supplier analytics surface patterns in nonconformities and late responses so teams can act before issues spread. In facilities, sensor data paired with anomaly detection flags leaks, malfunctioning HVAC cycles, or outlier water use early, which cuts waste and maintenance costs. In product and portfolio work, scenario tools estimate lifecycle emissions or social impact under different design or financing choices, giving leaders a clearer basis for tradeoffs. These gains are real only when the models are grounded in clean data, tied to material topics, and used inside existing decisions with clear ownership [2, 4, 5].

Governance is the hinge. AI that touches sustainability reporting or stakeholder outcomes must meet explicit standards for data lineage, accuracy, and explainability. Teams should document what the model is for, which variables it uses, how it was validated, and who reviews its outputs before they influence plans, budgets, or public claims. Bias testing is not optional, particularly in social indicators where skewed inputs or labels can produce unfair results. Human oversight remains essential, both as a backstop for errors and as a way to capture context that data does not see. A practical approach is to start with a narrow use case that sits on high quality data, compare model outputs with current methods, and scale only after accuracy, stability, and control objectives are met. Assurance practices that already exist for high-stakes indicators can extend to AI components through simple checklists and periodic model reviews [4–6].

Integration matters as much as analytics. The most successful programs embed AI outputs in the same forums where operational and risk decisions are made, rather than creating a parallel sustainability dashboard. For example, an emissions intensity forecast belongs in the monthly production review where managers can adjust shifts or maintenance windows, not only in the annual report team's workspace. Likewise, a supplier risk score should trigger a standard corrective action workflow that assigns tasks, deadlines, and verification steps. Keeping AI inside familiar routines makes adoption easier and shortens the loop from insight to action. It also helps limit the indicator set to metrics that leaders will actually use, which reduces noise and the temptation to claim progress based on analytics that have not changed behavior [2, 4].

Finally, ethics sets the outer boundary. Clear statements about purpose, data rights, and accountability protect trust with employees, customers, and communities. When AI assists sustainability claims, organizations should be transparent about what is modeled versus measured, how uncertainties are handled, and what thresholds trigger human review. These choices reduce the risk of overclaiming or greenwashing and make external assurance simpler. Used this way, AI becomes an enablement layer that supports material KPIs, improves the speed and quality of decisions, and strengthens the credibility of reporting without replacing judgment or responsibility [2, 4–7].

6 Methods: Design Science Research Approach

The problem addressed here is not only to describe Corporate Social Responsibility and sustainability practices, but to organize them into an artifact that organizations can actually use. For that reason, the approach follows a Design Science Research logic. DSR is suitable when the aim is to produce something purposeful, grounded in existing knowledge, and open to refinement through application. It typically moves through linked activities: identify and justify the problem, define objectives for the solution, design and describe the artifact, demonstrate its use in a realistic context, and evaluate its utility against those objectives [6, 7].

Problem identification in the original work centered on three gaps. First, many organizations adopt elements of ISO 26000, GRI, or SASB, but do so in a fragmented way, which weakens implementation and makes reporting harder. Second, leadership and culture are often discussed as success factors yet are not explicitly built into the operational architecture. Third, the emergence of artificial intelligence for monitoring, optimization, and reporting introduces new opportunities and risks that are not always mapped to existing governance. The objective for the solution was therefore to produce a concise framework that integrates CSR principles, ESG measurement, leadership and culture, and AI enablement into a single structure that can be read by managers, compliance teams, and technology teams alike [1, 2, 4, 5].

The design activity drew on recognized standards and on the seven dimensions already identified in the source work: ethics, strategy, sustainable actions, AI with an ethical purpose, leadership and culture, sustainable performance, and feedback for continuous improvement. Each dimension was defined in terms of its intent, the processes it requires, the data it needs, and the roles that own or support it. To make these elements explicit across organizational layers, the structure was mapped using a Zachman perspective, which clarifies the what, how, where, who, when, and why of a system. This mapping helps reduce ambiguity and gives a common view to business, sustainability, and IT stakeholders [8].

Demonstration in DSR does not need to be a full deployment. In this case, the manuscript shows how the framework can be applied to different sectors by selecting material topics, assigning indicators, planning data collection, and placing AI components under governance. Evaluation is linked to utility. A framework like this is considered useful if it helps organizations choose a minimal, material KPI set, align responsibilities with capabilities, integrate sustainability into existing management routines, and document AI use in a way that supports assurance. Because the work is conceptual and qualitative, evaluation is formative rather than statistical, and the article closes by identifying where empirical testing is still required, especially in AI ethics and in cross sector generalization [6–8].

7 Results: Conceptual Framework

The artifact integrates Corporate Social Responsibility principles, ESG measurement, leadership and culture, and AI enablement into a single structure that can be read and used by managers, compliance teams, and technologists. It rests on seven working dimensions drawn from the source work: ethics, strategy, sustainable actions, AI with an ethical purpose, leadership and culture, sustainable performance, and feedback for continuous improvement. Each dimension is defined by a clear intent, a small set of routines, the data it needs, and the roles that own it, so the framework can be implemented without creating a parallel organization. The logic begins with ethics, where ISO 26000 themes and the organization's stated values shape policy choices, risk appetite, and stakeholder commitments. Those choices are translated into strategy by selecting material topics and setting targets that connect to the business model, which keeps attention on issues that matter for both stakeholders and performance [1, 2].

Sustainable actions are then specified as process changes, product and service design choices, supplier standards, and investment criteria. At this point, measurement becomes the hinge. A compact, material set of ESG indicators is selected and documented with definitions, baselines, and targets that align with GRI or SASB so that external transparency and internal management use the same language. Leadership and culture make these elements routine by assigning clear ownership for each indicator, embedding reviews in existing planning and operating cadences, and aligning recognition and accountability with the desired behaviors. This prevents sustainability from becoming a communications exercise and ensures that indicators inform decisions about budgets, capacity, and tradeoffs in the same rooms where commercial choices are made [2, 4].

AI with an ethical purpose operates as an enablement layer rather than a substitute for judgment. Its role is to improve forecasting, optimization, traceability, and monitoring where data quality and governance allow. The framework requires that any model linked to sustainability decisions be documented for purpose, variables, validation, ownership, and review thresholds, with explicit controls for bias, explainability, and data lineage. This approach keeps technology aligned with material topics, reduces the risk of overclaiming, and makes external assurance easier when indicators are used in public disclosures. Sustainable performance is defined as a balanced view of outcomes and the health of the routines that produce them, which means tracking impact metrics such as emissions intensity, injury rates, and supplier conformance alongside process metrics like audit closure, grievance resolution, and the percentage of decisions supported by verified data. Feedback for continuous improvement closes the loop by adjusting targets, reallocating effort, and refining data and models in response to evidence, regulation, and stakeholder expectations [2, 5].

To make responsibilities and data flows explicit, the framework is mapped using a Zachman perspective. At the scope and business model levels, the "what" is expressed as material topics and KPIs, the "why" as the strategic and ethical rationale, and the "who" as accountable executives and cross functional forums. At the system and technology levels, the "how" covers operating routines and AI components, the "where" locates data sources and controls across plants, suppliers, and markets, and the "when" sets the cadence for collection, review, and assurance. This mapping reduces ambiguity by showing how a single indicator, such as emissions intensity, traverses the cells from purpose to data source, from review forum to external disclosure, and from model output to human decision. Because the same pattern applies to safety, labor, or governance indicators, teams can scale without reinventing the architecture for each topic [8].

The framework is designed to be light enough for practical use. It favors a minimal viable set of indicators per material topic, relies on existing governance forums rather than creating new ones, and treats AI as modular so pilots can be switched off or replaced without disrupting assurance. Utility is judged by whether leaders can assign owners, connect indicators to decisions, and run short learning cycles that produce visible improvements in operations and reporting. The artifact acknowledges limits where empirical testing is still required, especially on AI ethics and on generalizing maturity profiles across industries, but it provides a coherent path that ties recognized guidance to measurable results through clear roles, disciplined routines, and technology used with purpose [1, 2, 4, 5, 8].

8 Implementation Guide

An effective rollout starts with materiality. The organization should identify the few environmental, social, and governance topics that matter most for its strategy and stakeholders, then translate those topics into a compact set of indicators with clear definitions, baselines, and targets aligned to ISO 26000 guidance and GRI or SASB for comparability [1, 2]. Materiality is not a one time exercise. It needs a simple cadence for review so the portfolio adapts to regulation, market shifts, and stakeholder feedback. Clear scoping notes and data boundaries reduce confusion during audits and public disclosure.

Governance comes next. Responsibilities must be explicit at two levels. Executives and a cross functional forum own priorities, tradeoffs, and escalation, while process owners manage data collection, quality checks, and corrective actions. The same forums used for planning and operating reviews should include these indicators so decisions about budgets, capacity, and scheduling reflect sustainability objectives, rather than creating a separate track that competes for attention [2, 4]. A light assurance plan identifies which indicators require independent verification and how often, focusing on those with the greatest decision or reputational impact.

Data readiness enables the indicators to work. For each metric, teams document sources, lineage, controls, and retention. They establish simple checks for completeness and accuracy and keep a short list of known issues with owners and closure dates. When AI is involved, the organization defines purpose, inputs, validation approach, and thresholds that trigger human review. Bias testing and explainability are recorded alongside accuracy results so models can be trusted and defended. Pilots begin with one or two high quality use cases, such as energy forecasting in a plant or anomaly detection in supplier declarations, and scale only when accuracy and control objectives are met [5–7].

Operations are where sustainability becomes routine. Indicators are integrated into monthly production, logistics, and risk reviews. Managers use them to adjust shifts, maintenance windows, routes, or supplier actions. Corrective action workflows assign tasks, deadlines, and verification steps when thresholds are breached. The rule is that every indicator must have a known decision home so there is no ambiguity about who acts and when. Training focuses on how to read the indicators, how to interpret uncertainty, and how to escalate issues quickly without blame.

A Zachman style mapping helps keep the architecture coherent as teams add topics or locations. At the scope and business layers, leaders record the why and what as ethical principles, strategy, material topics, and KPIs. At the system and technology layers, teams detail the how, where, and when as operating routines, data sources, controls, tools, and cadences. The who appears at each layer, from executive accountability to process owners and data stewards. This shared map reduces rework and speeds onboarding when new functions or vendors join the program [8].

Continuous improvement closes the loop. After each reporting cycle, teams compare targets with outcomes, review incident patterns, and adjust plans. They retire low value indicators, refine definitions that caused confusion, and reallocate effort toward the few levers that moved results. External feedback from investors, customers, or assurance bodies is logged with actions and due dates so the next cycle shows visible progress. Over time, this discipline creates a credible system where recognized guidance and reporting are supported by reliable data, clear roles, and AI used with purpose, which improves decisions and strengthens trust with stakeholders [1, 2, 4, 5, 8].

9 Discussion

The framework links CSR intent to day to day management by combining recognized guidance, material indicators, and clear ownership. The main benefit is focus. ISO 26000 helps define what responsible conduct means in practice, while GRI and SASB provide a common vocabulary for measurement. Together they reduce noise and make it easier to compare progress across units and time [1, 2]. When leaders embed a compact set of material KPIs into regular planning and operating reviews, sustainability work competes on equal terms with cost, quality, and delivery. This tends to raise decision quality and lowers the risk that CSR remains a communications exercise [2, 4].

There are tradeoffs. A small indicator set protects attention, yet it can miss emerging risks if materiality reviews are infrequent or narrow. Broad KPI portfolios promise coverage, but they fragment effort and complicate assurance. The practical answer is cadence. Organizations should revisit material topics on a schedule that matches regulatory change and market dynamics, then adjust the KPI set without expanding it beyond what teams can manage [2]. Another tradeoff sits between central control and local discretion. Central standards protect comparability and integrity. Local teams need room to tailor actions to their processes and geographies. A clear split of responsibilities, with corporate setting the why and what and local owners adapting the how and when, supports both consistency and relevance [1, 2, 4].

Externalities matter. A program can lower a firm's own emissions or labor incidents while shifting impacts upstream if supplier incentives are weak. The framework addresses this by treating supplier coverage and corrective action reliability as first class metrics, and by using assurance selectively where the risk to stakeholders is highest. Shared value thinking reinforces the point that credible CSR creates advantage when firms help partners improve rather than only policing them [3]. Boards and risk committees play a role here by aligning incentives and by reviewing unresolved tradeoffs where commercial targets and stakeholder commitments collide [4].

AI introduces both leverage and new responsibilities. Models that forecast energy demand, detect anomalies in supplier records, or estimate lifecycle impacts can raise speed and accuracy, but only when data quality, validation, and explainability are explicit. Without those controls, the risk of unfair outcomes or overclaiming rises, and the cost of remediation can exceed the benefit. The safest path is to deploy AI on a narrow set of well understood indicators, compare results with current methods, and scale once accuracy and governance objectives are met. Documentation of purpose, variables, validation, ownership, and review thresholds should sit with the indicator, not in a separate system, which simplifies audit and strengthens trust in reported results [5].

Cost and capability are common concerns. Building data pipelines, assurance routines, and basic AI pilots requires investment and skills that some organizations lack

The design choice to keep the architecture light helps. By using existing forums, selecting a minimal KPI set, and applying modular technology, firms can start small and grow without locking themselves into high fixed costs. The maturity profiles from the source work also help sequence effort. Teams can begin with ethics and strategy, secure a few material indicators with clear owners, pilot one AI use case with strong data, and expand only after the first loop of measure, learn, and adjust produces visible gains [1, 2, 5].

Boundary conditions are real. Highly regulated sectors may need more formal controls and slower change. Early stage firms may struggle to justify full assurance and should target a credible subset first. Geographies with weak data infrastructure will need to prioritize manual controls until sources mature. None of these conditions invalidate the framework. They guide pacing and scope. The core remains the same. Use recognized guidance to define what matters, measure only what leaders will use, embed ownership and cadence in the management system, and treat AI as an enablement layer with clear guardrails [1, 2, 4, 5].

10 Limitations and Future Work

This work is designed to be practical, which brings limits that readers should keep in mind. First, the framework is conceptual and was demonstrated through structured examples rather than a multi site field experiment. Utility has been judged formatively, based on fit to recognized guidance, decision routines, and assurance logic, not on statistical tests of performance change over time [6, 7]. Future studies should evaluate the framework in different sectors and sizes of organization, tracking before and after effects on a stable set of ESG and operational indicators aligned with GRI or SASB so results are comparable and auditable [2].

Second, measurement quality remains a constraint. Even when firms select material topics and define a compact KPI set, data lineage, boundary choices, and supplier coverage can distort results. External assurance helps, but it is uneven across regions and topics, and some indicators are still difficult to verify at scale. Research should examine which mixes of outcome and process metrics best predict real impact, and how often materiality should be refreshed in fast changing contexts so metrics stay decision ready without creating churn [1, 2].

Third, leadership and culture are treated here as design elements that can be built into routines. In practice, norms and incentives evolve slowly and vary across units and geographies. More evidence is needed on which governance patterns at board and executive levels sustain attention to sustainability when commercial pressure rises, and how inclusive practices influence adoption and learning in front line teams [4]. Comparative case work and survey research could map the conditions under which the same framework produces different trajectories.

Fourth, AI enablement is promising but fragile. Models depend on data quality, representativeness, and stable relationships. They also raise ethical questions that go beyond technical accuracy, including fairness, transparency, and recourse for those affected by automated decisions. Future work should document model governance in detail, test bias controls on social indicators, and compare human plus AI workflows with human only baselines to show when automation improves decisions without eroding trust [5]. Methods for explaining results to non technical stakeholders and for integrating model checks into existing assurance plans deserve specific attention.

Fifth, the Zachman mapping clarifies responsibilities and flows, yet it can feel abstract if teams do not see concrete examples that match their realities. Tooling and templates can help. Researchers and practitioners could co develop open assets that show, for a handful of common indicators such as emissions intensity, injury rates, or supplier conformance, exactly how to populate each cell from scope to technology and how to maintain the map as organizations grow [8].

Finally, generalizability and pacing are open questions. Highly regulated sectors may need deeper controls and slower cycles. Early stage firms may need credible subsets of the framework before full adoption. Geographies with weaker data infrastructure will rely on manual controls longer. Future work should specify minimal viable versions by context, identify the few design choices that most influence outcomes, and propose decision rules for when to add indicators, tighten assurance, or scale AI. Across these lines of work, the guiding principle remains the same. Anchor practice in recognized guidance, measure only what leaders will use, embed ownership and cadence in the management system, and treat AI as an enablement layer with clear guardrails [1–8].

References

1. International Organization for Standardization: ISO 26000:2010 — Guidance on Social Responsibility. ISO, Geneva (2018)
2. KPMG: The Time Has Come: The KPMG Survey of Sustainability Reporting 2020. KPMG International, Amstelveen (2020) Available at: https://home.kpmg/xx/en/home/insights/2020/11/the-time-has-come.html
3. Porter, M.E., Kramer, M.R.: Creating shared value: redefining capitalism and the role of business in society. Harv. Bus. Rev. **89**(1/2), 62–77 (2019)
4. OECD: OECD Corporate Governance Factbook 2021. OECD Publishing, Paris (2021). https://doi.org/10.1787/783b87df-en
5. Piteira, M., et al.: Ethics and artificial intelligence: a theoretical framework for a responsible use of AI. J. Bus. Ethics. **160**(4), 753–767 (2019). https://doi.org/10.1007/s10551-018-3921-0
6. Hevner, A.R., Chatterjee, S.: Design research in information systems: theory and practice. Springer, New York (2016). https://doi.org/10.1007/978-1-4419-5653-8
7. Peffers, K., Tuunanen, T., Gengatharen, D.C., Davis, A.M.: The design science research process: a model for producing and presenting information systems research. In: Proceedings of the 40th Hawaii International Conference on System Sciences, pp. 45–77 (2018). https://doi.org/10.1109/HICSS.2007.57
8. Zachman, J.A.: A framework for information systems architecture. IBM Syst. J. **26**(3), 276–292 (1987). https://doi.org/10.1147/sj.263.0276

Using Portable SDR Systems to Teach Spectrum Monitoring and RF Reconnaissance Concepts

Bryan Chanataxi-Nacimba(✉), Fernando Caicedo-Altamirano, Janneth Moreta-Changoluisa, Diego Alomoto-Robles, and David Rivas-Lalaleo

Department of Electrical Electronic and Telecommunication, Universidad de las Fuerzas Armadas ESPE, Sangolquí 171103, Ecuador
{bjchanataxi,fscaicedo,jemoreta1,drrivas}@espe.edu.edu

Abstract. Practical training in spectrum monitoring and radio frequency (RF) analysis is essential for communication engineering education. Traditional spectrum analyzers, while accurate, are prohibitively expensive and logistically impractical for widespread use in teaching laboratories or field exercises. This paper presents a reproducible, low-cost, and portable spectrum sensing kit built around the ADALM-Pluto SDR, a directional antenna, and a GNU Radio processing chain with an adaptive Cell-Averaging Constant False Alarm Rate (CA-CFAR) detector. A calibrated benchmarking protocol was implemented using a passive splitter to feed both the SDR and a high-end Agilent FieldFox analyzer simultaneously, ensuring fair comparisons with matched RBW/VBW and detector settings. Evaluation metrics included probability of detection, false-alarm rates, minimum detectable signal, dynamic range, and acquisition speed, with 95% confidence intervals over multiple runs. To connect technical validation with educational impact, key performance indicators (KPIs) were introduced—Sensing Efficacy per Dollar (SED), Power-per-Detection (PPD), and Mission-Adjusted Detection Efficacy (MADE)—allowing students to assess trade-offs between performance, cost, and portability.

Field experiments from 150 MHz to 2.4 GHz demonstrated that the SDR-based kit achieves reliable detection ($P_d > 0.9$ at moderate SNR) while reducing equipment cost by over an order of magnitude compared to commercial analyzers. Moreover, the modular design supports extensions to the 30–300 MHz VHF band using compact down-converters, enabling broader applicability in communication courses. These results position the proposed SDR kit as a viable, affordable, and pedagogically valuable platform for hands-on learning in spectrum monitoring, signal intelligence concepts, and modern RF communication systems.

Keywords: Software Defined Radio · Spectrum Monitoring · Tactical Communications · Interference Detection · Time-Frequency Analysis · GNU Radio

A. Rocha et al. (Eds.): ICITS 2026, LNNS 1966, pp. 705–717, 2027.
https://doi.org/10.1007/978-3-032-27538-7_62

1 Introduction

The electromagnetic spectrum is a foundational resource for modern communication systems, supporting wireless networks, broadcast services, and critical infrastructure. Beyond its strategic role in industry and defense, spectrum awareness has become increasingly relevant in academic programs, where future engineers must acquire practical skills in monitoring, detecting, and analyzing radio frequency (RF) signals [1]. Training in these areas enables students to understand interference, coexistence challenges, and spectrum management, which are central to courses in communications, wireless networks, and electronic engineering curricula.

Conventional spectrum analyzers, such as the Keysight FieldFox series, deliver high accuracy and dynamic range, but their prohibitive cost, weight, and limited availability make them impractical for widespread deployment in educational laboratories or classroom demonstrations [2]. As a result, many institutions restrict spectrum analysis training to theoretical exercises or shared access to a limited number of professional-grade devices, reducing opportunities for hands-on learning.

Software Defined Radios (SDRs) have emerged as a transformative alternative, offering reconfigurability, portability, and affordability by implementing radio functions in software rather than hardware [3,4]. Their use in academia has expanded rapidly, from simple energy detection experiments to advanced cognitive radio demonstrations [5,6]. Despite this progress, systematic benchmarking of SDRs against professional analyzers is rarely conducted in educational settings, and quantitative performance metrics are seldom introduced as part of the learning process.

This work addresses these gaps by presenting a reproducible, low-cost spectrum sensing kit tailored for educational use. The proposed platform leverages the ADALM-Pluto SDR and integrates an adaptive Cell-Averaging CFAR (CA-CFAR) detector implemented in GNU Radio. Its design enables not only hands-on experimentation but also rigorous evaluation of system performance using industry-inspired methodologies.

The main contributions of this work are the development of a calibrated, low-cost SDR monitoring kit with an integrated Cell-Averaging Constant False Alarm Rate (CA-CFAR) detector, and the implementation of a rigorous benchmarking protocol to validate the SDR system against commercial-grade analyzers under matched conditions. Furthermore, we introduce pedagogically oriented Key Performance Indicators (KPIs) (SED, PPD, MADE) to facilitate student analysis of performance, cost, and portability trade-offs, and detail strategies for broadening the frequency coverage to support essential VHF bands (30–300 MHz) for communication courses.

Field experiments (150 MHz to 2.4 GHz) demonstrate that the SDR-based kit achieves reliable detection performance ($P_d > 0.9$ at moderate SNR) with only 2–5% deviation relative to a commercial analyzer, while reducing system cost and weight by more than an order of magnitude. These results confirm that portable SDR platforms, combined with CFAR-based detection and rigorous calibration, are effective and reproducible teaching tools for communication engineering education [7,8]. The remainder of this paper is organized as follows: Section Methodology reviews related work on SDR-based spectrum monitoring and CFAR detection and presents the system design and benchmarking protocol; Section Experimental details the experimental setup; Section Results and Discussion discusses results including Pd/Pfa analysis, MDS, and educational KPIs.

2 Methodology

2.1 State of the Art

The evolution of Software Defined Radio (SDR) technologies has reshaped the way spectrum sensing is approached, not only in professional and defense contexts but also in engineering education. By shifting traditional RF functionalities to software, SDRs enable reconfigurable, adaptable, and affordable systems that can be deployed in laboratories, classrooms, and field exercises [3,6].

Recent research highlights the versatility of SDR in teaching and training environments, where students can explore multi-band signal processing, cognitive radio paradigms, and real-time spectrum occupancy mapping [4,5]. Educational implementations often emphasize hands-on learning by integrating SDR platforms with open-source tools such as GNU Radio or MATLAB, allowing learners to design and evaluate digital receivers, interference detection systems, and adaptive sensing schemes.

In addition, CFAR-based detection methods—originally developed for radar and defense applications—are increasingly applied in academic projects as a way to teach robust detection under low signal-to-noise ratios, bridging theory with practical experimentation [7]. Academic conferences such as IEEE EDUCON and ASEE have reported SDR-based curricula as effective approaches to link theoretical signal processing with experiential learning.

Despite these advances, most educational implementations remain limited to basic demonstrations of energy detection or spectrum visualization. Few studies provide calibrated comparisons against commercial analyzers, integrate statistical validation, or propose performance indicators that highlight the trade-offs between cost, portability, and accuracy. This gap motivates the present work, which introduces a reproducible SDR-based spectrum sensing kit explicitly designed for pedagogical use.

2.2 System Proposal Description

The proposed system demonstrates the feasibility of using a low-cost SDR solution for educational spectrum monitoring, with three distinctive contributions over prior teaching-oriented approaches:

- A **calibrated benchmarking protocol**, employing a passive 3 dB splitter to feed both the SDR and a professional spectrum analyzer simultaneously, ensuring fair comparisons and allowing students to contrast low-cost and high-end platforms under identical conditions.

- Integration of an **adaptive CA-CFAR detector** within GNU Radio and Python scripts, enabling autonomous operation and statistical evaluation of key metrics such as probability of detection (P_d), false alarm rate (P_{fa}), and minimum detectable signal (MDS) with confidence intervals.
- Introduction of **educational Key Performance Indicators (KPIs)**, including Sensing Efficacy per Dollar (SED), Power-per-Detection (PPD), and Mission-Adjusted Detection Efficacy (MADE), to help students understand how performance, cost, and portability interact in practical system design.

At the heart of the system is an ADALM-Pluto SDR device covering 325 MHz to 3.8 GHz, connected to a ruggedized laptop running a Linux-based environment with GNU Radio. The software chain implements FFT analysis, CA-CFAR detection, and spectrum occupancy mapping, with the flexibility to operate both interactively and autonomously. For instructional purposes, a compact downconverter is included to extend coverage to the 30–300 MHz VHF band, enabling students to experiment with a wider set of communication signals.

The antenna configuration includes a directional Yagi-Uda array (600 MHz–8 GHz) for UHF/SHF demonstrations, and a separate VHF antenna when downconversion is employed. This setup not only enhances sensitivity and suppresses interference but also provides a clear illustration for students of how antenna selection impacts system performance.

Figure 1 illustrates the system architecture, showing the integration between the SDR, analyzer, antenna, and processing terminal. Figure 2a presents the GNU Radio flowgraph, including the CA-CFAR block and visualization components. Figure 2b shows an example spectrogram obtained during field testing, highlighting common educational examples such as Wi-Fi emissions at 2.4 GHz and narrowband signals in other bands.

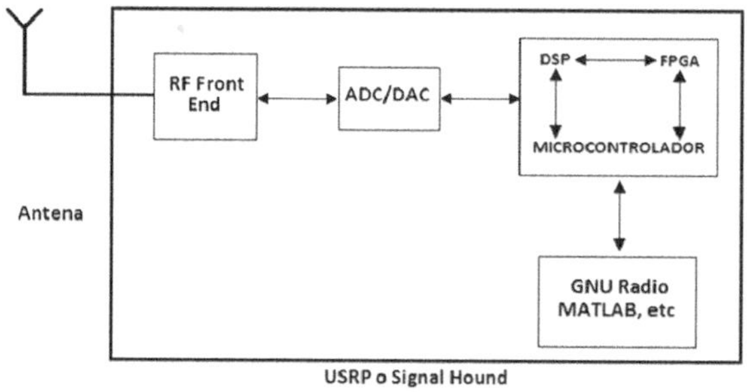

Fig. 1. System architecture integrating SDR, spectrum analyzer, and directional antenna via a calibrated power splitter.

2.3 System Architecture

The experimental platform was built upon the integration of the SDR Adalm Pluto with a directional antenna and an open-source signal processing environ-

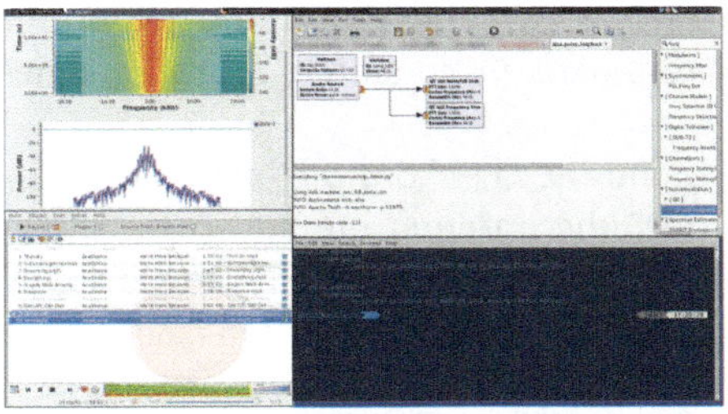

(a) GNU Radio flowgraph with FFT analysis, CA-CFAR detection, and real-time visualization.

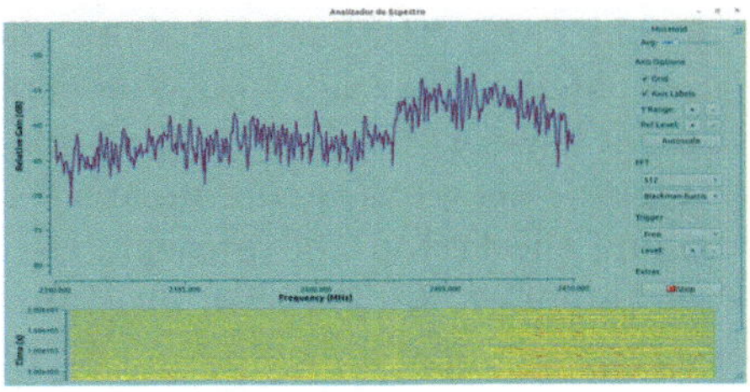

(b) Example spectrum capture during field test, illustrating Wi-Fi bands and narrowband signals, used here for teaching purposes.

Fig. 2. Flowchart in GNU Radio and example of spectrum capture.

ment. To ensure rigorous benchmarking and enhance educational value, its performance was evaluated against a high-end commercial spectrum analyzer, the Agilent FieldFox N9917A, under controlled conditions with a calibrated power splitter providing identical antenna input to both devices.

The SDR Adalm Pluto, developed by Analog Devices, is a compact and affordable software-defined radio module explicitly designed for education and experimentation. It supports full-duplex transmission and reception in the 325 MHz to 3.8 GHz frequency range, with a 20 MHz instantaneous bandwidth. The device features a 12-bit ADC/DAC, SMA connectors for RF I/O, and communicates via USB 2.0 with a host computer. Although originally conceived as an educational kit, when combined with open-source frameworks such as GNU Radio, it provides a versatile platform for laboratory courses and student projects, allowing learners to explore concepts ranging from basic spectrum visualization to advanced detection algorithms [9].

The Agilent FieldFox N9917A portable spectrum analyzer was employed as the reference instrument. Supporting frequency analysis from 30 kHz to 18 GHz, it offers high dynamic range and precision filtering for advanced signal characterization. Its rugged design, internal calibration, and Keysight's software integration make it a gold standard in professional RF measurements [10]. In the educational context, the FieldFox serves as an ideal benchmark against which students can validate SDR-based measurements, providing a meaningful comparison between professional and low-cost platforms.

To guarantee fairness and signal consistency during testing, both the SDR and the analyzer were connected to the same log-periodic dipole array (LPDA) antenna via a passive 3 dB power splitter. The antenna, designed for 600 MHz to 6 GHz operation, provides high gain and narrow beamwidth, enhancing the reception of weak or distant signals while suppressing off-axis interference.

Signal acquisition and visualization were performed using GNU Radio, which enabled real-time FFT analysis, CA-CFAR detection, and spectrogram visualization [11]. Standalone Python scripts were also developed to allow headless operation, automating data logging and snapshot collection during laboratory sessions or field campaigns. These scripts facilitated systematic measurement of probability of detection (P_d), false alarm rate (P_{fa}), and minimum detectable signal (MDS), thus allowing students to replicate professional methodologies within a low-cost experimental setup.

Figure 3 shows the experimental setup, including the Adalm Pluto, the Field-Fox, and the laptop and presents the log-periodic antenna employed for directional reception, while Fig. 4 illustrates the calibrated splitter configuration used for simultaneous comparisons.

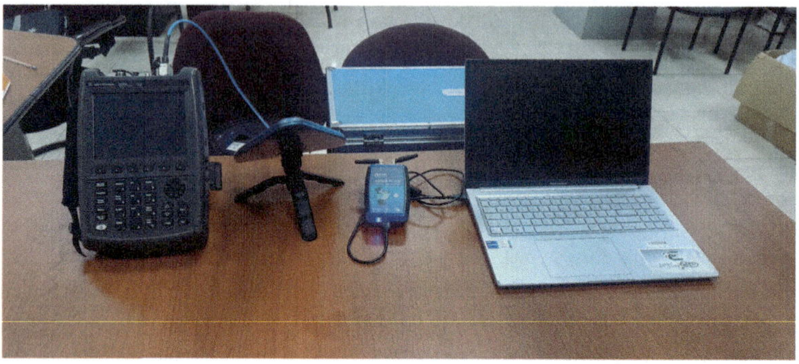

Fig. 3. Test scenario with all elements.

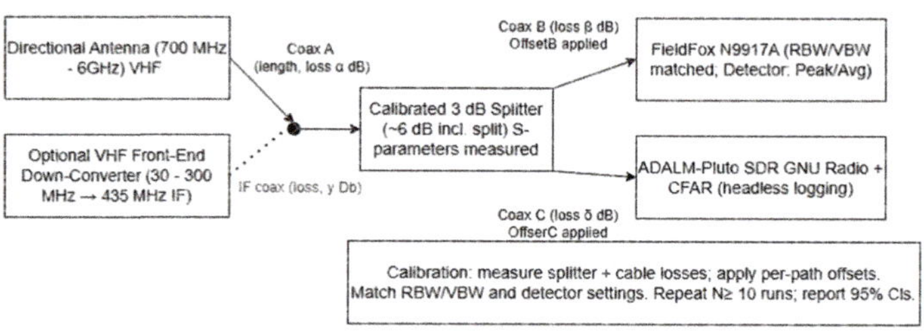

Fig. 4. Calibrated splitter configuration feeding both SDR and analyzer simultaneously to ensure fair comparison.

The technical comparison in Table 1 highlights key differences between the ADALM-Pluto SDR and the Agilent FieldFox N9917A. While the FieldFox provides superior measurement accuracy and broader frequency coverage, the SDR platform offers substantial advantages in affordability and accessibility, making it highly attractive for educational settings. This configuration demonstrates how the low-cost ADALM-Pluto can be effectively leveraged for educational spectrum analysis, allowing students to replicate industry-grade practices through hands-on experimentation, rigorous calibration protocols, and the flexibility of open-source tools and advanced detection algorithms.

3 Experimental

To evaluate the practical capabilities of the proposed SDR-based monitoring kit in an educational context, a series of controlled measurements were conducted

Table 1. Comparison between SDR Adalm Pluto and FieldFox N9917A

Specification	Adalm Pluto	FieldFox N9917A
Frequency Range	325 MHz–3.8 GHz	30 kHz–18 GHz
ADC Resolution	12 bits	14 bits
Instantaneous Bandwidth	20 MHz	Up to 100 MHz
Full Duplex	Yes	No
Weight	170 g	3.0 kg
Power Supply	USB powered	Internal battery
Software Compatibility	GNU Radio, MATLAB, Python	Keysight software
Approx. Cost (Ecuador)	\$900–\$1000	\$60,000–\$80,000

at the General Guillermo Rodríguez Lara Campus of the Universidad de las Fuerzas Armadas ESPE, located in Latacunga, Ecuador. The experiments were designed not only to validate system performance but also to serve as reproducible exercises for teaching spectrum analysis, detection theory, and measurement methodology in communication engineering courses.

3.1 Measurement Protocol

The experimental setup was designed to ensure fairness, reproducibility, and didactic clarity. Both the SDR Adalm Pluto and the Agilent FieldFox N9917A were connected to the same directional antenna through a passive 3 dB power splitter, ensuring that students could directly compare results obtained under identical reception conditions. This approach avoided variability due to reconnections or antenna reorientation, reinforcing best practices in experimental design.

The directional LPDA antenna was mounted on a fixed support, with stable orientation and elevation, covering the 600 MHz–6 GHz band. For VHF demonstrations (150–300 MHz), a dipole and a down-converter were introduced. Both instruments were configured with matched resolution bandwidths (RBW $\approx$ 1 kHz), video bandwidths (VBW), and sweep times, illustrating the importance of parameter alignment in fair benchmarking.

3.2 Frequency Coverage

The measurement campaign focused primarily on the 700 MHz–2.4 GHz range, which includes signals of everyday educational interest such as cellular, Wi-Fi, and telemetry. Additional experiments were carried out in the 150–300 MHz range to expose students to lower-frequency communications. For each band:

- **Adalm Pluto SDR:** Frequency sweep implemented via GNU Radio with dynamic center frequency shifts and 20 MHz instantaneous bandwidth.
- **FieldFox N9917A:** Built-in FFT sweep, calibrated as per manufacturer specifications.

Each sweep used overlapping 20 MHz windows to guarantee full coverage. Measurements were repeated at least 10 times per configuration, and results were averaged with 95% confidence intervals—allowing students to appreciate the role of statistical reliability in engineering evaluation.

3.3 Experimental Results

This protocol enabled the identification of active signal sources, interference patterns, and spectrum occupancy profiles at the test site. Figure 5 shows representative outputs obtained with the SDR, while Fig. 6 illustrates the joint configuration with both instruments connected.

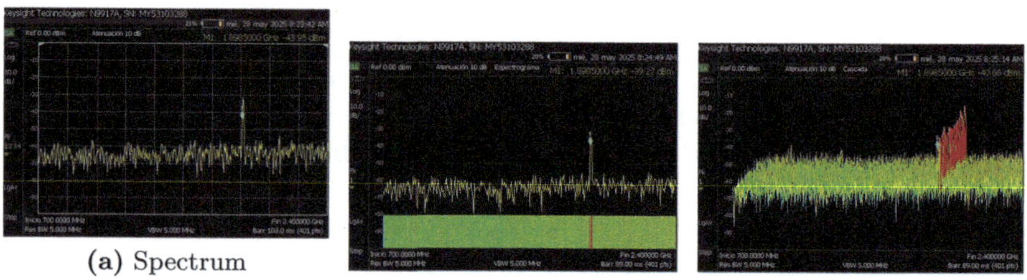

(a) Spectrum Measurement (b) Interference Detection (c) Time-Frequency Plot

Fig. 5. Measurement results obtained during educational field tests. Panel (a) shows a frequency-domain scan, (b) highlights interference events, and (c) presents a time-frequency waterfall plot used for teaching temporal dynamics.

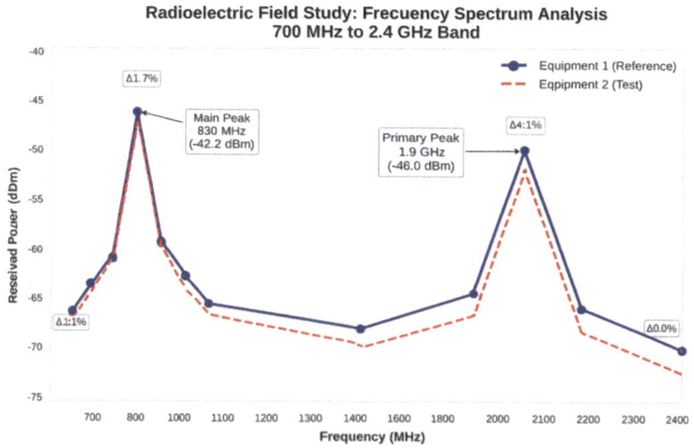

Fig. 6. Experimental configuration showing SDR and FieldFox connected via a calibrated splitter.

3.4 Quantitative Comparison

Table 2 summarizes the probability of detection (P_d) and false alarm rate (P_{fa}) for different SNR levels using the CA-CFAR detector in the SDR. The comparison with the FieldFox allowed students to directly evaluate the trade-offs between low-cost and professional instruments.

Table 2. Detection Performance of SDR vs. FieldFox (95% CI over 10 runs)

SNR (dB)	P_d SDR	P_d FieldFox	P_{fa} SDR	P_{fa} FieldFox
−5	0.62 ± 0.04	0.70 ± 0.03	0.11 ± 0.02	0.07 ± 0.01
0	0.81 ± 0.03	0.87 ± 0.02	0.08 ± 0.01	0.05 ± 0.01
5	0.93 ± 0.02	0.96 ± 0.01	0.05 ± 0.01	0.03 ± 0.01
10	0.98 ± 0.01	0.99 ± 0.01	0.04 ± 0.01	0.02 ± 0.01

Table 3 presents the minimum detectable signal (MDS). Although the FieldFox shows superior sensitivity, the SDR results were sufficient for classroom demonstrations of detection thresholds and noise floor estimation.

Table 3. Minimum Detectable Signal (MDS) Comparison

System	MDS (dBm)
Adalm Pluto SDR	−90 dBm
FieldFox N9917A	−110 dBm

Finally, Table 4 summarizes sweep performance. Here, students could contrast acquisition speed, dynamic range, and power deviation, reinforcing the importance of instrument specifications in real-world applications.

Table 4. Sweep Performance Metrics

Metric	Adalm Pluto SDR	FieldFox N9917A
Sweep Time (per 20 MHz)	0.35 s	0.12 s
Power Deviation (avg)	$\leq$ 5%	Baseline
Dynamic Range	65 dB	90 dB

These results demonstrate that while the commercial analyzer offers superior sensitivity and dynamic range, the SDR-based kit provides adequate detection capability and reproducibility at a fraction of the cost. More importantly, the experimental framework illustrates to students how to design fair tests, quantify uncertainty, and interpret system trade-offs—skills that are essential in both academic and professional engineering practice.

4 Results and Discussion

This section presents the quantitative evaluation of the SDR-based monitoring kit compared with the reference FieldFox analyzer, followed by an analysis of pedagogically adapted Key Performance Indicators (KPIs). Beyond technical validation, the discussion emphasizes the educational value of these experiments, showing how concepts such as detection probability, sensitivity, and efficiency trade-offs can be taught effectively through hands-on laboratory practice.

4.1 Detection Performance

The detection probability (P_d) as a function of SNR was evaluated using the CA-CFAR detector implemented in GNU Radio. Figure 7 shows the receiver operating characteristic (ROC) curves for both the SDR and the FieldFox. As expected, the reference analyzer exhibits superior sensitivity, but the SDR demonstrates acceptable performance ($P_d > 0.9$ for SNR ≥ 5 dB).

From an educational perspective, these curves provide students with a concrete example of how ROC analysis is applied in real RF systems, linking theory from detection courses to experimental practice.

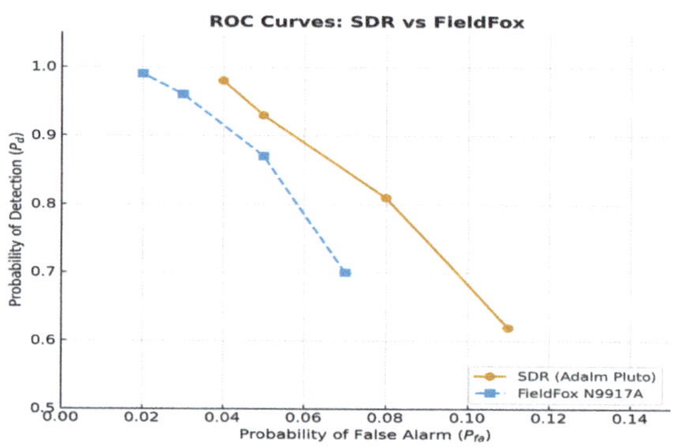

Fig. 7. ROC curves comparing SDR and FieldFox detection performance across SNR levels. Used in teaching to illustrate trade-offs between P_d and P_{fa}.

Table 2 provides numerical results for P_d and P_{fa}, confirming that the SDR delivers robust detection under moderate-to-high SNR conditions. This comparison helps students interpret how hardware limitations (e.g., ADC resolution) manifest in measurable detection metrics.

4.2 Minimum Detectable Signal (MDS)

Figure 8 illustrates the estimated MDS across different frequency bands for both systems. The FieldFox achieves −110 dBm sensitivity, while the SDR averages around −90 dBm, with performance degradation closer to its 3.8 GHz upper limit. In the VHF range (150–300 MHz, with down-converter), sensitivity remained within −85 dBm.

For educational use, these results allow students to explore the concept of noise floor and its relationship with signal detectability, encouraging discussion about trade-offs between equipment sophistication and affordability.

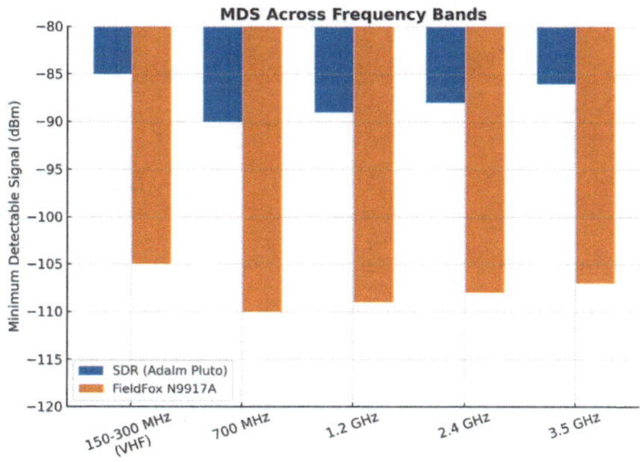

Fig. 8. Minimum detectable signal (MDS) across frequency bands for SDR and FieldFox. Students can use this as an example of sensitivity analysis.

4.3 Sweep Time and Power Deviation

The sweep performance results (Table 4) indicate that the SDR requires longer acquisition time per 20 MHz segment (0.35 s vs. 0.12 s). However, the power deviation remained within 5% of the FieldFox.

In a classroom context, this comparison highlights the trade-off between measurement speed and instrument cost. It provides a platform for students to critically evaluate whether high-end performance is always necessary for certain applications, such as lab exercises or prototype testing.

4.4 Educationally Adapted KPIs

To connect the technical outcomes with learning objectives, three KPIs were reinterpreted for pedagogical use:

- **Sensing Efficacy per Dollar (SED):** shows students how to analyze performance-to-cost ratios.
- **Power-per-Detection (PPD):** links detection efficiency to energy consumption, encouraging sustainability discussions.
- **Mission-Adjusted Detection Efficacy (MADE):** integrates P_d, sweep time, and portability, guiding students to think about multi-criteria evaluation.

Table 5 presents normalized KPI values (FieldFox baseline = 1). The SDR significantly outperforms in SED and PPD, demonstrating that while professional analyzers excel in sensitivity, low-cost SDR platforms can offer superior accessibility and efficiency in educational labs.

Table 5. Educational KPI Comparison (Normalized to FieldFox = 1)

KPI	SDR (Adalm Pluto)	FieldFox N9917A
SED	12.5	1.0
PPD	8.7	1.0
MADE	2.4	1.0

4.5 Discussion

The results confirm that while the FieldFox remains the gold standard for sensitivity and dynamic range, the SDR-based kit provides sufficiently accurate results for teaching and research training. More importantly, the analysis framework—including ROC curves, MDS evaluation, and KPIs—offers students practical exposure to methods they would otherwise encounter only theoretically.

By integrating low-cost SDRs into structured experimental protocols, universities can democratize access to advanced RF education, enabling large groups of students to gain hands-on experience. This strengthens the bridge between classroom learning and real-world engineering practice.

Acknowledgment. The authors would like to express their sincere gratitude to the Universidad de las Fuerzas Armadas ESPE for providing institutional support and access to laboratory and field testing facilities. This work was made possible through the contributions of the internal research and development projects *"Entrenamiento de Salto de Banda"* and *"Automatización de Polígono de Tiro"*, whose infrastructure and resources facilitated the experimental validation of the proposed system.

AI Use Declaration. Artificial Intelligence (AI) tools, specifically large language models, were employed only for language editing and translation support. All data, analyses, and results presented originate from real measurements with SDR hardware and professional equipment, ensuring the scientific integrity of the study.

References

1. Zhao, Q., Song, M., Zhang, X.: A survey of dynamic spectrum access: signal processing, networking, and regulatory policy. IEEE Signal Process. Mag. **35**(3), 36–53 (2018)
2. Keysight Technologies. Keysight FieldFox N9917A Microwave Analyzer Datasheet (2020). https://www.keysight.com

3. Mitola, J., Maguire, G.Q.: Cognitive radio: making software radios more personal. IEEE Pers. Commun. **6**(4), 13–18 (1999)
4. Al-Hourani, A., Gomez, J.: Low-cost SDR platforms for real-time spectrum monitoring in IoT and tactical networks. In: IEEE Military Communications Conference (MILCOM), pp. 1234–1240. IEEE (2022)
5. Yang, X., Chen, Y., Li, J.: Cognitive radio networks for tactical communications: recent advances and open challenges. IEEE Commun. Mag. **59**(6), 86–92 (2021)
6. Li, W., Li, X.: Software-defined radio: a survey of recent research and applications. IEEE Access **7**, 117241–117260 (2019)
7. Alvarez, M., Ruiz, P., Torres, L.: CFAR-based spectrum sensing for portable SDR platforms: performance and trade-offs. In: IEEE International Symposium on Dynamic Spectrum Access Networks (DySPAN), pp. 45–50. IEEE (2023)
8. Panwar, R., Singh, S.: Portable SDR systems for military spectrum reconnaissance: a comparative study. IEEE Access **10**, 45321–45335 (2022)
9. Analog Devices. Adalm-Pluto: active learning module for SDR (2021). https://www.analog.com/en/design-center/evaluation-hardware-and-software/eval-adalm-pluto.html
10. Keysight Technologies. Fieldfox n9917a data sheet (2020). https://www.keysight.com/us/en/assets/7018-05919/data-sheets/5992-3662.pdf
11. GNU Radio Foundation. Gnu radio documentation (2021). https://wiki.gnuradio.org/

Instructional Designer Competencies: A Systematic Literature Review

Margarida Sampaio(✉) and Guilhermina Lobato Miranda

Instituto de Educação, Universidade de Lisboa, Lisbon, Portugal
{asampaio1,gmiranda}@edu.ulisboa.pt

Abstract. The concepts of Instructional Design and *Instructional Designer* have only recently emerged in the academic literature in Portugal. This research aims to enhance understanding in this field by contributing to the existing body of knowledge. Instructional Design is a field that is constantly evolving, mainly due to scientific and technological advances. The recent introduction of Artificial Intelligence in various contexts, including education, is an example of this process. This article reflects the initial results of a systematic literature review on the Instructional Designer's Competency Profile who acts in academic and corporate e-Learning settings. The literature review followed the protocol and recommendations suggested by the Preferred Reporting Items for Systematic Reviews and Meta-Analyses (PRISMA) methodology. The initial search resulted in 863 records obtained from the Educational Resources Information Centre (ERIC), Elton B. Stephens Company (EBSCO), Jstor, Scopus, Springer-Link, Web of Science and Wiley databases. After eliminating duplicate records and applying the defined inclusion and exclusion criteria 80 records were obtained, which constituted our *corpus* for analysis. So far, the findings indicate that there are several ways to name this professional, he comes from different initial backgrounds and he doesn't always have specific training to practice this activity. Some skills were identified as essential. Highlights include learning and instruction theories and knowledge models, training projects management, digital technologies knowledge, and their application to specific contexts, and communication strategies mastery aimed to different audiences.

Keywords: Competencies · Instructional Designer · Systematic Literature Review

1 Introduction

The development of information and communication technologies (ICT) has a major impact on the field of Instructional Design (ID). The advent of mobile technologies or recent phenomena such as the introduction of Artificial Intelligence (AI) in education bring new challenges to professionals in this already challenging area. The Instructional Designer encounters various challenges in the process of creating content, such as the need to respond to a wide range of learner characteristics, adapt to constant technological evolution, and need to motivate and maintain learner engagement.

A. Rocha et al. (Eds.): ICITS 2026, LNNS 1966, pp. 718–730, 2027.
https://doi.org/10.1007/978-3-032-27538-7_63

ID involves analyzing what should be taught and learned, defining how it should be taught and evaluating the learning process [14]. Sims and Koszalka [40] define ID as "an intentional activity that results in a combination of strategies, activities and resources that facilitate learning" and the Instructional Designer as "someone with the skills to design the instruction" (p. 570).

The dynamic nature of the field of ID makes it difficult to identify and characterize the professionals who work in this area, which is why their competencies have long been the subject of numerous studies [19, 35, 36, 46]. There are also several organizations that have paid attention to this issue to define the Instructional Designers activity's competencies and frameworks. The Association for Talent Development (ATD), the International Board of Standards for Training, Performance and Instruction (IBSTPI) and the Association for Educational Communications and Technology (AECT) are some examples of such organizations.

The aim of this systematic literature review is to identify the main competencies that Instructional Designers must possess to fulfil the several roles assigned to them in the different contexts in which they can act. Two other objectives of the research are: to identify the discrepancies between the competencies identified in the literature and those perceived by practicing Instructional Designers and to explore models for developing specific competencies for these professionals.

2 Theoretical Framework

The concepts of competence and competency have different meanings for different authors. This terminological confusion often reflects conflation of distinct concepts and inconsistent usage of terms as much as different cultural traditions [53]. However, it always include knowledge of a given field and the ability to apply it, in a flexible way, to several situations. Connolly and Bruner [8] refer competence as an operative knowledge. Gilbert and Parlier [12] consider that this concept only makes sense in an action and according to the objectives being pursued, taking advantage of the elements of the situation to adapt to its evolving conditions. Richey et al. [33] bring to us the concept of competency as "a knowledge, skill or attitude that makes it possible to perform effectively the activities of a given task or function, according to professional standards" (p. 26). Professionals express these competencies as standards, a predefined level of quality or achievement [25].

Smith and Ragan [43] define ID as a "systematic process of translating the principles of learning and instruction into the planning of teaching materials, activities and assessments" (p. 24). For Richey et al. [33] ID is "a science and art of defining detailed specifications for the development, evaluation and support of situations that facilitate learning and performance" (p. 5). The Instructional Designer's role is not restricted to integrating technology into learning processes, or applying ID models, using learning theories or endeavoring to promote engagement in courses. These professionals are involved in a variety of other initiatives, such as mapping training needs, managing projects, researching and implementing emerging technologies, running business areas or managing professional development events [18]. Due to the variety of roles, they take on, Instructional Designers are also referred as Learning Designers, Learning

Experience Designers, Educational Technologists, Learning Technologists, Instructional Systems Designers, Learning & Development Professionals, among others [32, 39].

In addition to these empirical studies, different professional associations, such as IBSTPI [17], have produced knowledge and published standard competency models for professionals in the field of ID and Educational Technologies [37]. In this context, a model is an organization of competencies, which offers a conceptual framework based on a certain function and provides information on the knowledge, skills and tasks that are required to perform certain activities [19]. The existence of models and standards is fundamental to establishing a field of intervention.

In the literature we found references to Instructional Designer competencies that are valued in the various contexts and industries in which they work, such as knowledge of ID, theories and models of learning and instruction, information organization and assessment [6, 11, 15, 20, 30, 35, 51], communication skills [23, 35, 36, 46, 50] and the ability of these professionals to adapt to new circumstances [23, 47].

Regarding the academic background, we found that many had no access to formal learning in the field of ID [9]. Bean [5] refers to these professionals, who have become Instructional Designers without any formal education, as "Accidental Instructional Designers". This situation is very common in corporate contexts [5]. In the literature we found several references to formal education and the development of Instructional Designers' competencies [20, 39]. As in other areas of activity, it is essential to promote the continuous training of these professionals [49], ensuring an adequate response to pedagogical demands in a changing evolving context.

3 Methodology

In the literature review we followed the protocol and recommendations provided in the Preferred-Reporting of Items for Systematic Reviews and Meta-Analyses Statement (PRISMA) to ensure quality and rigor in the systematic review [21, 29]. This methodology was selected because of its transparency, validity, adaptability and because it can be applied to a wide variety of disciplines, contexts and subjects, including the social sciences and education [38]. Despite being a 2009 framework, the PRISMA methodology has been regularly updated and is globally recognized and formally endorsed by more than 170 publishing organizations [44]. The systematic review process helped refine the research questions, map the scope of relevant studies, assess and evaluate selected works, and establish evidence to guide future research [13].

3.1 Purpose of the Study

At this research stage, the purpose of the study is to analyze the Instructional Designer's competencies found in scientific literature.

3.2 Initial Screening

The diverse terminology associated with the field of Instructional Design and, more specifically, the role of the Instructional Designer presented a challenge in selecting appropriate search terms. This selection was informed by the thesaurus of the Educational Resources Information Center (ERIC) database and the terminology identified in the literature review. We used a variety of combinations related with the main synonyms for "Instructional Designer" (subject of study), synonyms for e-Learning (context of study), synonyms and words related to competencies (what is to be studied) and synonyms for ID (field of the study).

We began our search by consulting the reference (online) bibliographic databases for the subject [31], namely: ERIC, Elton B. Stephens Company (EBSCO), Jstor, Scopus, Springer-Link, Web of Science and Wiley. This search resulted in 863 records. In addition, we used the Google Scholar and LearnTechLib databases, as well as a direct web search to find some bibliography referenced in the records obtained. Figure 1 shows a summary of the distribution of records among the databases consulted.

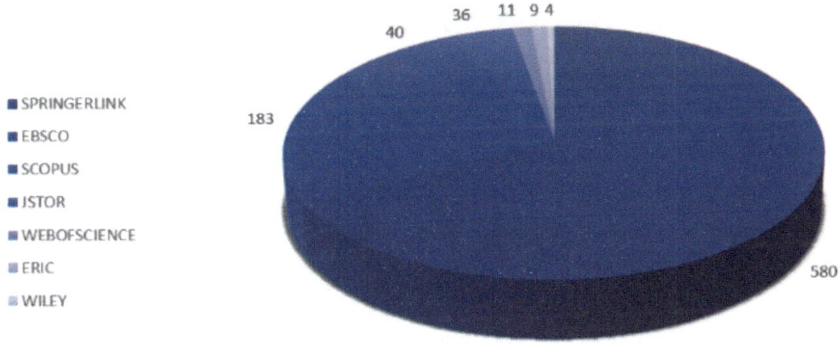

Fig. 1. Number of Records Obtained from the Databases.

3.3 Selection Criteria

In the literature review, we defined the inclusion criteria as scientific articles published in academic journals between 2018 and 2023, peer-reviewed, in english language and with full text available. Regarding the context, we included studies on Instructional Designers working in different settings, namely academic and corporate, and studies on the competencies of these education professionals and models of competency and development.

To detail the reasons that led us to adopt the inclusion criteria, we have: (i) We only considered peer-reviewed scientific articles - this option was because these were studies validated by the scientific community; (ii) Including only the english language was due to the small number of records in portuguese and portuguese-brazilian languages obtained in the defined time. To maximize the number of records for our study, it was essential to adopt english language, the predominant one of scientific discourse. Furthermore, this field has been extensively studied in the United States; (iii) Access to the full text, through the online bibliographic databases made available by our academic services was another criteria because it enabled us to assess their relevance; (iv) With regard to the time frame defined, five years, since the field of ID is in constant evolution and technological and social change permeable we couldn't have a very broad time period. Research should be based on recent data (the state of the art). The five-year period includes two pandemic years, which conditioned scientific production between 2020 and 2021 and made us extend the time frame of the research. On the other hand, the pandemic context has led to a growing demand for ID professionals and their skills in academic and corporate settings [28], to support emergency remote teaching and the transition from face-to-face learning contexts to online learning experiences.

Eliminating duplicate records and applying the defined selection criteria resulted in 60 records. Reading the full text of these records led us to identify 20 additional references. The final *corpus* of analysis consisted of 80 records.

A schematic representation [16] of the search carried out so far is shown in Fig. 2.

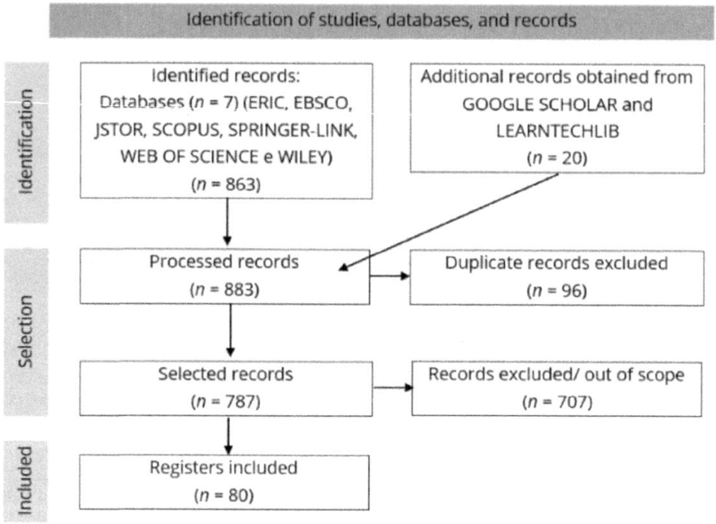

Fig. 2. PRISMA Flow Diagram.

4 Findings

All empirical studies included in this review were published in peer review journals. In this section, we summarize our first findings under the following dimensions: (1) Competency Models, (2) Origin of Articles, (3) Temporal Distribution, (4) Geographical Distribution and (5) Methodological Approach.

4.1 Instructional Designer Competency Models

Considering the scope of the study, we analyzed the frequency with which the Competency Models for the Instructional Designer were referenced. We found citations for the following models: International Board of Standards for Training, Performance and Instruction (IBSTPI), Association for Talent Development (ATD), Association for Educational Communications and Technology (AECT), International Society for Performance Improvement (ISPI), International Society for Technology in Education (ISTE), University Professional and Continuing Education Association (UPCEA) and Online Learning Consortium (OLC). IBSTPI was the most cited Competency Model, probably because it is a model validated at a global level by Instructional Designers working in different industries and contexts, covering the entire content design process and the different roles that the Instructional Designer can take on [33]. The frequency of citations ranges from 128 (the highest) and 17 (the lowest). Figure 3 shows the distribution of references for each of the Models.

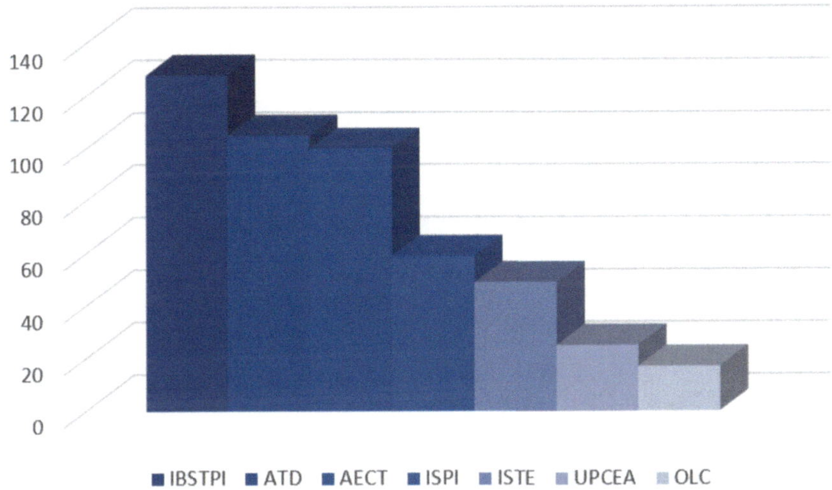

Fig. 3. Distribution of the Records by Competency Models.

4.2 Origin of the Records

Regarding the origin of the records, there is a wide variety of publications, with Techtrends (14 articles) leading, followed by the Journal of Computing in Higher Education (10 articles), Educational Technology Research and Development (9 articles) and the Online Learning Journal (9 articles). This distribution can be seen in Fig. 4. The publications with the highest number of records are all indexed in Scopus and are positioned in the 1st Quartile (Q1) in the field of Education (rankings for 2022).

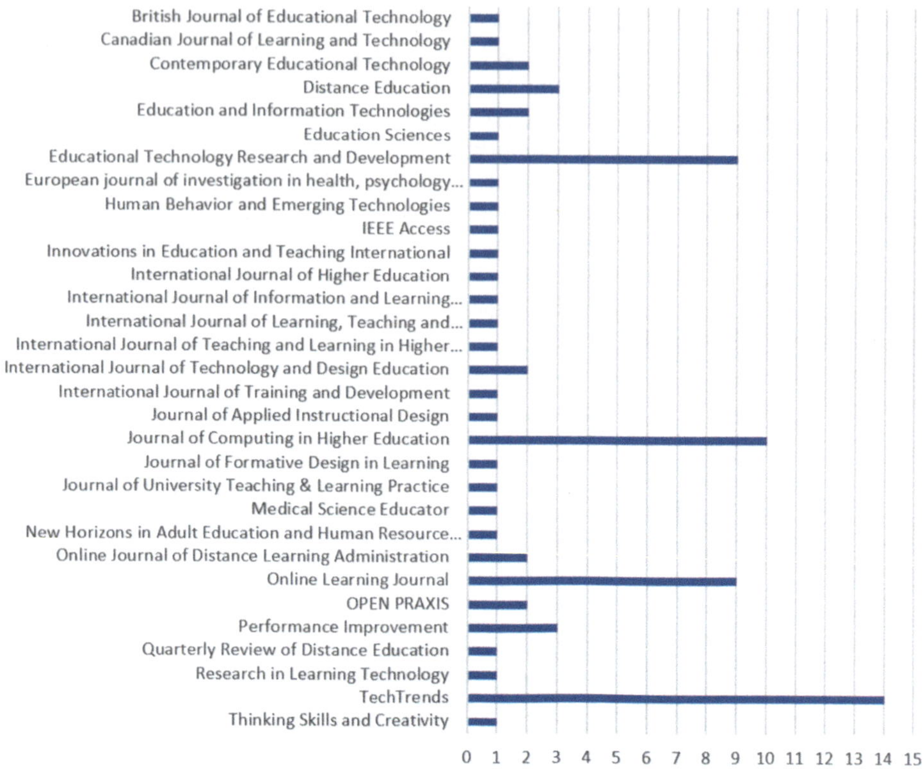

Fig. 4. Records Distribution by Journal.

4.3 Temporal Distribution

Concerning the time distribution of the records, as shown in Fig. 5, we found that the interest in the topic of Instructional Designer competencies grew in 2019. The literature also addresses this growing interest in the role of Instructional Designers [41]. In 2020, probably due to the pandemic context, we saw a decrease in scientific production in the field of ID. In 2022, interest in our study's subject increased once again, likely due to the growing recognition of this professional's role in adapting contents for digital environments [4]. It should be noted that the 2023 results only reflect one semester.

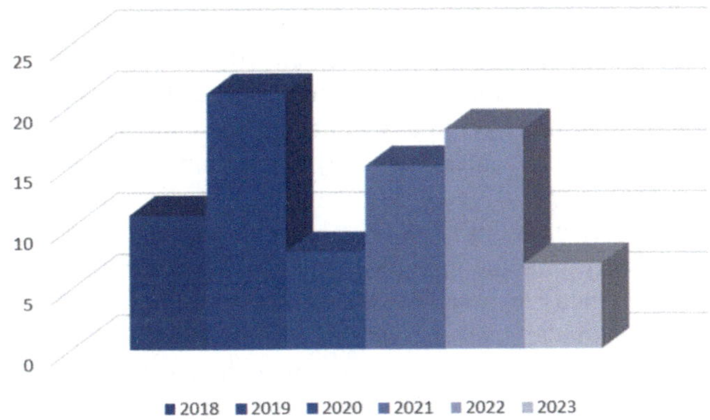

Fig. 5. Distribution of Articles by Research Time (2018–2023).

4.4 Geographical Distribution

Regarding the geographical location of the studies, we found a large concentration in the United States, with 68 records. This high number of records is probably related to the history of research in the field of ID. Figure 6 shows the geographical distribution of the studies.

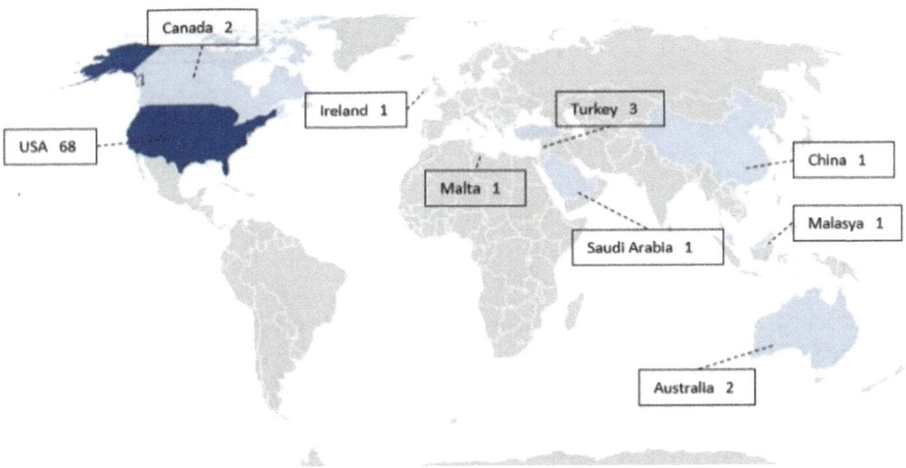

Fig. 6. Distribution of the Studies by Geographic Regions.

4.5 Methodological Approach

Relating to the methodological approach, we found that the studies in our *corpus* of analysis tended to adopt a qualitative approach (62%; 50 studies), focusing on understanding the phenomena and not aiming to generalize the results. Regarding research methods, we identified several approaches, including Design Based Research [54], Phenomenology [27], Action Research [42], Ethnographic Research [10], Survey Research [2], Grounded Theory [3], Case Study [26], Multiple Case Study [45] and Quasi-Experimental Study [48]. We also found some diversity in data collection techniques, such as Questionnaire Surveys [24], Semi-Structured Interviews [55], Document Analysis [7], Focus Groups [11], Observation [51] and Field Notes [54]. Some studies presented data collection techniques combined [34], assimilating evidence from multiple sources and ensuring triangulation [22]. The diversity of methodologies and data collection instruments allowed us to identify competencies in Instructional Designers from multiple angles. Figure 7 provides an overview of the methodological approach used to study Instructional Designer's competencies. N/A (Not Applicable) refers to systematic literature reviews [1] and articles without methodological references [52].

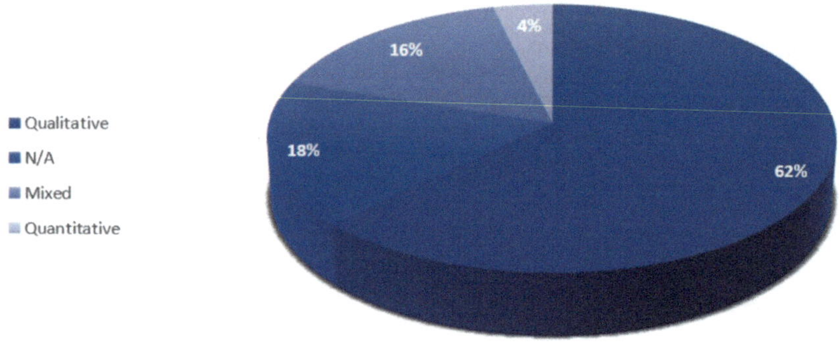

Fig. 7. Types of Studies.

5 Conclusion

This article presents the first results of a systematic literature review on the Instructional Designer's Competency Profile working in academic and corporate e-learning settings. We analyzed 80 references, examining the Instructional Designer Competency Model discussed, the origin of the articles, the temporal and geographical distribution, and the methodological approach adopted.

The Instructional Designer is an educational professional who designs instruction, someone who has mastered how people learn and is able to define strategies to promote the learning process. The simplicity of this definition has nothing to do with the diversity and complexity of the tasks they carry out. This diversity is at the root of the wide variety of professional titles by which they are known - Instructional Designer, Learning Designer, Learning Experience Designer, Educational Technologist, Educational Designer, Training Manager, Learning/Training Project Manager, e-Learning Developer, e-Learning Consultant, among others. This diversity can also be seen in the context in which they work.

The study of the essential competencies of Instructional Designers is part of the growing demand for these professionals in the expansion of digital learning environments. Studies in this field contribute to the understanding of the competencies needed for these professionals to ensure that they can fulfil the roles assigned to them, that they are able to adapt to emerging technologies, that they are able to communicate effectively, that they are aligned with theories of instruction and learning and to regulate their professional practice.

Instructional Design is an evolving field translated into a need to constantly update the competencies of the professionals who work in it. The knowledge of these competencies enables development plans to be drawn up and provides important insights for the initial training and the support of the knowledge acquisition process of the Instructional Designers.

Acknowledgments. This work was supported by National Funds through FCT-Portuguese Foundation for Science and Technology, I.P., under the scope of UIDEF - Unidade de Investigação e Desenvolvimento em Educação e Formação, UIDB/04107/2020, https://doi.org/10.54499/UIDB/04107/2020.

References

1. Abuhassna, H., Alnawajha, S.: Instructional design made easy! Instructional design models, categories, frameworks, educational context, and recommendations for future work. European J. Investig. Health Psychol. Educ. **13**(4), 715–735 (2023)
2. Anderson, M.C., Love, L.M., Haggar, F.L.: Looking beyond the physician educator: the evolving roles of instructional designers in medical education. Med. Sci. Educ. **29**(2), 507–513 (2019)
3. Baldwin, S.J., Ching, Y.-H., Friesen, N.: Online course design and development among college and university instructors: an analysis using grounded theory. Online Learn. **22**(2), 157–172 (2018)
4. Bao, W.: COVID-19 and online teaching in higher education: a case study of Peking University. Hum. Behavior Emerg. Technol. **2**(2), 113–115 (2020)
5. Bean, C.: The Accidental Instructional Designer: Learning Design for the Digital Age, 1st edn. ASTD Press, USA (2014)
6. Bond, J., Dirkin, K.: What models are instructional designers using today? J. Appl. Instruct. Des. **9**(2), 127–140 (2020)
7. Chongwony, L., Gardner, J.L., Tope, A.: Instructional design leadership and management competencies: job description analysis. Online J. Dist. Learn. Administration. **23**(1) (2020)
8. Connolly, K., Bruner, J.: The Growth of Competence. Academic, London (1974)
9. Cox, S., Osguthorpe, R.T.: How do instructional design professionals spend their time? TechTrends. **45-47** (2003)
10. DeVaughn, P., Stefaniak, J.: An exploration of the challenges instructional designers encounter while conducting evaluations. Perform. Improv. Q. **33**(4), 443–470 (2021)
11. Fiock, H., et al.: Instructional design learners make sense of theory: A collaborative autoethnography. Educ. Technol. Res. Dev. **70**(1), 31–57 (2022)
12. Gilbert, P., Parlier, M.: La compétence: Du ‘mot-valise’ au concept opératoire. Actualité de la formation permanente. **116**, 14–18 (1992)
13. Gough, D., Oliver, S., Thomas, J.: An Introduction to Systematic Reviews. Sage, London (2017)

14. Gustafson, K.L.: International Encyclopedia of Educational Technology. Pergamon, USA (1996)
15. Sh!ft Homepage.:https://www.shiftelearning.com/blog/adult-learning-theories-instructional-design. Last accessed 23 Mar 2025
16. Haddaway, N.R., Page, M.J., Pritchard, C.C., McGuinness, L.A.: PRISMA2020: an R package and shiny app for producing PRISMA 2020-compliant flow diagrams, with interactivity for optimized digital transparency and open synthesis. Campbell Syst. Rev. **18**(2), e1230 (2022)
17. International Board of Standards for Training, Performance, and Instruction Homepage.: https://ibstpi.org/competency-sets-services/instructional-designer-competencies/. Last accessed 23 Mar 2025
18. Kenny, R., Zhang, Z., Schwier, R., Campbell, K.: A review of what instructional designers do: questions answered and questions not asked. Canadian J. Learn. Technol/ La revue canadienne de l'apprentissage et de la technologie. **31**(1) (2005)
19. Koszalka, T.A., Russ-Eft, D.F., Reiser, R.: Instructional Designer Competencies: the Standards, 4th edn. Information Age Publishing, North Carolina (2013)
20. Kumar, S., Ritzhaupt, A.: What do instructional designers in higher education really do? Int. J. E-Learning. **16**(4), 371–393 (2017)
21. Liberati, A.: The PRISMA statement for reporting systematic reviews and meta-analyses of studies that evaluate health care interventions: explanation and elaboration. Ann. Intern. Med. **151**(4), 65–94 (2009)
22. Lincoln, Y.S., Guba, E.G.: Naturalistic Inquiry. Sage, Thousand Oaks (1985)
23. Lowell, V.L., Moore, R.L.: Developing practical knowledge and skills of online instructional design students through authentic learning and real-world activities. TechTrends. **64**(4), 581–590 (2020)
24. Lowenthal, P.R., Lomellini, A.: Accessible online learning: A preliminary investigation of educational technologists' and faculty members' knowledge and skills. TechTrends. **67**(2), 384–392 (2023)
25. Martin, F., Ritzhaupt, A.D.: Standards and competencies for instructional design and technology professionals. In: McDonald, J.K., West, R.E. (eds.) Design for Learning: Principles, Processes, and Praxis, pp. 249–258. EdTech Books, USA (2020)
26. McDonald, J.K., Jackson, B.D., Hunter, M.B.: Understanding distinctions of worth in the practices of instructional design teams. Educ. Technol. Res. Dev. **69**(3), 1641–1663 (2021)
27. Muljana, P.S., Luo, T.: Pursuing professional learning by using social media: how do instructional designers apply self-regulated learning? Res. Learn. Technol. **31** (2023)
28. North, C., Shortt, M., Bowman, M.A., Akinkuolie, B.: How instructional design is operationalized in various industries for job-seeking learning designers: engaging the talent development capability model. TechTrends: linking research & practice to improve learning. TechTrends. **65**(5), 713–730 (2021)
29. Page, M., et al.: The PRISMA 2020 statement: an updated guideline for reporting systematic reviews. BMJ. **372**(71), 1–9 (2021)
30. Ramani, S.: Background and skill set of instructional designers questioned in the OPM-University business model. Int. J. Teach. Learn. Higher Educ. **34**(1), 25–35 (2022)
31. Randolph, J.: A guide to writing the dissertation literature review. Pract. Assess. Res. Eval. **14**(13), 1–13 (2009)
32. Raynis, M.: Analysis of instructional design job announcements. Curr. Issues Emerg. eLearning. **4**(1), 166–223 (2018)
33. Richey, R.C., Fields, D.C., Foxon, M.: Instructional Design Competencies: the Standards, 3rd edn. ERIC Clearinghouse on Information and Technology, New York (2001)
34. Ritzhaupt, A.D., Martin, F., Pastore, R., Kang, Y.: Development and validation of the educational technologist competencies survey (ETCS): knowledge, skills, and abilities. J. Comput. High. Educ. **30**, 3–33 (2018)

35. Ritzhaupt, A.D., Kumar, S.: Knowledge and skills needed by instructional designers in higher education. Perform. Improv. Q. **28**(3), 51–69 (2015)
36. Ritzhaupt, A.D., Martin, F.: Development and validation of the educational technologist multimedia competency survey. Educ. Technol. Res. Dev. **62**(1), 13–33 (2014)
37. Ritzhaupt, A.D., Martin, F., Daniels, K.: Multimedia competencies for an educational technologist: A survey of professionals and job announcement analysis. J. Educ. Multimed. Hypermed. **19**(4), 421–449 (2010)
38. Shadiev, R., Liu, T., Hwang, W.Y.: Review of research on mobile-assisted language learning in familiar, authentic environments. Br. J. Educ. Technol. **51**(3), 709–720 (2020)
39. Sharif, A., Cho, S.: 21st-century instructional designers: bridging the perceptual gaps between identity, practice, impact and professional development. RUSC. Univ. Knowl. Soc. J. **12**(3), 72–85 (2015)
40. Sims, R.C., Koszalka, T.: Competencies for the new-age instructional designer. In: Spector, J.M., Merrill, M.D., Merrienboer, J., Driscoll, M.P. (eds.) Handbook of Research on Educational Communications and Technology, pp. 569–575. Taylor and Francis, New York (2008)
41. Slade, C., McGrath, D., Greenaway, R.: Professionalization in academic development: exploring learning designer roles in a changing higher education sector. In: Advance HE Teaching & Learning Conference 2018: teaching in the spotlight: learning from Global Communities, UK (2018)
42. Slagter van Tryon, P.J., McDonald, J., Hirumi, A.: Preparing the next generation of instructional designers: A cross-institution faculty collaboration. J. Comput. High. Educ. **30**(1), 125–153 (2018)
43. Smith, P.L., Ragan, T.J.: Instructional design. Wiley, New York (1999)
44. Sønderlund, A.L., Hughes, E., Smith, J.: The efficacy of learning analytics interventions in higher education: A systematic review. Br. J. Educ. Technol. **50**(5), 2594–2618 (2019)
45. Stefaniak, J., Baaki, J., Hoard, B., Stapleton, L.: The influence of perceived constraints during needs assessment on design conjecture. J. Comput. High. Educ. **30**(1), 55–71 (2018)
46. Sugar, W.A., Moore, R.L.: Documenting current instructional design practices: towards a typology of instructional designer activities, roles, and collaboration. J. Appl. Instruct. Des. **5**(1), 51–59 (2015)
47. Sugar, W., Hoard, B., Brown, A., Daniels, L.: Identifying multimedia production competencies and skills of instructional design and technology professionals: an analysis of recent job postings. J. Educ. Technol. Syst. **40**(3), 227–249 (2012)
48. Toker, S.: The progress of 21st-century skills throughout instructional design projects: A quasi-experimental comparison of rapid prototyping and dick and Carey models. Educ. Inf. Technol. **27**(2), 1959–1992 (2022)
49. Villachica, S.W., Marker, A., Taylor, K.: But what do they really expect? Employer perceptions of the skills of entry-level instructional designers. Perform. Improv. Q. **22**(4), 33–51 (2010)
50. Wakefield, J., Warren, S., Mills, L.: Traits, skills, & competencies aligned with workplace demands: what today's instructional designers need to master. In: Resta, P. (ed.) Proceedings of SITE 2012–Society for Information Technology & Teacher Education International Conference, pp. 3126–3132. Association for the Advancement of Computing in Education (2012) https://www.learntechlib.org/noaccess/40070/
51. Wang, X., Chen, Y., Ritzhaupt, A.D., Martin, F.: Examining competencies for the instructional design professional: an exploratory job announcement analysis. Int. J. Train. Dev. **25**(2), 95–123 (2021)
52. Wilson, B.G., Ozyer, A.: The role of graduate programs in fostering IDT identities: reflections on an emerging profession. J. Comput. High. Educ. **31**(3), 557–572 (2019)
53. Winterton, J., Delamare-Le Deist, F., Stringfellow, E.: Prototype Typology of Knowledge. Skillsand Competences. CEDEFOP, Thessaloniki (2005)

54. Xu, M., Yang, X., Stefaniak, J.: A design-based research study exploring pre-service teachers' instructional design decision-making for technology integration. TechTrends. **66**(6), 968–979 (2022)
55. Zhu, M., Basdogan, M., Bonk, C.J.: A case study of the design practices and judgments of novice instructional designers. Contemp. Educ. Technol. **12**(2), 1–19 (2020)

The Use of Photography, Through the Smartphone, for the Development of Creative Competence in the Education

Silvia Cáceres-Bazalar and María Chávez-Chuquimango(✉)

Universidad Peruana de Ciencias Aplicadas, Lima 15023, Peru
maria.chavez@upc.pe

Abstract. The present study aims to analyze the perception of secondary school's teachers at a public school located in Lima, Peru, on the use of photography through the smartphone as a tool for the development of creative competence in regular basic education. A qualitative approach with an ethnographic design was applied, in which 11 teachers were interviewed. For data processing, the thematic analysis was used, highlighting the importance of including photography in the educational curriculum as a strategy to promote creativity, strengthen interdisciplinary learning and enrich the school environment. Likewise, the findings were the following: visual narrative, responsible use of technology, critical perception, among others.

Keywords: Photography · Smartphone · Visual Education

1 Introduction

Currently, the exploration of visual teaching and learning encourages the process of activating knowledge through imaginative and creative thinking, with the aim of promoting photography as an integral tool for the visual exercise and the construction of learning [1, 2]. In fact, it is important that photography be contemplated from two different approaches: as a contemporary artistic expression in educational, training and research contexts; and as a reflection on the artistic learning process where the student can develop spontaneously [3, 4]. Consequently, those who develop a creative personality trait tend to face more complex challenges, which generates a greater interest in acquiring new knowledge [5, 6]

For example, in the regular basic education setting, between the ages of 13 and 14, formal education involves only the students and their efforts to make themselves heard [7]. Often, schools do not focus on children and adolescents as individuals, but on their role as "pupils", overlooking their knowledge, emotions, aspirations and feelings [8]. Encouraging creativity in the educational setting contributes significantly to the enrichment of academic and creative learning. This practice aims to develop various aspects that boost professional competencies [9]. It also contributes to the integration of creative methods and tools, such as photography and other art forms, to stimulate the comprehensive development of students, preparing them to face the challenges of the professional

A. Rocha et al. (Eds.): ICITS 2026, LNNS 1966, pp. 731–741, 2027.
https://doi.org/10.1007/978-3-032-27538-7_64

and academic environment with more complete and versatile skills [10]. Therefore, the openness of arts education should facilitate the design of educational projects aimed at forming emancipated, supportive, sensitive, critical and creative individuals, capable of taking control of their lives, creating their own narratives and feeling active agents in shaping their world and managing their environment [1]. The way to provide personal development and learning becomes under the activity of visual creativity [11]. To this end, photography, as a means of constructing and extracting complex meanings, encourages students to adopt a habitual practice of creativity [12].

Thus, photography and art education are closely interrelated, offering a valuable contribution to visual development, critical thinking, and skill execution in students [13]. Likewise, art education documents rural, academic, and technoscientific evidence for its objective development [14]. The relationship of photography in art education offers an innovative and contemporary approach that aims to provide new creative expectations on traditional teaching methods [15]. This development not only enriches learning processes and educational experiences but also transforms creative competencies in the student [16]. In today's professional environment, increased attention to creative work is required to foster such competence [17]. However, this competence is not widely developed due to the lack of recognition in the curricula, generating an imbalance between the creative competence and its respective needs [18]. Therefore, in promoting for educational development, digital technologies through mobile devices such as smartphone offer applicable opportunities for education in developing students' creative competence [19]. Since, with the use of mobile devices, students can capture, explore, and experiment with images in an immediate and accessible way [20].

At the international level, arts-based research indicates the exploration on the use of photography as a pedagogical tool to deepen teachers' methodology and initial practices to promote visual art [3]. Another study mentions that promoting photography as an artistic tool contributes to the development of creativity and visual literacy using mobile devices [19]. Likewise, the implementation of mobile devices in art teaching brings improvements in the academic field and a more concrete work in the methodological introduction to the student [21].

At the national level, the General Education Law No. 28044, the National Directorate of Initial and Primary Education (DINEIP) and the National Directorate of Secondary and Higher Technological Education (DINESST) have taken responsibility for developing the National Curricular Design of Regular Basic Education, seeking to integrate and articulate the curricular proposals of the Initial, Primary and Secondary Education levels, where learning is strengthened in terms of competence, capacity and attitude [22, 23]. At the same time, it should be adapted to the social, cultural and linguistic diversity of each region of the country [22].

Therefore, this study aims to analyze the perception of secondary school teachers at a public school, located in Lima (Peru), on the use of photography through the smartphone for the development of creative competence in their students. Thus, the use of photography, through the smartphone, represents pedagogical innovation between teaching and learning [24]. Likewise, photography as a means of artistic expression combines subjectivity and discipline in the student [20]. In addition, understanding teachers' perceptions and experiences provides a comprehensive view on the effectiveness in school methodology [25]. Therefore, this study focuses on answering the following research question, how do secondary education teachers perceive the use of photography as an artistic education tool, through the *smartphone*, for the development of creative competence in a public school, located in Lima (Peru)?

2 Methodology

This work is positioned in the interpretative paradigm, because it is based on understanding and interpreting the interactions and perceptions that encompass a certain social group [26]. Similarly, it is qualitative in nature, because it will allow understanding the perception of the interviewees regarding the research problem [26]. The design is ethnographic, because it seeks to understand and examine the ideas of a certain group of people with common interests to support the analysis of the study [27].

Eleven secondary school teachers from a public school located in Lima were selected due to their proximity to students in regular basic education [28]. The teachers at the public school were selected because there is a need to motivate students to develop their creative and artistic side to express their emotions and feelings; since 2024, mental health problems have increased in schoolchildren and university students. Also, it is sought to explore the social behavior in students using technologies such as smartphone [29]. Since, a recent study [30], indicates that most adolescents and young people use photography, through mobile devices, to express their emotions, feelings, etc. On the other hand, this study revolves around the exploration of creative competence in educational settings, where the development of activities and use of tools to strengthen this competence in students is suggested [31]. For the selection of the participants, convenience sampling was used, which allowed the selection of the interviewees and their availability to participate in this study [32].

Semi-structured interviews were applied as an instrument, with twenty-four questions divided into three parts: a) Elements of creative competence using photography and b) Changes generated using photography, through the smartphone, in art education and c) Use of the smartphone for the development of creative competence.

Prior to the start of the interviews, the instrument was validated through three pilot interviews to ensure the understanding of the questions by the participants. Between September and October 2024, a total of 11 semi-structured interviews were conducted. Interviews were conducted in virtual mode only, through the Zoom platform, with a duration between 18 to 78 min. In addition, all participants were provided with an information sheet addressing aspects related to the research, confidentiality and the handling of their personal data.

The data analysis was carried out using thematic analysis, which allows identifying and examining patterns within the qualitative data, allowing an exhaustive exploration of the data collected [33]. This analysis was based on the following steps: a) Transcription of the interviews in Word format. (b) Detailed and repeated reading to acquire a deeper understanding of the information. (c) Thematic coding and categorization to identify themes and categories. (d) Writing of results (Table 1).

Table 1. Characteristics of the participants

Participant	Sex	Age	Duration
E01	Male	48	26:52
E02	Female	47	28:38
E03	Female	39	18:36
E04	Female	64	38:50
E05	Female	42	38:50
E06	Female	33	38:50
E07	Female	41	38:00
E08	Female	30	53:17
E09	Female	34	37:40
E10	Male	56	78:36
E11	Female	30	24:57

Source: Own elaboration

3 Results and Discussion

A. What elements of creative competence are identified through the use of photography?

Visual Narrative: According to the comments of the interviewees, involving the use of photography with students allows them to communicate their ideas, tell stories through graphic elements and fosters the development of their visual narrative. Likewise, by experimenting and exploring symbols, images and visual metaphors, students strengthen their ability to express complex concepts. This type of activity not only stimulates creativity but also teaches students to express thoughts and emotions visually and to interpret abstract concepts. Expressing their identity through symbolic images strengthens students' ability to construct meaning and communicate indirectly, a fundamental skill for creating complex visual narratives in the future. According to Díaz et al. [34], the era of communication and image communicates the evident need for visual development to be critical and responsible consumers, capable of decoding visual messages in each cultural context. On the other hand, Capistran [35], indicates that regular teaching at the level of regular basic education helps to improve the level of educational inclusion, emotional intelligence and the development of awareness through art and visual narrative, predominating the importance of teamwork, the democracy of their own decisions and promoting coexistence and collective work in students from different cultures [36]. In conclusion, the intervention of photography generates the creation of visual narratives, which fosters creativity, emotional expression and development of skills to understand and communicate complex concepts.

> "It is very important to know how to handle the creative side, we talk about soft skills that I consider fundamental for the development and learning of a young

person who is in full growth, especially his personality and, in fact, it is the part of the development of creativity" (P11).

"The competencies are going to go from lower to higher. With photography, education is going to vary according to the creativity that each one has and under what approach. Because an education of children, adolescents or young people is not the same." (P09)

Critical Perception: For the interviewees, critical perception is based on the ability to analyze and evaluate information and situations in a reflective manner. In the educational context, it is observed that students who participate in creative activities, such as the design of environmental care campaigns, develop critical thinking skills. Students identify problems and propose solutions, which implies a deep analysis of their environment. This critical capacity is essential for students to go beyond consuming information and become agents of change in their communities. Fostering critical perception through creative activities can help students develop a sense of social responsibility and a greater understanding of the problems they face. Students not only consume visual information but also engage in a process of critical reflection on the content they are creating. According to Briones [36], critical perception encourages students in decision making, project creation, self-efficacy, and collaborative teamwork. On the other hand, another study [37] mentions that creative competence is a support for the development of creativity, personal and professional talent. This enhances school management through critical judgment work [30]. In conclusion, critical perception and reflective thinking are fundamental skills that are significantly developed through creative educational activities. These projects not only allow students to identify and analyze social problems but also turn them into active agents of change from their own environment by questioning, reflecting and proposing innovative solutions.

"As a teacher, I told the students that they should interpret an image from their perspective. So, the exciting thing for them, is to be in control of their activity and to have the freedom to create and support their idea according to their critical judgment" (P08).

B. What are the changes generated by the use of photography, through the smartphone, in art education?

Responsible Use of Technology: According to interviewees, it is essential to promote the responsible use of technology in education. The use of cameras and smartphones can be a valuable pedagogical tool. However, if not properly managed, these devices can become sources of distraction for students. Therefore, it is crucial to establish rules and limits that ensure that their use is focused and intended for learning. This action encourages critical reflection on the content that constantly creates, publishes and consumes online. In this way, influencing the responsible use of technology not only improves educational outcomes, but also contributes to the development of a critical awareness of its impact on personal life, society, and particularly the academic environment. According to Alonzo-Yaranga et al. [28], the importance of quality education is the socioeconomic development and advancement of society at all levels, providing students with technological tools, where they develop skills to grow intellectually, emotionally and socially. However, a study [38], mentions that smartphones have been identified as possible distractors in situations that demand concentration and attention. In conclusion, the responsible use of technology in the educational environment is essential to maximize its pedagogical benefits and avoid distractions that may affect the learning process. The use of smartphones can be a valuable tool for developing technical skills, such as capturing images, and encouraging critical reflection on the content that students create, publish and consume online. However, it is essential to establish clear rules that regulate their use, ensuring that the devices are focused on learning and not on distracting activities.

> "In some activities, I ask students to participate in building concepts and they use their cell phones to search for images that represent the topic. In that way, they do field work with the use of technology, where they not only capture images but also investigate and search for information for class projects." (P07)

Autonomous Learning. According to interviewees, photography encourages autonomous learning by allowing students to choose their own themes and approaches. Inviting students to select images that are meaningful to them and that reflect their perspective of the world, thus giving them a sense of ownership over their learning process. This approach promotes both self-exploration and critical reflection by encouraging students to think carefully about the message they wish to convey through their images. In addition, participation in individual and team projects helps them develop collaborative and communication skills, elements that enrich their educational experience. According to Antunes & Roque [39] mention that, for the development and management of a dynamic and collaborative educational class, professionals trained with new technologies and learning methodologies are required, where the teacher is perceived as a learner and mediator during the process. In addition, another study [19], indicates that photography provides autonomy to the student's development, since, through his lens, he captures the images he wants to communicate, which he implements in his academic activities. In conclusion, photography not only fosters students' creativity and personal expression, but also promotes autonomous learning, allowing them to make decisions about their academic projects. Therefore, while photography and technologies offer educational opportunities, it is essential to accompany them with a reflective and prepared approach to overcome the challenges presented in the current educational context.

> "When I worked in elementary education, I had the challenge of using photography as a tool to achieve concentration and patience in students. Since, I assigned them activities that they had to validate through images. Then, these group works allowed them to delegate, focus and concentrate to know what kind of image they wanted to capture to present the concepts seen in the session" (P04).

C. Use of the smartphone for the development of creative competence

Identity Building: Interviewees emphasize that photography goes beyond simply capturing images. It is about finding the right angle to communicate a message. In this process, students' sensitivity and emotions are stimulated, helping them to manage their feelings. Through photography, students can explore their identity and express their emotions, which contributes to their personal and emotional development. In addition, it is noted that experiences captured in images can enrich their learning and their understanding of the world around them. According to Surva et al. [40], identity exploration provides positive impacts in areas such as personal development, social cohesion, empowerment and community self-determination. This is related to the use of photography, where students explore their identity through images and generate a creative process when sharing their messages [30]. In conclusion, photography not only acts as a technical tool, but also to promote the personal, educational and emotional development of students. Through this process, educators highlight how photography allows students to explore their identity and share it through artistic works.

> "Many schools have installed all kinds of courses that help enhance certain soft skills. For example, through photography, students explore, investigate and share messages through images. This allows them to get to know the student's personal look, that is, their identity". (P11)

Use of Technology in Learning: The interviewees mentioned that, by using photography, through the smartphone, technical and social skills can be developed in students. Since, the student can capture and edit images easily, which is essential to enrich and streamline the visualization of educational content. However, they stress the importance of using technology responsibly, avoiding distractions and focusing attention on learning. Likewise, students develop social skills by communicating their ideas and identity through artistic works such as photographic projects. Therefore, relationship between photography and education enhances the technical and social skills of students, because it gives them the opportunity to participate in creative projects that can be presented and shared in the classroom, promoting collaborative and dynamic learning. According to Ramos and Izquierdo [41], the development of technical skills that students achieve with the use of photography, not only allows them to express themselves, but also to use it as an educational support tool in their projects. Likewise, a study [15], indicates that, at present, it is vital to integrate technology into the academic process to achieve autonomous, interactive and innovative learning. In conclusion, the use of smartphones transforms education, because students use them to learn, socialize and express themselves through the content they share and generate.

> "Nowadays, students use applications, search engines, among others, to carry out their activities through their cell phones. We cannot push technology away; we have to use it responsibly." (P05)

4 Conclusions

The results obtained highlight the relevance and potential of photography, as a tool for artistic expression, to promote critical thinking, creativity and reflection in the teaching-learning process, especially in the context of regular basic education.

Throughout the interviews, it was evident that teachers recognize the importance of photography as a medium that not only enriches the academic experience but also allows students to explore and express their emotions, thoughts and perspectives visually. The ability of smartphones to capture, edit and share images immediately is considered a valuable resource that promotes greater interaction and participation of students in their learning, opening spaces for creativity and autonomous thinking.

Likewise, the integration of photography in regular basic education is perceived as an innovative approach that transforms the traditional teaching methodology, offering new ways of approaching creative competencies. Teachers pointed out that, by using mobile devices, students not only develop technical skills related to photography, but also expand their cognitive, emotional and social competencies. This generates a more dynamic and participatory environment, favoring the inclusion of different ways of learning and adapting to the diverse needs of students.

On the other hand, the study has two limitations: a) there was a small sample of interviews, considering that the school had a small number of teachers; b) being a qualitative study, it is not possible to generalize the results obtained.

Finally, it is suggested that future studies should include an analysis of educational policies that integrate the use of digital technologies in the school curriculum, especially in arts education. Photography, as an accessible and versatile medium, has the potential to transform the educational process, allowing students to develop key competencies for their integral formation and their future participation in an increasingly visual and digital world.

Acknowledgments. A la Dirección de Investigación de la Universidad Peruana de Ciencias Aplicadas por el apoyo brindado para realización de este trabajo de investigación a través del incentivo UPC-EXPOST-2025-1.

Disclosure of Interests The authors have no competing interests to declare that are relevant to the content of this article.

References

1. Castro-Fernández, B., López-Facal, R.: Emotional cartography of everyday heritage via photography. Sustainability (Switzerland). **13**, 1–27 (2021). https://doi.org/10.3390/su13094764
2. Janiaski, F., Fagundes, T.: Clases de arte en Rondonópolis: Una construcción dialógica sobre el ensenando arte en la escuela. Revista Brasileira de Educacao. **28**, 1–18 (2023). https://doi.org/10.1590/S1413-24782023280096
3. Mesías-Lema, J., Ramon, R.: La fotografía en la investigación educativa basada en las artes. IJABER. International journal of arts-based Educ. Res. **1**, 7–22 (2021). https://doi.org/10.17979/ijaber.2021.1.1.7618
4. Oliver-Barceló, M., Ferrer-Ribot, M.: Educación artística y transformación social. Análisis de los currículums iberoamericanos de educación infantil. Artseduca. **32**, 35–46 (2022). https://doi.org/10.6035/artseduca.6307
5. Ramos, N.: El aprendizaje-servicio para el desarrollo de la motivación en educación artística. Estudio de caso en el área artística de Educación secundaria. Artseduca. **31**, 79–93 (2021). https://doi.org/10.6035/artseduca.5814
6. Zhou, K.: The influence of creative personality and goal orientation on innovation performance. Front. Psychol. **12**, 1–10 (2021). https://doi.org/10.3389/fpsyg.2021.634951

7. Pereira, S., Fillol, J., Moura, P.: El aprendizaje de los jóvenes con medios digitales fuera de la escuela: De lo informal a lo formal. Comunicar. Revista científica de educomunicación. **27**, 41–50 (2019). https://doi.org/10.3916/C58-2019-04
8. Fernández-Rodrigo, L., Gutiérrez-Ujaque, D.: Promoción de la resiliencia y educación artística en centros de máxima complejidad: vínculos y orientaciones desde la literatura. Revista Complutense de Educación. **33**, 237–246 (2021). https://doi.org/10.5209/rced.73990
9. Gargallo, B., Sahuquillo, P., Verde, I., Almerich, G.: Volver ¿Qué ocurre cuando los profesores utilizan métodos centrados en el aprendizaje? Efectos en los enfoques de aprendizaje, en las capacidades del alumno y en su percepción del entorno de aprendizaje. Revista de Educación. **1**, 163–197 (2018). https://doi.org/10.4438/1988-592X-RE-2018-382-396
10. Kawar, L.: Aprender la gobernanza de la migración: la representación y recepción de las normas internacionales en los programas de desarrollo de capacidades. Rev. Eur. Migr. Int. **40**, 41–64 (2024). https://doi.org/10.4000/remi.25320
11. Guervós, J.: Procesos cognitivos, discurso y control social. Pragmalinguística. **31**, 503–519 (2023). https://doi.org/10.25267/Pragmalinguistica.2023.i31.21
12. Lafaire, A., Kuismin, A., Moisander, J., Grünbaum, L.: Interspace for empathy: engaging with work-related uncertainty through artistic intervention in management education. Cult. Organ. **28**, 227–244 (2022). https://doi.org/10.1080/14759551.2022.2029442
13. Silveira, H.: Utilizando la fotografía en la historia de la ciencia y de la tecnología. Fotocinema, Revista científica de cine y fotografía. **1**, 113–138 (2023). https://doi.org/10.24310/fotocinema.28.2024.17609
14. Susperregui, J.: Ligeramente fuera de foco. Análisis de la ambigüedad en la fotografía de Robert Capa. Commun. Soc. **37**, 109–124 (2023)
15. Caeiro-Rodríguez, M., Murillo-Ligorred, V., Revilla, A., Ramos-Vallecillo, N.: Ecosistema de la Educación Artística en Red: una revisión de los antecedentes, las posibilidades y las perspectivas en la era digital. Revista Complutense de Educación. **33**, 679–690 (2021). https://doi.org/10.5209/rced.76499
16. Hernández, C.: Tareas propias de su sexo: costura y educación de la mujer en la España del siglo XIX. Domesticidad, profesión artística y construcción de género. Historia y Memoria de la Educación. **17**, 343–372 (2023). https://doi.org/10.5944/hme.17.2023.32419
17. Elkheir, Z., Abdul, A.: Aplicaciones de aprendizaje móvil diseño de conceptos y desafíos: encuesta. Res. J. Appl. Sci. Eng. Technol. **10**, 438–442 (2015). https://doi.org/10.19026/rjaset.10.2509
18. Estrada-Araoz, E., et al.: Satisfacción laboral y compromiso organizacional en docentes peruanos de educación básica regular: Un estudio correlacional. Salud, Ciencia y Tecnologia - Serie de Conferencias. **3**, 1–9 (2024). https://doi.org/10.56294/sctconf2024709
19. Pérez, J., Bedoya, C.: La fotografía como herramienta para el desarrollo de la creatividad y la alfabetidad visual: investigación en el aula. Kepes. **16**, 377–404 (2020). https://doi.org/10.17151/kepes.2019.16.20.14
20. López-Noguero, F., Romero-Díaz, T., Gallardo-López, J.: Smartphone como herramienta de enseñanza-aprendizaje en educación superior en Nicaragua. Revista Iberoamericana de Educación a Distancia. **26**, 307–330 (2023). https://doi.org/10.5944/ried.26.1.34016
21. Redifer, J., Bae, C., Zhao, Q.: Self-efficacy and performance feedback: impacts on cognitive load during creative thinking. Learn. Instr. **71**, 1–11 (2021). https://doi.org/10.1016/j.learninstruc.2020.101395
22. Martens, M., Miranda, L., Montané, A., Burga, E., Rivas, M. del C.: Currículo Nacional de la educación básica., https://www.minedu.gob.pe/curriculo/pdf/curriculo-nacional-de-la-educacion-basica.pdf, (2016)
23. Sota, J., et al.: Diseño curricular nacional de educación básica regular., https://www.minedu.gob.pe/normatividad/reglamentos/DisenoCurricularNacional.pdf, (2005)

24. Pascuas-Rengifo, Y., García-Quintero, J., Mercado-Varela, M.: Dispositivos móviles en la educación: tendencias e impacto para la innovación. Revista Politécnica. **16**, 97–109 (2020). https://doi.org/10.33571/rpolitec.v16n31a8
25. Giménez-Dasí, M., Quintanilla, L.: "Competencia" social, "competencia" emocional: Una propuesta para intervenir en educación infantil. Infancia y Aprendizaje. **32**, 359–373 (2009). https://doi.org/10.1174/021037009788964222
26. Creswell, J., Creswell, J.: Diseño de la investigación, vol. 1, pp. 1–320. SAGE Publications (2022)
27. Gallardo, L., Toledo, M.: Etnografía para proyectos arquitectónicos: inclusión de la perspectiva del habitante. Arquitetura revista. **16**, 1–20 (2020). https://doi.org/10.4013/arq.2020.162.02
28. Alonzo-Yaranga, L., et al.: Intervención en Liderazgo Directivo: Estrategias y Resultados en Educación Pública de Tambopata. Salud, Ciencia y Tecnologia - Serie de Conferencias. **3**, 1–8 (2024). https://doi.org/10.56294/sctconf2024778
29. Evaristo, E.: Gestión curricular para mejorar la practica docente en la comprensión de textos del área de comunicación en la institución educativa., https://repositorio.umch.edu.pe/handle/20.500.14231/2837, (2018)
30. Ávata, M., Ponce, F., Cordero, F., Palacios, F.: Una gestión educativa innovadora permite la construcción de procesos de enseñanza inclusivos, de calidad y creatividad. Sapienza. **3**, 310–333 (2022). https://doi.org/10.51798/sijis.v3i2.335
31. Falcon, L., Serguei, F.: Escritura Creativa y podcasting en competencias comunicativas de estudiantes de secundaria, Ica 2022., https://repositorio.ucv.edu.pe/handle/20.500.12692/112932, (2023)
32. Otero-Prevost, L., et al.: Propuesta para obtener el tamaño de muestra óptimo de plagas con exceso de ceros. Rev Mex De Cienc Agric. **15**, 1–12 (2024). https://doi.org/10.29312/remexca.v15i1.3618
33. Braun, V., Clarke, V.: Supporting best practice in reflexive thematic analysis reporting in palliative medicine: a review of published research and introduction to the reflexive thematic analysis reporting guidelines. Palliat. Med. **38**, 608–616 (2024). https://doi.org/10.1177/02692163241234800
34. Díaz, Á., Dacuña, I., Ferreira, C., Mesías, J.: Género, inclusión y narrativa visual en monstruo rosa: comunicar para educar en la diversidad. CIC. Cuadernos de Información y Comunicación. **26**, 175–190 (2021). https://doi.org/10.5209/ciyc.75809
35. Capistrán, R.: Escuelas musicales: Buenas prácticas docentes en centros de Primaria y Secundaria que educan a través de la música. Octaedro. Revista Electrónica de LEEME. **1**, 139–143 (2022). https://doi.org/10.7203/LEEME.49.21751
36. Briones, C.: Programa de empoderamiento emocional a personas de 40 a 45 años de edad con rezago estudiantil Yaguachi-Ecuador., https://repositorio.ucv.edu.pe/handle/20.500.12692/128240, (2023)
37. Romero-García, C., Suárez-Ortega, M.: El desarrollo de competencias emprendedoras desde la acción profesional orientadora en la universdad. Bordón. Revista de Pedagogía. **76**, 145–167 (2024). https://doi.org/10.13042/Bordon.2024.101120
38. Centella-Centeno, D., Vidal, J., Arcata, E., Huamani, V.: Detrás de la pantalla: Investigando la Adicción a Redes Sociales y el Estrés como factores explicativos de la Distracción por Teléfonos Inteligentes en jóvenes y adultos. PsiqueMag. **13**, 14–22 (2024). https://doi.org/10.18050/psiquemag.v13i1.2789
39. Antunes, F., Roque, M.: Aprendizajes en proceso de co-creación: eficacia de una experiencia educativa en la enseñanza superior. European Public Soc. Innov. Rev. **9**, 1–18 (2024). https://doi.org/10.31637/epsir-2024-601

40. Surya, R., et al.: Exploración del impacto de la identidad social y la dinámica de equipo en el comportamiento de búsqueda de ayuda en materia de salud mental entre atletas universitarios: un estudio cualitativo. Retos. **58**, 426–434 (2024)
41. Ramos, D., Izquierdo, J.: Ejercicios para el desarrollo de habilidades ortográficas en estudiantes de las carreras técnicas. Salud, Ciencia y Tecnología. **2**, 1–8 (2022)

Photographic Language and Peruvian Gastronomy in the Digital Era

Génesis Quispe-Coronel and María Chávez-Chuquimango(✉)

Universidad Peruana de Ciencias Aplicadas, Lima 15023, Peru
maria.chavez@upc.pe

Abstract. This study focuses on highlighting how the use of photographic language is used to promote the consumption of Peruvian gastronomy in social networks. This is due to the need to know how to communicate through images in the gastronomic field, since there has been a lack of studies in this area. For this purpose, a qualitative approach was applied, with an ethnographic design, conducting 15 interviews with specialists with more than 5 years of experience in the gastronomic industry. The following findings were identified: use of narrative, color, composition, framing and lighting.

Keywords: Gastronomy · Photographic Language · Social Networks

1 Introduction

Gastronomic photography proposes multisensory effects, which makes it a multimedia phenomenon that is present in the consumption behavior of society, as well as in corporate and advertising communication [1, 2]. In addition, active advertising by companies and brands are given through digital marketing and communication strategies, where content is created by professionals and consumers distributing on social networks such as Instagram [2, 3]. The public no longer only promotes data in the communication process, it is also an essential piece of reproduction and generation of content [4, 11, 12]

The main purpose of representing gastronomy through social networks by experts and users is the passion for sharing and communicating to the public about the various Peruvian dishes. Since, content about food preparation and plating that reaches consumers is disseminated [5, 13, 14]. These images promoted on social networks such as Instagram, Pinterest and Twitter have been used as a critique to explore and share knowledge. Likewise, they have become ubiquitous in daily visual interactions. This is due to the use of digital communication strategies, which promote the consumption of products [6, 15, 16].

At the international level, it is possible to observe cases on the use of photographic language to promote the consumption of gastronomy through Instagram. First, Tifentale & Manovich [17] present us with three types of photography that exist on Instagram: informal, professional and designed. Each of these types of images, are related to the photographic culture of the 20th century onwards. The photographs professional and

A. Rocha et al. (Eds.): ICITS 2026, LNNS 1966, pp. 742–752, 2027.
https://doi.org/10.1007/978-3-032-27538-7_65

designed are examples of what Alise Tifentale calls "competitive photography", the authors of these photographs bet on an aesthetic of "good photography" taking into account a correct composition and visual language, being aware of the quality and order of content when posting on Instagram; unlike the "informal photographs", which do not have any of these parameters and is based on "*Lifestyle*" looking at the American culture [17, 18]. Secondly, the case is presented in Spain, where they generate a great importance to food *influencers* with respect to Instagram. This exemplifies a characteristic phenomenon of food, which is to relate and share food with the aim of generating an emotional bond capable of creating loyalty among fans with gastronomy, where they get to promote a consumption behavior through the photographs that they publish without the need to compose with a correct photographic language and visual structure [19, 20]. Thirdly, in the United States, a Maru/Matchbox* study conducted with people aged 18 to 34 concluded that 69% of *millennials* take a photo or record a video of their food before eating it, which generates that restaurants use this trend to build an "Instagram-compatible" environment, from the decoration to the lighting and the way the food or cocktail is presented with the intention that customers post photos on their social networks [8, 21].

At the national level, cases are observed on the use of photographic language to promote the consumption of Peruvian gastronomy through Instagram. First, a study conducted in a restaurant in Trujillo is presented, it is identified that it is necessary to have a relationship between the use of gastronomic photography and the formation of brand identity among consumers, taking into account the elements of photographic language, such as composition, in order to communicate a clear message, striking and concise [22] Likewise, the case is presented at the photographic level and *influencers* such as Cholo Mena and his level of credibility before the cases he exposes in social networks, he shows Peruvian food as a cultural heritage, which is spread mostly on Instagram, because it is the most visual social network [23]. Finally, there is the case of the development of a Peruvian creative company, which comments that to carry out a correct management of the restaurant at the communicational and marketing level, it is necessary to emphasize the visual identity starting from the brand manual, photographic style, photographic language, art direction and the process of conceptualization of a photo shoot, it is important to present a good brand image having clear the concept to communicate assertively to the target audience [24].

Therefore, this study aims to analyze the perceptions of the use of photographic language to promote the consumption of Peruvian gastronomy through social networks such as Instagram, according to the perceptions of specialists in culinary photography. In a recent article [25], it is shown in a panoramic way how the evolution of food production in Peru has been and the potential that exists in the country by producing organic food of high quality, variety and diversity that have allowed restaurants to flourish. In addition, in 2024, Peru was recognized as one of the 10 best gastronomic destinations in the world by National Geographic magazine, which highlights the richness, diversity and creativity of Peruvian cuisine [26]. To communicate assertively and generate consumption motives within Peruvian gastronomy, it is necessary to understand the concept and generate a correct communicational tone for each restaurant, which is necessary within the role of marketing, branding and visual identity within the gastronomic field to achieve effective communication [27, 28]. In a research work conducted for the Catholic University of Salta, it is stated that there is an effect of advertising photography in the sales environment, these increase as the product is frequently displayed through photographs in social networks. This is thanks to employ a correct composition where the concept of the restaurant is channeled through professional photographers [29, 30]. Finally, this study seeks to answer the following question: What elements of photographic language are used to promote the consumption of Peruvian gastronomy in social networks?

2 Methodology

The present work is positioned in the interpretative paradigm with a qualitative approach, because it seeks to know the views, perceptions, experiences and meanings of individuals within a specific and dynamic context [31]. The design is ethnographic, since it allows us to know and analyze the thoughts of a group of people with similar characteristics who belong to a specific social context [32]. For this purpose, 15 specialists, with an age range between 30 and 50 years, from the gastronomic field in Lima, Peru, were interviewed. They have experience in culinary art, which is related to gastronomic photographic language, visual identity and the management of gastronomic photography [33]. Recall that Peruvian gastronomy is considered cultural heritage and identity from 2012 to the present day before UNESCO [34]. Likewise, the use of photographic language in gastronomic businesses has generated a dialogue on its exhibition in social networks, because the visual identity of brands is built on these platforms [35, 36]. In addition, the gastronomic category is developed within areas such as marketing, advertising and brand image, where photography has the role of persuading and capturing the attention of users [37]. The elements of photographic language allow the photo to contain an equivalent meaning between the structure of visual language and verbal language [38].

On the other hand, convenience sampling was performed, and data was collected through semi-structured interviews, since this flexible instrument allows for an in-depth conversation that facilitates the discovery of key information and personal experiences by the interviewees [39]. The instrument has 21 questions focused on the use of photographic language to promote the consumption of Peruvian gastronomy from social networks such as Instagram. Preliminarily, the instrument was validated through three pilot interviews to ensure the understanding of the questions. In addition, Please check and confirm if the inserted citations of Fig. 1 is correct. If not, please suggest an alternate citation. interviewee was given an information sheet that informed them about aspects related to the research, confidentiality, and the handling of the information (Table 1).

During the months of August and October 2024, a total of 15 semi-structured interviews were conducted, lasting between 40 and 60 min. These were conducted virtually and in person using the Zoom platform. For data processing, the thematic analysis was applied in four stages: (a) Transcription of the interviews in Word format; (b) Meticulous reading of the information; (c) Coding and thematic categorization; (d) Construction of the results [40].

Table 1. Characteristics of the participants

Participant	Sex	Age	Duration
E01	Male	37	45:38
E02	Male	25	32:17
E03	Male	38	48:06
E04	Male	31	25:04
E05	Male	30	15:02
E06	Male	25	50:82
E07	Male	38	58:14
E08	Male	32	47:35
E09	Male	45	41:59
E10	Male	39	37:22
E11	Male	30	46:00
E12	Male	35	61:00
E13	Male	24	23:55
E14	Male	26	28:36
E15	Male	28	19:42

Source: Own elaboration

3 Results and Discussion

This section answers the research questions posed for this study.

A. What elements of photographic language are used to promote the consumption of Peruvian gastronomy on the social network Instagram?

Lighting: According to the interviewees' comments, lighting is one of the most important elements in gastronomic photography because it influences the perception of the food and the product within the visual appeal. The specialists mentioned that proper lighting highlights the details, texture and colors of each dish, making them look more appetizing and natural. This finding coincides with another study [41], commenting that it is important to have a reflection before a photographic shot, so it is necessary to have deep knowledge and know how to illuminate, either with natural light or studio light. Lacking this basic knowledge weakens the professional process of the photographic session. Likewise, an article [42] comments that gastronomy can become ephemeral if it does not have the support of photography and audiovisual content. For this to stand out, it is important to use photographic elements accurately, such as lighting, basic for the photographic and audiovisual world.

"Lighting is the main thing. If you don't know how to light, you're not going to achieve anything with that plate of food. For example, I'm from the old school, I think your photo must come out as is. What you do in digital retouching should be basic, but you must know how to communicate your dish through an image" (P09).

In addition, the interviewees mentioned that lighting helps to manage shadows and reflections in an appropriate way, allowing the dishes not to look flat, which highlights their quality and provokes an emotional response in the viewer. This finding coincides with the following study [43], which comments on the need to develop creativity and sensitivity with respect to photography, because lighting allows lights and shadows to play an important role. With this, a creative, attractive and voluminous image is obtained, which is ideal in most cases. In conclusion, it is considered necessary to use lighting correctly to have an attractive image and highlight details, textures and colors. This can be done if creative photo shoots are developed. This will help the brand and the gastronomic industry, allowing the dishes not to be ephemeral or to be written in letters and in the memories of the diners, but to be recorded correctly in the form of images.

"A clean, minimalist image perhaps? Which is very fashionable. The colors and the lighting. I think by fulfilling these points in the right way, the image will be able to capture and persuade the consumer." (P01)

Framing: According to those interviewed, framing is another important photographic element because it determines the way the dish is presented and directs the viewer's attention. Specialists comment that a good framing can highlight the most important details and the art of plating by the chef. Alaei et al. [44] comment that image processing is becoming faster and faster, so good photographs with creative framing are needed to attract attention and be retained in the memory. In addition, Guamán-Guevara [45], examines the general perception of gastronomy and comments that plating or presentation is an attribute that should be considered in gastronomic businesses, which should highlight freshness, authenticity and flavor. The purpose is to have differentiating attributes, which develop positioning strategies, branding and sensory marketing, so it is necessary to use good photography and video to highlight these attributes.

"They capture the users' attention, when you show the true essence of the image and talk about the product as the protagonist (…) and overall, with the final dishes, you show a clean, clear, well-framed and story-driven photograph." (P08)

On the other hand, the framing also influences the perception of the size, proportion and context of the dish, i.e. what it seeks to convey through a story; and with it, an emotion. This finding coincides with Barrientos-Báez et al. [46], because it comments that the new digital paradigm is applied in the audiovisual and the new transmedia narratives. Both represent a new communicational phenomenon, being one of its bases the framing, which allows transmitting a story. Also, other authors [47], comment that brands have become vehicles of communication through stories, emotional and positive dialogues that they share with their audience. Therefore, it is necessary to use the framing correctly to have a clear image that highlights the most important details and can capture the viewer's attention. This can be achieved if the framing is developed with creativity.

Therefore, it is necessary for the photographer to work hand in hand with the chef to clearly understand the concept of each dish to tell a story.

> "All this is achieved if the image has a clear message, a purpose, a story, good framing, lighting, etc." (P14).

Narrative: According to the comments of the interviewees, narrative or *storytelling* is essential in gastronomic photography because it tells a story that connects emotionally with the viewer, either through sensations or memories. This finding coincides with Kaufman [48],who comments on the importance of *storytelling* or narrative in commercial communication, since it has a wide and varied range of applications where storytelling adds value and generates bonds with the customer, which provides better performance at both personal and organizational levels.

Likewise, a study [49], comments that there is a great innovation in the form of communication, where narrative and creativity stand out, which allows creating high impact images and connecting emotionally with the recipients.

> "With the images, I can transmit more familiar emotions, that take you back to a memory. That has helped me, and I perceive it with my clients. For example, when they identify with a dish, they enjoy it, they look at it, they photograph it. I feel happy, it's not just a dish, it's a journey of emotions." (P12)

Likewise, the interviewees commented that, through elements such as the setting, arrangement of objects and details, it is possible to evoke memories and transmit the origin of the dish, generating an emotional bond with the consumer. This finding coincides with other results [50], which comment on the existence of *visual* storytelling, where photos and videos are created through art direction, colors, archetypes and a good story, which generates a connection with the public alluding to their memories, sensations and attraction to visit the restaurant, bar or cafeteria. A recent case [51], indicates that the most attractive segment for visual narratives in gastronomic businesses is *foodies*, as they have a strong involvement with specialty food products and connect quickly with the visual narratives exposed in social networks by gastronomic businesses. Finally, this analysis concludes that, it is necessary to create and use the element of narrative or *storytelling* in gastronomic photography because of the innovative way to communicate, the desire to differentiate between brands and penetrate the consumer's mind. In addition, as there is a niche in the market called *foodies*, it opens new opportunities to create and generate content to promote gastronomic businesses.

> "One thing is to do your artistic gastronomic photography; and another thing is to dedicate yourself to gastronomic photography as a sales line. Because we do commercial photography, and there is a client in the middle, which we must satisfy to promote the restaurant's sales. So, you need to have the elements of photography as a concept, the narrative and what you want to say." (P09)

Color: According to the interviewees, the product should be highlighted by its colors because they are gastronomic photographs, such as fruits, vegetables and colorful products. In addition, a good image of the product attracts and captures the user's attention, so that it is consumed, photographed and disseminated in social networks. This finding coincides with a study [52], which reports on the Gestalt theory that influences visual appeal through photographic composition such as colors and shapes that persuade users. Which causes the consumer to photograph and disseminate in social networks. In addition, along the lines of Öztürk [53], he comments that the growth of culinary culture is booming; and, therefore, the awareness of gastronomic creativity is increasing both in the kitchen itself and in communication. Therefore, there is a great demand for photo shoots shared by culinary businesses on their digital platforms.

> "The color leads me to have a previous reference of the place. Through them, I can tell if it's a very hipster coffee shop, a very serious coffee shop for office people, a very casual coffee shop for neighborhood people. It can convey a lot of things to me through the colors." (P05)

Likewise, the interviewees commented that color highlights the quality, freshness and details of the product, since it activates the senses such as taste and sight by stimulating users. This finding coincides with Velasco et al. [54], who comment on the relationship between taste and color that exists in neuromarketing, reflected in gastronomic images, which are intertwined with memories. In addition, Kontukoski et al. [55], comment on the existence of associations between color and sensations with dishes. These arouse emotions that are rooted in the previous cultural and gastronomic experiences of each person. Since, color is not only related to aesthetics, but it can also awaken special meanings in a person's behavior.

As a result, together with the specialists and interviewees, it was concluded that color is a photographic element that is highly used through psychology and neuromarketing. It is rooted in the psychological processes and behavior of people because of their memories. It is important to use them properly in the photographic compositions according to the colors of the inputs to show the concept that each brand wants to develop, to generate an action in the consumer.

> "If it's a super friendly, super jovial brand, you're going to use a lot of colors and decorative elements. And if it is a more sober brand, you will use more neutral colors. Therefore, I feel that it will depend a lot on the audience you want to reach" (P14).

4 Conclusions

It is evident that the way of perceiving the use of photographic language to promote the consumption of Peruvian gastronomy through social networks is based on two important factors: the correct use of photographic language and visual identity. However, there is a lack in the level of demand for compliance with these, since not all gastronomic businesses work with professionals who have theoretical and practical knowledge of both factors. This decline has been improving, but it is still necessary to put more emphasis on these issues that encompass photography, video, advertising and marketing oriented to gastronomy.

Likewise, this study indicates that the most important elements of photographic language are lighting, framing, narrative and color. This is because Peruvian gastronomy is mainly based on the product and the story behind it. Peruvian products present different textures, shapes and colors that make it special. On the other hand, this study has limitations such as the reduced sample since there are not many professionals dedicated to visual communication in this field. In addition, the lack of publications on photography and gastronomy. Finally, for future studies, we recommend a quantitative analysis of the use of photography in the culinary field, the level of knowledge about the importance of photography in gastronomy and, in addition, to broaden the cultural context of this field. Finally, studies aligned with photography, gastronomy and tourism could be considered for future research.

Acknowledgments. A la Dirección de Investigación de la Universidad Peruana de Ciencias Aplicadas por el apoyo brindado para realización de este trabajo de investigación a través del incentivo UPC-EXPOST-2025-1.

Disclosure of Interests The authors have no competing interests to declare that are relevant to the content of this article.

References

1. Múnera, B.: Fotografía y comunicación: un espejo, una imagen, un ojo. Análisis. **53** (2021). https://doi.org/10.15332/21459169.6314
2. Schwarz, H.: Estudio Courret - Historia de la fotografía en Lima. Municipalidad Metropolitana de Lima, Lima (2017)
3. Gambetti, A., Han, Q.: Camera eats first: exploring food aesthetics portrayed on social media using deep learning. Int. J. Contemp. Hosp. Manag. **34**, 3300–3331 (2022). https://doi.org/10.1108/IJCHM-09-2021-1206
4. Hernández, Ó.: El proyecto fotográfico de las Naciones Unidas en América Latina (1945-1980). Una primera caracterización. Revista de historia de América., 269–303 (2023). https://doi.org/10.35424/RHA.165.2023.3372
5. Leal, M., Casais, B., Proença, J.: Tourism co-creation in place branding: the role of local community. Tour. Rev. **77**, 1322–1332 (2022). https://doi.org/10.1108/TR-12-2021-0542/FULL/XML
6. Caicedo, P., Gavilanes, E., Domínguez, F.: Innovación al montaje de platos y bebidas del cantón Ambato según la cocina sensorial. Revista Científica Arbitrada Multidisciplinaria PENTACIENCIAS. **5**, 426–438 (2023). https://doi.org/10.59169/pentaciencias.v5i4.678
7. Claux, D.: "Eating the barrio": gastronomy and urban change in Santa Cruz, Lima. Íconos - Revista de Ciencias Sociales., 37–58 (2024). https://doi.org/10.17141/iconos.79.2024.6041
8. Thaís, B..: Fotografía Gastronómica: análisis de la comunicación digital entre Brasil y España, https://rua.ua.es/dspace/bitstream/10045/109496/1/Fotografia_Gastronomica_analisis_de_la_comunicacion_digi_Krug__Barbara_Thais.pdf, (2020)
9. Aznar Íñiguez, C., Segarra Saavedra, J.: "Health foodies" como prescriptores saludables, gastronómicos y publicitarios. BiD: textos universitaris de biblioteconomia i documentació. (2022). https://doi.org/10.1344/bid2022.48.08
10. Cankul, D., Ari, O.P., Okumus, B.: The current practices of food and beverage photography and styling in food business. J. Hosp. Tour. Technol. **12**, 287–306 (2020). https://doi.org/10.1108/JHTT-02-2020-0052

11. Jarrín, D.: Mercadotecnia en redes sociales de los restaurantes de cocina fusión más importantes de Quito. Revista Enfoques. **6**, 329–340 (2022). https://doi.org/10.33996/revistaenfoques.v6i24.145
12. Vilela, Weinberg Jefman, T.M.: La cocina posmoderna de MasterChef Brasil: Social TV y medios que se difunden en Twitter. Sessões do Imaginário., https://revistaseletronicas.pucrs.br/ojs/index.php/famecos/issue/view/1017. (2015)
13. Rubalcava, C., Sánchez, Y., Sánchez, M.: La conciencia de marca en redes sociales: impacto en la comunicación boca a boca. Estudios Gerenciales., 313–320 (2019). https://doi.org/10.18046/j.estger.2019.152.3108
14. Tichý, J., Rosenlacher, P., Slavíková, B.: Creating of effective product photography form the perspective of neuromarketing. EMI. **10**, 1353–1805 (2018)
15. Del-Moral, M., Bellver, M., Guzmán, A.: Evaluación de la potencialidad creativa de aplicaciones móviles creadoras de relatos digitales para Educación Primaria. Ocnos. Revista de estudios sobre lectura. **18**, 7–20 (2019). https://doi.org/10.18239/ocnos_2019.18.1.1866
16. Taylor, F.: Seeing with photography collective. Photogr. Cult. **12**, 393–402 (2019). https://doi.org/10.1080/17514517.2019.1641288
17. Tifentale, A., Manovich, L.: Lev Manovich - Competitive Photography and the Presentation of the Self, http://manovich.net/index.php/projects/competitive-photography-and-the-presentation-of-the-self, (2016)
18. Navarro, M., Herrero, L.: uso de Instagram por parte de los Influencers Gastronómicos. VISUAL REVIEW. International Visual Culture Review / Revista Internacional de Cultura Visual. **14**, 1–18 (2023). https://doi.org/10.37467/revvisual.v10.4608
19. Amaya, S., Silva, G., Rincón, Y.: Impacto de los influencers en las estrategias promocionales del sector gastronómico de la ciudad de Medellín. Fonseca, J. Commun. (2021). https://doi.org/10.14201/fjc-v22-22695
20. Cava, A.: cultura del gusto. La alimentación entre representaciones y conexiones. Barataria. Revista Castellano-Manchega de Ciencias Sociales., 121–133 (2019). https://doi.org/10.20932/barataria.v0i26.462
21. Sabogal, L., Rojas, S.: Evolución del marketing experiencial: una aproximación teórica a su definición - Revisión Sistemática de Literatura. Entramado. **16**, 94–107 (2019). https://doi.org/10.18041/1900-3803/entramado.1.6140
22. Cepeda, M., Flórez, W.: La fotografía gastronómica y la identidad de marca en el consumidor de un restaurante gourmet. Trujillo - 2021. (2023)
23. Quispe, N.: Credibilidad de los foodies influencers en redes sociales. El caso del Cholo Mena (2023)
24. Martínez, R.: La importancia de la educación ambiental ante la problemática actual. Revista Electrónica Educare. **14**, 97–111 (2010). https://doi.org/10.15359/ree.14-1.9
25. Vidal, M., Pando, G.: Gastronomía: Cadena alimentaria de consumo en el Perú. Cultura, Ciencia y Tecnología., 15–27 (2019)
26. Peruano, E.: Perú en el top 10 de destinos gastronómicos del mundo según National Geographic, (2024)
27. Da Silva, D., Costa, C., Amorim, G., Roberto, J., Pinto, J.: Proposta de implementação de branding: um estudo de caso na empresa Tá na Arena. Revista de Gestão e Secretariado (Management and Administrative Professional Review). **14**, 18254–18274 (2023). https://doi.org/10.7769/gesec.v14i10.3045
28. Bordelon, L., Ferreira, S.: Tourist photographs and destination imagery on social media: reading the Stellenbosch winelands through a tourist lens. Tour. Rev. Int. **21**, 317–329 (2017). https://doi.org/10.3727/154427217X15022104437765
29. Digdanian, A.: El efecto de la fotografía publicitaria en la venta de un producto gastronómico, (2017)

30. Yadav, M.S., Pavlou, P.A.: Marketing in Computer-Mediated Environments: research synthesis and new directions. J. Mark. **78**, 20–40 (2014). https://doi.org/10.1509/jm.12.0020
31. Creswell, J., Creswell, J.: Research Design: Qualitative, Quantitative, and Mixed Methods Approaches, 6th edn. SAGE Publications (2022)
32. Lara, E.: La fotografía como documento histórico - artístico y etnográfico: una epistemología. Revista de Antropología Experimental. **5**, 1–28 (2005). https://doi.org/10.4067/S0719-26812016000200008
33. Akram, H., Abdelrady, A.H., Al-Adwan, A.S., Ramzan, M.: Teachers' perceptions of technology integration in teaching-learning practices: A systematic review. Front. Psychol. **13** (2022). https://doi.org/10.3389/fpsyg.2022.920317
34. Guardia, S.: Gastronomía Peruana Patrimonio Cultural de la Humanidad. Cátedra UNESCO Patrimonio Cultural y Turismo Sostenible de la Universidad de San Martín de Porres, Perú (2020)
35. Del Prete, A., Redon Pantoja, S.: Las redes sociales virtuales: Espacios de socialización y definición de identidad. Psicoperspectivas. Individuo y Sociedad. **19** (2020). https://doi.org/10.5027/psicoperspectivas-Vol19-Issue1-fulltext-1834
36. Trujillo, J.M., Portilla, B.d.R.: La influencia del contenido visual en el marketing digital para el turismo gastronómico. Economía Negocios. **6**, 69–77 (2024). https://doi.org/10.33326/27086062.2024.2.1972
37. Acosta, Y.: Del bodegón al Porn Food. Fotografia gastronómica. Reseña de Josep Àngel Guimerà Orts. Fotocinema. Revista científica de cine y fotografía. (2012). https://doi.org/10.24310/Fotocinema.2012.v0i4.5888
38. Concha, J.: Imagen fotográfica y Lenguaje. AISTHESIS. **3**, 264–273 (2001)
39. Del Toro Cabrera, A.: concepción de cliente y su repercusión en la calidad del servicio en el restaurante. Res Non Verba Revista Científica. **9**, 1–16 (2019). https://doi.org/10.21855/resnonverba.v9i2.214
40. Braun, V., Clarke, V.: Using thematic analysis in psychology. Qual. Res. Psychol. **3**, 77–101 (2006). https://doi.org/10.1191/1478088706qp063oa
41. Martín, D., Martín, J.: La importancia de la reflexión antes del disparo fotográfico. Un análisis estadístico de las preferencias de los usuarios. In: Libro de Actas - III Congreso Internacional de Investigación en Artes Visuales ANIAV 2017 GLOCAL, pp. 1–8. Universitat Politècnica València, Valencia (2017). https://doi.org/10.4995/ANIAV.2017.4568
42. Santamarina, A.: El arte cinematográfico y el arte gastronómico: dos caminos paralelos que transitan en torno al tiempo. Fotocinema. Revista Científica de Cine y Fotografía., 1–60 (2016). https://doi.org/10.24310/Fotocinema.2016.v0i12.6042
43. Pérez, J., Bedoya, C.: La fotografía como herramienta para el desarrollo de la creatividad y la alfabetidad visual: investigación en el aula. Kepes. **16**, 377–404 (2019). https://doi.org/10.17151/kepes.2019.16.20.14
44. Alaei, A., Bui, V., Doermann, D., Pal, U.: Document image quality assessment: a survey. ACM Comput. Surv. **56**, 1–36 (2024). https://doi.org/10.1145/3606692
45. Guamán-Guevara, F.: Atributos de la comida tradicional de la ciudad Latacunga. Evidencia empírica con futura aplicación en marketing sensorial. RECIMUNDO. **8**, 91–99 (2024). https://doi.org/10.26820/recimundo/8.(especial).octubre.2024.91-99
46. Barrientos - Báez, A., Cadevilla, D., Blanco, M.: Estrategias de promoción de la serie española La Peste (2018): Cine, gastronomía y social media. Rev. Cienc. Soc. **27**, 1–14 (2021). https://doi.org/10.31876/rcs.v27i3.36759
47. Rogel, C., Marcos, M.: Branded content audiovisual, la clave de la comunicación aspiracional de las marcas. Comunicación Revista Internacional de Comunicación Audiovisual Publicidad y Literatura. **1**, 114–135 (2024). https://doi.org/10.12795/Comunicacion.2024.v22.i01.07
48. Kaufman, B.: Stories that SELL, stories that TELL. J. Bus. Strateg. **24**, 11–15 (2003). https://doi.org/10.1108/02756660310508155

49. Carrascoza, J.: Lágrimas na chuva: a publicidade no futuro e o futuro da publicidade. Galáxia (São Paulo)., 207–222 (2020). https://doi.org/10.1590/1982-25532020345012
50. Mendieta, G.: Rol del visual-storytelling en la publicidad digital de restobares limeños. Lanzamiento de Sastrería Martínez, Caso (2022). https://doi.org/10.19083/tesis/667172
51. Calvo-Porral, C., Lévy-Mangin, J.-P.: From "foodies" to "cherry-pickers": A clustered-based segmentation of specialty food retail customers. J. Retail. Consum. Serv. **43**, 278–284 (2018). https://doi.org/10.1016/j.jretconser.2018.04.010
52. Lee, K.-H., Wang, Z., Syazreema, N.: Uncovering food experience from social media. In: Handbook on Tourism and Social Media. Edward Elgar Publishing (2022). https://doi.org/10.4337/9781800371415.00042
53. Öztürk, B.: Art of food: systematic literature review of culinary creativity. Int. J. Gastron. Food Sci. **36**, 1–200 (2024). https://doi.org/10.1016/j.ijgfs.2024.100940
54. Velasco, C., Barbosa, F., Spence, C., Olier, J.: The taste of colours. Food Qual. Prefer. **112**, 1–150 (2023). https://doi.org/10.1016/j.foodqual.2023.105009
55. Kontukoski, M., Paakki, M., Thureson, J., Uimonen, H., Hopia, A.: Imagined salad and steak restaurants: consumers' colour, music and emotion associations with different dishes. Int. J. Gastron. Food Sci. **4**, 1–11 (2016). https://doi.org/10.1016/j.ijgfs.2016.04.001

Integrating Artificial Intelligence and Self-reports for Emotional Recognition and Management in Primary and Secondary Education

Eduardo Ñaupa Yapura(✉), Antony Elio Ayansi Huisa, Anthony Dennis Oyola Saloma, Daniel Hachircana Méndez, Yadira Melina Zapata Ttito, and Diana Salas Sanchez

Fabrication Laboratory, Continental University, Cusco, Peru
{60859960,77021318,75389246,41547290,76763410, dsalas}@continental.edu.pe

Abstract. This study presents the development and validation of ALI, a technological system based on artificial intelligence designed to recognize and manage emotional states in school-aged students. The system combines computer vision techniques, micro expression analysis through the Facial Action Coding System (FACS), and self-reported emotional records collected from primary and secondary students in Peruvian schools. Data was obtained from three groups (6th grade primary, 2nd grade A secondary, and 2nd grade B secondary) across multiple school days, integrating both automated recognition and student self-assessment. Results indicate differentiated emotional patterns by educational stage: primary students exhibited a balanced distribution between positive and negative emotions with medium-high intensity (M = 5.17, SD = 2.74), whereas secondary students showed greater volatility and predominance of negative emotions such as sadness, anger, and disgust, with progressive increases in emotional intensity throughout the week. The ALI system achieved over 75% accuracy in Action Unit detection and demonstrated strong agreement with self-report data (68% overlap). These findings highlight the potential of combining artificial intelligence with psychological self-reports for early detection of emotional risks, allowing tailored interventions by educators and school psychologists. The study contributes to the growing field of AI-supported socioemotional learning, offering a scalable and ethical framework for improving student well-being and academic performance.

Keywords: Emotional Recognition · Artificial Intelligence in Education · Self-Report Analysis · Socioemotional Learning

1 Introduction

1.1 Background and Motivation

Socioemotional well-being has become a central concern in educational research and practice, given its direct impact on learning, motivation, and long-term academic performance. According to the World Health Organization [1], nearly 14% of adolescents

A. Rocha et al. (Eds.): ICITS 2026, LNNS 1966, pp. 753–762, 2027.
https://doi.org/10.1007/978-3-032-27538-7_66

worldwide experience mental health problems, including anxiety and depression, which strongly affect their social and academic development. In Peru, the Ministry of Education [2] has reported that 1 in 5 students presents symptoms of school-related stress or demotivation, underscoring the urgent need for early detection and intervention strategies in schools. Within this context, emotions are not only outcomes of the learning process but also critical drivers of engagement, retention, and well-being [3].

Furthermore, current evidence shows that educational systems are progressively incorporating emotional education as a strategic pillar, recognizing that the development of cognitive abilities alone is insufficient for achieving sustainable learning outcomes [4]. Schools that promote emotional literacy, empathy, and resilience tend to foster stronger academic performance and healthier interpersonal relationships [5]. This connection reinforces the argument that socioemotional development should be considered as important as traditional academic subjects, positioning it as a cornerstone of twenty-first century education [6].

1.2 Problem Statement

Despite the recognized importance of emotions in education, psychologists and teachers still face significant challenges in detecting emotional risks in real time. Traditional self-report measures provide useful insights but are often limited by subjectivity, delayed reporting, and underreporting of negative emotions. The absence of objective, real-time tools hamper the ability to intervene promptly [7]. Against this backdrop, this research addresses the following guiding question: Can an AI-based system for emotional recognition and management in school settings complement self-reports to provide valid, actionable insights for intervention?

This problem is particularly relevant in contexts with large classrooms or resource-limited schools, where individualized attention is challenging [8]. In such settings, educators may overlook subtle emotional cues, and interventions often occur only after issues have escalated. Thus, the development of technological systems capable of capturing and processing emotional information instantaneously becomes a powerful complement to the human capacity for observation, potentially reducing the lag between emotional need and professional support [8].

1.3 Theoretical Framework: Emotions in Learning

The theoretical foundation of this study is grounded in the work of Paul Ekman [9], who identified six basic universal emotions joy, sadness, anger, disgust, and surprise that are expressed consistently across cultures through facial cues. The Facial Action Coding System [10] (FACS) provides a structured method to decode these micro expressions into Action Units (AUs), linking specific muscle movements to emotional states. In educational settings, emotions play a pivotal role in influencing attention, memory consolidation, and motivation [11]. While self-report methods are widely used, they remain vulnerable to biases and may not capture the immediacy of emotional responses in classroom interactions.

The incorporation of emotion research into pedagogy has expanded the understanding of how affect and cognition intertwine. Studies show that emotions can act as both facilitators and barriers to learning, with positive effects linked to increased creativity and problem-solving, and negative effects sometimes sharpening focus but often undermining sustained engagement [12]. This dual nature of emotions highlights the complexity of assessing them, reinforcing the need for triangulated methods such as combining FACS-based detection with self-reports to achieve a fuller and more accurate understanding of students' emotional states.

1.4 Technology and Education Convergence

The integration of artificial intelligence into education has emerged as a transformative trend, offering opportunities to personalize learning, enhance student monitoring, and support teachers in decision-making. Recent systematic reviews, such as that of Vistorte [13], have emphasized the role of AI in assessing emotions within learning environments. Building on this momentum, the ALI system was developed as a novel technological contribution. It integrates computer vision, micro expression analysis, and student self-reports, generating real-time insights into students' emotional states and offering actionable data for intervention.

This convergence with education is also reshaping the responsibilities of teachers and psychologists. While AI can provide immediate and objective measurements, human expertise remains essential for interpretation, contextualization, and ethical application [14]. The success of systems like ALI depends not only on their technical precision but also on their integration into pedagogical practices that prioritize student agency and well-being. Therefore, the intersection of advanced technology with educational psychology represents both an opportunity and a challenge: to design solutions that enhance human decision-making without replacing it.

1.5 Research Gap and Contribution

Although international literature has explored AI-driven emotion recognition, there is a notable gap in validated, real-time systems applied in Latin American schools, where cultural, social, and infrastructural contexts differ from those in high-income countries [15]. This study contributes to closing that gap by:

- Validating an AI-based emotional recognition system (ALI) in real classrooms.
- Comparing automated detection with student self-reports to strengthen validity.
- Identifying developmental differences in emotional expression between primary and secondary students.
- Exploring contextual effects, such as subject-specific emotional responses and end-of-week intensity patterns.
- Proposing an ethical framework for responsible deployment, emphasizing privacy and data protection.

The general objective of this research is to evaluate the feasibility and validity of integrating the ALI system with self-reports for emotion recognition in school-aged students. The specific objectives are:

- To analyze patterns of emotional expression across educational stages.

- To assess the concordance between AI detection and student self-reports.
- To identify contextual variables (subjects, days of the week) influencing emotions.
- To propose an ethical framework for the deployment of AI-driven emotional recognition systems in schools.

2 Methodology

2.1 Participants and Data Collection Protocol

The study was conducted in two private educational institutions in southern Peru. Ethical approval was obtained from school authorities, ensuring compliance with confidentiality and data protection standards. Informed consent was secured from parents, teachers, and students prior to participation. No videos were stored; only anonymized facial landmark points and emotional reports were recorded to safeguard student privacy.

The research followed a two-phase protocol:

- Initial Phase: A pilot assessment was carried out with approximately 300 secondary school students in social sciences classes, allowing calibration of the system and refinement of the instruments.
- Effective Phase: A sample of 90 students (ages 11–15) from primary (6th grade) and secondary (2nd grade A and B) levels was evaluated. Data collection integrated two complementary sources: (1) automated recognition through the ALI system and (2) student self-reported emotional diaries.

2.2 Ethical Considerations

The study adhered to the ethical guidelines for research with minors, emphasizing voluntary participation, anonymization of data, and the right to withdraw at any stage. The protocol was designed to avoid psychological harm by limiting recording sessions to short intervals and ensuring that results were used exclusively for educational and psychological support.

2.3 Emotional Coding Framework (FACS)

The ALI system employed the Facial Action Coding System (FACS) developed by Ekman and Friesen (1978) to decode facial microexpressions into Action Units (AUs). Each AU corresponds to specific muscle movements associated with basic emotions: joy, sadness, anger, fear, disgust, and surprise. For example, AU12 (lip corner puller) indicates joy, whereas AU4 (brow lowering) signals anger or sadness.

2.3.1 Technologies for Emotional Recognition

The system was built on open-source and widely validated tools:

- OpenCV: for image acquisition and preprocessing.
- Dlib: for facial landmark detection and geometric mapping.
- Mediapipe: for real-time face detection with high accuracy.
- FER+ dataset: for training deep learning models in emotion classification.

These technologies enabled robust detection of facial expressions in classroom environments with minimal latency, see Fig. 1.

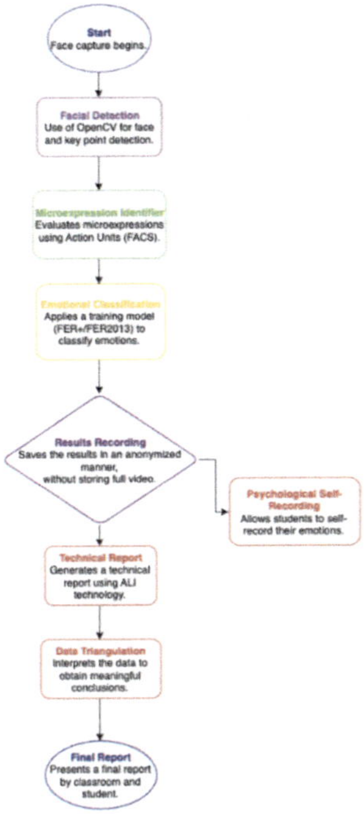

Fig. 1. Digital workflow.

2.4 Technical Approach and Micro expression Flow

The system followed a structured technical pipeline divided into three main stages:

Preparation	Emotional processing flow	Integration and reporting
Video captured through IP cameras (minimum 720p, 30fps) under controlled lighting conditions.	Face detection and landmark extraction using OpenCV, Dlib, and Mediapipe.	Alerts were generated for school psychologists when recurrent negative emotions were detected (e.g., stress, sadness).
Facial landmark detection based on the FACS framework.	Action Unit Activation Analysis to map muscle movements to emotions.	Reports were produced at both individual and group levels, triangulating system outputs with student self-reports to enhance validity and provide actionable insights for educators.
	Neural network–based classification to assign probability scores for each emotion (e.g., sadness 70%, joy 42%).	

(*continued*)

(*continued*)

Preparation	Emotional processing flow	Integration and reporting
	Contextual recording that associates emotional states with time of day, subject, and classroom activity.	

3 Results

3.1 Sample Characteristics

The effective phase of the study included a sample of 90 students aged between 11 and 15 years, distributed as follows: 30 students from 6th grade of primary education, and 60 students from 2nd grade of secondary education (30 from group A and 30 from group B). Emotional states were captured during regular classes across three consecutive school days (Wednesday–Friday). Two complementary data sources were analyzed: (1) self-reported diaries of emotions and (2) automated recognition outputs from the ALI system.

3.2 Descriptive Statistics of Self-reported Emotions

6th Grade Primary: The distribution of emotions was balanced, with surprise (20%), anger (20%), and sadness (20%) being the most reported, followed by joy (16.7%). Negative emotions such as disgust and fear appeared less frequently (13.3% each). The average emotional intensity was 5.17 (SD = 2.74) on a 1–9 scale, reflecting a medium-high expressivity with several high-intensity episodes (scores of 8 and 9).

2nd Grade A Secondary: Emotional oscillations were more pronounced. On Wednesday, sadness predominated (26.7%), but on Thursday, joy peaked at 36.7%, with anger as a secondary response (20%). By Friday, negative emotions dominated again, with fear (23.3%), anger (20%), and disgust (20%) as the most frequent. Intensities were high and variable, ranging from mild (2–3) to very strong (8–9), evidencing emotional instability typical of early adolescence.

2nd Grade B Secondary: A more consistent pattern of negative emotions emerged. On Wednesday, sadness reached 30%, while joy was limited to 20%. On Thursday, disgust (33.3%) and fear (20%) predominated, with a decline in joy (13.3%). On Friday, surprise (26.7%) and sadness (23.3%) were the most frequent. The average intensity rose progressively from 4.80 (Wednesday) to 5.37 (Friday), indicating a trend toward experiencing more intense negative emotions as the week advanced.

3.3 Automated Recognition with ALI System

The ALI system, based on FACS micro expression analysis, reported the following general distribution of emotions across all sessions:

- Sadness: 51.1%
- Discomfort/anger-related expressions: 23.1%
- Happiness: 17.3%
- Disgust and surprise: 3.5% each
- Fear: 1.5%

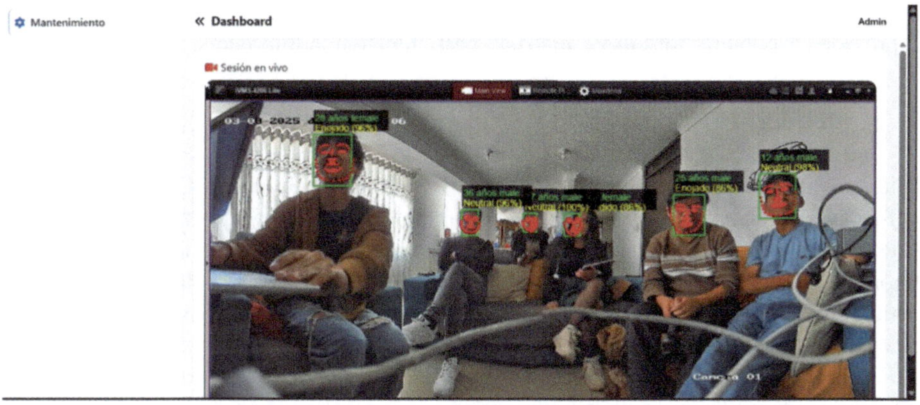

Fig. 2. Real-time capture of the ALI system recognizing multiple emotions during a simulated session.

The system achieved >75% accuracy in Action Unit detection (e.g., AU1, AU4, AU12) and showed a 68% overlap with self-reported emotions, reinforcing its validity as a complementary tool for psychological evaluation (Fig. 2).

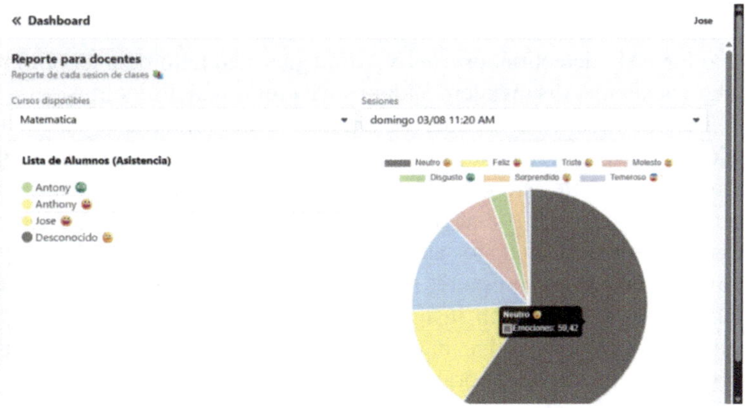

Fig. 3. Visual report from the ALI system on the emotional analysis of students during a mathematics class.

When triangulating both data sources, a clear distinction between primary and secondary students was observed:

- Primary education presented a more balanced emotional profile, combining positive and negative emotions with moderate intensity.
- Secondary education showed higher levels of emotional volatility (group A) or persistent negative emotional dominance (group B), with intensities that tended to increase toward the end of the week.

These findings highlight the developmental differences in emotional regulation across age groups and the potential of integrating AI-based recognition with self-reports to provide early alerts for emotional risk, particularly in adolescence (Fig. 3).

4 Conclusion

This study demonstrates that the integration of artificial intelligence systems with student self-reports provides a reliable, innovative, and interdisciplinary approach for the recognition and management of emotions in primary and secondary education. By combining automated recognition through the ALI system with psychological self-assessments, the research introduces a methodological triangulation that strengthens validity and opens new perspectives for applied socioemotional learning. The ALI system, grounded in the Facial Action Coding System (FACS), achieved over 75% accuracy in Action Unit detection and a 68% agreement with self-reported emotions, confirming its robustness as a complementary tool for psychological evaluation and educational intervention.

The results highlight developmental differences in emotional regulation, evidencing the nuanced ways in which emotions manifest across educational stages. Primary students exhibited a more balanced and spontaneous emotional profile, where positive and negative emotions were distributed in equilibrium and expressed with medium intensity. In contrast, secondary students especially those in grade 2B showed a predominance of negative emotions such as sadness, anger, and disgust, coupled with a progressive increase in intensity toward the end of the week. These patterns align with psychological theories of adolescence as a stage of heightened volatility, and underscore the critical need for early detection, preventive strategies, and tailored interventions to mitigate the risk of academic disengagement and socioemotional difficulties.

Beyond descriptive trends, the findings reveal important contextual determinants of emotional expression. The association between specific academic subjects and particular emotional states suggests that emotions are not random occurrences but are linked to the cognitive and social demands of classroom contexts. Furthermore, the escalation of emotional intensity as the week progresses illustrates a cumulative effect of academic and social pressures, which may contribute to stress and reduced well-being if left unaddressed. Such insights provide an empirical foundation for the design of adaptive, context-sensitive interventions that align emotional support strategies with subject-specific and temporal patterns.

From an ethical standpoint, this research demonstrates that technological innovation in education must be accompanied by strong ethical safeguards. The protocol ensured informed consent, anonymization of facial landmarks, and avoidance of video storage, thereby addressing the risks of emotional labeling, stigmatization, or invasive surveillance. The ALI system was intentionally framed as a supportive and non-substitutive tool: its purpose is not to replace the human judgment of psychologists or teachers, but rather to empower professionals with objective, real-time data that enhances their capacity to respond promptly and effectively to students' emotional needs.

The study's contributions extend to three levels:

- Theoretical: It enriches the growing field of AI in education by demonstrating how computational models of emotion, when integrated with psychological self-report measures, can produce more accurate and reliable indicators of student well-being.
- Practical: It offers a replicable framework for schools to monitor socioemotional dynamics, generate early alerts for risk, and adapt teaching practices to promote healthier classroom climates.
- Technological-Ethical: It provides evidence that AI-driven systems can be implemented responsibly in sensitive domains like education, provided that safeguards for privacy, cultural adaptation, and human oversight are respected.

In conclusion, the ALI system represents a scalable, ethical, and impactful framework for socioemotional learning in schools. Its integration of artificial intelligence and psychological methods offers a promising pathway for improving both academic performance and socioemotional development. Future research should aim to (a) conduct longitudinal studies that capture emotional dynamics across academic years, (b) adapt and validate the system in diverse cultural and neurodivergent contexts, and (c) refine algorithms to reduce potential biases and enhance cross-cultural generalizability. By doing so, the field can ensure that AI-driven emotional recognition evolves as an inclusive and equitable technology, capable of fostering educational environments where well-being and learning outcomes reinforce each other.

References

1. W. H. Organization: World Mental Health Report: Transforming Mental Health for all. World Health Organization (2022)
2. Con Bienestar, C.A., et al.: "Monitoreo de la afectación del bienestar socioemocional docente en el contexto post pandemia por COVID-19. Informe de Evidencia sobre Educación y Salud Mental, N° 5 (2024)
3. Tan, J., Mao, J., Jiang, Y., Gao, M.: The influence of academic emotions on learning effects: a systematic review. Int. J. Environ. Res. Public Health. **18**(18), 9678 (2021)
4. Dubovi, I.: Cognitive and emotional engagement while learning with VR: the perspective of multimodal methodology. Comput. Educ. **183**, 104495 (2022)
5. Nandan, A., Ranjana, M.: Emotional Intelligence in Children: Building Resilience and Empathy at Home and in School
6. Pondiscio, R.: The Unexamined rise of Therapeutic Education: how Social-emotional Learning Extends K-12 Education's reach into Students' Lives and Expands Teachers' Roles. American Enterprise Institute (2021)
7. Schiepe-Tiska, A., Dzhaparkulova, A., Ziernwald, L.: A mixed-methods approach to investigating social and emotional learning at schools: teachers' familiarity, beliefs, training, and perceived school culture. Front. Psychol. **12**, 518634 (2021)
8. Srinivasa, K.G., Kurni, M., Saritha, K.: Harnessing the power of AI to education. In: Learning, Teaching, and Assessment Methods for Contemporary Learners: Pedagogy for the Digital Generation, pp. 311–342. Springer (2022)

9. Kim, H.: Paul Ekman and the search for the isolated face in the 1960s. J. Hist. Behav. Sci. **60**(4), e22322 (2024)
10. Luo, C., Song, S., Xie, W., Shen, L., Gunes, H.: Learning multi-dimensional edge feature-based au relation graph for facial action unit recognition. arXiv preprint arXiv:2205.01782. (2022)
11. Tyng, C.M., Amin, H.U., Saad, M.N.M., Malik, A.S.: The influences of emotion on learning and memory. Front. Psychol. **8**, 235933 (2017)
12. Nash, P., Schlösser, A.: Working with schools in identifying and overcoming emotional barriers to learning. Educ. Stud. **41**(1–2), 143–155 (2015)
13. Vistorte, A.O.R., et al.: Integrating artificial intelligence to assess emotions in learning environments: a systematic literature review. Front. Psychol. **15**, 1387089 (2024)
14. Farnsworth, B.: How to measure emotions and feelings (and the difference between them). April. **14**, 591–632 (2020)
15. Wu, Y., Zhang, L., Gu, Z., Lu, H., Wan, S.: Edge-AI-driven framework with efficient mobile network design for facial expression recognition. ACM Trans. Embed. Comput. Syst. **22**(3), 1–17 (2023)

Author Index

A
Aguilar, Hugo Gonzalez 44, 552
Alexandra, Nasimba Yautibug Lizbeth 582
Almeida Aguilar, María Alejandrina 559
Alomoto-Robles, Diego 705
Alves, Nelson 120
Arcos, Renato 458
Armas-Herrera, Reinaldo 414, 473
Atencio-Iturry, Ivette Eneida 518
Au-Yong-Oliveira, Manuel 278

B
Balk, Roosa 97
Barzev, Iliyan 540
Bayas, Carolina Paola Jaramillo 495
Becerra, Mónica Hinojosa 156
Benavides-Montenegro, Gerardo 436
Borges, Ana Patrícia 169
Borissova, Daniela 16, 540
Bravo-Quezada, Omar G. 62

C
Cáceres-Bazalar, Silvia 731
Córdova, Rodolfo 429
Cabrera-Paucar, Edwin José 74
Cahuate, Jessica 27
Caicedo-Altamirano, Fernando 705
Calle-Landázuri, Doris 506
Canedo, Edna Dias 403
Carrasco, Franklin 429
Carreira, Dário 169
Carvache-Franco, Mauricio 317
Carvache-Franco, Orly 317
Carvache-Franco, Wilmer 317
Carvalho, Suzete 278
Castro, Santiago 484
Castro, Verónica 27
Ccosi, Bryan Antony Quinta 269
Chávez-Chuquimango, María 731, 742
Chanataxi-Nacimba, Bryan 705
Chiguano, Moisés Pallo 484
Chindongo, Bupe 446
Chiquito, Roxana 208, 220
Cooper, Kenny Joseph 381
Cordóva-Vera, Karen Stephany 131
Cornejo, Ángel 27
Corona Ferreira, Arturo 559
Costa, Tiago 297
Costa, Victor 169
Cruces Gutiérrez, Eduardo 559
Cusco-V, Victor 131
Custodio Izquierdo, Carlos Arturo 559

D
De la Torre-Guzmán, Javier 232
de Mendonca, Fabio Lucio L. 403
Dhar, Bablu Kumar 290
Dionísio, António 694

E
El Khori El Khouri, Sandra 637
Encalada Ochoa, Carlos Francisco 637
Espín-Lopez, José Luis 414, 473
Espinoza Vidaurre, Sam M. 518

F
Fernández-Mora, Lady 108
Filho, Geraldo Pereira Rocha 403
Flórez-Villamil, Daniela 108
Franco, Glendamira Serrano 360

G
Gambetta Quelopana, Renza L. 518
García, Víctor Manuel Zamudio 360
García, José 381
García-Acosta, Mishell Alejandra 414, 473
Gasca-Hurtado, Gloria Piedad 3
Gavilanez, María Merchán 436
Gómez Ramos, José Luis 559
Gómez-Velásquez, Natalia 3
Gonçalves, Bruno F. 181

A. Rocha et al. (Eds.): ICITS 2026, LNNS 1966, pp. 763–765, 2027.
https://doi.org/10.1007/978-3-032-27538-7

Gonzales-Inca, E. 679
González Zacarías, Carlos 559
Guaigua-Albarracín, Gustavo 232
Guamán, Darío 614
Guamán, Mario 614
Guevara-Arteaga, Adriana 341
Gutiérrez, César Ariel Quesada 657
Gutierrez-Aguilar, O. 679

H
Hänninen, Vihtori 306
Hernández-Rodríguez, Juan David 108
Herrera, Reinaldo 569
Higuerey-Gómez, Angel Alexander 414, 473
Huisa, Antony Elio Ayansi 753

I
Iriarte-Solano, Margoth 74
Irias, José 381

J
Jadira, Vallejo Burbano Alejandra 582
Jaramillo, Rubén 614
Jawad, Daniyal 257
Jiménez-González, Ledys L. 62, 657
Joel, Toapanta Vizuete Kevin 582
Juca-Aulestia, Marcelo 74

K
Khokhar, Danish 257

L
Lancheros-Cuesta, Diana Janeth 350
Laque, Luis Daniel Haro 269
Lara, Adrian 589
Laura-De La Cruz, Kevin Mario 52
Le Fur, Eric 245
Lekova, Anna 16
Liyanage, Gayani Hawpe 257
Londoño-Proaño, Cristián 392
Loo, Luis 381
Loza, Ricardo 669
Luján-Mora, Sergio 436

M
Méndez, Daniel Hachircana 753
Machado, Carmen 694
Maldonado, Selvin 381
Mansilla-Lopez, Juan 669
Manzur-Vera, Gabriela 52
Marcão, Ricardo 290, 297, 694
Marcillo, Fabricio 429
Martinho, Filipa 290, 297, 694
Mena, Liliana Marisol Jumbo 74
Méndez-Beltrán, María Andrea 108
Merizalde-Torres, Karen 506
Mesquita, Elza 181
Miguéis, Vera 120
Minchenkova, Aleksandra 317
Miranda, Aldenis E. 381
Miranda, Guilhermina Lobato 718
Montalvo, William 331
Moodley, Tevin 446
Morán-Velasco, Jonathan M. 62
Moreta-Changoluisa, Janneth 705

N
Núñez-Naranjo, Aracelly 131
Naranjo, William Ortega 495
Nascimento, Rodrigo 290
Navarro-Alfaro, Jonathan 589
Nolasco-Mamani, Marco Antonio 518
Nova-Arévalo, Nestor Armando 108
Nuñez, S. Hinojosa 679
Núñez-Naranjo, Aracelly 317
Nze, Georges Daniel Amvame 403

O
Ocampo, Willian 429
Ochoa, Carlos Francisco Encalada 657
Orellana-Jaramillo, Pedro J. 62
Orozco, Fausto 429
Ortiz-Pabón, Efraín 108

P
Pérez, Lennis 27
Palacio Valdivieso, Gloria del Carmen 626
Palma-Moreira, Ana 278
Paththamesthrige, Ferdina 257
Patrício, Maria Raquel 181
Pavão, João 169
Peñarreta, Miguel Angel 569
Peña-Gómez, Sara Juliana 108
Perez, Christopher 669
Pérez, Mario 458
Pérez-Yufra, Domingo Nicolás 52
Pozo, Jairo 331

Pucha, María Elena Villafuerte 657
Puchaicela, Maribel 27
Puentes-Castro, Alexander 350
Punchihewa, Uthpala 257

Q
Quispe-Coronel, Génesis 742
Quispe-Vargas, Miliam 52
Quituizaca Songor, Ivanova 626

R
Ríos-Vera, K. 679
Rasheva-Yordanova, Katia 16
Reátegui, Ruth 192
Reis, Isabel 297
Restrepo-Tamayo, Luz Marcela 3
Rivas-Lalaleo, David 705
Rivera-Rogel, Diana 156
Rocha, Nelson Pacheco 169
Rodríguez, Roger Daniel Ponce 648
Rodrigues, Paulo Henrique B. 403
Rodriguez-Palumbo, Cynthia Karín 52
Romero, Ana C. 648
Romero, Julio Merchán 484
Ronceros, Marcos 669

S
Sabando-García, Ángel Ramón 414, 473
Sacoto-Cabrera, Jairo 369
Salamea-Palacios, Christian 369
Salazar-Jácome, Elizabeth 232, 506
Salazar-Paredes, Rolando 232
Salcedo-Quishpe, Miryam Johana 341
Saloma, Anthony Dennis Oyola 753
Sampaio, Margarida 718
Sanabria, María Eugenia Sandoval 648
Sanchez, Diana Salas 753
Santos, Vasco 290, 297, 694
Silva, Daniel Alves 403
Solorzano, Cristhian 208, 220
Stoev, Stefan 16
Suarez-Morales, Lizbeth 208, 220
Sugimoto, Masanori 85

T
Talavera, S. Hinojoza 679
Tandazo-Malla, Carlos 192
Tapia-Leon, Mariela 436
Teixeira, Ana Luísa 169
Tica, Estefany Nayeli Quispe 269
Tituaña, Lorena Elizabeth Muquis 495
Toasa, Renato M. 131, 458
Topon-Visarrea, Blanca 341
Torres-Carrión, Pablo Vicente 156
Torres-Diaz, Juan Carlos 156
Torres-Toukoumidis, Angel 484
Ttito, Yadira Melina Zapata 753
Tufiño, Rodrigo 331

U
Ugando-Peñate, Mikel 414, 473
Ulcuango, Luis 614
Ulloa, Andres 569
Uyaguari Morocho, Ana Isabel 637

V
Valarezo, Wellington Gerardo Abril 657
Valdivia, Ana Noelia Martinez 518
Valencia Berrezueta, Angela Gabriela 637
Vasquez-Avila, Juan 369
Vasquez-Peñafiel, María-Stefanie 317
Velásquez Rodríguez, Norma C. 518
Viñanzaca-Figueroa, Javier 369
Vilko, Jyri 97, 257, 306
Villa-Ochoa, Jhony Alexander 602

Y
Yánez, María José Carrera 141
Yapura, Eduardo Ñaupa 753

Z
Zabala-Cáceres, Samanta 506
Zapana-Aguilar, Pilar Valeria 52
Zumba, Aide 429

GPSR Compliance

The European Union's (EU) General Product Safety Regulation (GPSR) is a set of rules that requires consumer products to be safe and our obligations to ensure this.

If you have any concerns about our products, you can contact us on ProductSafety@springernature.com

In case Publisher is established outside the EU, the EU authorized representative is:

Springer Nature Customer Service Center GmbH
Europaplatz 3
69115 Heidelberg, Germany

Batch number: 10607216

Printed by Printforce, the Netherlands